W0111658

Progress in Mathematics
Volume 243

Series Editors
Hyman Bass
Joseph Oesterlé
Alan Weinstein

Studies in Lie Theory

Dedicated to A. Joseph
on his Sixtieth Birthday

Joseph Bernstein
Vladimir Hinich
Anna Melnikov
Editors

Birkhäuser
Boston • Basel • Berlin

Joseph Bernstein
Tel Aviv University
School of Mathematical Sciences
Ramat Aviv, Tel Aviv 69978
Israel

Vladimir Hinich
Anna Melnikov
University of Haifa
Department of Mathematics
Mount Carmel, Haifa 31905
Israel

Mathematics Subject Classifications (2000): 12A50, 13N10, 14L30, 14R20, 16K40, 16S30, 16S32, 16S36, 16S40, 16W35, 16W50, 17B35, 17B37, 17B56, 17B65, 17B69, 20C08, 20G10, 22D20, 22E35, 22E50, 32C38, 33C45, 35A27, 53D17

Library of Congress Control Number: 2005936729

ISBN-10: 0-8176-4342-7 e-ISBN: 0-8176-4478-4
ISBN-13: 978-0-8176-4342-3

Printed on acid-free paper.

©2006 Birkhäuser Boston *Birkhäuser*

All rights reserved. This work may not be translated or copied in whole or in part without the written permission of the publisher (Birkhäuser Boston, c/o Springer Science+Business Media Inc., 233 Spring Street, New York, NY 10013, USA), except for brief excerpts in connection with reviews or scholarly analysis. Use in connection with any form of information storage and retrieval, electronic adaptation, computer software, or by similar or dissimilar methodology now known or hereafter developed is forbidden.
The use in this publication of trade names, trademarks, service marks and similar terms, even if they are not identified as such, is not to be taken as an expression of opinion as to whether or not they are subject to proprietary rights.

Printed in the United States of America. (KeS/EB)

9 8 7 6 5 4 3 2 1

www.birkhauser.com

Anthony Joseph (shortly after joining the Weizmann Institute)

With Denise at Zichron Ya'acov, June 2002

Participants of the workshop on Representations of Lie algebras in honour of Anthony Joseph
First row from left to right: M. P. Malliavin, F. Fauquant-Millet, A. Braverman, G. Letzter, D. Joseph, A. Joseph, A. Melnikov, L. Makar-Limanov, V. Berkovich, V. Hinich;
Second row from left to right: J. Beck, D. Gaitsgory, D. Kazhdan, S. Arkhipov, O. Gabber, D. Todoric, R. Rentschler, N. Shomron, A. Elashvili, D. Vogan, M. Duflo, A. Stolin, G. Perets;
Third row from left to right: Ya. Greenstein, N. Papalexiou, C. Stroppel, B. Kostant, A. Kostant, I. Mirkovic, M. Finkelberg, V. Ginzburg, Yu. Bazlov, A. Retakh;
Last row from left to right: Sh. Zelikson, A. Juhasz, V. Vologodsky, B. Noyvert, E. Perelman, A. Vershik, R. Bezrukavnikov, M. Losik, J. Alev, V. Ostapenko, S. Khoroshkin, A. Reznikov.

Contents

Preface

This volume is dedicated to Anthony Joseph on the occasion of his 60th birthday. A conference entitled *Representations of Lie Algebras* was held in his honour at the Weizmann Institute, Rehovot, in July 2002. Subsequently, distinguished experts in representation theory and related areas were invited to contribute survey and research articles, which comprise this volume.

The focus here is on semisimple Lie algebras and quantum groups, the central subjects in representation theory to which the contribution of Tony Joseph is difficult to overestimate. For over three decades the impact of his work has been seminal and has changed the face of the subject.

The introductory part of the volume consists of a short note by Jacques Dixmier describing the beginnings of Tony's entry into mathematics, followed by the speech of Denise at the dinner honouring her husband. From Denise, the participants got a glimpse into another side of Tony's personality.

The scientific part of the volume begins with two surveys which give an overview of the central topics in representation theory to which Tony Joseph made his mark: the first, written by W. McGovern, describes Joseph's main input into the theory of primitive ideals in semisimple Lie algebras; the second, coauthored by D. Farkas and G. Letzter, is devoted to the study made by A. Joseph of quantized enveloping algebras. Thereafter, 16 research articles cover a number of different topics in representation theory.

J. Alev and F. Dumas study the invariants of the Weyl skew field $D_n(k)$ under the action of a subgroup G of $GL_n(k)$. The authors proved in some cases that the skew field of invariants is again a Weyl skew field $D_m(K)$ where K is a purely transcendental extension of k.

A. Beilinson presents a "spectral decomposition" of the category of Heisenberg modules (i.e., modules over a Heisenberg extension of a commutative Lie algebra of formal loops on a torus T). The "spectral parameters" form the moduli stack of $T^\vee$-local systems on Spec $k((t))$.

A. Braverman and P. Etingof study a generating function defined by certain equivariant integrals along a moduli space of framed G-bundles on P^2. They prove the conjecture of Nikita Nekrasov claiming that the leading term of asymptotics of this generating function is given by the instanton part of the Seiberg–Witten prepotential of the affine Toda system associated to the Langlands dual Lie algebra.

The paper of I. Cherednik is devoted to the study of polynomial representations of double affine Hecke algebras. It is proved that the quotient of such a representation by the radical of the duality pairing is irreducible if it is finite dimensional.

D. Gaitsgory and D. Kazhdan study representations of groups over two-dimensional local fields. Here it makes sense to consider representations in pro-vector spaces and work with the central extension of the original group. The authors construct the functor of "semiinfinite invariants" which pairs representations corresponding to two levels that sum up to the critical level.

A. Joseph studies a 20-year-old conjecture claiming that the tensor product of a Demazure module with a one-dimensional Demazure module admits a Demazure flag. This conjecture had already been proved for finite type. The present paper proves the result for the quantized universal enveloping algebra of a Kac–Moody Lie algebra with a simply-laced symmetrizable Cartan matrix in any characteristic.

M. Kashiwara, P. Schapira with their coauthors F. Ivorra and I. Waschkies construct a microlocalization functor

$$\mu_X : D^b(I(K_X)) \to D^b(I(K_{T^*X}))$$

from the derived category of Ind-sheaves of vector spaces on a C^∞-manifold X to the similar category constructed for the cotangent bundle of X. The classical microlocalization is expressed now as $\mu hom(A, B) = \mathrm{R}\mathcal{H}om(\mu_X(A), \mu_X(B))$.

D. Kazhdan and Y. Varshavsky study the endoscopic decomposition for supercuspidal level zero representations of a reductive group over a local nonarchimedean field.

A. Kirillov and L. Rybnikov introduce and study "odd analogues" of a special family of algebras ("family algebras") defined earlier by A. Kirillov.

B. Kostant and W. Wallach study a Poisson analog $J(n)$ of the Gelfand–Zetlin algebra. This is a maximal Poisson-commutative subalgebra of the algebra of polynomial functions on $\mathfrak{g} = M_n(\mathbb{C})$.

In the paper of S. Kumar and K.-H. Neeb the authors study a connection between the cohomology of an algebraic group with that of its Lie algebra. They prove an analog of the Van Est theorem and also study extensions of an algebraic group by an abelian algebraic group.

T. Levasseur and T. Stafford study the ring $D(X)$ of global differential operators on the "basic affine space" $X = G/U$ where G is a complex semisimple Lie group and U is a maximal unipotent subgroup. They prove that the cohomology $H(X, O_X)$ considered as a $D(X)$-module decomposes into a sum of non-isomorphic simple $D(X)$-modules indexed by the elements of the Weyl group.

G. Lusztig proves a remarkable identity in the Hecke algebra of type A generalizing an identity of Wallach in the group ring of the symmetric group.

L. Makar-Limanov studies centralizers of elements in a quantum space which is the $\mathbb{C}$-algebra generated by $x_1, \ldots, x_n$ subject to relations $x_i x_j = q_{ij} x_j x_i$ for $i < j$. In case the coefficients q_{ij} are "in general position," the centralizer of any non-constant element is a subalgebra of a polynomial ring in one variable. An example shows that the centralizer need not be integrally closed and that there is no upper bound on the number of generators of the centralizer.

M. Nazarov and A. Sergeev present a centralizer construction for the Yangian of the queer Lie superalgebra $\mathfrak{q}(N)$.

In his linguistically refreshing paper V. Schechtman presents a new proof of the theorem claiming that the vertex algebroid structures on a Lie algebroid T form a gerbe whose class coincides with the Chern–de Rham class of T. This had been proved earlier in a recent *Inventiones* paper by Gorbunov, Malikov and Schechtman.

Acknowledgments: The conference was supported by the Arthur and Rochelle Belfer Institute of Mathematics and Computer Science, the Maurice and Gabriella Goldshleger Conference Foundation, the Albert Einstein Minerva Center for Theoretical Physics at the Weizmann Institute of Science. Further support came from the TMR programme "Algebraic Lie Representations" of the European Union, and Minerva foundation, Germany.

The Editors would like to thank W. McGovern, D. Farkas and G. Letzter for their overview papers as well as all the authors of the research papers. The Editors further extend their thanks to Ann Kostant of Birkhäuser for her personal involvement in the project. Special thanks are due to Raanan Michael, the administrator of the Faculty of Mathematics and Computer Science of the Weizmann Institute, and his secretaries Michele Bensimon and Meira Hadar for being the most skillful, devoted and efficient team in all the organizational matters of the conference.

J. Bernstein
V. Hinich
A. Melnikov
Editors

Rehovot, June 2005

Publications of Anthony Joseph

1. (With C.A. Coulson), Spheroidal wave functions for the hydrogen atom, *Proc. Phys. Soc.*, **90** (1967), 887–893.
2. (With C.A. Coulson), A constant of the motion for the two-centre Kepler problem, *Int. J. Quant. Chem.*, **1** (1967), 337–347.
3. The influence of the mass on the bound states of molecular systems, *Proc. Phys. Soc.*, **91** (1967), 574–576.
4. The theory of local degeneracy, *Int. J. Quant. Chem.*, **1** (1967), 535–559.
5. On the determination of the exact number of bound states of a given potential, *Int. J. Quant. Chem.*, **1** (1967), 615–629.
6. The spin dipole, *Proc. Phys. Soc.*, **92** (1967), 847–852.
7. Self-adjoint ladder operators I, *Rev. Mod. Phys.*, **99** (1967), 829–837.
8. (With C.A. Coulson), Self-adjoint operators II, *Rev. Mod. Phys.*, **39** (1967), 838–849.
9. Self-adjoint ladder operators III, *Rev. Mod. Phys.*, **40** (1968), 845–871.
10. The theory of conditional invariance, *Proc. Roy. Soc.*, **A305** (1968), 405–427.
11. The hemispherical box: An example of virtual symmetry, *J. of Phys.*, **A2** (1969), 719.
12. (With A.I. Solomon), Global and infinitesimal non-linear chiral transformations, *J. Math. Phys.*, **11** (1970), 748.
13. Derivations of Lie brackets and canonical quantisation, *Commun. Math. Phys.*, **17** (1970), 210–232.
14. A second anticommutant theorem for symmetric ternary algebras, *Proc. Camb. Phil. Soc.*, **69** (1971), 25–52.
15. (With M.A. Jacobs, S. Nussinov and A.A. Rangwala), An improved upper bound on the imaginary part of elastic scattering, *Phys. Rev.*, **D2** (1970), 1970–1974.
16. A classification of local current algebras, *Commun. Math. Phys.*, **19** (1970), 106–118.
17. (With W.G. Sullivan), A commutant property of symmetric ternary algebras, *Proc. Camb. Phil. Soc.*, **70** (1971), 1–4.
18. Commuting polynomials in quantum canonical operators and realizations of Lie algebras, *J. Math. Phys.*, **13** (1972), 351–357.

19. Magnetic spin monopole from the Yang-Mills field, *Phys. Rev.*, **D5** (1972), 313–320.
20. Realizations of Lie algebras from the canonical anticommutation relations, *Nuovo Cimento*, **8A** (1972), 217–234.
21. On the dynamical solution to the Sugawara model, *Nucl. Phys.*, **B42** (1972), 418–436.
22. On the dynamical solution to the Sugawara model: II The Lie identities, *Ann. of Phys.*, **73** (1972), 417–438.
23. Recovery of the Lie identities of field theory from minimal conditions, *Nucl. Phys.*, **B43** (1972), 107–118.
24. An inequality relating cross sections to polarization from isospin conservation, *Phys. Letters*, **42B** (1972), 368.
25. Gelfand-Kirillov dimension for algebras associated with the Weyl algebra, *Ann. de l'Institut Henri Poincaré*, **17** (1972), 325–336.
26. Combinatorial analysis of the Lie identities in field theory, in *Statistical Mechanics and Field Theory*. Lectures given at the 1971 Haifa Summer School (Eds. R.N. Sen and C. Weil, Keter Publishing House, Jerusalem).
27. The Weyl algebra – semisimple and nilpotent elements, *Am. J. Math.*, **97** (1975), 597–615.
28. A characterization theorem for realizations of $s\ell(2)$, *Proc. Camb. Phil. Soc.*, **75** (1974), 119–131.
29. Proof of the Gelfand-Kirillov conjecture for solvable Lie algebras, *Proc. Amer. Math. Soc.*, **45** (1974), 1–10.
30. Symplectic structure in the enveloping algebra of a Lie algebra, *Bull. Soc. Math. France*, **102** (1974), 75–83.
31. Minimal realizations and spectrum generating algebras, *Commun. Math. Phys.*, **36** (1974), 325–338.
32. Realizations in classical and quantum mechanics, *Proc. of the Third Intern. Colloq. on Group Theoretical Methods in Physics* (Eds. H. Bacry and A. Grossman), 1974.
33. Infinite dimensional Lie algebras in mathematics and physics, *Proc. of the Third Intern. Colloq. on Group Theoretical Methods in Physics* (Eds. H. Bacry and A. Grossman), 1974.
34. A generalization of the Gelfand-Kirillov conjecture, *Amer. J. Math.*, **99** (1977), 1151–1165.
35. The algebraic method in representation theory, in *Group Theoretical Methods in Physics* (Ed. A. Janner), LN 50 (Physics), 95–109, Springer-Verlag, New York, 1976.
36. Second commutant theorems in enveloping algebras, *Amer. J. Math.*, **99** (1977), 1167–1192.
37. The minimal orbit in a simple Lie algebra and its associated maximal ideal, *Ann. Sci. Ec. Norm. Sup.*, **9** (1976), 1–30.
38. A preparation theorem for the prime spectrum of a semisimple Lie algebra, *J. of Algebra*, **48** (1977), 241–289.

39. A characteristic variety for the prime spectrum of a semisimple Lie algebra, in Non-commutative Harmonic Analysis (Ed. J. Carmona), LN 587, 102–118, Springer-Verlag, New York, 1977.
40. Sur les vecteurs de plus haut poids dans l'algèbre enveloppante d'une algèbre de Lie semisimple complexe, *Comptes Rendus A* **281**, (1975), 835–837.
41. Sur la classification des idéaux primitifs dans l'algèbre enveloppante d'une algèbre de Lie réductive, *Comptes Rendus A* **284** (1977), 425–427.
42. On the annihilators of the simple subquotients of the principal series, *Ann. Ec. Norm. Sup.*, **10** (1977), 419–440.
43. A wild automorphism of $Us\ell(2)$, *Math. Proc. Camb. Phil. Soc.*, **80** (1976), 61–64.
44. A generalization of Quillen's lemma and its application to the Weyl algebras, *Israel J. Math.*, **28** (1977), 177–192.
45. Gelfand-Kirillov dimension for the annihilators of simple quotients of Verma modules, *J. Lond. Math. Soc.*, **18** (1978), 50–60.
46. (With L.W. Small), An additivity principle for Goldie rank, *Israel J. Math.*, **31** (1978), 105–114.
47. Sur la classification des idéaux primitifs dans l'algèbre enveloppante de $s\ell(n+1,\mathbb{C})$, *Comptes Rendus Ser.*, A-287 (1978), 303–306.
48. W module structure in the primitive spectrum of the enveloping algebra of a semisimple Lie algebra, in Lecture Notes in Mathematics, **728**, Springer-Verlag, (1979), 116–135.
49. Dixmier's problem for Verma and principal series submodules, *J. Lond. Math. Soc.*, **20** (1979), 193–204.
50. Towards the Jantzen conjecture, *Compositio Math.*, **40** (1980), 35–67.
51. Towards the Jantzen conjecture II, *Compositio Math.*, **40** (1980), 69–78.
52. Towards the Jantzen conjecture III, *Compositio Math.*, **41** (1981), 23–30.
53. Goldie rank in the enveloping algebra of a semisimple Lie algebra I, *J. Algebra*, **65** (1980), 269–283.
54. Goldie rank in the enveloping algebra of a semisimple Lie algebra II, *J. Algebra*, **65** (1980), 284–306.
55. Kostant's problem, Goldie rank and the Gelfand-Kirillov conjecture, *Invent. Math.*, **56** (1980), 191–213.
56. On the Gelfand-Kirillov conjecture for induced ideals in the semisimple case, *Bull. Math. Soc. France*, **107** (1979), 139–159.
57. (With O. Gabber), On the Bernstein-Gelfand-Gelfand resolution and the Duflo sum formula, *Compos. Math.*, **43** (1981), 107–131.
58. (With O. Gabber), towards the Kazhdan-Lusztig conjecture, *Ann. Ec. Norm. Sup.*, **14** (1981), 261–302.
59. Goldie rank in the enveloping algebra of a semisimple Lie algebra III, *J. Algebra*, **73** (1981), 295–326.
60. The Enright functor in the Bernstein-Gelfand-Gelfand O category, *Invent. Math.*, **67** (1982), 423–445.
61. Completion functors in the O category, in Lecture Notes, **1020**, Springer-Verlag, 1983, 80–106.

62. On the classification of primitive ideals in the enveloping algebra of a semisimple Lie algebra, in Lecture Notes, **1024**, Springer-Verlag, 1983, 30–76.
63. Primitive ideals in enveloping algebras, ICM Proceedings, PWM – North Holland, Warsaw 1984, 403–414.
64. On the variety of a highest weight module, *J. Algebra*, **88** (1984), 238–278.
65. On the associated variety of a primitive ideal, *J. Algebra*, **93** (1985), 509–523.
66. (With J.T. Stafford), Modules of k-finite vectors over semisimple Lie algebras, *Proc. Lond. Math. Soc.*, **49** (1984), 361–384.
67. On the Demazure character formula, *Ann. Ec. Norm. Sup.*, **18** (1985), 389–419.
68. Three topics in enveloping algebras, Proceedings of Durham Symposium, 1983, (unpublished).
69. On the Demazure character formula II, *Compos. Math.*, **58** (1986), 259–278.
70. Kostant's problem and Goldie rank, in: Lecture Notes in Mathematics **880**, Springer-Verlag, Berlin, 1981, 249–266.
71. On the cyclicity of vectors associated with Duflo involutions, *Lecture Notes in Mathematics*, **1243**, 144–188, Springer-Verlag, Berlin, 1987.
72. A criterion for an ideal to be induced, *J. Algebra*, **110** (1987), 480–497.
73. On the multiplicities of the adjoint representation in simple quotients of an enveloping algebra of a semisimple Lie algebra, *Trans. Amer. Math. Soc.*, **316** (1989), 447–491.
74. Rings which are modules in the Bernstein-Gelfand-Gelfand O category, *J. Algebra*, **113** (1988), 110–126.
75. A sum rule for scale factors in the Goldie rank polynomials, *J. Algebra*, **118** (1988), 276–311.
76. A sum rule for scale factors in the Goldie rank polynomials, addendum, *J. Algebra*, **118** (1988), 312–321.
77. A surjectivity theorem for rigid highest weight modules, *Invent. Math.*, **92** (1988), 567–596.
78. The primitive spectrum of an enveloping algebra, *Astérisque*, **173–174** (1989), 13–53.
79. Characters for unipotent representations, *J. Algebra*, **130** (1990), 273–295.
80. On the characteristic polynomials for orbital varieties, *Ann. Ec. Norm. Sup.*, **22** (1989), 569–603.
81. Rings of $\mathfrak{b}$-finite endomorphisms of simple highest weight modules are Goldie, IMCP, Ring theory, 1989, The Weizmann Science Press, Jerusalem 1990, 124–134.
82. The surjectivity theorem, characteristic polynomials and induced ideals, in *The Orbit Method in Representation Theory*, Eds. M. Duflo, N.V. Pedersen, M. Vergne, Birkhäuser, Boston, 1990, 85–98.
83. (With T.J. Enright). An intrinsic analysis of unitarizable highest weight modules, *Math. Ann.*, **288** (1990), 571–594.
84. Annihilators and associated varieties of unitary highest weight modules, *Ann Ec. Norm. Sup.*, **25** (1992), 1–45.

85. Some ring theoretic techniques and open problems in enveloping algebras, Proceedings of Berkeley Conference, July 1989, Ed. S. Montgomery and L. Small, Springer-Verlag, Berlin 1992, p. 27–67.
86. (With G. Letzter). Local finiteness of the adjoint action for quantized enveloping algebras. *J. Algebra*, **153** (1992), 289–318.
87. (With G. Letzter). Separation of variables for quantized enveloping algebras. *Amer. J. of Math.*, **116** (1994), 125–177.
88. (With G. Perets and P. Polo). Sur l'équivalence de catégories de Beilinson et Bernstein, *Comptes Rendus*, Paris, **313** (1991), 705–709.
89. (With G. Letzter). Verma module annihilators for quantized enveloping algebras, *Ann. Ec. Norm. Sup.*, **28** (1995), 493–526.
90. Faithfully flat embeddings for minimal primitive quotients of quantized enveloping algebras, *Israel Math. Conf. Proc.*, **7** (1993), 79–106.
91. Idéaux premiers et primitifs de l'algèbre des fonctions sur un groupe quantique, *Comptes Rendus*, **316** (1993), 1139–1142.
92. On the prime and primitive spectra of the algebra of functions on a quantum group, *J. Algebra*, **169** (1994), 441–511.
93. (With G. Letzter). Rosso's form and quantized Kac-Moody algebras, *Math. Zeit.*, **222** (1996), 543–571.
94. *Enveloping Algebras: Problems Old and New*, in Progress in Mathematics **123**, Birkhäuser, Boston, 1994, 385–413.
95. Some Remarks on the R-matrix, *J. Algebra*, **180** (1996), 412–430.
96. Orbital varieties, Goldie rank polynomials and unitary highest weight modules, In B. Orsted and H. Schlichtkrull (eds.), *Algebraic and Analytic Methods in Representation Theory*, Academic Press, London 1997, 53–98.
97. Sur une conjecture de Feigin, *Comptes Rendus*, Paris, **320** (1995), 1441–1444.
98. Sur les idéaux génériques sur l'algèbre des fonctions sur un groupe quantique, *Comptes Rendus*, Paris, **321** (1995), 135–140.
99. Preservation of Coxeter structure under finite group action, Acts du Colloque, Reims 1995, Colloques et Seminaires du SMF, 1997, 185–219.
100. (With A. Braverman). The minimal realization from deformation theory, *J. Algebra*, **205** (1998), 13–36.
101. The orbital varieties of the minimal orbit, *Ann. Ec. Norm. Sup., Serie 4*, **31** (1998), 17–45.
102. On a Harish-Chandra homomorphism, *CRAS*, **324(I)** (1997), 759–764.
103. Sur l'annulateur d'un module de Verma, In B. Broer (ed.), *Representation Theories and Algebraic geometry*, Kluwer Academic Publishers, NATO Scientific Affairs, 1998, 237–300.
104. On the mock Peter-Weyl theorem and the Drinfeld double of a double, *J. reine angew. Math.*, **507** (1999), 37–56.
105. The admissibility of bounded modules for an affine Lie algebra, *Alg. Rept. Theory*, **3** (2000), 131–149.
106. A completion of the quantized enveloping algebra of a Kac-Moody algebra, *J. Algebra*, **214** (1999), 235–275.

107. On the Kostant-Parthasarathy-Ranga Rao-Varadarajan determinants, I. Injectivity and multiplicitie, *J. Algebra*, **241** (2001), 27–45.
108. (With G. Letzter). On the KPRV determinants, II. Construction of the KPRV determinants, *J. Algebra*, **241** (2001), 46–66.
109. (With G. Letzter and D. Todoric). On the KPRV determinants, III. Computation of the KPRV determinants, *J. Algebra*, **241** (2001), 67–88.
110. (With G. Letzter and S. Zelikson). On the Brylinski-Kostant filtration, *JAMS*, **14** (2000), 945–970.
111. (With F. Fauquant-Millet). Semi-invariants de l'algèbre enveloppante quantifiée d'une algèbre parabolique, *Transf. Groups.*, **6** (2001), 125–142.
112. On an affine KPRV determinant at $q = 1$, *Bull. Math. Sci.*, **125** (2001), 23–48.
113. (With J. Greenstein). A Chevalley-Kostant presentation of basic modules for $\mathfrak{sl}(2)^\wedge$ and the associated affine KPRV determinants at $q = 1$, *Bull. Math. Soc.*, **125** (2001), 85–108.
114. (With W. Borho). Sheets and topology of primitive spectra for semisimple Lie algebras, *J. Algebra*, **244** (2001), 76–167.
115. (With D. Todoric). On the quantum KPRV determinants for semisimple and affine Lie algebras, *Alg. Rept. Theory*, **5** (2002), 57–99.
116. (With G. Letzter). Evaluation of the quantum affine PRV determinant, *Math. Res. Lett.*, **2** (2002), no. 2–3, 307–322.
117. (With A. Melnikov). Quantization of Orbital Varieties in $\mathfrak{sl}_n$, Progr. in Math. Birkhäuser, **213** (2003), 165–196.
118. A Decomposition Theorem for Demazure Crystals, *J. Algebra*, **265** (2003), 562–578.
119. On the graded injectivity of the Conze embedding, *J. Algebra*, **265** (2003), 358–378.
120. (With W. Borho). Corrigendum to Sheets and topology of primitive spectra for semisimple Lie algebras, *J. Algebra*, **259** (2003), 310–311.

Communicated Preprints

121. (With V. Hinich). Orbital variety closures and the convolution product in Borel-Moore homology, to appear in *Selecta Mathematica, New Series.*
122. Modules with a Demazure flag, this volume.
123. (With F. Fauquant-Millet), Semi-centre de l'algèbre enveloppante d'une sous-algèbre parabolique d'une algèbre de Lie semi-simple, *Ann. Ec. Norm. Sup.*, **38**(2), (2005), 155–191.
124. The enigma of the missing invariants on the nilradical of a Borel, *Bull. Sci. Math.*, **128** (2004), 433–446.
125. Results and problems in enveloping algebras arising from quantum groups, to appear in volume dedicated to 70th birthday of A. N. Vershik.
126. Goldie rank ratios and quaternionic extensions, *J. Algebra* (to appear).

Book

Quantum Groups and Their Primitive Ideals, Springer-Verlag, Berlin 1995.

Edited Volumes

1. (With A. Connes, M. Duflo and R. Rentschler). *Operator Algebras, Unitary Representations, Enveloping Algebras and Invariant Theory*, Actes du colloque en l'honneur de Jacques Dixmier, Progr. in Math., Vol. 92 Birkhäuser, Boston 1990.
2. (With S. Shnider). *Quantum deformations of algebras and their representations*, Israel Mathematical Conference Proceedings Vol. 7, Bar-Ilan University, 1993.
3. (With F. Mignot, F. Murat, B. Prum and R. Rentschler). *First European Congress of Mathematics, Vol. I*, Progr. in Math. Vol. 119, Birkhäuser, Boston 1994.
4. (With F. Mignot, F. Murat, B. Prum and R. Rentschler). *First European Congress of Mathematics, Vol. II*, Progr. in Math. Vol. 120, Birkhäuser, Boston 1994.
5. (With F. Mignot, F. Murat, B. Prum and R. Rentschler). *First European Congress of Mathematics, Vol. III*, Progr. in Math. Vol. 121, Birkhäuser, Boston, 1994.
6. (With A. Melnikov and R. Rentschler). *Studies in Memory of Issai Schur*, Progr. in Math. Progr. in Math. Vol. 210, Birkhäuser, Boston, 2002.

Students of Anthony Joseph

S. Bamba
M. Hervé
P. Polo
G. Perets
E. Benlolo
V. Hinich
A. Melnikov
A. Zahid
A. El Alaoui
F. Fauquant-Millet
D. Todoric
M. Kébé
Sh. Zelikson
M. Gorelik
E. Lanzmann
J. Greenstein
Yu. Bazlov
E. Perelman

List of Summer Students

Anna Klugman (Melnikov), Technion
Gail Letzter, University of Chicago
Eti Gvirtz, Bar-Ilan University
Elise Benlolo, University of Haifa
Markus Stricker, ETH
Gadi Perets, Université Pierre et Marie Curie
Avital Frumkin, Hebrew University
Alexander Braverman, Tel Aviv University
Denis Gaitsgory, Tel Aviv University
Maria Gorelik, Moscow State University

Maxim Leenson, Moscow Independent University
Elena Perelman, St. Petersburg State University
Yuri Bazlov, St. Petersburg State University
Judita Preiss, Cambridge University
Tatiana Assing, Université Pierre et Marie Curie
Vera Zolataskai, Moscow State University
Shira Zerbiv, Technion
Zelijka Ljujic, University of Belgrade
Alexander Yom Dim, University of Haifa
Irina Bobkova, St. Petersburg State University

From Denise Joseph

First, I wish to thank all the guests who joined us tonight to celebrate Tony's 60th birthday and especially the scientists who came from abroad. They don't know how much we appreciate their coming to Israel in this difficult time and we thank them heartily for their courage, support and wish to separate science from politics. I also wish to thank Anna who worked so hard to make this conference a success, and all who also gave so much of their time to organize so many things.

I asked Tony what I should speak about and he said I should tell how difficult it is to be the wife of a mathematician, I would then have the sympathy of all the wives and husbands and persons living with mathematicians. But I shall not follow his advice.

Last week we went to the BA graduation ceremony of our son in the Faculty of Agriculture which is situated opposite the Weizmann Institute. On a big screen it was written, "60th birthday". Then we were shown a film of when the Faculty was inaugurated in 1942 with just one building and one department of Agriculture, and we could see how it developed so beautifully with a big campus and 14 departments. Tony said, "you imagine it is as old as me". I cannot speak of Tony's mathematical achievements although he has tried many times to explain to me on what subject he was working.

I know that Tony is not only hard on himself but also on his children and I am sure on his students. I was nicely surprised to find out when I met some of his students not only how highly they regard him but that they also like him.

Tony loves sport, tennis, skiing and windsurfing. I always admired his determination to be a good windsurfer although in the beginning he used to fall all the time. One day he said, "I am going to learn to play the flute". I was astonished since Tony had never played any musical instrument. I asked, "are you going to take lessons?" He replied, "of course not". He bought some books and after a week or two I was amazed to hear nice musical rhythms of known songs.

Tony likes to build and repair things in his spare time. I still remember when we moved some twenty years ago to our house he built a wooden table for the garden. The neighbour was so surprised; she said to me, "I thought people who use their brain are not good with their hands".

Let us raise our glasses and drink to the health happiness and long life — to Tony.

From Jacques Dixmier: A Recollection of Tony Joseph

When I saw Tony for the first time, he had published a number of papers about physics and even chemistry. He wanted to discuss my 1968 paper concerning the Weyl algebra A_1. He explained to me why physical arguments permitted a deeper understanding. But at some other points, he said "this is a typical mathematical idea". So maybe I played a little role in Tony's transition towards mathematics. If so, I think the mathematical world is greatly indebted to me!

After that, Tony often wrote to me (when he wasn't in Paris). In spite of my bad organization, I rediscovered the following old letter. Thank you, Tony, for your wonderful work, and my warmest encouragements for the future.

7/7/72.

Dear Professor Dixmier,

I must first thank you for the manuscripts you
have sent me. I have been working on some of the problems you
posed in your first paper on the Weyl algebra, in particular that
on ~~[illegible]~~ the question of the existence of endomorphisms which are not
automorphisms. Your lemma 7.3 is central to this question. By
slightly extending its validity and ~~[illegible]~~ through its systematic application,
I have been able to show that if Q, P are of degree less than
147, then they generate A1. Hence such an ~~every~~ endomorphism is an
automorphism. For polynomials of higher degree, I am able to partially
~~[illegible]~~ construct Q, P; but unfortunately I have been as yet unable to
exhibit an endomorphism. As far as I can see it is not possible to
push your leading term analysis further.

In another direction I have studied the question of the existence
of polynomial relations between commuting elements of the quotient field and
have obtained some partial results. There is a quite simple and explicit
method ~~[illegible]~~ computing the common divisor of two elements based on the fact

that the elements of the zero eigenspace ~~of the~~ ad $q_i p_i$: $i = 1, 2 .. n$, commute.
This is quite valuable in situations where one wishes to control the
degree of the denominator (or numerator).

I hope to spend a few days in Paris at IHES
with Professor Michel, ~~In that case~~ during the last week of
August. In that case I hope that I might be able to
drop in on you.

Yours sincerely,

Tony Joseph.

Part I

Survey and Review

The work of Anthony Joseph in classical representation theory

W. McGovern

Department of Mathematics
University of Washington
Seattle, WA 98195
USA

Over some three decades Joseph's groundbreaking work in classical representation theory has changed the face of the subject. Joseph has not only introduced many beautiful ideas of his own but also has shown a gift for reinterpreting the work of others in an entirely new light, establishing connections between quite disparate aspects of the subject. What follows is a survey of this part of Joseph's work, beginning with his papers on primitive ideals in the enveloping algebra of a semisimple Lie algebra. Such ideals were the primary focus of Joseph's work from the mid-1970s until the mid-1990s and he is responsible for the lion's share of the advances in their theory during that time.

Most of Dixmier's classic 1974 text on enveloping algebras of Lie algebras $\mathfrak{g}$ is devoted to the case where $\mathfrak{g}$ is solvable. What really put primitive ideals in enveloping algebras of semisimple Lie algebras on the map was Duflo's fundamental theorem that any such ideal is the annihilator of a very special kind of simple module, namely a highest weight module. Thus one can study such ideals via abstract ring theory or highest weight theory. Joseph used both of these tools in his first papers on primitive ideals. In [39, 41] he introduced the notion of the characteristic variety of a primitive ideal, using the idea behind the Harish-Chandra homomorphism, and used it to detect many inclusions and equalities among annihilators of simple highest weight modules.

In [42] he identified the annihilators of simple Harish-Chandra bimodules for complex groups as annihilators of explicit highest weight modules. In [38] he used localization techniques from noncommutative ring theory to decompose the primitive spectrum of an enveloping algebra as the union of spectra of more manageable rings; along the way he exhibited many subalgebras of a semisimple Lie algebra whose enveloping algebras have centers that are polynomial rings (see also [40]). He used these results in an unpublished preprint of 1976 to classify the primitive spectrum in types A_2 and B_2 before Duflo had proved his theorem. Later he used his work on characteristic varieties together with Duflo's theorem to classify primitive ideals in type A, via the well-known Robinson–Schensted bijection between permutations of n letters and pairs of standard Young n-tableaux of the same shape [50, 51, 47]. More recently

Barbasch–Vogan and Garfinkle have extended this classification to the other classical types, working with standard domino tableaux rather than standard Young tableaux.

The connection between primitive ideals and highest weight modules was further strengthened by the discovery of an equivalence between a large category of bimodules over an enveloping algebra, including all of its primitive quotients and the Bernstein–Gelfand–Gelfand category $\mathcal{O}$, which includes all simple highest weight modules. This discovery is due independently to Joseph, Bernstein–Gelfand, and Enright. Both Joseph and Bernstein–Gelfand used it to give an algebraic proof of Duflo's theorem ([49]; see also *Comp. Math.* **41** (1980), 245–285). Joseph started out by establishing a bijection that had been conjectured by Dixmier between ideals containing a fixed minimal primitive ideal and submodules of a Verma module. (Enright approached the category equivalence quite differently from Bernstein–Gelfand and Joseph: he introduced functors which essentially built up the bimodule structure in category $\mathcal{O}$ from scratch. Later Joseph studied these functors in [60, 61], establishing the connection between them and the techniques of [49].)

At about this time Joseph also refined his earlier work on inclusions and equalities among primitive ideals, showing (modulo a conjecture later proved by Vogan) that the set of primitive ideals of a fixed regular integral central character spans a vector space carrying a natural action of the Weyl group [48]; it is determined by the multiplicities of composition factors in Verma modules. This paper was one of the inspirations for Kazhdan and Lusztig's fundamental paper in *Inv. Math.* **53**, in which they formulated their famous conjecture for computing these multiplicities. In turn Gabber and Joseph related this conjecture to the hereditarity of the Jantzen filtration on Verma modules in [58]. This filtration was used to study composition factors of primitive quotients in [94]. Gabber and Joseph also studied natural questions arising from the category equivalence in [57].

Joseph first combined the tools of abstract ring theory and highest weight theory in [45], where he related the Gelfand–Kirillov dimension of a simple highest weight module L to that of U/I, I the annihilator of L in the enveloping algebra U. He did so again in an especially beautiful way in the papers [53, 54], which studied Goldie ranks of primitive quotients. He showed that the Goldie ranks of primitive quotients obtained from a fixed one P by translation functors are given by a polynomial function which moreover completely determines P once the central character of this quotient is given. In addition this polynomial is determined up to a multiplicative scalar by the composition factor multiplicities in Verma modules, which are given by the Kazhdan–Lusztig polynomials evaluated at $q = 1$. Goldie rank polynomials attached to the set of all primitive ideals with a fixed regular integral central character λ span a vector space which carries a natural action of the Weyl group. As a consequence the number of primitive ideals of central character λ equals the sum of the dimensions of certain Weyl group representations. Barbasch and Vogan later identified just which Weyl group representations occur in this way in *Math. Ann.* **259** (1982) and *J. Alg.* **80** (1983). In type A they all do, as was already clear from Joseph's earlier work; in general, exactly the special ones in Lusztig's sense do. Thus the picture of the primitive spectrum first introduced in [48] was brought to a very satisfactory culmination. In the course of his work on Goldie ranks Joseph also showed that they satisfy an additivity property which he generalized to a wider setting in joint work with Small [46].

It remained (and in fact still remains) to compute the multiplicative scalars in the Goldie rank polynomials. Joseph launched a program for doing this in [59] which was continued in [71, 75, 76]. In these papers an important topic is the relationship between a typical primitive quotient U/I and the Kostant ring of ad-finite maps from L to itself, where L is a simple highest weight module with annihilator I; Joseph had already studied this ring in [55, 70] and proved his Goldie rank theorems in a special case. In [59, 71, 75, 76] he studies numerical invariants attached to a Kostant ring and the quotient U/I that it contains and shows that their ratio is a positive integer. He goes on to establish beautiful relations among these ratios. They (and the multiplicative scalars) can be computed exactly if one can construct sufficiently many primitive quotients with known Goldie rank. It turns out that Goldie ranks are the easiest to compute (and the most interesting) when they are 1, or equivalently when the primitive quotient is completely prime.

Motivated by his earlier work in theoretical physics, Joseph had already introduced an important completely prime primitive quotient attached to the minimal nilpotent orbit in [37] (which he revisited with A. Braverman in [100], using deformation theory of Koszul algebras); now he proved a positivity property for Goldie rank polynomials which shows that completely prime primitive quotients are quite rare [80, 82]. This property was refined and extended in [114], where he and Borho extended Dixmier's old notion of a sheet of adjoint orbits to primitive ideals and showed that that there are only finitely many sheets of primitive ideals with Goldie rank bounded above by a fixed positive integer. This reviewer used Joseph's work extensively in *AMS Memoir* **519**, where he computed just which quotients by maximal ideals are completely prime in the classical cases. Joseph also computed the Goldie ranks of certain quotients by induced primitive ideals in [56], verifying for such quotients the Gelfand–Kirillov conjecture that their Goldie fields are always Weyl skew fields. He verified the same conjecture for type A in [55].

By 1982 the classification of the primitive spectrum was essentially complete, but still rather mysterious. The appearance of special Weyl group representations can be most satisfactorily explained by identifying them with the representations attached by Springer to certain nilpotent orbits (also called special). In turn the closures of these orbits arise as the associated varieties of primitive ideals. Borho and Brylinski had proved that such associated varieties are irreducible (and thus closures of single nilpotent orbits) in many special cases; Joseph proved it in general in [65], having earlier verified it for type A in [52].

He also gave a more direct construction of the Springer representation π attached to a nilpotent orbit $\mathcal{O}$ in [64], as follows. Intersect $\bar{\mathcal{O}}$ with the nilradical $\mathfrak{n}$ of a fixed Borel subalgebra of the ambient semisimple Lie algebra. One obtains an algebraic variety $\mathcal{V}$ with exactly $\dim \pi$ irreducible components, as was known from early work of Springer and Spaltenstein. Joseph looked at the leading terms of Hilbert–Samuel polynomials of the coordinate rings of these components as a function of the grading coming from the action of a Cartan subgroup. He showed that their numerators may be regarded as polynomials in the symmetric algebra of the dual of the Cartan subalgebra and proved that the span of these polynomials attached to the components of $\mathcal{V}$ is stable under the Weyl group W. He then conjectured that the resulting representation of W is isomorphic to the Springer representation π. Hotta proved this in 1982 by a fairly complicated computation; following work of Rossmann, Joseph later gave a more direct proof by realizing the polynomials in question as integrals which represent classes in a homology ring whose structure is well known [80]. The components of $\mathcal{V}$ are called orbital varieties; they form the basic building blocks of the associated variety of a highest weight module. Joseph also used earlier work of Steinberg on his famous triple variety to deduce in [64] that every orbital variety lying in $\mathfrak{n}$ arises by taking the intersection of $\mathfrak{n}$ with a translate of itself by some element of W, taking the saturation under the action of the Borel subgroup B, and finally passing to the closure. In later work with Hinich [121], Joseph showed that inclusions among orbital variety closures behave quite similarly to inclusions among primitive ideals, though there are also subtle differences between these two kinds of inclusions.

It turns out that simple highest weight modules, unlike primitive ideals, need not have irreducible associated varieties; Joseph gave the first counterexample in [64] and Tanisaki later gave other examples. Following the Kostant–Souriau program of geometric quantization, Joseph formulated a quantization subprogram of his own in [77], which essentially sought to invert the associated variety construction. More precisely, given an orbital variety V, Joseph asked first whether there is a simple highest weight module with associated variety V. If so, he called V weakly quantizable; the terminology comes from [101]. Whenever V is weakly quantizable, he asked more strongly whether the coordinate ring of V agrees as a module over a Cartan subalgebra with some highest weight module M up to a weight shift. If so, he called V strongly quantizable. Joseph's student Melnikov later showed that every orbital variety is weakly quantizable in type A (*C. R. Acad. Sci.* **316** (1993)).

Another student, Benlolo, exhibited two varieties in type A that are strongly quantizable, but only via nonsimple highest weight modules. Joseph himself showed that his quantization program fails in general by exhibiting orbital varieties of the minimal nilpotent orbit that are not even weakly quantizable [101]. Nevertheless both he and his students Benlolo, Melnikov, and Perelman have continued to investigate the quantizability of various orbital varieties and obtained many positive results (see e.g., [117]). In particular, orbital varieties lying in abelian nilradicals of maximal parabolic subalgebras all turn out to be strongly quantizable in an especially nice way, via unitary highest weight modules [84]. The unitary highest weight modules arising in this way are in some sense the most interesting ones; Enright and Joseph constructed all unitary highest weight modules starting from these in [83]. More generally, orbital varieties which are hypersurfaces in nilradicals of parabolic subalgebras have nice quantization properties.

In addition to his research papers on primitive ideals, Joseph has also written a number of expository papers on these and related topics: see [62, 63, 68, 78, 85, 94, 96]. He has also written on a variety of other topics in classical representation theory. His first papers in pure mathematics dealt with Weyl algebras, which he became interested in through his previous work in mathematical physics [25, 27, 44]. He also studied homomorphic images of subalgebras of the first Weyl algebra in [43]. He proved the

Gelfand–Kirillov conjecture mentioned above for solvable Lie algebras in [29], building on his work in [30], and generalized it in [34]. He studied double commutants in enveloping algebras in [36]. In [67, 69] he ventured into finite-dimensional representation theory, proving the Demazure character formula for $\mathfrak{b}$-submodules generated by extremal weight vectors in finite-dimensional irreducible $\mathfrak{g}$-modules whose highest weight is sufficiently far from the walls ($\mathfrak{b}$ a Borel subalgebra); later Andersen and Kumar proved this formula in general and Kumar and Mathieu deduced the PRV conjecture from it.

The Demazure operators, which appeared for the first time in Joseph's work in these papers, reappeared beautifully and unexpectedly in [101]. They appear yet again in a very recent paper on crystal bases [118], which established the existence of Demazure flags for certain tensor products; see also [122]. Joseph continued to study finite-dimensional modules in [110]; here he, Letzter, and Zelikson gave a proof valid for all integral weights that the jump polynomials attached to Brylinski's filtration of a finite-dimensional module can be computed from Lusztig's q-polynomials. In [66] Joseph and Stafford studied the homological properties of primitive quotients, showing that such quotients behave well at regular central character but can be quite pathological otherwise.

In [72] Joseph revisited the induced ideals he had studied in [56] from a very different point of view, studying the conditions under which they are generated by minimal primitive ideals together with one copy of the adjoint representation. (He was led to study this question by previous unpublished work of R. Brylinski on it.) Along the way he derived an interesting and surprisingly difficult formula for the multiplicity of the adjoint representation in any quotient of the enveloping algebra by a maximal ideal [73]. He studied modules in category $\mathcal{O}$ with a ring structure in [74], finding that such modules have a remarkably rigid structure. These modules arose again in [77], where he studied very general rings of differential operators and developed a criterion for such a ring to be a quotient of the enveloping algebra. This paper was motivated by work of Levasseur and Stafford and recovered their results as a special case. Such rings of differential operators are not in general Noetherian; Joseph showed that they are nevertheless Goldie in [81].

In [79] Joseph studied unipotent representations of complex groups, which had been defined (purely algebraically) by Barbasch and Vogan five years earlier. He gave a new proof of the Barbasch–Vogan character formulas for them, using some fundamental calculations of Lusztig. In [88] he, Perets, and Polo gave an elementary ring-theoretic proof of the Beilinson–Bernstein equivalence of categories between modules over an enveloping algebra and D-modules, using just the Hilbert Nullstellensatz and the translation principle. Their construction carries over to the quantum case [90], which Lunts and Rosenberg interpreted as a Beilinson–Bernstein equivalence of categories in a nonabelian framework.

Following Polo, Joseph studied Galois theory in the setting of an enveloping algebra in [99]. Given a finite group G of automorphisms of the quotient U/I of the enveloping algebra by a minimal primitive ideal of regular integral central character, he asked whether the ring $(U/I)^G$ of G-invariants could be another quotient U'/I' of the same type. He was able to show under these conditions that the Lie algebras corresponding to the enveloping algebras U, U' must have the same Coxeter diagram, but he could not quite show that their Dynkin diagrams are the same. In 1997 he returned to the Harish-Chandra homomorphism, which he had used in one of his first papers on primitive ideals ([39]). This time he worked with the extended Harish-Chandra homomorphism, defined on the space of invariant differential operators on a semisimple Lie algebra, and showed that it is graded surjective [102]. This gave an affirmative answer to a question of Wallach, a weaker version of which had been answered affirmatively by Levasseur and Stafford.

Joseph's most recent work in classical representation theory is largely motivated by his work in quantized enveloping algebras, described elsewhere in this volume. He gave a new proof of an old result of Duflo that the annihilator of a Verma module is generated by its intersection with the center of the enveloping algebra in [103]; his approach here generalizes to the super and quantum cases. He studied simple weight modules over an affine Kac–Moody algebra in [105], classifying all such modules whose weights have bounded norm and showing in particular that any such module has finite-dimensional weight spaces. In an important series of papers [107, 108, 109] he studied analogues of the classical PRV determinants (studied also by Kostant) for generalized Verma modules. He gave formulas for their characters in [107], worked with G. Letzter to interpret their zeros and give formulas for the degrees in which these zeros occur in [108], and finally gave an explicit formula for the determinants themselves in [109], working with Letzter and Todoric. Working with his student Greenstein, he modified an old construction of Kostant and Chevalley in the semisimple case to show that any integrable highest weight module for affine $\mathfrak{sl}_2$ admits a realization as a subalgebra of a suitable Clifford algebra [113]. This has no analog for $\mathfrak{g}$ semisimple and does not extend to any other affine Lie algebra.

Although Joseph's research interests have largely shifted to quantized enveloping algebras in the last few years, much of what he does with them is motivated by and has ramifications for classical representation theory (as indicated above). Representation theory has benefited immensely from his many contributions. I wish him the best in the years to come.

Quantized representation theory following Joseph

Daniel R. Farkas and Gail Letzter

Department of Mathematics
Virginia Tech
Blacksburg, VA 24061
USA

A quantum group . . . is at present a purely mythical being . . . A. Joseph

We are ignorant of the meaning of the dragon . . . but there is something in the dragon's image that fits man's imagination and this accounts for the dragon's appearance in different places . . . L. Borges

At the beginning of his lecture for the Microprogram on Noncommutative Rings held at the Mathematical Sciences Research Institute in July of 1989, Tony Joseph remarked that he had finally seen an application of quantum groups — he had passed a Volkswagen Quantum GL5 on his way to the talks. Not all successful marriages begin with love at first sight.

We quickly set notation. Assume that $\mathfrak{g}$ is a finite-dimensional semisimple Lie algebra of rank n. Fix a root system with simple roots $\alpha_1, \ldots, \alpha_n$. Let Q (and let P) denote its root lattice (resp. weight lattice). Assume that q is always an indeterminate. The quantized enveloping algebra $U_q(\mathfrak{g})$, *which we shall always shorten to* U, is the $\mathbf{C}(q)$-algebra generated by symbols x_i, y_i, and $t_i^{\pm 1}$ for $1 \leq i \leq n$, subject to the associative algebra and Hopf algebra relations found in [Jbook]. (Other practitioners use E_i instead of x_i, F_i instead of y_i, and K_i instead of t_i.) The group generated by all t_i is denoted T.

Write τ for the isomorphism that sends the additive group Q to the multiplicative group T via $\alpha_i \mapsto t_i$. Enlarge T to a group $\check{T}$ isomorphic to P by extending τ in the obvious way. The "simply connected" quantized enveloping algebra $\check{U}$ is generated by U and $\check{T}$. For many purposes, $\check{U}$ is a more suitable algebra than U and the scalar field should be replaced with its algebraic closure. We will be blithely sloppy in this review and use all such algebras interchangeably. The reader, concerned whether the dropped ornamentation of U is required for precise descriptions of Joseph's contributions, is welcome to suffer through the subtleties and complications on his or her own.

The moral dual to U is the quantized function algebra (or coordinate ring) $R_q[G]$ where G is the Lie group associated to $\mathfrak{g}$. Again, we abbreviate notation and simply write R. It is the subalgebra of the (finite) Hopf dual of U generated by the coordinate functions of the simple highest weight modules $L(\lambda)$ for $\lambda \in P^+$.

Precise descriptions for much of the material we cover here can be found in Joseph's monograph, *Quantum Groups and Their Primitive Ideals*. Ken Brown has written a featured review of the book for *Math Reviews*. We warmly recommend reading his essay; it provides background and insights that complement our own idiosyncratic retrospective.

1 Local Finiteness

Much of the classical representation theory for the enveloping algebra of a finite-dimensional semisimple Lie algebra begins with the observation that the algebra is a direct sum of finite-dimensional simple modules under the adjoint action. This proposition is indispensable in the analysis of primitive ideals. In his elegant lectures at MSRI in 1989 (which made the world of quantum groups accessible to ring theorists), Paul Smith pointed out the disappointing fact that the quantized enveloping algebra is not a locally finite module with respect to the quantum adjoint action ([Sm]). However, at this conference, Smith jotted down for Joseph an intriguing example of a finite-dimensional ad $U_q(\mathfrak{sl}\,2)$ module inside $U_q(\mathfrak{sl}\,2)$. (A careful description of this example appears in [Lg1].) This calculation became the starting point for the portrait of the "locally finite part" of the quantized enveloping algebra, painted in a series of joint papers with Gail Letzter ([J86], [J87], [J89], [J93]).

Set

$$F(U) = \{a \in U \mid \dim(\operatorname{ad} U)a < \infty\}.$$

Since the action of U on finite-dimensional modules is completely reducible, $F(U)$ can be written as a direct sum of finite-dimensional simple ad U modules. Using Hopf algebra arguments, it follows that $F(U)$ is a subalgebra of U.

Smith's example was parlayed into the fundamental observation that $F(U)$ is "large" inside U ([J86]). The first step in making this assertion plausible is a computation establishing

$$F(U) \cap T = \tau(-2P^+).$$

Notice the immediate consequence that $F(U) \cap T$ is a multiplicative monoid but not a group; unfortunately, this means that $F(U)$ is not a Hopf subalgebra of U. This is one of several shortcomings of the locally finite part that cause serious complications in Joseph's papers in that there is not a good representation theory for $F(U)$ that easily lifts to all of U. On the bright side, a calculation shows that $x_i\tau(-2\lambda)$ lies in $F(U)$ for λ sufficiently large in P^+. This suggests that one can recover x_i from the locally finite part by inverting members of $F(U) \cap T$. Indeed, if $T_\diamond$ denotes the group generated by this intersection, then

Theorem 1. *U is a finitely generated free module over $F(U)T_\diamond$ and generators can be chosen to be a transversal of coset representatives for $T_\diamond$ in T.*

The extension U over $F(U)$ has enough integrality properties to allow one to compare prime ideals in the two rings. (See [J89], Section 6.1, for an illustration.)

The classical adjoint action behaves compatibly with respect to the degree function on the enveloping algebra. The close quantum analogs to degree for the quantized enveloping algebra (see for example [DeK], Section 1.7) are not $\operatorname{ad} U$ invariant. Joseph introduces a different filtration that *is* ad-invariant; information gleaned from this choice does not appear in the classical set-up.

Recall that $L(\lambda)$ denotes the finite-dimensional simple U module of highest weight λ for a dominant integral weight λ. Viewed through the new filtration, the locally finite part completely reduces ([J87]):

$$F(U) = \bigoplus_{\lambda \in P^+} (\operatorname{ad} U)\tau(-2\lambda). \tag{1}$$

The representation of $F(U)$ on the module $L(\lambda)$ induces an isomorphism between $(\operatorname{ad} U)\tau(-2\lambda)$ and $\operatorname{End} L(\lambda)$ due to a positive definite property of Rosso's form (J93]). As a consequence, one obtains a module decomposition formula.

Theorem 2. $F(U) = \bigoplus_{\lambda \in P^+} \operatorname{End} L(\lambda)$.

Since this decompositon looks like the Peter–Weyl Theorem for the quantized function algebra, Joseph refers to it as the "mock Peter–Weyl theorem"; it is an example of a quantum phenomenon that does not occur in the classical situation. The mockification was explained by Caldero, who showed how the Rosso form can be used to define an $\operatorname{ad} U$ module isomorphism from $F(U)$ onto the quantized function algebra ([Ca1]). Joseph exploits this idea to initiate his study of quantum Kac–Moody algebras ([J104], [J106]). He shows that in this greater generality, there is an $\operatorname{ad} U$ module isomorphism from the integral part of the quantized enveloping algebra into the direct sum of the endomorphism algebras of the highest weight simple integrable modules $L(\lambda)$. Working backwards, he then lifts elements from the endomorphism algebras to form a suitable completion of the integrable part, thus obtaining a mock Peter–Weyl theorem in this context ([J106]). The locally finite machinery also leads to a classification of bicovariant differential calculi in [BaS] and [J104].

Theorem 2 leads to a powerful description of the center Z of U. The invariants of $\operatorname{End} L(\lambda)$ are scalars, so one can lift the identity endomorphism to an invariant (and hence central) element z_λ of U. It follows that $\{z_\lambda \mid \lambda \in P^+\}$ is a basis for Z. (Up to a scalar, z_λ is the unique central element of

$$\tau(-2\lambda) + (\operatorname{ad} U_+)\tau(-2\lambda) \tag{2}$$

where U_+ is the augmentation ideal of U.) Furthermore, Z is a polynomial ring in the z_ω where ω runs over the fundamental weights in P^+.

There are other approaches to the description of the center. (See [Ba1] for a comparison of several methods that produce the central generators.) One consequence of the derivation found in [J86] is a transparent computation of the image of z_λ under the Harish-Chandra map; it produces an expression closely related to the character formula for $L(\lambda)$. A second application is found in Letzter's theory of quantum symmetric pairs ([Lg2]). Earlier results identified zonal spherical functions with orthogonal polynomials by means of the radial components of certain central elements. Using the placement of z_λ in $(\operatorname{ad} U)\tau(-2\lambda)$ and its nice image under the Harish-Chandra map, these radial components are calculated in a uniform manner with a relative minimum of computation.

The local finiteness philosophy is adapted to the quantized function algebra in Joseph's comprehensive paper [J92]. In order to avoid losing information by localizing, he looks at a family of locally finite subalgebras of R obtained by twisting the adjoint action via appropriate automorphisms of the quantized function algebra. (These twisted adjoint maps have proved fruitful, see [Le].) Playing the action of R on itself off the action of U on R, Joseph shows that R is a finite module over the direct limit $F(R)$ of these locally finite subalgebras. In various precise senses, the relation of $F(R)$ to R parallels that of $F(U)$ to U. This apparatus is used in [J92] to establish the orbit yoga for R: a natural bijection between its primitive ideals and the symplectic leaves of the associated semisimple Lie group. More precisely, Joseph completes a program of Hodges and Levasseur ([HoL1], [HoL2]), based on earlier work of Soibelman ([So]) classifying the unitary irreducible representations of R. He shows how the prime and primitive ideals of R break into strata parametrized by pairs of elements from the Weyl group. The description is refined in [J98]. A reader-friendly geometric rendition of the stratification picture can be found in the lecture notes of Brown and Goodearl ([BrG]).

The consequences of [J92] are legion. For example, R is a noetherian domain in which all prime ideals are completely prime. Descriptions of the dual R^* and the Hopf algebra automorphisms of R can be deduced as well as applications to R-matrices ([J95]). Ideas generated in [J92] have been exploited by a veritable crowd of mathematicians, including Brown, Cauchon, Goodearl, Gorelik, Hodges, E. Letzter, Levasseur, and Yakimov.

2 Geometry

There is a genre of mysteries in which the clue to the crime is an item that is missing rather than one that is there. This device lurks behind some of Joseph's fundamental work on quantized enveloping algebras; geometry that appears to be crucial in the classical setting is absent here. Noncommutative algebraic geometry is not sufficiently developed to fill the gap. Indeed, some of the directions that Joseph takes have inspired researchers who are developing the foundations of this new subject.

The Separation of Variables Theorem, due to Kostant, and the Verma Module Annihilator Theorem, due to Duflo, require the fact that the ideal in the associated symmetric algebra for $\mathfrak{g}$ generated by the nonconstant invariant homogeneous polynomials is prime. The argument provided by Kostant ([Ko1]) involves nontrivial algebraic geometry. Thus it was quite surprising when [J87] proved

Theorem 3. *There exists an* $\operatorname{ad} U$ *submodule* H *of* $F(U)$ *such that the multiplication map*

$$H \otimes Z \to F(U)$$

is an $\operatorname{ad} U$ *module isomorphism.*

The argument rests on delicate manipulations of the filtration associated with $F(U)$ (including Theorem 2) and the critical intervention of the Common Basis Theorem to prove that the corresponding map at the graded level is injective. ("Common Basis" is Joseph's diplomatic name for the Canonical/Crystal Basis/Path circle of results.) This tour-de-force turned out to be less astonishing after Bernstein and Lunts ([BeL]) presented a proof of the original Separation of Variables Theorem that also avoided algebraic geometry via a filtration argument. More recently, Baumann ([Ba2]) has given a slightly different proof of Theorem 3, again using Common Basis. His methods combined with the analysis of [J87] generalize to the affine case (cf. [J115], [J116]).

A good choice for the quantum harmonics H is proposed in [J93]. With respect to the positive definite Rosso form, graded harmonics comprise the orthogonal complement to the nonconstant top symbols of $(\operatorname{ad} U)\tau(\lambda)$ as λ varies over $-2P^+$. This is further explored in [Ba2] and [Ca3]. In particular, Caldero specializes the quantum picture to gain new information for classical harmonics.

Joseph's idea for circumventing geometry in a proof of the Verma Module Annihilator Theorem is to compose a brilliant fugue played by two determinants. The PRV determinant was introduced by Parthasarathy, Ranga Rao, and Varadarajan ([PRV] and generalized by Kostant ([Ko2]) in order to prove the irreducibility and unitarity of certain principal series representations. The Shapovalov determinant first appeared in 1972 ([Sh]); it has appeared frequently since then because of its utility in determining when Verma-like modules are simple (and possibly unitary) in a variety of contexts. (See [Ja] for a notable example.)

We briefly review the quantum versions of each determinant. Let $L(\mu)$ be a simple highest weight module (for $\mu \in P^+$) and let

$$H(\mu) = L(\mu)_1 \oplus \cdots \oplus L(\mu)_m$$

denote the isotypical component of $L(\mu)$ in H. Choose a basis $v_1, \ldots, v_m$ for the zero weight space of $L(\mu)$ and write v_{ij} for the copy of v_j in $L(\mu)_i$. The PRV determinant P^μ is the determinant of the matrix $(\phi(v_{ij}))_{i,j}$ where ϕ is the quantum Harish-Chandra map sending U to the group algebra of T. Consider a character Λ on T and extend it to a scalar-valued algebra homomorphism of the group algebra. Finally, set $L(\Lambda)$ to be the simple highest weight U module of highest weight Λ. As in the classical case, one can show that

$$\Lambda(P^\mu) = 0 \text{ if and only if } \operatorname{Ann}_{H(\mu)} L(\Lambda) = 0.$$

In other words, $\operatorname{Ann}_H L(\Lambda) = 0$ if and only if $\Lambda(P^\mu) = 0$ for all μ with $H(\mu) \neq 0$.

As to the Shapovalov determinant, set U^- to be the algebra generated by $y_1, \ldots, y_n$. If κ is the Chevalley antiautomorphism of U, then the Shapovalov form on U^- sends (a, b) to $\phi(\kappa(a)b)$. The Shapovalov determinant S^η is the determinant of this form when restricted to the $-\eta$ weight space of U^- for $\eta \in Q^+$. The construction can be extended to the Verma module $M(\Lambda)$ for a character Λ of T, making $\Lambda(S^\eta)$ the determinant of the Shapovalov form on the $\Lambda q^{-\eta}$ weight space. This time, the vanishing property asserts that $M(\Lambda)$ is simple if and only if $\Lambda(S^\eta) = 0$ for all $\eta \in Q^+$.

Both determinants can be completely factored as a product of Laurent linear polynomials. (Here an element of the group algebra of T is "Laurent linear" provided it is a linear combination of t and t^{-1} for some $t \in T$.) Miraculously, the set of factors (up to scalars) that appear in at least one of the PRV determinants is the same as the set of factors that appear in at least one of the Shapovalov determinants. Hence $\Lambda(S^\eta) = 0$ for all η if and only if $\Lambda(P^\mu) = 0$ for all μ. Consequently,

$$\text{if } M(\Lambda) \text{ is simple then } \mathrm{Ann}_H\, M(\Lambda) = 0. \tag{3}$$

Now the Verma module $M(\Lambda)$ contains a simple Verma module $M(\Lambda')$. Assertion (3) and Theorem 3 imply that

$$\mathrm{Ann}_{F(U)}\, M(\Lambda') = H\,\mathrm{Ann}_Z\, M(\Lambda') = F(U)\mathrm{Ann}_Z\, M(\Lambda').$$

Since all submodules of a Verma module admit the same central character, $M(\Lambda)$ and $M(\Lambda')$ are annihilated by the same central elements. Thus

$$\mathrm{Ann}_{F(U)}\, M(\Lambda) \subseteq \mathrm{Ann}_{F(U)}\, M(\Lambda') = F(U)\mathrm{Ann}_Z\, M(\Lambda') = F(U)\mathrm{Ann}_Z\, M(\Lambda).$$

With a bit of extra work and Theorem 1, the desired result is obtained.

Theorem 4. *The annihilator in U of the Verma module $M(\Lambda)$ is generated by its intersection with the center.*

After some traveling back and forth between U and $F(U)$, the quantum analog of Duflo's Theorem ([Du]) is obtained as a corollary.

Theorem 5. *The set of primitive ideals of U coincides with the set of annihilators of the simple highest weight modules.*

The development of two-determinant technology is the first step in Joseph's program to extend annihilator theorems to other types of enveloping algebras. We focus here on quantized enveloping algebras for the Kac–Moody case and discuss applications outside the quantum realm in the next section. The Shapovalov determinant is factored in [Jbook] (chapter 3.4), completing work in [J93]. Unexpected factors give rise to a new family of Verma modules that does not have a counterpart either for the quantized enveloping algebra of a semisimple Lie algebra or for the ordinary enveloping algebra. Even the definition of quantum PRV determinants for Kac–Moody algebras is problematic. Joseph and his student, D. Todoric, present a definition ([J115]) based on his Harish-Chandra map for a completed quantized enveloping algebra ([J106]) and a weak separation of variables proposition. Later, the quantum affine PRV determinant is factored ([J116]), this time into an infinite product. Thus it is not clear how to evaluate it under a character Λ, leaving the goal of a Verma module annihilator theorem in the distance.

Inspired by Kostant's comment in [Ko2], the scope of the PRV determinant can be expanded by replacing the role of a Borel subalgebra with that of a parabolic. [J115] defines the analogous quantum KPRV determinant in a natural way, using a construction not immediately available in the classical context.

It turns out that the distinction between avoiding geometry and promoting geometry may not exist. In [J90], Joseph identifies the quantization of the algebras of functions and of differential operators on the big open Bruhat cell (determined by the longest element of the Weyl group) and its translates. The core difficulty is to understand his candidate for the quantization of the coordinate ring of the basic affine space that lifts the flag variety. The justification of these constructions is a Beilinson–Bernstein equivalence of categories in the spirit of Hodges–Smith ([HoS]) as completed in [J88]. Joseph gingerly sticks a big toe into the waters of noncommutative algebraic geometry: "The appropriate framework to discuss equivalence of categories in the present noncommutative setup seems to be provided by the Artin–van den Bergh–Schelter–Tate theory."

While offering the geometric challenge, [J90] only records the first step. Joseph proves that cousins of quantum coordinate rings, including $F(U)$, are noetherian. Ultimately, the argument produces filtrations with "Hilbert basis properties", courtesy of the Common Basis Theorem. His student, M. Gorelik, describes the prime and primitive spectra of these translated function algebras in [Go1]. The skew fields of fractions of closely related rings are characterized in [J97], an echo of Joseph's early studies of the Gelfand–Kirillov conjecture. More on this theme is found in [Ca2].

Although even definitions of such objects as the noncommutative ring of differential operators remain fluid, Joseph's papers are a fertile source of insights and problems for those mathematicians building the foundations of noncommutative algebraic geometry. Lunts and Rosenberg ([LuR]) redo [J90] in a much broader geometric context but make clear that they have been inspired by Joseph's body of work. (Also see [BaK].)

3 Trickle Down Economics

When the algebra community embraced quantum groups, optimists expected that the new subject would ultimately elucidate long standing questions about representations of semisimple Lie algebras. (See the introduction of [Jbook].) Ideas generated by Joseph and his collaborators have, indeed, percolated through quantum technicalities, contributing to both classical and cutting edge representation theory.

The two-determinant strategy for proving Verma module annihilator theorems has been put to advantage in many settings. Joseph has shown how to reprove Duflo's classical result in this fashion ([J103]). Gorelik and Lanzmann employ this strategy to establish a conjecture of Musson for Lie superalgebras. These two students of Joseph determine precisely which Verma modules of orthosymplectic superalgebras have annihilators generated by their intersection with the center ([GoL1]). A true annihilator theorem is obtained in their follow-up paper [GoL2]. Gorelik further pursues the annihilator theme for basic classical Lie superalgebras in [Go2].

New meets old in [J102]. Recall the description of the center that came out of the mock Peter–Weyl theorem. Exploiting equations (1) and (2), one sees that there is a projection $\mathcal{L}$ from $F(U)$ onto Z that sends a to the unique central element contained in $a + (\operatorname{ad} U_+)a$. (If $a \in U_+$ then $\mathcal{L}(a) = 0$.) When restricted to the Weyl group invariants in the locally finite part of the group algebra for T, this function can be regarded as a sort of inverse to the Harish-Chandra map. Indeed, it is easy to compute the Harish-Chandra map of the restricted $\mathcal{L}(a)$ using the quantum trace.

Joseph adapts this to the classical environment by constructing an injection $\mathcal{L}$: $S(\mathfrak{h})^W \to S(\mathfrak{g})^G$ where $\mathfrak{g}$ is a complex semisimple Lie algebra with adjoint group G, Cartan subalgebra $\mathfrak{h}$, and Weyl group W. Extending this injection to a doubling of $\mathcal{L}$ (with the help of Common Basis combinatorics and a result of Kumar [Ku]), he obtains a multivariable version of Chevalley's theorem. In turn, this proves the normality of the commuting variety and settles a question asked by Wallach: the Harish-Chandra homomorphism for the space of invariant differential operators on $\mathfrak{g}$ is graded surjective.

References

All articles that include Joseph as an author are designated [Jn] where n is the number recorded in his list of publications. His monograph *Quantum Groups and Their Primitive Ideals* is referred to as [Jbook].

[BaK] E. Backelin and K. Kremnitzer, *Quantum flag varieties, equivariant quantum $\mathcal{D}$-modules, and localization of quantum groups*, arXiv:math. QA/0401108.

[Ba1] P. Baumann, *On the center of quantized enveloping algebras*, J. Algebra. **203** (1998), 244–260.

[Ba2] P. Baumann, *Another proof of Joseph and Letzter's separation of variables theorem for quantum groups*, Transform. Groups **5** (2000), 3–20.

[BaS] P. Baumann, F. Schmidt, *Classification of bicovariant calculi on quantum groups*, Comm. Math. Physics **194** (1998), 71–86.

[BeL] J. Bernstein and V. A. Lunts, *A simple proof of Kostant's theorem that $U(\mathfrak{g})$ is free over its center*, Amer. J. Math., **118** (1996), 979–987.

[BrG] K. Brown and K. R. Goodearl, *Lectures on Algebraic Quantum Groups*, Adv. Courses Math. CRM Barcelona, Birkhäuser, Basel, 2002.

[Ca1] P. Caldero, *Éléments ad-finis de certains groupes quantiques*, C. R. Acad. Sci. Paris (I), **316** (1993), 327–329.

[Ca2] P. Caldero, *On the Gelfand-Kirillov conjecture for quantum algebras*, Proc. Amer. Math. Soc., **128** (2000), 943–951.

[Ca3] P. Caldero, *On harmonic elements for semi-simple Lie algebras*, Adv. Math., **166** (2002), 73–99.

[DeK] C. De Concini and V. G. Kac, *Representations of quantum groups at roots of* 1 in: *Operator Algebras, Unitary Representations, Enveloping Algebras, and Invariant Theory*, Prog. in Math. **92**, Birkhäuser, Boston 1990, 471–506.

[Du] M. Duflo, *Sur la classification des idéaux primitifs dans l'algèbre enveloppante d'une algèbre de Lie semi-simple*, Ann. Math., **105** (1977), 107–120.

[Go1] M. Gorelik, *The prime and the primitive spectra of a quantum Bruhat cell translate*, J. Algebra, **227** (2000), 211–253.

[Go2] M. Gorelik, *Annihilation theorem and separation theorem for basic classical Lie superalgebras*, J. Amer. Math. Soc., **15** (2002), 113–165.

[GoL1] M. Gorelik and E. Lanzmann, *The annihilation theorem for the completely reducible Lie superalgebras*, Invent. Math., **137** (1999), 651–680.

[GoL2] M. Gorelik and E. Lanzmann, *The minimal primitive spectrum of the enveloping algebra of the Lie superalgebra $osp(1, 2l)$*, Adv. Math., **154** (2000), 333–336.

[HoL1] T. J. Hodges and T. Levasseur, *Primitive ideals of $C_q[SL(3)]$*, Comm. Math. Phys., **156** (1993), 581–605.

[HoL2] T. J. Hodges and T. Levasseur, *Primitive ideals of $C_q[SL(n)]$*, J. Algebra, **156** (1993), 581–605.

[HoS] T. J. Hodges and S. P. Smith, *Sheaves of noncommutative algebras and the Beilinson-Bernstein equivalence of categories*, Proc. Amer. Math. Soc., **93** (1985), 379–386.

[Ja] J.-C. Jantzen, *Kontravariante Formen auf induzierten Darstellungen halbeinfacher Lie-Algebren*, Math. Ann. **226** (1977), 53–65.

[Ko1] B. Kostant, *Lie group representations on polynomial rings*, Amer. J. Math., **85** (1963), 327–404.

[Ko2] B. Kostant, *On the existence and irreducibility of certain series of representations*, in: *Lie Groups and Their Representations*, Proc. Summer School Bolyai János Math. Soc., Halsted, New York 1975, 231–329.

[Ku] S. Kumar, *A refinement of the PRV conjecture*, Invent. Math., **97** (1989), 305–311.

[Le] E. Letzter, *Remarks on the twisted adjoint representation of $R_q[G]$*, Comm. Algebra, **27** (1999), 1889–1893.

[Lg1] G. Letzter, *Representation theory for quantized enveloping algebras* in: *Algebraic Groups and Their Generalizations: Quantum and Infinite Dimensional Methods*, Proc. Symp. Pure Math. **56**, Part 2, Amer. Math. Soc., Providence 1994, 63–80.

[Lg2] G. Letzter, *Quantum zonal spherical functions and Macdonald polynomials*, Adv. Math., **189** (2004), no. 1, 88–147.

[LuR] V. A. Lunts and A. L. Rosenberg, *Localization for quantum groups*, Selecta Math. NS, **5** (1999), 123–159.

[PRV] K. P. Parthasarathy, R. Ranga Rao, V. S. Varadarajan, *Representations of complex semisimple Lie groups and Lie algebras*, Ann. Math., **85** (1967), 383–429.

[Sh] N. N. Shapovalov, *On a bilinear form on the universal enveloping algebra of a complex semisimple Lie algebra*, Funct. Anal. Appl., **6** (1972), 65–70.

[Sm] S. P. Smith, *Quantum groups: an introduction and survey for ring theorists* in: *Noncommutative Rings*, MSRI Publ. **24**, Springer, Berlin 1992, 131–178.

[So] Y. S. Soibelman, *The algebra of functions on a compact quantum group and its representations*, Leningrad Math. J., **2** (1991), 161–178.

Part II

Research Articles

Opérateurs différentiels invariants et problème de Noether

Jacques Alev et François Dumas[1–2]

[1] Université de Reims-Champagne-Ardenne
Laboratoire de Mathématiques (UMR 6056 du CNRS)
B.P. 1039, 51687 Reims Cedex, France
jacques.alev@univ-reims.fr
[2] Université Blaise Pascal (Clermont-Ferrand 2)
Laboratoire de Mathématiques (UMR 6620 du CNRS)
63177, Aubière, France
Francois.Dumas@math.univ-bpclermont.fr

Dédié à Anthony Joseph, à l'occasion de son soixantième anniversaire

Summary. Let $\mathcal{G}$ be a group and $\rho : \mathcal{G} \to GL(V)$ a representation of $\mathcal{G}$ in a vector space V of dimension n over a commutative field k of characteristic zero. The group $\rho(\mathcal{G})$ acts by automorphisms on the algebra of regular functions $k[V]$, and this action can be canonically extended to the Weyl algebra $A_n(k)$ of differential operators over $k[V]$ and then to the skewfield of fractions $D_n(k)$ of $A_n(k)$. The problem studied in this paper is to determine sufficient conditions for the subfield of invariants of $D_n(k)$ under this action to be isomorphic to a Weyl skewfield $D_m(K)$ for some integer $0 \leq m \leq n$ and some purely transcendental extension K of k. We obtain such an isomorphism in two cases: (1) when ρ splits into a sum of representations of dimension one, (2) when ρ is of dimension two. We give some applications of these general results to the actions of tori on Weyl algebras and to differential operators over Kleinian surfaces.

Subject Classifications: 16S32, 13A50, 16K40, 17B35

Introduction

0.1.

Toute représentation ρ d'un groupe $\mathcal{G}$ dans un k-espace vectoriel V de dimension n induit sur l'algèbre $S = k[V]$ des fonctions régulières une action d'un sous-groupe $G \simeq \rho(\mathcal{G})$ du groupe des automorphismes de k-algèbre de S. Celle-ci se prolonge de façon canonique en une action par automorphismes sur l'algèbre de Weyl $A_n(k) = \operatorname{Diff} S$ des opérateurs différentiels sur S. Au niveau du corps commutatif $k(V) = \operatorname{Frac} S$, tout un pan de l'étude des fonctions rationnelles invariantes sous G

gravite autour du problème classique, dit problème de Noether, consistant à déterminer sous quelles conditions le corps $k(V)^G$ est extension transcendante pure du corps de base k. Nous renvoyons par exemple aux articles [5], [10], [11], [14], [16], [19], [20] ou [22], et à leurs bibliographies pour des références sur ce point.

La question discutée dans cet article peut être vue comme une extension du problème de Noether au prolongement de l'action de G sur le corps de fractions $D_n(k)$ de l'algèbre de Weyl $A_n(k)$. Il s'agit cette fois de déterminer des conditions suffisantes pour que le sous-corps d'invariants $D_n(k)^G$ soit isomorphe à un corps de Weyl $\mathcal{D}_{m,t}(k) = D_m(K)$, avec m un entier tel que $0 \leq m \leq n$ et K une extension transcendante pure de degré de transcendance t sur k. Les travaux fondateurs de I. M. Gelfand et A. A. Kirillov ont en effet mis en évidence que ce sont ces corps de Weyl $\mathcal{D}_{m,t}(k)$ qui jouent pour l'étude de l'équivalence birationnelle des algèbres non commutatives intervenant en théorie de Lie le rôle des extensions transcendantes pures de la théorie commutative classique.

0.2.

La première section de l'article présente avec précision le problème évoqué ci-dessus, indique quelques éléments de réponse généraux, et rappelle un résultat démontré en [1] sur les invariants des corps de fractions des extensions de Ore, qui peut être considéré comme une forme non commutative d'un théorème classique de K. Miyata (voir [16]). Par la description qu'il donne du corps des fractions des invariants d'une algèbre de polynômes sous l'action d'un groupe stabilisant le corps des coefficients, le théorème de Miyata permet de donner une réponse positive au problème de Noether dans des situations commutatives où l'on sait triangulariser l'action. C'est le cas lorsque la représentation considérée est de dimension 2 ou 3 (W. Burnside, E. Noether; voir par exemple [11]). C'est aussi le cas des travaux de E. B. Vinberg (voir [22] ou [11]) sur les actions de groupes algébriques résolubles connexes, où le théorème de Lie–Kolchin assure la triangularisation nécessaire à l'application itérative du théorème de Miyata.

Pour des algèbres non commutatives de polynômes (en l'occurence des extensions de Ore itérées), le problème supplémentaire qui se pose est, une fois que l'on a obtenu le corps des invariants comme un corps de fractions d'une extension de Ore itérée, de le reconnaître comme un corps de Weyl. Une situation où l'on sait, en petite dimension et de façon élémentaire, mener à bien cette reconnaissance est celle de l'article [1]; on y démontre que pour tout groupe fini G d'automorphismes de toute extension de Ore R en deux variables, le sous-corps des invariants sous G du corps des fractions de R est un corps de Weyl D_1 ou son analogue quantique. Dans le cadre différent du problème de Noether non commutatif tel qu'on l'a formulé en 0.1, on parvient également à obtenir une réponse positive dans deux situations qui font l'objet des parties suivantes.

0.3.

On montre dans la deuxième partie que, pour tout groupe $\mathcal{G}$, et pour toute représentation de $\mathcal{G}$ qui est somme directe de n représentations de dimension 1, il existe un entier $0 \leq s \leq n$ unique tel que $D_n(k)^G \simeq \mathcal{D}_{n-s,s}(k)$. En particulier, si k est algébriquement clos, on a $D_n(k)^G \simeq D_n(k)$ pour toute représentation de tout groupe abélien fini. L'exemple de l'action naturelle du tore $\mathbb{T}_n = (k^*)^n$ sur l'algèbre de Weyl $A_n(k)$, déjà considérée pour d'autres problèmes par exemple en [18], montre que toutes les valeurs de s entre 0 et n peuvent être atteintes, pour différents choix de sous-groupes $\mathcal{G}$ de $\mathbb{T}_n$.

0.4.

Le théorème central de la troisième partie établit que, pour tout groupe $\mathcal{G}$ et toute représentation de $\mathcal{G}$ de dimension 2, il existe deux entiers positifs m, t vérifiant $1 \leq m + t \leq 2$ satisfaisant $D_2(k)^G \simeq \mathcal{D}_{m,t}(k)$.

L'application principale que l'on développe concerne la représentation standard de dimension 2 des sous-groupes finis de $\mathrm{SL}(2, \mathbb{C})$. Si G est un tel sous-groupe agissant par automorphismes sur $S = k[V]$ en prolongeant son action linéaire naturelle sur $V = \mathbb{C}^2$, l'algèbre S^G est l'algèbre des fonctions régulières sur la surface de Klein associée à G. Il est bien connu que les surfaces de Klein sont classées suivant la classification en types A_n, D_n, E_6, E_7 et E_8 des sous-groupes finis de $\mathrm{SL}(2, \mathbb{C})$. Par ailleurs, l'algèbre de Weyl $A_1(\mathbb{C})$ peut être considérée comme une déformation non commutative de S, de dimension 2, sur laquelle G agit par automorphismes. On a montré en [1] que pour cette action, on a $D_1(\mathbb{C})^G \simeq D_1(\mathbb{C})$. La problématique abordée ici est différente, puisqu'il s'agit de considérer le prolongement canonique de l'action de G sur S à l'algèbre $\mathrm{Diff}(S) = A_2(\mathbb{C})$, qui est de dimension 4. Comme G ne contient pas de pseudoreflexion non triviale, il résulte du théorème 5 de [13] que l'algèbre $A_2(\mathbb{C})^G$ n'est autre que l'algèbre des opérateurs différentiels sur la surface de Klein correspondante. L'application du théorème principal de cette partie montre que l'on a alors $\mathrm{Frac}\,(\mathrm{Diff}\,(S^G)) = D_2(\mathbb{C})^G \simeq D_2(\mathbb{C})$. En outre, la méthode permet de calculer explicitement des générateurs de $D_2(\mathbb{C})^G$, dont la détermination repose sur les propriétés de certaines dérivations hamiltoniennes pour la structure de Poisson standard sur S.

Dans tout l'article, k désigne un corps commutatif de caractéristique nulle.

1. Une extension du problème de Noether pour les algèbres de Weyl

1.1. Action sur une algèbre de Weyl canoniquement associée à une représentation d'un groupe

1.1.1. Données et notations

(i) Soit $\mathcal{G}$ un groupe. Considérons une représentation $\rho : \mathcal{G} \to GL(V)$ de $\mathcal{G}$ dans un k-espace vectoriel V de dimension finie n. Notons $G = \rho(\mathcal{G})$.

(ii) Notons $S = k[V]$ l'algèbre des fonctions régulières sur V, et $\mathrm{End}_k S$ la k-algèbre des endomorphismes k-linéaires de S. Le plongement canonique $\mu : S \to \mathrm{End}_k S$ consistant à identifier tout élément s à la multiplication μ_s par s dans S est un morphisme d'algèbres.

(iii) Considérons dans $\mathrm{End}_k S$ la sous-algèbre $\mathrm{Diff}\, S$ des opérateurs différentiels de S, et dans $\mathrm{Diff}\, S$ le sous-espace vectoriel $\mathrm{Der}_k S$ formé des k-dérivations de S. Si $(q_1, \ldots, q_n)$ est une base de V^* sur k, on a $S = k[q_1, \ldots, q_n]$. Une S-base de $\mathrm{Der}_k S$ est $(\partial_{q_1}, \partial_{q_2}, \ldots, \partial_{q_n})$, où ∂_{q_i} désigne la dérivation par rapport à q_i pour tout $1 \leq i \leq n$. On a alors que $\mathrm{Diff}\, S$ est la sous-algèbre de $\mathrm{End}_k S$ engendrée par $\mu_{q_1}, \ldots, \mu_{q_n}, \partial_{q_1}, \ldots, \partial_{q_n}$. L'algèbre $\mathrm{Diff}\, S = \mathrm{Diff}\, k[q_1, \ldots, q_n]$ est appelée l'algèbre de Weyl d'indice n sur k. On la note $A_n(k)$.

On synthétise dans la proposition suivante quelques propriétés bien connues de cette action.

Proposition 1.1. *Avec les données et notations ci-dessus:*

(i) L'action de G par automorphismes sur V se prolonge canoniquement en une action par automorphismes sur l'algèbre des fonctions régulières $S = k[V]$.
(ii) L'application $G \times \mathrm{End}_k S \to \mathrm{End}_k S$ définie par $(g, \varphi) \mapsto g.\varphi = g\varphi g^{-1}$ détermine une action de G par automorphismes sur $\mathrm{End}_k S$ prolongeant l'action canonique de G sur S en rendant le morphisme μ covariant.
(iii) L'action par automorphismes de G sur $\mathrm{End}_k S$ définie ci-dessus se restreint en une action par automorphismes de G sur la sous-algèbre $A_n(k) = \mathrm{Diff}\, S$.
(iv) Pour toute base $(q_1, \ldots, q_n)$ de V^*, l'action ci-dessus de G sur $A_n(k)$ se restreint en une action par automorphismes sur l'espace vectoriel $V = k\partial_{q_1} \oplus \cdots \oplus k\partial_{q_n}$, où l'algèbre symétrique $S(V)$ est identifiée à l'algèbre des opérateurs différentiels à coefficients constants sur V, et cette restriction est l'action définie au départ par la représentation ρ.

1.1.2. Définition et notation

Pour toute représentation ρ de dimension finie n d'un groupe $\mathcal{G}$, l'action du groupe $G = \rho(\mathcal{G})$ par automorphismes sur l'algèbre de Weyl $A_n(k)$ définie à la proposition précédente sera dite canoniquement associée à ρ. On note

$$A_n(k)^{\mathcal{G},\rho} = A_n(k)^G = \{f \in A_n(k)\,;\; g(f) = f,\ \forall\, g \in G\}$$

la sous-algèbre des invariants de $A_n(k)$ sous cette action.

Le lemme suivant est bien connu. Il met en évidence que, contrairement à l'algèbre commutative d'invariants S^G, l'algèbre $A_n(k)^{\mathcal{G},\rho}$ ne peut pas se réduire au corps de base k.

Lemme 1.2. *Avec les données et notations ci-dessus, la dérivation d'Euler:*

$$\omega = \mu_1\partial_1 + \mu_2\partial_2 + \cdots + \mu_n\partial_n$$

appartient toujours à l'algèbre d'invariants $A_n(k)^{\mathcal{G},\rho}$.

1.1.3. Notations

Introduisons la notation usuelle p_i pour désigner (pour tout $1 \leq i \leq n$) la dérivée ∂_{q_i} par rapport à q_i dans l'algèbre $S = k[q_1, \dots, q_n]$. L'algèbre de Weyl $A_n(k)$ apparaît donc comme l'algèbre engendrée sur k par $q_1, \dots, q_n, p_1, \dots, p_n$ avec les relations canoniques:

$$[p_i, q_i] = 1, \quad [p_i, q_j] = [p_i, p_j] = [q_i, q_j] = 0 \text{ pour } i \neq j.$$

D'après la proposition 1.1, pour tout groupe $\mathcal{G}$ et toute représentation ρ de dimension n de $\mathcal{G}$, le groupe $\rho(\mathcal{G})$ est isomorphe à un sous-groupe G d'automorphismes linéaires de $A_n(k)$ stabilisant les k-espaces vectoriels $V = kp_1 \oplus \cdots \oplus kp_n$ et $V^* = kq_1 \oplus \cdots \oplus kq_n$, avec

$$[g(p_i), q_j] = [p_i, g^{-1}(q_j)] \text{ pour tous } g \in G, 1 \leq i, j \leq n,$$

ou encore,

$$g(p_i) = \sum_{j=1}^{n} \partial_{q_i}(g^{-1}(q_j))p_j \text{ pour tous } g \in G, 1 \leq i \leq n.$$

Le groupe G apparaît en particulier comme un sous-groupe du groupe symplectique $\mathrm{Sp}(2n, k)$ isomorphe à un sous-groupe de $\mathrm{GL}(n, k)$. Remarquons que, dans cette construction, le groupe G ne dépend (à conjugaison près) que de la classe de ρ pour l'équivalence des représentations de $\mathcal{G}$. Rappelons enfin qu'avec ces notations, l'élément

$$w = q_1 p_1 + q_2 p_2 + \cdots + q_n p_n$$

de $A_n(k)$ est invariant sous l'action de G.

1.2. Prolongement du problème de Noether aux algèbres de Weyl

1.2.1. Définitions et notations

(i) Pour tout entier $n \geq 1$, et tout corps commutatif K extension de k, on note $D_n(K)$ le corps de fractions de l'algèbre de Weyl $A_n(K)$. Dans le cas particulier où K est une extension transcendante pure $k(z_1, \dots, z_t)$ avec t entier naturel, on note $\mathcal{D}_{n,t}(k)$ le corps $D_n(K) = D_n(k(z_1, \dots, z_t))$. Pour englober dans cette notation le cas commutatif, on convient de noter $\mathcal{D}_{0,t}(k) = k(z_1, \dots, z_t)$ pour tout $t \geq 0$. Les corps $\mathcal{D}_{n,t}(k)$ où $n, t \geq 0$ sont appelés corps de Weyl. Remarquons que, dans cette notation, l'indice t désigne le degré de transcendance sur k du centre de $\mathcal{D}_{n,t}(k)$, qui n'est autre que $k(z_1, \dots, z_t)$.

(ii) Tout automorphisme de $A_n(k)$ se prolonge de façon unique en un automorphisme de $D_n(k)$. Donc, pour tout sous-groupe G du groupe des automorphismes de l'algèbre $A_n(k)$ provenant d'une représentation ρ de dimension finie n d'un groupe $\mathcal{G}$ suivant la construction détaillée ci-dessus en 1.1, on peut considérer le sous-corps des invariants:

$$D_n(k)^{\mathcal{G},\rho} = D_n(k)^G = \{f \in D_n(k)\,;\; g(f) = f, \ \forall\, g \in G\}.$$

1.2.2. Problème de Noether non commutatif

Dans toute la problématique initiée par l'article [7] de Gelfand et Kirillov (voir [8], [9]), les corps de Weyl jouent dans l'étude de l'équivalence birationnelle pour les algèbres non commutatives intervenant en théorie de Lie (algèbres enveloppantes, leurs quotients primitifs,...) un rôle comparable à celui des extensions transcendantes pures dans la théorie commutative classique. En relation avec le problème de Noether en théorie des invariants, on peut alors de façon significative formuler la question suivante, que l'on appelera dans la suite problème de Noether non commutatif.

(*) Soit ρ une représentation d'un groupe $\mathcal{G}$, de dimension finie n sur k. Le sous-corps $D_n(k)^{\mathcal{G},\rho}$ est-il isomorphe à un corps de Weyl $\mathcal{D}_{m,t}(k)$ pour certains entiers naturels m et t ?

Rappelons que le problème de Noether classique concerne le prolongement de l'action de $G = \rho(\mathcal{G})$ par automorphismes sur l'algèbre $k[V]$ des fonctions régulières en une action par automorphismes sur le corps commutatif $k(V) = \operatorname{Frac} k[V]$ des fonctions rationnelles, (la question étant alors de savoir si le corps d'invariants $k(V)^G$ est extension transcendante pure du corps de base k, voir par exemple [5], [11], [19], [20], [10], [14]).

1.2.3. Remarques

(i) Les sous-corps commutatifs maximaux du corps de Weyl $D_n(k)$ sont de degré de transcendance au plus n (voir [15], corollary 6.6.18). Comme un corps de Weyl $\mathcal{D}_{m,t}(k)$ contient effectivement des sous-corps commutatifs de degré de transcendance $m + t$, on déduit qu'une réponse positive au problème (*) ci-dessus n'est possible que pour des valeurs m, t telles que $m + t \leq n$, et donc en particulier $m \leq n$.

(ii) A notre connaissance, un des seuls éléments de réponse au problème (*) ci-dessus apparaissant dans la littérature est le théorème 3.3 de [2], qui démontre que $D_n(k)^G \simeq D_n(k)$ pour tout groupe G fini et abélien d'automorphismes linéaires de $A_n(k)$. Plus généralement, dans le cas d'un groupe fini, on peut apporter au problème de Noether non commutatif la précision suivante.

Proposition 1.3. *Soit $\mathcal{G}$ un groupe fini. Soit ρ une représentation de $\mathcal{G}$ de dimension finie n sur k. Si le sous-corps $D_n(k)^{\mathcal{G},\rho}$ est isomorphe à un corps de Weyl $\mathcal{D}_{m,t}(k)$, alors $m = n$ et $t = 0$, c'est-à-dire que $D_n(k)^{\mathcal{G},\rho} \simeq D_n(k)$.*

Preuve. On suppose que ρ est une représentation d'un groupe fini $\mathcal{G}$, de dimension finie n sur k, telle que $D_n(k)^{\mathcal{G},\rho}$ soit isomorphe à un corps de Weyl $\mathcal{D}_{m,t}(k)$ pour certains entiers $m, t \geq 0$. Pour alléger, on note $D = D_n(k)$ et $Q = D_n(k)^{\mathcal{G},\rho}$. On a donc: $\mathcal{D}_{m,t}(k) \simeq Q \subset D \simeq D_n(k)$. Parce que le sous-groupe $G = \rho(\mathcal{G})$ de $\operatorname{Aut} D$ est fini par hypothèse, on sait d'après le lemme 2.18 de [17] que $[D : Q]$ est fini et $\leq |G|$. La question de savoir en général si $[D : Q] < +\infty$ implique GK-trdeg $Q =$ GK-trdeg D reste à notre connaissance encore ouverte, où GK-trdeg désigne le degré de transcendance de Gelfand et Kirillov (voir [7] ou [12]). En revanche, J. Zhang a prouvé en [23] que $[D : Q] < +\infty$ implique $\mathrm{Ld}\, D = \mathrm{Ld}\, Q$, où Ld désigne le degré de transcendance inférieur (lower transcendence degree). Ce nouvel invariant dimensionnel introduit en [23], (voir aussi [12]), coïncide avec le GK-trdeg pour une large classe d'algèbres classiques, dont les algèbres de Weyl. On a donc ici $2m + t = 2n$. Par ailleurs, la remarque (i) du 1.2.3 ci-dessus assure que $m + t \leq n$. Les deux conditions impliquent $m = n$ et $t = 0$. □

1.3. Invariants des corps de fractions des extensions de Ore

1.3.1. Rappels et notations

(i) Soient A un anneau (non nécessairement commutatif), σ un automorphisme de A, et δ une σ-dérivation de A. On note $A[x\,;\,\sigma, \delta]$ l'anneau des polynômes de Ore en une variable à coefficients dans A, dont le produit est défini à partir de la relation de commutation $xa = \sigma(a)x + \delta(a)$ pour tout $a \in A$, (voir par exemple [15]). Comme d'habitude, on simplifie cette notation en $A[x\,;\,\sigma]$ lorsque δ est nulle, et en $A[x\,;\,\delta]$ lorsque σ est l'identité. Lorsque A est noethérien intègre, de corps de fractions $K = \mathrm{Frac}\, A$, il en est de même des anneaux $R = A[x\,;\,\sigma, \delta]$ et $S = K[x\,;\,\sigma, \delta]$, et l'on note $\mathrm{Frac}\, R = \mathrm{Frac}\, S = K(x\,;\,\sigma, \delta)$.

(ii) L'algèbre de Weyl $A_n(k)$ peut être vue comme l'anneau de polynômes de Ore obtenu en itérant n fois la construction ci-dessus à partir de l'anneau commutatif $A = k[q_1, q_2, \ldots, q_n]$:

$$A_n(k) = k[q_1, q_2, \ldots, q_n][p_1\,;\,\partial_1][p_2\,;\,\partial_2]\ldots[p_n\,;\,\partial_n],$$

et son corps de fractions $D_n(k)$ comme le corps de fonctions rationnelles non commutatif:

$$D_n(k) = k(q_1, q_2, \ldots, q_n)(p_1\,;\,\partial_1)(p_2\,;\,\partial_2)\ldots(p_n\,;\,\partial_n),$$

où chaque dérivation ∂_i est définie sur $k(q_1, q_2, \ldots, q_n)$ par $\partial_i(q_j) = \delta_{i,j}$, et s'annule en les p_j précédents $(j < i)$. Par ailleurs, pour $1 \leq i \leq n$, considérons dans $A_n(k)$ l'élément

$$w_i = q_i p_i,$$

qui vérifie

$$p_i w_i - w_i p_i = p_i, \quad w_i q_i - q_i w_i = q_i, \quad [p_i, w_j] = [q_i, w_j] = [w_i, w_j] = 0 \text{ si } j \neq i.$$

On en déduit deux autres façons de voir le corps $D_n(k)$ comme corps de fractions d'anneaux de polynômes de Ore itérés. D'une part:

$$D_n(k) = k(q_1, q_2, \ldots, q_n)(w_1\,;\,d_1)(w_2\,;\,d_2)\ldots(w_n\,;\,d_n),$$

avec d_i la dérivation eulérienne $d_i = q_i\partial_i$ pour tout $1 \leq i \leq n$. D'autre part:

$$D_n(k) = k(w_1, w_2, \ldots, w_n)(p_1\,;\,\sigma_1)(p_2\,;\,\sigma_2)\cdots(p_{n-1}\,;\,\sigma_{n-1})(p_n\,;\,\sigma_n),$$

où chaque automorphisme σ_i est défini sur $k(w_1, w_2, \ldots, w_n)$ par $\sigma_i(w_j) = w_j + \delta_{i,j}$, et fixe les p_j précédents $(j < i)$.

(iii) Pour tout groupe G d'automorphismes d'un anneau S, on note S^G le sous-anneau des invariants de S sous G. Lorsque S admet un corps de fractions $D = \text{Frac}\, S$, les éléments de G se prolongent de façon unique en des automorphismes de D, et D^G est un sous-corps de D contenant $\text{Frac}\,(S^G)$.

Le théorème rappelé ci-dessous est un argument clef des preuves des résultats suivants.

Théorème 1.4. *Soient K un corps non nécessairement commutatif, σ un automorphisme et δ une σ-dérivation de K. Notons $S = K[x\,;\,\sigma,\delta]$ et $D = \text{Frac}\, S = K(x\,;\,\sigma,\delta)$. Soit G un groupe d'automorphismes d'anneau de S tel que $g(K) \subseteq K$ pour tout $g \in G$.*

(i) Si $S^G \subseteq K$, alors $S^G = D^G = K^G$.
(ii) Si $S^G \nsubseteq K$, alors, pour tout $u \in S^G$ de degré ≥ 1 minimum parmi les degrés des éléments de S^G de degré ≥ 1, il existe σ' un automorphisme de K^G et δ' une σ'-dérivation de K^G tels que $S^G = K^G[u\,;\,\sigma',\delta']$ et $D^G = \text{Frac}\,(S^G) = K^G(u\,;\,\sigma',\delta')$.

Preuve. Une démonstration détaillée figure en [1]. □

1.3.2. Remarques

(i) Ce théorème étend aux algèbres de polynômes non commutatives définies par extension de Ore un résultat de T. Miyata ([16]) en théorie classique des invariants. On trouvera en [11] et à la remarque 1.3 de [2] des références bibliographiques sur ce théorème de Miyata et certaines de ses nombreuses applications.
(ii) Parmi ces applications figurent toutes les situations où le théorème de Lie–Kolchin permet de triangulariser l'action du groupe (voir en particulier l'article [22] de E.B. Vinberg). On conclut alors que le corps des fonctions rationnelles invariantes est extension transcendante pure de k simplement en appliquant un nombre fini de fois le théorème de Miyata. Pour le prolongement de l'action aux opérateurs différentiels considéré en (*), une telle application itérative du théorème 1.4 ci-dessus permet de même de prouver que $D_n(k)^{\mathcal{G},\rho}$ est le corps de fractions d'une extension itérée de Ore. Le problème difficile (qui ne se pose pas dans la situation commutative) est de reconnaître ce corps $D_n(k)^{\mathcal{G},\rho}$ comme un corps de Weyl. C'est ce que l'on parvient à faire dans les deux situations particulières qui font l'objet des deux sections suivantes.

2. Cas d'une somme directe de représentations de dimension 1

2.1. Résolution du problème de Noether non commutatif pour une somme directe de représentations de dimension 1

Le résultat central de cette section est le théorème suivant.

Théorème 2.1. *Soit $\mathcal{G}$ un groupe. Soit ρ une représentation de $\mathcal{G}$ qui est somme directe de n représentations de dimension 1 de $\mathcal{G}$. Alors, il existe un entier $0 \leq s \leq n$ unique tel que $D_n(k)^{\mathcal{G},\rho} \simeq \mathcal{D}_{n-s,s}(k)$.*

Preuve. L'unicité de s est claire puisque il s'agit du degré de transcendance sur k du centre de $\mathcal{D}_{n-s,s}(k)$. La preuve de l'existence de s procède par récurrence sur n.

1) Supposons d'abord que $n = 1$. Avec les notations du paragraphe 1.2, on considère donc une représentation ρ de $\mathcal{G}$ de dimension 1. Le groupe $G = \rho(\mathcal{G})$ opère alors sur l'algèbre de Weyl $A_1(k) = k[q_1][p_1\,;\,\partial_1]$ par des automorphismes de la forme

$$g(q_1) = \chi_1(g)q_1, \quad g(p_1) = \chi_1(g)^{-1}p_1, \qquad \text{pour tout } g \in G$$

où χ_1 est un caractère $G \to k^*$. Comme en 1.3.1.(ii), considérons dans $A_1(k)$ l'élément $w_1 = q_1p_1$, qui est invariant sous l'action de G. Considérons dans le corps $D_1(k) = k(w_1)(p_1, \ \sigma_1)$ la sous-algèbre $S_1(k) = k(w_1)[p_1, \ \sigma_1]$. Il est clair que $\operatorname{Frac} S_1(k) = D_1(k)$. Tout $g \in G$ fixe w_1 et agit sur p_1 par $g(p_1) = \chi_1(g)p_1$. On est donc dans les conditions d'application du théorème 1.4. Ou bien $S_1(k)^G \subseteq k(w_1)$, alors $D_1(k)^G = S_1(k)^G = k(w_1)^G = k(w_1)$; on conclut dans ce cas que $D_1(k)^G \simeq \mathcal{D}_{1-s,s}(k)$ avec $s = 1$. Ou bien $S_1(k)^G \nsubseteq k(w_1)$, et alors $S_1(k)^G$ est une extension de Ore de la forme $k(w_1)[u\,;\,\sigma', \delta']$ avec σ' un automorphisme et δ' une σ'-dérivation de $k(w_1)$, et u un polynôme en p_1 à coefficients dans $k(w_1)$, tel que $g(u) = u$ pour tout $g \in G$, et de degré ≥ 1 minimal parmi les degrés des éléments possédant cette propriété. Vu la forme de l'action de G sur p_1, il est clair que l'on peut choisir un tel u sous la forme d'une puissance $u = p_1^a$ avec a entier ≥ 1, et donc $\sigma' = \sigma_1^a$ et $\delta' = 0$. En résumé $D_1(k)^G = \operatorname{Frac} S_1(k)^G = k(w_1)(p_1^a\,;\,\sigma_1^a)$. Ce corps est encore engendré par $x = p_1^a$ et $y = a^{-1}w_1p_1^{-a}$, qui vérifient $xy - yx = 1$. On conclut que $D_1(k)^G \simeq D_1(k) = \mathcal{D}_{1-s,s}(k)$ avec $s = 0$.

2) Supposons maintenant, par hypothèse de récurrence, que le théorème est vrai pour toute représentation de dimension $\leq n-1$ de tout groupe sur tout corps de caractéristique nulle. Considérons un groupe $\mathcal{G}$ quelconque et $\rho : \mathcal{G} \to GL(V)$ une représentation de dimension n, que l'on suppose être somme directe de représentations de dimension 1. En reprenant les notations des paragraphes 1.1 et 1.3.1.(ii), le sous-groupe $G = \rho(\mathcal{G})$ de $GL(V)$ opère donc sur $A_n(k)$ par automorphismes de la forme

$$g(q_i) = \chi_i(g)q_i, \quad g(p_i) = \chi_i(g)^{-1}p_i, \qquad \text{pour tout } g \in G \text{ et tout } 1 \leq i \leq n,$$

où $\chi_1, \chi_2, \ldots, \chi_n$ sont des caractères $G \to k^*$. Donc

$$g(w_i) = w_i, \qquad \text{pour tout} \quad g \in G \quad \text{et tout} \quad 1 \leq i \leq n.$$

Dans le corps de Weyl

$$D_n(k) = k(w_1, w_2, \ldots, w_n)(p_1\,;\,\sigma_1)(p_2\,;\,\sigma_2)\cdots(p_{n-1}\,;\,\sigma_{n-1})(p_n\,;\,\sigma_n),$$

considérons les sous-corps:

$$L = k(w_n),$$

$$\begin{aligned} K &= k(w_1, w_2, \ldots, w_n)(p_1 ; \sigma_1)(p_2 ; \sigma_2) \cdots (p_{n-1} ; \sigma_{n-1}) \\ &= k(w_n)(w_1, w_2, \ldots, w_{n-1})(p_1 ; \sigma_1)(p_2 ; \sigma_2) \cdots (p_{n-1} ; \sigma_{n-1}) \\ &\simeq D_{n-1}(L), \end{aligned}$$

ainsi que la sous-algèbre $S_n(k) = K[p_n ; \sigma_n]$, qui vérifie $\operatorname{Frac} S_n(k) = D_n(k)$. D'après l'hypothèse de récurrence appliquée à la restriction de l'action de G par L-automorphismes sur $A_{n-1}(L)$, il existe un entier $0 \leq s \leq n-1$ tel que $D_{n-1}(L)^G \simeq \mathcal{D}_{n-1-s,s}(L) \simeq \mathcal{D}_{n-(s+1),s+1}(k)$. Puisque K est stable par G, on peut appliquer le théorème 1.4 à l'anneau de polynômes de Ore $S_n(k) = K[p_n ; \sigma_n]$. Deux cas sont donc possibles.

Premier cas: $S_n(k)^G = K^G$. On obtient directement que: $D_n(k)^G = \operatorname{Frac}(S_n(k)^G) = K^G \simeq D_{n-1}(L)^G \simeq \mathcal{D}_{n-(s+1),s+1}(k)$, ce qui montre le résultat voulu.

Second cas: il existe un polynôme u de degré ≥ 1 en p_n, à coefficients dans K, tel que $g(u) = u$ pour tout $g \in G$. En choisissant u de degré ≥ 1 minimum parmi les degrés en p_n des éléments de $S_n(k)^G$ n'appartenant pas à K, il existe alors un automorphisme σ' et une σ'-dérivation δ' de K^G tels que: $S_n(k)^G = K^G[u ; \sigma', \delta']$ et $D_n(k)^G = \operatorname{Frac} S_n(k)^G = K^G(u ; \sigma', \delta')$.

Développons $u = f_m p_n^m + \cdots + f_1 p_n + f_0$ où m entier ≥ 1 et $f_i \in K^G$ pour tout $0 \leq i \leq m$. Il est clair, vu la forme de l'action de G sur p_n, que $f_m p_n^m$ est lui-même invariant sous G. On développe f_m dans le corps de séries de Laurent:

$$\overline{K} = k(w_1, w_2, \ldots, w_n)((p_1 ; \sigma_1))((p_2 ; \sigma_2)) \cdots ((p_{n-1} ; \sigma_{n-1})).$$

Comme l'action de G se prolonge à $\overline{K}$ en agissant diagonalement sur $p_1, \ldots, p_{n-1}$ et en fixant les w_i, on peut finalement choisir sans restriction un u monomial, c'est-à-dire de la forme

$$u = p_1^{a_1} \ldots p_n^{a_n} \text{ avec } (a_1, \ldots, a_n) \in \mathbb{Z}^n, \text{ et } a_n \geq 1.$$

Pour tout indice $1 \leq j \leq n$, la relation de commutation entre u et w_j est $uw_j = (w_j + a_j)u$. Introduisons donc les éléments:

$$w'_1 = w_1 - a_n^{-1} a_1 w_n, \quad w'_2 = w_2 - a_n^{-1} a_2 w_n, \ldots, \quad w'_{n-1} = w_{n-1} - a_n^{-1} a_{n-1} w_n,$$

de sorte que $w'_j u = u w'_j$ pour tout $1 \leq j \leq n-1$. Comme $\sigma_i(w'_j) = w'_j + \delta_{i,j}$ pour $1 \leq i, j \leq n-1$, le corps $F_{n-1} = k(w'_1, w'_2, \ldots, w'_{n-1})(p_1 ; \sigma_1)(p_2 ; \sigma_2) \cdots (p_{n-1} ; \sigma_{n-1})$ est isomorphe à $D_{n-1}(k)$. Plus précisément, F_{n-1} est le corps de fractions de l'algèbre $k[q'_1, \ldots, q'_{n-1}][p_1 ; \partial_{q'_1}] \ldots [p_{n-1} ; \partial_{q'_{n-1}}]$, où l'on a posé $q'_i = w_i p_i^{-1}$ pour tout $1 \leq i \leq n-1$. Cette dernière algèbre est isomorphe à l'algèbre de Weyl $A_{n-1}(k)$. En lui appliquant l'hypothèse de récurrence, il existe un entier $0 \leq s \leq n-1$ tel que $F_{n-1}^G \simeq \mathcal{D}_{n-1-s,s}(k)$. Il est clair par ailleurs par définition des w'_j que $k(w_n)(w'_1, w'_2, \ldots, w'_{n-1}) = k(w_n)(w_1, w_2, \ldots, w_{n-1})$; puisque w_n commute avec tout élément de F_{n-1}, on en déduit que $K = F_{n-1}(w_n)$. L'algèbre $S_n(k)^G = K^G[u ; \sigma', \delta']$ s'écrit alors sous la forme $S_n(k)^G = F_{n-1}^G(w_n)[u ; \sigma', \delta']$. Mais le générateur u commute avec chaque w'_j pour $0 \leq j \leq n-1$ comme on l'a vu plus haut, commute avec tous les p_i par définition, et vérifie avec w_n la relation $uw_n = (w_n + a_n)u$. Il suffit donc de poser $u' = a_n^{-1}u$ pour obtenir: $S_n(k)^G = F_{n-1}^G(w_n)[u' ; \sigma'']$, avec σ'' qui vaut l'identité sur F_n^G et satisfait: $\sigma''(w_n) = w_n + 1$. Il en résulte que Frac $S_n(k)^G \simeq D_1(F_{n-1}^G) \simeq D_1(\mathcal{D}_{n-1-s,s}(k)) \simeq \mathcal{D}_{n-s,s}(k)$, ce qui achève la preuve. □

2.2. Application aux groupes abéliens finis

Corollaire 2.2. *On suppose ici k algébriquement clos. Soit $\mathcal{G}$ un groupe abélien fini. Pour toute représentation ρ de G, de dimension finie n sur k, on a $D_n(k)^{\mathcal{G},\rho} \simeq D_n(k)$.*

Preuve. Comme $\mathcal{G}$ est abélien fini, toute représentation de $\mathcal{G}$ est somme directe de représentations de dimension 1. On applique alors le théorème 2.1 et la proposition 1.3. □

Le résultat de ce corollaire est prouvé, par une méthode directe, au théorème 3.3 de [2].

2.3. Application à l'action canonique de sous-groupes d'un tore

Le corollaire suivant montre en particulier que, si l'on ne suppose pas nécessairement que $\mathcal{G}$ est fini, toutes les valeurs possibles de s dans le théorème 2.1 peuvent être atteintes.

Corollaire 2.3. *Soit n un entier fixé ≥ 1. Soit ρ la représentation naturelle du tore $\mathbb{T}_n = (k^*)^n$ sur l'espace vectoriel k^n. Alors:*

(i) Pour tout sous-groupe $\mathcal{G}$ de $\mathbb{T}_n$, il existe un entier $0 \leq s \leq n$ unique tel que $D_n(k)^{\mathcal{G},\rho} \simeq \mathcal{D}_{n-s,s}(k)$.
(ii) Pour tout entier $0 \leq s \leq n$, il existe au moins un sous-groupe $\mathcal{G}$ de $\mathbb{T}_n$, tel que $D_n(k)^{\mathcal{G},\rho} \simeq \mathcal{D}_{n-s,s}(k)$.
(iii) En particulier, $s = n$ si $\mathcal{G} = \mathbb{T}_n$, et $s = 0$ si $\mathcal{G}$ est fini.

Preuve. Le point (i) est l'application directe du théorème 2.1. Pour le (ii), fixons un entier $0 \leq s \leq n$. Considérons dans $\mathbb{T}_n$ le sous-groupe:

$$G = \{\mathrm{Diag}\,(\alpha_1, \ldots, \alpha_s, 1, \ldots, 1) \ ; \ (\alpha_1, \ldots, \alpha_s) \in (k^*)^s\} \simeq \mathbb{T}_s,$$

opérant sur $A_n(k)$ par automorphismes:

$$q_i \mapsto \alpha_i q_i, \quad p_i \mapsto \alpha_i^{-1} p_i, \quad \text{pour tout } 1 \leq i \leq s,$$

$$q_i \mapsto q_i, \quad p_i \mapsto p_i, \quad \text{pour tout } s+1 \leq i \leq n.$$

Dans le corps $D_n(k) = k(w_1, w_2, \ldots, w_n)(p_1 ; \sigma_1)(p_2 ; \sigma_2) \cdots (p_n ; \sigma_n)$, introduisons le sous-corps $K = k(w_1, w_2, \ldots, w_n)(p_{s+1} ; \sigma_{s+1})(p_{s+2} ; \sigma_{s+2}) \cdots (p_n ; \sigma_n)$.

La sous-algèbre $S = K[p_1, ; \sigma_1] \cdots [p_s, ; \sigma_s]$ vérifie donc Frac $S = D_n(k)$. Il est clair que K est invariant par G. S'il existait dans S^G un polynôme de degré ≥ 1 en l'un au moins des $p_1, \dots, p_s$, il existerait en particulier (vu la forme de l'action de G) un monôme:

$$u = vp_1^{d_1} p_2^{d_2} \cdots p_s^{d_s} \quad v \in K,\ v \neq 0,\ d_1, \dots, d_s \in \mathbb{N},\ (d_1, \dots, d_s) \neq (0, \dots, 0)$$

invariant par G. On aurait alors $\alpha_1^{d_1} \alpha_2^{d_2} \cdots \alpha_s^{d_s} = 1$ pour tout $(\alpha_1, \alpha_2, \dots, \alpha_s) \in (k^*)^s$, ce qui est impossible. On conclut donc grâce au théorème 1.4 que $(\text{Frac } S)^G = S^G = K^G$, c'est-à-dire que $D_n(k)^G = K$. Il est clair que $K \simeq \mathcal{D}_{n-s,s}(k)$, ce qui achève la preuve du point (ii). Le point (iii) découle de la preuve du point (ii) ci-dessus et du corollaire 2.2 □

Les actions des tores $\mathbb{T}_n$ sur les algèbres de Weyl qui font l'objet de ce corollaire ont été entre autres considérées dans [18].

3. Cas d'une représentation de dimension 2

3.1. Résolution du problème de Noether non commutatif pour une représentation de dimension 2

3.1.1. Un résultat technique préliminaire

On commence par énoncer sous forme du lemme suivant un argument qui interviendra plusieurs fois dans la preuve du théorème principal.

Lemme 3.1. *Soit D une dérivation du corps commutatif de fractions rationnelles $k(x, y)$ de la forme $D = ax\partial_x + by\partial_y$, avec $(a, b) \in \mathbb{Z}^2$, $(a, b) \neq (0, 0)$. Alors:*

$$\text{Frac}\, k(x, y)[z\,;\ D] \simeq \mathcal{D}_{1,1}(k).$$

Preuve. Posons $A = k(x, y)[z\,;\ D]$. Notons d le pgcd de a et b. Désignons par a' et b' les entiers définis par $a = da'$ et $b = db'$. En posant $z' = d^{-1}z$ et $D' = d^{-1}D$, on a clairement $A = k(x, y)[z'\,;\ D']$. D'après le théorème de Bezout, il existe des entiers c, e tels que $a'e - b'c = 1$. Les éléments $x' = x^{-b'}y^{a'}$ et $y' = x^e y^{-c}$ vérifient donc $k(x, y) = k(x', y')$. On vérifie sans difficulté que $D'(x') = 0$ et $D'(y') = y'$. Donc Frac $A = k(x', y')(z'\,;\ D')$ avec $[z', x'] = 0$ et $[z', y'] = y'$. Un dernier changement de générateur $z'' = y'^{-1}z'$ ramène ces relations à $[z'', x'] = 0$ et $[z'', y'] = 1$. On conclut que Frac $A = k(x')(y')(z''\,;\ \partial_{y'}) \simeq D_1(k(x')) \simeq \mathcal{D}_{1,1}(k)$. □

3.1.2. Données et notations

Pour toute la suite, on fixe un groupe $\mathcal{G}$ et une représentation ρ de $\mathcal{G}$ de dimension 2 sur k. Le groupe $G = \rho(\mathcal{G})$ opère par automorphismes sur l'algèbre de Weyl $A_2(k)$. On note, pour $g \in G$ quelconque:

$$g(q_1) = \alpha_g q_1 + \beta_g q_2, \qquad g(q_2) = \gamma_g q_1 + \delta_g q_2,$$
$$g(p_1) = \frac{1}{\Delta_g}(\delta_g p_1 - \gamma_g p_2), \qquad g(p_2) = \frac{1}{\Delta_g}(-\beta_g p_1 + \alpha_g p_2),$$

où $\alpha_g, \beta_g, \gamma_g, \delta_g \in k$ tels que $\Delta_g = \alpha_g\delta_g - \beta_g\gamma_g \neq 0$.
Comme en 1.1.3, introduisons

$$w = q_1 p_1 + q_2 p_2,$$

qui est invariant sous l'action de G d'après le lemme 1.2.

3.1.3. Triangularisation de l'action

Suivant un procédé bien connu (qui semble remonter à W. Burnside, voir [11]), on commence par triangulariser l'action de G sur le corps commutatif $k(V) = k(q_1, q_2)$ en posant $v = q_1 q_2^{-1}$, de sorte que $k(V) = k(v, q_2)$ et que, avec les notations ci-dessus, on a pour tout $g \in G$:

$$g(v) = \frac{\alpha_g v + \beta_g}{\gamma_g v + \delta_g}, \quad g(q_2) = (\gamma_g v + \delta_g) q_2.$$

Comme en 1.3.1 (ii), on regarde $D_2(k)$ comme corps de fractions de l'extension de Ore:

$$D_2(k) = k(q_1, q_2)(w_1\,;\, d_1)(w_2\,;\, d_2),$$

avec $w_1 = q_1 p_1$, $w_2 = q_2 p_2$, $d_1 = q_1\partial_1$ et $d_2 = q_2\partial_2$. En notant d la dérivation de $k(q_1, q_2)(w_1\,;\, d_1)$ telle que $d(q_1) = q_1$, $d(q_2) = q_2$ et $d(w_1) = 0$, on a encore $D_2(k) = k(q_1, q_2)(w_1\,;\, d_1)(w\,;\, d)$, d'où

$$D_2(k) = k(v, q_2)(w_1\,;\, d_1)(w\,;\, d),$$

avec $d_1(v) = v$ et $d(v) = d(w_1) = 0$. Les générateurs v, q_2, w, w_1 du corps $D_2(k)$ vérifient donc les relations de commutation:

$$[v, q_2] = [w_1, q_2] = [w, v] = [w, w_1] = 0, \quad [w_1, v] = v, \quad [w, q_2] = q_2.$$

L'action de G sur v, q_2, w est décrite ci-dessus; pour w_1 on obtient:

$$\begin{aligned}
g(w_1) &= \frac{1}{\Delta_g}(\alpha_g q_1 + \beta_g q_2)(\delta_g p_1 - \gamma_g p_2)\\
&= \frac{1}{\Delta_g}(\alpha_g\delta_g q_1 p_1 - \beta_g\gamma_g q_2 p_2 - \alpha_g\gamma_g q_1 p_2 + \beta_g\delta_g q_2 p_1)\\
&= \frac{1}{\Delta_g}(\alpha_g\delta_g w_1 - \beta_g\gamma_g(w - w_1) - \alpha_g\gamma_g v(w - w_1) + \beta_g\delta_g v^{-1} w_1)\\
&= \frac{1}{\Delta_g}(\alpha_g + \beta_g v^{-1})(\gamma_g v + \delta_g) w_1 - \frac{1}{\Delta_g}\gamma_g(\beta_g + \alpha_g v) w.
\end{aligned}$$

En vue de simplifier la forme de l'action, on pose $x = (q_1q_2)^{-1}w_1 = v^{-1}q_2^{-2}w_1 = q_2^{-1}p_1$, de sorte que $g(x) = \frac{1}{\Delta_g}x - \frac{1}{\Delta_g}\gamma_g q_2^{-2}(\gamma_g v + \delta_g)^{-1}w$ pour tout $g \in G$. Il est clair que $D_2(k)$ est engendré sur k par v, q_2, w, x, et on calcule $[x, v] = q_2^{-2}$, $[x, q_2] = 0$, $[x, w] = 2x$.

Les résultats obtenus peuvent alors être synthétisés dans le lemme suivant:

Lemme 3.2. *Les données et notations sont celles de 3.1.2.*

(i) Les éléments $w = q_1p_1 + q_2p_2$, $v = q_1q_2^{-1}$ *et* $x = q_2^{-1}p_1$ *vérifient les relations de commutation:*

$$[q_2, v] = 0, \qquad [w, v] = 0, \qquad [x, v] = q_2^{-2},$$
$$[w, q_2] = q_2, \qquad [x, q_2] = 0,$$
$$[x, w] = 2x.$$

(ii) En notant:
- *d la dérivation du corps commutatif $k(v, q_2)$ telle que $d(v) = 0$ et $d(q_2) = q_2$,*
- *K le corps gauche $k(v, q_2)(w\,;\,d)$,*
- *σ' le k-automorphisme de K fixant v et q_2, et tel que $\sigma'(w) = w + 2$,*
- *d' la σ'-dérivation de K telle que $d'(v) = q_2^{-2}$, $d'(q_2) = 0$, $d'(w) = 0$,*
- *S la k-algèbre $K[x\,;\,\sigma', d']$,*

on a:

$$D_2(k) = \operatorname{Frac} S = K(x\,;\,\sigma',\, d') = k(v, q_2)(w\,;\,d)(x\,;\,\sigma',\, d').$$

(iii) L'action sur $D_2(k)$ d'un automorphisme quelconque g de G, défini par $\alpha_g, \beta_g, \gamma_g, \delta_g \in k^*$ avec $\Delta_g = \alpha_g\delta_g - \beta_g\gamma_g \neq 0$, est déterminée par son action sur ces générateurs:

$$g(v) = \frac{\alpha_g v + \beta_g}{\gamma_g v + \delta_g}, \qquad g(q_2) = (\gamma_g v + \delta_g)q_2,$$

$$g(w) = w, \qquad g(x) = \frac{1}{\Delta_g}x - \frac{1}{\Delta_g}\gamma_g q_2^{-2}(\gamma_g v + \delta_g)^{-1}w.$$

(iv) En particulier chacun des trois sous-corps $k(v) \subset k(v, q_2) \subset K = k(v, q_2)(w\,;\,d)$ est stable sous l'action de G.

Le point (iv), qui traduit la triangularisation de l'action de G sur $D_2(k)$, est à la base de la méthode utilisée pour résoudre ici le problème (*). Il va en effet permettre d'appliquer le théorème 1.4 à l'anneau de polynômes de Ore $S = K[x\,;\,\sigma', d']$. La preuve se scinde alors dès le départ en deux cas, dont le plus simple est celui traité au lemme suivant.

3.1.4. Cas où il n'existe pas dans S de polynômes de degré ≥ 1 en x invariants par G

On obtient alors directement une réponse positive au problème (*).

Lemme 3.3. *Les données et notations sont celles du lemme 3.2. On suppose de plus que $S^G \subset K$. Alors il existe deux entiers naturels m, t vérifiant $1 \leq 2m + t \leq 3$ tels que*

$$D_2(k)^G = S^G = K^G \simeq \mathcal{D}_{m,t}(k).$$

Preuve. En appliquant le théorème 1.4, on alors $D_2(k)^G = S^G = K^G$. Soit $A = k(v, q_2)[w\,;\,d]$, de sorte que $\operatorname{Frac} A = K$. Puisque $k(v, q_2)$ est stable par G, et comme w est lui-même invariant par G, (de degré 1 en w, donc nécessairement minimal), on réapplique le théorème 1.4 pour conclure que $K^G = \operatorname{Frac}(A^G) = k(v, q_2)^G(w\,;\,d)$. Il reste à déterminer $k(v, q_2)^G$. Posons pour cela $B = k(v)[q_2]$. Rappelons que G agit sur B par automorphismes de la forme $g(v) = \frac{\alpha_g v + \beta_g}{\gamma_g v + \delta_g}$, $g(q_2) = (\gamma_g v + \delta_g)q_2$. Deux cas peuvent alors de nouveau se présenter.

Premier cas: si $B^G \subset k(v)$, on a $k(v, q_2)^G = k(v)^G$ qui, d'après le théorème de Lüroth, est soit une extension transcendante pure monogène $k(t)$, soit réduit à k. On conclut respectivement que $D_2(k)^G = k(t)(w) \simeq \mathcal{D}_{0,2}(k)$ ou $D_2(k)^G = k(w) \simeq \mathcal{D}_{0,1}(k)$.

Second cas: sinon, il existe dans B^G des polynômes de degré ≥ 1 en q_2. Soit alors $q \in B^G$, de degré en q_2 non-nul minimal. Vu la forme de l'action de G sur B, on peut sans restriction prendre q monomial. Notons $q = sq_2^e$, avec $s \in k(v), s \neq 0$ et e entier ≥ 1. Il résulte du théorème 1.4 (sous la forme originale commutative de Miyata) que $k(v, q_2)^G = k(v)^G(q)$. Comme ci-dessus, le théorème de Lüroth conduit à $k(v, q_2)^G = k(t, q)$ si $k(v)^G = k(t)$ et $k(v, q_2)^G = k(q)$ si $k(v)^G = k$. Comme $[w, v] = 0$ (d'où $[w, t] = 0$) et $[w, q_2] = q_2$ (d'où $[w, q] = eq$), on déduit que

$$D_2(k)^G = k(t, q)(w\,;\,d) \text{ ou } D_2(k)^G = k(q)(w\,;\,d).$$

Il suffit de remplacer le générateur w par $(eq)^{-1}w$ pour conclure que $D_2(k)^G \simeq \mathcal{D}_{1,1}(k)$ ou $D_2(k)^G \simeq \mathcal{D}_{1,0}(k)$, ce qui achève la preuve. □

3.1.5. Cas où il existe dans S des polynômes de degré ≥ 1 en x invariants par G

Dans toute la suite, on supposera que S^G n'est pas inclus dans K^G.

On montre d'abord que l'on peut alors remplacer, dans l'extension de Ore $S = K[x\,;\,\sigma', d']$, le générateur x par un générateur y qui est vecteur propre de tous les automorphismes de G pour la représentation déterminant.

Lemme 3.4. *Les données, hypothèses et notations sont celles du lemme 3.2. On suppose de plus que S^G n'est pas inclus dans K^G.*

(i) Il existe une fraction rationnelle $b(v) \in k(v)$ *telle que*

$$g(b) = \frac{1}{\Delta_g}(\gamma_g v + \delta_g)^2 b + \frac{1}{\Delta_g}\gamma_g(\gamma_g v + \delta_g), \text{ pour tout } g \in G.$$

(ii) En posant $y = x + b(v)q_2^{-2}w = q_2^{-1}p_1 + b(v)q_2^{-2}w \in S$, *on a les relations de commutation:*

$$[q_2, v] = 0, \qquad [w, v] = 0, \qquad [y, v] = q_2^{-2},$$

$$[w, q_2] = q_2, \qquad [y, q_2] = b(v)q_2^{-1},$$

$$[y, w] = 2y.$$

(iii) En notant:

- *d la dérivation de* $k(v, q_2)$ *telle que* $d(v) = 0$ *et* $d(q_2) = q_2$,
- σ' *le k-automorphisme de K fixant v et* q_2, *et tel que* $\sigma'(w) = w + 2$,
- d'' *la* σ'*-dérivation de K telle que* $d''(v) = q_2^{-2}$, $d''(q_2) = b(v)q_2^{-1}$, $d''(w) = 0$,

on a

$$S = K[y\,;\, \sigma',\, d''] = k(v, q_2)(w\,;\, d)[y\,;\, \sigma',\, d''].$$

(iv) L'action sur S d'un automorphisme quelconque g de G, défini par $\alpha_g, \beta_g, \gamma_g, \delta_g \in k^*$ *avec* $\Delta_g = \alpha_g\delta_g - \beta_g\gamma_g \neq 0$, *est alors déterminée par son action sur ces générateurs:*

$$g(v) = \frac{\alpha_g v + \beta_g}{\gamma_g v + \delta_g}, \qquad g(q_2) = (\gamma_g v + \delta_g)q_2,$$

$$g(w) = w, \qquad g(y) = \frac{1}{\Delta_g}y.$$

Preuve. Désignons par n l'entier ≥ 1 qui est le minimum des degrés en x des éléments de S^G n'appartenant pas à K^G. Choisissons dans S^G un élément $u = u_n x^n + \cdots + u_1 x + u_0$ de degré n, avec $u_i \in K$, $u_n \neq 0$, $n \geq 1$.

On a vu au point (iv) du lemme 3.2 que l'action sur x d'un automorphisme quelconque $g \in G$ est de la forme

$$g(x) = \frac{1}{\Delta_g}(x + r_g), \quad \text{où l'on a posé } r_g = -\gamma_g(\gamma_g v + \delta_g)^{-1}q_2^{-2}w \in K.$$

On a donc

$$g(u) = g(u_n)\left(\frac{1}{\Delta_g}\right)^n (x + r_g)^n + g(u_{n-1})\left(\frac{1}{\Delta_g}\right)^{n-1}(x + r_g)^{n-1} + \cdots$$

$$+ g(u_1)\frac{1}{\Delta_g}(x + r_g) + g(u_0).$$

En rappelant que K est stable sous l'action de G, explicitons les coefficients dans K des termes de degré n et $n-1$ en x:

$$g(u) = g(u_n)\left(\frac{1}{\Delta_g}\right)^n x^n + \left[g(u_n)\left(\frac{1}{\Delta_g}\right)^n \sum_{i=0}^{n-1} \sigma'^i(r_g) + g(u_{n-1})\left(\frac{1}{\Delta_g}\right)^{n-1}\right] x^{n-1} + \cdots .$$

L'égalité $g(u) = u$ conduit donc, pour tout $g \in G$, aux égalités suivantes dans K:

$$g(u_n) = (\Delta_g)^n u_n,$$

$$g(u_{n-1}) = (\Delta_g)^{n-1} u_{n-1} + (\Delta_g)^{n-1} u_n \sum_{i=0}^{n-1} \gamma_g(\gamma_g v + \delta_g)^{-1} q_2^{-2}(w + 2i).$$

En multipliant les deux membres de la seconde par les inverses des deux membres de la première, et en notant que $\sum_{i=0}^{n-1}(w+2i) = nw + n(n-1)$, on obtient

$$g(u_n^{-1}u_{n-1}) = \left[\frac{1}{\Delta_g} u_n^{-1} u_{n-1} + \frac{1}{\Delta_g} n(n-1)\gamma_g(\gamma_g v + \delta_g)^{-1} q_2^{-2}\right] + \left[\frac{1}{\Delta_g} n\gamma_g(\gamma_g v + \delta_g)^{-1} q_2^{-2}\right] w.$$

L'élément $u_n^{-1}u_{n-1}$ de $K = k(v, q_2)(w\,;\, d)$ se développe dans le corps d'opérateurs pseudo-différentiels formels $K = k(v, q_2)((w^{-1}\,;\, -d))$, (voir par exemple [6]), en une certaine série de Laurent $u_n^{-1}u_{n-1} = \sum_{j>-\infty} \varphi_j w^{-j}$ avec $\varphi_j \in k(v, q_2)$ pour tout j. Puisque w est invariant sous G, on a $g(u_n^{-1}u_{n-1}) = \sum_{j>-\infty} g(\varphi_j) w^{-j}$, de sorte que l'identification des coefficients des termes en w dans l'égalité précédente conduit à

$$g(\varphi_{-1}) = \frac{1}{\Delta_g}\varphi_{-1} + \frac{1}{\Delta_g} n\gamma_g(\gamma_g v + \delta_g)^{-1} q_2^{-2}, \quad \text{pour tout } g \in G.$$

L'élément φ_{-1} de $k(v, q_2)$ se développe quant à lui dans $k(v)((q_2^{-1}))$ en une série $\varphi_{-1} = \sum_{l>-\infty} a_l q_2^{-l}$ avec $a_l \in k(v)$, de sorte que l'identification des coefficients des termes en q_2^{-2} dans l'égalité précédente conduit à

$$(\gamma_g v + \delta_g)^{-2} g(a_2) = \frac{1}{\Delta_g} a_2 + \frac{1}{\Delta_g} n\gamma_g(\gamma_g v + \delta_g)^{-1}, \quad \text{pour tout } g \in G.$$

On a ainsi trouvé dans $k(v)$ un élément $b = n^{-1}a_2$ qui vérifie

$$g(b) = \frac{1}{\Delta_g}(\gamma_g v + \delta_g)^2 b + \frac{1}{\Delta_g}\gamma_g(\gamma_g v + \delta_g), \quad \text{pour tout } g \in G.$$

Posons $y = x + bq_2^{-2}w$. D'une part, on a pour tout $g \in G$:

$$g(y) = g(x) + \left[\frac{1}{\Delta_g}(\gamma_g v + \delta_g)^2 b + \frac{1}{\Delta_g}\gamma_g(\gamma_g v + \delta_g)\right](\gamma_g v + \delta_g)^{-2} q_2^{-2} w$$
$$= \frac{1}{\Delta_g}x - \frac{1}{\Delta_g}\gamma_g q_2^{-2}(\gamma_g v + \delta_g)^{-1} w + \frac{1}{\Delta_g} bq_2^{-2} w + \frac{1}{\Delta_g}\gamma_g(\gamma_g v + \delta_g)^{-1} q_2^{-2} w$$
$$= \frac{1}{\Delta_g} y.$$

D'autre part, il est clair que y engendre S sur K, et on vérifie immédiatement à partir des relations entre v, q_2, w et x que l'on a: $[y, v] = q_2^{-2}$, $[y, q_2] = bq_2^{-1}$, $[y, w] = 2y$. Le reste du lemme en découle. □

Le but du lemme suivant est d'exprimer, grâce à deux applications successives du théorème 1.4, l'algèbre S^G comme une extension de Ore itérées en deux variables à coefficients dans le corps commutatif d'invariants $k(v, q_2)^G$.

Lemme 3.5. *Les données, hypothèses et notations sont celles du lemme 3.4. On désigne par n le minimum des degrés en x non-nuls des éléments de S^G n'appartenant pas à K.*

(i) On a $K^G = k(v, q_2)^G(w\,;\, d)$, où d désigne encore la restriction de d à $k(v, q_2)^G$.
(ii) Il existe un entier relatif m et une fraction rationnelle $f_m(v) \in k(v)$ tels que le monôme $f = f_m(v)q_2^m \in k(v, q_2)$ vérifie $g(f) = (\Delta_g)^n f$ pour tout $g \in G$.
(iii) L'élément $z = fy^n$ est invariant sous l'action de G.
(iv) En notant τ le k-automorphisme du corps gauche $K^G = k(v, q_2)^G(w\,;\, d)$ fixant tout élément de $k(v, q_2)^G$ et tel que $\tau(w) = w + 2n - m$, il existe une τ-dérivation D de K^G vérifiant $D(w) = 0$ telle que:

$$S^G = k(v, q_2)^G(w\,;\, d)[z\,;\, \tau,\ D] \quad et \quad D_2(k)^G = k(v, q_2)^G(w\,;\, d)(z\,;\, \tau,\ D).$$

De plus la restriction de D à $k(v, q_2)^G$ est nulle lorsque $n \geq 2$.

Preuve. Rappelons d'abord que l'élément non-nul u_n de K considéré au début de la preuve du lemme 3.4 vérifie $g(u_n) = (\Delta_g)^n u_n$ pour tout $g \in G$. Cet élément u_n de $K = k(v, q_2)(w\,;\, d)$ se développe dans le corps d'opérateurs pseudo-différentiels formels $K = k(v, q_2)((w^{-1}\,;\, -d))$ en une certaine série de Laurent $u_n = \sum_{j \geq j_0} \psi_j w^{-j}$ avec $\psi_j \in k(v, q_2)$ pour tout j et $\psi_{j_0} \neq 0$. Comme w est invariant sous l'action de G, on a en particulier $g(\psi_{j_0}) = (\Delta_g)^n \psi_{j_0}$ pour tout $g \in G$. L'élément non-nul ψ_{j_0} de $k(v, q_2)$ se développe lui-même dans $k(v)((q_2))$ en une série de Laurent: $\psi_{j_0} = \sum_{i \geq m} f_i q_2^i$ avec $f_i \in k(v)$, $f_m \neq 0$. Et donc en particulier $g(f_m)(\gamma_g v + \delta_g)^m = (\Delta_g)^n f_m$ pour tout $g \in G$.

Posons alors $f = f_m q_2^m \in k(v, q_2)$. Il vérifie par construction $f \neq 0$ et $g(f) = (\Delta_g)^n f$ pour tout $g \in G$. Introduisons enfin $z = fy^n \in S$. Par construction, $g(z) = z$ pour tout $g \in G$. Par ailleurs, il est clair que le degré en x de z dans S est égal à n, puisque le degré en x de y est égal à 1. Comme n est précisément la valeur minimale des degrés des polynômes de $S = K[x\,;\,\sigma', d']$ n'appartenant pas à K et invariants sous G, le théorème 1.4 montre qu'il existe un automorphisme τ et une τ-dérivation D de K^G tels que

$$S^G = K^G[z\,;\,\tau, D] \text{ et } D_2(k)^G = \text{Frac}\, S^G = K^G(z\,;\,\tau, D).$$

En appliquant le théorème 1.4 dans l'algèbre $S_0 = k(v, q_2)[w\,;\,d]$, dont le corps de fractions est K, il est clair (puisque w est invariant sous G) que $K^G = k(v, q_2)^G(w\,;\,d)$, d'où

$$S^G = k(v, q_2)^G(w\,;\,d)[z\,;\,\tau, D] \text{ et } D_2(k)^G = k(v, q_2)^G(w\,;\,d)(z\,;\,\tau, D).$$

On peut expliciter la relation de commutation entre z et w. En effet, comme $yw = (w+2)y$, on a $y^n w = (w+2n)y^n$, donc $fy^n w = (fw+2nf)y^n = (wf - d(f))y^n + 2nz = wz + (2n - d(f)f^{-1})z$. Rappelons que la dérivation d de $k(v, q_2)$ vérifie $d(v) = 0$ et $d(q_2) = q_2$, de sorte que $d(f) = d(f_m q_2^m) = m f_m q_2^m = mf$. On a donc finalement:

$$zw - wz = (2n - m)z.$$

On peut donner quelques précisions sur les valeurs prises par τ et D sur $K^G \cap k(v, q_2) = k(v, q_2)^G$. Rappelons d'abord que la restriction de la σ'-dérivation d'' à $k(v, q_2)$ est une (vraie) dérivation, définie par $d''(v) = q_2^{-2}$ et $d''(q_2) = b(v)q_2^{-1}$. Pour toute fraction rationnelle $a \in k(v, q_2)$, on a $ya = ay + d''(a)$, d'où par une récurrence évidente: $y^n a = ay^n + nd''(a)y^{n-1} + \cdots$, où le reste désigné par les points de suspension est de degré $\leq n-2$ en y. En multipliant à gauche par l'élément $f \in k(v, q_2)$, on obtient

$$za = az + nd''(a)fy^{n-1} + \cdots \quad \text{pour tout } a \in k(v, q_2),$$

où le reste $nd''(a)fy^{n-1} + \cdots$ est dans $k(v, q_2)[y\,;\,d'']$. Si $a \in k(v, q_2)^G$, on a par ailleurs $za = \tau(a)z + D(a)$. On tire de ces deux relations que

$$(\tau(a) - a)z + D(a) = nd''(a)fy^{n-1} + \cdots \quad \text{pour tout} a \in k(v, q_2)^G.$$

Donc en identifiant les coefficients dans l'égalité $(\tau(a) - a)fy^n + D(a) = nd''(a)fy^{n-1} + \cdots$ ainsi obtenue, on déduit que la restriction de τ à $k(v, q_2)^G$ est l'identité, de sorte que la restriction de D à $k(v, q_2)^G$ est une (vraie) dérivation. Si de plus $n \geq 2$, l'identification implique aussi que $d''(a) = 0$ pour tout $a \in k(v, q_2)^G$, et donc la restriction de D à $k(v, q_2)^G$ est nulle. Ce qui achève la preuve. □

Le lemme précédant établissant que $D_2(k)^G$ est le corps des fractions d'une extension de Ore itérée, il reste à l'identifier comme un corps de Weyl. Le raisonnement se scinde en deux cas suivant que $n \geq 2$ ou que $n = 1$.

3.1.6. Cas où $n \geq 2$

Lemme 3.6. *Les données, hypothèses et notations sont celles du lemme 3.5. On suppose de plus que n* ≥ 2. *Alors* $D_2(k)^G \simeq \mathcal{D}_{0,2}(k)$, *ou* $D_2(k)^G \simeq \mathcal{D}_{1,0}(k)$, *ou* $D_2(k)^G \simeq \mathcal{D}_{1,1}(k)$.

Preuve. D'après le lemme 3.5, z commute avec tout élément de $k(v, q_2)^G$. On note donc que $D_2(k)^G = k(v, q_2)^G(w\,;\,d)(z\,;\,\tau)$. De plus, il résulte des calculs faits ci-dessus que $d''(a) = 0$ pour tout $a \in k(v)^G$. Puisque la restriction à $k(v, q_2)$ de d'' est $q_2^{-2}\partial_v + b(v)q_2^{-1}\partial_{q_2}$, l'intersection avec $k(v)$ du noyau de d'' est réduite à k. Donc $k(v)^G = k$. Comme dans la preuve du lemme 3.3, on discute alors suivant les cas possibles pour le corps commutatif $k(v, q_2)^G$. Rappelons que, pour tout $g \in G$, on a

$$g(v) = \frac{\alpha_g v + \beta_g}{\gamma_g v + \delta_g}, \qquad g(q_2) = (\gamma_g v + \delta_g)q_2.$$

Le groupe G opère donc par automorphismes sur la sous-algèbre $B = k(v)[q_2]$, ce qui permet d'appliquer le théorème 1.4 (sous la forme commutative originale de Miyata, [16]). On distingue deux cas:

Premier cas: on suppose que $B^G \subset k(v)^G$. Donc $k(v)(q_2)^G = k(v)^G$. On a vu ci-dessus que $k(v)^G = k$. On obtient ainsi $D_2(k)^G = k(w)(z\,;\,\tau)$. Puisque τ est défini par $\tau(w) = w+2n-m$, on conclut que $D_2(k)^G = k(w, z) \simeq \mathcal{D}_{0,2}(k)$ lorsque $m = 2n$, et que $D_2(k)^G \simeq \mathcal{D}_{1,0}(k)$ lorsque $m \neq 2n$.

Second cas: on suppose que $B^G \not\subset k(v)^G$. Notons e le minimum des degrés en q_2 non-nuls des éléments de B^G. Soit q un élément de B de degré e invariant sous G. Vu la forme de l'action de G sur q_2, on peut sans restriction supposer que q est un monôme de la forme $q = sq_2^e$, avec $s \in k(v)$, $s \neq 0$. L'application du théorème 1.4 conduit alors à $k(v, q_2)^G = k(v)^G(q)$, c'est-à-dire $k(v, q_2)^G = k(q)$. On en déduit que $D_2(k)^G = k(q)(w\,;\,d)(z\,;\,\tau)$. Rappelons que la dérivation d de $k(v, q_2)$ est définie par $d(v) = 0$ et $d(q_2) = q_2$. Donc $d(q) = d(sq_2^e) = seq_2^e = eq$. En résumé:

$$[w, q] = eq, \quad [z, q] = 0, \quad [z, w] = (2n - m)z.$$

Comme l'entier e est non-nul, on conclut en appliquant le lemme 3.1 à la dérivation $(2n - m)z\partial_z + eq\partial_q$ de $k(z, q)$ que $D_2(k)^G \simeq \mathcal{D}_{1,1}(k)$. □

3.1.7. Cas où $n = 1$

Lemme 3.7. *Les données, hypothèses et notations sont celles du lemme 3.5. On suppose de plus que $n = 1$. Alors $D_2(k)^G \simeq \mathcal{D}_{0,2}(k)$, ou $D_2(k)^G \simeq \mathcal{D}_{1,0}(k)$, ou $D_2(k)^G \simeq \mathcal{D}_{1,1}(k)$ ou $D_2(k)^G \simeq \mathcal{D}_{2,0}(k)$.*

Preuve. On peut expliciter les valeurs prises par D sur $k(v, q_2)$. En effet, rappelons (voir le lemme 3.5) que $z = fy$, où $f = f_m q_2^m$ pour un certain entier $m \in \mathbb{Z}$ et une fraction $f_m \in k(v)$ non-nulle satisfaisant la condition $g(f_m) = \Delta_g(\gamma_g v + \delta_g)^{-m} f_m$ pour tout $g \in G$. Donc $D = fd''$, où la dérivation d'' de $k(v, q_2)$ est définie (voir le lemme 3.4) par $d''(v) = q_2^{-2}$ et $d''(q_2) = bq_2^{-1}$, pour un certain $b \in k(v)$. En rappelant le lemme 3.5, on peut donc voir aussi le corps $D_2(k)^G$ comme corps de fractions d'un anneau de polynômes de Ore tordu uniquement par dérivations, sous la forme

$$D_2(k)^G = k(v, q_2)^G(z\,;\,D)(w\,;\,d),$$

où D est la restriction à $k(v,q_2)^G$ de la dérivation de $k(v,q_2)$ définie par $D(v) = f_m q_2^{m-2}$ et $D(q_2) = f_m b q_2^{m-1}$, et d est la dérivation de $k(v,q_2)^G(z\,;\,D)$ telle que $d(z) = (m-2)z$ et dont la restriction à $k(v,q_2)^G$ est définie par $d(v) = 0$ et $d(q_2) = q_2$.

Comme dans la preuve du lemme 3.6, on considère alors deux cas suivant la nature des invariants sous G de l'algèbre commutative $B = k(v)[q_2]$.

Premier cas: on suppose que $B^G \subset k(v)^G$. Donc $k(v)(q_2)^G = k(v)^G$. D'après le théorème de Lüroth, on a $k(v)^G = k$ ou $k(v)^G = k(t)$, extension transcendante pure de k.

Lorsque $k(v)^G = k$, on obtient $D_2(k)^G = k(z)(w\,;\,d)$ avec $d = (m-2)\partial_z$, d'où $D_2(k)^G = k(z,w) \simeq \mathcal{D}_{0,2}(k)$ si $m = 2$, et $D_2(k)^G \simeq D_1(k) \simeq \mathcal{D}_{1,0}(k)$ si $m \neq 2$.

Lorsque $k(v)^G = k(t)$, on remarque d'abord que, puisque $t \in k(v)$, on a: $D(t) = \partial_v(t)D(v) = \partial_v(t) f_m q_2^{m-2}$. Comme $D(t)$ doit appartenir à $k(v)(q_2)^G = k(t) \subset k(v)$, et comme ni $\partial_v(t)$ ni f_m ne peuvent être nuls dans $k(v)$, on a forcément $m = 2$, ce qui implique que $d(z) = 0$. Par ailleurs, $d(t) = 0$ puisque d est nulle sur $k(v)$. On obtient donc dans ce cas que $D_2(k)^G = k(t)(z\,;\,D)(w)$. Puisque $D(t) \neq 0$, on pose $z' = D(t)^{-1}z$ pour réécrire $D_2(k)^G$ sous la forme $k(w)(t)(z'\,;\,\partial_t) \simeq D_1(k(w))$ et conclure $D_2(k)^G \simeq \mathcal{D}_{1,1}(k)$.

Second cas: on suppose que $B^G \not\subset k(v)^G$. Comme on l'a vu au début du second cas de la preuve du lemme 3.6, on a alors $k(v,q_2)^G = k(v)^G(q)$, où q est un monôme de la forme $q = sq_2^e$, avec $s \in k(v)$ non-nulle et e un entier ≥ 1. Il est clair qu'alors $d(q) = eq$, puisque $d(v) = 0$ et $d(q_2) = q_2$. On calcule ensuite:

$$\begin{aligned} D(q) = D(s)q_2^e + seD(q_2)q_2^{e-1} &= \partial_v(s)D(v)q_2^e + sef_m b q_2^{e+m-2} \\ &= (\partial_v(s) + ensb) f_m q_2^{e+m-2}. \end{aligned}$$

Cet élément $D(q) = [z,q] \in k(v,q_2)^G = k(v)^G(q)$ se développe dans $k(v)^G((q))$ en une série de Laurent $D(q) = \sum_{j>-\infty} h_j q^j$ avec $h_j \in k(v)^G$. Puisque $q^j = s^j q_2^{ej}$ pour tout j, on déduit de l'égalité obtenue ci-dessus que $D(q)$ est un monôme $h_r q^r$, où l'entier r vérifie $e + m - 2 = er$, et $h_r = s^{-r}(\partial_v(s) + ensb) f_m$. On convient de noter simplement dans la suite $c = h_r \in k(v)^G$. On a donc pour résumer:

$$D_2(k)^G = k(v)^G(q)(z\,;\,D)(w\,;\,d),$$

avec $D(q) = cq^r$ et $d(q) = eq$, $c \in k(v)^G$, e entier ≥ 1, r entier tel que $e+m-2 = er$.

D'après le théorème de Lüroth, on a $k(v)^G = k$ ou $k(v)^G = k(t)$, extension transcendante pure de k. D'où la disjonction ci-dessous en deux sous-cas.

Premier sous-cas: on suppose $k(v)^G = k$. Donc $D_2(k)^G = k(q)(z\,;\,D)(w\,;\,d)$, avec les relations:

$$[z,q] = D(q) = cq^r, \quad [w,q] = d(q) = eq, \quad [w,z] = d(z) = (m-2)z.$$

Remarquons que, puisque $c \in k(v)^G$, on a ici $c \in k$. Lorsque $c = 0$, on obtient $D_2(k)^G = k(q,z)(w\,;\,d)$ avec $d = e\partial_q + (m-2)\partial_z$, et il suff it d'appliquer le lemme

3.1 pour conclure que $D_2(k)^G \simeq \mathcal{D}_{1,1}(k)$. Lorsque $c \neq 0$, posons $z' = c^{-1}q^{1-r}z$, de façon à avoir $[z', q] = q$. On calcule: $d(z') = d(q^{1-r})c^{-1}z + c^{-1}q^{1-r}d(z) = (1-r)eq^{1-r}c^{-1}z + c^{-1}q^{1-r}(m-2)z = (e - re + m - 2)z'$. Mais on a vu plus haut que $e + m - 2 - re = 0$, de sorte que, en posant $D' = c^{-1}q^{1-r}D$, on obtient $D_2(k)^G = k(q)(z' ; D')(w ; d)$, avec les relations:

$$[z', q] = D'(q) = q, \quad [w, q] = d(q) = eq, \quad [w, z'] = d(z') = 0.$$

On pose enfin $w' = w - ez'$, de sorte que $[w', q] = 0$, avec $[w', z'] = 0$ et $[z', q] = q$, d'où $D_2(k)^G = k(w', q)(z' ; q\partial_q)$, et on conclut que $D_2(k)^G = D_1(k(w')) \simeq \mathcal{D}_{1,1}(k)$.

Second sous-cas: on suppose $k(v)^G = k(t)$, extension transcendante pure de k. Il est clair que $d(t) = 0$ puisque d est nulle sur $k(v)^G$. On calcule $D(t) = \partial_v(t)D(v) = \partial_v(t)f_m q_2^{m-2}$. Mais $q = sq_2^e$ et $m - 2 = e(r-1)$, d'où $q_2^{m-2} = s^{1-r}q^{r-1}$, et donc $D(t) = \partial_v(t)f_m s^{1-r}q^{r-1}$. Posons pour simplifier $a = \partial_v(t)f_m s^{1-r}$. Remarquons que $a \neq 0$ dans $k(v)$, car f_m et s sont non-nuls (voir le début de cette preuve) et $t \notin k$. De plus, $a = D(t)q^{1-r} = [z, t]q^{1-r}$ est invariant sous G, donc appartient à $k(t)$. Ainsi

$$D_2(k)^G = k(t, q)(z ; D)(w ; d),$$

avec

$$\begin{aligned} &[z, t] = D(t) = aq^{r-1}, && [z, q] = D(q) = cq^r, && \text{où } a, c \in k(t), a \neq 0,\\ &[w, t] = d(t) = 0, && [w, q] = d(q) = eq, && [w, z] = d(z) = (m-2)z, \end{aligned}$$

et en rappelant que les entiers e, m, r vérifient $e \geq 1$ et $e + m - 2 - re = 0$.

Introduisons $z' = q^{1-r}z$ de sorte que $d(z') = (m-2)q^{1-r}z + (1-r)eq^{1-r}z = 0$. En notant $D' = q^{1-r}D$, on obtient

$$D_2(k)^G = k(t, q)(z' ; D')(w ; d),$$

avec

$$\begin{aligned} &[q, t] = 0, && [z', t] = D'(t) = a, && [w, t] = d(t) = 0,\\ & && [z', q] = D'(q) = cq, && [w, q] = d(q) = eq,\\ & && && [w, z'] = d(z') = 0. \end{aligned}$$

Les quatre éléments $t, q, w'' = \frac{1}{eq}w$ et $z'' = -\frac{c}{ea}w + \frac{1}{a}z'$ engendrent le corps $D_2(k)^G$ et vérifient:

$$\begin{aligned} &[q, t] = 0, && [z'', t] = 1, && [w'', t] = 0,\\ & && [z'', q] = 0, && [w'', q] = 1,\\ & && && [w'', z''] = 0. \end{aligned}$$

On conclut que $D_2(k)^G \simeq D_2(k) \simeq \mathcal{D}_{2,0}(k)$, ce qui achève la preuve du lemme.

□

On peut alors synthétiser les résultats obtenus sous la forme du théorème principal de cette section.

Théorème 3.8. *Soit $\mathcal{G}$ un groupe. Soit ρ une représentation de $\mathcal{G}$ de dimension 2 sur k. Alors, il existe deux entiers naturels m, t vérifiant* $1 \leq m+t \leq 2$, *tels que* $D_2(k)^{\mathcal{G},\rho} \simeq \mathcal{D}_{m,t}(k)$.

Preuve. Le résultat découle directement des lemmes 3.3, 3.7 et 3.6. □

Corollaire 3.9. *Pour toute représentation ρ de dimension 2 d'un groupe fini $\mathcal{G}$, on a* $D_2(k)^{\rho,\mathcal{G}} \simeq D_2(k)$.

Preuve. On applique le théorème 3.8 et la proposition 1.3. □

3.2. Application aux opérateurs différentiels invariants sur les surfaces de Klein

3.2.1. Données, notations, et synthèse

On prend ici $k = \mathbb{C}$. On fixe un sous-groupe fini G de $\mathrm{SL}(2, \mathbb{C})$ agissant naturellement sur $\mathbb{C}^2$ et on considère le prolongement canonique de cette action par automorphismes sur l'algèbre de Weyl $A_2(\mathbb{C}) = \mathbb{C}[q_1, q_2][p_1\,;\,\partial_1][p_2\,;\,\partial_2]$. On note toujours S la sous-algèbre $\mathbb{C}[q_1, q_2]$ et $D_2(\mathbb{C})$ le corps non commutatif $\mathrm{Frac}A_2(\mathbb{C})$. Comme expliqué au paragraphe 0.4 de l'introduction, le théorème 5 de [13] montre que $A_2(\mathbb{C})^G$ n'est autre que l'algèbre des opérateurs différentiels sur la surface de Klein S^G. Le corollaire 3.9 assure que $D_2(\mathbb{C})^G \simeq D_2(\mathbb{C})$. L'objet de ce qui suit est d'illustrer par quelques calculs explicites les raisonnements effectués en toute généralité dans la preuve du théorème 3.8.

Commençons par synthétiser les différentes étapes de la méthode.

(i) Un automorphisme quelconque $g \in G$ agit sur $A_2(\mathbb{C})$ par

$$g(q_1) = \alpha_g q_1 + \beta_g q_2, \qquad g(q_2) = \gamma_g q_1 + \delta_g q_2,$$
$$g(p_1) = \delta_g p_1 - \gamma_g p_2, \qquad g(p_2) = -\beta_g p_1 + \alpha_g p_2,$$

où $\alpha_g, \beta_g, \gamma_g, \delta_g \in \mathbb{C}$ avec ici $\Delta_g = \alpha_g\delta_g - \beta_g\gamma_g = 1$.

(ii) La triangularisation de l'action de G sur $\mathrm{Frac}S = \mathbb{C}(q_1, q_2) = \mathbb{C}(v, q_2)$ est donnée par

$$v = q_1 q_2^{-1}, \quad g(v) = \frac{\alpha_g v + \beta_g}{\gamma_g v + \delta_g 0}, \quad g(q_2) = (\gamma_g v + \delta_g) q_2.$$

Parce que G est fini, il existe $t \in \mathbb{C}(v)$ tel que $\mathbb{C}(v)^G = \mathbb{C}(t)$, et $\mathbb{C}(v)[q_2]^G \not\subset \mathbb{C}(v)[q_2]$. On a noté e le minimum des degrés en q_2 non-nuls des éléments de $\mathbb{C}(v)[q_2]^G$, et montré qu'il existe $s(v) \in \mathbb{C}(v)$ non-nul tel que l'élément $q = sq_2^e$ vérifie $\mathbb{C}(v)[q_2]^G = \mathbb{C}(t)[q]$. En résumé:

$$t, s \in \mathbb{C}(v) \text{ non-nuls, } e \geq 1 \text{ entier, } \mathbb{C}(v)^G = \mathbb{C}(t),$$
$$\mathbb{C}(q_1, q_2)^G = \mathbb{C}(v, q_2)^G = \mathbb{C}(t, q).$$

(iii) Toujours parce que G est fini, on ne peut pas être dans le cadre du lemme 3.3, et c'est donc le lemme 3.4 qui s'applique, permettant de déterminer une fraction rationnelle $b(v)$ et un générateur y; mais un point crucial est qu'ici $\Delta_g = 1$ pour tout $g \in G$, ce qui rend inutile le nouveau changement de variable $z = fy^n$ du lemme 3.5. Ce dernier s'applique donc directement avec $n = 1, m = 0, f = f_m = 1, z = y$ et $D = d''$. Et comme $n = 1$, on est ensuite dans le cas d'application du lemme 3.7 et non du lemme 3.6. En résumé, on obtient:

$$D_2(\mathbb{C})^G = \mathbb{C}(v, q_2)^G(z\,;\, D)(w\,;\, d) = \mathbb{C}(t, q)(z\,;\, D)(w\,;\, d),$$

avec
(1) $w = q_1p_1 + q_2p_2$,
(2) $z = q_2^{-1}p_1 + b(v)q_2^{-2}w$, où $b(v) \in \mathbb{C}(v)$ vérifie $g(b) = (\gamma_g v + \delta_g)^2 b + \gamma_g(\gamma_g v + \delta_g)$, pour tout $g \in G$.
(3) D est la restriction à $\mathbb{C}(t, q) = \mathbb{C}(v, q_2)^G$ de la dérivation de $\mathbb{C}(v, q_2)$ définie par $D(v) = q_2^{-2}$ et $D(q_2) = b(v)q_2^{-1}$,
(4) d est la restriction à $\mathbb{C}(t, q)(z\,;\, D)$ de la dérivation de $\mathbb{C}(v, q_2)(z\,;\, D)$ définie par $d(v) = 0$, $d(q_2) = q_2$, et $d(z) = -2z$.

(iv) Il est alors possible de reconnaître $D_2(\mathbb{C})^G$ comme un corps de Weyl $D_2(\mathbb{C})$ en déterminant explicitement des générateurs P_1, P_2, Q_1, Q_2 du corps d'invariants satisfaisant les relations canoniques. Rappelons le procédé décrit pour cela au second sous-cas du second cas de la preuve du lemme 3.7. On vérifie que $D(q)$ est de la forme

$$D(q) = c(t)q^r, \text{ où } r \text{ entier tel que } e - 2 = re, \text{ et } c(t) \in \mathbb{C}(t).$$

On vérifie que $D(t)$ est de la forme

$$D(t) = a(t)q^{j-r}, \text{ pour l'entier } r \text{ ci-dessus et}$$

$$a(t) = \partial_v(t)s^{1-r} \in \mathbb{C}(t), a(t) \neq 0.$$

On pose $P_1 = -\frac{c(t)}{ea(t)}w + \frac{q^{1-r}}{a(t)}z,\ Q_1 = t,\ P_2 = \frac{1}{eq}w,\ Q_2 = q$.
On a $D_2(\mathbb{C})^G = \mathbb{C}(Q_1, Q_2)(P_1\,;\, \partial_{Q_1})(P_2\,;\, \partial_{Q_2})$.

(v) En résumé, si l'on excepte l'étape (iv) ci-dessus, qui se limite à une vérification calculatoire automatique, la méthode se ramène à deux problèmes:
- le premier de théorie classique des invariants commutatifs est la détermination de t et q comme au (ii) ci-dessus,
- le second, spécifique au problème de Noether non commutatif, est la détermination d'un générateur z, c'est-à-dire d'une fraction rationnelle $b(v) \in \mathbb{C}(v)$, satisfaisant la condition (2) du point (iii) ci-dessus. On donne dans la suite une méthode de construction explicite d'un tel z.

3.2.2. Méthode de détermination d'un générateur z

L'algèbre commutative $S = \mathbb{C}[q_1, q_2]$ est munie d'une structure d'algèbre de Poisson pour le crochet de Poisson défini à partir des relations $\{q_1, q_2\} = 1$ et $\{q_1, q_1\} = \{q_2, q_2\} = 0$, c'est-à-dire défini par

$$\{s,t\} = \partial_1(s)\partial_2(t) - \partial_2(s)\partial_1(t) \quad \text{pour tous } s, t \in S.$$

Pour tout $s \in S$, l'application $\sigma_s : t \mapsto \{s,t\}$ est une dérivation de S, dite dérivation hamiltonienne associée à s. L'action naturelle de $\mathrm{SL}(2,\mathbb{C})$ est compatible avec le crochet de Poisson, et donc toute dérivation hamiltonienne associée à un élément de S^G est invariante pour le prolongement canonique de cette action à $A_2(\mathbb{C})$. Par l'introduction de dérivations logarithmiques hamiltoniennes associés à certains éléments homogènes de S^G, on en déduit au lemme suivant la construction de dérivations $z \in A_2(\mathbb{C})^G$ satisfaisant les conditions voulues.

Lemme 3.10. *Les données hypothèses et notations sont celles de 3.2.1. Si f est un élément non-nul de S^G homogène de degré $k \geq 2$, alors:*

(i) l'élément $z = -\frac{1}{k}f^{-1}\sigma_f$ de $D_2(\mathbb{C})$ est invariant sous l'action de G;
(ii) l'élément $b = -\frac{1}{k}f^{-1}\partial_1(f)q_2$ de $\mathbb{C}(q_1,q_2)$ appartient à $\mathbb{C}(v)$, et il vérifie

$$z = q_2^{-1}p_1 + b(v)q_2^{-2}w \text{ et } g(b) = (\gamma_g v + \delta_g)^2 b(v) + \gamma_g(\gamma_g v + \delta_g) \text{ pour tout } g \in G.$$

Preuve. Le point (i) est clair puisque $f \in S^G$, et donc $\sigma_f \in A_2(\mathbb{C})^G$. Pour (ii), notons $f = \sum \lambda_i q_1^{a_i} q_2^{k-a_i}$ avec $\lambda_i \in \mathbb{C}$ et $1 \leq a_i \leq k$. On a $f = (\sum \lambda_i v^{a_i})q_2^k$, en rappelant que $v = q_1 q_2^{-1}$. Par ailleurs $\partial_1(f) = \sum \lambda_i a_i q_1^{a_i-1} q_2^{k-a_i} = (\sum \lambda_i a_i v^{a_i-1})q_2^{k-1}$. D'où $b \in \mathbb{C}(v)$. Puisque f est homogène de degré k, on a $(q_1\partial_1 + q_2\partial_2)(f) = kf$, donc

$$\partial_2(f) = q_2^{-1}kf - q_2^{-1}q_1\partial_1(f) = k[q_2^{-1} + q_1 q_2^{-2} b(v)]f.$$

Il en résulte que $\sigma_f = \partial_1(f)p_2 - \partial_2(f)p_1 = -kb(v)q_2^{-1}fp_2 - k[q_2^{-1} + q_1 q_2^{-2} b(v)]fp_1$. L'élément z défini en (i) vérifie donc

$$z = b(v)q_2^{-1}p_2 + [q_2^{-1} + q_1 q_2^{-2} b(v)]p_1 = q_2^{-1}p_1 + b(v)q_2^{-2}w,$$

en rappelant que par définition $w = q_1 p_1 + q_2 p_2$. Le calcul de $g(b)$ s'en déduit en utilisant les égalités $g(z) = z$, $g(w) = w$, $g(q_2) = \gamma_g q_1 + \delta_g q_2$ et $g(p_1) = \delta_g p_1 - \gamma_g p_2$. □

La classification des sous-groupes finis de $\mathrm{SL}(2,\mathbb{C})$ en types A_n, D_n, E_6, E_7, E_8 est un résultat classique (voir par exemple [21] ou [1]). Pour tout sous-groupe fini G de $\mathrm{SL}(2,\mathbb{C})$, S^G est la $\mathbb{C}$-algèbre commutative engendrée par 3 polynômes homogènes f_1, f_2, f_3 explicitement déterminés suivant les types (voir par exemple [3] ou [4]):

- type A_n : $f_1 = q_1 q_2,\ f_2 = q_1^n,\ f_3 = q_2^n$,
- type D_n : $f_1 = q_1^2 q_2^2,\ f_2 = q_1^{2n} + (-1)^n q_2^{2n},\ f_3 = q_1^{2n+1} q_2 - (-1)^n q_1 q_2^{2n+1}$,
- type E_6 : $f_1 = q_1 q_2^5 - q_1^5 q_2,\ f_2 = q_1^8 + 14 q_1^4 q_2^4 + q_2^8$,
  $f_3 = q_1^{12} - 33 q_1^8 q_2^4 - 33 q_1^4 q_2^8 + q_2^{12}$,
- type E_7 : $f_1 = q_1^8 + 14 q_1^4 q_2^4 + q_2^8,\ f_2 = q_1^{10} q_2^2 - 2 q_1^6 q_2^6 + q_1^2 q_2^{10}$,
  $f_3 = q_1^{17} q_2 - 34 q_1^{13} q_2^5 + 34 q_1^5 q_2^{13} - q_1 q_2^{17}$,

• type E_8 : $f_1 = q_1^{11}q_2 + 11q_1^6q_2^6 - q_1q_2^{11}$,
$f_2 = q_1^{20} - 228q_1^{15}q_2^5 + 494q_1^{10}q_2^{10} + 228q_1^5q_2^{15} + q_2^{20}$,
$f_3 = q_1^{30} + 522q_1^{25}q_2^5 - 10005q_1^{20}q_2^{10} - 10005q_1^{10}q_2^{20} - 522q_1^5q_2^{25} + q_2^{30}$.

L'application du lemme 3.10 en choisissant par exemple $f = f_1$ permet de déterminer explicitement $b(v)$ et z.

Proposition 3.11. *Les données, hypothèses et notations sont celles de 3.2.1. Une fraction rationnelle $b(v) \in \mathbb{C}(v)$ telle que l'élément $z = q_2^{-1}p_1 + b(v)q_2^{-2}w$ satisfasse les conditions du point (iii) de 3.2.1 est donnée suivant le type de G par*

type	A_n	D_n	E_6	E_7	E_8
$b(v)$	$-\dfrac{1}{2v}$	$-\dfrac{1}{2v}$	$-\dfrac{1-5v^4}{6(v-v^5)}$	$-\dfrac{8v+56v^5}{8(v^8+14v^4+1)}$	$-\dfrac{11v^{10}+66v^5-1}{12(v^{11}+11v^6-v)}$

Preuve. Pour chaque type, on choisit le générateur f_1 de S^G donné dans la liste ci-dessus. En appelant k son degré total, on calcule $b(v) = -\frac{1}{k}f^{-1}\partial_1(f)q_2$ et on applique le lemme 3.10. □

La proposition ci-dessus résolvant la seconde des questions posées au point (v) de 3.2.1, la méthode résumée en 3.2.1 permet en théorie de calculer explicitement des générateurs P_1, Q_1, P_2, Q_2 de $D_2(\mathbb{C})^G$ en fonction des générateurs p_1, q_1, p_2, q_2 de $D_2(\mathbb{C})$ donnés au départ. On donne ci-dessous un exemple d'un tel calcul pour le type A_n.

3.2.3. Cas où G est du type A_{n-1}

On suppose ici que $G = \langle\gamma\rangle$ est cyclique d'ordre $n \geq 2$, où γ agit par

$$\gamma : q_1 \mapsto \omega q_1, \quad q_2 \mapsto \omega^{-1}q_2, \quad \text{avec } \omega = \exp\frac{2i\pi}{n}.$$

On a $\gamma(v) = \omega^2 v$. On peut expliciter les calculs résumés précédemment en 3.2.1, dont on reprend ci-dessous toutes les notations:

Premier cas: n est impair. Posons $n = 2p+1$, $(p \geq 0)$. On a $\mathbb{C}(v)^G = \mathbb{C}(t)$ pour $t = v^n$. L'élément $q = v^{p+1}q_2$ de $\mathbb{C}(v)[q_2]$ est invariant par G, de degré en q_2 égal à 1 donc minimal; donc $e = 1$ et $s(v) = v^{p+1}$. Conformément à la proposition 3.11, on prend $b(v) = -\frac{1}{2v}$. Donc $D(q_2) = -\frac{1}{2v}q_2^{-1}$ et $D(v) = q_2^{-2}$ permettent de calculer $D(q) = \frac{n}{2}tq^{-1}$ et $D(t) = nt^2q^{-2}$, d'où $r = -1$, $c(t) = \frac{n}{2}t$ et $a(t) = nt^2$. Par ailleurs, le calcul de $z = q_2^{-1}p_1 + b(v)q_2^{-2}w$ conduit à $z = \frac{1}{2}q_2^{-1}p_1 - \frac{1}{2}q_1^{-1}p_2$. On conclut que les quatre éléments:

$$Q_2 = q = q_1^{p+1}q_2^{-p}$$

$$
\begin{aligned}
Q_1 &= t = q_1^{2p+1} q_2^{-2p-1} \\
P_2 &= q^{-1} w = q_1^{-p} q_2^{p} p_1 + q_1^{-p-1} q_2^{p+1} p_2 \\
P_1 &= -c(t)a(t)^{-1} w + q^2 a(t)^{-1} z \\
&= -\frac{p}{2p+1} q_1^{-2p} q_2^{2p+1} p_1 - \frac{p+1}{2p+1} q_1^{-2p-1} q_2^{2p+2} p_2
\end{aligned}
$$

vérifient

$$\operatorname{Frac} S^G = \mathbb{C}(q_1, q_2)^G = \mathbb{C}(Q_1, Q_2) \text{ et } D_2(\mathbb{C})^G = \mathbb{C}(Q_1, Q_2)(P_1 ; \partial_{Q_1})(P_2 ; \partial_{Q_2}).$$

Second cas: n'est pair. Posons $n = 2p$, $(p \geq 1)$. On détermine successivement $t = v^p$ et $q = vq_2^2$, d'où $e = 2$ et $s(v) = v$. Avec toujours $D(q_2) = -\frac{1}{2v} q_2^{-1}$ et $D(v) = q_2^{-2}$, on calcule $D(q) = 0$ et $D(t) = pt2q^{-1}$, d'où $r = 0$, $c(t) = 0$ et $a(t) = pt$. A partir de $z = \frac{1}{2} q_2^{-1} p_1 - \frac{1}{2} q_1^{-1} p_2$, on obtient que les quatre éléments:

$$
\begin{aligned}
Q_2 &= q = q_1 q_2 \\
Q_1 &= t = q_1^p q_2^{-p} \\
Q_1 &= t = q_1^p q_2^{-p} \\
P_2 &= \frac{1}{2} q^{-1} w = \frac{1}{2}(q_2^{-1} p_1 + q_1^{-1} p_2) \\
P_1 &= -\frac{1}{2} c(t)a(t)^{-1} w + q a(t)^{-1} z = \frac{1}{2p} q_1^{1-p} q_2^{p} p_1 - \frac{1}{2p} q_1^{-p} q_2^{p+1} p_2
\end{aligned}
$$

vérifient

$$\operatorname{Frac} S^G = \mathbb{C}(q_1, q_2)^G = \mathbb{C}(Q_1, Q_2) \text{ et } D_2(\mathbb{C})^G = \mathbb{C}(Q_1, Q_2)(P_1 ; \partial_{Q_1})(P_2 ; \partial_{Q_2}).$$

Le premier cas où l'on ne peut pas choisir pour appliquer le lemme 3.11 un élément $f_1 \in S^G$ monomial, ce qui rend un peu plus complexes les calculs, est celui du type E_6, que l'on traite ci-dessous à titre d'exemple.

3.2.4. Cas où G est du type E_6

On suppose ici que G est le groupe tétraédral binaire. Ce groupe G, d'ordre 24, est engendré par γ, μ et η agissant par

$$
\begin{aligned}
&\gamma : q_1 \mapsto i q_1, && q_2 \mapsto -i q_2, \\
&\mu : q_1 \mapsto i q_2, && q_2 \mapsto i q_1, \\
&\eta : q_1 \mapsto \frac{1}{\sqrt{2}}(\zeta^7 q_1 + \zeta^7 q_2), && q_2 \mapsto \frac{1}{\sqrt{2}}(\zeta^5 q_1 + \zeta q_2),
\end{aligned}
$$

avec $i=\exp\frac{2i\pi}{4}$ et $\zeta=\exp\frac{2i\pi}{8}$. L'action de G sur $\mathbb{C}(c,q_2)$ est donc définie par

$$\gamma(v)=-v,\qquad \mu(v)=\frac{1}{v},\qquad \eta(v)=i\frac{v+1}{v-1},$$
$$\gamma(q_2)=-iq_2,\qquad \mu(q_2)=ivq_2,\qquad \eta(q_2)=\frac{\zeta}{\sqrt{2}}(-v+1)q_2.$$

On définit dans $\mathbb{C}(v)$ les éléments:

$$u=\frac{v^2+1}{v^2-1}=\frac{q_1^2+q_2^2}{q_1^2-q_2^2}\quad \text{et}$$
$$h=\frac{u^2+j}{u^2+j^2}=\frac{jv^4+2(1-j^2)v^2+j}{v^4+2(j-j^2)v^2+1}$$
$$=\frac{jq_1^4+2(1-j^2)q_1^2q_2^2+jq_2^4}{q_1^4+2(j-j^2)q_1^2q_2^2+q_2^4}.$$

Ils vérifient par construction:

$$\gamma(u)=u,\qquad \mu(u)=-u,\qquad \eta(u)=\frac{2v}{v^2+1},$$
$$\gamma(h)=h,\qquad \mu(h)=h,\qquad \eta(h)=jh.$$

On a $\mathbb{C}(v)^\gamma=\mathbb{C}(v^2)$. Mais $\mathbb{C}(v^2)=\mathbb{C}(u)$ car $v^2=\frac{u+1}{u-1}$. Or $\mu(u)=-u$, donc $\mathbb{C}(v)^{\langle\gamma,\mu\rangle}=\mathbb{C}(u^2)$, qui n'est autre que $\mathbb{C}(h)$ car $u^2=\frac{j-j^2h}{h-1}$. Puisque $\eta(h)=jh$, on pose

$$t=h^3=\left(\frac{u^2+j}{u^2+j^2}\right)^3=\left(\frac{jv^4+2(1-j^2)v^2+j}{v^4+2(j-j^2)v^2+1}\right)^3$$
$$=\left(\frac{jq_1^4+2(1-j^2)q_1^2q_2^2+jq_2^4}{q_1^4+2(j-j^2)q_1^2q_2^2+q_2^4}\right)^3,$$

et on conclut que $\mathbb{C}(v)^G=\mathbb{C}(u^2)^\eta=\mathbb{C}(h)^\eta=\mathbb{C}(t)$.

On détermine ensuite $q=s(v)q_2^e$. Comme on doit avoir $\gamma(q)=q$ avec $\gamma(v)=-v$ et $\gamma(q_2)=-iq_2$, on a forcément e pair. Le degré minimum cherché est $e=2$ car l'élément

$$s(v)=vuh(h-1)=(j-j^2)\frac{(v^2+1)(jv^4+2(1-j^2)v^2+j)}{(v^2-1)(v^4+2(j-j^2)v^2+1)^2},$$

vérifie $\gamma(s)=-s$, $\mu(s)=-v^{-2}s$ et $\eta(s)=-2i(v-1)^{-2}s$, conditions qui traduisent que $q=s(v)q_2^2$ est invariant par G. Conformément à la proposition 3.11, on prend $b(v)=\frac{5v^4-1}{6v(1-v^4)}$, que l'on reporte dans $D(q_2)=b(v)q_2^{-1}$ pour calculer, avec $D(v)=$

q_2^{-2}, la valeur de $D(q) = D(s)q_2^2 + 2s(v)b(v)$. Après un calcul élémentaire mais fastidieux, on obtient $D(q) = \frac{2}{j-j^2}(1-2t)$, d'où $r = 0$ et $c(t) = \frac{2}{j-j^2}(1-2t)$. Puisque $r = 0$, la fraction $D(q)$ est donnée par $D(q) = a(t) = \partial_v(t)s$. On obtient après calcul $a(t) = \frac{6}{j^2-j}t(t-1)$. Suivant les formules données au point (iv) du paragraphe 3.2.1, on conclut que

$$Q_2 = s(v)q_2^2,$$
$$Q_1 = t$$
$$P_2 = \frac{1}{2s(v)}q_2^{-2}q_1p_1 + \frac{1}{2s(v)}q_2^{-1}p_2,$$
$$P_1 = \left[\left(\frac{2s(v)b(v) - c(t)}{2a(t)}\right)q_1 + \left(\frac{s(v)}{a(t)}\right)q_2\right]p_1 + \left(\frac{2s(v)b(v) - c(t)}{2a(t)}\right)q_2p_2,$$

vérifient

$$\operatorname{Frac} S^G = \mathbb{C}(q_1, q_2)^G = \mathbb{C}(Q_1, Q_2) \text{ et } D_2(\mathbb{C})^G = \mathbb{C}(Q_1, Q_2)(P_1\,;\,\partial_{Q_1})(P_2\,;\,\partial_{Q_2}).$$

On peut évidemment exprimer P_1, Q_1, P_2, Q_2 uniquement en fonction des générateurs p_1, q_1, p_2, q_2 de départ en utilisant les formules explicites définissant $b(v), s(v), t, a(t), c(t)$ données ci-dessus.

Bibliographie

[1] J. Alev and F. Dumas, Invariants du corps de Weyl sous l'action de groupes finis, *Commun. Algebra* **25** (1997), 1655–1672.

[2] ——, *Sur les invariants des algèbres de Weyl et de leurs corps de fractions*, Lectures Notes Pure and Applied Math. **197** (1998), 1–10.

[3] J. Alev and Th. Lambre, Comparaison de l'homologie de Hochschild et de l'homologie de Poisson pour une déformation des surfaces de Klein, in *Algebra and Operator Theory* (Tashkent, 1997), Kluwer Acad. Publ., Dordrecht, 1998, 25–38.

[4] L. Chiang, H. Chu and M. Kang, Generation of invariants, *J. Algebra* **221** (1999), 232–241.

[5] I. V. Dolgachev, Rationality of fields of invariants, *Proc. Symposia Pure Math.* **46** (1987), 3–16.

[6] K. R. Goodearl and R. B. Warfield, *An Introduction to non commutative Noetherian Rings*, Cambridge University Press, London, 1985.

[7] I. M. Gelfand and A. A. Kirillov, Sur les corps liés aux algèbres enveloppantes des algèbres de Lie, *Inst. Hautes Etudes Sci. Publ. Math.* **31** (1966), 509–523.

[8] A. Joseph, A generalization of the Gelfand-Kirillov conjecture, *Amer. J. Math.* **99** (1977), 1151–1165.

[9] ——, Coxeter structure and finite group action, in *Algèbre non commutative, groupes quantiques et invariants* (Reims, 1995), 185–219, Sémin. Congr. 2, Soc. Math. France, Paris, 1997.

[10] P. I. Katsylo, Rationality of fields of invariants of reducible representations of the group SL_2, *Vestnik Moskov. Univ. Ser. I Mat. Mekh.* **39** (1984), 77–79.

[11] M. Kervaire et T. Vust, Fractions rationnelles invariantes par un groupe fini, in *Algebraische Transformationsgruppen und Invariantentheorie*, D.M.V. Sem. **13** Birkhäuser, Basel, (1989), 157–179.
[12] G.R. Krause and T.H. Lenagan, *Growth of algebras and Gelfand–Kirillov dimension. Revised version.* Graduate Studies in Mathematics, **22** American Mathematical Society, 2000.
[13] Th. Levasseur, Anneaux d'opérateurs différentiels, in *Séminaire d'Algèbre P. Dubreil - M.-P. Malliavin (1980)*, Lecture Notes in Math. **867**, Springer, Berlin-New York, 1981, 157–173.
[14] T. Maeda, On the invariant field of binary octavics, *Hiroshima Math. J.* **20** (1990), 619–632.
[15] J. C. Mc Connell and J. C. Robson, *Non commutative Noetherian Rings*, Wiley, Chichester, 1987.
[16] T. Miyata, Invariants of certain groups I, *Nagoya Math. J.* **41** (1971), 68–73.
[17] S. Montgomery, *Fixed Rings of Finite Automorphism Groups of Associative Rings*, Lecture Notes in Math. **818**, Springer-Verlag, Berlin, 1980.
[18] I. M. Musson, Actions of tori on Weyl algebras, *Commun. Algebra* **16** (1988), 139–148.
[19] D. Saltman, Noether's problem over an algebraically closed field, *Invent. Math.* **77** (1984), 71–84.
[20] ——, Groups acting on fields: Noether's problem, in *Group Actions on Rings* (Brunswick, Maine, 1984), Contemp. Math., **43**, Amer. Math. Soc., Providence, RI, 1985, 267–277.
[21] T. A. Springer, *Invariant Theory*, Lectures Notes in Maths **585**, Springer-Verlag, Berlin, 1977.
[22] E. B. Vinberg, Rationality of the field of invariants of a triangular group, *Vestnik Mosk. Univ. Mat.* **37** (1982), 23–24.
[23] J. Zhang, On lower transcendence degree, *Adv. Math.* **139** (1998), 157–193.

Langlands parameters for Heisenberg modules

A. Beilinson

University of Chicago
Department of Mathematics
Chicago, Illinois 60637
USA
sasha@math.uchicago.edu

To Tony Joseph on his 60th birthday

Summary. We describe a "spectral decomposition" of the category of representations of a Heisenberg Lie algebra whose parameter space is the moduli space of the de Rham local systems for the dual torus on the formal punctured disc, and the fibers are equal to the category of representations of the (twisted) lattice Heisenberg algebra.

Subject Classification: 17B69

Introduction

For an abelian category $\mathcal{A}$ it is often important to consider its "spectral decomposition" over some space S of spectral parameters. If S is set, then this is a direct product decomposition $\mathcal{A} = \prod_{s \in S} \mathcal{A}_s$. If S is an affine scheme, $S = \mathrm{Spec}\mathcal{O}(S)$, then this is a structure of $\mathcal{O}(S)$-category on $\mathcal{A}$. Notice that there is a universal affine S equal to Spec $Z(\mathcal{A})$, where $Z(\mathcal{A})$ is the Bernstein center of $\mathcal{A}$ (which is the endomorphism ring of the identity endofunctor of $\mathcal{A}$). If S is allowed to be a space of more general nature, then this is no longer true.

For example, the category of $\mathcal{D}$-modules on a smooth variety X can be seen as the category of $\mathcal{O}$-modules on the (non-algebraic) stack S equal to the quotient of X modulo the action of universal formal groupoid. Thus it has a spectral decomposition over S. As Fourier transform shows, a given abelian category may have different presentations of this type.

Spectral decompositions may provide a clue to the anticipated de Rham version of the local Langlands theory. Namely, let k be a field of characteristic 0, and $K \simeq k((t))$ a local k-field, so Spec$K$ is a formal punctured disc. Let G be a reductive group over k and $G^\vee$ its Langlands dual. Denote by $G(K)$ the group ind-scheme of formal loops (so for a test k-algebra R one has $G(K)(R) := G(K \hat{\otimes} R)$). A first approximation to the de Rham version of a smooth representation of a p-adic group is a "geometric" abelian category $\mathcal{A}$ equipped with a $G(K)$-action and a rigidification of the corresponding Lie algebra action. The Langlands philosophy suggests looking for a natural spectral decomposition of $\mathcal{A}$ over the moduli space $\mathcal{LS}_{G^\vee}$ of de Rham $G^\vee$-local systems on $\operatorname{Spec} K$.[1] Probably, to define it one needs to elevate the $G(K)$-action on $\mathcal{A}$ to a certain structure of chiral algebra origin.

In this article we consider, in an ad hoc manner, a toy example when G is a torus T and $\mathcal{A}$ is the category of Heisenberg modules, i.e., representations of a Heisenberg extension of the Lie algebra $\mathfrak{t}(K)$ of some integral level κ (here $\mathfrak{t}$ is the Lie algebra of T). We show that it admits a natural spectral decomposition over $\mathcal{LS}_{T^\vee}$ whose fibers are equivalent to the category of representations of a lattice vertex algebra of level κ (if κ is non-degenerate, then this is a semisimple category with $|\det\kappa|$ non-isomorphic irreducibles). As was noticed by E. Frenkel, the picture fits into the general pattern of [DVVV].

In the classical situation ($\kappa = 0$) it has a simple geometric meaning. Namely, the Heisenberg modules are the same as $\mathcal{O}^!$-modules on the space of connections $\mathcal{C}$ on a $T^\vee$-bundle over $\operatorname{Spec} K$, and our spectral decomposition corresponds to the evident projection $\mathcal{C} \twoheadrightarrow \mathcal{LS}_{T^\vee}$.

Conjecturally, the story admits a generalization to the case when G is any reductive group and $\mathcal{A}$ is the category of representations of the corresponding Kac–Moody algebra of some negative integral level (on which $G(K)$ acts by conjugation). We hope to return to this subject elsewhere.

V. Drinfeld was first to highlight the idea of spectral decomposition in the geometric theory of automorphic forms. I am grateful to him, D. Gaitsgory, and D. Kazhdan for stimulating discussions. The work was partially supported by NSF grant DMS-0100108.

1. The space of $T^\vee$-local systems

1.1.

As in the introduction, k is our base field of characteristic 0, so "scheme" means "k-scheme", etc. For the language of ind-schemes the reader is referred to [D] 6.3 or [BD2] 7.11. In fact, every ind-scheme S that appears below is an ind-affine reasonable $\aleph_0$-ind-scheme. Such S amounts to a topological commutative algebra $\mathcal{O}(S)$ whose topology admits a base formed by a sequence of ideals $I_1 \supset I_2 \supset \dots$ such that $\mathcal{O}(S) = \varprojlim \mathcal{O}(S)/I_n$ and each $I_a/I_n \subset \mathcal{O}(S)/I_n$ is a finitely generated ideal. We write $S = \operatorname{Spf} \mathcal{O}(S) := \cup \operatorname{Spec} \mathcal{O}(S)/I_n$, so for a commutative algebra R the set of R-points $S(R)$ is the set of all continuous morphisms of algebras $\mathcal{O}(S) \to R$. An *$\mathcal{O}^!$-module on S* is the same as a discrete $\mathcal{O}(S)$-module M of its sections; so $M = \cup M^{I_n}$ where $M^{I_n} \subset M$ is the submodule of elements killed by I_n (which are sections supported on $\operatorname{Spec} \mathcal{O}(S)/I_n$). Such M's form an abelian k-category $\mathcal{M}(S)$.

[1] Notice that $\mathcal{LS}_{G^\vee}$ has no global functions other than constants, so the usual Bernstein center of $\mathcal{A}$ does not help.

From now on T is our torus, $\Gamma := \mathrm{Hom}(\mathbb{G}_m, T)$ the corresponding lattice, so $T = \mathbb{G}_m \otimes \Gamma$. Let $\Gamma^\vee := \mathrm{Hom}(T, \mathbb{G}_m) = \mathrm{Hom}(\Gamma, \mathbb{Z})$ be the dual lattice, $T^\vee := \mathbb{G}_m \otimes \Gamma^\vee$ the dual torus. The corresponding Lie algebras are $\mathfrak{t} := k \otimes \Gamma$ and $\mathfrak{t}^\vee = \mathfrak{t}^* = k \otimes \Gamma^\vee$.

We fix a local field $K \simeq k((t))$; let $O \simeq k[[t]] \subset K$ be the ring of integers, $\mathfrak{m} \subset O$ its maximal ideal. For a commutative algebra R we write $K_R := K \hat{\otimes} R \simeq R((t))$, $O_R := O \hat{\otimes} R \simeq R[[t]]$, $\mathfrak{m}_R := \mathfrak{m} \hat{\otimes} R \simeq R[[t]] \simeq tR[[t]]$.

Any affine scheme Y of finite type defines an ind-scheme of formal loops $Y(K)$ of the above kind; one has $Y(K)(R) := Y(K_R)$. It contains a subscheme $Y(O)$, $Y(O)(R) := Y(O_R)$; there is a canonical morphism $Y(O) \to Y$ defined by the projection $O_R \to O_R/\mathfrak{m}_R = R$. So we have a group ind-scheme $T(K)$ with the Lie algebra $\mathfrak{t}(K) = \mathfrak{t} \otimes K$, and so on.

1.2.

Let $\mathcal{LS}_{T^\vee} = \mathcal{LS}_{T^\vee}(X)$ be the moduli stack of de Rham $T^\vee$-local systems on $X :=$ Spec K. By definition, an R-point of $\mathcal{LS}_{T^\vee}$ is a pair $(\mathfrak{F}, \nabla)_R = (\mathfrak{F}_R, \nabla)$ where $\mathfrak{F}_R$ is a $T^\vee$-torsor on $X_R :=$ Spec K_R and ∇ is an R-relative continuous connection on $\mathfrak{F}_R$. We assume that $\mathfrak{F}_R$ is trivial étale locally on Spec R.

Here are some convenient descriptions of $\mathcal{LS}_{T^\vee}$:

Let $\omega(K)$ be the space of 1-forms on X. This is naturally a commutative group ind-scheme: $\omega(K)(R)$ is the space of 1-forms on X_R relative to R. Thus $\omega(K) = \mathrm{Spf}\,\hat{\mathrm{Sym}} K$ where $\hat{\mathrm{Sym}} K := \varprojlim \mathrm{Sym}(K/\mathfrak{m}^n)$ and K is identified with the space of continuous linear functionals on $\omega(K)$ via a canonical pairing $K \times \omega(K) \to k$, $f, \phi \mapsto \mathrm{Res}\, f\phi$. The ind-scheme $\mathcal{C}$ of connections on the trivialized $T^\vee$-torsor on X identifies canonically with $\mathfrak{t}^\vee \otimes \omega(K)$, $\nu \mapsto \nabla_\nu := \partial_t + \nu$, so

$$\mathcal{O}(\mathcal{C}) = \hat{\mathrm{Sym}}\,\mathfrak{t}(K) := \varprojlim \mathrm{Sym}(\mathfrak{t} \otimes (K/\mathfrak{m}^n)). \tag{1.2.1}$$

The group ind-scheme $T^\vee(K)$ of automorphisms of a $T^\vee$-bundle acts on $\mathcal{C}$; the corresponding gauge action on $\mathfrak{t}^\vee \otimes \omega(K)$ is $g(\nu) = \nu + d\log(g)$. Thus

$$\mathcal{LS}_{T^\vee} = \mathcal{C}/T^\vee(K) = \mathfrak{t}^\vee \otimes \omega(K)/T^\vee(K). \tag{1.2.2}$$

Set $T^\vee(\mathfrak{m}) := \mathrm{Ker}(T^\vee(O) \twoheadrightarrow T^\vee)$ and $\Phi := T^\vee(K)/T^\vee(\mathfrak{m})$. Thus $T^\vee = T^\vee(O)/T^\vee(\mathfrak{m})$ is a subgroup of Φ and $\Phi/T^\vee = Q \times \Gamma^\vee$ where Q is the formal group whose Lie algebra equals $\mathfrak{t}^\vee \otimes (K/O)$. Set $\omega(K)^- := \omega(K)/\omega(O)$, $\bar{\mathcal{C}} := \mathfrak{t}^\vee \otimes \omega(K)^-$; one has $\bar{\mathcal{C}} = \mathrm{Spf}(\mathrm{Sym}\,\mathfrak{t}(O)) := \varinjlim \mathrm{Spec}(\mathrm{Sym}\,\mathfrak{t}(O/\mathfrak{m}^n))$ (see (1.2.1)). Since $d\log$ yields an isomorphism $T^\vee(\mathfrak{m}) \xrightarrow{\sim} \mathfrak{t}^\vee \otimes \omega(O)$, the group scheme $T^\vee(\mathfrak{m})$ acts freely along the fibers of the projection $\mathcal{C} \to \bar{\mathcal{C}}$, i.e., one has $\bar{\mathcal{C}} = \mathcal{C}/T^\vee(\mathfrak{m})$ and

$$\mathcal{LS}_{T^\vee} = \bar{\mathcal{C}}/\Phi. \tag{1.2.3}$$

There is a canonical decomposition $\omega(K)^- = \omega(K)^{irr} \times k$ where the projection $\omega(K)^- \to k$ is the residue map, and $k \hookrightarrow \omega(K)^-$ is the subspace of forms with pole of order one. Tensoring it by $\mathfrak{t}^\vee$, we get $\bar{\mathcal{C}} = \mathcal{C}^{irr} \times \mathfrak{t}^\vee$. Since the morphism $d\log : \Phi \to \mathfrak{t}^\vee \otimes \omega(K)^-$ kills $T^\vee$, it yields a morphism $Q \times \Gamma \to (\mathfrak{t}^\vee \otimes \omega(K)^{irr}) \times \mathfrak{t}^\vee$.

This is an embedding compatible with the product decomposition; it identifies Q with the formal completion of $\mathfrak{t}^\vee \otimes \omega(K)^{irr}$ at 0, and $\Gamma^\vee \hookrightarrow \mathfrak{t}^\vee$ is the usual embedding. Therefore, by (1.2.3), we have a canonical projection

$$\mathcal{LS}_{T^\vee} \to (\mathcal{C}^{irr}/Q) \times (\mathfrak{t}^\vee/\Gamma^\vee) \tag{1.2.4}$$

which makes $\mathcal{LS}_{T^\vee}$ a $T^\vee$-gerbe over $(\mathcal{C}^{irr}/Q) \times (t^\vee/\Gamma^\vee)$. Any splitting of the extension[2] $0 \to T^\vee \to \Phi \to Q \times \Gamma \to 0$ yields a trivialization of this gerbe, i.e., an identification

$$\mathcal{LS}_{T^\vee} \tilde{\to} (\mathcal{C}^{irr}/Q) \times (t^\vee/\Gamma^\vee) \times BT^\vee. \tag{1.2.5}$$

1.3.

We define an *$\mathcal{O}^!$-module on $\mathcal{LS}_{T^\vee}$* as an $\mathcal{O}^!$-module on $\mathcal{C}$ equivariant with respect to the action of the group ind-scheme $T^\vee(K)$. Thus this is a vector space M equipped with a discrete $\mathcal{O}(\mathcal{C})$-module structure and a $T^\vee(K)$-action[3] so that the obvious compatibilities are satisfied. The $\mathcal{O}^!$-modules on $\mathcal{LS}_{T^\vee}$ form an abelian k-category $\mathcal{M}(\mathcal{LS}_{T^\vee})$.

Since $T^\vee(\mathfrak{m})$ acts freely along the fibers of the projection of $\mathcal{C} \to \bar{\mathcal{C}}$, we see that $\mathcal{M}(\mathcal{LS}_{T^\vee})$ identifies canonically with the category of Φ-equivariant $\mathcal{O}^!$-modules on $\bar{\mathcal{C}}$. The equivalence assigns to M as above the module of $T^\vee(\mathfrak{m})$-invariants $M^{T^\vee(\mathfrak{m})}$ equipped with the induced actions of $\Phi := T^\vee(K)/T^\vee(\mathfrak{m})$ and the algebra $\mathcal{O}(\bar{\mathcal{C}})$.

Denote by $\mathcal{V}(M)$ the vector space of coinvariants of the action of the group ind-scheme Φ on $M^{T^\vee(\mathfrak{m})}$. Since $T^\vee(O)/T^\vee(\mathfrak{m}) = T^\vee$ is reductive, its coinvariants are the same as invariants, so one has

$$\mathcal{V}(M) := (M^{T^\vee(\mathfrak{m})})_\Phi = (M^{T(O)})_{Q\times\Gamma^\vee}. \tag{1.3.1}$$

The functor $M \mapsto \mathcal{V}(M)$ is *right* exact.

Remark. According to (1.2.5), an object of $\mathcal{M}(\mathcal{LS}_{T^\vee})$ can be thought of as a $\mathfrak{t}^\vee$-family of $\mathcal{D}$-modules on $\mathcal{C}^{irr}$ equipped with an action of the $\Gamma^\vee$-translations along $\mathfrak{t}^\vee$ and a Γ-grading. Then $\mathcal{V}$ is the middle de Rham cohomology functor along the $\mathcal{C}^{irr}$-variables, followed by taking the $\Gamma^\vee$-coinvariants and the 0 component of the grading.

Problem. The definitions of $\mathcal{M}(\mathcal{LS}_{T^\vee})$ and $\mathcal{V}$ look quite ad hoc for they use two specific structures available on the stack $\mathcal{LS}_{T^\vee}$. It would be very nice to find a right general geometric setting where they belong, which would cover, in particular, the case of moduli space of de Rham G-local systems on X for an arbitrary group G.

---

[2] For $T = \mathbb{G}_m$ such a splitting amounts to a choice of 1-jet of coordinate (which is a non-zero element of $\mathfrak{m}/\mathfrak{m}^2$); such a choice yields a splitting for any T.

[3] A rule that assigns to any test algebra R an action of the group $T^\vee(K_R)$ on $M \otimes R$ compatible with morphisms of R's.

1.4.

This subsection will not be used in the sequel. The category $\mathcal{M}(\mathcal{LS}_{T^\vee})$ has a following "Mellin transform" description which can be considered as a manifestation of the de Rham version of the local geometric class field theory.[4]

To formulate it, consider the ind-scheme $T(K)$. A *$\mathcal{D}$-module on $T(K)$* is an $\mathcal{O}^!$-module on $T(K)$ equivariant with respect to the translation action of the Lie algebra $\mathfrak{t}(K)$ of $T(K)$. Precisely, this is a vector space M equipped with $\mathcal{O}(T(K))$-module structure and a action of the commutative Lie algebra $\mathfrak{t}(K)$ that are compatible via the translation action of $\mathfrak{t}(K)$ on $\mathcal{O}(T(K))$; both $\mathcal{O}(T(K))$- and $\mathfrak{t}(K)$-actions are assumed to be continuous (the topology on M is discrete). These objects form an abelian k-category $\mathcal{M}(T(K), \mathcal{D})$.

Proposition. *There is a canonical equivalence of categories*

$$\mathcal{M}(\mathcal{LS}_{T^\vee}) \xrightarrow{\sim} \mathcal{M}(T(K), \mathcal{D}). \tag{1.4.1}$$

Proof. The equivalence is a Fourier–Laumon transform [L] combined with the local self-duality of Contou–Carrère [CC]. Namely, let M be a vector space. Let us show that the two kind of structures on M — a structure of $\mathcal{O}^!$-module on $\mathcal{LS}_{T^\vee}$ and of $\mathcal{D}$-module on $T(K)$ — actually do not differ.

The first structure consists of compatible $\mathcal{O}(\mathcal{C})$- and $T^\vee(K)$-actions, the second one of compatible $\mathcal{O}(T(K))$- and $\mathfrak{t}(K)$-actions. Now a $\mathcal{O}(\mathcal{C})$-action is the same as a $\mathfrak{t}(K)$-action by (1.2.1). Also a $T^\vee(K)$-action is the same as an $\mathcal{O}(T(K))$-action. Indeed, by [CC], the group ind-scheme $\mathbb{G}_m(K)$ is Cartier self-dual in a canonical way, so $T(K) = \Gamma \otimes \mathbb{G}_m(K)$ is Cartier dual to $T^\vee(K) = \Gamma^\vee \otimes \mathbb{G}_m(K)$, which means that the corresponding topological Hopf algebras are mutually dual (see [BD1] 3.10.12, 3.10.13), hence the assertion.

It remains to check that if one pair of structures is compatible, then such is the corresponding other pair. We leave it as an exercise to the reader. □

2. The Heisenberg modules and the spectral decomposition

2.1.

The basic material on lattice vertex (or chiral) algebras and their representations can be found in textbooks [K], [FBZ], [L], or [BD1]. We follow [BD1] 3.10 since the exposition in *loc. cit.* highlights the structures we need. Notice that [BD1] deals with chiral algebras on an algebraic curve, while we consider chiral algebras on $X :=$ Spec K and Spec R-families of such algebras (which are K_R-modules with an extra structure). The results and constructions from [BD1] 3.10 remain valid in this setting.

[4] The next proposition was discovered by V. Drinfeld and the author about 10 years ago. There is a companion global statement.

Let A be a lattice chiral algebra on X for the lattice Γ and $\kappa : \Gamma \times \Gamma \to \mathbb{Z}$ the corresponding bilinear symmetric form (so the isomorphism class of A is uniquely determined by κ, see [BD1] 3.10.4).

Our A is a Γ-graded chiral algebra. The component A^0 is a chiral subalgebra of A identified naturally with the twisted chiral enveloping algebra $U(\mathfrak{t}_{\mathcal{D}})^\kappa$ of a Heisenberg extension $\mathfrak{t}_{\mathcal{D}}^\kappa$ of the commutative Lie* algebra $\mathfrak{t}_{\mathcal{D}}$ (see [BD1] 3.10.9). We have the corresponding topological Heisenberg Lie algebra $\mathfrak{t}(K)^\kappa := h(\mathfrak{t}_{\mathcal{D}}^\kappa)$ which is a central k-extension of the commutative Lie algebra $\mathfrak{t}(K) = K \otimes \Gamma$ with the commutator pairing $(f_1 \otimes \gamma_1, f_2 \otimes \gamma_2) \mapsto \kappa(\gamma_1, \gamma_2)\mathrm{Res} f_1 df_2$.

2.2.

Let $(\mathfrak{F}, \nabla)_R$ be a Spec R-family of de Rham $T^\vee$-local systems as in 1.2. The Γ-grading amounts to a $T^\vee$-action on A, so, as in [BD1] 3.4.17, one gets the twisted form $A(\mathfrak{F}, \nabla)_R$ of the chiral R-algebra $A_R := A \otimes_K K_R$. This is a Spec R-family of lattice chiral algebras in the evident way.

Since the $T^\vee$-action on A^0 is trivial, the 0-component $A(\mathfrak{F}, \nabla)_R^0 = A^0(\mathfrak{F}, \nabla)_R$ of $A(\mathfrak{F}, \nabla)_R$ does not feel the twist, i.e., it equals A_R^0. Therefore one has a canonical identification

$$\alpha : U(\mathfrak{t}_{\mathcal{D}})_R^\kappa = A_R^0 \xrightarrow{\sim} A(\mathfrak{F}, \nabla)_R^0. \tag{2.2.1}$$

The above construction is compatible with the base change, so $A(\mathfrak{F}, \nabla)_R$ together with the base change identifications form a $\mathcal{LS}_{T^\vee}$-family of chiral algebras which we denote by $A^\natural$.

In particular, we have a family of chiral algebra $A_{\mathcal{C}}^\natural$ parametrized by the ind-scheme $\mathcal{C}$ (the moduli space of connections on the trivial torsor, see 1.2) equivariant with respect to the $T^\vee(K)$-action on $\mathcal{C}$.

2.3.

Below "A-module" means "chiral A-module supported at the closed point of Spec O" (see [BD1] 3.6.2); for a vector space M we refer to an A-module structure on M as an A-action on M. Same for $A(\mathfrak{F}, \nabla)_R$-modules, and so on.

Variant: Let $S = \mathrm{Spf}\, \mathcal{O}(S)$, $\mathcal{O}(S) = \varprojlim \mathcal{O}(S)/I_n$, be an ind-scheme as in 1.1, and $(\mathfrak{F}, \nabla)_S$ be an S-family of local systems. For an $\mathcal{O}^!$-module M on S an $A(\mathfrak{F}, \nabla)_S$-*action on* M is same as a datum of mutually compatible $A(\mathfrak{F}, \nabla)_{\mathcal{O}(S)/I_n}$-actions on M^{I_n} (or $A(\mathfrak{F}, \nabla)_{\mathcal{O}(S)/I_n}$-actions on M^{I_ℓ} for $n \geq \ell$).

We define an $A^\natural$-*module on* $\mathcal{LS}_{T^\vee}$ as a $T^\vee(K)$-equivariant $\mathcal{O}^!$-module M on $\mathcal{C}$ (see 1.3) equipped with an $A_{\mathcal{C}}^\natural$-action such that this action is compatible with the $T^\vee(K)$-actions on $A_{\mathcal{C}}^\natural$ and M. Such objects form an abelian category $\mathcal{M}(\mathcal{LS}_{T^\vee}, A^\natural)$.

2.4.

An $U(\mathfrak{t}_{\mathcal{D}})^\kappa$-module is the same as a discrete $\mathfrak{t}(K)^\kappa$-module such that the central element $1 \in k \subset \mathfrak{t}(K)^\kappa$ acts as identity (see [BD1] 3.7.22). We denote by $\mathcal{M}(\mathfrak{t}(K))^\kappa$ the category of $\mathfrak{t}(K)^\kappa$-modules as above.

The morphism of chiral $\mathcal{C}$-algebras $\alpha : U(\mathfrak{t}_{\mathcal{D}})^{\kappa}_{\mathcal{C}} \to A^{\natural}_{\mathcal{C}}$ (see (2.2.1)) is compatible with the $T^{\vee}(K)$-action. For $M \in \mathcal{M}(\mathcal{LS}_{T^{\vee}}, A^{\natural})$ it defines an action of $\mathfrak{t}(K)^{\kappa}$ on M which commutes with the $T^{\vee}(K)$-action. Therefore $\mathfrak{t}(K)^{\kappa}$ acts on the vector space $\mathcal{V}(M)$ (see 1.3). We have defined a functor

$$\mathcal{V} : \mathcal{M}(\mathcal{LS}_{T^{\vee}}, A^{\natural}) \to \mathcal{M}(\mathfrak{t}(K))^{\kappa}. \tag{2.4.1}$$

Theorem. *This is an equivalence of categories.*

This equivalence is the promised "spectral decomposition" of $\mathcal{M}(\mathfrak{t}(K))^{\kappa}$ over the moduli space of local systems $\mathcal{LS}_{T^{\vee}}$.

Remark. Choose any extension of A to a lattice chiral algebra on the disc Spec $O \supset$ Spec K. It yields a splitting of the Heisenberg extension $\mathfrak{t}(K)^{\kappa}$ over $\mathfrak{t}(O) \subset \mathfrak{t}(K)$. Therefore every character $\chi : \mathfrak{t}(O) \to k$ yields the induced $\mathfrak{t}(K)^{\kappa}$-module $V_{\chi} \in \mathcal{M}(\mathfrak{t}(K))^{\kappa}$. The corresponding $A^{\natural}$-module is supported at the point of $\mathcal{LS}_{T^{\vee}}$ equal to the class of $\chi \in \mathfrak{t}(O)^{*} = \mathfrak{t}^{\vee} \otimes \omega(K)^{-} = \bar{\mathcal{C}}$ in $\bar{\mathcal{C}}/\Phi$ (see (1.2.3)).

Proof of the Theorem. This takes the rest of the article.

2.5.

Let $\mathcal{T}^{\vee}_X = \operatorname{Spec} F^{\ell}_X := \mathcal{J}T^{\vee}_X$ be the jet $\mathcal{D}_X$-scheme of the group X-scheme $T^{\vee}_X = T^{\vee} \times X$. As in [BD1] 3.10.1, our A carries a canonical $\mathcal{T}^{\vee}_X$-action. The $T^{\vee}$-action on A, which defines the Γ-grading, is the restriction of this action to the constant group $\mathcal{D}_X$-subscheme $T^{\vee}_X \hookrightarrow \mathcal{T}^{\vee}_X$. The $\mathcal{T}^{\vee}_X$-action yields an action on A of the group ind-scheme $T^{\vee}(K)$ (= the ind-scheme of horizontal sections of $\mathcal{T}^{\vee}_X$).

The Hopf $\mathcal{D}_X$-algebra F^{ℓ}_X is also Γ-graded. Consider the 0-component $F^{\ell 0}_X$, so Spec $F^{\ell 0}$ is the quotient group $\mathcal{D}_X$-scheme $\mathcal{T}^{\vee}_X / T^{\vee}_X$. This is a vector group $\mathcal{D}_X$-scheme: as in [BD1] 3.10.9, there is a canonical isomorphism of Hopf $\mathcal{D}_X$-algebras $\operatorname{Sym} \mathfrak{t}^{\ell}_{\mathcal{D}} \xrightarrow{\sim} F^{\ell 0}_X$. It identifies the group ind-scheme of horizontal sections of $\mathcal{T}^{\vee}_X / T^{\vee}_X$ with $\mathfrak{t}^{\vee} \otimes \omega(K)$. Denote the projection $\mathcal{T}^{\vee}_X \twoheadrightarrow \mathcal{T}^{\vee}_X / T^{\vee}_X \xrightarrow{\sim} \operatorname{Spec} \operatorname{Sym} \mathfrak{t}^{\ell}_{\mathcal{D}}$ by $d \log$; the corresponding morphism of ind-schemes of horizontal sections $T^{\vee}(K) \to \mathfrak{t}^{\vee} \otimes \omega(K)$ is $g \mapsto d \log(g) = dg/g$.

The $\mathcal{T}^{\vee}_X$-action on $A^0 \subset A$ is trivial on $T^{\vee}_X$, hence it yields the action of the group $\mathcal{D}_X$-scheme $\mathcal{T}^{\vee}_X / T^{\vee}_X$. By [BD1] 3.10.9, the corresponding action on $A^0 = U(\mathfrak{t}_{\mathcal{D}})^{\kappa}$ of the group ind-scheme of horizontal sections $\mathfrak{t}^{\vee} \otimes \omega(K)$ comes from its evident action on the Heisenberg extension: $\mathfrak{t}^{\vee} \otimes \omega(K) = \operatorname{Hom}(\mathfrak{t}_{\mathcal{D}}, \omega) \xrightarrow{\sim} \operatorname{Aut}(\mathfrak{t}^{\kappa}_{\mathcal{D}})$.

2.6.

We see that for any $(\mathfrak{F}, \nabla)_R$ the twisted algebra $A(\mathfrak{F}, \nabla)_R$ coincides with the twist of A_R by the induced $\mathcal{D}_X$-scheme $\mathcal{T}^{\vee}_X$-torsor, which is the same as the twist of A_R by the $T^{\vee}(K)$-torsor of sections of $\mathfrak{F}_R$ (see Remark (iv) in [BD1] 3.4.17). Thus any section (i.e., trivialization) s of $\mathfrak{F}_R$ yields an isomorphism of chiral R-algebras $\beta_s : A_R \tilde{\to} A(\mathfrak{F}, \nabla)_R$.

In particular, since $\mathcal{C}$ is the space of connections on the *trivialized* $T^\vee$-torsor, one has a canonical identification of the $\mathcal{C}$-families of chiral algebras

$$\beta : A_{\mathcal{C}} \tilde{\to} A^\natural_{\mathcal{C}}. \tag{2.6.1}$$

By construction, it is compatible with the $T^\vee(K)$-actions (here $T^\vee(K)$ acts on $A_{\mathcal{C}}$ via the tensor product of the above action on A and the action on $\mathcal{C}$).

Let $\mathfrak{a}$ be a canonical automorphism of the chiral $\mathcal{C}$-algebra $A^0_{\mathcal{C}}$ whose value $\mathfrak{a}_\nu$ at a point $\nu \in \mathcal{C} = \mathfrak{t}^\vee \otimes \omega(K)$ is the action of ν on A^0.

Lemma. *One has*[5] $\beta^{-1}\alpha = \mathfrak{a} \in \mathrm{Aut}(A^0_{\mathcal{C}})$.

Proof. Let s be the distinguished section of the trivialized $T^\vee$-torsor $\mathfrak{F}$ on X, and $\mathcal{J}s$ the corresponding horizontal section of $\mathcal{J}\mathfrak{F}$. Any point $\nu \in \mathcal{C}$ yields a horizontal embedding $i_\nu : (\mathfrak{F}, \nabla_\nu) \hookrightarrow \mathcal{J}\mathfrak{F}$. For $a \in A$ one has $\beta(a) = \mathcal{J}s \cdot a$; if $a \in A^0$, then $\alpha(a) = i_\nu(s) \cdot a$. Hence $\beta^{-1}\alpha(a) = (i_\nu(s)/\mathcal{J}s)a$. The image by $d\log$ of the section $i_\nu(s)/\mathcal{J}s$ of $T^\vee_X$ is a horizontal section of Spec Sym $\mathfrak{t}^\ell_{\mathcal{D}}$ equal to ν; we are done. □

2.7.

Let $T(K)^\kappa$ be the Heisenberg $\mathbb{G}_m$-(super)extension of the group ind-scheme $T(K)$ that corresponds to A (see [BD1] 3.10.14). Therefore the structure of an A-module on a vector space amounts to a $T(K)^\kappa$-module structure, i.e., a $T(K)^\kappa$-action on it such that $\mathbb{G}_m \subset T(K)^\kappa$ acts by homotheties.

The $T^\vee(K)$-action on A yields, by transport of structure, a $T^\vee(K)$-action on the extension $T(K)^\kappa$ which can be described as follows (see loc. cit.). Write $T^\vee(K) = \Gamma^\vee \otimes \mathbb{G}_m(K)$ and $\mathrm{Aut}(T(K)^\kappa) = \mathrm{Hom}(T(K), \mathbb{G}_m) = \Gamma^\vee \otimes \mathrm{Hom}(\mathbb{G}_m(K), \mathbb{G}_m)$; our action is the tensor productt of $id_{\Gamma^\vee}$ and the Contou–Carrère self-duality isomorphism $\mathbb{G}_m(K) \tilde{\to} \mathrm{Hom}(\mathbb{G}_m(K), \mathbb{G}_m)$.

Consider the semidirect product $T^\vee(K) \ltimes T(K)^\kappa$ with respect to this action. This is a central $\mathbb{G}_m$-(super)extension of $T^\vee(K) \times T(K) = (T^\vee \times T)(K)$ split over $T^\vee(K)$; denote it by $(T^\vee \times T)(K)^{h_\kappa}$. We see that $(T^\vee \times T)(K)^{h_\kappa}$ is a Heisenberg extension for the symmetric bilinear form h_κ on $\Gamma^\vee \times \Gamma$, $h_\kappa(\gamma_1^\vee + \gamma_1, \gamma_2^\vee + \gamma_2) = \gamma_1^\vee(\gamma_2) + \gamma_2^\vee(\gamma_1) + \kappa(\gamma_1, \gamma_2)$ (see [BD1] 3.10.13).

2.8.

Denote by $\mathcal{M}((T^\vee \times T)(K))^{h_\kappa}$ the category of $(T^\vee \times T)(K)^{h_\kappa}$-modules. Take any $M \in \mathcal{M}((T^\vee \times T)(K))^{h_\kappa}$ and consider it as a $T^\vee(K)$-module via the canonical embedding $T^\vee(K) \hookrightarrow (T^\vee \times T)(K)^{h_\kappa}$; set

$$\mathcal{V}(M) := (M^{T^\vee(O)})_{Q \times \Gamma^\vee}. \tag{2.8.1}$$

(recall that $Q \times \Gamma^\vee = T^\vee(K)/T^\vee(O)$ is a group ind-finite ind-scheme),

[5] The arrow α was defined in (2.2.1).

Lemma. *The functor* $\mathcal{V} : \mathcal{M}((T^\vee \times T)(K))^{h_\kappa} \to \mathcal{V}ect$ *is an equivalence of categories.*

Proof. The bilinear form h_κ is non-degenerate over $\mathbb{Z}$, hence $\mathcal{M}((T^\vee \times T)(K))^{h_\kappa}$ is a semisimple category having a single irreducible object. It remains to show that for irreducible M one has $\dim \mathcal{V}(M) = 1$. Choose any splittting $T(O) \to T(K)^\kappa$; an irreducible M is induced from the trivial representation of the subgroup $T^\vee(O) \times T(O)$, and the computation of the dimension is immediate. □

2.9.

Recall that an object of $\mathcal{M}(\mathcal{LS}_{T^\vee}, A^\natural)$ is a vector space M equipped with an $A^\natural_{\mathcal{C}}$-module structure and a $T^\vee(K)$-action which are compatible via the $T^\vee(K)$-action on $A^\natural_{\mathcal{C}}$.

Lemma. *This structure on* M *amounts to commuting* $(T^\vee \times T)(K)^{h_\kappa}$*- and* $\mathfrak{t}(K)^\kappa$*-module structures.*

Proof. The structure on M can be rewritten in the following equivalent ways:

(i) By (2.6.1), this amounts to *an* A*-module structure, a commuting discrete* $\mathcal{O}(\mathcal{C})$*-module structure, and a* $T^\vee(K)$*-action* compatible with the module structures (here $T^\vee(K)$ acts both on A and $\mathcal{O}(\mathcal{C})$).

(ii) Replacing the A-module structure by the corresponding $T(K)^\kappa$-action, we see that our structure amounts to *discrete* $\mathcal{O}(\mathcal{C})$*-module and* $(T^\vee \times T)(K)^{h_\kappa}$*-module structures* which are compatible (here $(T^\vee \times T)(K)^{h_\kappa}$ acts on $\mathcal{O}(\mathcal{C})$ via the projection to $T^\vee(K)$).

(iii) A discrete $\mathcal{O}(\mathcal{C})$-module structure is the same as an action of the commutative topological Lie algebra $\mathfrak{t}(K)$ of its generators (see (1.2.1)). Therefore our structure amounts to *a* $(T^\vee \times T)(K)^{h_\kappa}$*-module structure and an action of the Lie algebra* $\mathfrak{t}(K)$ such that for every $a \in \mathfrak{t}(K)$, $g^\vee \in T^\vee(K)$, $g \in T(K)$, $m \in M$, and a lifting $q \in (T^\vee \times T)(K)^{h_\kappa}$ of $(g^\vee, g)$, one has

$$qaq^{-1}(m) = a(m) - \operatorname{Res}(a, d \log g^\vee)m. \tag{2.9.1}$$

Notice that the action of $(T^\vee \times T)(K)^{h_\kappa}$ yields an action on M of its Lie algebra $\mathfrak{t}^\vee(K) \ltimes \mathfrak{t}(K)^\kappa$. For $\tilde{a} \in \mathfrak{t}(K)^\kappa$ that lifts $a \in \mathfrak{t}(K)$ consider an operator on M

$$m \mapsto \tilde{a} \star m := (\kappa(a), \tilde{a})(m) + a(m). \tag{2.9.2}$$

Here $\kappa(a) \in \mathfrak{t}^\vee(K)$ is the image of a by $\kappa : \mathfrak{t} \to \mathfrak{t}^\vee$, so $(\kappa(a), \tilde{a}) \in \mathfrak{t}^\vee(K) \ltimes \mathfrak{t}(K)^\kappa$, and the first part of (2.9.2) is the action of this Lie algebra element on M; the second part of (2.9.2) is the action on M of $a \in \mathfrak{t}(K)$.

Compatibility (2.9.1) implies that (2.9.2) is a $\mathfrak{t}(K)^\kappa$-action on M that commutes with the $(T^\vee \times T)(K)^{h_\kappa}$-action, so we arrive to the datum from the statement of the lemma. Conversely, for any $\mathfrak{t}(K)^\kappa$-action $\star$ on M that commutes with a $(T^\vee \times T)(K)^{h_\kappa}$-action the operators $m \mapsto a(m)$ recovered from (2.9.2) form an action of the commutative Lie algebra $\mathfrak{t}(K)$ that satisfies (2.9.1), i.e., we have recovered the structure from (iii), and we are done. □

2.10.

Now we can finish the proof of the theorem. By the lemma from 2.9, $\mathcal{M}(\mathcal{L}\mathcal{S}_{T^\vee}, A^\natural)$ is the category of vector spaces equipped with commuting $(T^\vee \times T)(K)^{h_\kappa}$- and $\mathfrak{t}(K)^\kappa$-actions. By the lemma from 2.8, the functor $\mathcal{V}$ identifies it with the category of vector spaces equipped with an $\mathfrak{t}(K)^\kappa$-action, i.e., with $\mathcal{M}(\mathfrak{t}(K))^\kappa$. We get an equivalence

$$\mathcal{M}(\mathcal{L}\mathcal{S}_{T^\vee}, A^\natural) \xrightarrow{\sim} \mathcal{M}(\mathfrak{t}(K))^\kappa. \tag{2.10.1}$$

To finish the proof, let us show that it coincides with (2.4.1). Indeed, both functors assign to $M \in \mathcal{M}(\mathcal{L}\mathcal{S}_{T^\vee}, A^\natural)$ the same vector space; all we need to check is that the two $\mathfrak{t}(K)^\kappa$-actions on it coincide. As follows from the lemma in 2.6, for $\tilde{a}$ as above its action on $\mathcal{V}(M)$ from (2.4.1) comes from the operator $m \mapsto \tilde{a}(m) + a(m)$ on M (commuting with the $T^\vee(K)$-action). The action from (2.10.1) comes from operator (2.9.2). The two operators differ by an operator from the Lie algebra of $T^\vee(K)$, which dies on $\mathcal{V}(M)$. We are done. □

References

[BBE] A. Beilinson, S. Bloch, H. Esnault, $\mathcal{E}$-factors for Gauss–Manin determinants, *Moscow Mathematical Journal* **2** (2002) no. 3, pp. 477–532.

[BD1] A. Beilinson, V. Drinfeld, *Chiral Algebras*, Colloquium Publications, Vol. 51, Amer. Math. Soc., Providence, RI, 2004.

[BD2] A. Beilinson, V. Drinfeld, *Quantization of Hitchin's hamiltonians and Hecke eigensheaves*, an unfinished version is available at http://www.math.uchicago.edu/~arinkin/langlands/

[CC] C. E. Contou–Carrère, Jacobienne locale, groupe de bivecteurs de Witt universel et symbole local modéré, *C. R. Acad. Sci. Paris, Série* I **318** (1994) pp. 743–746.

[DVVV] R. Dijkgraaf, C. Vafa, E. Verlinde, H. Verlinde, The operator algebra of orbifold models, *Comm. Math. Phys.* **123** (1989) pp. 485–526.

[Dr] V. Drinfeld, Infinite-dimensional vector bundles in algebraic geometry, to appear in *Unity of Mathematics*, in Honor of I. M. Gelfand, Birkhäuser, Boston, MA.

[FBZ] E. Frenkel, D. Ben–Zvi, *Vertex Algebras and Algebraic Curves*, Mathematical Surveys and Monographs, Vol. 88, Amer. Math. Soc., Providence, RI, 2001.

[K] V. Kac, *Vertex Algebras for Beginners*, University Lecture Series, Vol. 10, Amer. Math. Soc., Providence, RI, 1998.

[LL] J. Lepowsky, H. Li, *Introduction to Vertex Operator Algebras and Their Representations*, Progress in Mathematics, Vol. 227, Birkhäuser, Boston, MA, 2004.

Instanton counting via affine Lie algebras II: From Whittaker vectors to the Seiberg–Witten prepotential

A. Braverman[1] and P. Etingof[2]

[1] Department of Mathematics
Brown University
151 Thayer street
Providence, RI
USA
and
Einstein Institute of Mathematics
Edmond J. Safra Campus
Givat Ram
The Hebrew University of Jerusalem
Jerusalem 91904, Israel
braval@math.brown.edu

[2] Department of Mathematics
Massachusetts Institute of Technology
77 Mass. Ave.
Cambridge, MA 02139
USA
etingof@math.mit.edu

Dedicated to A. Joseph on the occasion of his 60th birthday

Summary. Let G be a simple simply connected algebraic group over $\mathbb{C}$ with Lie algebra $\mathfrak{g}$. Given a parabolic subgroup $P \subset G$, in [1] the first author introduced a certain generating function $Z^{\text{aff}}_{G,P}$. Roughly speaking, these functions count (in a certain sense) framed G-bundles on $\mathbb{P}^2$ together with a P-structure on a fixed (horizontal) line in $\mathbb{P}^2$. When $P = B$ is a Borel subgroup, the function $Z^{\text{aff}}_{G,B}$ was identified in [1] with the Whittaker matrix coefficient in the universal Verma module over the affine Lie algebra $\check{\mathfrak{g}}_{\text{aff}}$ (here we denote by $\mathfrak{g}_{\text{aff}}$ the affinization of $\mathfrak{g}$ and by $\check{\mathfrak{g}}_{\text{aff}}$ the Lie algebra whose root system is dual to that of $\mathfrak{g}_{\text{aff}}$).

For $P = G$ (in this case we shall write $\mathcal{Z}^{\text{aff}}_G$ instead of $\mathcal{Z}^{\text{aff}}_{G,P}$) and $G = SL(n)$ the above generating function was introduced by Nekrasov (see [7]) and studied thoroughly in [5] and [8]. In particular, it is shown in *loc. cit.* that the leading term of certain asymptotic of $\mathcal{Z}^{\text{aff}}_G$ is given by the (instanton part of the) *Seiberg–Witten prepotential* (for $G = SL(n)$). The prepotential is defined using the geometry of the (classical) periodic Toda integrable system. This result was conjectured in [7].

The purpose of this paper is to extend these results to arbitrary G. Namely, we use the above description of the function $\mathcal{Z}^{\mathrm{aff}}_{G,B}$ to show that the leading term of its asymptotic (similar to the one studied in [7] for $P = G$) is given by the instanton part of the prepotential constructed via the Toda system attached to the Lie algebra $\check{\mathfrak{g}}_{\mathrm{aff}}$. This part is completely algebraic and does not use the original algebro-geometric definition of $\mathcal{Z}^{\mathrm{aff}}_{G,B}$. We then show that for fixed G these asymptotic are the same for *all* functions $\mathcal{Z}^{\mathrm{aff}}_{G,P}$.

1. Introduction

1.1. The partition function

This paper has grown out of a (still unsuccessful) attempt to understand the following object. Let K be a simple[1] simply connected compact Lie group and let d be a non-negative integer. Denote by $\mathcal{M}^d_K$ the moduli space of (framed) K-instantons on $\mathbb{R}^4$ of second Chern class $-d$. This space can be naturally embedded into a larger *Uhlenbeck space* $\mathcal{U}^d_K$. Both spaces admit a natural action of the group K (by changing the framing at ∞) and the torus $(S^1)^2$ acting on $\mathbb{R}^4$ after choosing an identification $\mathbb{R}^4 \simeq \mathbb{C}^2$. Moreover, the maximal torus of $K \times (S^1)^2$ has a unique fixed point on $\mathcal{U}^d_K$. Thus we may consider (see [1], [5] or [7] for precise definitions) the *equivariant integral*

$$\int\limits_{\mathcal{U}^d_K} 1^d$$

of the unit $K \times (S^1)^2$-equivariant cohomology class (which we denote by 1^d) over $\mathcal{U}^d_K$; the integral takes values in the field $\mathcal{K}$ of fractions of the algebra $\mathcal{A} = H^*_{K\times(S^1)^2}(pt)$.[2] Note that $\mathcal{A}$ is canonically isomorphic to the algebra of polynomial functions on $\mathfrak{k}\times\mathbb{R}^2$ (here $\mathfrak{k}$ denotes the Lie algebra of K) which are invariant with respect to the adjoint action of K on $\mathfrak{k}$. Thus each $\int_{\mathcal{U}^d_K} 1^d$ may be naturally regarded as a rational function of $a \in \mathfrak{k}$ and $(\varepsilon_1, \varepsilon_2) \in \mathbb{R}^2$.

Now consider the generating function

$$\mathcal{Z} = \sum_{d=0}^{\infty} Q^d \int\limits_{\mathcal{U}^d_K} 1^d.$$

It can (and should) be thought of as a function of the variables Q and $a, \varepsilon_1, \varepsilon_2$ as before. In [7] it was conjectured that the first term of the asymptotic in the limit $\lim_{\varepsilon_1,\varepsilon_2\to 0} \ln \mathcal{Z}$ is closely related to the *Seiberg–Witten prepotential* of K. For $K = SU(n)$ this conjecture has been proved in [8] and [5]. Also in [7] an explicit combinatorial expression for $\mathcal{Z}$ has been found.

---

[1] In this paper by a simple Lie (or algebraic) group we mean a group whose Lie algebra is simple.

[2] In this paper we always consider cohomology with complex coefficients.

1.2. Algebraic version

In [1] the first author has defined some more general partition functions containing the function $\mathcal{Z}_K$ as a special case. Let us recall that definition. First, it will be convenient for us to make the whole situation completely algebraic.

Namely, let G be a complex simple algebraic group whose maximal compact subgroup is isomorphic to K. We shall denote by $\mathfrak{g}$ its Lie algebra. Let also $\mathbf{S} = \mathbb{P}^2$ and denote by $\mathbf{D}_\infty \subset \mathbf{S}$ the "straight line at ∞"; thus $\mathbf{S}\backslash\mathbf{D}_\infty = \mathbb{A}^2$. It is well known that $\mathcal{M}_K^d$ is isomorphic to the moduli space $\mathrm{Bun}_G^d(\mathbf{S}, \mathbf{D}_\infty)$ of principal G-bundles on $\mathbf{S}$ of second Chern class $-d$ endowed with a trivialization on $\mathbf{D}_\infty$. When it does not lead to confusion we shall write Bun_G instead of $\mathrm{Bun}_G(\mathbf{S}, \mathbf{D}_\infty)$. The algebraic analog of $\mathcal{U}_K^d$ has been constructed in [2]; we denote this algebraic variety by $\mathcal{U}_G^d$. This variety is endowed with a natural action on $G \times (\mathbb{C}^*)^2$.

1.3. Parabolic generalization of the partition function

Let $\mathbf{C} \subset \mathbf{S}$ denote the standard horizontal line. Choose a parabolic subgroup $P \subset G$. Let $\mathrm{Bun}_{G,P}$ denote the moduli space of the following objects:

1) A principal G-bundle $\mathcal{F}_G$ on $\mathbf{S}$;
2) A trivialization of $\mathcal{F}_G$ on $\mathbf{D}_\infty \subset \mathbf{S}$;
3) A reduction of $\mathcal{F}_G$ to P on $\mathbf{C}$ compatible with the trivialization of $\mathcal{F}_G$ on $\mathbf{C} \cap \mathbf{D}_\infty$.

Let us describe the connected components of $\mathrm{Bun}_{G,P}$. Let M be the Levi group of P. Denote by $\check{M}$ the *Langlands dual* group of M and let $Z(\check{M})$ be its center. We denote by $\Lambda_{G,P}$ the lattice of characters of $Z(\check{M})$. Also let $\Lambda_{G,P}^{\mathrm{aff}} = \Lambda_{G,P} \times \mathbb{Z}$ be the lattice of characters of $Z(\check{M}) \times \mathbb{C}^*$. Note that $\Lambda_{G,G}^{\mathrm{aff}} = \mathbb{Z}$.

The lattice $\Lambda_{G,P}^{\mathrm{aff}}$ contains a canonical semigroup $\Lambda_{G,P}^{\mathrm{aff,pos}}$ of positive elements (see [2] and [1]). It is not difficult to see that the connected components of $\mathrm{Bun}_{G,P}$ are parameterized by the elements of $\Lambda_{G,P}^{\mathrm{aff,pos}}$:

$$\mathrm{Bun}_{G,P} = \bigcup_{\theta_{\mathrm{aff}} \in \Lambda_{G,P}^{\mathrm{aff,pos}}} \mathrm{Bun}_{G,P}^{\theta_{\mathrm{aff}}}.$$

Typically, for $\theta_{\mathrm{aff}} \in \Lambda_{G,P}^{\mathrm{aff}}$ we shall write $\theta_{\mathrm{aff}} = (d, \theta)$ where $\theta \in \Lambda_{G,P}$ and $d \in \mathbb{Z}$.

Each $\mathrm{Bun}_{G,P}^{\theta_{\mathrm{aff}}}$ is naturally acted on by $P \times (\mathbb{C}^*)^2$; by embedding M into P we get an action of $M \times (\mathbb{C}^*)^2$ on $\mathrm{Bun}_{G,P}^{\theta_{\mathrm{aff}}}$. In [2] we define for each $\theta_{\mathrm{aff}} \in \Lambda_{G,P}^{\mathrm{aff,pos}}$ a certain Uhlenbeck scheme $\mathcal{U}_{G,P}^{\theta_{\mathrm{aff}}}$ which contains $\mathrm{Bun}_{G,P}^{\theta_{\mathrm{aff}}}$ as a dense open subset. The scheme $\mathcal{U}_{G,P}^{\theta_{\mathrm{aff}}}$ still admits an action of $M \times (\mathbb{C}^*)^2$.

We want to do some equivariant intersection theory on the spaces $\mathcal{U}_{G,P}^{\theta_{\mathrm{aff}}}$. For this let us denote by $\mathcal{A}_{M\times(\mathbb{C}^*)^2}$ the algebra $H^*_{M\times(\mathbb{C}^*)^2}(pt, \mathbb{C})$. Of course this is just the algebra of M-invariant polynomials on $\mathfrak{m} \times \mathbb{C}^2$. Also let $\mathcal{K}_{M\times(\mathbb{C}^*)^2}$ be its field of fractions. We can think about elements of $\mathcal{K}_{M\times(\mathbb{C}^*)^2}$ as rational functions on $\mathfrak{m}\times\mathbb{C}^2$ that are invariant with respect to the adjoint action.

Let $T\subset M$ be a maximal torus. Then one can show that $(\mathcal{U}^{\theta_{\text{aff}}}_{G,P})^{T\times(\mathbb{C}^*)^2}$ consists of one point. This guarantees that we may consider the integral $\int_{\mathcal{U}^{\theta_{\text{aff}}}_{G,P}} 1^{\theta_{\text{aff}}}_{G,P}$ where $1^{\theta_{\text{aff}}}_{G,P}$ denotes the unit class in $H^*_{M\times(\mathbb{C}^*)^2}(\mathcal{U}^{\theta_{\text{aff}}}_{G,P},\mathbb{C})$. The result can be thought of as a rational function on $\mathfrak{m}\times\mathbb{C}^2$ that is invariant with respect to the adjoint action of M. Define

$$\mathcal{Z}^{\text{aff}}_{G,P}=\sum_{\theta\in\Lambda^{\text{aff}}_{G,P}}\mathfrak{q}_{\text{aff}}^{\theta_{\text{aff}}}\int\limits_{\mathcal{U}^{\theta_{\text{aff}}}_{G,P}}1^{\theta_{\text{aff}}}_{G,P} \tag{1.1}$$

(we refer the reader to Section 2 of [1] for a detailed discussion of equivariant integration). One should think of $\mathcal{Z}^{\text{aff}}_{G,P}$ as a formal power series in $\mathfrak{q}_{\text{aff}}\in Z(\check{M})\times\mathbb{C}^*$ with values in the space of ad-invariant rational functions on $\mathfrak{m}\times\mathbb{C}^2$. Typically, we shall write $\mathfrak{q}_{\text{aff}}=(\mathfrak{q},Q)$ where $\mathfrak{q}\in Z(\check{M})$ and $Q\in\mathbb{C}^*$. Also we shall denote an element of $\mathfrak{m}\times\mathbb{C}^2$ by $(a,\varepsilon_1,\varepsilon_2)$ (note that for general P (unlike in the case $P=G$) the function $\mathcal{Z}^{\text{aff}}_{G,P}$ is not symmetric with respect to switching ε_1 and ε_2). Here is the main result of this paper.

Theorem 1.4. *Let $P\subset G$ be a parabolic subgroup as a above.*

1. *There exists a function $\mathcal{F}^{\text{inst}}\in\mathbb{C}(a)[[Q]]$ such that*

$$\lim_{\varepsilon_1\to 0}\lim_{\varepsilon_2\to 0}\varepsilon_1\varepsilon_2\ln\mathcal{Z}^{\text{aff}}_{G,P}=\mathcal{F}^{\text{inst}}(a,Q). \tag{1.2}$$

   In particular, the above limit does not depend on $\mathfrak{q}$ and it is the same for all P.
2. *The function $\mathcal{F}^{\text{inst}}(a,Q)$ is equal to the instanton part of the Seiberg–Witten prepotential of the affine Toda system associated with the Langlands dual Lie algebra $\check{\mathfrak{g}}_{\text{aff}}$ (see Section 3 for the explanation of these words).*

Since the function $\mathcal{Z}^{\text{aff}}_G$ is symmetric in ε_1 and ε_2, Theorem 1.4 implies the following result:

Corollary 1.5. *The function $\varepsilon_1\varepsilon_2\ln\mathcal{Z}^{\text{aff}}_G$ is regular when both ε_1 and ε_2 are set to 0. Moreover, one has*

$$(\varepsilon_1\varepsilon_2\ln\mathcal{Z}^{\text{aff}}_G)|_{\varepsilon_1=\varepsilon_2=0}=\mathcal{F}^{\text{inst}}.$$

Corollary 1.5 was conjectured by N. Nekrasov in [7] (in fact [7] contains only the formulation for $G=SL(n)$ but the generalization to other groups is straightforward). For $G=SL(n)$ Nekrasov's conjecture was proved in [5] and [8]. Also, more recently, this conjecture was proved in [9] for all classical groups. These papers, however, utilize methods that are totally different from ours. In particular, in our approach the existence of the partition functions $\mathcal{Z}^{\text{aff}}_{G,P}$ for $P\neq G$ (in particular, for P being the Borel subgroup) plays a crucial role.

In fact, we are going to prove the following slightly stronger version of Theorem 1.4.

Theorem 1.6. 1. *Theorem 1.4 holds for* $P = B$.
2. *For every parabolic subgroup* $P \subset G$ *one has*

$$\lim_{\varepsilon_2 \to 0} \varepsilon_2 (\ln \mathcal{Z}^{\mathrm{aff}}_{G,P} - \ln \mathcal{Z}^{\mathrm{aff}}_{G}) = 0.$$

1.7. Plan of the proof

Let us explain the idea of the proof of Theorem 1.6. The second part is in fact rather routine so let us explain the idea of the proof of the first part.

The "Borel" partition function $\mathcal{Z}^{\mathrm{aff}}_{G,B}$ was realized in [1] as the Whittaker matrix coefficient in the universal Verma module over the Lie algebra $\check{\mathfrak{g}}_{\mathrm{aff}}$. As a corollary one gets that the function $\mathcal{Z}^{\mathrm{aff}}_{G,B}$ is an eigenfunction of *the non-stationary Toda hamiltonian* associated with the affine Lie algebra $\check{\mathfrak{g}}_{\mathrm{aff}}$ (see Corollary 3.7 from [1] for the precise statement; we use [3] as our main reference about Toda hamiltonians).

It turns out that this is all that we have to use in order to prove Theorem 1.6(1). Namely, in this paper (see Sections 2 and 3) we introduce the notion of the Seiberg–Witten prepotential (more precisely, its instanton part) for a very general class of non-stationary Schrödinger operators in such a way that by the definition it is equal to some asymptotic (in some sense) of the universal eigenfunction of this operator (we were unable to find such a definition in the literature). Usually the prepotential is attached to a classical completely integrable system (our main references on the definition of the Seiberg–Witten prepotential are [6] and [7]). We show that in the integrable case our definition of the prepotential coincides with the one from *loc. cit.*

2. Schrödinger operators and the prepotential: the one-dimensional case

2.1. Schrödinger operators

Let $x \in \mathbb{C}$ and let $U(x)$ be a trigonometric polynomial in x—i.e., a polynomial in e^x and e^{-x}. Let also $\hbar$ and Q be formal variables. We want to study the eigenvalues of the Schrödinger operator

$$T = \hbar^2 \frac{d^2}{dx^2} + QU(x).$$

More precisely, for each $a \in \mathbb{C}$ let W_a denote the space $e^{\frac{ax}{\hbar}} \mathbb{C}(\hbar)[e^x, e^{-x}][[Q]]$ with the natural action of the algebra of linear differential operators in x. Then we would like to look for eigenfunctions of T in W_a. After conjugating T with $e^{\frac{ax}{\hbar}}$ the operator T turns into the operator

$$\hbar^2 \frac{d^2}{dx^2} + 2\hbar a \frac{d}{dx} + QU(x) + a^2.$$

Let

$$T^a = \hbar^2 \frac{d^2}{dx^2} + 2\hbar a \frac{d}{dx} + QU(x).$$

We now want to look for eigenfunctions of T^a in W_0 (this problem is obviously equivalent to finding eigenfunctions of T in W_a). In fact, we want them to depend nicely on a, so we set $W = \mathbb{C}(a, \hbar)[e^x, e^{-x}][[Q]]$ and we want to look for eigenfunctions of T^a (considered now as a differential operator with coefficients in $\mathbb{C}(a, \hbar)[[Q]]$) in W.

Proposition 2.2. 1. *There exist $\psi \in W$ and $b \in Q\mathbb{C}(a, \hbar)[[Q]]$ such that*

$$T^a \psi = b\psi \tag{2.1}$$

and such that $\psi = 1 + O(Q)$. Moreover, under such conditions b is unique and ψ is unique up to multiplication by an element of $1 + Q\mathbb{C}(a, \hbar)[[Q]]$.

2. *Let $\phi = \hbar \ln \psi$ (note that ϕ is defined uniquely up to adding an element of $Q\mathbb{C}(a, \hbar)[[Q]]$). Then ϕ is regular at $\hbar = 0$ provided this is true for its constant term.* [3]
3. *The limit $\upsilon(a, Q) := \lim_{\hbar \to 0} b(a, \hbar, Q)$ exists in $\mathbb{C}(a)[[Q]]$.*

Proof. Let us prove the first assertion. Let us write

$$\psi = \sum_{n=0}^{\infty} \psi_n Q^n \qquad \text{and} \qquad b = \sum_{n=0}^{\infty} b_n Q^n.$$

Note that $\psi_0 = 1$ and thus automatically $b_0 = 0$. Thus the equation (2.1) becomes

$$\hbar^2 \psi_n'' + 2\hbar a \psi_n' + U(x)\psi_{n-1} = \sum_{i=0}^{n-1} b_{n-i}\psi_i, \tag{2.2}$$

which should be valid for each $n > 0$ (here and in what follows the prime denotes the derivative of a function with respect to x). It is enough for us to prove that the system of equations (2.2) has a unique solution if we require that for all $n > 0$ the constant term of the function ψ_n is equal to 0.

Equation (2.2) is equivalent to

$$\hbar^2 \psi_n'' + 2\hbar a \psi_n' = -U(x)\psi_{n-1} + \sum_{i=0}^{n-1} b_{n-i}\psi_i, \tag{2.3}$$

where the left-hand side is just the differential operator $D = \hbar \frac{d^2}{dx^2} + 2\hbar a \frac{d}{dx}$ applied to ψ_n, and the right-hand side only depends on the ψ_i's with $i < n$.

Let us now argue by induction on n. By the induction hypothesis we assume that ψ_i and b_i have already been uniquely determined for all $i < n$. Note that the operator D has the following properties (whose verification is left to the reader):

[3] By the "constant term" we shall always mean the constant term of a trigonometric polynomial. The reader should not confuse this with the notion of "free term" by which we always mean the coefficient of the 0-th power of the variable in a formal power series.

1) $\ker D$ consists of constant (i.e., independent of x) functions.
2) $\operatorname{im} D$ consists of all trigonometric polynomials whose constant term is equal to 0.

Observe now that the coefficient of b_n in the RHS of (2.3) is $\psi_0 = 1$. Thus property 2) above determines b_n uniquely—it has to be chosen so that the constant term of the RHS is equal to 0. If b_n is chosen in this way, then there exists some ψ_n satisfying (2.3). *A priori* such ψ_n is defined uniquely up to adding a constant trigonometric polynomial, but the requirement that the constant term of ψ_n is equal to 0 determines ψ_n uniquely.

Let us prove the second and third assertions (this is a standard WKB argument which we include for the sake of completeness). Let us write

$$\phi = \hbar \ln \psi$$

(the logarithm is taken in the sense of formal power series in Q; this makes sense because $\psi_0 = 1$).

Let us rewrite (2.1) in terms of ϕ. We get

$$(\phi')^2 + \hbar\phi'' + 2a\phi' + QU(x) = b. \tag{2.4}$$

Let us now look for a solution ϕ of the form

$$\phi = \sum_{n=1}^{\infty} \phi_n Q^n.$$

Then (2.4) is equivalent to the following system of equations:

$$\hbar\phi_1'' + 2a\phi_1' = b_1 - U(x) \tag{2.5}$$

and

$$\hbar\phi_n'' + 2a\phi_n' = b_n - \sum_{i=1}^{n-1} \phi_i'\phi_{n-i}' \tag{2.6}$$

for all $n > 1$.

Without loss of generality we may assume that the constant term of all ϕ_n is equal to zero. We need to show that under such conditions all ϕ_n and b_n are regular when $\hbar = 0$. Let us prove by induction in k that the statement is valid for $n \leq k$. If $k = 0$, the statement is clear, so let $k > 0$; we need to prove the statement for $n = k$. By the induction assumption we may assume that $\sum_{i=1}^{n-1} \phi_i'\phi_{n-i}'$ is regular at $\hbar = 0$. Arguing as before, we see that if (2.6) has a solution, then b_n has to be equal to the constant term of $\sum_{i=1}^{n-1} \phi_i'\phi_{n-i}'$ for $n > 1$ and of $U(x)$ for $n = 1$, and thus it is also regular at $\hbar = 0$. Thus the right-hand side of (2.6) is regular at $\hbar = 0$. This immediately implies that the same is true for ϕ_n. □

2.3. Explicit calculation of $\lim_{\hbar\to 0} b$ via periods

We now want to explain how to evaluate the function $\lim_{\hbar\to 0} b(a,\hbar,Q) = \upsilon(a,Q)$ using period integrals on a certain algebraic curve. More precisely, we are going to express a as a (multivalued) function of υ and Q which will be written in terms of such periods. Let φ denote the limit of ϕ as $\hbar \to 0$. Then we have the equation

$$(\varphi')^2 + 2a\varphi' = \upsilon - QU(x). \tag{2.7}$$

In other words, φ' satisfies a quadratic equation. Thus we may write

$$\varphi' = -a + \sqrt{a^2 + \upsilon - QU(x)}.$$

This is an equality of formal power series in Q. The square root is chosen in such a way that the right-hand side is equal to 0 when $Q = 0$ (note we automatically have $\upsilon = 0$ when $Q = 0$).

Recall, however, that φ was a trigonometric polynomial. This implies that

$$\int_0^{2\pi i} \varphi' dx = 0.$$

This is equivalent to the equation

$$2\pi i a = \int_0^{2\pi i} \sqrt{a^2 + \upsilon - QU(x)} dx. \tag{2.8}$$

Set $w = e^x$ and recall that $U(x) = P(w)$ for some polynomial P in w and w^{-1}. Set also $u = a^2 + \upsilon$ and consider the algebraic curve $C = C_u$ which is the projectivization of the affine curve given by the equation

$$z^2 + QP(w) = u.$$

We claim that we may write a locally as a function $a(u,Q)$ of u and Q. Namely, first of all $a_0 := a(u,0)$ must satisfy $a_0^2 = u$. Let us locally choose one of the square roots. Then the function a is found as a series $a_0 + a_1 Q + \cdots + a_n Q^n + \cdots$, where a_i with $i > 0$ are found recursively.

Note that when $Q = 0$ the above curve breaks into two components corresponding to $z = \pm a_0$. Let $A = A_{u,Q}$ denote the one-dimensional cycle in C satisfying the following conditions:

1) The projection of A to the w-plane is an isomorphism between A and the unit circle.
2) A depends continuously on Q and when $Q = 0$ it lies in the component of C corresponding to $z = a_0$.

Such a cycle is unique at least for small values of Q. Thus the equation (2.8) becomes equivalent to

$$a = \frac{1}{2\pi i} \oint_A z \frac{dw}{w}. \tag{2.9}$$

Note that $z\frac{dw}{w}$ is a well-defined meromorphic differential on C. Note also that C and A depend only on Q and u; thus we may think of (2.9) as expressing a as a function of u and Q.

2.4. Eigenfunctions of non-stationary Schrödinger operators

Let us now change our problem a little. Introduce one more variable κ and define new operators

$$\mathcal{L} = T - \kappa Q \frac{\partial}{\partial Q} \qquad \text{and} \qquad \mathcal{L}^a = T^a - \kappa Q \frac{\partial}{\partial Q}.$$

Let us now look for solutions of the equation

$$\mathcal{L}^a \Psi = 0 \tag{2.10}$$

where $\Psi \in \mathbb{C}(a, \hbar, \kappa)[e^x, e^{-x}][[Q]]$ (we shall denote this space by $W(\kappa)$). Of course this equation is equivalent to the equation

$$\mathcal{L}(e^{\frac{ax}{\hbar}} \Psi) = a^2 e^{\frac{ax}{\hbar}} \Psi.$$

More precisely, we want to look for the asymptotic of these eigenfunctions when both $\hbar$ and κ go to 0.

Proposition 2.5. 1. *There exists unique solution Ψ of (2.10) in $W(\kappa)$ such that $\Psi = 1 + O(Q)$.*

2. *This solution Ψ takes the form*

$$\Psi = e^{\frac{\Phi}{\kappa} + g} \tag{2.11}$$

where $g \in Q\mathbb{C}(a, \hbar)[e^x, e^{-x}][[\kappa, Q]]$ and $\Phi \in Q\mathbb{C}(a, \hbar)[[\kappa, Q]]$.

3. *One has*

$$\hbar Q \frac{\partial \Phi}{\partial Q} = b.$$

4. *The limit*

$$\mathcal{F}^{\text{inst}} = \lim_{\hbar \to 0} \hbar \Phi(a, \hbar, Q)$$

exists and one has

$$Q \frac{\partial \mathcal{F}^{\text{inst}}}{\partial Q} = v. \tag{2.12}$$

Remark. We will explain the origin of the notation a little later.

Proof. Let us first prove (1). Let us write

$$\Psi = \sum_{n=0}^{\infty} \Psi_n Q^n, \quad \Psi_0 = 1.$$

Then (2.10) becomes equivalent to the sequence of equations:

$$\hbar^2 \Psi_n'' + 2\hbar a \Psi_n' - \kappa n \Psi_n = U(x)\Psi_{n-1}. \tag{2.13}$$

Let D_n denote the differential operator $\hbar^2 \frac{d^2}{dx^2} + 2\hbar a \frac{d}{dx} - \kappa n$. Then it is easy to see that D_n is invertible when acting on $\mathbb{C}(a, \hbar, \kappa)[e^x, e^{-x}]$ (it is diagonal in the basis given by the functions $\{e^{kx}\}_{k\in\mathbb{Z}}$ with non-zero eigenvalues). Thus by induction we get a unique solution for each Ψ_n, $n \geq 1$.

Let $F = \ln \Psi$. First, we claim that κF is regular when $\kappa = 0$. This is proved exactly in the same way as part (2) of Proposition 2.2 and we leave it to the reader. Let us now write

$$F = \sum_{n=-1}^{\infty} F_n \kappa^n.$$

We want to compute F_{-1}.

Equation (2.10) is equivalent to the equation

$$\hbar^2((F')^2 + F'') + 2a\hbar F' + QU(x) = \hbar\kappa Q \frac{\partial F}{\partial Q}. \tag{2.14}$$

Decomposing this in a power series in κ and looking at the coefficient of κ^{-2} we see that $\Phi = F_{-1}$ satisfies the equation $(\Phi')^2 = 0$; in other words Φ is indeed independent of x.

Let us now look at the free term (in κ) in the above identity (it is easy to see that the coefficient of κ^{-1} is automatically 0 on both sides). We get the equation

$$\hbar^2(F_0')^2 + \hbar^2 F_0'' + 2a\hbar F_0' + QU(x) = \hbar Q \frac{\partial \Phi}{\partial Q}. \tag{2.15}$$

Note now that (2.15) is basically the same equation as (2.4) if we set $b = \hbar Q \frac{\partial \Phi}{\partial Q}$ and $F_0 = \hbar^{-1}\phi$. Since obviously $F_0|_{Q=0} = 0$, the uniqueness statement from Proposition 2.2(1) implies (3). Now (4) is equivalent to Proposition 2.2(3). □

Definition 2.6. The function $\mathcal{F}^{\text{inst}}(a, Q)$ is called *the instanton part* of the prepotential.

Remark. In the context of integrable systems one is usually interested in the *full* Seiberg–Witten prepotential $\mathcal{F}$ which is defined as the sum of $\mathcal{F}^{\text{inst}}$ and $\mathcal{F}^{\text{pert}}$; here $\mathcal{F}^{\text{pert}}$ is called the perturbative part of the prepotential and it is usually given by some simple formula. We do not know if there is a canonical choice of $\mathcal{F}^{\text{pert}}$ in our generality. However, we may observe that in all the known cases $\mathcal{F}^{\text{pert}}$ satisfies the equation

$$Q\frac{\partial\mathcal{F}^{\text{pert}}}{\partial Q}=a^2.$$

This fixes $\mathcal{F}^{\text{pert}}$ uniquely up to adding a function which is independent of Q. Note that if we now define $\mathcal{F}=\mathcal{F}^{\text{inst}}+\mathcal{F}^{\text{pert}}$ (for any choice of $\mathcal{F}^{\text{pert}}$ satisfying the above equation) then the equation (2.12) gets simplified: it is now equivalent to

$$Q\frac{\partial\mathcal{F}}{\partial Q}=u. \tag{2.16}$$

3. Schrödinger operators in higher dimensions and integrable systems

We now want to generalize the results of the previous section to the higher dimensional situation.

3.1. The setup

In this section we are going to work with the following general setup. Let $\mathfrak{h}$ be a finite dimension vector space over $\mathbb{C}$ and let $\Lambda\subset\mathfrak{h}$ be a lattice. We denote by H the algebraic torus whose lattice of co-characters is Λ (analytically one may think of H as $\mathfrak{h}/2\pi i\Lambda$ by means of the map $x\mapsto e^x$); we let $\mathbb{C}[H]$ denote the algebra of polynomial functions on H; we might think of elements of $\mathbb{C}[H]$ as trigonometric polynomials on $\mathfrak{h}$. We assume that $\mathfrak{h}$ is endowed with a non-degenerate bilinear form $\langle\cdot,\cdot\rangle$ which takes integral values on Λ.

Let $\mathcal{K}$ denote the field of rational functions on $\mathfrak{h}^*\times\mathbb{C}^2$ (typically, we denote an element in $\mathfrak{h}^*\times\mathbb{C}^2$ by $(a,\hbar,\kappa)$ with $a\in\mathfrak{h}^*$; so, sometimes we shall write $\mathbb{C}(a,\hbar,\kappa)$ instead of $\mathcal{K}$). Let Q be another indeterminate. We are going to be interested in the space $W(\kappa):=\mathcal{K}[H][[Q]]$; its elements are power series in Q whose coefficients lie in $\mathcal{K}$.

Let Δ denote the Laplacian on $\mathfrak{h}$ (or H) corresponding to the bilinear form fixed above. Fix now any $P\in\mathbb{C}[H]$. We shall denote by U the corresponding function on $\mathfrak{h}$ given by the formula

$$U(x)=P(e^x).$$

Now, following the previous section we define the operators

$$T=\hbar^2\Delta+QU(x);\qquad T^a=\hbar^2\Delta+2\hbar\langle\nabla,a\rangle+QU(x).$$

Here for a function ψ we denote by $\nabla\psi$ its differential in the $\mathfrak{h}$-direction. Similarly, we define

$$\mathcal{L}=T-\kappa Q\frac{\partial}{\partial Q}\qquad\text{and}\qquad\mathcal{L}^a=T^a-\kappa Q\frac{\partial}{\partial Q}.$$

Here $a\in\mathfrak{h}^*$. Note that as before for a fixed a the operator $T^a+\langle a,a\rangle$ is formally conjugate to T, and the operator $\mathcal{L}^a+\langle a,a\rangle$ is formally conjugate to $\mathcal{L}$.

As before, we set $W = \mathbb{C}(a, \hbar)[H][[Q]]$. Then with such notations Propositions 2.2 and 2.5 hold as stated in the current situation as well. The proofs are just word-by-word repetitions of those from the one-dimensional situation.

However, generalizing the results of Section 2.3 turns out to be a little bit more tricky. In order to do this we need to make some integrability assumptions.

3.2. Integrability

Let us denote by $\mathcal{D}$ the subalgebra of the algebra of differential operators on H with coefficients in $\mathbb{C}[\hbar, Q]$ consisting of all differential operators of the form $\sum \hbar^i D_i$ (where D_i is a differential operator on H with coefficients in $\mathbb{C}[Q]$) such that the order of D_i is $\leq i$. It is clear that $\mathcal{D}/\hbar\mathcal{D}$ is canonically isomorphic to $\mathcal{O}(T^*H) \otimes \mathbb{C}[Q] = \mathcal{O}(T^*H \times \mathbb{C})$ (here T^*H denotes the cotangent bundle to H and $\mathcal{O}(T^*H)$ is the algebra of regular functions on it). Note that $T^*H = H \times \mathfrak{h}^*$. We shall denote the resulting map from $\mathcal{D}$ to $\mathcal{O}(T^*H) \otimes \mathbb{C}[Q]$ by σ and call it the symbol map.

Similarly we let $\mathcal{D}^a = \mathcal{D} \otimes \mathcal{O}(\mathfrak{h}^*)$; we have $\mathcal{D}^a/\hbar\mathcal{D}^a \simeq \mathcal{O}(T^*H \times \mathbb{C} \times \mathfrak{h}^*)$. We let $\sigma^a : \mathcal{D}^a \to \mathcal{O}(T^*H \times \mathbb{C} \times \mathfrak{h}^*)$ denote the corresponding symbol map.

From now on we want to change our point of view a little bit and think about T^a as a differential operator on H rather than on $\mathfrak{h}$. Note that if we do so, then T^a lies in $\mathcal{D}^a$.

We now assume that in addition to the above data we are given the following:

a) An affine algebraic variety S such that $\dim S = \dim H$;
b) A finite morphism $\pi : \mathfrak{h}^* \to S$;
c) An injective homomorphism $\eta : \mathcal{O}(S) \to \mathcal{D}$.

These data must satisfy the following conditions:

1) T lies in the image of η; we let $C \in \mathcal{O}(S)$ denote the (unique) function for which $\eta(C) = T$.
2) $\eta|_{Q=0}$ is equal to the composition of $\pi^* : \mathcal{O}(S) \to \mathcal{O}(\mathfrak{h}^*)$ with the natural embedding $\mathcal{O}(\mathfrak{h}^*) \to \mathcal{D}$ which sends every function $h \in \mathcal{O}(\mathfrak{h}^*)$ that is homogeneous of degree d to $\hbar^d D_h$ where D_h is the differential operator with constant coefficients corresponding to h.

In this case we shall say that T is *integrable*. Note that if $\dim H = 1$, then T is automatically integrable.

Let $\eta^a : \mathcal{O}(S) \to \mathcal{D}^a$ denote the composition of η with the conjugation by $e^{\frac{\langle a,x\rangle}{\hbar}}$. Note that $T^a = \eta^a(C) - \langle a, a\rangle$.

Let also $p : T^*H \times \mathbb{C} \to S$ denote the morphism such that for every $f \in \mathcal{O}(S)$ we have

$$p^*(f) = \sigma \circ \eta(f).$$

This morphism represents the classical integrable system, which is the classical limit of the quantum integrable system defined by η.

3.3. Computation of $\lim_{\hbar\to 0} b$ via periods in the integrable case

We now want to explain how to generalize the results Section 2.3 to our multidimensional situation in the integrable case.

First, the operator T^a has simple spectrum in W; therefore the function ψ which is an eigenfunction of T^a is automatically an eigenfunction of every operator of the form $\eta^a(f)$ ($f \in \mathcal{O}(S)$). More precisely, we get a homomorphism $\mathbf{b}: \mathcal{O}(S) \to \mathcal{O}(\mathfrak{h}^*)[[Q]]$ such that for each $f \in \mathcal{O}(S)$ we have

$$\eta^a(f)(\psi)(t, a, Q, \hbar) = f(\mathbf{b}(a, Q, \hbar))\psi(t, a, Q, \hbar).^4$$

Note that $b = \mathbf{b}^*(C) - \langle a, a\rangle$. It is easy to see that the limit $\lim_{\hbar\to 0} \mathbf{b}^*(C)$ exists; we denote it by $\mathbf{u}$. By the definition $\mathbf{u}$ is a map from $\mathfrak{h}^* \times \Sigma$ to S where Σ denotes the formal disc with coordinate Q. It is clear that $\mathbf{u}|_{Q=0} = \pi$.

Let us now look at the function $\varphi = \lim_{\hbar\to 0}(\hbar \ln \psi)$. Then we have

$$p(d\varphi(t, a, Q) + a, Q) = \mathbf{u}(a, Q). \tag{3.1}$$

On the other hand, for any $\lambda \in \Lambda$ considered as a morphism $\lambda : \mathbb{C}^* \to H$ we must have

$$\oint \lambda^* d\varphi = 0 \tag{3.2}$$

where $\oint$ denotes the integral over the unit circle in $\mathbb{C}^*$. Let us think of $d\varphi$ as a morphism $H \to T^*H$ (which depends on a and Q). We denote by α the canonical one-form on T^*H. Let also L_λ denote the image of the unit circle under λ. Then (3.2) is equivalent to

$$\oint_{L_\lambda} (d\varphi)^*\alpha = 0. \tag{3.3}$$

We can now again write a locally as a function of $\mathbf{u}$ and Q; $a = a(\mathbf{u}, Q)$. To do this we must make a (local) choice of $a_0 := a_0(\mathbf{u}, Q)$. Note that a_0 must satisfy

$$\pi(a_0) = \mathbf{u}$$

and therefore choosing a_0 amounts to choosing a local branch of π. Let now λ be as above. Then we denote by $A_{\lambda,\mathbf{u},Q}$ the unique 1-dimensional cycle in T^*H such that:

1) the projection of $A_{\lambda,\mathbf{u},Q}$ to H is equal to L_λ;
2) $A_{\lambda,\mathbf{u},Q} \subset p^{-1}(\mathbf{u})$;
3) $A_{\lambda,\mathbf{u},Q}$ depends continuously on Q and for $Q = 0$ it lies in the above chosen branch of π.[5]

---

[4] Here $t \in H$ (i.e., we think about ψ as a function on H rather than on $\mathfrak{h}$).

[5] More precisely, this means the following: for $Q = 0$ the map p is equal to the composition of the natural projection $T^*H \to \mathfrak{h}^*$ and $\pi : \mathfrak{h}^* \to S$. Thus for every $\mathbf{u}$ we have $p|_{Q=0}^{-1}(\mathbf{u}) = H \times \pi^{-1}(\mathbf{u})$. We require that A_λ lie in the product of H and the corresponding branch of π.

Then (3.1) says that for every $\lambda \in \Lambda$ we have

$$\langle a, \lambda \rangle = \frac{1}{2\pi i} \oint_{A_{\lambda,\mathbf{u},Q}} \alpha. \tag{3.4}$$

3.4. Some variants

Let us choose a closed cone $\mathfrak{h}_+^* \subset \mathfrak{h}^*$ which is integral with respect to Λ (i.e., given by finitely many inequalities given by elements of Λ). We assume also that $a \in \mathfrak{h}_+^*$ implies that $-a \notin \mathfrak{h}_+^*$ for $a \neq 0$ (i.e., 0 is an extremal point of $\mathfrak{h}_+$). Set $\Lambda_+^\vee = \Lambda^\vee \cap \mathfrak{h}_+^*$. We denote by $\widehat{W}$ the corresponding completion of W; by the definition it consists of all formal sums

$$\sum c_\gamma e^{\langle \gamma, x\rangle}$$

where $\gamma \in \Lambda^\vee$ and such that for each $\check{\lambda} \in \Lambda^\vee$ the set

$$\{\gamma \in \check{\lambda} - \Lambda_+ | \text{ such that } c_\gamma \neq 0\}$$

is finite.

It is easy to see that the results of this section generalize immediately to the situation when the initial Schrödinger operator T takes the form

$$T = \hbar^2 \Delta + \mathbf{U}(Q, x)$$

where $\mathbf{U} \in \mathbb{C}[H][[Q]]$ subject to the following condition:

- The function $\mathbf{U}(0, x)$ is a linear combination of $e^{\langle \check{\lambda}, x\rangle}$ with $\check{\lambda} \in \mathfrak{h}_+^*$, $\check{\lambda} \neq 0$.

In this case the eigenfunctions ψ and Ψ should be elements of respectively $\widehat{W}$ and $\widehat{W}(\kappa)$.

The above condition guarantees in particular that 0 is an eigenvalue of T^a on $\widehat{W}$. The definition of the prepotential goes through in this case without any changes.

Here is the basic example of the above situation. Let

$$\mathfrak{h} = \{(x_1, \ldots, x_n) \in \mathbb{C}^n\}/\mathbb{C}(1, \ldots, 1) \qquad \Lambda = \{(x_1, \ldots, x_n) \in \mathbb{Z}^n\}/\mathbb{Z}(1, \ldots, 1).$$

Clearly,

$$\mathfrak{h}^* = \left\{(a_1, \ldots, a_n) \middle| \sum a_i = 0\right\}$$

and we set

$$\mathfrak{h}_+^* = \{(a_1, \ldots, a_n) \in \mathfrak{h}^* | \; a_1 + a_2 + \cdots + a_k \geq 0 \text{ for each } 1 \leq k \leq n\}.$$

Let

$$\mathbf{U}(Q, x) = 2(e^{x_1 - x_2} + e^{x_2 - x_3} + \cdots + e^{x_{n-1} - x_n} + Qe^{x_n - x_1})$$

be the periodic Toda potential. It is clear that the condition above is satisfied and therefore we may speak of the corresponding prepotential. In the next section we explain its connection with the standard physical definition of the prepotential.

The periodic Toda potential is equal to the Toda potential defined by the affine Lie algebra $\widehat{sl_n}$ (see for example [3]). One can easily see that the Toda potential for any affine Lie algebra (see [3]) satisfies our conditions and thus the corresponding prepotential is well defined.

Note also that the operator

$$\hbar^2\Delta + \mathbf{U}(Q,x)$$

turns into the operator

$$\hbar^2\Delta + 2Q^{1/n}(e^{x_1-x_2} + \cdots + e^{x_{n-1}-x_n} + e^{x_n-x_1})$$

after the change of variables

$$x_j \mapsto x_j + \frac{j\ln Q}{n}.$$

Thus when computing the prepotential we may deal with the latter operator (a similar statement is true for all affine Lie algebras).

Remark. The variable Q that we are using is connected with the variable Λ (which is commonly used by physicists—see [6], [7] etc.) by the formula

$$Q = \Lambda^{2n}.$$

Remark. It is not difficult to check that if U is equal to the Toda potential for $\widehat{sl_n}$, then our definition of $\mathcal{F}^{\text{inst}}$ coincides with the one usually given by physicists (see Chapter 2 of [6]). Let us give a very brief sketch of the proof of this result (details will appear in a subsequent publication in a more general setting). Namely, in this case (2.12) becomes equivalent to the *renormalization group equation* (Proposition 2.10 of [6]). Note that in the original (Seiberg–Witten) definition of the prepotential a is expressed in terms of periods of some family of curves over $\mathfrak{h}^*/W \times \mathbb{C}$ where $W = S_n$ is the Weyl group of sl_n (here the second factor is the line with coordinate Q). These curves are called the *Seiberg–Witten curves* (see Section 2.1 of [6]). In our case, a is expressed via periods on the fibers of the map $p_Q : T^*H \to \mathfrak{h}^*/W$ (note that in this case we have $S = \mathfrak{h}^*/W$). However, it is well known (see [4]) that the fibers of the map p_Q are open pieces in the Jacobians of the Seiberg–Witten curves; thus periods of a regular one-form over these fibers are equal to the periods of a certain meromorphic one-form over the curves themselves and it is not difficult to check that we get exactly the same periods as we need. Some generalization of this fact will be considered in much more detail in a further publication.

4. Proof of Nekrasov's conjecture

In this section we want to prove Theorem 1.6 (and thus also Theorem 1.4). The first part of Theorem 1.6 is an immediate corollary of Corollary 3.7 from [1] combined with the definition of $\mathcal{F}^{\text{inst}}$ given by definition 2.6. Thus it remains to prove the second part of Theorem 1.6. The proof is based on the following result.

Theorem 4.1. *Let $P \subset G$ be a parabolic and let $(d,\theta) \in \Lambda^{\mathrm{aff},+}_{G,P}$. Then one of the following is true:*

a) Both $\int_{\mathcal{U}^{d,\theta}_{G,P}} 1$ and $\int_{\mathcal{U}^d_G} 1$ are 0.
b) $\int_{\mathcal{U}^d_G} 1 \neq 0$ and the ratio

$$\frac{\int\limits_{\mathcal{U}^{d,\theta}_{G,P}} 1}{\int\limits_{\mathcal{U}^d_G} 1}$$

is regular when $\varepsilon_2 \to 0$.

Let us first explain why Theorem 4.1 implies Theorem 1.6. First of all, we claim that for any $d \le d'$ we have

$$\int\limits_{\mathcal{U}^{d'}_G} 1 = A_d \int\limits_{\mathcal{U}^d_G} 1$$

where A_d is a regular function on $\mathfrak{h} \times \mathbb{C}^2$ (in particular, it is regular when $\varepsilon_2 \to 0$). Indeed, according to [2] there exists a closed $G \times (\mathbb{C}^*)^2$-equivariant embedding $\mathcal{U}^d \stackrel{i_d}{\to} \mathcal{U}^{d'}$. Since $\mathcal{U}_{d'}$ is contractible we have $H^*_{G\times(\mathbb{C}^*)^2}(\mathcal{U}^{d'}_G) = \mathcal{A}_{G\times(\mathbb{C}^*)^2}$. Thus it follows that the direct image $(i_d)_*1$ of the equivariant unit cohomology class is equal to some $A_d \in \mathcal{A}_{G\times(\mathbb{C}^*)^2}$.

Now it follows from Theorem 4.1 that the ratio

$$\frac{\mathcal{Z}^{\mathrm{aff}}_{G,P}(\mathfrak{q}, Q, a, \varepsilon_1, \varepsilon_2)}{\mathcal{Z}^{\mathrm{aff}}_G(Q, a, \varepsilon_1, \varepsilon_2)}$$

is regular when $\varepsilon_2 \to 0$. This means that

$$\lim_{\varepsilon_2 \to 0} \varepsilon_2 (\ln \mathcal{Z}^{\mathrm{aff}}_{G,P} - \ln \mathcal{Z}^{\mathrm{aff}}_G) = 0.$$

This is the statement of Theorem 1.6(2).

Thus to complete the proof we need to prove Theorem 4.1. The proof is based on the following general lemma.

Lemma 4.2. *Let L_1 and L_2 be two algebraic tori and let $L = L_1 \times L_2$. We let $\mathfrak{l}_1$ and $\mathfrak{l}_2$ denote the corresponding Lie algebras. We shall denote a typical element in $\mathfrak{l}$ by (l_1, l_2), where $l_i \in \mathfrak{l}_i$.*

Let $\pi : X \to Y$ be a morphism of L-varieties. Assume that:

1) both X^L and Y^L are proper;
2) the natural map $X^{L_1} \to Y^{L_1}$ is proper.

Then if $\int_Y 1$ is zero, then $\int_X 1$ is also zero (here we consider both integrals in L-equivariant cohomology). If $\int_Y 1 \neq 0$, then the ratio

$$\frac{\int_X 1}{\int_Y 1}$$

(where the integral is taken in L-equivariant cohomology) is regular when $l_2 \to 0$.

Lemma 4.2 is an easy corollary of the definition of the above integrals given in Section 2 of [1] and we leave the proof to the reader.

4.3. End of the proof

We now want to apply Lemma 4.2 to the case when $X = \mathcal{U}_{G,P}^{d,\theta}$, $Y = \mathcal{U}_G^d$, $L_1 = T \times \mathbb{C}^*$ where the $\mathbb{C}^*$ factor corresponds to ε_1 and $L_2 = \mathbb{C}^*$ corresponding to ε_2. To avoid confusion in the notation we shall denote the "first" (i.e., horizontal) copy of $\mathbb{C}^*$ by $\mathbb{C}_1^*$ and the other copy by $\mathbb{C}_2^*$. We need to show that the map $(\mathcal{U}_{G,P}^{d,\theta})^{T\times\mathbb{C}_1^*} \to (\mathcal{U}_G^d)^{T\times\mathbb{C}_1^*}$ is proper. In fact, we claim that the following stronger statement is true:

Lemma 4.4. *The map* $(\mathcal{U}_{G,P}^{d,\theta})^{\mathbb{C}_1^*} \to (\mathcal{U}_G^d)^{\mathbb{C}_1^*}$ *is an isomorphism.*

Proof. This is an easy corollary of Theorem 10.2 of [2]. In *loc. cit* a natural stratification of $\mathcal{U}_{G,P}$ is described and it follows immediately that

$$(\mathcal{U}_{G,P}^{d,\theta})^{\mathbb{C}_1^*} = (\mathcal{U}_G^d)^{\mathbb{C}_1^*} = \mathrm{Sym}^d(\mathbf{X}\backslash\{\infty\})$$

(recall that $\mathbf{X}$ denotes the "vertical" axis in $\mathbb{P}^2$). □

Acknowledgments

A.B. would like to thank M. Finkelberg, A. Gorsky, A. Marshakov, A. Mironov, I. Krichever, H. Nakajima, N. Nekrasov, A. Okounkov and Y. Oz for very valuable discussions on the subject. The work of A.B. was partially supported by the NSF grant DMS-0300271. The work of P.E. was partially supported by the NSF grant DMS-9988796 and the CRDF grant RM1-2545-MO-03.

References

[1] A. Braverman, *Instanton counting via affine Lie algebras I. Equivariant J-functions of (affine) flag manifolds and Whittaker vectors*, to appear in the proceedings of the "Workshop on algebraic structures and moduli spaces", Montréal, 2003.

[2] A. Braverman, M. Finkelberg and D. Gaitsgory, Uhlenbeck spaces via affine Lie algebras, math.AG/0301176; in *Unity of Mathematics* (dedicated to I. M. Gelfand's 90th birthday, Harvard University, 2003), Prog. in Math., vol. 224, pp. 17–135.

[3] P. Etingof, *Whittaker functions on quantum groups and q-deformed Toda operators*, in: Differential topology, infinite-dimensional Lie algebras, and applications, 9–25, Amer. Math. Soc. Transl. Ser. 2, **194**, Amer. Math. Soc., Providence, RI, 1999.

[4] I. Krichever, *Algebraic curves and nonlinear difference equations* (Russian), *Uspekhi Mat. Nauk* 33 (1978), no. 4, **202**, 215–216.
[5] H. Nakajima and K. Yoshioka, *Instanton counting on blow-ups I*, math.AG/0306198.
[6] H. Nakajima and K. Yoshioka, *Lectures on instanton counting, Algebraic structures and moduli spaces*, 31–101, CRM Proc. Lecture Notes, 38, Amer. Math. Soc., Providence, RI, 2004.
[7] N. Nekrasov, *Seiberg–Witten prepotential from instanton counting*, Adv. Theor. Math. Phys. 7 (2003), no. 5, 831–864.
[8] N. Nekrasov and A. Okoun'kov, *Seiberg–Witten theory and random partitions*, in *Unity of Mathematics*, (dedicated to I. M. Gelfand's 90th birthday, Harvard University, 2003), Prog. Math., vol. 244, pp. 525–596.
[9] N. Nekrasov and S. Shadchin, *ABCD of instantons*, hep-th/0404225.

Irreducibility of perfect representations of double affine Hecke algebras

Ivan Cherednik*

Department of Mathematics
UNC Chapel Hill
Chapel Hill, North Carolina 27599
USA
chered@math.unc.edu

Dedicated to A. Joseph on his 60th birthday

Summary. It is proved that the quotient of the polynomial representation of the double affine Hecke algebra by the radical of the duality pairing is always irreducible apart from the roots of unity provided that it is finite dimensional. We also find necessary and sufficient conditions for the radical to be zero, a generalization of Opdam's formula for the singular parameters such that the corresponding Dunkl operators have multiple zero-eigenvalues.

Subject Classification: 20C08

In the paper we prove that the quotient of the polynomial representation of the double affine Hecke algebra (DAHA) by the radical of the duality pairing is always irreducible (apart from the roots of unity) provided that it is finite dimensional. We also find necessary and sufficient conditions for the radical to be zero, which is a q-generalization of Opdam's formula for the *singular* k-parameters with the multiple zero-eigenvalue of the corresponding Dunkl operators.

Concerning the terminology, *perfect modules* in the paper are finite dimensional possessing a non-degenerate *duality pairing*. The latter induces the canonical *duality anti-involution* of DAHA. Actually, it suffices to assume that the pairing is *perfect*, i.e., identifies the module with its dual as a vector space, but we will stick to the finite-dimensional case.

We also assume that perfect modules are *spherical*, i.e., quotients of the polynomial representation of DAHA, and invariant under the *projective action of* $PSL(2,\mathbb{Z})$. We do not impose the semisimplicity in contrast to [C3]. The irreducibility theorem in this paper is stronger and at the same time the proof is simpler than that in [C3].

The irreducibility follows from the projective $PSL(2,\mathbb{Z})$-action which readily results from the τ_--invariance. The latter always holds if q is not a root of unity. At

* Partially supported by NSF grant DMS-0200276.

roots of unity, it is true for special k only. We do not give in the paper necessary and sufficient conditions for the τ_--invariance as q is a root of unity. Generally, it is not difficult to check (if it is true).

The polynomial representation has the canonical duality paring. It is defined in terms of the difference-trigonometric Dunkl operators, similar to the rational case where the differential-rational operators are used, and involves the evaluation at $q^{-\rho_k}$ instead of the value at zero. The quotient of the polynomial representation by the radical Rad of this pairing is a universal *quasi-perfect* representation. By the latter, we mean a DAHA-module with a non-degenerate but maybe non-perfect duality paring.

The polynomial representation, denoted by $\mathcal{V}$ in the paper, is quasi-perfect and irreducible for generic values of the DAHA-parameters q, t. It is also Y-semisimple, i.e., there exists a basis of eigenvectors of the Y-operators, and has the simple Y-spectrum for generic q, t.

The radical Rad is nonzero when q is a root of unity or as $t = \varsigma\, q^k$ for *special fractional* k and proper roots of unity ς.

We give an example of *reducible* $\mathcal{V}$ which has no radical (B_n). The complete list will be presented in the next paper.

Semisimplicity. Typical examples of Y-semisimple perfect representations are the *non-symmetric Verlinde algebras*, generalizing the Verlinde algebras. The latter describe the fusion of the integrable representations of the Kac–Moody algebras, and, equivalently, the reduced category of representations of quantum groups at roots of unity. The third interpretation is via factors/subfactors. Generally, these algebras appear in terms of the vertex operators (coinvariants) associated with Kac–Moody or Virasoro-type algebras.

There are at least two important reasons to drop the semisimplicity constraint:

First, it was found recently that the fusion procedure for a certain Virasoro-type algebra leads to a non-semisimple variant of the Verlinde algebra. As a matter of fact, there are no general reasons to expect semisimplicity in the *massless conformal field theory.* The positive definite inner product in the Verlinde algebra, which guarantees the semisimplicity, is given in terms of the masses of the points/particles.

Second, non-semisimple representations of DAHA are expected to appear when the whole category of representations of *Lusztig's quantum group* at roots of unity is considered. Generally, non-spherical representations could be necessary. However the anti-spherical (Steinberg-type) representations, which are spherical constructed for t^{-1} in place of t, are expected to play an important role.

The simplest *non-semisimple* example at roots of unity (A_1) is considered at the end of the paper in detail.

Concerning the necessary and sufficient condition for the radical of $\mathcal{V}$ to be nonzero, it readily follows from the evaluation formula for the nonsymmetric Macdonald polynomials [C2]. This approach does require the q, t-setting because the evaluation formula collapses in the limit. See [DO], Section 3.2.

The method from [O2] (see also [DJO] and [J]) based on the shift operator is also possible, and even becomes simpler with q, t than in the rational/trigonometric case. It will be demonstrated in the next paper. The definition of the radical of the polynomial representation is due to Opdam in the rational case. See, e.g., [DO]. In the q, t-case, the radical was introduced in [C1, C2].

Rational limit. Interestingly, the quotient of $\mathcal{V}$ by the radical is always irreducible for the *rational DAHA*. The justification is immediate and goes as follows.

This quotient has the zero-eigenvalue (no other eigenvalues appear in the rational setting) of multiplicity one. Any its proper submodule will generate at least one additional zero-eigenvector, which is impossible.

The DAHA and its rational degeneration are connected by exp–log maps of some kind [C4], but these maps are of analytic nature in the infinite dimensional case and cannot be directly applied to the polynomial representation.

Generally, the q, t-methods are simpler in many aspects than those in the rational degeneration thanks to the existence of the Macdonald polynomials and their analytic counterparts. It is somewhat similar to the usage of the unitary invariant scalar product in the theory of *compact* Lie groups vs. the abstract theory of Lie algebras. The q, t-generalization of Opdam's formula for singular k and the theory of perfect representations are typical examples in favor of the q, t-setting. However, with the irreducibility of the universal quasi-perfect quotient of the polynomial representation, it is the other way round.

My guess is that it happens because the q, t-polynomial representation contains more information than could be seen after the rational degeneration. I mean mainly the semisimplicity which does not exist in the rational theory and can be incorporated only if the rational DAHA is extended by the "first jet" towards q (not published).

It must be mentioned here that the rational theory is for complex reflection groups. The q, t-theory is mainly about the crystallographic groups. Not all complex reflection groups have affine extensions.

I thank A. Garsia, E. Opdam, and N. Wallach for useful discussions. I would like to thank UC at San Diego and IML (Luminy) for their kind invitations.

1. Affine Weyl groups

Let $R = \{\alpha\} \subset \mathbb{R}^n$ be a root system of type $A, B, \ldots, F, G$ with respect to a euclidean form (z, z') on $\mathbb{R}^n \ni z, z'$, W the **Weyl group** generated by the reflections s_α, R_+ the set of positive roots ($R_- = -R_+$), corresponding to (fixed) simple roots $\alpha_1, \ldots, \alpha_n$, Γ the Dynkin diagram with $\{\alpha_i, 1 \le i \le n\}$ as the vertices.

We will also use the dual roots (coroots) and the dual root system

$$R^\vee = \{\alpha^\vee = 2\alpha/(\alpha, \alpha)\}.$$

The root lattice and the weight lattice are

$$Q = \oplus_{i=1}^n \mathbb{Z}\alpha_i \subset P = \oplus_{i=1}^n \mathbb{Z}\omega_i,$$

where $\{\omega_i\}$ are fundamental weights: $(\omega_i, \alpha_j^\vee) = \delta_{ij}$ for the simple coroots $\alpha_i^\vee$.

Replacing $\mathbb{Z}$ by $\mathbb{Z}_\pm = \{m \in \mathbb{Z}, \pm m \ge 0\}$ we obtain $Q_\pm, P_\pm$. Note that $Q \cap P_+ \subset Q_+$. Moreover, each ω_j has all non-zero coefficients (sometimes rational) when expressed in terms of $\{\alpha_i\}$. Here and further see [B].

The form will be normalized by the condition $(\alpha,\alpha)=2$ for the *short* roots. Thus, $\nu_\alpha \stackrel{\mathbf{def}}{=\!=} (\alpha,\alpha)/2$ is either 1, or $\{1,2\}$, or $\{1,3\}$.

We will use the notation ν_{lng} for the long roots ($\nu_{\mathrm{sht}}=1$).

Let $\vartheta \in R^\vee$ be the **maximal positive coroot**. Considered as a root (it belongs to R because of the choice of normalization) it is maximal among all short positive roots of R.

Setting $\nu_i = \nu_{\alpha_i}$, $\nu_R = \{\nu_\alpha, \alpha \in R\}$, one has

$$\rho_\nu \stackrel{\mathbf{def}}{=\!=} (1/2)\sum_{\nu_\alpha=\nu}\alpha = \sum_{\nu_i=\nu}\omega_i, \text{ where } \alpha \in R_+,\ \nu \in \nu_R. \tag{1.1}$$

Note that $(\rho_\nu, \alpha_i^\vee)=1$ as $\nu_i=\nu$. We will call ρ_ν **partial** ρ.

Affine roots. The vectors $\tilde\alpha = [\alpha, \nu_\alpha j] \in \mathbb{R}^n \times \mathbb{R} \subset \mathbb{R}^{n+1}$ for $\alpha \in R, j \in \mathbb{Z}$ form the **affine root system** $\tilde R \supset R$ ($z \in \mathbb{R}^n$ are identified with $[z,0]$). We add $\alpha_0 \stackrel{\mathbf{def}}{=\!=} [-\vartheta, 1]$ to the simple roots for the maximal short root ϑ. The corresponding set $\tilde R$ of positive roots coincides with $R_+ \cup \{[\alpha,\nu_\alpha j],\ \alpha \in R,\ j>0\}$.

We complete the Dynkin diagram Γ of R by α_0 (by $-\vartheta$ to be more exact). The notation is $\tilde\Gamma$. One can obtain it from the completed Dynkin diagram for $R^\vee$ from [B] reversing the arrows. The number of laces between α_i and α_j in $\tilde\Gamma$ is denoted by m_{ij}.

The set of the indices of the images of α_0 by all the automorphisms of $\tilde\Gamma$ will be denoted by O ($O=\{0\}$ for E_8, F_4, G_2). Let $O' = r \in O, r \neq 0$. The elements ω_r for $r \in O'$ are the so-called minuscule weights: $(\omega_r, \alpha^\vee) \le 1$ for $\alpha \in R_+$.

Given $\tilde\alpha = [\alpha, \nu_\alpha j] \in \tilde R$, $b \in B$, let

$$s_{\tilde\alpha}(\tilde z) = \tilde z - (z,\alpha^\vee)\tilde\alpha,\quad b'(\tilde z) = [z, \zeta - (z,b)] \tag{1.2}$$

for $\tilde z = [z,\zeta] \in \mathbb{R}^{n+1}$.

The **affine Weyl group** $\widetilde W$ is generated by all $s_{\tilde\alpha}$ (we write $\widetilde W = \langle s_{\tilde\alpha}, \tilde\alpha \in \tilde R_+\rangle$). One can take the simple reflections $s_i = s_{\alpha_i}$ $(0 \le i \le n)$ as its generators and introduce the corresponding notion of the length. This group is the semidirect product $W \ltimes Q'$ of its subgroups $W = \langle s_\alpha, \alpha \in R_+\rangle$ and $Q' = \{a', a \in Q\}$, where

$$\alpha' = s_\alpha s_{[\alpha,\nu_\alpha]} = s_{[-\alpha,\nu_\alpha]} s_\alpha \text{ for } \alpha \in R. \tag{1.3}$$

The **extended Weyl group** $\widehat W$ generated by W and P' (instead of Q') is isomorphic to $W \ltimes P'$:

$$(wb')([z,\zeta]) = [w(z), \zeta - (z,b)] \text{ for } w \in W, b \in P. \tag{1.4}$$

From now on, b and b', P and P' will be identified.

Given $b \in P_+$, let w_0^b be the longest element in the subgroup $W_0^b \subset W$ of the elements preserving b. This subgroup is generated by simple reflections. We set

$$u_b = w_0 w_0^b \in W,\ \pi_b = b(u_b)^{-1} \in \widehat W,\ u_i = u_{\omega_i}, \pi_i = \pi_{\omega_i}, \tag{1.5}$$

where w_0 is the longest element in W, $1 \le i \le n$.

The elements $\pi_r \stackrel{\text{def}}{=\!=} \pi_{\omega_r}, r \in O'$ and $\pi_0 = \text{id}$ leave $\tilde{\Gamma}$ invariant and form a group denoted by Π, which is isomorphic to P/Q by the natural projection $\{\omega_r \mapsto \pi_r\}$. As to $\{u_r\}$, they preserve the set $\{-\vartheta, \alpha_i, i > 0\}$. The relations $\pi_r(\alpha_0) = \alpha_r = (u_r)^{-1}(-\vartheta)$ distinguish the indices $r \in O'$. Moreover (see e.g., [C3]):

$$\widehat{W} = \Pi \ltimes \widetilde{W}, \quad \text{where } \pi_r s_i \pi_r^{-1} = s_j \text{ if } \pi_r(\alpha_i) = \alpha_j, \ 0 \le j \le n. \tag{1.6}$$

Setting $\widehat{w} = \pi_r \widetilde{w} \in \widehat{W}$, $\pi_r \in \Pi, \widetilde{w} \in \widetilde{W}$, the length $l(\widehat{w})$ is by definition the length of the reduced decomposition $\widetilde{w} = s_{i_l} \cdots s_{i_2} s_{i_1}$ in terms of the simple reflections $s_i, 0 \le i \le n$.

The length can be also defined as the cardinality $|\lambda(\widehat{w})|$ of

$$\lambda(\widehat{w}) \stackrel{\text{def}}{=\!=} \tilde{R}_+ \cap \widehat{w}^{-1}(\tilde{R}_-) = \{\tilde{\alpha} \in \tilde{R}_+, \ \widehat{w}(\tilde{\alpha}) \in \tilde{R}_-\}, \ \widehat{w} \in \widehat{W}.$$

Reduction modulo W. The following proposition is from [C2]. It generalizes the construction of the elements π_b for $b \in P_+$.

Proposition 1.1. *Given $b \in P$, there exists a unique decomposition $b = \pi_b u_b$, $u_b \in W$ satisfying one of the following equivalent conditions:*

(i) $l(\pi_b) + l(u_b) = l(b)$ and $l(u_b)$ is the greatest possible,
(ii) $\lambda(\pi_b) \cap R = \emptyset$. Moreover, $u_b(b) \stackrel{\text{def}}{=\!=} b_- \in P_- = -P_+$ is a unique element from P_- which belongs to the orbit $W(b)$. □

For $\tilde{\alpha} = [\alpha, \nu_\alpha j] \in \tilde{R}_+$, one has

$$\begin{aligned}
\lambda(b) = \{\tilde{\alpha}, \ &(b, \alpha^\vee) > j \ge 0 \text{ if } \alpha \in R_+, \\
&(b, \alpha^\vee) \ge j > 0 \text{ if } \alpha \in R_-\},
\end{aligned} \tag{1.7}$$

$$\begin{aligned}
\lambda(\pi_b) = \{\tilde{\alpha}, \ \alpha \in R_-, \ &(b_-, \alpha^\vee) > j > 0 \text{ if } u_b^{-1}(\alpha) \in R_+, \\
&(b_-, \alpha^\vee) \ge j > 0 \text{ if } u_b^{-1}(\alpha) \in R_-\},
\end{aligned} \tag{1.8}$$

$$\lambda(u_b) = \{\alpha \in R_+, \ (b, \alpha^\vee) > 0\}. \tag{1.9}$$

2. Double Hecke algebras

By m, we denote the least natural number such that $(P, P) = (1/m)\mathbb{Z}$. Thus $m = 2$ for D_{2k}, $m = 1$ for B_{2k} and C_k, otherwise $m = |\Pi|$.

The double affine Hecke algebra depends on the parameters $q, t_\nu, \nu \in \{\nu_\alpha\}$. The definition ring is $\mathbb{Q}_{q,t} \stackrel{\text{def}}{=\!=} \mathbb{Q}[q^{\pm 1/m}, t^{\pm 1/2}]$ formed by the polynomials in terms of $q^{\pm 1/m}$ and $\{t_\nu^{\pm 1/2}\}$. We set

$$\begin{aligned}
&t_{\tilde{\alpha}} = t_\alpha = t_{\nu_\alpha}, \ t_i = t_{\alpha_i}, \ q_{\tilde{\alpha}} = q^{\nu_\alpha}, \ q_i = q^{\nu_{\alpha_i}}, \\
&\text{where } \tilde{\alpha} = [\alpha, \nu_\alpha j] \in \tilde{R}, \ 0 \le i \le n.
\end{aligned} \tag{2.1}$$

It will be convenient to use the parameters $\{k_\nu\}$ together with $\{t_\nu\}$, setting

$$t_\alpha = t_\nu = q_\alpha^{k_\nu} \ \text{ for } \nu = \nu_\alpha, \ \text{ and } \rho_k = (1/2)\sum_{\alpha>0} k_\alpha \alpha.$$

For pairwise commutative $X_1, \ldots, X_n$,

$$X_{\tilde b} = \prod_{i=1}^{n} X_i^{l_i} q^j \ \text{ if } \tilde b = [b, j],\ \widehat{w}(X_{\tilde b}) = X_{\widehat{w}(\tilde b)}, \tag{2.2}$$
$$\text{where } b = \sum_{i=1}^{n} l_i \omega_i \in P,\ j \in \frac{1}{m}\mathbb{Z},\ \widehat{w} \in \widehat{W}.$$

We set $(\tilde b, \tilde c) = (b, c)$ ignoring the affine extensions.

Later $Y_{\tilde b} = Y_b q^{-j}$ will be needed. Note the negative sign of j.

Definition 2.1. The double affine Hecke algebra $\mathcal{HH}$ is generated over $\mathbb{Q}_{q,t}$ by the elements $\{T_i,\ 0 \le i \le n\}$, pairwise commutative $\{X_b,\ b \in P\}$ satisfying (2.2), and the group Π, where the following relations are imposed:

(0) $(T_i - t_i^{1/2})(T_i + t_i^{-1/2}) = 0,\ 0 \le i \le n;$
(i) $T_i T_j T_i \ldots = T_j T_i T_j, \ldots,\ m_{ij}$ factors on each side;
(ii) $\pi_r T_i \pi_r^{-1} = T_j$ if $\pi_r(\alpha_i) = \alpha_j$;
(iii) $T_i X_b T_i = X_b X_{\alpha_i}^{-1}$ if $(b, \alpha_i^\vee) = 1,\ 0 \le i \le n;$
(iv) $T_i X_b = X_b T_i$ if $(b, \alpha_i^\vee) = 0$ for $0 \le i \le n;$
(v) $\pi_r X_b \pi_r^{-1} = X_{\pi_r(b)},\ r \in O'.$ □

Given $\widetilde{w} \in \widetilde{W}, r \in O,$ the product

$$T_{\pi_r \widetilde{w}} \overset{\mathbf{def}}{=\!=} \pi_r \prod_{k=1}^{l} T_{i_k}, \ \text{ where } \widetilde{w} = \prod_{k=1}^{l} s_{i_k}, l = l(\widetilde{w}), \tag{2.3}$$

does not depend on the choice of the reduced decomposition (because $\{T\}$ satisfy the same "braid" relations as $\{s\}$ do). Moreover,

$$T_{\hat v} T_{\widehat w} = T_{\hat v \widehat w} \ \text{ whenever } l(\hat v \widehat w) = l(\hat v) + l(\widehat w) \ \text{ for } \hat v, \widehat w \in \widehat W. \tag{2.4}$$

In particular, we arrive at the pairwise commutative elements

$$Y_b = \prod_{i=1}^{n} Y_i^{l_i} \ \text{ if } b = \sum_{i=1}^{n} l_i \omega_i \in P, \ \text{ where } Y_i \overset{\mathbf{def}}{=\!=} T_{\omega_i}, \tag{2.5}$$

satisfying the relations

$$T_i^{-1} Y_b T_i^{-1} = Y_b Y_{\alpha_i}^{-1} \ \text{ if } (b, \alpha_i^\vee) = 1,$$
$$T_i Y_b = Y_b T_i \ \text{ if } (b, \alpha_i^\vee) = 0,\ 1 \le i \le n. \tag{2.6}$$

The Demazure–Lusztig operators are defined as follows:

$$T_i = t_i^{1/2} s_i + (t_i^{1/2} - t_i^{-1/2})(X_{\alpha_i} - 1)^{-1}(s_i - 1),\ 0 \le i \le n, \tag{2.7}$$

and obviously preserve $\mathbb{Q}[q, t^{\pm 1/2}][X]$. We note that only the formula for T_0 involves q:

$$T_0 = t_0^{1/2} s_0 + (t_0^{1/2} - t_0^{-1/2})(qX_\vartheta^{-1} - 1)^{-1}(s_0 - 1),$$
$$\text{where } s_0(X_b) = X_b X_\vartheta^{-(b,\vartheta)} q^{(b,\vartheta)},\ \alpha_0 = [-\vartheta, 1]. \tag{2.8}$$

The map sending T_j to the formula in (2.7), and $X_b \mapsto X_b$ (see (2.2)), $\pi_r \mapsto \pi_r$ induces a $\mathbb{Q}_{q,t}$-linear homomorphism from $\mathcal{HH}$ to the algebra of linear endomorphisms of $\mathbb{Q}_{q,t}[X]$. This $\mathcal{HH}$-module, which will be called the **polynomial representation**, is faithful and remains faithful when q, t take any non-zero complex values assuming that q is not a root of unity.

The images of the Y_b are called the **difference Dunkl operators**. To be more exact, they must be called difference-trigonometric Dunkl operators, because there are also difference-rational Dunkl operators.

The polynomial representation is the $\mathcal{HH}$-module induced from the one dimensional representation $T_i \mapsto t_i^{1/2}$, $Y_i \mapsto Y_i^{1/2}$ of the affine Hecke subalgebra $\mathcal{H}_Y = \langle T, Y\rangle$. Here the PBW-Theorem is used: for arbitrary nonzero q, t, any element $H \in \mathcal{HH}$ has a unique decomposition in the form

$$H = \sum_{w \in W} g_w f_w T_w,\ g_w \in \mathbb{Q}_{q,t}[X],\ f_w \in \mathbb{Q}_{q,t}[Y]. \tag{2.9}$$

The definition of DAHA and the polynomial representation are compatible with the **intermediate subalgebras** $\mathcal{HH}^\flat \subset \mathcal{HH}$ with P replaced by any lattice $B \ni b$ between Q and P. Respectively, Π is changed to the image $\Pi^\flat$ of B/Q in Π. From now on, we take X_a, Y_b with the indices $a, b \in B$. We will continue using the notation $\mathcal{V}$ for the B-polynomial representation:

$$\mathcal{V} = \mathbb{Q}_{q,t}[X_b] = \mathbb{Q}_{q,t}[X_b, b \in B].$$

We also set $\widehat{W}^\flat = B \cdot W \subset \widehat{W}$, and replace m by the least $\tilde{m} \in \mathbb{N}$ such that $\tilde{m}(B, B) \subset \mathbb{Z}$ in the definition of the $\mathbb{Q}_{q,t}$.

Automorphisms. The following **duality anti-involution** is of key importance for the various duality statements:

$$\phi : X_b \mapsto Y_b^{-1},\ T_i \mapsto T_i\ (1 \le i \le n). \tag{2.10}$$

It preserves q, t_ν and their fractional powers.

We will also need the automorphisms of $\mathcal{HH}^\flat$ (see [C2], [C3]):

$$\tau_+ : X_b \mapsto X_b,\ Y_r \mapsto X_r Y_r q^{-\frac{(\omega_r,\omega_r)}{2}},\ \pi_r \mapsto q^{-(\omega_r,\omega_r)} X_r \pi_r,$$
$$\tau_+ : Y_\vartheta \mapsto q^{-1} X_\vartheta T_0^{-1} T_{s_\vartheta},\ T_0 \mapsto q^{-1} X_\vartheta T_0^{-1},\ \text{and} \tag{2.11}$$
$$\tau_- \stackrel{\text{def}}{=} \phi\tau_+\phi,\ \sigma \stackrel{\text{def}}{=} \tau_+\tau_-^{-1}\tau_+ = \tau_-^{-1}\tau_+\tau_-^{-1}, \tag{2.12}$$

where $r \in O'$. They fix $T_i\ (i \geq 1),\ t_\nu,\ q$ and fractional powers of $t_\nu,\ q$. Note that $\tau_- = \sigma\tau_+\sigma^{-1}$.

In the definition of $\tau_\pm$ and σ, we need to add $q^{\pm 1/(2m)}$ to $\mathbb{Q}_{q,t}$.

The automorphism τ_- acts trivially on $\{T_i(i \geq 0),\ \pi_r,\ Y_b\}$. Hence it naturally acts in the polynomial representation $\mathcal{V}$. The automorphism τ_+ and therefore σ do not act in $\mathcal{V}$. The automorphism σ sends X_b to Y_b^{-1} and is associated with the Fourier transform in the DAHA theory.

Actually, all these automorphisms act in the central extension of the *elliptic braid group* defined by the relations of $\mathcal{HH}$, where the quadratic relation is dropped. The central extension is by the fractional powers of q.

The elements $\tau_\pm$ generate the projective $PSL(2,\mathbb{Z})$, which is isomorphic to the braid group B_3 due to Steinberg.

3. Macdonald polynomials

This definition is due to Macdonald (for $k_{\text{sht}} = k_{\text{lng}} \in \mathbb{Z}_+$), who extended in [M] Opdam's nonsymmetric polynomials introduced in the differential case in [O1] (Opdam mentions Heckman's contribution in [O1]). The general case was considered in [C2].

We continue using the same notation X, Y, T for these operators acting in the polynomial representation. The parameters q, t are generic in the following definition.

Definition 3.1. The **nonsymmetric Macdonald polynomials** $\{E_b, b \in P\}$ are unique (up to proportionality) eigenfunctions of the operators

$$\{L_f \overset{\mathbf{def}}{=\!=} f(Y_1,\ldots,Y_n),\ f \in \mathbb{Q}[X]\}$$

acting in $\mathbb{Q}_{q,t}[X]$:

$$L_f(E_b) = f(q^{-b_\sharp})E_b, \quad \text{where } b_\sharp \overset{\mathbf{def}}{=\!=} b - u_b^{-1}(\rho_k), \tag{3.13}$$

$$X_a(q^b) = q^{(a,b)} \text{ for } a, b \in P,\ u_b = \pi_b^{-1}b, \tag{3.14}$$

where u_b is from Proposition 1.1.

They satisfy

$$E_b - X_b \in \oplus_{c \succ b}\mathbb{Q}(q,t)X_c,\ \langle E_b, X_c\rangle_\circ = 0 \text{ for } P \ni c \succ b, \tag{3.15}$$

where we set $c \succ b$ if

$$c_- - b_- \in B \cap Q_+ \text{ or } c_- = b_- \text{ and } c - b \in B \cap Q_+.$$

The following **intertwiners** are the key in the theory:

$$\Psi_i = \tau_+(T_i) + \frac{t_i^{1/2} - t_i^{-1/2}}{Y_{\alpha_i}^{-1} - 1},\ i \geq 0,\ \ P_r = \tau_+(\pi_r),\ r \in O', \tag{3.16}$$

$$\Psi_{\widehat{w}} = P_r\Psi_{i_l}\ldots\Psi_{i_1} \text{ for reduced decompositions } \widehat{w} = \pi_r s_{i_l}\ldots s_{i_1}.$$

Note the formulas

$$\tau_+(T_0) = X_0^{-1}T_0^{-1},\ X_0 = qX_\vartheta^{-1},\ \tau_+(\pi_r) = q^{-(\omega_r,\omega_r)/2}X_r\pi_r.$$

The products $\Psi_{\widehat{w}}$ do not depend on the choice of the reduced decomposition, intertwine Y_b, and transform the E-polynomials correspondingly. Namely, for $\widehat{w} \in \widehat{W}$,

$$\Psi_{\widehat{w}}Y_b = Y_{\widehat{w}(b)}\Psi_{\widehat{w}},\ \text{where } Y_{[b,j]} \stackrel{\mathbf{def}}{=\!=} Y_b q^{-j}, \qquad (3.17)$$
$$E_b = \mathrm{Const}\,\Psi_{\widehat{w}}(E_c)\ \text{ for }\ \mathrm{Const} \neq 0,\ b = \widehat{w}((c)),$$

provided that $\pi_b = \widehat{w}\pi_c$ and $l(\pi_b) = l(\widehat{w}) + l(\pi_c)$.

Here we use the **affine action** of $\widehat{W}$ on $z \in \mathbb{R}^n$:

$$(wb)((z)) = w(b+z),\ w \in W, b \in P,$$
$$s_{\tilde{\alpha}}((z)) = z - ((z,\alpha)+j)\alpha,\ \tilde{\alpha} = [\alpha, \nu_\alpha j] \in \tilde{R}. \qquad (3.18)$$

The definition of the E-polynomials and the action of the intertwiners are compatible with the transfer to the intermediate subalgebras $\mathcal{HH}^\flat$. Recall that the B-polynomial representation is

$$\mathcal{V} = \mathbb{Q}_{q,t}[X_b] \stackrel{\mathbf{def}}{=\!=} \mathbb{Q}_{q,t}[X_b, b \in B].$$

We note that the Ψ-intertwiners were introduced by Knop and Sahi in the case of GL_n.

The coefficients of the Macdonald polynomials are rational functions in terms of q_ν, t_ν. The following evaluation formula holds:

$$E_b(q^{-\rho_k}) = q^{(\rho_k,b_-)} \prod_{[\alpha,j]\in\lambda'(\pi_b)} \Big(\frac{1 - q_\alpha^j t_\alpha X_\alpha(q^{\rho_k})}{1 - q_\alpha^j X_\alpha(q^{\rho_k})}\Big), \qquad (3.19)$$
$$\lambda'(\pi_b) = \{[\alpha, j] \mid [-\alpha, \nu_\alpha j] \in \lambda(\pi_b)\}. \qquad (3.20)$$

Explicitly, (see (1.8)),

$$\lambda'(\pi_b) = \{[\alpha, j] \mid \alpha \in R_+, \qquad (3.21)$$
$$-(b_-,\alpha^\vee) > j > 0 \text{ if } u_b^{-1}(\alpha) \in R_-,$$
$$-(b_-,\alpha^\vee) \geq j > 0 \text{ if } u_b^{-1}(\alpha) \in R_+\}.$$

Formula (3.19) is the Macdonald **evaluation conjecture** in the nonsymmetric variant from [C2].

Note that one has to consider only long α (resp., short) if $k_{\text{sht}} = 0$ (resp., $k_{\text{lng}} = 0$) in the λ'-set.

We have the following **duality formula** for $b, c \in P$:

$$E_b(q^{c_\sharp})E_c(q^{-\rho_k}) = E_c(q^{b_\sharp})E_b(q^{-\rho_k}),\ b_\sharp = b - u_b^{-1}(\rho_k). \qquad (3.22)$$

See [C2]. The proof is based on the anti-involution ϕ from (2.10).

The action of τ_-. The automorphism τ_+ is a *formal* conjugation by the Gaussian $\gamma(q^z) = q^{(z,z)/2}$ where we set $X_b(q^z) = q^{(b,z)}$. We treat γ as an element in a completion of the polynomial representation with the extended action of $\mathcal{HH}$. Actually, only the W-invariance of γ and the relations

$$\omega_j(\gamma) = q^{(\omega_i,\omega_i)/2} X_i^{-1}\gamma \text{ for } j = 1,\ldots,n$$

are needed here. For instance, one can (formally) take

$$\gamma_x \stackrel{\mathbf{def}}{=\!=} \sum_{b\in B} q^{-(b,b)/2} X_b. \tag{3.23}$$

Applying σ and using that $\tau_- = \sigma\tau_+^{-1}\sigma^{-1}$, we obtain that the automorphism τ_- in $\mathcal{V}$ is proportional to the *multiplication* by

$$\gamma_y \stackrel{\mathbf{def}}{=\!=} \sum_{b\in B} q^{(b,b)/2} Y_b \tag{3.24}$$

provided that $|q| < 1$. We use that $\mathcal{V}$ is a union of finite-dimensional spaces preserved by the Y-operators. This observation is convenient, although not absolutely necessary, to check the following proposition.

Proposition 3.2. *i) For generic q, t or for any q, t provided that the polynomial E_b for $b \in B$ is well defined,*

$$\tau_-(E_b) = q^{-\frac{(b_-,b_-)}{2}+(b_-,\rho_k)} E_b \text{ for } P_- \ni b_- \in W(b). \tag{3.25}$$

ii) For arbitrary q, t,

$$\begin{aligned} \tau_-(T_i) = T_i,\ \tau_-(\Psi_i) = \Psi_i \text{ for } i > 0, \\ \tau_-(\tau_+(\pi_r)) = q^{(\omega_r,\omega_r)/2} Y_r \tau_+(\pi_r),\ \tau_-(\tau_+(T_0)) = \tau_+(T_0)^{-1} Y_0, \\ \tau_-(\Psi_0) = \Psi_0 Y_0 = Y_0^{-1}\Psi_0,\ Y_0 = q^{-1}Y_\vartheta^{-1}. \end{aligned} \tag{3.26}$$

iii) If q is not a root of unity for arbitrary t_ν, then τ_- preserves an arbitrary Y-submodule of $\mathcal{V}$.

Proof. The first two claims are straightforward. As for (iii), since q is generic one can assume that $0 < q < 1$ and define $\widetilde{\tau_-}$ as the operator of multiplication by $C^{-1}\gamma_y$ using (3.24) and taking

$$C = \sum_{b\in B} q^{(b,b)/2} Y_b(1) = \sum_{b\in B} q^{(b,b)/2} q^{(b,\rho_k)}.$$

Then $\widetilde{\tau_-}$ coincides with τ_- for generic k, when all E-polynomials exist and the X-spectrum of $\mathcal{V}$ is simple, due to (i). This gives the coincidence for any k. □

4. The radical

Following [C1, C2], we set

$$\{f, g\} = \{L_{\iota(f)}(g(X))\} = \{L_{\iota(f)}(g(X))\}(q^{-\rho_k}) \text{ for } f, g \in \mathcal{V}, \tag{4.1}$$
$$\iota(X_b) = X_{-b} = X_b^{-1},\ \iota(z) = z \text{ for } z \in \mathbb{Q}_{q,t},$$

where L_f is from Definition 3.1. It induces the $\mathbb{Q}_{q,t}$-linear anti-involution ϕ of $\mathcal{HH}^\flat$ from (2.10).

Lemma 4.1. *For arbitrary non-zero* q, t_{sht}, t_{lng},

$$\{f, g\} = \{g, f\} \text{ and } \{H(f), g\} = \{f, H^\phi(g)\},\ H \in \mathcal{HH}^\flat. \tag{4.2}$$

The quotient $\mathcal{V}'$ *of* $\mathcal{V}$ *by the radical* Rad $\stackrel{\text{def}}{=}$ Rad $\{\ ,\ \}$ *of the pairing* $\{\ ,\ \}$ *is an* $\mathcal{HH}^\flat$*-module such that*

a) all Y*-eigenspaces of* $\mathcal{V}'$ *are zero or one-dimensional,*
b) $E(q^{-\rho_k}) \neq 0$ *if the image* E' *of* E *in* $\mathcal{V}'$ *is a non-zero* Y*-eigenvector.*

The radical Rad *is the greatest* $\mathcal{HH}^\flat$*-submodule in the kernel of the map* $f \mapsto \{f, 1\} = f(q^{-\rho_k})$.

Proof. Formulas (4.2) are from Theorem 2.2 of [C2]. Concerning the rest, let us recall the argument from [C3]. Since Rad$\{\ ,\ \}$ is a submodule, the form $\{\ ,\ \}$ is well defined and non-degenerate on $\mathcal{V}'$. For any pullback $E \in \mathcal{V}$ of $E' \in \mathcal{V}'$, $E(q^{-\rho_k}) = \{E, 1\} = \{E', 1'\}$. If E' is a Y-eigenvector in $\mathcal{V}'$ and $E(q^{-\rho_k})$ vanishes, then

$$\{Q_{q,t}[Y_b](E'), \mathcal{H}_Y^\flat(1')\} = 0 = \{E', \mathcal{V} \cdot \mathcal{H}_Y^\flat(1')\}.$$

Therefore $\{E', \mathcal{V}'\} = 0$, which is impossible. □

In the following lemma, q is generic, but t_ν are not supposed generic. The Macdonald polynomials E_b always exist for $b = b^o$, satisfying the conditions

$$q^{-a_\sharp} \neq q^{-b^o_\sharp} \text{ for all } a \succ b^o. \tag{4.3}$$

We call such b^o **primary.** Sufficiently big b are primary.

Lemma 4.2. *i) A* Y*-eigenvector* $E \in \mathcal{V}$ *belongs to* Rad *if and only if* $E(q^{-\rho_k}) = 0$. *The equality* $E(q^{-\rho_k}) = 0$ *automatically results in the equalities*

$$E(q^{-b^o_\sharp}) = 0 \text{ for all } b^o \in B^\star \stackrel{\text{def}}{=} \{b^o \in B \mid E_{b^o}(q^{-\rho_k}) \neq 0\}. \tag{4.4}$$

ii) Let us assume that the radical is nonzero. Then for any constant $C > 0\,(1 \le i \le n)$*, there exists primary* b^o *such that* $(\alpha_i, b^o) > C$ *and* $E_{b^o}(q^{-\rho_k}) = 0$*, i.e.,* $E_{b^o} \in$ Rad.

Proof. The first claim follows from Lemma 4.1. If $E \in \text{Rad}$ and there is no such b^o for certain C, then the number of common zeros of the translations $c(E)$ of E for *any* number of $c \in B$ is infinite, which is impossible because the degree of E is finite.

□

We come to the following theorem generalizing the description of singular k from [O2].

Theorem 4.3. *Assuming that q is generic, the radical vanishes if and only if $E_{b^o}(q^{-\rho_k}) \neq 0$ for all sufficiently big primary b^o, i.e., if the product in the right-hand side of (3.19) is non-zero for all $b \in B$ with sufficiently big (b, α_i) for $i > 0$.* □

We can define **quasi-perfect representations** as $\mathcal{HH}^\flat$-modules which have a non-degenerate form $\{\ ,\ \}$ satisfying (4.2). Then the greatest quasi-perfect quotient of the polynomial represntation is $\mathcal{V}/\text{Rad}$. Indeed, any quasi-perfect quotient V of $\mathcal{V}$ supplies it with a form $\{f, g\}_V = \{f', g'\}$ for the images f', g' of f, g in V. Then a proper linear combination $\{\ ,\ \}_o$ of $\{\ ,\ \}$ and $\{\ ,\ \}_V$ will satisfy $\{1, 1\}_o = 0$, which immediately makes it zero identically.

5. The irreducibility

In this section q, t are arbitrary non-zero, including roots of unity.

Theorem 5.1. *i) If the quotient $\mathcal{V}'$ of the polynomial representation $\mathcal{V}$ by the radical* $\text{Rad}\{\ ,\ \}$ *is finite dimensional and τ_--invariant, then it is an irreducible $\mathcal{HH}^\flat$-module. The radical is always τ_--invariant if q is not a root of unity.*

ii) At roots of unity, the τ_--invariance holds when the radical is $\mathcal{HH}^\flat$-generated by linear combinations $\sum c_b E_b$ (provided that E_b exist) where the summations are over b with coinciding $q^{-(b_-,b_-)/2-(b_-,\rho_k)}$ from (3.25).

Proof. Using $\phi\tau_-\phi = \tau_+$, the relation

$$\{\tau_+ f, g\} = \{f, \tau_- g\} \text{ for } f, g \in \mathcal{V}'$$

defines the action of τ_+ in $\mathcal{V}'$ and therefore the action of σ there satisfying

$$\tau_+\tau_-^{-1}\tau_+ = \sigma = \tau_-^{-1}\tau_+\tau_-^{-1}.$$

The pairing $\{f, g\}_\sigma \overset{\text{def}}{=\!=} \{\sigma f, g\} = \{f, \sigma^{-1} g\}$ corresponds to the *anti-involution* $\heartsuit = \sigma \cdot \phi = \phi \cdot \sigma^{-1}$ of $\mathcal{HH}^\flat$, sending

$$\heartsuit : T_i \mapsto T_i,\ \pi_r \mapsto \pi_r,\ Y_b \mapsto Y_b,\ X_b \mapsto T_{w_0}^{-1} X_{\varsigma(b)} T_{w_0} \tag{5.1}$$

for $0 \le i \le n,\ b \in B$.

It holds in either direction, from f to g and the other way round, but the form $\{f, g\}_\sigma$, generally speaking, could be non-symmetric. Actually it is symmetric, but we do not need it for the proof.

Using this *non-degenerate* pairing, we proceed as follows. Any proper $\mathcal{HH}^\flat$-submodule $\mathcal{V}''$ of $\mathcal{V}'$ contains at least one Y-eigenvector e'', so we can assume that $\mathcal{V}'' = \mathcal{HH}^\flat e''$. The corresponding eigenvalue cannot coincide with that of 1 thanks to the previous lemma. Therefore $\{1', \mathcal{V}''\}_\sigma = 0$ for the image $1'$ of 1 in $\mathcal{V}'$, and the orthogonal complement of $\mathcal{V}''$ in $\mathcal{V}'$ is a *proper* $\mathcal{HH}^\flat$-submodule of $\mathcal{V}'$ containing $1'$, which is impossible.

Using Proposition 3.2, we obtain (ii). □

Let us check that the pairing $\{f, g\}_\sigma$ is symmetric. First,

$$\begin{aligned}&\{\tau_+(1'), 1'\} = \{1', \tau_-(1')\} = \{1', 1'\} = 1 \Rightarrow\\ &\quad \{1', 1'\}_\sigma = \{\sigma(1'), 1'\} = \{\tau_+(1'), \tau_-(1')\} = \{\tau_+(1'), 1'\} = 1.\end{aligned}$$

Then, $\{1', f\}_\sigma - \{f, 1'\}_\sigma = \{(\sigma - \sigma^{-1})(1'), f\} = \{(1 - \sigma^{-2})(1'), f\}_\sigma$. However σ^{-2} coincides with T_{w_o} up to proportionality in *irreducible* $\mathcal{HH}$-modules where σ acts (see [C3]). Thus $(1 - \sigma^{-2})(1')$ is proportional to $1'$ and must be zero in $\mathcal{V}'$ due to the calculation above. We obtain that $1'$ is in the radical of the pairing $\{f, g\}_\sigma - \{g, f\}_\sigma$, which makes this difference identically zero since $1'$ is a generator.

The quotient $\mathcal{V}'$ *is not* τ_--invariant if q is a root of unity and k are *generic*. In this case (see [C2, C3]), all E_b and $\mathcal{E}_b = E_b/E_b(q^{-\rho_k})$ are well defined. The radical is linearly generated by the differences $\mathcal{E}_b - \mathcal{E}_c$ when

$$u_b = u_c,\ b_- = c_- \mod NA \cap B \ \text{ for } (A, B) = \mathbb{Z},\ q^N = 1.$$

The polynomials E_b and $\mathcal{E}_b$ are τ_--eigenvectors. Their eigenvalues are $q^{-(b_-,b_-)/2-(b_-,\rho_k)}$. Therefore τ_- does not preserve the radical.

An example of reducible $\mathcal{V}'$. For the root system $B_n (n > 2)$, let

$$n \ge l > n/2 + 1,\ r = 2(l-1),\ k_{\text{lng}} = -\frac{s}{r},\ l, s \in \mathbb{N},\ (s, r) = 1.$$

We will assume that k_{sht} is generic.

Then Theorem 4.3 readily gives that the radical is zero. Indeed, the numerator of the formula from (3.19) is nonzero for *all* b because

$$q_\alpha^j t_\alpha X_\alpha(q^{\rho_k}) = q_\alpha^{j+k_\alpha+(\alpha^\vee,\rho_k)} \neq 1 \ \text{ for any } \alpha \in R_+, j > 0, \tag{5.2}$$

and the denominator is non-zero because

$$q_\alpha^j X_\alpha(q^{\rho_k}) = q_\alpha^{j+(\alpha^\vee,\rho_k)} \neq 1 \ \text{ for any } \alpha \in R_+, j > 0. \tag{5.3}$$

We use that $(\alpha^\vee, \rho_k)$ involves k_{sht} unless α belongs to the root subsystem A_{n-1} formed by $\epsilon_l - \epsilon_m$ in the notation of [B].

Thus all Macdonald polynomials E_b are well defined and the Y-action in $\mathcal{V}$ is semisimple. The semisimplicity results from (5.3).

The following relation holds:

$$q_\alpha^j t_\alpha^{-1} X_\alpha(q^{\rho_k}) = q_\alpha^{j-k_\alpha+(\alpha^\vee,\rho_k)} = 1 \ \text{ for } \alpha = \epsilon_l,\ j = 2(l-1)s \tag{5.4}$$

in the notation from [B]. Indeed, $(\alpha^\vee, \rho_k) = k_{\text{sht}} + 2(l-1)k_{\text{lng}}$. Let

$$\tilde{\alpha}^\bullet = [-\alpha, \nu_\alpha j] = [-\epsilon_l, 2(l-1)s].$$

Here α is short, so $\nu_\alpha = 1$.

Proposition 5.2. *The polynomial representation has a proper submodule $\mathcal{V}^\bullet$ which is the linear span of E_b for b such that $\lambda(\pi_b)$ contains $\tilde{\alpha}^\bullet$. The quotient $\mathcal{V}/\mathcal{V}^\bullet$ is irreducible.*

Proof. This statement follows from the Main Theorem of [C3]. It is easy to check it directly using the intertwiners from (3.17). Indeed, given b, the linear span $\sum_{\widehat{w}} \Psi_{\widehat{w}}(E_b)$ is an $\mathcal{HH}^\flat$-submodule of $\mathcal{V}$ when *all* $\widehat{w} \in \widehat{W}$ are taken, not only the ones satisfying $l(\widehat{w}\pi_b) = l(\widehat{w}) + l(\pi_b)$. If π_b contains $\tilde{\alpha}^\bullet$ but $\widehat{w}\pi_c$ does not, then $\Psi_{\widehat{w}}(E_b) = 0$ because the product $\Psi_{\widehat{w}}\Psi_{\pi_b}(1)$ can be transformed using the homogeneous Coxeter relations to get the combination

$$\cdots(\tau_+(T_i) - t_i^{1/2})(\tau_+(T_i) + t_i^{-1/2})\cdots(1)$$

somewhere. This combination is identically zero. □

6. A non-semisimple example

Let us consider the case of A_1 assuming that $q^{1/2}$ is a primitive $2N$-th root of unity. We set $t = q^k$,

$$B = P = \mathbb{Z},\ Q = 2\mathbb{Z},\ X = X_{\omega_1},\ Y = Y_{\omega_1},\ T = T_1.$$

Thus the E-polynomials will be numbered by integers, and $Y(E_m) = q^{\lambda_m}E_m$ for

$$\lambda_m = -m_\sharp,\ m_\sharp \stackrel{\mathbf{def}}{=} (m + \operatorname{sgn}(m)k)/2,\ \operatorname{sgn}(0) = -1,$$

provided that E_m exists. The λ_m are called weights of E_m.

Note that $\pi = sp$ in the polynomial representation $\mathcal{V} = \mathbb{Q}_{q,t}[X, X^{-1}]$ for $s(f(X)) = f(X^{-1})$, $p(f(X)) \stackrel{\mathbf{def}}{=} f(q^{1/2}X)$. The definition ring is $\mathbb{Q}_{q,t} = \mathbb{Q}[q^{\pm 1/4}, t^{\pm 1/2}]$, where $q^{1/4}$ is use to introduce of $\tau_\pm$. Otherwise $q^{1/2}$ is sufficient.

We will need the following lemma, which is similar to the considerations from [CO].

Let $\widehat{V}_0 = \mathbb{Q}_{q,t}$, $\widehat{V}_1 = \mathbb{Q}_{q,t}X, \ldots,$

$$\widehat{V}_{-m} = B_m\widehat{V}_m,\ \widehat{V}_{m+1} = A_{-m}\widehat{V}_{-m}, \ldots,$$

where $m > 0$, $A_{-m} = q^{m/2}X\pi$, B_m is the restriction of the intertwiner $t^{1/2}(T + \frac{t^{1/2}-t^{-1/2}}{Y^{-2}-1})$ to $\widehat{V}_m$ provided that $q^{2\lambda_m} \neq 1$ for $\lambda_m = -m/2 - k/2$. If $q^{2\lambda_m} = 1$ and the denominator of B_m becomes infinity, then we set $B_m = t^{1/2}T,\ \ \widehat{V}_{-m} = \widehat{V}_m + T\widehat{V}_m$.

Lemma 6.1. *i) The space $\widehat{V}_{\pm m}$ is one-dimensional or two-dimensional. In the latter case, it is the Jordan 2-block satisfying $(Y - q^{\pm\lambda_m})^2 \widehat{V}_{\pm m} = \{0\}$. If $\dim \widehat{V}_{-m} = 1$, then $\dim \widehat{V}_{m+1} = 1$ and the generators are*

$$E_{-m} = B_{m-1} \cdots B_1 A_0(1), \;\; E_{m+1} = A_{-m} E_{-m}.$$

If $\dim \widehat{V}_{-m} = 2$, then $\dim \widehat{V}_{m+1} = 2$ and the E-polynomials E_{-m}, E_{m+1} do not exist, although these spaces contain the E-polynomials of smaller degree.

ii) Let us assume that either $q^{2\lambda_m} = t$ or $q^{2\lambda_m} = t^{-1}$. Then $\dim \widehat{V}_{-m} = 1$ and this space is generated by E_{-m}. If $\widehat{V}_m$ is one-dimensional, then respectively $(T + t^{-1/2})E_{-m} = 0$ or $(T - t^{1/2})E_{-m} = 0$. If $\dim \widehat{V}_m = 2$, then respectively

$$(T + t^{-1/2})E_{-m} \;\; or \;\; (T - t^{1/2})E_{-m}$$

is non-zero and proportional to the (unique) E-polynomial which is contained in the space $\widehat{V}_m$. □

We are going to apply the lemma to *integral* k. In the range $0 < k < N/2$, the corresponding perfect representation is Y-semisimple. Using the reduction modulo N (see [CO]), it suffices to consider the interval $-N/2 \le k < 0$.

Proposition 6.2. *i) For integral k such that $-N/2 \le k < 0$, the quotient $V_{2N+4|k|} \stackrel{\text{def}}{=\!=} \mathcal{V}/\mathrm{Rad}$ by the radical of the pairing $\{\, ,\, \}$ is an irreducible $\mathcal{HH}$-module of dimension $2N + 4|k|$.*

ii) The polynomials E_m exist and $E_m(q^{-k/2}) \ne 0$ for the sequences:

$$\begin{aligned} m &= \{0, 1, -1, \ldots, -|k| + 1, |k|\}, \\ m &= \{-2|k|, 2|k| + 1, \ldots, -N + 1, N\}, \\ m &= \{-N, N + 1, \ldots, -N - |k| + 1, N + |k|\}, \end{aligned}$$

respectively with $2|k|$, $2(N - 2|k|)$, and $2|k|$ elements. They do not exist for $2|k| + 2|k|$ indices

$$\begin{aligned} m &= \{-|k|, |k| + 1, \ldots, -2|k| + 1, 2|k|\}, \\ m &= \{-N - |k|, N + |k| + 1, \ldots, -N - 2|k| + 1, N + 2|k|\}. \end{aligned}$$

iii) The Y-semisimple component of $V_{2N+4|k|}$ of dimension $2N - 4|k|$ is linearly generated by E_m for

$$\{m = -2|k|, \; 2|k| + 1, \; -2|k| - 1, \; \ldots, \; -N + 1, \; N\}.$$

The corresponding Y-weights are

$$\left\{\lambda = \frac{|k|}{2}, \frac{-|k| - 1}{2}, \frac{|k| + 1}{2}, \ldots, \frac{N - 1 - |k|}{2}, \frac{|k| - N}{2}\right\}.$$

iv) *The rest of* $V_{2N+4|k|}$ *is the direct sum of* $4|k|$ *Jordan 2-blocks of the total dimension* $8|k|$. *There are two series of the corresponding (multiple) weights* λ :

$$\left\{\frac{-|k|}{2}, \frac{|k|-1}{2}, \ldots, \frac{-1}{2}, \frac{0}{2}\right\}, \left\{\frac{N-|k|}{2}, \frac{|k|-N-1}{2}, \ldots, \frac{N-1}{2}, \frac{-N}{2}\right\}.$$

Proof. We will use the chain of the spaces of generalized eigenvectors

$$\widehat{V}_0 = \mathbb{Q}_{q,t}, \ \widehat{V}_1 = \mathbb{Q}_{q,t}X, \ \widehat{V}_{-1}, \ \ldots, \ \widehat{V}_m, \ \ldots$$

from Lemma 6.1. Recall that $m > 0$. The following holds:

0) the spaces $\widehat{V}_{\pm m}$ are all one-dimensional from 0 to $m = |k|$, i.e., in the sequence $V_0, \ldots, V_{-|k|+1}, V_{|k|}$;
1) the intertwiner B_m becomes infinity at $m = |k|$ ($B_{|k|} = t^{1/2}T$) and $\dim\widehat{V}_m = 2$ in the range $|k| < m \leq 2|k|$;
2) the intertwiner B_m kills $1 \in \widehat{V}_m$ at $m = 2|k|$, and after this $\dim\widehat{V}_m = 1$ for $2|k| < m \leq N$;
3) B_m is proportional to $(T+t^{-1/2})$ at $m = N$, $E_{-N} = X^N + X^{-N}$, and $\dim\widehat{V}_m = 1$ as $N < m \leq N + |k|$;
4) the intertwiner B_m becomes infinity again at $m = N + |k|$, and afterwards $\dim\widehat{V}_m = 2$ when $N + |k| < m \leq N + 2|k|$;
5) B_m kills E_{-N} at $m = N + 2|k|$, and $B_m(\widehat{V}_m)$ is generated by $E_{-N-2|k|}$ of same Y-eigenvalue as E_N.

Concerning step (5), the polynomials $E_{-N-2|k|}$ and E_N both exist, there evaluations are nonzero, and the difference

$$E = E_N/E_N(q^{-k/2}) - E_{-N-2|k|}/E_{-N-2|k|}(q^{-k/2})$$

belongs to the radical Rad, i.e., becomes zero in $V_{2N+4|k|}$.

Note that $(T + t^{1/2})E = 0$, which is important to know to continue the decomposition of $\mathcal{V}$ further. It follows the same lines.

We see that step (5) is the first step which produces no new elements in $V_{2N+4|k|}$. Namely,

$$B_{N+2|k|}(\widehat{V}_{N+2|k|}) \ = \ \mathbb{Q}_{q,t}E_N \ \text{ in } \ V_{2N+4|k|},$$

and we can stop here.

The lemma gives that between (2) and (3), the polynomials E_m exist, their images linearly generate the Y-semisimple part of V. It is equivalent to the inequalities $E_m(q^{-\rho_k}) \neq 0$ because they have different Y-eigenvalues.

Apart from (2)-(3), there will be Jordan 2-blocks with respect to Y. Let us check it.

First, we obtain the 2-dimensional irreducible representation of $\mathcal{H}_Y = \langle T, Y, \pi\rangle$ in the corresponding $\widehat{V}$-space at step (1). Then we apply invertible intertwiners to this space (the weights will go back) and eventually will obtain the two-dimensional $\widehat{V}$-space for the starting weight $\lambda = -|k|/2$. Note that $E_0 = 1$ is not from the Y-semisimple component of $V_{2N+4|k|}$. It belongs to a Jordan 2-block.

Second, the intertwiner (2) makes the last space one-dimensional and Y-semisimple (the corresponding eigenvalue is simple in $V_{2N+4|k|}$). It will remain one-dimensional until (3). After step (3), we obtain the Jordan blocks. The steps (4)–(5) are parallel to (1)–(2). □

The above consideration readily results in the irreducibility of the module $V_{2N+4|k|}$. Indeed, Lemma 6.1, (ii) gives that if a submodule of $V_{2N+4|k|}$ contains at least one simple Y-eigenvector then it contains the image of 1 and the whole space. Step (5) guarantees that it is always the case, because we can obtain E_N beginning with an arbitrary Y-eigenvector.

The irreducibility and the existence of the projective $PSL(2,\mathbb{Z})$-action in $V_{2N+4|k|}$ also follow from Theorem 5.1, (ii) because the radical is generated by E which is a linear combination of the E-polynomials with the coinciding τ_--eigenvalues.

References

[B] N. Bourbaki, *Groupes et algèbres de Lie*, Ch. 4–6, Hermann, Paris, 1969.

[C1] I. Cherednik, Macdonald's evaluation conjectures and difference Fourier transform, *Inventiones Math.* **122** (1995), 119–145.

[C2] —, Nonsymmetric Macdonald polynomials, *IMRN* **10** (1995), 483–515.

[C3] —, Double affine Hecke algebras and difference Fourier transform, *Inventiones Math.* **152** (2003), 213–303.

[C4] — , Diagonal coinvariants and Double Affine Hecke algebras, *IMRN* **16** (2004), 769–791.

[CO] — and V. Ostrik, From Double Hecke Algebras to Fourier Transform, *Selecta Math. New Ser.* **9** (2003), 1022–182.

[DO] C.F. Dunkl, and E.M. Opdam, Dunkl operators for complex reflection groups, *Proc. London Math. Soc.* (3), **86** (2003), 70–108.

[DJO] C.F. Dunkl, and M. de Jeu, and E.M. Opdam, Singular polynomials for finite reflection groups, *Trans. Amer. Math. Soc.* **346** (1994), 237–256.

[J] M. de Jeu, *The Dunkl operators*, Thesis (1993), 1–92.

[M] I. Macdonald, *Affine Hecke algebras and orthogonal polynomials*, Séminaire Bourbaki **47**:797 (1995), 01–18.

[O1] E. Opdam, Harmonic analysis for certain representations of graded Hecke algebras, *Acta Math.* **175** (1995), 75–121.

[O2] — Dunkl operators, Bessel functions and the discriminant of a finite Coxeter group, *Compositio Mathematica* **85** (1993), 333–373.

Algebraic groups over a 2-dimensional local field: Some further constructions

Dennis Gaitsgory[1] and David Kazhdan[2]

[1] Department of Mathematics
The University of Chicago
5734 University Ave.
Chicago, IL 60637
USA
gaitsgde@math.uchicago.edu
[2] Einstein Institute of Mathematics
The Hebrew University of Jerusalem
Givat Ram, Jerusalem 91904
Israel
kazhdan@math.huji.ac.il

Dedicated to A. Joseph on his 60th birthday

Summary. In [GK] we developed a framework to study representations of groups of the form $G((t))$, where G is an algebraic group over a local field K. The main feature of this theory is that natural representations of groups of this kind are not on vector spaces, but rather on pro-vector spaces.

In this paper we present some further constructions related to this theory. The main results include: 1) General theorems insuring representability of covariant functors, 2) Study of the functor of semi-invariants, which is an analog of the functor of semi-infinite cohomology for infinite-dimensional Lie algebras, 3) Construction of representations from the moduli space of G-bundles on algebraic curve over K.

Subject Classification: 22D20

Introduction

0.1.

Let $\mathbf{K}$ be a local field, G a split reductive group over $\mathbf{K}$, and $G((t))$ the corresponding loop group, regarded as a group ind scheme. In [GK] we suggested a categorical framework in which one can study representations of the group $G((t))(\mathbf{K}) = G(\mathbf{K}((t)))$.

The main point is that $\mathbb{G} := G((t))(\mathbf{K})$ admits no interesting representations on vector spaces, and we have to consider pro-vector spaces instead. In more detail, we regard $\mathbb{G}$ as a group-like object in the category $\mathbb{S}\text{et} := \text{Ind}(\text{Pro}(\text{Ind}(\text{Pro}(\mathcal{S}et_0))))$, where $\mathcal{S}et_0$ denotes the category of finite sets. We observe that $\mathbb{S}\text{et}$ has a natural pseudo-action on the category $\mathbb{V}\text{ect} = \text{Pro}(\text{Vect})$ of pro-vector spaces, and we define the category $\text{Rep}(\mathbb{G})$ to consist of pairs $(\mathbb{V}, \rho)$, where $\mathbb{V} \in \mathbb{V}\text{ect}$, and ρ is an action map $\mathbb{G} \times \mathbb{V} \to \mathbb{V}$ in the sense of the above pseudo-action, satisfying the usual properties.

In [GK] several examples of objects of $\text{Rep}(\mathbb{G})$ were considered. One such example is the principal series representation Π, considered by M. Kapranov in [Ka]. Combining the results of [Ka] and the formalism of adjoint functors developed in [GK] we showed that the endomorphism algebra of Π could be identified with the Cherednik double affine Hecke algebra.

Another example is the "left regular" representation, corresponding to functions on $\mathbb{G}$, with respect to the action of $\mathbb{G}$ on itself by left translations, denoted $M(\mathbb{G})$. The main feature of $M(\mathbb{G})$ is that the right action develops an anomaly: instead of the action of $\mathbb{G}$ we obtain an action of the Kac–Moody central extension $\widehat{\mathbb{G}}_0$ of $\mathbb{G}$ by means of the multiplicative group $\mathbf{G}_m$, induced by the adjoint action of G on its Lie algebra.

0.2.

In the present paper we continue the study of the category $\text{Rep}(\mathbb{G})$. It is natural to subdivide the contents into three parts:

In the first part, which consists of Sections 1 and 2, we prove some general results about representability of various covariant functors on the category $\text{Rep}(\mathbb{G})$. These results are valid when $\mathbb{G}$ is replaced by an arbitrary group-like object on $\mathbb{S}\text{et}$. We also introduce the pro-vector space of distributions on an object of $\mathbb{S}\text{et}$ with values in a pro-vector space; this notion is used to construct actions on invariants and coinvariants of representations of $\mathbb{G}$.

The second part occupies Sections 3, 4, and 5. We study representations of a central extension $\widehat{\mathbb{G}}$ of $\mathbb{G}$ by means of $\mathbf{G}_m$ with a fixed central character $c : \mathbf{G}_m \to \mathbb{C}^*$; the corresponding category is denoted $\text{Rep}_c(\widehat{\mathbb{G}})$, and $(\widehat{\mathbb{G}}', c')$ denotes the opposite extension with its central character, see [GK], Sect. 5.9.

Our goal here is to study the functor of semi-invariants

$$\overset{\frac{\infty}{2}}{\underset{\mathbb{G}}{\otimes}} : \text{Rep}_c(\widehat{\mathbb{G}}) \times \text{Rep}_{c'}(\widehat{\mathbb{G}}') \to \mathbb{V}\text{ect},$$

which couples the categories of representations at opposite levels. The motivation for the existence of such functor is provided by the semi-infinite cohomology functor on the category of representations of a Kac–Moody Lie algebra.

The construction of $\overset{\frac{\infty}{2}}{\underset{\mathbb{G}}{\otimes}}$ presented here follows the categorical interpretation of semi-infinite cohomology, developed by L. Positselsky (unpublished).

We use the functor of semi-invariants to prove the main result of this paper, Theorem 3.3. This theorem describes for any quasi pro-unipotent subgroup $\mathbb{H}$ of $\mathbb{G}$

(Sect. 2.6) the ring of endomorphisms of the functor $\mathrm{Coinv}_{\mathbb{H}} : \mathrm{Rep}(\mathbb{G}) \to \mathbb{V}\mathrm{ect}$, as the algebra of endomorphisms of a certain object in the category of representations of $\widehat{\mathbb{G}}_0$.

In particular, we obtain a functorial interpretation of the double affine (Cherednik) algebra in terms of the category $\mathrm{Rep}(\mathbb{G})$, as the algebra of endomorphisms of the functor of coinvariants with respect to the maximal quasi pro-unipotent subgroup of $\mathbb{G}$.

The third part consists of Sections 7 and 8, preceded by some preliminaries in Sect. 6. We construct some more examples of objects of $\mathrm{Rep}(\mathbb{G})$, this time using the moduli stack of bundles on an algebraic curve X over $\mathbf{K}$, when we think of the variable t as a local coordinate near some point $\mathbf{x} \in X$.

In particular, we show in Theorem 7.9 that in this way one naturally produces a pro-vector space, endowed with an action of $\mathbb{G} \times \mathbb{G}$, such that the space of bi-coinvariants with respect to the maximal quasi pro-unipotent subgroup $\mathbf{I}^{00}$ of $\mathbb{G}$ is a bi-module over Cherednik's algebra, isomorphic to the regular representation of this algebra.

0.3. Notation

We keep the notations introduced in [GK]. In particular, for a category $\mathcal{C}$ we denote by $\mathrm{Ind}(\mathcal{C})$ (resp., $\mathrm{Pro}(\mathcal{C})$) its ind- (resp., pro-) completion.

For a filtering set I and a collection A_i of objects of $\mathcal{C}$ indexed by I, we will denote by $\text{“}\varinjlim_I\text{”}\, A_i$ the resulting object of $\mathrm{Ind}(\mathcal{C})$ and by $\varinjlim_I A_i := \mathrm{limInd}(\text{“}\varinjlim_I\text{”}\, A_i) \in \mathcal{C}$ the inductive limit of the latter, if it exists. The notation for inverse families is similar.

As was mentioned above $\mathcal{S}et_0$ denotes the category of finite sets. We use the shorthand notation $\mathbf{Set} = \mathrm{Ind}(\mathrm{Pro}(\mathcal{S}et_0))$ and $\mathbb{S}\mathrm{et} = \mathrm{Ind}(\mathrm{Pro}(\mathbf{Set}))$. We denote by Vect_0 the category of finite-dimensional vector space, $\mathrm{Vect} \simeq \mathrm{Ind}(\mathrm{Vect}_0)$ is the category of vector spaces, and $\mathbb{V}\mathrm{ect} := \mathrm{Pro}(\mathrm{Vect})$ is the category of pro-vector spaces.

0.4. A correction to [GK]

As was pointed out by A. Shapira, Lemma 2.13 of [GK] is wrong. Namely, he explained to us a counterexample of a pro-vector space $\mathbb{V}$, acted on by a discrete set X (thought of as an object of $\mathbf{Set}$), such that the action of every element of X on $\mathbb{V}$ is trivial, whereas the action of X on $\mathbb{V}$ in the sense of the pseudo-action of $\mathbf{Set} \subset \mathbb{S}\mathrm{et}$ on $\mathbb{V}\mathrm{ect}$ is non-trivial. Namely, $\mathbb{V} = \text{“}\varprojlim_{n\in\mathbb{N}}\text{”}\, \mathrm{Funct}_c(\mathbb{Z}^{\geq n})$ and $X = \mathbb{N}$, such that $i \in \mathbb{N}$ acts on each $\mathrm{Funct}_c(\mathbb{Z}^{\geq n})$ by

$$\begin{cases} f(x_n, x_{n+1}, \ldots) \mapsto f(x_n, x_{n+1}, \ldots, x_i+1, \ldots) - f(x_n, x_{n+1}, \ldots, x_i, \ldots), \ i \geq n \\ f(x_n, x_{n+1}, \ldots) \mapsto 0, \ i < n. \end{cases}$$

However, we have the following assertion. Let $\mathbb{G}$ be as in [GK], Sect. 1.12 let and $\Pi_1 = (\mathbb{V}_1, \rho_1)$, $\Pi_2 = (\mathbb{V}_2, \rho_2)$ be two objects of $\mathrm{Rep}(\mathbb{G}, \mathbb{V}\mathrm{ect})$. Assume that $\mathbb{V}_1$ is strict as a pro-vector space, i.e., that it can be represented as $\text{“}\varprojlim\text{”}\, \mathbf{V}_1^i$, where the maps in the inverse systesm $\mathbf{V}_1^j \to \mathbf{V}_1^i$ are surjective. Let $\phi : \mathbb{V}_1 \to \mathbb{V}_2$ be a map in $\mathbb{V}$ect, which intertwines the actions of the set $G(\mathbf{F}) = \mathbb{G}^{\mathrm{top}}$ on $\mathbb{V}_1$ and $\mathbb{V}_2$.

Lemma 0.5. *Under the above circumstances, the map ϕ is a map in* $\mathrm{Rep}(\mathbb{G}, \mathbb{V}\mathrm{ect})$.

Proof. We will prove a more general assertion when we do not require $\mathbb{V}_1$ and $\mathbb{V}_2$ to be representations of $\mathbb{G}$ on $\mathbb{V}$ect, but just objects of endowed with an action of $\mathbb{G}$, regarded as an object of $\mathbb{S}$et. We claim that a map $\mathbb{V}_1 \to \mathbb{V}_2$ compatible with a pointwise action of $\mathbb{G}^{\mathrm{top}}$ is compatible with an action of $\mathbb{G}$ as an object of $\mathbb{S}$et, under the assumption that $\mathbb{V}_1$ is strict.

We represent $\mathbb{G}$ as "$\underset{\longrightarrow}{\lim}$"$\mathbb{X}_k$, $\mathbb{X}_k \in \mathrm{Pro}(\mathbf{Set})$, and for each k, $\mathbb{X}_k \simeq$ "$\underset{\longleftarrow}{\lim}$"$\mathbf{X}_k^l$, such the maps $(\mathbf{X}_k^{l'})^{\mathrm{top}} \to (\mathbf{X}_k^l)^{\mathrm{top}}$ are surjective. The assertion of the lemma reduces immediately to the case when $\mathbb{V}_2 = \mathbf{W} \in$ Vect, and $\mathbb{G}$ is replaced by $\mathbb{X}_k$. In this case

$$\mathcal{H}om(\mathbb{X}_k \otimes \mathbb{V}_2, \mathbf{W}) \simeq \underset{\underset{i}{\longrightarrow}}{\lim}\, \mathcal{H}om(\mathbb{X}_k \otimes \mathbf{V}_2^i, \mathbf{W}).$$

However, by the assumption on the inverse system $\{\mathbf{V}_1^i\}$, for every i the map

$$\mathrm{Hom}((\mathbb{X}_k)^{\mathrm{top}} \times \mathbf{V}_1^i, \mathbf{W}) \to \mathrm{Hom}((\mathbb{X}_k)^{\mathrm{top}} \times \mathbb{V}_1, \mathbf{W})$$

is injective. This reduces us to the case when $\mathbb{V}_1 = \mathbf{V}$ is an object of Vect. The rest of the proof proceeds as in Lemma 2.13 of [GK]. □

1. The pro-vector space of distributions

1.1.

Let $\mathbb{X}$ be an object of $\mathbb{S}$et and $\mathbb{V} \in \mathbb{V}$ect. Consider the covariant functor on $\mathbb{V}$ect that assigns to $\mathbb{W}$ the set of actions $\mathbb{X} \times \mathbb{V} \to \mathbb{W}$. We claim that this functor is representable. We will denote the representing object by $\mathbb{D}\mathrm{istr}_c(\mathbb{X}, \mathbb{V}) \in \mathbb{V}$ect; its explicit construction is given below. It is clear from the definition that covariant functor $\mathbb{V} \to \mathbb{D}\mathrm{istr}_c(\mathbb{X}, \mathbb{V})$ is right exact.

We begin with some preliminaries of categorical nature:

Lemma 1.2. *The category* $\mathbb{V}$ect *is closed under inductive limits.*

Proof. Since $\mathbb{V}$ect is abelian, it is enough to show that it is closed under direct sums.

Let $\mathbb{V}^\kappa$ be a collection of pro-vector spaces, $\mathbb{V}^\kappa \simeq$ "$\underset{\longleftarrow}{\lim}$" $\mathbf{V}^\kappa_{i^\kappa}$ with i^κ running over a filtering set I^κ. Consider the set $\prod_\kappa I^\kappa$, whose elements can be thought of as families $\{\varphi(\kappa) \in I^\kappa, \forall \kappa\}$. This set is naturally filtering, and

$$\bigoplus_\kappa \mathbb{V}^\kappa \simeq \text{“}\underset{\longleftarrow}{\lim}\text{”} \left(\bigoplus_\kappa \mathbf{V}^\kappa_{\phi(\kappa)}\right),$$

where the inverse system is taken with respect to $\prod_\kappa I^\kappa$. □

1.3.

Let us now describe explicitly the pro-vector space $\mathbb{D}\mathrm{istr}_c(\mathbb{X}, \mathbb{V})$.

If X is a finite set and V is a finite-dimensional vector space, let $\mathrm{Distr}_c(X, V)$ be the set of V-valued functions on X, thought of as distributions. If $\mathbf{X}^0 \in \mathrm{Pro}(Set_0)$ equals "$\varprojlim$" X_i with $X_i \in Set_0$ and V is as above, set

$$\mathbb{D}\mathrm{istr}_c(\mathbf{X}^0, V) = \text{“}\varprojlim\text{”}\ \mathrm{Distr}_c(X_i, V) \in \mathbb{V}\mathrm{ect}.$$

Set also $\mathrm{Distr}(\mathbf{X}^0, V) = \varprojlim \mathrm{Distr}_c(X_i, V) \in \mathrm{Vect}$, i.e.,

$$\mathrm{Distr}(\mathbf{X}^0, V) = \mathrm{limProj}\ \mathbb{D}\mathrm{istr}_c(\mathbf{X}^0, V).$$

If $\mathbf{X}$ is an object of $\mathbf{S}$et equal to "$\varinjlim$"$\mathbf{X}^j$, $\mathbf{X}^j \in \mathrm{Pro}(Set_0)$ and $\mathbf{V} \in \mathrm{Vect}$ is "$\varinjlim$" V_m with $V_m \in \mathrm{Vect}_0$, set

$$\mathbb{D}\mathrm{istr}_c(\mathbf{X}, V) = \varinjlim_{j,m} \mathbb{D}\mathrm{istr}_c(\mathbf{X}^j, V_m) \in \mathbb{V}\mathrm{ect},$$

where the inductive limit is taken in $\mathbb{V}$ect. Set also

$$\mathrm{Distr}_c(\mathbf{X}, V) = \varinjlim_{j,m} \mathrm{Distr}_c(\mathbf{X}^j, V_m) \in \mathrm{Vect}.$$

When V is finite-dimensional, the latter is the vector space, which is the topological dual of the topological vector space $\mathrm{Funct}^{lc}(\mathbf{X}, V^*)$ of locally constant functions on $\mathbf{X}$ with values in V^*. Note that $\mathrm{Distr}_c(\mathbf{X}, V)$ is not isomorphic to $\mathrm{limProj}\ \mathbb{D}\mathrm{istr}_c(\mathbf{X}, V)$ even if V is finite-dimensional.

For $\mathbb{X}^0 \in \mathrm{Pro}(\mathbf{S}\mathrm{et})$ equal to "$\varprojlim$"$\mathbf{X}_l$ with $\mathbf{X}_l \in \mathbf{S}$et and $\mathbb{V}$ is a pro-vector space equal to "$\varprojlim$"$\mathbf{V}_n$, set

$$\mathbb{D}\mathrm{istr}_c(\mathbb{X}^0, \mathbb{V}) = \varprojlim_{l,n} \mathbb{D}\mathrm{istr}_c(\mathbf{X}_l, \mathbf{V}_n) \in \mathbb{V}\mathrm{ect}.$$

Finally, for $\mathbb{X} \in \mathbb{S}\mathrm{et}$ equal to "$\varinjlim$"$\mathbb{X}^k$ and $\mathbb{V} \in \mathbb{V}\mathrm{ect}$, set

$$\mathbb{D}\mathrm{istr}_c(\mathbb{X}, \mathbb{V}) = \varinjlim_{k} \mathbb{D}\mathrm{istr}_c(\mathbb{X}^k, \mathbb{V}).$$

Lemma-Construction 1.4. *For* $\mathbb{D}\mathrm{istr}_c(\mathbb{X}, \mathbb{V}) \in \mathbb{V}\mathrm{ect}$ *constructed above, there exists a natural isomorphism*

$$\mathrm{Hom}_{\mathbb{V}\mathrm{ect}}(\mathbb{D}\mathrm{istr}_c(\mathbb{X}, \mathbb{V}), \mathbb{W}) \simeq \mathcal{H}om(\mathbb{X} \otimes \mathbb{V}, \mathbb{W}).$$

Proof. By the definition of both sides, we can assume that $\mathbb{X} \in \mathrm{Pro}(\mathbf{S}\mathrm{et})$ and $\mathbb{W} = \mathbf{W} \in \mathrm{Vect}$. We have the following (evident) sublemma:

Sublemma 1.5. *If* $\mathbb{U} = \varprojlim \mathbb{U}_m$, *where the projective limit is taken in the category* $\mathbb{V}\mathrm{ect}$, *then for any* $\mathbb{X} \in \mathrm{Pro}(\mathbf{Set})$ *and* $\mathbf{W} \in \mathrm{Vect}$,

$$\mathcal{H}om(\mathbb{X} \otimes \mathbb{U}, \mathbf{W}) \simeq \varinjlim \mathcal{H}om(\mathbb{X} \otimes \mathbb{U}_m, \mathbf{W}).$$

The sublemma implies that we can assume that $\mathbb{V} = \mathbf{V} \in \mathrm{Vect}$. By applying again the construction of $\mathbb{D}\mathrm{istr}_c(\mathbb{X}, \mathbf{V})$, we reduce the assertion of the lemma further to the case when $\mathbb{X} = \mathbf{X} \in \mathbf{Set}$, i.e., we have to show that

$$\mathrm{Hom}_{\mathbb{V}\mathrm{ect}}(\mathbb{D}\mathrm{istr}_c(\mathbf{X}, \mathbf{V}), \mathbf{W}) \simeq \mathcal{H}om(\mathbf{X} \otimes \mathbf{V}, \mathbf{W}).$$

By the construction of $\mathbb{D}\mathrm{istr}_c(\mathbf{X}, \mathbf{V})$ and the definition of the action, we can assume that $\mathbf{X} \in \mathrm{Pro}(\mathcal{S}et_0)$ and $\mathbf{V}$ is finite-dimensional. In this case the assertion is evident. □

Remark. For fixed $\mathbb{X}$ and $\mathbb{V}$ as above we can also consider the contravariant functor on $\mathbb{V}\mathrm{ect}$, given by $\mathbb{W} \mapsto \mathcal{H}om(\mathbb{X} \times \mathbb{W}, \mathbb{V})$. It is easy to see that this functor is ind-representable, but Lemma 1.8 shows that it is not in general representable. We will denote the resulting object of $\mathrm{Ind}(\mathbb{V}\mathrm{ect})$ by $\mathbb{F}\mathrm{unct}(\mathbb{X}, \mathbb{V})$.

1.6.

Let now $\mathbb{X}$, $\mathbb{Y}$ be two objects of $\mathbb{S}\mathrm{et}$. The associativity constraint of the pseudo-action of $\mathbb{S}\mathrm{et}$ and $\mathbb{V}\mathrm{ect}$ gives rise to a map

$$\mathbb{D}\mathrm{istr}_c(\mathbb{X} \times \mathbb{Y}, \mathbb{V}) \to \mathbb{D}\mathrm{istr}_c(\mathbb{X}, \mathbb{D}\mathrm{istr}_c(\mathbb{Y}, \mathbb{V})). \tag{1}$$

Let us now recall the following definition from [GK], Sect. 2.10:
An object $\mathbb{X} \in \mathbb{S}\mathrm{et}$ is said to satisfy condition (**) if it can be represented as "$\varinjlim$" $\mathbb{X}_k$ with each $\mathbb{X}_k \in \mathrm{Pro}(\mathbf{Set})$ being weakly strict. We remind (see [GK], Sect. 1.10) that an object $\mathbb{X}' \in \mathrm{Pro}(\mathbf{Set})$ is said to be weakly strict if it can be represented as "$\varinjlim$" $\mathbf{X}'_i$, $\mathbf{X}'_i \in \mathbf{Set}$, such that the transition maps $\mathbf{X}'_i \to \mathbf{X}'_j$ are *weakly surjective*; in the case of interest when all $\mathbf{X}'_i$'s are locally compact, the latter condition means that the map of topological spaces $\mathbf{X}'^{top}_i \to \mathbf{X}'^{top}_j$ has dense image.

As was shown in [GK], Sect. 2.12, if G is an algebraic group over $\mathbf{K}$, then the corresponding object $\mathbb{G} \in \mathbb{S}\mathrm{et}$ satisfies condition (**).

Proposition 1.7. *If* $\mathbb{X} \in \mathbb{S}\mathrm{et}$ *satisfies condition (**), then the map in* (1) *is surjective.*[1]

This map is not in general an isomorphism. To construct a counter-example, it suffices to take $\mathbb{V} = \mathbb{C}$—the 1-dimensional vector space, and $\mathbb{Y}$ a discrete set $Y \in \mathcal{S}et \simeq \mathrm{Ind}(\mathcal{S}et_0)$, regarded as an object of $\mathbb{S}\mathrm{et}$ by means of $\mathcal{S}et_0 \to \mathrm{Pro}(\mathbf{Set})$.

[1] We are grateful to Alon Shapira who discovered an error in the previous version of the paper, where the (**) assumption on $\mathbb{X}$ was omitted.

Proof. We need to show that for a pro-vector space $\mathbb{W}$, the map

$$\mathcal{H}om(\mathbb{X} \otimes \mathbb{D}istr_c(\mathbb{Y}, \mathbb{V}), \mathbb{W}) \to \mathcal{H}om((\mathbb{X} \times \mathbb{Y}) \otimes \mathbb{V}, \mathbb{W}) \tag{2}$$

is injective. We will repeatedly use the facts that the functor limInd : Ind(Vect) $\to$ Vect is exact and the functor limProj : Pro(Vect) $\to$ Vect is left-exact.

By assumption, $\mathbb{X}$ can be written as "$\varinjlim$" $\mathbb{X}_k$ with $\mathbb{X}_k \in$ Pro(**Set**) being weakly strict. Set also $\mathbb{W} =$ "$\varprojlim$" $\mathbf{W}_j$, $\mathbf{W}_j \in$ Vect. Both sides of (2) are projective limits over k and j of the corresponding objects with $\mathbb{X}$ replaced by $\mathbb{X}_k$ and $\mathbb{W}$ replaced by $\mathbf{W}_j$. So, we can assume that $\mathbb{X}$ is a weakly strict object of Pro(**Set**) and $\mathbb{W} = \mathbf{W} \in$ Vect.

Let us write now $\mathbb{Y} =$ "$\varinjlim$" $\mathbb{Y}_{k'}$ with $\mathbb{Y}_{k'} \in$ Pro(**Set**), in which case $\mathbb{D}istr_c(\mathbb{Y}, \mathbb{V}) \simeq \varinjlim \mathbb{D}istr_c(\mathbb{Y}_{k'}, \mathbb{V})$, and

$$\mathcal{H}om((\mathbb{X} \times \mathbb{Y}) \otimes \mathbb{V}, \mathbb{W}) \simeq \varprojlim \mathcal{H}om((\mathbb{X} \times \mathbb{Y}_{k'}) \otimes \mathbb{V}, \mathbb{W}).$$

Lemma 1.8. *If* $\mathbb{U} = \varinjlim \mathbb{U}_m$, *the inductive limit taking place in* $\mathbb{V}$ect, *then for an object* $\mathbb{X} \in \mathbb{S}$et, *satisfying condition (**), and* $\mathbb{W} \in \mathbb{V}$ect, *the natural map*

$$\mathcal{H}om(\mathbb{X} \otimes \mathbb{U}, \mathbb{W}) \to \varprojlim \mathcal{H}om(\mathbb{X} \otimes \mathbb{U}_m, \mathbb{W})$$

is injective. If $\mathbb{X} \in$ **Set**, *then this map is an isomorphism.*

Proof. As above, we can assume that $\mathbb{W} = \mathbf{W} \in$ Vect, and $\mathbb{X}$ is a weakly strict object of Pro(**Set**). Assume first that $\mathbb{X} = \mathbf{X} \in$ **Set**. In this case the assertion of the lemma follows from the description of inductive limits in $\mathbb{V}$ect given in Lemma 1.2.

Thus, let $\mathbb{X}$ be represented as "$\varprojlim$" $\mathbf{X}_l$, $\mathbf{X}_l \in$ **Set**, with the transition maps $\mathbf{X}_{l'} \to \mathbf{X}_l$ being weakly surjective. Then

$$\mathcal{H}om(\mathbb{X} \otimes \mathbb{U}, \mathbf{W}) \simeq \varinjlim_l \mathcal{H}om(\mathbf{X}_l \otimes \mathbb{U}, \mathbf{W}) \simeq \varinjlim_l \varprojlim_m \mathcal{H}om(\mathbf{X}_l \otimes \mathbb{U}_m, \mathbf{W}),$$

and

$$\varprojlim_m \mathcal{H}om(\mathbb{X} \otimes \mathbb{U}_m, \mathbf{W}) \simeq \varprojlim_m \varinjlim_l \mathcal{H}om(\mathbf{X}_l \otimes \mathbb{U}_m, \mathbf{W}).$$

However, by the assumption, the transition maps $\mathcal{H}om(\mathbf{X}_l \otimes \mathbb{U}_m, \mathbf{W}) \to \mathcal{H}om(\mathbf{X}_{l'} \otimes \mathbb{U}_m, \mathbf{W})$ are injective. Therefore, the natural map

$$\varinjlim_l \varprojlim_m \mathcal{H}om(\mathbf{X}_l \otimes \mathbb{U}_m, \mathbf{W}) \to \varprojlim_m \varinjlim_l \mathcal{H}om(\mathbf{X}_l \otimes \mathbb{U}_m, \mathbf{W})$$

is injective. □

Hence, we are reduced to the case when $\mathbb{Y}$ is also an object of Pro(**Set**). Using Sublemma 1.5, we reduce the assertion further to the case when $\mathbb{V} = \mathbf{V} \in$ Vect and $\mathbb{Y} = \mathbf{Y} \in$ **Set**.

If $\mathbb{X} = \text{“}\underset{\longleftarrow}{\lim}\text{”}\mathbf{X}_l$, then both sides of (2) are inductive limits over l of the corresponding objects with $\mathbb{X}$ replaced by $\mathbf{X}_l$. Thus, from now on we will assume that $\mathbb{X} = \mathbf{X} \in \mathbf{Set}$, and we have to show that the map

$$\mathcal{H}om(\mathbf{X} \otimes \mathbb{D}\mathrm{istr}_c(\mathbf{Y}, \mathbf{V}), \mathbf{W}) \to \mathcal{H}om((\mathbf{X} \times \mathbf{Y}) \otimes \mathbf{V}, \mathbf{W}) \tag{3}$$

is injective, where on the left-hand side $\mathcal{H}om$ is understood in the sense of the pseudo-action of $\mathbf{Set} \subset \mathbb{S}\mathrm{et}$ on $\mathbb{V}\mathrm{ect}$.

By applying Lemma 1.8, we reduce the assertion to the case when $\mathbf{Y} \in \mathrm{Pro}(\mathcal{S}et_0)$ and $\mathbf{V}$ is finite-dimensional. It is clear that when $\mathbf{Y}$ belongs to $\mathcal{S}et_0$, the map in (3) is an isomorphism. Consider now the case when $\mathbf{Y} = \text{“}\underset{\longleftarrow}{\lim}\text{”}\, Y_i$ with $Y_i \in \mathcal{S}et_0$ and $\mathbf{X} = \text{“}\underset{\longrightarrow}{\lim}\text{”}\, \mathbf{X}_n$ with $\mathbf{X}_n \in \mathrm{Pro}(\mathcal{S}et_0)$. Then, by Sublemma 1.5

$$\begin{aligned}\mathcal{H}om(\mathbf{X} \otimes \mathbb{D}\mathrm{istr}_c(\mathbf{Y}, \mathbf{V}), \mathbf{W}) &\simeq \underset{\underset{i}{\longrightarrow}}{\lim}\, \mathcal{H}om(\mathbf{X} \otimes \mathbb{D}\mathrm{istr}_c(Y_i, \mathbf{V}), \mathbf{W}) \\ &\simeq \underset{\underset{i}{\longrightarrow}}{\lim}\, \underset{\underset{n}{\longleftarrow}}{\lim}\, \mathcal{H}om(\mathbf{X}_n \otimes \mathbb{D}\mathrm{istr}_c(Y_i, \mathbf{V}), \mathbf{W}) \simeq \underset{\underset{i}{\longrightarrow}}{\lim}\, \underset{\underset{n}{\longleftarrow}}{\lim}\, \mathcal{H}om((\mathbf{X}_n \times Y_i) \otimes \mathbf{V}, \mathbf{W}).\end{aligned}$$

We also have an identification

$$\begin{aligned}\mathcal{H}om((\mathbf{X} \times \mathbf{Y}) \otimes \mathbf{V}, \mathbf{W}) &\simeq \underset{\underset{n}{\longleftarrow}}{\lim}\, \mathcal{H}om((\mathbf{X}_n \times \mathbf{Y}) \otimes \mathbf{V}, \mathbf{W}) \\ &\simeq \underset{\underset{n}{\longleftarrow}}{\lim}\, \underset{\underset{i}{\longrightarrow}}{\lim}\, \mathcal{H}om((\mathbf{X}_n \times Y_i) \otimes \mathbf{V}, \mathbf{W}).\end{aligned}$$

Since Y_i are finite sets, we can assume that the transition maps $Y_{i'} \to Y_i$ are surjective. Therefore, the map

$$\underset{\underset{i}{\longrightarrow}}{\lim}\, \underset{\underset{n}{\longleftarrow}}{\lim}\, \mathcal{H}om((\mathbf{X}_n \times Y_i) \otimes \mathbf{V}, \mathbf{W}) \to \underset{\underset{n}{\longleftarrow}}{\lim}\, \underset{\underset{i}{\longrightarrow}}{\lim}\, \mathcal{H}om((\mathbf{X}_n \times Y_i) \otimes \mathbf{V}, \mathbf{W})$$

is injective. □

1.9.

As an application of Proposition 1.7, we will prove the following result.

Let $\rho : \mathbb{X} \times \mathbb{V} \to \mathbb{W}$ be an action map. We can consider $\ker(\rho)$ and $\mathrm{coker}(\rho)$ as functors on $\mathbb{V}\mathrm{ect}$:

$$\ker(\rho)(\mathbb{U}) = \{\phi : \mathbb{U} \to \mathbb{V} \mid \rho \circ \phi = 0\} \text{ and } \mathrm{coker}(\rho)(\mathbb{U}) = \{\psi : \mathbb{W} \to \mathbb{U} \mid \psi \circ \rho = 0\}.$$

As in [GK], Proposition 2.8, one shows that $\mathrm{coker}(\rho)$ is always representable, and $\ker(\rho)$ is representable if condition (**) is satisfied.

Corollary 1.10. *Let $\mathbb{Y} \times \mathbb{V} \to \mathbb{V}$ and $\mathbb{Y} \times \mathbb{W} \to \mathbb{W}$ be actions commuting in the natural sense with ρ. Then, if $\mathbb{Y}$ satisfies (**), we have an action of $\mathbb{Y}$ on* $\mathrm{coker}(\rho)$, *and if $\mathbb{X}$ satisfies condition (**), we have an action of $\mathbb{Y}$ on* $\ker(\rho)$.

This corollary will be used when $\mathbb{V} = \mathbb{W}$, and both $\mathbb{X} = \mathbb{G}$ and $\mathbb{Y} = \mathbb{H}$ are group-like objects in $\mathbb{S}$et, whose actions on $\mathbb{V}$ commute. In this case we obtain that $\mathbb{G}$ acts on both invariants and coinvariants of $\mathbb{H}$ on $\mathbb{V}$.

Proof. Let us first prove the assertion about the cokernel. Note that coker(ρ) is isomorphic to the cokernel of the map $\mathbb{D}\mathrm{istr}_c(\mathbb{X}, \mathbb{V}) \to \mathbb{W}$ obtained from ρ. We need to show that the composition

$$\mathbb{D}\mathrm{istr}_c(\mathbb{Y}, \mathbb{W}) \to \mathbb{W} \to \mathrm{coker}(\rho)$$

factors through $\mathbb{D}\mathrm{istr}_c(\mathbb{Y}, \mathrm{coker}(\rho))$. By the right-exactness of the functor $\mathbb{D}\mathrm{istr}_c(\mathbb{Y}, \cdot)$,

$$\mathbb{D}\mathrm{istr}_c(\mathbb{Y}, \mathrm{coker}(\rho)) \simeq \mathrm{coker}\big(\mathbb{D}\mathrm{istr}_c(\mathbb{Y}, \mathbb{D}\mathrm{istr}_c(\mathbb{X}, \mathbb{V})) \to \mathbb{D}\mathrm{istr}_c(\mathbb{Y}, \mathbb{W})\big),$$

and it is enough to show that the composition

$$\mathbb{D}\mathrm{istr}_c(\mathbb{Y}, \mathbb{D}\mathrm{istr}_c(\mathbb{X}, \mathbb{V})) \to \mathbb{W} \to \mathrm{coker}(\rho)$$

vanishes.

However, using Proposition 1.7, we can replace $\mathbb{D}\mathrm{istr}_c(\mathbb{Y}, \mathbb{D}\mathrm{istr}_c(\mathbb{X}, \mathbb{V}))$ by $\mathbb{D}\mathrm{istr}_c(\mathbb{Y} \times \mathbb{X}, \mathbb{V})$, and the required assertion follows from the commutative diagram:

$$\begin{array}{ccccc}
\mathbb{D}\mathrm{istr}_c(\mathbb{X}, \mathbb{V}) & \xrightarrow{\rho} & \mathbb{W} & \longrightarrow & \mathrm{coker}(\rho) \\
\uparrow & & \uparrow & & \\
\mathbb{D}\mathrm{istr}_c(\mathbb{Y} \times \mathbb{X}, \mathbb{V}) & \xrightarrow{\rho} & \mathbb{D}\mathrm{istr}_c(\mathbb{Y}, \mathbb{W}). & &
\end{array}$$

The proof for ker(ρ) is similar. We have to show that the composition

$$\mathbb{D}\mathrm{istr}_c(\mathbb{X}, \mathbb{D}\mathrm{istr}_c(\mathbb{Y}, \ker(\rho))) \to \mathbb{D}\mathrm{istr}_c(\mathbb{X}, \mathbb{V}) \to \mathbb{W}$$

vanishes. Using Proposition 1.7, it is sufficient to show that the composition

$$\mathbb{D}\mathrm{istr}_c(\mathbb{X} \times \mathbb{Y}, \ker(\rho)) \to \mathbb{D}\mathrm{istr}_c(\mathbb{X}, \mathbb{V}) \to \mathbb{W}$$

vanishes, which follows from the assumption. □

2. Existence of certain left adjoint functors

2.1.

In what follows $\mathbb{G}$ will be group-like object in $\mathbb{S}$et satisfying assumption (**). Following [GK], we will denote by Rep$(\mathbb{G})$ the category of representations of $\mathbb{G}$ on $\mathbb{V}$ect.

Proposition 2.2. *The forgetful functor* Rep$(\mathbb{G}) \to \mathbb{V}$ect *admits a left adjoint.*

Proof. We have to prove for any $\mathbb{W} \in \mathbb{V}\mathrm{ect}$ the representability of the functor on $\mathrm{Rep}(\mathbb{G})$ given by $\Pi = (\mathbb{V}, \rho) \mapsto \mathrm{Hom}_{\mathbb{V}\mathrm{ect}}(\mathbb{W}, \mathbb{V})$. This functor obviously commutes with projective limits in $\mathrm{Rep}(\mathbb{G})$; so, by Proposition 1.2 of [GK] (with Ind replaced by Pro), it is enough to show that it is pro-representable.

Consider the category of pairs (Π, α), where $\Pi = (\mathbb{V}, \rho)$ is an object in $\mathrm{Rep}(\mathbb{G})$ and $\alpha : \mathbb{W} \to \mathbb{V}$ is a map in $\mathbb{V}\mathrm{ect}$. For any such pair we obtain an action map $\mathbb{G} \times \mathbb{W} \to \mathbb{V}$, and hence a map $\mathbb{D}\mathrm{istr}_c(\mathbb{G}, \mathbb{W}) \to \mathbb{V}$. Since for an object of $\mathbb{V}\mathrm{ect}$ the class of its quotient objects is clearly a set, the subclass of those (Π, α), for which the above map $\mathbb{D}\mathrm{istr}_c(\mathbb{G}, \mathbb{W}) \to \mathbb{V}$ is surjective, is also a set. This set is naturally filtered, and let us denote it by $A(\mathbb{W})$; it is endowed with a functor to $\mathrm{Rep}(\mathbb{G})$ given by $(\Pi, \alpha) \mapsto \Pi$.

We claim that $\underset{(\Pi,\alpha)\in A(\mathbb{W})}{\underleftarrow{\lim}} \Pi$ is the object on Pro(Rep), which pro-represents our functor.

Indeed, for $\Pi' = (\mathbb{V}', \rho') \in \mathrm{Rep}(\mathbb{G})$, the map

$$\mathrm{Hom}_{\mathrm{Pro}(\mathrm{Rep}(\mathbb{G}))}\left(\underset{(\Pi,\alpha)\in A(\mathbb{W})}{\underleftarrow{\lim}} \Pi, \Pi'\right) = \underset{(\Pi,\alpha)\in A(\mathbb{W})}{\underrightarrow{\lim}} \mathrm{Hom}_{\mathrm{Rep}(\mathbb{G})}(\Pi, \Pi') \\ \to \mathrm{Hom}_{\mathbb{V}\mathrm{ect}}(\mathbb{W}, \mathbb{V}')$$

is evident. Vice versa, given a map $\mathbb{W} \to \mathbb{V}'$ consider the induced map $\mathbb{D}\mathrm{istr}_c(\mathbb{G}, \mathbb{W}) \to \mathbb{V}'$, and let $\mathbb{U}$ be its image. We claim that the action map $\mathbb{G} \times \mathbb{U} \to \mathbb{V}'$ factors through $\mathbb{U}$; this would mean that $\Pi := (\mathbb{U}, \rho'|_{\mathbb{U}})$ is a sub-object of Π', and we obtain a morphism from $\underset{(\Pi,\alpha)\in A(\mathbb{W})}{\underleftarrow{\lim}} \Pi$ to Π'.

Consider the commutative diagram:

$$\begin{array}{ccc} \mathbb{D}\mathrm{istr}_c(\mathbb{G} \times \mathbb{G}, \mathbb{W}) & \longrightarrow & \mathbb{D}\mathrm{istr}_c(\mathbb{G}, \mathrm{Distr}_c(\mathbb{G}, \mathbb{W})) \\ {\scriptstyle\mathrm{mult}}\downarrow & & \downarrow \\ \mathbb{D}\mathrm{istr}_c(\mathbb{G}, \mathbb{W}) & & \mathbb{D}\mathrm{istr}_c(\mathbb{G}, \mathbb{U}) \\ \downarrow & & \downarrow \\ \mathbb{U} & \longrightarrow & \mathbb{V}'. \end{array}$$

We need to show that the image of the vertical map $\mathrm{Distr}_c(\mathbb{G}, \mathbb{U}) \to \mathbb{V}'$ is contained in $\mathbb{U}$. Since, by construction, the morphism $\mathbb{D}\mathrm{istr}_c(\mathbb{G}, \mathbb{W}) \to \mathbb{U}$ is surjective, and the functor $\mathbb{D}\mathrm{istr}_c(\mathbb{G}, \cdot)$ is right-exact, it suffices to show that the image of the composed vertical map is contained in $\mathbb{U}$.

However, by Proposition 1.7, it is sufficient to check that the composed map

$$\mathrm{Distr}_c(\mathbb{G} \times \mathbb{G}, \mathbb{W}) \to \mathbb{V}'$$

has its image contained in $\mathbb{U}$, but this follows from the above diagram. □

2.3.

Let us now derive some corollaries of Proposition 2.2. We will denote the left adjoint constructed above by $\mathbb{V} \mapsto \mathrm{Free}(\mathbb{V}, \mathbb{G})$.

Corollary 2.4. *Let* $\mathbb{G}_1 \to \mathbb{G}_2$ *be a homomorphism of group-objects of* $\mathbb{S}\mathrm{et}$. *Then the natural forgetful functor* $\mathrm{Rep}(\mathbb{G}_2) \to \mathrm{Rep}(\mathbb{G}_1)$ *admits a left adjoint.*

Proof. Let Π_1 be an object of $\mathrm{Rep}(\mathbb{G}_1)$. The functor on $\mathrm{Rep}(\mathbb{G}_2)$ given by $\Pi \mapsto \mathrm{Hom}_{\mathbb{G}_1}(\Pi_1, \Pi)$ commutes with projective limits. Therefore, by Lemma 1.2 of [GK] it suffices to show that it is pro-representable.

Let $\mathbb{V}_1$ be the pro-vector space underlying Π_1. We have an injection $\mathrm{Hom}_{\mathbb{G}_1}(\Pi_1, \Pi) \hookrightarrow \mathrm{Hom}_{\mathbb{V}\mathrm{ect}}(\mathbb{V}_1, \mathbb{V})$, where $\mathbb{V}$ is the pro-vector space underlying Π.

By Proposition 2.2 we know that the functor $\Pi \mapsto \mathrm{Hom}_{\mathbb{V}\mathrm{ect}}(\mathbb{V}_1, \mathbb{V})$ is representable. Therefore, the assertion of the proposition follows from Proposition 1.4 of [GK]. □

We will denote the resulting functor $\mathrm{Rep}(\mathbb{G}_1) \to \mathrm{Rep}(\mathbb{G}_2)$ by $\Pi \mapsto \mathrm{Coind}_{\mathbb{G}_1}^{\mathbb{G}_2}(\Pi)$ and call it the coinduction functor.

Corollary 2.5. *The category* $\mathrm{Rep}(\mathbb{G})$ *is closed under inductive limits.*

Remark. Note that if $\mathbb{G} = \mathbf{G}$ is a group-object in $\mathbf{S}\mathrm{et}$, then the proof of Lemma 1.2 shows that the category $\mathrm{Rep}(\mathbf{G}, \mathbb{V}\mathrm{ect})$ is closed under inductive limits. Moreover, the forgetful functor $\mathrm{Rep}(\mathbf{G}, \mathbb{V}\mathrm{ect}) \to \mathbb{V}\mathrm{ect}$ commutes with inductive limits.

For an arbitrary $\mathbb{G} \in \mathbb{S}\mathrm{et}$, the latter fact is not true, and we need to resort to Proposition 2.2 even to show the existence of inductive limits. We will always have a surjection from the inductive limit of underlying pro-vector spaces to the pro-vector space, underlying the inductive limit.

Proof. Let $\Pi_i = (\mathbb{V}_i, \rho_i)$ be a filtering family of objects of $\mathrm{Rep}(\mathbb{G})$. Consider the covariant functor F on $\mathrm{Rep}(\mathbb{G})$ given by

$$\Pi \mapsto \varprojlim \mathrm{Hom}_{\mathrm{Rep}(\mathbb{G})}(\Pi_i, \Pi).$$

Consider also the functor F' that sends $\Pi = (\mathbb{V}, \rho)$ to $\varprojlim \mathrm{Hom}_{\mathbb{V}\mathrm{ect}}(\mathbb{V}_i, \mathbb{V})$.

By Proposition 2.2 and Lemma 1.2, the functor F' is representable. Hence, by Proposition 1.4 of [GK], we conclude that F is pro-representable. Since F obviously commutes with projective limits in $\mathrm{Rep}(\mathbb{G})$, it is representable by Lemma 1.2 of [GK]. □

2.6. Inflation

Let us call a group-object $\mathbf{H}$ of $\mathbf{S}\mathrm{et}$ quasi-unipotent if it can be presented as "$\varinjlim$" $\mathbf{H}_i$, where $\mathbf{H}_i$ are group-objects of $\mathrm{Pro}(\mathcal{S}et_0)$ and transition maps being homomorphisms, see [GK].

Let us call a group-object $\mathbb{H} \in \text{Pro}(\mathbf{S}\text{et})$ quasi-pro-unipotent if it can be presented as "$\underleftarrow{\lim}$" $\mathbf{H}^l$, where $\mathbf{H}^l$ are quasi-unipotent group-objects of $\mathbf{S}$et, and the transition maps $\mathbf{H}^{l'} \to \mathbf{H}^l$ being weakly surjective homomorphisms; see [GK], Sect. 1.10.

According to Lemma 2.7 of [GK], if $\mathbb{H}$ is quasi-pro-unipotent, the functor of $\mathbb{H}$-coinvariants

$$\text{Coinv}_{\mathbb{H}} : \text{Rep}(\mathbb{H}, \mathbb{V}\text{ect}) \to \mathbb{V}\text{ect}$$

is exact.

Proposition 2.7. *If* $\mathbb{H}$ *is quasi-pro-unipotent, the functor* Coinv$_{\mathbb{H}}$ *admits a left adjoint.*

We will refer to the resulting adjoint functor as "inflation", and denote it by $\mathbb{V} \mapsto \text{Inf}^{\mathbb{H}}(\mathbb{V})$.

2.8. Proof of Proposition 2.7

Let us first take $\mathbf{H}$ to be a quasi-unipotent group-object of $\mathbf{S}$et, isomorphic to "$\underrightarrow{\lim}$" $\mathbf{H}_i$, where $\mathbf{H}_i$ are group-objects in $\text{Pro}(\mathcal{S}\text{et}_0)$.

Let us show that for a vector space $\mathbf{V}$, the functor $\text{Rep}(\mathbf{H}, \text{Vect}) \to \text{Vect}$ given by $\Pi \mapsto \text{Hom}(\mathbf{V}, \Pi_{\mathbf{H}})$ is pro-representable.

For an index i, consider the object $\text{Coind}^{\mathbf{H}}_{\mathbf{H}_i}(\mathbf{V}) \in \text{Rep}(\mathbf{H}, \mathbb{V}\text{ect})$, where $\mathbf{V}$ is regarded as a trivial representation of $\mathbf{H}_i$, and Coind is as in Corollary 2.4. Using Proposition 2.4 of [GK], we obtain that $\text{Coind}^{\mathbf{H}}_{\mathbf{H}_i}(\mathbf{V})$ is a well-defined object of $\text{Pro}(\text{Rep}(\mathbf{H}, \text{Vect}))$, which pro-represents the functor $\Pi \mapsto \Pi^{\mathbf{H}_i}$.

Note that if $\mathbf{H}$ is locally compact, and $\mathbf{H}_i \subset \mathbf{H}$ is open, then $\text{Coind}^{\mathbf{H}}_{\mathbf{H}_i}(\mathbf{V})$ belongs in fact to $\text{Rep}(\mathbf{H}, \text{Vect})$, and is isomorphic to the space of compactly supported $\mathbf{V}$-valued distributions on $\mathbf{H}/\mathbf{H}_i$, i.e., to the ordinary compact induction.

Since $\mathbf{H}_i$ is compact, we have $\Pi^{\mathbf{H}_i} \simeq \Pi_{\mathbf{H}_i}$. Therefore, for $j > i$ we have natural maps

$$\text{Coind}^{\mathbf{H}}_{\mathbf{H}_j}(\mathbf{V}) \to \text{Coind}^{\mathbf{H}}_{\mathbf{H}_i}(\mathbf{V}).$$

Therefore, we can consider the object

$$\underleftarrow{\lim}\, \text{Coind}^{\mathbf{H}}_{\mathbf{H}_j}(\mathbf{V}) \in \text{Pro}(\text{Rep}(\mathbf{H}, \text{Vect})),$$

where the projective limit is taken in the category $\text{Pro}(\text{Rep}(\mathbf{H}, \text{Vect}))$.

For $\Pi \in \text{Rep}(\mathbf{H}, \text{Vect})$ we have

$$\text{Hom}\left(\underleftarrow{\lim}\, \text{Coind}^{\mathbf{H}}_{\mathbf{H}_i}(\mathbf{V}), \Pi\right) \simeq \underrightarrow{\lim}\, \text{Hom}(\mathbf{V}, \Pi_{\mathbf{H}_i}).$$

Since $\Pi_{\mathbf{H}} \simeq \underrightarrow{\lim}\, \Pi_{\mathbf{H}_i}$, the RHS of the above expression is not in general isomorphic to $\text{Hom}(\mathbf{V}, \Pi_{\mathbf{H}})$, except when $\mathbf{V}$ is finite dimensional. In the latter case we set $\text{Inf}^{\mathbf{H}}(\mathbf{V}) := \underleftarrow{\lim}\, \text{Coind}^{\mathbf{H}}_{\mathbf{H}_i}(\mathbf{V})$.

For general $\mathbf{V}$, isomorphic to $\varinjlim \mathbf{V}_k$ with $\mathbf{V}_k \in \mathrm{Vect}_0$, we set

$$\mathrm{Inf}^{\mathbf{H}}(\mathbf{V}) = \varinjlim \mathrm{Inf}^{\mathbf{H}}(\mathbf{V}_k),$$

where the inductive limit is taken in Pro(Rep($\mathbf{H}$, Vect)), see Lemma 1.2.

Now, the existence (and construction) of the functor $\mathrm{Inf}^{\mathbb{H}}$ follows from Proposition 2.4 of [GK]. Namely, if $\mathbb{H} =$ "$\varprojlim$" $\mathbf{H}^l$ with $\mathbf{H}^l$ being group-objects in $\mathbf{S}$et as above, and $\mathbb{V} =$ "$\varprojlim$" $\mathbf{V}_m$, we set

$$\mathrm{Inf}^{\mathbb{H}}(\mathbb{V}) = \varprojlim_{l,m} \mathrm{Inf}^{\mathbf{H}^l}(\mathbf{V}_m),$$

where the projective limit is taken in the category Rep($\mathbb{H}$, $\mathbb{V}$ect) $\simeq$ Pro(Rep($\mathbb{H}$, Vect)), and each $\mathrm{Inf}^{\mathbf{H}^l}(\mathbf{V}_m)$ is regarded as a representation of $\mathbb{H}$ via $\mathbb{H} \to \mathbf{H}^l$.

3. The functor of coinvariants

3.1.

From now on we will assume that the group-like object $\mathbb{G}$ is obtained from a split reductive group G over $\mathbf{K}$, as in [GK], Sect. 2.12. More generally, we will consider a central extension $\widehat{G}$ of $G((t))$ as in Sect. 2.14 of [GK], and denote by $\mathrm{Rep}_c(\widehat{\mathbb{G}})$ the category of representations of $\widehat{\mathbb{G}}$ at level c.

Let $\mathbb{H}$ be a quasi-pro-unipotent group-object in Pro($\mathbf{S}$et). Let $\mathbb{H} \to \mathbb{G}$ be a homomorphism, and we will assume that we are given a splitting of the induced extension $\widehat{\mathbb{G}}|_{\mathbb{H}}$. In particular, we have the forgetful functor $\mathrm{Rep}_c(\widehat{\mathbb{G}}) \to \mathrm{Rep}(\mathbb{H}, \mathbb{V}\mathrm{ect})$.

Consider the functor

$$\mathrm{Rep}_c(\widehat{\mathbb{G}}) \to \mathbb{V}\mathrm{ect},$$

given by $\Pi \mapsto \mathrm{Coinv}_{\mathbb{H}}(\Pi)$. Let $E(\mathbb{G}, \mathbb{H})_c$ denote the algebra of endomorphisms of this functor.

Remark. One can regard $E(\mathbb{G}, \mathbb{H})_c$ as an analogue of the Hecke algebra of a locally compact subgroup with respect to an open compact subgroup. Indeed, if $\mathbf{G}$ is a locally compact group-like object in $\mathbf{S}$et and $\mathbf{H} \subset \mathbf{G}$ is open and compact, the corresponding Hecke algebra, which by definition is the algebra of $\mathbf{H}$-bi-invariant compactly supported functions on $\mathbf{G}$, can be interpreted both as the algebra of endomorphisms of the representation $\mathrm{Coind}_{\mathbf{H}}^{\mathbf{G}}(\mathbb{C})$, where $\mathbb{C}$ is the trivial representation and as the algebra of endomorphisms of the functor $\Pi \mapsto \mathrm{Coinv}_{\mathbf{H}}(\Pi) : \mathrm{Rep}(\mathbf{G}, \mathrm{Vect}) \to \mathrm{Vect}$.

3.2.

Recall now the representation $M_c(\mathbb{G})$, introduced in Sect. 5.6 of [GK]. According to the main theorem of *loc.cit.*, the structure of $\widehat{\mathbb{G}}$-representation on $M_c(\mathbb{G})$ extends naturally to a structure of $\widehat{\mathbb{G}} \times \widehat{\mathbb{G}}'$-representation, where $\widehat{\mathbb{G}}'$ is the group-object of $\mathbb{S}$et corresponding to the central extension $\widehat{G}'$ of $G((t))$, the latter being the Baer sum of $\widehat{G}$ and the canonical extension $\widehat{G}_0$, corresponding to the adjoint action of G on its Lie algebra. The action of $\widehat{\mathbb{G}}'$ of $M_c(\mathbb{G})$ has central character c', given by the formula in Sect. 5.9 of [GK].

In what follows we will call objects of $\mathrm{Rep}_{c'}(\widehat{\mathbb{G}}')$ "representations at the opposite level" to that of $\mathrm{Rep}_c(\widehat{\mathbb{G}})$. We will refer to the $\widehat{\mathbb{G}}'$-action on $M_c(\mathbb{G})$ as the "right action".

Using Corollary 1.10, by taking $\mathbb{H}$-coinvariants with respect to $\mathbb{H}$ mapping to $\widehat{\mathbb{G}}'$, we obtain an object of $\mathrm{Rep}_c(\widehat{\mathbb{G}})$ which we will denote by $M_c(\mathbb{G}, \mathbb{H})$. By construction, we have a natural map

$$E(\mathbb{G}, \mathbb{H})_{c'} \to \mathrm{End}_{\mathbb{V}\mathrm{ect}}(M_c(\mathbb{G}, \mathbb{H})).$$

However, since the $\widehat{\mathbb{G}}$ and $\widehat{\mathbb{G}}'$ actions on $M_c(\mathbb{G})$ commute, from Lemma 0.5 we obtain that endomorphisms of $M_c(\mathbb{G}, \mathbb{H})$, resulting from the above map, commute with the $\widehat{\mathbb{G}}$-action.

Hence, we obtain a map

$$E(\mathbb{G}, \mathbb{H})_{c'} \to \mathrm{End}_{\mathrm{Rep}_c(\widehat{\mathbb{G}})}(M_c(\mathbb{G}, \mathbb{H})). \tag{4}$$

We will prove the following theorem:

Theorem 3.3. *The map in* (4) *is an isomorphism.*

3.4.

Let us consider a few examples. Suppose first that the group $\mathbb{H}$ is trivial. As a corollary of Theorem 3.3 we obtain:

Theorem 3.5. *The algebra* $E(\mathbb{G})_c$ *of endomorphisms of the forgetful functor* $\mathrm{Rep}_c(\widehat{\mathbb{G}}) \to \mathbb{V}\mathrm{ect}$ *is isomorphic to the algebra of endomorphisms of the object* $M_c(\mathbb{G}) \in \mathrm{Rep}_{c'}(\widehat{\mathbb{G}}')$.

Let now $\mathbb{H}$ be a *thick* subgroup of $\mathbf{G}[[t]]$ (see [GK], Sect. 2.12). Note that in this case, the object $M_c(\mathbb{G}, \mathbb{H})$ is isomorphic to the induced representation $i_{\mathbb{H}}^{\widehat{\mathbb{G}}}(\mathbb{C})$ of [GK], Sect. 3.3, where $\mathbb{C}$ is the trivial 1-dimensional representation of $\mathbb{H}$.

In particular, let us take $\mathbb{H}$ to be $\mathbf{I}^{00}$, the subgroup of $\mathbf{I}$ equal to the kernel of the natural map $\mathbf{I} \to \mathbf{T} \to \Lambda$, where $I \subset G[[t]]$ is the Iwahori subgroup and Λ is the lattice of cocharacters of T, regarded as a quotient of $\mathbf{T}$ by its maximal compact subgroup.

The corresponding induced representation $i_{\mathbb{H}}^{\widehat{\mathbb{G}}}(\mathbb{C})$ is isomorphic to Kapranov's representation, denoted in Sect. 4 of [GK] by $\mathbb{V}_c$. Assume now that G is semisimple and simply-connected. In this case it follows from Corollary 4.4 of [GK] that the algebra $\mathrm{End}(\mathbb{V}_c)$ is isomorphic to the Cherednik algebra $\ddot{\mathsf{H}}_{q,c'}$. From Theorem 3.3 we obtain:

Corollary 3.6. *The Cherednik algebra* $\ddot{\mathsf{H}}_{q,c'}$ *is isomorphic to the algebra of endomorphisms of the functor* $\Pi \to \mathrm{Coinv}_{\mathbf{I}^{00}}(\Pi) : \mathrm{Rep}_c(\widehat{\mathbb{G}}) \to \mathbb{V}\mathrm{ect}$.

3.7.

Note that by combining Proposition 2.7 and Corollary 2.4, we obtain that the above functor $\mathrm{Coinv}_{\mathbb{H}} : \mathrm{Rep}_c(\widehat{\mathbb{G}}) \to \mathbb{V}\mathrm{ect}$ admits a left adjoint:

$$\mathbb{V} \mapsto \mathrm{Coind}_{\mathbb{H}}^{\widehat{\mathbb{G}}}(\mathrm{Inf}^{\mathbb{H}}(\mathbb{V})).$$

Of course, the algebra of endomorphisms of this functor is isomorphic to $E(\mathbb{G}, \mathbb{H})_c^o$.

Consider now the functor $\mathrm{Rep}_c(\widehat{\mathbb{G}}) \to \mathrm{Vect}$ obtained by composing $\mathrm{Coinv}_{\mathbb{H}}$ with the functor $\mathrm{limProj} : \mathbb{V}\mathrm{ect} \to \mathrm{Vect}$. Let $\overline{E}(\mathbb{G}, \mathbb{H})_c$ be the algebra of endomorphisms of this latter functor. We have a natural map $E(\mathbb{G}, \mathbb{H})_c \to \overline{E}(\mathbb{G}, \mathbb{H})_c$.

Proposition 3.8.

(a) The map $E(\mathbb{G}, \mathbb{H})_c \to \overline{E}(\mathbb{G}, \mathbb{H})_c$ *is injective.*
(b) The algebra $\overline{E}(\mathbb{G}, \mathbb{H})_c^o$ *is isomorphic to* $\mathrm{End}_{\mathrm{Rep}_c(\widehat{\mathbb{G}})}\left(\mathrm{Coind}_{\mathbb{H}}^{\widehat{\mathbb{G}}}(\mathrm{Inf}^{\mathbb{H}}(\mathbb{C}))\right)$.

We do not know under what conditions on $\mathbb{H}$ one might expect that the above map $E(\mathbb{G}, \mathbb{H})_c \to \overline{E}(\mathbb{G}, \mathbb{H})_c$ is an isomorphism.

Proof. To prove the first assertion of the proposition, note that by Theorem 3.3, the evaluation map $E(\mathbb{G}, \mathbb{H})_c \to \mathrm{End}_{\mathbb{V}\mathrm{ect}}(\mathrm{Coinv}_{\mathbb{H}}(M_c(\mathbb{G})))$ is injective.

By construction, the pro-vector space $M_c(\mathbb{G})$ can be represented as a countable inverse limit with surjective restriction maps. Hence, by Proposition 2.5 of [GK], $\mathrm{Coinv}_{\mathbb{H}}(M_c(\mathbb{G})) \in \mathbb{V}\mathrm{ect}$ will also have this property. We have:

Lemma 3.9. *For any pro-vector space, which can be represented as a countable inverse limit with surjective restriction maps, the morphism* $\mathrm{limProj}(\mathbb{V}) \to \mathbb{V}$ *is surjective.*

This lemma implies that the map $\mathrm{End}_{\mathbb{V}\mathrm{ect}}(\mathbb{V}) \to \mathrm{End}_{\mathrm{Vect}}(\mathrm{limProj}(\mathbb{V}))$ is injective.

To prove the second assertion, we must analyze the endomorphism algebra of the functor $\mathrm{Vect} \to \mathrm{Rep}_c(\widehat{\mathbb{G}})$ given by

$$\mathbf{V} \mapsto \mathrm{Coind}_{\mathbb{H}}^{\widehat{\mathbb{G}}}(\mathrm{Inf}^{\mathbb{H}}(\mathbf{V})).$$

However, as every left adjoint, this functor commutes with inductive limits. Therefore, it is enough to consider its restriction to the subcategory Vect_0. This implies the proposition. □

4. The functor of semi-invariants

4.1.

Our method of proof of Theorem 3.3 in based on considering the functor of $\mathbb{G}$-semi-invariants

$$\overset{\frac{\infty}{2}}{\underset{\mathbb{G}}{\otimes}} : \mathrm{Rep}_{c'}(\widehat{\mathbb{G}}') \times \mathrm{Rep}_c(\widehat{\mathbb{G}}) \to \mathbb{V}\mathrm{ect},$$

where c and c' are opposite levels. The construction of this functor mimics the construction of the semi-infinite cohomology functor for associative algebras by L. Positselsky, [Pos].

For $\Pi_c \in \mathrm{Rep}_c(\widehat{\mathbb{G}})$, $\Pi_{c'} \in \mathrm{Rep}_{c'}(\widehat{\mathbb{G}}')$ consider the pro-vector spaces

$$\Pi_{c'} \otimes \Pi_c \text{ and } \Pi_{c'} \otimes M_c(\mathbb{G}) \otimes \Pi_c.$$

We consider the former as acted on by the diagonal copy of $\mathbf{G}[[t]]$, and the latter by two mutually commuting copies of $\mathbf{G}[[t]]$: one acts diagonally on $\Pi_{c'} \otimes M_c(\mathbb{G})$ via the *left* $\widehat{\mathbb{G}}$-action on $M_c(\mathbb{G})$; the other copy acts diagonally on $M_c(\mathbb{G}) \otimes \Pi_c$ via the *right* action. Consider the object

$$(\Pi_{c'} \otimes M_c(\mathbb{G}) \otimes \Pi_c)_{\mathbf{G}[[t]]\times\mathbf{G}[[t]]}.$$

We will construct two natural maps

$$(\Pi_{c'} \otimes \Pi_c)_{\mathbf{G}[[t]]} \rightrightarrows (\Pi_{c'} \otimes M_c(\mathbb{G}) \otimes \Pi_c)_{\mathbf{G}[[t]]\times\mathbf{G}[[t]]}. \tag{5}$$

To construct the first map recall from Lemma 5.8 of [GK] that

$$(M_c(\mathbb{G}) \otimes \Pi_c)_{\mathbf{G}[[t]]} \simeq i^{\widehat{\mathbb{G}}}_{\mathbf{G}[[t]]}\left(r^{\widehat{\mathbb{G}}}_{\mathbf{G}[[t]]}(\Pi_c)\right). \tag{6}$$

Since $\mathbb{G}/\mathbf{G}[[t]]$ is ind-compact, the functor $i^{\widehat{\mathbb{G}}}_{\mathbf{G}[[t]]}$ is isomorphic to the induction functor, $\widetilde{i}^{\widehat{\mathbb{G}}}_{\mathbf{G}[[t]]}$. Therefore, we obtain a morphism of $\widehat{\mathbb{G}}$-representations

$$\Pi_c \to i^{\widehat{\mathbb{G}}}_{\mathbf{G}[[t]]}\left(r^{\widehat{\mathbb{G}}}_{\mathbf{G}[[t]]}(\Pi_c)\right) \simeq (M_c(\mathbb{G}) \otimes \Pi_c)_{\mathbf{G}[[t]]} \tag{7}$$

by adjunction from the identity map $r^{\widehat{\mathbb{G}}}_{\mathbf{G}[[t]]}(\Pi_c) \to r^{\widehat{\mathbb{G}}}_{\mathbf{G}[[t]]}(\Pi_c)$.

The first map in (5) comes from (7) by tensoring with $\Pi_{c'}$ and taking $\mathbf{G}[[t]]$-coinvariants.

To construct the second map in (5) we will use the following observation. Let $\widetilde{M}_c(\mathbb{G})$ be a representation of $\widehat{\mathbb{G}} \times \widehat{\mathbb{G}}'$, obtained from the representation $M_{c'}(\mathbb{G})$ of $\widehat{\mathbb{G}}' \times \widehat{\mathbb{G}}$, by flipping the roles of $\widehat{\mathbb{G}}$ and $\widehat{\mathbb{G}}'$. We have:

Proposition 4.2.

(1) We have a natural isomorphism of $\widehat{\mathbb{G}} \times \widehat{\mathbb{G}}'$*-representations* $\widetilde{M}_c(\mathbb{G}) \simeq M_{c'}(\mathbb{G})$.
(2) The resulting two morphisms

$$M_c(\mathbb{G}) \rightrightarrows \big(M_c(\mathbb{G}) \otimes M_c(\mathbb{G})\big)_{\mathbf{G}[[t]]}$$

one, coming from (7)*, and the other from interchanging the roles of* c *and* c'*, coincide.*

Remark. It will follow from the proof that statement (2) of the proposition fixes the isomorphism of statement (1) uniquely.

The proof will be given in Sect. 5. Using this proposition we construct the second map in (5) by simply interchanging the roles of c and c'.

4.3.

For $\Pi_c, \Pi_{c'}$ as above, we set $\Pi_{c'} \overset{\frac{\infty}{2}}{\underset{\mathbb{G}}{\otimes}} \Pi_c$ to be the equalizer (i.e., the kernel of the difference) of the two maps in (5). Note that since the functor of $\mathbf{G}[[t]]$-coinvariants is only right-exact, the resulting functor $\overset{\frac{\infty}{2}}{\underset{\mathbb{G}}{\otimes}}$ is a priori neither right nor left exact.

Suppose now that Π_c is not only a representation of $\widehat{\mathbb{G}}$, but carries an additional commuting action of some group-object $\mathbb{H} \in \mathbb{S}\text{et}$, which satisfies condition (**). In this case it follows from Corollary 1.10 that $\Pi_{c'} \overset{\frac{\infty}{2}}{\underset{\mathbb{G}}{\otimes}} \Pi_c$ is an object of $\text{Rep}(\mathbb{H})$.

The key assertion describing the behavior of the functor of semi-invariants is the following:

Proposition 4.4. *For $M_c(\mathbb{G})$, regarded as an object of $\text{Rep}_c(\widehat{\mathbb{G}})$, we have a natural isomorphism $\Pi_{c'} \overset{\frac{\infty}{2}}{\underset{\mathbb{G}}{\otimes}} M_c(\mathbb{G}) \simeq \Pi_{c'}$. Moreover, this isomorphism is compatible with the $\widehat{\mathbb{G}}$-actions.*

Proof. Consider the following general set-up. Let $\mathcal{C}_1$ and $\mathcal{C}_2$ be two abelian categories, $\mathsf{G} : \mathcal{C}_1 \to \mathcal{C}_2$ be a functor, and $\mathsf{F} : \mathcal{C}_2 \to \mathcal{C}_1$ its right adjoint. By composing with $\mathsf{F} \circ \mathsf{G}$ on the left and on the right, the adjunction map $\text{Id}_{\mathcal{C}_1} \to \mathsf{F} \circ \mathsf{G}$ gives rise to two maps

$$\mathsf{F} \circ \mathsf{G} \rightrightarrows \mathsf{F} \circ \mathsf{G} \circ \mathsf{F} \circ \mathsf{G}, \tag{8}$$

such that $\text{Id}_{\mathcal{C}_1}$ maps to their equalizer.

Lemma 4.5. *Assume that the functor G is exact and faithful. Then the map*

$$\text{Id}_{\mathcal{C}_1} \to \text{Equalizer}\left(\mathsf{F} \circ \mathsf{G} \rightrightarrows \mathsf{F} \circ \mathsf{G} \circ \mathsf{F} \circ \mathsf{G}\right)$$

is an isomorphism.

Proof. By assumption o G, it is enough to show that

$$\mathsf{G} \to \text{Equalizer}\left(\mathsf{G} \circ \mathsf{F} \circ \mathsf{G} \rightrightarrows \mathsf{G} \circ \mathsf{F} \circ \mathsf{G} \circ \mathsf{F} \circ \mathsf{G}\right)$$

is an isomorphism, but this happens for any pair of adjoint functors. □

We apply this lemma to $\mathcal{C}_1 = \text{Rep}_{c'}(\widehat{\mathbb{G}}')$, $\mathcal{C}_2 = \text{Rep}(\mathbf{G}[[t]], \mathbb{V}\text{ect})$ with $\mathsf{F} = i^{\widehat{\mathbb{G}}}_{\mathbf{G}[[t]]}$, $\mathsf{G} = r^{\widehat{\mathbb{G}}}_{\mathbf{G}[[t]]}$. To prove the proposition it is sufficient to show that for $\Pi_{c'} \in \text{Rep}_{c'}(\widehat{\mathbb{G}}')$ the terms and maps in (5) are equal to the corresponding ones in (8).

First, by (6) and Proposition 4.2(1), for $\Pi_{c'}$ as above, $\mathsf{F} \circ \mathsf{G}(\Pi_{c'})$ is indeed isomorphic to $(\Pi_{c'} \otimes M_c(\mathbb{G}))_{\mathbf{G}[[t]]}$. Furthermore, by applying the functor $\mathsf{F} \circ \mathsf{G}$ to the adjunction map $\Pi_{c'} \to \mathsf{F} \circ \mathsf{G}(\Pi_{c'})$ we obtain the second of the two maps from (5).

Let us now calculate the adjunction map $\mathrm{Id}_{\mathrm{Rep}_{c'}(\widehat{\mathbb{G}}')} \to i^{\widehat{\mathbb{G}}}_{\mathbf{G}[[t]]} \circ r^{\widehat{\mathbb{G}}}_{\mathbf{G}[[t]]}$ applied to

$$\mathsf{F} \circ \mathsf{G}(\Pi_{c'}) \simeq (\Pi_{c'} \otimes M_c(\mathbb{G}))_{\mathbf{G}[[t]]}.$$

By construction, it is obtained from the adjunction map

$$M_c(\mathbb{G}) \to \mathsf{F} \circ \mathsf{G}(M_c(\mathbb{G})) \simeq (M_c(\mathbb{G}) \otimes M_c(\mathbb{G}))_{\mathbf{G}[[t]]}$$

by tensoring with $\Pi_{c'}$ and taking $G[[t]]$-coinvariants. Therefore, by Proposition 4.2(2), it coincides with the first map from (5). □

Remark. Note that by Proposition 4.2(2), the two identifications $M_c(\mathbb{G}) \overset{\frac{\infty}{2}}{\underset{\mathbb{G}}{\otimes}} M_c(\mathbb{G}) \simeq M_c(\mathbb{G})$, one coming from Proposition 4.4 applied to $\Pi_{c'} = M_c(\mathbb{G})$ and the other from interchanging the roles of c and c' as in Proposition 4.2(1), coincide.

4.6. Proof of Theorem 3.3

Let $\Pi_{c'}$ be an object of $\mathrm{Rep}_{c'}(\widehat{\mathbb{G}}')$, and let Π_c be an object of $\mathrm{Rep}_c(\widehat{\mathbb{G}})$, carrying an additional commuting action of a group-object $\mathbb{H} \in \mathbb{S}\mathrm{et}$, which is quasi-pro-unipotent. Then, using Corollary 1.10 and the fact that the functor $\mathrm{Coinv}_{\mathbb{H}}$ is exact (Lemma 2.7 of [GK]), we obtain an isomorphism:

$$\Pi_{c'} \overset{\frac{\infty}{2}}{\underset{\mathbb{G}}{\otimes}} (\Pi_c)_{\mathbb{H}} \simeq \left(\Pi_{c'} \overset{\frac{\infty}{2}}{\underset{\mathbb{G}}{\otimes}} \Pi_c \right)_{\mathbb{H}}.$$

Applying this for $\Pi_c = M_c(\mathbb{G})$, we obtain a functorial isomorphism:

$$\Pi_{c'} \overset{\frac{\infty}{2}}{\underset{\mathbb{G}}{\otimes}} M_c(\mathbb{G}, \mathbb{H}) \simeq (\Pi_{c'})_{\mathbb{H}}. \tag{9}$$

Therefore, we obtain a map

$$\mathrm{End}_{\mathrm{Rep}_c(\widehat{\mathbb{G}})}(M_c(\mathbb{G}, \mathbb{H})) \to E(\mathbb{G}, \mathbb{H})_{c'}. \tag{10}$$

The fact that the composition

$$\mathrm{End}_{\mathrm{Rep}_c(\widehat{\mathbb{G}})}(M_c(\mathbb{G}, \mathbb{H})) \to E(\mathbb{G}, \mathbb{H})_{c'} \to \mathrm{End}_{\mathrm{Rep}_c(\widehat{\mathbb{G}})}(M_c(\mathbb{G}, \mathbb{H}))$$

is the identity map follows from the remark following the proof of Proposition 4.4.

Therefore, to finish the proof of the theorem it suffices to show that the map of (4) is injective. For that note that for any $\Pi_{c'} \in \mathrm{Rep}_{c'}(\widehat{\mathbb{G}}')$ we have an injection $\Pi_{c'} \hookrightarrow (\Pi_{c'} \otimes M_c(\mathbb{G}))_{\mathbf{G}[[t]]}$ (coming from the above adjunction $\mathrm{Id}_{\mathrm{Rep}_{c'}(\widehat{\mathbb{G}}')} \to i^{\widehat{\mathbb{G}}}_{\mathbf{G}[[t]]} \circ r^{\widehat{\mathbb{G}}}_{\mathbf{G}[[t]]}$) and a surjection $\Pi_{c'} \otimes M_c(\mathbb{G}) \twoheadrightarrow (\Pi_{c'} \otimes M_c(\mathbb{G}))_{\mathbf{G}[[t]]}$ of objects of $\mathrm{Rep}_{c'}(\widehat{\mathbb{G}}')$.

Lemma 4.7. *Suppose an element $\alpha \in E(\mathbb{G}, \mathbb{H})_{c'}$ annihilates $(\Pi_{c'})_{\mathbb{H}}$ for some $\Pi_{c'} \in \mathrm{Rep}_{c'}(\widehat{\mathbb{G}}')$. Then α annihilates all objects of the form $(\mathbb{V} \otimes \Pi_{c'})_{\mathbb{H}}$ for $\mathbb{V} \in \mathbb{V}\mathrm{ect}$.*

Proof. Suppose that $\mathbb{V} = \text{"}\varprojlim\text{"}\mathbf{V}_i$, $\mathbf{V}_i \in \text{Vect}$. Then $\mathbb{V} \otimes \Pi_{c'} \simeq \varprojlim (\mathbf{V}_i \otimes \Pi_{c'})$, where the projective limit is taken in the category $\mathbb{V}\text{ect}$.

Using Corollary 2.6 of [GK], we have: $(\mathbb{V} \otimes \Pi_{c'})_{\mathbb{H}} \simeq \varprojlim (\mathbf{V}_i \otimes \Pi_{c'})_{\mathbb{H}}$. This shows that we can assume that $\mathbb{V}$ is a *vector space*, which we will denote by $\mathbf{V}$.

Let us write $\mathbf{V} = \varinjlim \mathbf{V}_i$, where $\mathbf{V}_i \in \text{Vect}_0$.

Sublemma 4.8. *For* $\mathbf{V} = \varinjlim \mathbf{V}_i$ *and* $\mathbb{W} \in \mathbb{V}\text{ect}$ *the natural map*

$$\varinjlim (\mathbf{V}_i \otimes \mathbb{W}) \to (\varinjlim \mathbf{V}_i) \otimes \mathbb{W}$$

is surjective.

Therefore, we have a surjection

$$\varinjlim (\mathbf{V}_i \otimes \Pi_{c'}) \twoheadrightarrow \mathbf{V} \otimes \Pi_{c'},$$

and, hence, a surjection on the level of coinvariants. Since by assumption, α annihilates every $(\mathbf{V}_i \otimes \Pi_{c'})_{\mathbb{H}}$, and the functor $\text{Coinv}_{\mathbb{H}}$ commutes with inductive limits (see Corollary 1.10), we obtain that α annihilates also $\left(\varinjlim (\mathbf{V}_i \otimes \Pi_{c'})\right)_{\mathbb{H}}$. Hence, by the above, it annihilates also $(\mathbf{V} \otimes \Pi_{c'})_{\mathbb{H}}$. □

Using this lemma and the exactness of the functor of $\mathbb{H}$-coinvariants, we obtain that any $\alpha \in \ker(E(\mathbb{G}, \mathbb{H})_{c'} \to \text{End}(M_c(\mathbb{G}, \mathbb{H}))$ annihilates all $(\Pi_{c'} \otimes M_c(\mathbb{G}))_{\mathbb{H}}$, and hence $(\Pi_{c'})_{\mathbb{H}}$ for any $\Pi_{c'}$.

Remark. Note that the same argument proves the following more general assertion. Let $\mathbb{H}_1$, and $\mathbb{H}_2$ be two quasi-pro-unipotent groups endowed with homomorphisms to $\widehat{\mathbb{G}}$. Then the space of natural transformations between the functors $\text{Coinv}_{\mathbb{H}_1}, \text{Coinv}_{\mathbb{H}_2} : \text{Rep}_c(\widehat{\mathbb{G}}) \to \mathbb{V}\text{ect}$ is isomorphic to $\text{Hom}_{\text{Rep}_{c'}(\widehat{\mathbb{G}}')}(M_c(\mathbb{G}, \mathbb{H}_1), M_c(\mathbb{G}, \mathbb{H}_2))$.

5. Proof of Proposition 4.2

5.1.

We will repeatedly use the following construction:

Let $Z_1 \to Z_2$ be a map of schemes of finite type over $\mathbf{K}$, such that Z_1 is a principal bundle with respect to a smooth unipotent group-scheme H on Z_2. Let $\mathcal{L}$ be the line bundle on Z_2, given by $z \mapsto \det(\mathfrak{h}_z)$, where $\mathfrak{h}_z$ is the fiber at $z \in Z_2$ of the sheaf of Lie algebras corresponding to H. Let $\widehat{Z}_1$ be the total space of the pullback of the resulting G_m-torsor to Z_1.

Lemma 5.2. *Under these circumstances we have a natural map*

$$\left(\text{Funct}_c^{lc}(\widehat{\mathbf{Z}}_1) \otimes \mathbb{C}\right)_{\mathbf{G}_m} \to \text{Funct}_c^{lc}(\mathbf{Z}_2),$$

where $\mathbf{G}_m$ *acts on* $\mathbb{C}$ *via the standard character* $\mathbf{G}_m \to \mathbb{Z} \overset{1 \mapsto q}{\to} \mathbb{C}^*$.

5.3.

Let us recall the construction of $M_c(\mathbb{G})$, following [GK], Sect. 5. To simplify the exposition, we will first assume that $c = 1$, in which case we will sometimes write $M(\mathbb{G})$ instead of $M_c(\mathbb{G})$.

Consider the set of pairs (i, Y), where Y is a sub-scheme of $G((t))$, stable under the right action of the congruence subgroup G^i. Note that in this case the quotient Y/G^i is a scheme of finite type over $\mathbf{K}$.

The above set is naturally filtered: $(i, Y) < (i', Y')$ if $i' \geq i$ and $Y \subset Y'$. Note also that $Y/G^{i'} \to Y/G^i$ is a principal bundle with respect to the group $G^i/G^{i'}$.

Let $\mathbf{Y}/\mathbf{G}^i$ denote the object of $\mathbf{S}$et, corresponding to the scheme Y/G^i. Consider the vector space $\mathbf{V}(i, Y) := \mathrm{Funct}_c^{lc}(\mathbf{Y}/\mathbf{G}^i) \otimes \mu(\mathbf{G}[[t]]/\mathbf{G}^i)$, see [GK], Sect. 3.2, where for a locally compact group $\mathbf{H}$, we denote by $\mu(\mathbf{H})$ the space of left-invariant Haar measures on it.

Whenever $(i, Y) < (i', Y')$, we have a natural map $\mathbf{V}(i', Y') \to \mathbf{V}(i, Y)$. It is defined as the composition of the restriction map $\mathrm{Funct}_c^{lc}(\mathbf{Y}'/\mathbf{G}^{i'}) \to \mathrm{Funct}_c^{lc}(\mathbf{Y}/\mathbf{G}^{i'})$, followed by the map

$$\mathrm{Funct}_c^{lc}(\mathbf{Y}/\mathbf{G}^{i'}) \otimes \mu(\mathbf{G}^i/\mathbf{G}^{i'}) \to \mathrm{Funct}_c^{lc}(\mathbf{Y}/\mathbf{G}^i),$$

coming from Lemma 5.2, using $\mu(\mathbf{G}[[t]]/\mathbf{G}^{i'}) \simeq \mu(\mathbf{G}[[t]]/\mathbf{G}^i) \otimes \mu(\mathbf{G}^i/\mathbf{G}^{i'})$.

We have

$$M(\mathbb{G}) = \text{“}\lim_{\substack{\longleftarrow \\ (i,Y)}}\text{”}\, \mathbf{V}(i, Y),$$

as a pro-vector space.

Let us now describe the action of $\mathbb{G} \times \widehat{\mathbb{G}}_0$ on $M(\mathbb{G})$. For our purposes it would suffice to do so on the level of groups of $\mathbf{K}$-valued points of the corresponding group-indschemes.

For $\mathbf{g} \in G((t))(\mathbf{K})$ acting on $M(\mathbb{G})$ *on the left*, we define $\mathbf{V}(i, Y) \to \mathbf{V}(i, \mathbf{g} \cdot Y)$ to be the natural map. In this way we obtain an action of $\mathbf{g}$ on the entire inverse system.

To define the right action, for (i, Y) as above, let j be a large enough integer, so that $\mathrm{Ad}_{\mathbf{g}^{-1}}(G^j) \subset G^i$. Then the right multiplication by $\mathbf{g}$ defines a map of schemes,

$$Y/G^j \to Y \cdot \mathbf{g}/G^i,$$

such that the former is a principal $G^i/\mathrm{Ad}_{\mathbf{g}^{-1}}(G^j)$-bundle over the latter.

A lift of $\mathbf{g}$ to a point $\widehat{\mathbf{g}}$ of the central extension $\widehat{G}_0$ defines an identification $\mu(\mathbf{G}[[t]]/\mathbf{G}^j) \simeq \mu(\mathbf{G}[[t]]/\mathrm{Ad}_{\mathbf{g}^{-1}}(G^j))$. Hence, by Lemma 5.2, we obtain a map

$$\mathbf{V}(j, Y) \to \mathbf{V}(i, Y \cdot \mathbf{g}),$$

and hence an action of $\widehat{\mathbf{g}}$ on the inverse system.

5.4.

Let now $\widehat{\mathbb{G}}$ and c be general. We modify the above construction as follows. For each $Y \subset G((t))$ as above, let $\widehat{Y}$ be its pre-image in $\widehat{G}$. Set

$$\mathbf{V}_c(j, Y) := \left(\mathrm{Funct}_c^{lc}(\mathbf{G}^j \backslash \widehat{\mathbf{Y}}) \otimes \mathbb{C}\right)_{\mathbf{G}_m} \otimes \mu(\mathbf{G}[[t]]/\mathbf{G}^j),$$

where $\mathbf{G}_m$ acts naturally on $\widehat{\mathbf{Y}}$ and by the character c on $\mathbb{C}$. We have

$$M_c(\mathbb{G}) = \text{“}\lim_{\overleftarrow{(j,Y)}}\text{”}\, \mathbf{V}_c(j, Y),$$

and the action of $\widehat{\mathbb{G}} \times \widehat{\mathbb{G}}'$ is described in the same way as above.

By definition, the representation $\widetilde{M}_{c'}(\mathbb{G})$ is the same as $M_{c'}(\mathbb{G})$, viewed as a representation of $\widehat{\mathbb{G}} \times \widehat{\mathbb{G}}' \simeq \widehat{\mathbb{G}}' \times \widehat{\mathbb{G}}$. Explicitly it can be written as follows. Consider the set of pairs (j, Y), where $Y \subset G((t))$ is stable under the action of G^j *on the left*; let $\widehat{Y}'$ be the preimage of Y in $\widehat{G}'$. We have

$$\widetilde{M}_{c'}(\mathbb{G}) = \text{“}\lim_{\overleftarrow{(j,Y)}}\text{”}\, \widetilde{\mathbf{V}}_{c'}(j, Y),$$

where

$$\widetilde{\mathbf{V}}_{c'}(j, Y) := \left(\mathrm{Funct}_c^{lc}(\mathbf{G}^j \backslash \widehat{\mathbf{Y}}') \otimes \mathbb{C}\right)_{\mathbf{G}_m} \otimes \mu(\mathbf{G}[[t]]/\mathbf{G}^j),$$

where $\mathbf{G}_m$ acts naturally on $\widehat{\mathbf{Y}}$ and by the character c' on $\mathbb{C}$. In this presentation, the *right* action of $\widehat{\mathbb{G}}'$ is defined in an evident fashion, and the *left* action of $\widehat{\mathbb{G}}$ is defined as in the case of the right action of $\widehat{\mathbb{G}}_0$ on $M(\mathbb{G})$.

5.5.

We shall now construct the sought-after map $\widetilde{M}_{c'}(\mathbb{G}) \to M_c(\mathbb{G})$. Let us mention that when G is the multiplicative group G_m the sought-after isomorphism amounts to simply the inversion on the group.

For a pair (i, Y) as in the definition of $M_c(\mathbb{G})$, there exists an integer j large enough so that $\mathrm{Ad}_{y^{-1}}(G^j) \subset G^i$ for $y \in Y(\overline{\mathbf{K}})$. In particular, over Y/G^i we obtain a group-scheme, denoted $G_Y^{i,j}$, whose fiber over $y \in Y$ is $G^i / \mathrm{Ad}_{y^{-1}}(G^j)$, and we have a map

$$G^j \backslash Y \to Y/G^i, \tag{11}$$

such that the former scheme is a principal $G_Y^{i,j}$-bundle over the latter.

Note that the fiber of $\widehat{Y}$ over a given point $y \in Y$ identifies with $\det(\mathrm{Ad}_y(\mathfrak{g}[[t]]), \mathfrak{g}[[t]])$, where $\mathfrak{g}$ is the Lie algebra of G. Hence, we obtain a natural map

$$\widetilde{\mathbf{V}}_{c'}(j, Y) \to \mathbf{V}_c(i, Y)$$

from Lemma 5.2. Thus, we obtain a map $\widetilde{M}_{c'}(\mathbb{G}) \to M_c(\mathbb{G})$, and from the construction, it is clear that this map respects the action of $\widehat{G}(\mathbf{K}) \times \widehat{G}'(\mathbf{K})$. Now Lemma 0.5 implies that the constructed map is a morphism of $\widehat{\mathbb{G}} \times \widehat{\mathbb{G}}'$-representations.

The map in the opposite direction: $M_c(\mathbb{G}) \to \widetilde{M}_{c'}(\mathbb{G})$ is constructed similarly, and by the definition of the transition maps giving rise to the inverse systems $M_c(\mathbb{G})$ and $\widetilde{M}_{c'}(\mathbb{G})$, it is clear that both compositions $M_c(\mathbb{G}) \to \widetilde{M}_{c'}(\mathbb{G}) \to M_c(\mathbb{G})$ and $\widetilde{M}_{c'}(\mathbb{G}) \to M_c(\mathbb{G}) \to \widetilde{M}_{c'}(\mathbb{G})$ are the identity maps.

This proves point (1) of Proposition 4.2.

5.6.

Following [GK], let us denote by $M(\mathbf{G}[[t]])$ the pro-vector space

$$\text{``}\varprojlim\text{''}\ \mathrm{Funct}_c^{lc}(\mathbf{G}[[t]]/\mathbf{G}^i) \otimes \mu(\mathbf{G}[[t]]/\mathbf{G}^i),$$

where the transition maps are given by fiber-wise integration. This space carries an action of the group $\mathbf{G}[[t]] \times \mathbf{G}[[t]]$. The convolution product defines an isomorphism

$$(M(\mathbf{G}[[t]]) \otimes M(\mathbf{G}[[t]])_{\mathbf{G}[[t]]} \simeq M(\mathbf{G}[[t]]), \tag{12}$$

where $\mathbf{G}[[t]]$ acts diagonally.

By construction, as a representation of $\widehat{\mathbb{G}}$ under the left action, $M_c(\mathbb{G})$ identifies with $i^{\widehat{\mathbb{G}}}_{\mathbf{G}[[t]]}(M(\mathbf{G}[[t]]))$. Therefore,

$$\mathrm{Hom}_{\mathrm{Rep}_c(\widehat{\mathbb{G}})}(\widetilde{M}_{c'}(\mathbb{G}), M_c(\mathbb{G})) \simeq \mathrm{Hom}_{\mathbf{G}[[t]]}(\widetilde{M}_{c'}(\mathbb{G}), M(\mathbf{G}[[t]])). \tag{13}$$

The map $\widetilde{M}_{c'}(\mathbb{G}) \to M_c(\mathbb{G})$ constructed above corresponds to the natural restriction morphism $\widetilde{M}_{c'}(\mathbb{G}) \to M(\mathbf{G}[[t]])$.

Remark. From the latter description it is not immediately clear why this map is compatible with the right $\widehat{\mathbb{G}}'$-action.

Note also that the map $\widetilde{M}_c(\mathbb{G}) \to M_{c'}(\mathbb{G})$ can be described by a similar adjunction property with respect to the right $\widehat{\mathbb{G}}'$-action.

Let us prove now point (2) of Proposition 4.2. For any Π, which is a representation of $\widehat{\mathbb{G}} \times \widehat{\mathbb{G}}'$ at levels (c, c') we have

$$\mathrm{Hom}_{\widehat{\mathbb{G}} \times \widehat{\mathbb{G}}'}\left(\Pi, (M_c(\mathbb{G}) \otimes \widetilde{M}_{c'}(\mathbb{G}))_{\mathbf{G}[[t]]}\right) \simeq \mathrm{Hom}_{\mathbf{G}[[t]] \times \mathbf{G}[[t]]}(\Pi, M(\mathbf{G}[[t]]),$$

with the isomorphism being given by the restriction map

$$(M_c(\mathbb{G}) \otimes \widetilde{M}_{c'}(\mathbb{G}))_{\mathbf{G}[[t]]} \to (M(\mathbf{G}[[t]]) \otimes M(\mathbf{G}[[t]]))_{\mathbf{G}[[t]]},$$

followed by the map of (12).

Let us apply this to $\Pi = M_c(\mathbb{G})$. It is clear that both maps appearing in Proposition 4.2(2), correspond under the above isomorphism to the restriction map $M_c(\mathbb{G}) \to M(\mathbf{G}[[t]])$. Therefore, these two maps coincide.

6. Distributions on a stack

6.1.

First, let $\mathbf{X}$ be a locally compact object of $\mathbf{S}$et. Recall that $\mathbb{F}\mathrm{unct}^{lc}(\mathbf{X})$ denotes the corresponding (strict) object in $\mathbb{V}\mathrm{ect}$ (see [GK], Sect. 3.2), and $\mathrm{Funct}^{lc}(\mathbf{X}) = \mathrm{limProj}\ \mathbb{F}\mathrm{unct}^{lc}(\mathbf{X})$. The vector space $\mathrm{Distr}_c(\mathbf{X})$ introduced in Sect. 1.3 identifies with $\mathrm{Hom}_{\mathbb{V}\mathrm{ect}}(\mathbb{F}\mathrm{unct}^{lc}(\mathbf{X}), \mathbb{C})$, or, which is the same, with the space of linear functionals $\mathrm{Funct}^{lc}(\mathbf{X}) \to \mathbb{C}$, continuous in the topology of projective limit.

Suppose now that $\mathbf{X} = X(\mathbf{K})$, where X is a *smooth* algebraic variety over $\mathbf{K}$. In this case we can introduce the subspace $\mathrm{Distr}_c^{lc}(\mathbf{X})$ of locally constant distributions on $\mathbf{X}$ (see e.g., [GK], Sect. 5.1).

Indeed, it is well known that a choice of a top differential form ω on X defines a measure $\mu(\omega)$ on $\mathbf{X}$, i.e., a functional on the space $\mathrm{Funct}_c^{lc}(\mathbf{X})$. For $\omega' = \omega \cdot f$, where f is an invertible function on X, we have $\mu(\omega') = \mu(\omega) \cdot |f|$. Hence, the subset of elements in $\mathrm{Distr}_c(\mathbf{X})$, which can be (locally) written as $\mu(\omega) \cdot g$, where g is a locally constant function on $\mathbf{X}$ with compact support, is independent of the choice of ω. This subset is by definition $\mathrm{Distr}_c^{lc}(\mathbf{X})$.

Although the following is well known, we give a proof for the sake of completeness.

Proposition 6.2. *Let $f : X_1 \to X_2$ be a smooth map between smooth varieties over* $\mathbf{K}$. *Then*

(1) The push-forward map $\mathrm{Distr}_c(\mathbf{X}_1) \to \mathrm{Distr}_c(\mathbf{X}_2)$ *sends* $\mathrm{Distr}_c^{lc}(\mathbf{X}_1)$ *to* $\mathrm{Distr}_c^{lc}(\mathbf{X}_2)$.

(2) If $X_1(\mathbf{K}) \to X_2(\mathbf{K})$ is surjective, then $f_! : \mathrm{Distr}_c^{lc}(\mathbf{X}_1) \to \mathrm{Distr}_c^{lc}(\mathbf{X}_2)$ is also surjective.

Proof. Statement (1) is local in the analytic, and a fortiori in the Zariski topology on $\mathbf{X}_1$. Therefore, we can assume that our morphism f factors as $X_1 \overset{f'}{\to} X_2 \times Z \overset{f''}{\to} X_2$, where Z is another smooth variety, with f' being étale, and f'' being the projection on the first factor.

Since an étale map induces a local isomorphism in the analytic topology, it is clear that $f'_!$ maps $\mathrm{Distr}_c^{lc}(\mathbf{X}_1)$ to $\mathrm{Distr}_c^{lc}(\mathbf{X}_2 \times \mathbf{Z})$. From the definition of $\mathrm{Distr}_c^{lc}(\cdot)$, it is clear that

$$\begin{array}{ccc} \mathrm{Distr}_c^{lc}(\mathbf{Z}_1) \otimes \mathrm{Distr}_c^{lc}(\mathbf{Z}_2) & \xrightarrow{\sim} & \mathrm{Distr}_c^{lc}(\mathbf{Z}_1 \times \mathbf{Z}_2) \\ \downarrow & & \downarrow \\ \mathrm{Distr}_c(\mathbf{Z}_1) \otimes \mathrm{Distr}_c(\mathbf{Z}_2) & \longrightarrow & \mathrm{Distr}_c(\mathbf{Z}_1 \times \mathbf{Z}_2). \end{array} \tag{14}$$

So the map $f''_! : \mathrm{Distr}_c^{lc}(\mathbf{X}_2 \times \mathbf{Z}) \to \mathrm{Distr}_c(\mathbf{X}_2)$ can be identified with

$$\mathrm{Distr}_c^{lc}(\mathbf{Z}) \otimes \mathrm{Distr}_c^{lc}(\mathbf{X}_2) \overset{\int \times \mathrm{id}}{\to} \mathrm{Distr}_c^{lc}(\mathbf{X}_2),$$

implying assertion (1) of the proposition.

We will prove a slight strengthening of assertion (2). Note that since f is smooth, the image of $\mathbf{X}_1$ in $\mathbf{X}_2$ is open, and hence, also closed in the analytic topology. We will show that f_1 maps $\mathrm{Distr}_c^{lc}(\mathbf{X}_1)$ surjectively onto the subspace of $\mathrm{Distr}_c^{lc}(\mathbf{X}_2)$, consisting of distributions, supported on the image.

The assertion is local in the analytic topology on $\mathbf{X}_2$. Let $x_2 \in X_2(\mathbf{K})$ be a point, and let $x_1 \in X_1(\mathbf{K})$ be its pre-image. Then the local factorization of f as $f'' \circ f'$ as above makes the assertion manifest. □

6.3.

In what follows we will need a relative version of the above notions. For a smooth morphism $g : X \to Z$ let ω_{rel} be a relative top differential form on X. It defines a relative measure $\mu(\omega_{\mathrm{rel}}) : \mathrm{Funct}_c^{lc}(\mathbf{X}) \to \mathrm{Funct}_c^{lc}(\mathbf{Z})$. As in the absolute situation, by multiplying $\mu(\omega_{\mathrm{rel}})$ by locally constant functions on $\mathbf{X}$, whose support is proper over $\mathbf{Z}$, we obtain a pro-vector sub-space inside $\mathrm{Hom}_{\mathrm{Funct}^{lc}(\mathbf{Z})}(\mathrm{Funct}_c^{lc}(\mathbf{X}), \mathrm{Funct}_c^{lc}(\mathbf{Z}))$, which we will denote by $\mathbb{D}\mathrm{istr}_c^{lc}(\mathbf{X}/\mathbf{Z})$. Note that when $X = X' \times Z$, we have: $\mathbb{D}\mathrm{istr}_c^{lc}(\mathbf{X}/\mathbf{Z}) \simeq \mathrm{Distr}_c^{lc}(\mathbf{X}') \otimes \mathbb{F}\mathrm{unct}^{lc}(\mathbf{Z})$ (the tensor product being taken in the sense of $\mathbb{V}$ect). We will denote by $\mathrm{Distr}_c^{lc}(\mathbf{X}/\mathbf{Z})$ the vector space $\mathrm{limProj}\ \mathbb{D}\mathrm{istr}_c^{lc}(\mathbf{X}/\mathbf{Z})$.

When $f : X_1 \to X_2$ is a smooth map of schemes smooth over Z, as in Proposition 6.2 we have a push-forward map $f_! : \mathbb{D}\mathrm{istr}_c^{lc}(\mathbf{X}_1/\mathbf{Z}) \to \mathbb{D}\mathrm{istr}_c^{lc}(\mathbf{X}_2/\mathbf{Z})$, which is surjective if $f : X_1(\mathbf{K}) \to X_2(\mathbf{K})$ is; moreover, in this case the map $f_! : \mathrm{Distr}_c^{lc}(\mathbf{X}_1/\mathbf{Z}) \to \mathrm{Distr}_c^{lc}(\mathbf{X}_2/\mathbf{Z})$ is also easily seen to be surjective. In the particular case when $X_2 = Z$ we obtain a map $\int : \mathbb{D}\mathrm{istr}_c^{lc}(\mathbf{X}/\mathbf{Z}) \to \mathbb{F}\mathrm{unct}^{lc}(\mathbf{Z})$.

If Y is another scheme over Z, consider the Cartesian diagram

$$\begin{array}{ccc} X \underset{Z}{\times} Y & \xrightarrow{f'} & X \\ {\scriptstyle g'}\downarrow & & \downarrow{\scriptstyle g} \\ Y & \xrightarrow{f} & Z. \end{array} \tag{15}$$

We have a pullback map $f^* : \mathbb{D}\mathrm{istr}_c^{lc}(\mathbf{X}/\mathbf{Z}) \to \mathbb{D}\mathrm{istr}_c^{lc}(\mathbf{X} \underset{\mathbf{Z}}{\times} \mathbf{Y}/\mathbf{Y})$.

Suppose now that the scheme Z is itself smooth, and X is smooth over Z as above. In this case the spaces $\mathrm{Distr}_c^{lc}(\mathbf{X})$ and $\mathrm{Distr}_c^{lc}(\mathbf{Z})$ are well defined, and we have an isomorphism

$$\mathrm{Distr}_c^{lc}(\mathbf{X}) \simeq \mathrm{Distr}_c^{lc}(\mathbf{X}/\mathbf{Z}) \underset{\mathrm{Funct}^{lc}(\mathbf{Z})}{\otimes} \mathrm{Distr}_c^{lc}(\mathbf{Z}).$$

If in the situation of (15) Y is also smooth over $\mathbf{Z}$, and $\xi_Y \in \mathrm{Distr}_c^{lc}(\mathbf{Y})$, $\xi_{X/Z} \in \mathrm{Distr}_c^{lc}(\mathbf{X}/\mathbf{Z})$, consider the element $f^*(\xi_{X/Z}) \otimes \xi_Y \in \mathrm{Distr}_c^{lc}(\mathbf{X} \underset{\mathbf{Z}}{\times} \mathbf{Y})$. We have

$$f'_!(f^*(\xi_{X/Z}) \otimes \xi_Y) = \xi_{X/Z} \otimes f_!(\xi_Y) \in \mathrm{Distr}_c^{lc}(\mathbf{X}), \text{ and} \tag{16}$$

$$g'_!(f^*(\xi_{X/Z}) \otimes \xi_Y) = f^*(g_!(\xi_{X/Z})) \cdot \xi_Y \in \mathrm{Distr}_c^{lc}(\mathbf{Y}). \tag{17}$$

Finally, let us assume that both maps f and g induce surjections on the level of $\mathbf{K}$-valued points.

Lemma 6.4. *The maps $f_!$, $g_!$ induce an isomorphism*

$$\mathrm{Distr}_c^{lc}(\mathbf{Z}) \simeq \mathrm{coker}\left(\mathrm{Distr}_c^{lc}(\mathbf{X} \underset{\mathbf{Z}}{\times} \mathbf{Y}) \overset{(f_!', -g_!')}{\longrightarrow} \mathrm{Distr}_c^{lc}(\mathbf{X}) \oplus \mathrm{Distr}_c^{lc}(\mathbf{Y})\right).$$

Proof. Let $(\xi_X, \xi_Y) \in \mathrm{Distr}_c^{lc}(\mathbf{X}) \oplus \mathrm{Distr}_c^{lc}(\mathbf{Y})$ be an element such that $f_!(\xi_X) = g_!(\xi_Y)$. We need to find an element $\xi' \in \mathrm{Distr}_c^{lc}(\mathbf{X}_1 \underset{\mathbf{Z}}{\times} \mathbf{Y})$, such that $f_!'(\xi') = \xi_X$ and $g_!(\xi') = \xi_Y$. Using Lemma 6.2, we can assume that $\xi_X = 0$.

Let $\xi_{X/Z}$ be an element in $\mathrm{Distr}_c^{lc}(\mathbf{X}/\mathbf{Z})$, such that $\int \xi = 1 \in \mathrm{Funct}(\mathbf{Z})$. Then $\xi' := f^*(\xi_{X/Z}) \otimes \xi_Y$ satisfies our requirements, by (16). □

Let now X and Y be smooth varieties, and $f : Z \times X \to Y$ a map, such that the corresponding map $f' : Z \times X \to Z \times Y$ is smooth.

Lemma-Construction 6.5. *Under the above circumstances we have a natural action map*

$$\mathbf{Z} \times \mathrm{Distr}_c^{lc}(\mathbf{X}) \to \mathrm{Distr}_c^{lc}(\mathbf{Y}).$$

Proof. Consider the map

$$f_!' : \mathbb{D}\mathrm{istr}_c^{lc}(\mathbf{Z} \times \mathbf{X}/\mathbf{Z}) \to \mathbb{D}\mathrm{istr}_c^{lc}(\mathbf{Z} \times \mathbf{Y}/\mathbf{Z}).$$

By composing it with $\cdot \otimes 1 : \mathrm{Distr}_c^{lc}(\mathbf{X}) \to \mathbb{D}\mathrm{istr}_c^{lc}(\mathbf{Z} \times \mathbf{X}/\mathbf{Z})$ we obtain a map

$$\mathrm{Distr}_c^{lc}(\mathbf{X}) \to \mathrm{Distr}_c^{lc}(\mathbf{Y}) \otimes \mathbb{F}\mathrm{unct}^{lc}(\mathbf{Z}).$$

The latter is, by definition, the same as an action map $\mathbf{Z} \times \mathrm{Distr}_c^{lc}(\mathbf{X}) \to \mathrm{Distr}_c^{lc}(\mathbf{Y})$. □

6.6.

Let $\mathcal{Y}$ be an algebraic stack. We will say that $\mathcal{Y}$ is **K**-admissible (or just admissible) if there exists a smooth covering $Z \to \mathcal{Y}$, such that for any map $X \to \mathcal{Y}$, the corresponding map of schemes

$$X \underset{\mathcal{Y}}{\times} Z \to X$$

is surjective on the level of **K**-points.

If $\mathcal{Y}$ is admissible, a covering $Z \to \mathcal{Y}$ having the above property will be called admissible. It is clear that the class of admissible coverings is closed under Cartesian products. It is also clear that if $\mathcal{Y}$ is admissible, and $\mathcal{Y}' \to \mathcal{Y}$ is a representable map, then $\mathcal{Y}'$ is also admissible.

Lemma 6.7. *Suppose that $\mathcal{Y}$ is a stack, which is locally in the Zariski topology has the form Z/G, where Z is a scheme, and G is an affine algebraic group. Then $\mathcal{Y}$ is admissible.*

Proof. First, we can assume that $G = GL_n$. Indeed, by assumption, there is an embedding $G \to GL_n$, and consider the scheme $Z' := Z \underset{G}{\times} GL_n$. Then $\mathcal{Y} = Z'/GL_n$.

Now the assertion follows from Hilbert's 90: for $y \in \mathcal{Y}(\mathbf{K})$ its pre-image in Z is a GL_n-torsor, which is necessarily trivial. □

From now on, we will assume that $\mathcal{Y}$ is admissible. Assume in addition that $\mathcal{Y}$ is smooth. We will now define the space, denoted, $\mathrm{Distr}_c^{lc}(\mathbf{Y})$, of locally constant compactly supported distributions on $\mathbf{Y}$.

Namely, given two admissible coverings $Z_1, Z_2 \to \mathcal{Y}$ we define

$$\mathrm{Distr}_c^{lc}(\mathbf{Y}) := \mathrm{coker}\left(\mathrm{Distr}_c^{lc}(\mathbf{Z}_1 \underset{\mathbf{Y}}{\times} \mathbf{Z}_2) \to \mathrm{Distr}_c^{lc}(\mathbf{Z}_1) \oplus \mathrm{Distr}_c^{lc}(\mathbf{Z}_2)\right).$$

Lemma 6.4, combined with Proposition 6.2(2), implies that $\mathrm{Distr}_c^{lc}(\mathbf{Y})$ is well-defined, i.e., is independent of the choice of Z_1, Z_2.

If $f : \mathcal{Y}_1 \to \mathcal{Y}_2$ is a smooth representable map of (smooth admissible) stacks, from Proposition 6.2(1) we obtain that there exists a well-defined map $f_! : \mathrm{Distr}_c^{lc}(\mathbf{Y}_1) \to \mathrm{Distr}_c^{lc}(\mathbf{Y}_2)$.

Assume now that $\mathcal{Y} = Z/G$, where G is an algebraic group acting on Z. By Lemma 6.5, we have an action of $\mathbf{G}$ on the vector space $\mathrm{Distr}_c^{lc}(\mathbf{Z})$. From Lemma 6.4 we obtain

Corollary 6.8. *For $\mathcal{Y}$ as above,*

$$\mathrm{Distr}_c^{lc}(\mathbf{Y}) \simeq \mathrm{Coinv}_{\mathbf{G}}(\mathrm{Distr}_c^{lc}(\mathbf{Z})).$$

6.9. Relative version

Assume now that $\mathcal{Y}$ is a stack, endowed with a smooth map to a scheme Z. For a pair of admissible coverings $X_1, X_2 \to \mathcal{Y}$, we define the pro-vector space $\mathbb{D}\mathrm{istr}_c^{lc}(\mathcal{Y}/Z)$ as

$$\mathrm{coker}\left(\mathbb{D}\mathrm{istr}_c^{lc}(\mathbf{X}_1 \underset{\mathbf{Y}}{\times} \mathbf{X}_2/\mathbf{Z}) \to \mathbb{D}\mathrm{istr}_c^{lc}(\mathbf{X}_1/\mathbf{Z}) \oplus \mathbb{D}\mathrm{istr}_c^{lc}(\mathbf{X}_2/\mathbf{Z})\right).$$

A relative version of Lemma 6.4 shows that this is well-defined, i.e., independent of the choice of X_1 and X_2.

Finally, the assertion of Lemma–Construction 6.5 remains valid, where Z is a scheme, $\mathcal{Y}, \mathcal{Y}'$ are smooth stacks, and the map $f : Z \times \mathcal{Y} \to \mathcal{Y}'$ is such that the corresponding map $f' : Z \times \mathcal{Y} \to Z \times \mathcal{Y}'$ is smooth and representable.

7. Induction via the moduli stack of bundles

7.1.

Let X be a (smooth complete) algebraic curve over $\mathbf{K}$, $\mathbf{x} \in \mathbf{X}$ a rational point, and let t be a coordinate near $\mathbf{x}$.

If G be a split reductive group, let Bun_G denote the moduli stack of principal G-bundles on X. For $i \in \mathbb{Z}$, let $\mathrm{Bun}_G^{i,\mathbf{x}}$ denote the stack classifying bundles equipped with a trivialization on the i-th infinitesimal neighbourhood of $\mathbf{x}$. By construction, $\mathrm{Bun}_G^{i,\mathbf{x}}$ is a principal $G[[t]]/G^i$-bundle over Bun_G.

If $\mathcal{Y} \subset \mathrm{Bun}_G$ is an open substack of finite type, we let $\mathcal{Y}^{i,\mathbf{x}}$ denote its pre-image in $\mathrm{Bun}_G^{i,\mathbf{x}}$. The following is well-known:

Lemma 7.2. *For any $\mathcal{Y} \subset \mathrm{Bun}_G$ of finite type and i large enough, the stack $\mathcal{Y}^{i,\mathbf{x}}$ is a scheme of finite type.*

For $\mathcal{Y}$ as above, we let $\mathcal{Y}^{\infty,\mathbf{x}}$ denote the object of $\mathrm{Pro}(Sch^{ft})$ equal to "$\underleftarrow{\lim}$" $\mathcal{Y}^{i,\mathbf{x}}$. We let $\mathrm{Bun}_G^{\infty,\mathbf{x}}$ denote the object

$$\text{“}\underset{\mathcal{Y}}{\underrightarrow{\lim}}\text{”}\, \mathcal{Y}^{\infty,\mathbf{x}} \in \mathrm{Ind}(\mathrm{Pro}(Sch^{ft})).$$

Another basic fact is that $G((t))$, viewed as a group-object of $\mathrm{Ind}(\mathrm{Pro}(Sch^{ft}))$, acts on $\mathrm{Bun}_G^{\infty,\mathbf{x}}$ in the sense of the tensor structure on $\mathrm{Ind}(\mathrm{Pro}(Sch^{ft}))$.

7.3.

By Lemma 6.7, the stacks $\mathcal{Y}^{i,\mathbf{x}}$ are admissible. Set $\mathbf{W}^i_{\mathcal{Y}} = \mathrm{Distr}_c^{lc}(\mathbf{Y}^{i,\mathbf{x}})$. For $\mathcal{Y}_1 \hookrightarrow \mathcal{Y}_2$ we have a natural push-forward map on the level of distributions $\mathbf{W}^i_{\mathcal{Y}_1} \to \mathbf{W}^i_{\mathcal{Y}_2}$. Set

$$\mathbf{W}^i := \underset{\mathcal{Y}}{\underrightarrow{\lim}}\, \mathbf{W}^i_{\mathcal{Y}} \in \mathrm{Vect}.$$

For a fixed $\mathcal{Y}$ and $j > i$ we have a smooth representable map of stacks $\mathbf{Y}^{j,\mathbf{x}} \to \mathbf{Y}^{i,\mathbf{x}}$; hence we obtain a map $\mathbf{W}^j_{\mathcal{Y}} \to \mathbf{W}^i_{\mathcal{Y}}$ and, finally, a map $\mathbf{W}^j \to \mathbf{W}^i$.

We define the pro-vector space

$$\mathbb{W}_{X,\mathbf{x}} := \text{“}\underset{i}{\underleftarrow{\lim}}\text{”}\, \mathbf{W}^i.$$

Now we are ready to state:

Theorem 7.4. *The pro-vector space $\mathbb{W}_{X,\mathbf{x}}$ carries a natural action of the group $\mathbb{G}$, such that* $\mathrm{Coinv}_{\mathbf{G}^i}(\mathbb{W}_{X,\mathbf{x}}) \simeq \mathbf{W}^i$.

Note that by construction we have:

Corollary 7.5. *The $\mathbb{G}$-representation $\mathbb{W}_{X,\mathbf{x}}$ is admissible.*

Indeed, the coinvariants $\mathrm{Coinv}_{\mathbf{G}^i}(\mathbb{W}_{X,\mathbf{x}}) \simeq \mathbf{W}^i$ all belong to Vect.

7.6. Proof of Theorem 7.4

Let $G((t)) = \text{“}\varinjlim_k\text{”}\, Z_k$ with $Z_k = \text{“}\varprojlim_l\text{”}\, Z_k^l$, where Z_k^l are schemes of finite type.

To define an action

$$\mathbb{G} \times \mathbb{W}_{X,\mathbf{x}} \to \mathbb{W}_{X,\mathbf{x}}$$

we need to give for every k and i a map

$$\mathbf{Z}_k^l \times \mathbf{W}^j \to \mathbf{W}^i$$

defined for j and l sufficiently large.

For k and i as above let j be such that $\mathrm{Ad}_{Z_k}(G^j) \subset G^i$. The action of $G((t))$ on $\mathrm{Bun}_G^{\infty,\mathbf{x}}$ yields a map of stacks $Z_k \times \mathrm{Bun}_G^{j,\mathbf{x}} \to \mathrm{Bun}_G^{i,\mathbf{x}}$, which factors through Z_k^l for some l. Moreover, for every sub-stack $\mathcal{Y} \subset \mathrm{Bun}_G$ of finite type, there exists another sub-stack $\mathcal{Y}'$ of finite type, such that we have a map

$$Z_k^l \times \mathcal{Y}^{j,\mathbf{x}} \to \mathcal{Y}'^{j,\mathbf{x}}.$$

We claim that for $i, j, k, l, \mathcal{Y}, \mathcal{Y}'$ as above, we have a map

$$\mathbf{Z}_k^l \times \mathbf{W}_{\mathcal{Y}}^j \to \mathbf{W}_{\mathcal{Y}'}^j. \tag{18}$$

This follows from the stack-theoretic version of Lemma–Construction 6.5, see Sect. 6.9. The fact that the resulting action map $\mathbb{G} \times \mathbb{W}_{X,\mathbf{x}} \to \mathbb{W}_{X,\mathbf{x}}$ respects the group law on $\mathbb{G}$ is a straightforward verification.

To compute $\mathrm{Coinv}_{\mathbf{G}^i}(\mathbb{W}_{X,\mathbf{x}})$ note that $G[[t]]$, and hence all G^i, act on each $\mathrm{Bun}_G^{j,\mathbf{x}}$ individually.

Hence,

$$\mathrm{Coinv}_{\mathbf{G}^i}(\mathbb{W}_{X,\mathbf{x}}) \simeq \text{“}\varprojlim_{j\geq i}\text{”}\, \mathrm{Coinv}_{\mathbf{G}^i/\mathbf{G}^j}(\mathbf{W}^j).$$

We claim that for $j \geq i$, $\mathrm{Coinv}_{\mathbf{G}^i/\mathbf{G}^j}(\mathbf{W}^j) \simeq \mathbf{W}^i$. Indeed, since each $\mathcal{Y}^{j,\mathbf{x}}$ is stable under G^i/G^j, we have:

$$\mathrm{Coinv}_{\mathbf{G}^i/\mathbf{G}^j}(\mathbf{W}^j) \simeq \varinjlim_{\mathcal{Y}} \mathrm{Coinv}_{\mathbf{G}^i/\mathbf{G}^j}(\mathbf{W}_{\mathcal{Y}}^j) \simeq \varinjlim_{\mathcal{Y}} \mathbf{W}_{\mathcal{Y}}^i \simeq \mathbf{W}^i,$$

where the middle isomorphism follows from Corollary 6.8.

7.7. Variants and generalizations

Recall that the stack Bun_G is endowed with a canonical line bundle $\mathcal{L}_{\mathrm{Bun}_G}$, with the basic property that the $\mathbb{G}$-action on $\mathrm{Bun}_G^{\mathbf{x},\infty}$ extends to an action of a central extension $\widehat{\mathbb{G}}$ on the pull-back of $\mathcal{L}_{\mathrm{Bun}_G}$ to $\mathrm{Bun}_G^{\mathbf{x},\infty}$.

By the same token, we consider now a representation $\widehat{\mathbb{W}}_{X,\mathbf{x}}$ of $\widehat{\mathbb{G}}$, and for every $c : \mathbf{G}_m \to \mathbb{C}^*$ the object

$$\mathbb{W}_{X,\mathbf{x},c} := (\widehat{\mathbb{W}}_{X,\mathbf{x}} \otimes \mathbb{C})_{\mathbf{G}_m} \in \mathrm{Rep}_c(\widehat{\mathbb{G}}).$$

Note that instead of a single point $\mathbf{x}$ we could have considered any finite collection $\overline{\mathbf{x}} = \mathbf{x}_1, \ldots, \mathbf{x}_n$ of rational points. By repeating the construction we obtain a pro-vector space $\mathbb{W}_{X,\overline{\mathbf{x}}}$, acted on by the product $\prod_k \widehat{\mathbb{G}}_{\mathbf{x}_k}$, where each $\widehat{\mathbb{G}}_{\mathbf{x}_k}$ identifies with $\widehat{\mathbb{G}}$ once we identify the local ring of $\mathbf{X}$ at $\mathbf{x}_k$ with $\mathbf{F}$.

Again, for a choice of a character $c : \mathbf{G}_m \to \mathbb{C}^*$, we obtain a representation of $\prod_k \widehat{\mathbb{G}}_{\mathbf{x}_k}$, denoted $\mathbb{W}_{X,\overline{\mathbf{x}},c}$, such that the center $\mathbf{G}_m^k$ acts via the multiplication map $\mathbf{G}_m^k \to \mathbf{G}_m$.

7.8.

From now on we will suppose that X is isomorphic to the projective line P^1, and the number of points is two, which we will denote by $\mathbf{x}_1$, and $\mathbf{x}_2$, respectively. Assume also that G is semi-simple and simply connected.

Consider the representation $\widehat{\mathbb{W}}_{P^1,\mathbf{x}_1,\mathbf{x}_2,c}$ of $\widehat{\mathbb{G}}_{\mathbf{x}_1} \times \widehat{\mathbb{G}}_{\mathbf{x}_2}$. Let us take its coinvariants with respect to $\mathbf{I}^{00}_{\mathbf{x}_1} \subset \widehat{\mathbb{G}}_{\mathbf{x}_1}$. By Corollary 1.10, on the resulting pro-vector space we will have an action of $\widehat{\mathbb{G}}_{\mathbf{x}_2}$; we will denote this representation by Π_c^{thick}, i.e.,

$$\Pi_c^{\mathrm{thick}} = \mathrm{Coinv}_{\mathbf{I}^{00}_{\mathbf{x}_1}}(\widehat{\mathbb{W}}_{P^1,\mathbf{x}_1,\mathbf{x}_2,c}).$$

By Theorem 3.3, the algebra $\ddot{\mathsf{H}}_{q,c'}$ acts on Π_c^{thick} by endomorphisms. Consider now

$$\mathbf{U}_c := \mathrm{Coinv}_{\mathbf{I}^{00}_{\mathbf{x}_1} \times \mathbf{I}^{00}_{\mathbf{x}_2}}(\widehat{\mathbb{W}}_{P^1,\mathbf{x}_1,\mathbf{x}_2,c}) \simeq \mathrm{Coinv}_{\mathbf{I}^{00}_{\mathbf{x}_2}}(\Pi_c^{\mathrm{thick}}).$$

By Corollary 7.5, this is a vector space, endowed with two commuting actions of $\ddot{\mathsf{H}}_{q,c'}$. We have:

Theorem 7.9. *There exists a canonically defined vector $\mathbf{1}_{\mathbf{U}_c} \in \mathbf{U}_c$, which freely generates $\mathbf{U}_c$ under each of the two $\ddot{\mathsf{H}}_{q,c'}$-actions.*

8. Proof of Theorem 7.9

8.1.

Let W_{aff} be the affine Weyl group corresponding to G. Since G was assumed simply connected, W_{aff} is a Coxeter group.

If α is a simple affine root, let $I_\alpha \subset \widehat{G}$ denote the corresponding sub-minimal parahoric; let $N(I_\alpha)$ denote the (pro)-unipotent radical of I_α, and $M_\alpha := I_\alpha / N(I_\alpha)$ the Levi quotient.

By definition, M_α is a reductive group of semi-simple rank 1, with a distinguished copy of G_m in its center; we will denote by M'_α the quotient M_α / G_m. Let B_α denote the Borel subgroup of M_α, and $\mathbf{B}^0_\alpha$ the kernel of

$$\mathbf{B}_\alpha \to \mathbf{T} \to \Lambda.$$

Let Π_c^α be the quotient of the principal series representation of $\mathbf{M}_\alpha$, given by the condition that $\mathbf{G}_m \subset \mathbf{M}_\alpha$ acts by the character c, i.e.,

$$\Pi_c^\alpha = \left(\mathrm{Funct}_c^{lc}(\mathbf{M}_\alpha/\mathbf{B}_\alpha^0) \otimes \mathbb{C}\right)_{\mathbf{G}_m}. \tag{19}$$

Let us denote by $\dot{\mathsf{H}}_{q,c}^\alpha$ the corresponding affine Hecke algebra of M_α, i.e., the algebra of endomorphisms of the functor $\mathrm{Coinv}_{\mathbf{B}_\alpha^0} : \mathrm{Rep}(\mathbf{M}_\alpha, \mathrm{Vect})_c \to \mathrm{Vect}$, or which is the same, the algebra of endomorphisms of Π_c^α as a $\mathbf{M}_\alpha$-representation. It is well known that $\mathbf{U}_\alpha := \mathrm{Coinv}_{\mathbf{B}_\alpha^0}(\Pi_c^\alpha)$, as a bi-module over $\dot{\mathsf{H}}_{q,c}^\alpha$, is isomorphic to the regular representation. In a sense, Theorem 7.9 generalizes this result to the affine case.

The functor $\Pi \mapsto \mathrm{Coinv}_{\mathbf{I}^{00}}(\Pi)$ on $\mathrm{Rep}_c(\widehat{\mathbb{G}})$ can be factored into two steps. We first apply the functor

$$r_{\mathbf{M}^\alpha}^{\widehat{\mathbb{G}}} : \mathrm{Rep}_c(\widehat{\mathbb{G}}) \to \mathrm{Rep}_c(\mathbf{I}_\alpha, \mathbb{V}\mathrm{ect}) \xrightarrow{\mathrm{Coinv}_{\mathbf{N}_\alpha}} \mathrm{Rep}_c(\mathbf{M}_\alpha, \mathbb{V}\mathrm{ect}),$$

where the first arrow is the forgetful functor, and then apply

$$\mathrm{Coinv}_{\mathbf{B}_\alpha^0} : \mathrm{Rep}_c(\mathbf{M}_\alpha, \mathbb{V}\mathrm{ect}) \to \mathbb{V}\mathrm{ect}.$$

In particular, endomorphisms of the latter functor map to endomorphisms of the composition. As a result, we obtain the canonical embedding $\dot{\mathsf{H}}_{q,c'}^\alpha \to \ddot{\mathsf{H}}_{q,c}$.

Recall also that the group-algebra $\mathbb{C}[\Lambda]$ is canonically a subalgebra in $\ddot{\mathsf{H}}_{q,c}$, contained in each $\dot{\mathsf{H}}_{q,c'}^\alpha$.

8.2.

The strategy of the proof of Theorem 7.9 will be as follows. We will endow the vector space $\mathbf{U}_c$ with an increasing filtration

$$\mathbf{U}_c = \underset{w \in W_{\mathrm{aff}}}{\cup} \mathbf{U}_w$$

with $\mathbf{U}_{w_1} \subset \mathbf{U}_{w_2}$ if and only if $w_1 \leq w_2$ in the Bruhat order. This filtration will be stable under the action of $\mathbb{C}[\Lambda] \subset \ddot{\mathsf{H}}_{q,c}$ with respect to both actions of the latter on $\mathbf{U}_c$.

The subquotients

$$\mathbf{U}^w := \mathbf{U}_w / \underset{w' < w}{\cup} \mathbf{U}_{w'}$$

will be free Λ-modules of rank 1 (with respect to each of the actions of $\ddot{\mathsf{H}}_{q,c}$). In particular, for $w = 1$, the space $\mathbf{U}^1 \simeq \mathbf{U}_1$ will contain a canonical element $\mathbf{1}_{\mathbf{U}^1} \in \mathbf{U}^1$, which generates $\mathbf{U}^1$ under each of the Λ-actions. This will be the element $\mathbf{1}_{\mathbf{U}_c}$ of Theorem 7.9.

Moreover, the following crucial property will be satisfied. Suppose that w is an element of W_{aff}, and s_α is a simple affine reflection, such that $s_\alpha \cdot w > w$ (resp., $w \cdot s_\alpha > w$). Then the subquotient

$$\mathbf{U}_{s_\alpha \cdot w} / \bigcup_{w' < s_\alpha \cdot w, w' \neq w} \mathbf{U}_{w'}, \left(\text{resp., } \mathbf{U}_{w \cdot s_\alpha} / \bigcup_{w' < w \cdot s_\alpha, w' \neq w} \mathbf{U}_{w'} \right) \tag{20}$$

is stable under the action of $\dot{\mathsf{H}}^\alpha_{q,c}$, embedded into the first (resp., second) copy of $\ddot{\mathsf{H}}_{q,c}$, and as a $\dot{\mathsf{H}}^\alpha_{q,c}$-module, it is isomorphic to $\dot{\mathsf{H}}^\alpha_{q,c} \underset{\mathbb{C}[\Lambda]}{\otimes} \mathbf{U}^w$.

The existence of a filtration with the above properties clearly implies the assertion of the theorem.

8.3.

Let I^0 denote the (pro)-unipotent radical of I; we have $\mathbf{I}^{00}/\mathbf{I}^0 \simeq \mathbf{T}^0$, where $\mathbf{T}^0 \subset \mathbf{T}$ is the maximal compact subgroup of $\mathbf{T}$.

Consider the scheme $\mathcal{G}_G := \mathrm{Bun}_G^{\infty, \mathbf{x}_1, \mathbf{x}_2} / I_{\mathbf{x}_2}$, called the thick Grassmannian of G. By definition, it classifies principal G-bundles on $X = P^1$, endowed with a trivialization at the formal neighbourhood of $\mathbf{x}_1$ and a reduction to B of their fiber at $\mathbf{x}_2$. Consider also the base affine space $\widetilde{\mathcal{G}}_G := \mathrm{Bun}_G^{\infty, \mathbf{x}_1, \mathbf{x}_2} / I^0_{\mathbf{x}_2}$, which is a principal T-bundle over $\mathcal{G}_G$. The loop group $G((t))$, where t is the coordinate near $\mathbf{x}_1$ acts naturally on both $\mathcal{G}_G$ and $\widetilde{\mathcal{G}}_G$.

It is well known that $\mathcal{G}_G$ can be written as a union of open sub-schemes $\mathcal{G}_{G,w}$, $w \in W_{\mathrm{aff}}$, each being stable under the action of $I_{\mathbf{x}_1} = I \subset G((t))$, such that $\mathcal{G}_{G,w_1} \subset \mathcal{G}_{G,w_2}$ if and only if $w_1 < w_2$ in the Bruhat order. Let us denote by $\mathcal{G}_G^w$ the locally closed subscheme $\mathcal{G}_{G,w} - \bigcup_{w' < w} \mathcal{G}_{G,w'}$, and by $\widetilde{\mathcal{G}}_{G,w}$, $\widetilde{\mathcal{G}}_G^w$ the corresponding sub-schemes in $\widetilde{\mathcal{G}}$. It is well known that the group I^0 (resp., I) acts transitively on each $\mathcal{G}_G^w$ (resp., $\widetilde{\mathcal{G}}_G^w$) with finite-dimensional unipotent stabilizers. Choosing a point in each $\widetilde{\mathcal{G}}_G^w$, we will denote by N_w its stabilizer in I, or, which is the same, the stabilizer in I^0 of the projection of this point to $\mathcal{G}_G^w$.

Consider the stack

$$\mathrm{Bun}_G^{\mathbf{x}_1, \mathbf{x}_2} := \mathrm{Bun}_G^{\infty, \mathbf{x}_1, \mathbf{x}_2} / (I^0_{\mathbf{x}_1} \times I^0_{\mathbf{x}_2}) \simeq \widetilde{\mathcal{G}}_G / I^0.$$

By definition, it classifies G-bundles on $X = P^1$ with a reduction to the maximal unipotent at $\mathbf{x}_1$ and $\mathbf{x}_2$, and it carries a natural action of the group $T \times T$. From the above discussion, we obtain that $\mathrm{Bun}_G^{\mathbf{x}_1, \mathbf{x}_2}$ can be canonically written as a union of open substacks of finite type

$$\mathrm{Bun}_G^{\mathbf{x}_1, \mathbf{x}_2} = \bigcup_{w \in W_{\mathrm{aff}}} \mathcal{Y}_w$$

with $\mathcal{Y}_{w_1} \subset \mathcal{Y}_{w_2}$ if and only if $w_1 \leq w_2$.

Consider the locally-closed sub-stack $\mathcal{Y}^w := \mathcal{Y}_w - \bigcup_{w'<w} \mathcal{Y}_{w'}$. We obtain that $\mathcal{Y}^w$ is isomorphic to $T \times (\mathrm{pt}/N_w)$, where N_w is as above. The first copy of T acts via multiplication on the first factor, and the action of the second copy is twisted by the projection of w to the finite Weyl group, acting by automorphisms on T.

We will denote by $\widehat{\mathcal{Y}}_w$, $\widehat{\mathcal{Y}}^w$ the pull-back of the total space of the G_m-torsor corresponding to $\mathcal{L}_{\mathrm{Bun}_G}$ to these sub-stacks.

8.4.

We have:

$$\mathbf{U}_c \simeq \left(\varinjlim_{w} \mathrm{Distr}_c^{lc}(\widehat{\mathcal{Y}}_w) \otimes \mathbb{C}\right)_{\mathbf{T}^0 \times \mathbf{T}^0 \times \mathbf{G}_m}.$$

Set

$$\mathbf{U}_w := (\mathrm{Distr}_c^{lc}(\widehat{\mathcal{Y}}_w) \otimes \mathbb{C})_{\mathbf{T}^0 \times \mathbf{T}^0 \times \mathbf{G}_m}. \tag{21}$$

We claim that each $\mathbf{U}_w$ maps injectively into $\mathbf{U}_c$; and the images of $\mathbf{U}_w$ define a filtration with the required properties. One thing is clear, however: by construction, $\mathbf{U}_w$ carries an action of $\Lambda \times \Lambda$, and its map to $\mathbf{U}_c$ is compatible with this action.

8.5.

To proceed we need to introduce some more notation. Let Z be a smooth scheme, and let $\mathcal{L}$ be a line bundle on Z. Let $\overset{\circ}{\mathcal{L}}$ denote the total space of the corresponding G_m-torsor over Z. We will denote by $\mathrm{Distr}_c^{lc}(Z)_{\mathcal{L}}$ the space

$$\left(\mathrm{Distr}_c^{lc}(\overset{\circ}{\mathbf{L}}) \otimes \mathbb{C}\right)_{\mathbf{G}_m},$$

where $\mathbf{G}_m$ acts on $\mathbb{C}$ via the standard character $\mathbf{G}_m \to \mathbb{Z} \overset{1\mapsto q}{\longrightarrow} \mathbb{C}^*$.

Let now $Z_1 \subset Z$ be a smooth closed subscheme, and let Z_2 be its complement. We have:

Lemma 8.6. *There exists a natural short exact sequence:*

$$0 \to \mathrm{Distr}_c^{lc}(Z_2)_{\mathcal{L}} \to \mathrm{Distr}_c^{lc}(Z)_{\mathcal{L}} \to \mathrm{Distr}_c^{lc}(Z_1)_{\mathcal{L}\otimes\mathcal{L}_n} \to 0,$$

where $\mathcal{L}_n$ is the top power of the normal bundle to Z_1 inside Z.

Proof. Note that by definition we have

$$\mathrm{Distr}_c^{lc}(Z)_{\mathcal{L}_0} \simeq \mathrm{Funct}_c^{lc}(Z),$$

where $\mathcal{L}_0$ is the inverse of the line bundle of top forms on Z. The assertion of the lemma follows now from the fact that for any $Z_1 \subset Z$ we have a short exact sequence for the corresponding spaces of locally constant functions with compact support:

$$0 \to \mathrm{Funct}_c^{lc}(Z_2) \to \mathrm{Funct}_c^{lc}(Z) \to \mathrm{Funct}_c^{lc}(Z_1) \to 0.$$

□

For each $w < w'$, the open embedding $\mathcal{Y}_w \hookrightarrow \mathcal{Y}_{w'}$ can be covered by an open embedding of schemes $Z_w \hookrightarrow Z_{w'}$, such that $\mathcal{Y}_{w'} = Z_{w'}/N$, $\mathcal{Y}_w = Z_w/N$with N being a unipotent algebraic group. Therefore, by Lemma 6.8 and the exactness of the functor $\text{Coinv}_{\mathbf{N}}$, the map $\mathbf{U}_w \to \mathbf{U}_{w'}$ is an embedding. Hence, $\mathbf{U}_w \to \mathbf{U}_c$ is also an embedding.

Moreover, we claim that from Lemma 8.6 we obtain a (non-canonical) isomorphism

$$\mathbf{U}^w \simeq (\text{Distr}_c^{lc}(\mathcal{Y}^w))_{\mathbf{T}^0 \times \mathbf{T}^0}, \tag{22}$$

compatible with the $\Lambda \times \Lambda$-action.

Indeed, *a priori*, $\mathbf{U}^w \simeq \left((\text{Distr}_c^{lc}(\mathcal{Y}^w))_{\mathcal{L}}\right)_{\mathbf{T}^0 \times \mathbf{T}^0}$ for a certain $T \times T$-equivariant line bundle $\mathcal{L}$ on $\mathcal{Y}^w$. However, from the description of $\mathcal{Y}^w$ as $T \times (\text{pt}/N_w)$, this line bundle is (non-canonically) trivial. Note, however, that this line bundle is canonically trivial for $w = 1$.

Now, the same description of $\mathcal{Y}^w$ implies that $\text{Distr}_c^{lc}(\mathcal{Y}^w) \simeq \text{Funct}_c^{lc}(\mathbf{T})$, with the first action of $\mathbf{T}$ being given by multiplication, and the second action is twisted by w. This implies that $\mathbf{U}^w \simeq (\text{Funct}_c^{lc}(\mathbf{T}))_{\mathbf{T}^0 \times \mathbf{T}^0} \simeq \mathbb{C}[\Lambda]$.

8.7.

We will now study the subquotient $\mathbf{U}_{s_\alpha \cdot w} / \underset{w' < s_\alpha \cdot w, w' \neq w}{\cup} \mathbf{U}_{w'}$, where s_α is a simple affine reflection such that $s_\alpha \cdot w > w$. (The case $w \cdot s_\alpha > w$ is analyzed similarly.)

Note first of all that for any $w' \in W_{\text{aff}}$, we have $I_\alpha \cdot \mathcal{G}_G^{w'} \subset \mathcal{G}_G^{w'} \cup \mathcal{G}_G^{s_\alpha \cdot w'}$. Hence, the open subset $\mathcal{G}_{G, s_\alpha \cdot w}$ is I_α-stable, and so is the union $\underset{w' < s_\alpha \cdot w, w' \neq w}{\cup} \mathcal{G}_G^{w'}$. Therefore, the subquotient in (20) is indeed $\dot{\mathsf{H}}^\alpha_{q,c'}$-stable.

We will consider two additional stacks. One is $'\mathcal{Y} := \text{Bun}_G^{\infty, \mathbf{x}_1, \mathbf{x}_2} / (N(I_\alpha)_{\mathbf{x}_1} \times I^0_{\mathbf{x}_2})$, on which we have an action of M'_α. We will denote by pr the projection

$$'\mathcal{Y} \xrightarrow{pr} {'\mathcal{Y}}/N_\alpha \simeq \text{Bun}_G^{\mathbf{x}_1, \mathbf{x}_2},$$

where $N_\alpha := B_\alpha \cap I^0$.

Another stack is the quotient

$$''\mathcal{Y} := \text{Bun}_G^{\infty, \mathbf{x}_1, \mathbf{x}_2} / (M'_\alpha \times I^0).$$

The stack $''\mathcal{Y}$ can be written as a union of open sub-stacks $''\mathcal{Y}_w$ numbered by left cosets $\{1, s_\alpha\} \backslash W_{\text{aff}}$; we will denote by $''\mathcal{Y}^w$ the corresponding locally closed sub-stacks. Let also $'\mathcal{Y}_w$ and $'\mathcal{Y}^w$ denote the pre-images of the corresponding sub-stacks in $'\mathcal{Y}$, and $'\widehat{\mathcal{Y}}_w$, $'\widehat{\mathcal{Y}}^w$ the total spaces of the G_m-torsors, corresponding to the pull-backs of the line bundle $\mathcal{L}_{\text{Bun}_G}$.

If w is an element of W_{aff} we have:

$$'\mathcal{Y}^w = pr^{-1}(\mathcal{Y}^w \cup \mathcal{Y}^{s_\alpha \cdot w}). \tag{23}$$

Using Lemma 8.6, the subquotient (20) is isomorphic to

$$\left(\mathrm{Distr}_c^{lc}('\widehat{\mathcal{Y}}^w)\otimes\mathbb{C}\right)_{(\mathbf{B}^0_\alpha)_{\mathbf{x}_1}\times(\mathbf{T}^0)_{\mathbf{x}_2}\times\mathbf{G}_m}.$$

The vector space $\left(\mathrm{Distr}_c^{lc}('\widehat{\mathcal{Y}}^w)\otimes\mathbb{C}\right)_{(\mathbf{T}^0)_{\mathbf{x}_2}\times\mathbf{G}_m}$ is naturally a representation of the group $\mathbf{M}_\alpha$. We claim that as such,

$$\left(\mathrm{Distr}_c^{lc}('\widehat{\mathcal{Y}}^w)\otimes\mathbb{C}\right)_{(\mathbf{T}^0)_{\mathbf{x}_2}\times\mathbf{G}_m}\simeq\Pi_c^\alpha. \tag{24}$$

Clearly, the above isomorphism implies our assertion about the action of $\dot{\mathsf{H}}^\alpha_{q,c'}$ on the subquotient in (20).

8.8.

To prove (24) let us observe that $''\mathcal{Y}\simeq{}'\mathcal{Y}/M'_\alpha$ and that $''\mathcal{Y}^w\simeq\mathrm{pt}/N_{w,\alpha}$, where $N_{w,\alpha}$ is a unipotent group, so that the Cartesian product

$$\mathrm{pt}\underset{''\mathcal{Y}^w}{\times}{}'\mathcal{Y}^w$$

is isomorphic to M'_α, and the action of $N_{w,\alpha}$ on M'_α comes from a surjective homomorphism $N_{w,\alpha}\to N_\alpha$ and the action of the latter on M'_α by right multiplication. Hence,

$$\mathrm{Distr}_c^{lc}('\widehat{\mathcal{Y}}^w)\simeq\left(\mathrm{Distr}_c^{lc}(\mathbf{M}_\alpha)\right)_{\mathbf{N}_\alpha}\simeq\left(\mathrm{Distr}_c^{lc}(\mathbf{M}_\alpha/\mathbf{N}_\alpha)\right),$$

implying (24).

Acknowledgements

We would like to thank A. Shapira for pointing out two mistakes in the previous version of the paper.

The research of D.G. is supported by a long-term fellowship at the Clay Mathematics Institute and a grant from DARPA. He would also like to thank the Einstein Institute of Mathematics of the Hebrew University of Jerusalem and IHES, where this work was written. The research of D.K. is supported by an ISF grant.

References

[GK] D. Gaitsgory, D. Kazhdan, Representations of algebraic groups over a 2-dimensional local field, math.RT/0302174, GAFA **14** (2004), 535–574.

[Ka] M. Kapranov, Double affine Hecke algebras and 2-dimensional local fields, *JAMS* **14** (2001), 239–262.

[Ka1] M. Kapranov, The elliptic curve in the S-duality theory and Eisenstein series for Kac–Moody groups, math.AG/0001005.

[Pos] Positselsky, *Private communications.*

Modules with a Demazure flag

Anthony Joseph*

The Donald Frey Professorial Chair
Department of Mathematics
The Weizmann Institute of Science
Rehovot 76100
Israel
Anthony.joseph@weizmann.ac.il

To Denise

Summary. A Demazure module can be described as the space of global sections of a suitable line bundle on a Schubert variety. A problem posed by the author in 1985 was to show that the tensor product of a one-dimensional Demazure module with an arbitrary one admits a Demazure flag, that is, a filtration whose quotients are Demazure modules. This was shown by P. Polo (who called such filtrations "excellent") in a large number of cases including positive characteristic and by O. Mathieu for all semisimple algebraic groups first in zero characteristic and later in arbitrary characteristic.

This paper settles this question in the context of a Kac–Moody algebra with symmetric simply-laced Cartan datum and in arbitrary characteristic. The method combines the corresponding "combinatorial excellent filtration" established independently by P. Littelmann et al. and the author with the globalization techniques of G. Lusztig and M. Kashiwara. In principle the method applies to an arbitrary symmetrizable Kac–Moody algebra; but for technical reasons it is necessary to use a positivity result of Lusztig which applies to only the simply-laced case.

Subject Classification: 17B37

1. Introduction

Global (or canonical) bases have been of tremendous interest for their own sake exhibiting a deep combinatorial structure as exemplified by their multiplication properties and by their $q \to 0$ limit leading to crystal bases. Besides this, global bases are indispensable for certain purposes. An example of this is the common basis theorem

* Work supported by European Community RTN network "Liegrits", Grant No. MRTN-CT-2003-505078.

[J2, 6.2.19] which leads to the separation theorem [J2, 7.3.8] and quantum PRV determinants [J2, 8.2] as well as to a simple proof of Richardson's theorem (see [B] for example, which nevertheless gives a slightly different proof though also using global bases).

This work is a further example where the use of global bases is essential. It also motivates two further questions concerning their structure (see 3.9, 5.18, Remark 1 and 5.23).

1.1.

The aim of this paper is to settle in some generality a question I posed [J1, 5.8] twenty years ago. In a slightly modified form (see [J4, introduction] for details) it asks if the tensor product of a one-dimensional Demazure module by an arbitrary Demazure module admits a Demazure flag, that is, a filtration whose quotients are Demazure modules. In this van der Kallen [V1] gave an interesting criterion for the existence of a Demazure flag based on the annihilator formula for extremal vectors given in [J1, 3.4], and generalized by Polo to positive characteristic. From this Polo [P] was able to positively answer my question in most cases including nearly all positive characteristic. At a similar time Mathieu [M1] transformed van der Kallen's criterion using a result of Bott and the Whitehead lemma (concerning semisimplicity in zero characteristic) to a criterion involving the vanishing of certain higher sheaf cohomologies, which he verified using some particular Frobenius splittings. This latter criterion and the resulting proof was valid only in zero characteristic. However Mathieu later modified his proof to include all characteristics. A trick of Donkin then recovers the assertion over $\mathbb{Z}$. (See [M5, V2]).

1.2.

Whilst the annihilator formula holds for an arbitrary symmetrizable Kac–Moody Lie algebra, van der Kallen's criterion fails on account of imaginary roots. Thus these methods do not go over to the Kac–Moody case.

1.3.

Recently I gave [J4] a purely combinatorial version of the existence of a Demazure flag at the level of crystals. (This is even valid in the non-symmetrizable case, see [J5, Sect. 19].) It turned out that Littelmann [L4] had done something similar (see also [LLM, 2.4]); but neither of us were able to recover the module theoretic version, although Littelmann [L4, Thm. 4] did obtain the existence of a Weyl flag for the full tensor product. The basis of the latter was standard monomial theory which Littelmann himself established in complete generality [L3, Sect. 6]. Weyl flags were first studied by Donkin [Do1,2] and Wang [W]. They have applications to invariant theory. Mathieu resolved the semisimple case in full generality — see [M3] and references therein. Later, alternative proofs for the semisimple case were given by Paradowski [Pa] and Kaneda [Ka] notably using canonical bases. Of course, the existence of a Weyl flag (for tensor product) is much easier than the existence of a Demazure flag, the former being trivial in zero characteristic.

1.4.

In this paper I prove the existence of a Demazure flag for the tensor products described in 1.1 with respect to the Lusztig form of the quantized enveloping algebra of a Kac–Moody algebra $\mathfrak{g}$ with simply-laced symmetric Cartan datum in the sense of [Lu1, 2.1.3] defined over $\mathbb{Z}[q, q^{-1}]$. Setting $q = 1$ gives the corresponding result over the Kostant form of the enveloping algebra and hence for its specialization over an arbitrary field. In particular, the result is valid for a split semisimple algebraic group with symmetric simply-laced Cartan datum over an arbitrary field. The proof depends heavily on the corresponding combinatorial result as well as on Kashiwara's globalization technology. A key point is that for fundamental weights occurring in the first factor the filtration respects the global basis (on the second factor). Moreover in this sense it is canonical depending only on a lifting of a natural partial order (see 5.5). Once it is realized that this should be so the proof is rather natural though still very difficult. (Moreover it is not yet known if the filtration respects the global basis when arbitrary dominant weights occur in the first factor (see 5.23).) As a bonus one obtains a Weyl flag for the full tensor product (over $\mathbb{Z}[q, q^{-1}]$). Our analysis should cover an arbitrary symmetrizable Kac–Moody Lie algebra; but for the present at a certain point a positivity result of Lusztig [Lu2, 22.1.7] is used and this is why the restriction to symmetric simply-laced Cartan datum is needed. Henceforth we shall just say that $\mathfrak{g}$ is simply-laced.

2. Notation and background

2.1.

Our analysis depends heavily on Kashiwara's work on crystals and their globalization [K1,2]. For our own convenience we follow the exposition of this work given in [J2, Chaps. 5, 6] adopting the notation there with only small changes. Here we need to extend [J2, 6.3.8] replacing $\mathbb{Q}(q)$ by $\mathbb{Z}[q, q^{-1}]$. The possibility for doing this was indicated to me by Kashiwara who also makes a brief reference to this generalization in [K2, Remark 3.2.6]. During the writing of this manuscript a paper by Ryom–Hansen appeared in which he also wrote down a proof [R, Lemma 3.3] of 3.8. However for the sake of completion we have retained our analysis which also serves as a preamble.

2.2.

Let q be an indeterminate and set $K = \mathbb{Q}(q)$. Let ψ be the $\mathbb{Q}$-linear automorphism of K sending q to q^{-1}. Set $A = \mathbb{Z}[q, q^{-1}]$, $A^+ = \mathbb{Q}[q]_0$ (localization at $q = 0$), $A^- = \psi(A^+) = \mathbb{Q}[q^{-1}]_\infty$ (localization at $q = \infty$). For each integer $n \geq 0$ set $[n]_q = (q^n - q^{-n})/(q - q^{-1})$, $[n]_q^! = [n]_q[n-1]_q \cdots [1]_q$, $\left[\begin{smallmatrix} n \\ m \end{smallmatrix}\right]_q = [n]_q^!/[n-m]_q^![m]_q^!$, for m integer $0 \leq m \leq n$.

Let $U_q(\mathfrak{g})$ be the Hopf algebra over K with generators $e_i, f_i, t_i, t_i^{-1} : i = 1, 2, \ldots, \ell$, satisfying the relations given for example in [J2, 5.1.1] and corresponding to a symmetrizable Kac–Moody algebra $\mathfrak{g}$ with Cartan subalgebra $\mathfrak{h}$. Recall that

$\mathfrak{h}^*$ admits a non-degenerate symmetric form (,) defined through the symmetrizability. We shall often replace the subscript by $\alpha = \alpha_i$ which is the simple root corresponding to i. Set $\pi = \{\alpha_i\}_{i=1}^{\ell}$ and let $P(\pi)$ (or simply P) denote the $\mathbb{Z}$ lattice generated by a choice of fundamental weights. We may assume that $(\mu, \lambda) \in \mathbb{Z}$, for all $\mu, \lambda \in P$. Let P^+ denote the dominant elements of P. Given $\lambda \in P^+$ one has (by definition) $\alpha^\vee(\lambda) \in \mathbb{N}$ for every simple coroot $\alpha^\vee = 2\alpha/(\alpha, \alpha)$.

Set $q_i = q^{\frac{1}{2}(\alpha_i,\alpha_i)}$ and define the divided powers $e_i^{(n)} = e_i^n/[n]_{q_i}^!$, $f_i^{(n)} = f_i^n/[n]_{q_i}^!$. We shall generally omit the q_i subscript (even though it depends on i).

Let $U_q^{\mathbb{Z}}(\mathfrak{n}^+)$ (resp. $U_q^{\mathbb{Z}}(\mathfrak{n}^-)$) denote the A subring of $U_q(\mathfrak{g})$ generated by the $e_i^{(n)}$ (resp. $f_i^{(n)}$) : $n \in \mathbb{N}, i = 1, 2, \dots, \ell$. For short we may denote them by $\mathcal{E}$ and $\mathcal{F}$, respectively.

In order to separate roots it is necessary to enlarge $U_q(\mathfrak{g})$ slightly. Thus we view P as a multiplicative group T by defining a map $\tau : P \to T$ satisfying $\tau(\alpha_i) = t_i$, $\forall i$, $\tau(\lambda + \mu) = \tau(\lambda)\tau(\mu)$, $\forall\ \lambda, \mu \in P$ and we enlarge $U_q(\mathfrak{g})$ so that it contains T. Let $U_q^{\mathbb{Z}}(\mathfrak{g})$ be the A subring of $U_q(\mathfrak{g})$ generated by $U_q^{\mathbb{Z}}(\mathfrak{n}^\pm)$, and T. This is the Lusztig quantum analogue of the Kostant $\mathbb{Z}$ form for $U(\mathfrak{g})$. It specializes to the latter at $q = 1$, $\tau(\mu) = 1$, $\forall\ \mu \in P$. A weight submodule M of $U_q^{\mathbb{Z}}(\mathfrak{g})$ is an A submodule on which T acts by a character. We shall only consider characters of the form $\tau(\mu) \mapsto q^{(\mu,\lambda)} : \lambda \in P$. Then the corresponding weight submodule is denoted by M_λ.

For each $\alpha \in \pi$, let $U_q(\mathfrak{s}_\alpha)$ denote the subalgebra of $U_q(\mathfrak{g})$ generated by e_α, f_α over KT and let $U_q^{\mathbb{Z}}(\mathfrak{s}_\alpha)$ be its corresponding Lusztig form.

2.3.

For each $\lambda \in P^+$, let $V(\lambda)$ denote the simple highest weight module with highest weight λ. It is integrable and so for each $\alpha \in \pi$ a direct sum of simple finite-dimensional $U_q(\mathfrak{s}_\alpha)$ modules. Each such submodule V has a one dimensional highest (resp. lowest) weight space V^{e_α} (resp. V^{f_α}) and the Kashiwara operator $\tilde{f}_\alpha$ (resp. $\tilde{e}_\alpha$) on V and hence on $V(\lambda)$ is defined by setting $(\tilde{f}_\alpha^n - f_\alpha^{(n)})|_{V^{e_\alpha}} = 0$ (resp. $(\tilde{e}_\alpha^n - e_\alpha^{(n)})|_{V^{f_\alpha}} = 0$). This coincides with the definition in [J2, 5.1.2]. Indeed if $u_n \in V^{e_\alpha}$ has t_α eigenvalue q^n, then $f_\alpha^{(n)} u_n \in V^{f_\alpha} \setminus \{0\}$ and $e_\alpha^{(m)} f_\alpha^{(n)} u_n = f_\alpha^{(n-m)} u_n$, for all $m : 0 \leq m \leq n$. Then $\tilde{e}_\alpha f_\alpha^{(n-m)} u_n = \tilde{e}_\alpha e_\alpha^{(m)} f_\alpha^{(n)} u_n = e_\alpha^{(m+1)} f_\alpha^{(n)} u_n = f_\alpha^{(n-m-1)} u_n$, as required. Let $\tilde{\mathcal{E}}$ (resp. $\tilde{\mathcal{F}}$) denote the monoid generated by the $\tilde{e}_\alpha$ (resp. $\tilde{f}_\alpha$) : $\alpha \in \pi$.

2.4.

Fix a generator u_λ of $V(\lambda)$ of weight λ. Set $L(\lambda) = A^+\tilde{\mathcal{F}}u_\lambda$. A seemingly innocent but in fact very difficult result of Kashiwara is that $L(\lambda)$ is $\tilde{\mathcal{E}}$ stable. Moreover the images of the distinct $\tilde{f}u_\lambda : \tilde{f} \in \tilde{\mathcal{F}}$ form a $\mathbb{Q}$ basis $B(\lambda)$ of $L(\lambda)/qL(\lambda)$ called the crystal basis for $V(\lambda)$ [J2, Thm. 5.4.27]. There is an induced action of $\tilde{e}_\alpha, \tilde{f}_\alpha : \alpha \in \pi$ on $B(\lambda)$ which gives rise to a combinatorial description of $V(\lambda)$. This combinatorics behaves particularly well for tensor product [J2, Thm. 5.1.12]. An alternative description of this combinatorial structure has been given by Littelmann [L1] but it is unlikely that his globalization technique [L3] will work here. In particular, Littelmann's monomial basis [L3, L5], although explicit, depends on a choice of reduced decomposition and none of these choices can be made to satisfy the conclusion of 3.5 simultaneously for all $\alpha \in \pi$. Possibly Littelmann's standard monomial basis [L3, Sect. 6] for the dual module might be better.

2.5.

For the present work we rely heavily on Kashiwara's global basis. It is obtained as follows.

Set $V^{\mathbb{Z}}(\lambda) = U_q^{\mathbb{Z}}(\mathfrak{n}^-)u_\lambda$. It is a $U_q^{\mathbb{Z}}(\mathfrak{g})$ module. Set $L^{\mathbb{Z}}(\lambda) = V^{\mathbb{Z}}(\lambda) \cap L(\lambda)$. Then (see [J2, 6.2.2]) $L^{\mathbb{Z}}(\lambda)/qL^{\mathbb{Z}}(\lambda)$ is a free $\mathbb{Z}$ module with basis $B(\lambda)$.

Recall 2.2 and extend ψ to a ring homomorphism of $U_q(\mathfrak{n}^-)$ by setting $\psi(f_\alpha) = f_\alpha,\ \forall \alpha \in \pi$.

Define a $\mathbb{Q}$-linear isomorphism on $V(\lambda)$, by $\psi(au_\lambda) = \psi(a)u_\lambda,\ \forall a \in U_q(\mathfrak{n}^-)$. Since $\psi(f_\alpha^{(n)}) = f_\alpha^{(n)}, \forall n \in \mathbb{N}, \alpha \in \pi$, it follows that ψ restricts to a $U_q^{\mathbb{Z}}(\mathfrak{n}^-)$ module isomorphism of $V^{\mathbb{Z}}(\lambda)$ to itself. One has

$$[e_\alpha, f_\alpha^{(n)}] = f_\alpha^{(n-1)} \frac{(q_\alpha^{-(n-1)} t_\alpha - q_\alpha^{(n-1)} t_\alpha^{-1})}{q_\alpha - q_\alpha^{-1}},$$

and moreover the last factor becomes $[\alpha^\vee(\gamma) - (n-1)]_{q_\alpha}$ on a vector of weight γ. It follows that ψ commutes with e_α. Consequently ψ is also a $U_q^{\mathbb{Z}}(\mathfrak{n}^+)$ module isomorphism. Set $L(\lambda)^- = \psi(L(\lambda))$ and $L^{\mathbb{Z}}(\lambda)^- = \psi(L^{\mathbb{Z}}(\lambda)) = L(\lambda)^- \cap V^{\mathbb{Z}}(\lambda)$. One has

$$V(\lambda) \cong K \otimes_A V^{\mathbb{Z}}(\lambda) \cong K \otimes_{A^+} L(\lambda) \cong K \otimes_{A^-} L(\lambda)^-. \qquad (*)$$

By construction $L(\lambda)$ is $\tilde{\mathcal{F}}$ stable. Fix $\alpha \in \pi$. After Kashiwara [J2, 5.1.7] the splitting of $V(\lambda)$ into a direct sum of simple $U_q(\mathfrak{s}_\alpha)$ modules results (rather remarkably) in a direct sum decomposition of $L(\lambda)$ whose direct summands tensored over K are simple $U_q(\mathfrak{s}_\alpha)$ modules. (One may add that the global basis only respects this splitting up to "triangularity" based on dimension — see 3.8.) In particular, $L(\lambda) = \sum_{n\in\mathbb{N}} A^+ \tilde{f}_\alpha^n L(\lambda)^{e_\alpha}$. Consequently, $L(\lambda)^- = \psi(L(\lambda)) = A^- \sum_{n\in\mathbb{N}} \psi(f_\alpha^{(n)} L(\lambda)^{e_\alpha}) = A^- \sum_{n\in\mathbb{N}} f_\alpha^{(n)} \psi(L(\lambda))^{e_\alpha} = A^- \sum_{n\in\mathbb{N}} \tilde{f}_\alpha^n \psi(L(\lambda))^{e_\alpha}$, which shows that $L(\lambda)^-$ is also $\tilde{\mathcal{F}}$ stable. Similarly $L(\lambda)^-$ is $\tilde{\mathcal{E}}$ stable. By contrast $V^{\mathbb{Z}}(\lambda)$ is neither $\tilde{\mathcal{E}}$ nor $\tilde{\mathcal{F}}$ stable. Curiously, the limiting module $V^{\mathbb{Z}}(\infty)$, which identifies with $U_q^{\mathbb{Z}}(\mathfrak{n}^-)$ and so is $\mathcal{F}$ stable, is also $\tilde{\mathcal{E}}$ and $\tilde{\mathcal{F}}$ stable [J2, 6.1.7]. However although the corresponding A^+ lattice $L(\infty)$ is both $\tilde{\mathcal{E}}$ and $\tilde{\mathcal{F}}$ stable and splits as above, it is no longer true that $L(\infty)^- := \psi(L(\infty))$ is $\tilde{\mathcal{E}}$ or $\tilde{\mathcal{F}}$ stable. This is because Kashiwara's truncated version of e_α (namely e'_α in the notation of [J2, 5.3.1]) no longer admits a commutator with $f_\alpha^{(n)}$ invariant under ψ. We emphasize these points as they may not be recognized by the casual (or even not so casual) reader of [K1].

Recall that $L^{\mathbb{Z}}(\lambda)^- = \psi(L^{\mathbb{Z}}(\lambda))$. A further remarkable theorem of Kashiwara (see [J2, 6.2.3]) is that $E := V^{\mathbb{Z}}(\lambda) \cap L(\lambda) \cap L(\lambda)^- = L^{\mathbb{Z}}(\lambda) \cap L^{\mathbb{Z}}(\lambda)^-$ maps isomorphically to $L^{\mathbb{Z}}(\lambda)/qL^{\mathbb{Z}}(\lambda)$. This and (∗) above makes $(V^{\mathbb{Z}}(\lambda), L(\lambda), L(\lambda)^-)$ a balanced triple for $V(\lambda)$ in the language of Kashiwara [K3, Sect. 2]. Let G_λ denote the inverse map. This provides (see [J2, 6.2.8]) a global A basis $G_\lambda(b) : b \in B(\lambda)$ of $V^{\mathbb{Z}}(\lambda)$. It coincides with Lusztig's canonical basis [Lu1,2]. Moreover this construction mimics that of Lusztig [Lu1, Sect. 3] given earlier in the semisimple case.

2.6.

Recall [J2, 5.1.3] that the coproduct on $U_q(\mathfrak{g})$ satisfies

$$\Delta(e_\alpha) = e_\alpha \otimes t_\alpha^{-1} + 1 \otimes e_\alpha, \ \Delta(f_\alpha) = f_\alpha \otimes 1 + t_\alpha \otimes f_\alpha, \ \forall\, \alpha \in \pi.$$

For Lemma 5.13 some readers may prefer to now compute $\Delta(f_\alpha^{(n)})$. In any case one has to be particularly careful about q factors.

3. Properties of the global basis

3.1.

We first extend in sections 3.2–3.8 the results of Kashiwara described in [J2, 6.3.4–6.3.8] to be valid over A. The possibility for doing this was indicated in [K2, Remark 3.2.6]. We give details for completion and as a preliminary to sections 3.9–3.14. One may also consult [R] which makes this extension in a slightly different manner.

3.2.

Fix $\lambda \in P^+$ and write simply $B = B(\lambda)$, $V = V(\lambda)$, $V^{\mathbb{Z}} = V^{\mathbb{Z}}(\lambda)$, $L = L(\lambda)$, $L^- = L(\lambda)^-$, $L^{\mathbb{Z}} = L^{\mathbb{Z}}(\lambda)$, $L^{\mathbb{Z}-} = L^{\mathbb{Z}}(\lambda)^-$. Following 2.3 we use a subscript to denote a weight subspace of these modules. Obviously G_λ respects weight space decomposition. We shall often omit the subscript λ where it is understood. Fix $\alpha \in \pi$ and set $e = e_\alpha$, $f = f_\alpha$, $\tilde{e} = \tilde{e}_\alpha$, $\tilde{f} = \tilde{f}_\alpha$.

3.3.

Given $b \in B \subset L/qL$, choose a representative $\hat{b} \in L$. Then $\hat{b} - G(b) \in qL$ and similarly $\tilde{f}^n\hat{b} - G(\tilde{f}^n b) \in qL$. Yet L is $\tilde{f}$ stable, so we conclude that

$$\tilde{f}^n G(b) - G(\tilde{f}^n b) \in qL, \ \forall\, n \in \mathbb{N}.$$

Now suppose $eG(b) = 0$. Then $f^{(n)}G(b) = \tilde{f}^n G(b)$. Since $G(b) \in E = V^{\mathbb{Z}} \cap L \cap L^-$, $V^{\mathbb{Z}}$ is $f^{(n)}$ stable and $L \cap L^-$ is $\tilde{f}^n$ stable, we obtain $\tilde{f}^n G(b) \in E$. Consequently,

$$\tilde{f}^n G(b) - G(\tilde{f}^n b) \in E \cap qL = 0.$$

Combined with a similar argument for $\tilde{e}$ we obtain the

Lemma. *For all $b \in B$ and all $n \in \mathbb{N}$ one has*

(i) $f^{(n)}G(b) = \tilde{f}^n G(b) = G(\tilde{f}^n b)$, *given* $eG(b) = 0$,
(ii) $e^{(n)}G(b) = \tilde{e}^n G(b) = G(\tilde{e}^n b)$, *given* $fG(b) = 0$.

3.4.

The result in 3.3 is a consequence of E being defined by the balanced triple ($V^{\mathbb{Z}}$, L, L^-) for V, combined with $V^{\mathbb{Z}}$ (resp. L, L^-) being stable under the $e^{(n)}, f^{(n)}$ (resp. $\tilde{e}^n, \tilde{f}^n$) : $n \in \mathbb{N}$. Suppose that $eG(b) = 0$. Then $\tilde{e}b = 0$ and we set $B' = \{\tilde{f}^i b : i \in \mathbb{N}\}$. By 3.3(i) the A (resp. A^+, A^-) submodules generated by the $G(b') : b' \in B'$ form a balanced subtriple $(W^{\mathbb{Z}}, M, M^-)$ for the $U_q(\mathfrak{s}_\alpha)$ submodule W of V generated by $G(b)$. Then $(V^{\mathbb{Z}}/W^{\mathbb{Z}}, L/M, L^-/M^-)$ is a balanced triple for V/W through the remaining elements $G(b'') : b'' \in B \setminus B'$ of the global basis. (This is obtained in a wider context, though over $\mathbb{Q}$, in [K3, Lemma 2.2.2].) We conclude that if $eG(b) = 0 \mod W$, then similarly 3.3(i) holds mod W. A similar assertion holds if $fG(b) = 0$ and moreover the submodule factored out can be assumed to be the same in both cases. This eventually gives a finer version of Kashiwara's [J2, Lemma 6.3.4], described as follows. It is also due to Kashiwara.

3.5.

Following Kashiwara (see J2, 5.2.1) one may define maps $\varepsilon_\alpha, \varphi_\alpha : B \to \mathbb{N} : \alpha \in \pi$ through $\varepsilon_\alpha(b) = \max\{n | \tilde{e}_\alpha^n b \neq 0\}$, $\varphi_\alpha(b) = \max\{n | \tilde{f}_\alpha^n b \neq 0\}$. If $b \in B$, we define $\{\tilde{e}_\alpha^m b, \tilde{f}_\alpha^n b : m, n \in \mathbb{N}\}$ to be the α-string through b. Since $\tilde{e}_\alpha \tilde{f}_\alpha b = b$ if $\tilde{f}_\alpha b \neq 0$ and $\tilde{f}_\alpha \tilde{e}_\alpha b = b$ if $\tilde{e}_\alpha b \neq 0$, it is stable by $\tilde{e}_\alpha, \tilde{f}_\alpha$ and has $\varepsilon_\alpha(b) + \varphi_\alpha(b) + 1$ elements. We call $\ell_\alpha(b) := \varepsilon_\alpha(b) + \varphi_\alpha(b)$ its length. If b has weight ξ, then $\varphi_\alpha(b) - \varepsilon_\alpha(b) = \alpha^\vee(\xi)$. Note in particular that

$$\ell_\alpha(b) = \alpha^\vee(wt\ b) \ \text{ given } \ \varepsilon_\alpha(b) = 0. \tag{$*$}$$

As before we fix $\alpha \in \pi$ and omit the α-subscript. For each $r \in \mathbb{N}$, set $I^r(B) = \{b \in B | \varepsilon(b) + \varphi(b) = r\}$ and $W^r(B) = \bigcup_{s \geq r} I^s(B)$. Given $\xi \in \lambda - \mathbb{N}\pi$, set $B_{\xi+\mathbb{Z}\alpha} = \bigcup_{n\in\mathbb{Z}} B_{\xi+n\alpha}$ and $W^r(B_{\xi+\mathbb{Z}\alpha}) = B_{\xi+\mathbb{Z}\alpha} \cap W^r(B)$.

Similarly consider $V^{\mathbb{Z}}$. Set $V_{\xi+\mathbb{Z}\alpha} = \bigoplus_{n\in\mathbb{Z}} V_{\xi+n\alpha}$ (resp. $V^{\mathbb{Z}}_{\xi+\mathbb{Z}\alpha} = \bigoplus_{n\in\mathbb{Z}} V^{\mathbb{Z}}_{\xi+n\alpha} = V_{\xi+\mathbb{Z}\alpha} \cap V^{\mathbb{Z}}$). It is an $U_q(\mathfrak{s}_\alpha)$ (resp. $U_q^{\mathbb{Z}}(\mathfrak{s}_\alpha)$) submodule of V (resp. $V^{\mathbb{Z}}$). The former is a finite direct sum of finite-dimensional simple $U_q(\mathfrak{s}_\alpha)$ modules. Take the direct sum $W^r(V_{\xi+\mathbb{Z}\alpha})$ of those of dimension $\geq r+1$ up to the maximal dimension $s+1$ and set $W^r(V^{\mathbb{Z}}_{\xi+\mathbb{Z}\alpha}) = V^{\mathbb{Z}}_{\xi+\mathbb{Z}\alpha} \cap W^r(V_{\xi+\mathbb{Z}\alpha})$. Since every simple submodule of $V_{\xi+\mathbb{Z}\alpha}/W^r(V_{\xi+\mathbb{Z}\alpha})$ has dimension $\leq r$, it follows that the image of $V^{\mathbb{Z}}_{\xi+\mathbb{Z}\alpha}$ in this quotient has weights $\xi + n\alpha$ satisfying $|\alpha^\vee(\xi + n\alpha)| \leq (r-1)$. Thus the increasing family of $U_q^{\mathbb{Z}}(\mathfrak{s}_\alpha)$ submodules $W^s(V^{\mathbb{Z}}_{\xi+\mathbb{Z}\alpha}) \subset W^{s-1}(V^{\mathbb{Z}}_{\xi+\mathbb{Z}\alpha}) \subset \cdots$, of $V^{\mathbb{Z}}_{\xi+\mathbb{Z}\alpha}$ with union $V^{\mathbb{Z}}_{\xi+\mathbb{Z}\alpha}$, has successive quotients $W^r(V^{\mathbb{Z}}_{\xi+\mathbb{Z}\alpha})/W^{r+1}(V^{\mathbb{Z}}_{\xi+\mathbb{Z}\alpha})$ with weights $\xi + n\alpha$ satisfying $|\alpha^\vee(\xi + n\alpha)| \leq r$.

Lemma. *For all $r, k \in \mathbb{N}$, $b \in I^r(B_{\xi+\mathbb{Z}\alpha})$ one has*

(i) $f^{(k)}G(b) = \begin{bmatrix} \varepsilon(b) + k \\ k \end{bmatrix} G(\tilde{f}^k b) \mod W^{r+1}(V^{\mathbb{Z}}_{\xi+\mathbb{Z}\alpha})$

(ii) $e^{(k)}G(b) = \begin{bmatrix} \varphi(b) + k \\ k \end{bmatrix} G(\tilde{e}^k b) \mod W^{r+1}(V^{\mathbb{Z}}_{\xi+\mathbb{Z}\alpha})$

(iii) $W^r(V^{\mathbb{Z}}_{\xi+\mathbb{Z}\alpha}) = \bigoplus_{b\in W^r(B_{\xi+\mathbb{Z}\alpha})} AG(b)$.

Proof. Take $b \in I^r(B)$ satisfying $\varepsilon(b) = 0$. Then $\alpha^\vee(wt\ b) = r$. Consequently $eG(b) \in W^{r+1}(V_{\xi+\mathbb{Z}\alpha})$. Moding out by this submodule, 3.3 applies and we obtain $f^{(n)}G(b) - G(\tilde{f}^n b) \in W^{r+1}(V_{\xi+\mathbb{Z}\alpha}) \cap V^{\mathbb{Z}}_{\xi+\mathbb{Z}\alpha}$. Hence (i). The proof of (ii) is similar. For (iii) recall that $\{G(b) : b \in B\}$ is an A basis for $V^{\mathbb{Z}}$. It follows by weight space decomposition that $G(b) : b \in B_{\xi+\mathbb{Z}\alpha}$ is an A basis for $V^{\mathbb{Z}}_{\xi+\mathbb{Z}\alpha}$. Consider $b \in B_{\xi+\mathbb{Z}\alpha}$ belonging to an α-string of length $t \geq r$. By (i), (ii) it follows that the $U_q^{\mathbb{Z}}(\mathfrak{s}_\alpha)$ module generated by $G(b)$ has a weight vector $\xi+n\alpha$ satisfying $\alpha^\vee(\xi+n\alpha) = t$. Consequently $G(b)$ belongs to $W^r(V^{\mathbb{Z}}_{\xi+\mathbb{Z}\alpha})$. Hence (iii). □

3.6.

For each $\alpha \in \pi$, set $\mathcal{E}_\alpha = \sum_{n\in\mathbb{N}} Ae^{(n)}$, $\mathcal{F}_\alpha = \sum_{n\in\mathbb{N}} Af^{(n)}$. As before we fix $\alpha \in \pi$; but we do not drop the subscript on $\mathcal{E}_\alpha$ and $\mathcal{F}_\alpha$. Similarly we set $\tilde{\mathcal{E}}_\alpha = \bigcup_{n\in\mathbb{N}} \tilde{e}^n_\alpha$, $\tilde{\mathcal{F}}_\alpha = \bigcup_{n\in\mathbb{N}} \tilde{f}^n_\alpha$.

3.7.

Take $s \in \mathbb{N}$ and consider $W_s(V^{\mathbb{Z}}_{\xi+\mathbb{Z}\alpha}) := V^{\mathbb{Z}}_{\xi+\mathbb{Z}\alpha}/W^{s+1}(V^{\mathbb{Z}}_{\xi+\mathbb{Z}\alpha})$. By 3.5(ii), $W_s(V^{\mathbb{Z}}_{\xi+\mathbb{Z}\alpha})$ is a $U_q^{\mathbb{Z}}(\mathfrak{s}_\alpha)$ quotient of $V^{\mathbb{Z}}_{\xi+\mathbb{Z}\alpha}$ with A basis $\{G(b)|b \in B_{\xi+\mathbb{Z}\alpha} \setminus W^{s+1}(B_{\xi+\mathbb{Z}\alpha})\}$. This has no α-strings of length $> s$ and by taking s smaller if necessary we can assume it to have an α-string of length s. Then s has the same meaning as in 3.5. Let N be an $\mathcal{E}_\alpha$ submodule of $W_s(V^{\mathbb{Z}}_{\xi+\mathbb{Z}\alpha})$. We shall say that N admits a global basis if we can write

$$N = \bigoplus_{b\in B(N)} AG(b) \qquad (*)$$

for some $B(N) \subset B$. Here and in 3.8 we assume N as above.

Lemma. $\tilde{e}B(N) \subset B(N) \cup \{0\}$.

Proof. Take $b \in B(N)$. If $\tilde{e}b \neq 0$, then by 3.5(ii) the expansion of $eG(b)$ in the global basis for $W_s(V^{\mathbb{Z}}_{\xi+\mathbb{Z}\alpha})$ has $G(\tilde{e}b)$ as a non-zero coefficient. Hence the assertion. □

3.8.

Set $M = U_q^{\mathbb{Z}}(\mathfrak{s}_\alpha)N = \sum_{n\geq 0} f^{(n)}N = \mathcal{F}_\alpha N$ and $B(M) = \tilde{\mathcal{F}}_\alpha B(N) \setminus \{0\}$. Since M is a submodule of $W_s(V^{\mathbb{Z}}_{\xi+\mathbb{Z}\alpha})$ which is free of finite rank, it follows that M admits a highest weight space of weight $\eta \in \xi + \mathbb{Z}\alpha$ satisfying $\alpha^\vee(\eta) =: s \in \mathbb{N}$, by our choice of s. Factoring out by the $U_q^{\mathbb{Z}}(\mathfrak{s}_\alpha)$ module $I^s(M)$ it generates, repeating the procedure and taking inverse images we obtain an increasing family of $U_q^{\mathbb{Z}}(\mathfrak{s}_\alpha)$ submodules of M with union M and whose successive quotients $W^r(M)/W^{r+1}(M)$ have weights η satisfying $\alpha^\vee(\eta) \leq r$. Set $W^r(N) = N \cap W^r(M)$. Set $B(M) = \tilde{\mathcal{F}}_\alpha(B(N))$ which as noted in 3.5 is a union of α-strings. Define $W^r(B(M))$ as in 3.5 and set $W^r(B(N)) = B(N) \cap W^r(B(M))$.

Proposition. *For all $r \in \mathbb{N}$ one has*

$$W^r(N) = \bigoplus_{b \in W^r(B(N))} AG(b) \tag{i}$$

$$W^r(M) = \bigoplus_{b \in W^r(B(M))} AG(b). \tag{ii}$$

Proof. These are proved by decreasing induction on r. Since $M = \mathcal{F}_\alpha N$, it follows that its highest weight space M_η equals N_η. Also $B(M)_\eta = B(N)_\eta$ by 3.7. Since G respects weights, the hypothesis 3.7(∗) on N implies that

$$M_\eta = \bigoplus_{b \in B(M)_\eta} AG(b).$$

Since $eM_\eta = 0$, $f^{(n)}$ and $\tilde{f}^n$ coincide on M_η. Hence by 3.3(i) we obtain

$$W^s(M) = \mathcal{F}_\alpha M_\eta = \bigoplus_{b \in \tilde{\mathcal{F}}_\alpha(B(M)_\eta)} AG(b). \tag{$*$}$$

Moreover each $b \in \tilde{\mathcal{F}}_\alpha(B(M)_\eta) \subset B(M)$ lies in an α-string of length s and so $\tilde{\mathcal{F}}_\alpha(B(M)_\eta) \subset W^s(B(M))$. By the hypothesis 3.7(∗) on N and (∗) we further obtain

$$W^s(N) = N \cap W^s(M) = \bigoplus_{b \in \tilde{\mathcal{F}}_\alpha(B(M)_\eta) \cap B(N)} AG(b). \tag{$**$}$$

Set $N' = N/W^s(N)$, $M' = M/W^s(M)$. Then N' is an $\mathcal{E}_\alpha$ submodule of M' satisfying $M' = \mathcal{F}_\alpha N'$. In turn M' is a $U_q^{\mathbb{Z}}(\mathfrak{s}_\alpha)$ submodule of $W_{s-1}(V^{\mathbb{Z}}_{\xi+\mathbb{Z}\alpha})$. By the hypothesis on N and (∗∗) we may write

$$N' = \bigoplus_{b \in B(N')} AG(b),$$

where $B(N') = B(N) \setminus \tilde{\mathcal{F}}_\alpha(B(M)_\eta)$. In particular, $B(N')_\eta = \emptyset$. Setting $B(M') = \tilde{\mathcal{F}}_\alpha B(N')$, we obtain $B(M')_\eta = \emptyset$. Since $B(M') \subset \tilde{\mathcal{F}}_\alpha B(N) \subset B_{\xi+\mathbb{Z}\alpha}$, every α-string of length s in $B(M')$ must have an element of weight η. This forces $\tilde{\mathcal{F}}_\alpha(B(M)_\eta) = W^s(B(M))$. Consequently (i), (ii) are proved for $r = s$. Repeating the argument with N' which is an $\mathcal{E}_\alpha$ submodule of $W_{s-1}(V^{\mathbb{Z}}_{\xi+\mathbb{Z}\alpha})$ gives the general case. □

3.9.

Take $b \in B_\xi(\mu)$ and let $N_b(\mu)$, or simply N_b or N, be the smallest $\mathcal{E}_\alpha$ submodule of $V^{\mathbb{Z}}_{\xi+\mathbb{Z}\alpha}$ containing $G_\mu(b)$ and admitting a global basis. Thus N satisfies the hypotheses of 3.7 and we set $M_b(\mu) = U_q^{\mathbb{Z}}(\mathfrak{s}_\alpha)N_b(\mu)$, or simply M_b or M. From 3.7 and 3.8 we obtain

$$B(N) \supset \tilde{\mathcal{E}}_\alpha b \cup \{b' \in B(M) | \varepsilon_\alpha(b') = 0\}. \tag{$*$}$$

Unfortunately little else seems to be known, although it is an interesting and perhaps important problem to determine $B(N)$. Indeed the technical difficulty of our main result would be significantly reduced if we could prove that

$$B(N) \supset \{b' \in B(M) | \varepsilon_\alpha(b') \leq \varepsilon_\alpha(b)\}. \tag{$**$}$$

Indeed through the string property (4.7, 4.8) the conclusion of 5.11 would become almost immediate. Such a result is very natural from the point of view of Littelmann's monomial basis (which approximates the global basis [L5, Prop. 10.4], [G, Sects. 4,5]). Unfortunately($**$) is false. For example take $\pi = \{\alpha, \beta\}$ of type A_2. Then by [Lu3], every global basis element for $U_q^{\mathbb{Z}}(\mathfrak{n}^-)$ is a monomial and so a Littelmann monomial for *some choice of reduced decomposition.* Given $\lambda \in P^+$, let b_λ be the unique element of $B(\lambda)$ of weight λ. For simplicity of notation we write $\tilde{f} b_\lambda : \tilde{f} \in \tilde{\mathcal{F}}$ simply as $\tilde{f}$, and $f u_\lambda : f \in \mathcal{F}$ as f. Then for example $G(\tilde{f}_\beta \tilde{f}_\alpha^n) = f_\beta f_\alpha^{(n)}$, $\forall\, n \in \mathbb{N}$. (This may also be deduced from 3.5 and [J2, 6.2.9].) Now $\tilde{e}_\alpha(\tilde{f}_\beta \tilde{f}_\alpha^n) = \tilde{f}_\beta \tilde{f}_\alpha^{n-1}$, for all $n \geq 2$, whilst $\tilde{e}_\alpha(\tilde{f}_\beta \tilde{f}_\alpha) = 0$.

Now take $\lambda \in P^+(\pi)$ with $s := \alpha^\vee(\lambda)$, $t := \beta^\vee(\lambda)$ sufficiently large. Take $b = \tilde{f}_\beta \tilde{f}_\alpha^n : s \geq n \geq 1$ in the above. Then

$$B(N) = \{\tilde{f}_\beta \tilde{f}_\alpha^m : m \leq n\},$$

whilst $B(M)$ consists of the α-string $\tilde{f}_\beta \tilde{f}_\alpha^i : 1 \leq i \leq s$ and the α-string $\tilde{f}_\alpha^i \tilde{f}_\beta : 0 \leq i \leq (s+1)$. From this we see that equality holds in ($*$). This precludes ($**$) if $n \geq 2$.

For a second example we note that $G_\lambda(\tilde{f}_\alpha \tilde{f}_\beta^2 \tilde{f}_\alpha) = f_\alpha f_\beta^{(2)} f_\alpha u_\lambda$. In this case taking $b = \tilde{f}_\alpha \tilde{f}_\beta^2 \tilde{f}_\alpha$ we obtain

$$B(N) = \{\tilde{f}_\alpha \tilde{f}_\beta^2 \tilde{f}_\alpha, \tilde{f}_\beta^2 \tilde{f}_\alpha\} \cup \{\tilde{f}_\alpha \tilde{f}_\beta^2, \tilde{f}_\beta^2\}, \tag{$***$}$$

whilst $B(M)$ consists of the α-strings generated by $b_1 = \tilde{f}_\beta^2 \tilde{f}_\alpha$, $b_2 = \tilde{f}_\beta^2$ (here both elements are annihilated by $\tilde{e}_\alpha$). Thus in this case ($**$) holds. Curiously for $n \geq 2$ one also has $G_\lambda(\tilde{f}_\beta \tilde{f}_\alpha^n \tilde{f}_\beta) = f_\beta f_\alpha^{(n)} f_\beta u_\lambda$. Thus $e_\alpha G_\lambda(\tilde{f}_\beta \tilde{f}_\alpha^n \tilde{f}_\beta) = [\alpha^\vee(\lambda) - (n-2)]_{q_\alpha} G_\lambda(\tilde{f}_\beta \tilde{f}_\alpha^{n-1} \tilde{f}_\beta)$, for $n > 2$, so there is a contribution to the longer string just for $n = 1, 2$.

One may ask if this is a general phenomenon. We say that the global basis satisfies the no-gap hypothesis if whenever $G(b')$ does not appear in the expansion of $e_\alpha G(b)$, then for all $i \in \mathbb{N}$, $G(\tilde{f}_\alpha^i b')$ does not appear in the expansion of $e_\alpha G(\tilde{f}_\alpha^i b)$. Such a property would be enough to extend our theorem (5.22) to the general case (see Remark 1 of 5.18).

One may remark that the essence of the difficulty in passing from the crystal to the global basis (both of which are canonical) is that there are more monomial identities between the crystal operators than between the divided powers. By contrast the latter admit, also non-monomial identities (for example Serre relations). Consequently the relationship between the monomials formed from the crystal operators and (the sums of) monomials of divided powers describing the corresponding global basis element must get increasingly complex. This makes the proof of our main result more difficult.

3.10.

Set $n = \varepsilon_\alpha(b)$ and $b_0 = \tilde{e}_\alpha^n b$, which is just the unique highest element in the α-string S_0 containing b. Obviously $N_{b_0} \subset N_b$ and the inclusion is strict unless $n = 0$. Recall 3.8.

Lemma. $B(M_{b_0}) \coprod \{0\} = \tilde{\mathcal{F}}_\alpha B(N_{b_0}) = \tilde{\mathcal{F}}_\alpha B(N_b) = B(M_b) \coprod \{0\}$.

Proof. Suppose the inclusion $B(M_{b_0}) \subset B(M_b)$ is strict. Let S be an α-string of minimal length of $B(M_b)$ not occurring in $B(M_{b_0})$. Recalling 3.8, factor M_b (resp. M_{b_0}) by the submodule defined by all α-strings of length $\geq \ell(S)$ excluding S. This reduces us to the case $B(M_b) = B(M_{b_0}) \coprod S$. Write $B(M_{b_0})$ as a disjoint union of α-strings $S_i : i = 0, 1, 2, \ldots, m$ of increasing length, and let b_i be the unique highest element of S_i. Then by construction $e_\alpha G(b_m) = 0$, so by 3.3(i) the A submodule generated by the $G(b) : b \in S_m$ is a submodule of both M_{b_0} and M_b. Factoring out by this submodule and continuing in this fashion a contradiction is reached. □

3.11.

Fix $b \in B_\xi$ and let N_b^-, or simply N^-, be the smallest $\mathcal{F}_\alpha$ submodule of $V_{\xi+\mathbb{Z}\alpha}^{\mathbb{Z}}$ containing $G(b)$ and admitting a global basis. The symmetry with respect to $\tilde{e}, \tilde{f}$ (resp. e, f) interchange as described in 3.2–3.4 implies results analogous to 3.7 and 3.8. In particular, $B(N^-) \cup \{0\}$ is $\tilde{\mathcal{F}}_\alpha$ stable and $M^- := U_q^{\mathbb{Z}}(\mathfrak{s}_\alpha)N^-$ admits a global basis parametrized by $B(M^-) := \tilde{\mathcal{E}}_\alpha B(N^-) \setminus \{0\}$. Now set $n' = \varphi_\alpha(b)$ and $b_0' = \tilde{f}_\alpha^{n'} b$, which is just the unique lowest weight element in the α-string S_0 containing b. As in 3.10 we obtain $B(M_{b_0'}^-) = B(M_b^-)$.

Lemma. $B(M_{b_0'}^-) = B(M_{b_0})$, *equivalently* $M_b^- = M_b$.

Proof. Let S be an α-string of minimal length which is either not in $B(M_{b_0'}^-)$ or not in $B(M_{b_0})$. As in 3.10, factor $M_{b_0'}^-$ (resp. M_{b_0}) by the submodule defined by all α-strings of length $\geq \ell(S)$ excluding S. This reduces us to the case $B(M_{b_0}) = B(M_{b_0'}) \coprod S$ (or vice versa). Using 3.3 the proof is completed exactly as in 3.10. □

3.12.

To overcome the difficulty evoked in 3.9 we shall use the compatibility of the global basis obtained by varying $\mu \in P^+(\pi)$. We recall this below.

Define an order relation on $P^+(\pi)$, through $\mu \succ \nu$ if $\mu - \nu \in P^+(\pi)$. Suppose $\mu' \succ \mu$. Then there is a surjection $\pi_{\mu',\mu} : V^{\mathbb{Z}}(\mu') \to V^{\mathbb{Z}}(\mu)$ of $\mathcal{F} = U_q^{\mathbb{Z}}(\mathfrak{n}^-)$ modules defined by $\pi_{\mu',\mu}(u_{\mu'}) = u_\mu$.

A deep result of Kashiwara is that $\pi_{\mu',\mu}$ is compatible with the respective global bases. (For a Cartan datum of finite type Lusztig [Lu1] previously obtained this result for the canonical basis). More precisely one may recall (see 4.3) that $B(\mu)$ can be viewed as a subset of $B(\mu')$. We use the convention that $G_\mu(b) = 0$ if $b \notin B(\mu)$. (This is justified by the fact that $G_{\mu'}(b)u_\mu = 0$.) Then by [J2, 6.2.9] (which is eventually shown to be valid for all k defined there) one has

$$\pi_{\mu',\mu}(G_{\mu'}(b)) = G_\mu(b), \ \forall\, b \in B(\mu'). \tag{$*$}$$

In $(*)$ we may also take $\mu' = \infty$, where $G_\infty(b) : b \in B(\infty)$ is the global basis for $U_q^{\mathbb{Z}}(\mathfrak{n}^-)$ with $B(\infty)$ being the direct limit of the $B(\mu) : \mu \in P^+(\pi)$ (see 4.3) and where $\pi_{\infty,\mu}$ is the $\mathcal{F}$ module map of $V^{\mathbb{Z}}(\infty) := U_q^{\mathbb{Z}}(\mathfrak{n}^-)$ onto $V^{\mathbb{Z}}(\mu)$ defined by $\pi_{\infty,\mu}(1) = u_\mu$. This means in particular that $G_\mu(b) = \pi_{\infty,\mu}(G_\infty(b)) = G_\infty(b)\pi_{\infty,\mu}(1) = G_\infty(b)u_\mu$.

By $(*)$ the expansion $fG_\mu(b) : b \in B(\mu), f \in \mathcal{F}$ as a sum (over A) of the $G_\mu(b')$, is independent of μ (except that some $G_\mu(b')$ are zero when $b \notin B(\mu)$) and is determined by the expansion of $fG_\infty(b) : b \in B(\infty), f \in \mathcal{F}$. This is not quite true of the expansion of $eG_\mu(b) : b \in B(\mu), e \in \mathcal{E}$ (and in particular for the $e_\alpha : \alpha \in \pi$) since we must eventually apply the $(t_\alpha - t_\alpha^{-1})/(q_\alpha - q_\alpha^{-1})$ factors to u_μ. Here if μ' is replaced by μ, the weights will be shifted by $\mu' - \mu$. This affects the coefficient, but not whether a given term appears (up to some accidental zeros). More precisely, if $G_\mu(b')$ occurs with a non-zero coefficient in $e_\alpha G_\mu(b)$, then this also holds for all $\mu' \succ \mu$ outside a Zariski closed subset. This means that if we take $b \in B_\xi(\mu)$ and define $N_b(\mu)$, or simply $N(\mu)$, as in 3.9, then the subset $B(N(\mu))$ of $B_{\xi+\mathbb{Z}\alpha}(\mu)$ it defines can be assumed increasing in μ — more precisely $B(N(\mu')) \supset B(N(\mu))$ for all $\mu' \succ \mu$ outside a Zariski closed subset. (Remarkably by 3.10, 3.11, this last proviso can be omitted.) Moreover we can write $\xi = \mu - \gamma$, for some $\gamma \in \mathbb{N}\pi$ and whilst $G_\mu(b)$ has weight ξ, $G_{\mu'}(b)$ has weight $\mu' - \gamma = \mu' - \mu + \xi$. Thus $|B(N(\mu'))|$ will be uniformly bounded from above. Precisely we have

$$|B(N(\mu'))| \leq \sum_{n\in\mathbb{N}} \dim\, U_q(\mathfrak{n}^-)_{-\gamma+n\alpha} < \infty \tag{$**$}$$

for all $\mu' \succ \mu$. In particular, the number of α-strings in M will reach a (finite) maximum (as a function of $\mu' \succ \mu$).

3.13.

Although the expansion of $eG_\mu(b)$ evoked in 3.12 depends on μ, this dependence vanishes on passing to the corresponding crystal. More precisely, the embedding of $B(\mu)$ in $B(\mu')$ (for $\mu \prec \mu'$) is, up to a shift of weights, a full embedding (in the language of Kashiwara [K2]), that is it commutes with the action of $\tilde{\mathcal{E}}$. (For more details see 4.3). Consequently, the elements of a given α-string in $B(M(\mu))$ with $M(\mu)$ defined as in 3.8, satisfying $\varepsilon_\alpha(b) = 0$, is independent of μ though some new α-strings may appear as μ increases, up to some finite bound.

The above may be illustrated by the second example of 3.9. Let ρ be the sum of the fundamental weights $\varpi_\alpha : \alpha \in \pi$. Then $B(N(\rho))$ is reduced to $\{\tilde{f}_\alpha \tilde{f}_\beta^2 \tilde{f}_\alpha, \tilde{f}_\beta^2 \tilde{f}_\alpha\}$ which is a complete α-string. On the other hand $B(N(2\rho))$ has the maximal possible size given in 3.9($***$), so in particular $B(N(2\rho)) = B(N(\mu))$, for all $\mu \succeq 2\rho$. The corresponding α-strings have lengths $\alpha^\vee(\mu - \alpha - 2\beta) = \alpha^\vee(\mu)$ and $\alpha^\vee(\mu) + 2$, which increases with μ. However the highest weight elements of the strings, namely $\tilde{f}_\beta^2 \tilde{f}_\alpha, \tilde{f}_\beta^2$, remain fixed.

We summarize the above conclusions in the following.

Proposition. *Fix $b \in B(\mu)$, $\mu \in P^+$ and $\alpha \in \pi$. Take $\mu' \in \mu + \mathbb{N}\varpi_\alpha$ and let $N(\mu')$ be the smallest $\mathcal{E}_\alpha$ stable submodule of $V^{\mathbb{Z}}(\mu')$ containing $G_{\mu'}(b)$ and admitting a global basis.*

(i) There exists a finite set $F \subset \mathbb{N}$ such that $B(N(\mu')) : \mu' \in \mu + (\mathbb{N} \setminus F)\varpi_\alpha$ is independent of μ' and contains $B(N(\mu))$.
(ii) $M(\mu') := \mathcal{F}_\alpha(N(\mu'))$ admits a global basis. Thus $B(M(\mu'))$ is defined, equals $\tilde{\mathcal{F}}_\alpha B(N(\mu'))$ and consists of finitely many α-strings, this number being independent of $\mu' \in \mu + (\mathbb{N} \setminus F)\varpi_\alpha$.
(iii) The length of each α-string in $B(M(\mu'))$ increases by n on passing to $B(M(\mu' + n\varpi_\alpha))$.
(iv) For each $b' \in B(M(\mu')) : \mu' \in \mu + \mathbb{N}\varpi_\alpha$ the decomposition of $f_\alpha G_{\mu'}(b')$ as a linear combination of the $G_{\mu'}(b'') : b'' \in B(M(\mu'))$ is the same as in the decomposition of $f_\alpha G_{\mu''}(b') : \mu'' \in \mu + \mathbb{N}\varpi_\alpha$ except that a given term $G_{\mu'}(b'')$ is equal to zero if $b'' \notin B(\mu')$.
(v) $B(N(\mu + n\varpi_\alpha))$ is increasing in n.

Proof. Clearly $B(N(\mu'))$ is just the set of all $G_{\mu'}(b')$ occurring in the decomposition of the $e_\alpha^{(n)} G_{\mu'}(b) : n \in \mathbb{N}$. Every such term can be non-zero only if $U_q(\mathfrak{n}^-)_{-\gamma+n\alpha} \neq 0$, where $\gamma = -wt\, b$ with b viewed as an element of $B(\infty)$. In particular, n is uniformly bounded by some $m \in \mathbb{N}$ from above. Since $G_{\mu'}(b') = G_\infty(b')u_{\mu'}$ the dependence of $e_\alpha^{(n)} G_{\mu'}(b')$ on μ' comes only through the evaluation of the $(t_\alpha - t_\alpha^{-1})/(q_\alpha - q_\alpha^{-1})$ on u_μ. If $G_\mu(b)$ involves k factors in the $f_\alpha^{(s)} : s \in \mathbb{N}$, the number of such evaluations is at most km. Moreover the evaluation depends linearly on $\mu' \in \mu + \mathbb{N}\varpi_\alpha$ and so they are all non-zero outside a finite set F. Finally the $B(\mu') : \mu' = \mu + r\varpi_\alpha : r \in \mathbb{N}$ are increasing in r whilst $\bigcup_{n\in\mathbb{N}} B(\mu')_{\mu'-\gamma+n\alpha} \subset \bigcup_{n\leq m} B(\infty)_{-\gamma+n\alpha}$ and hence the latter become stationary for r sufficiently large. All this establishes (i).

The first part of (ii) follows from 3.8. The second part from (i).

If $b' \in B(M(\mu'))$ satisfies $\varepsilon_\alpha(b') = 0$, then it generates an α-string of length $\alpha^\vee(wt\, b')$, where b' is viewed as an element of $B(\mu')$. Viewed as an element of $B(\mu' + n\varpi_\alpha)$, its weight increases by $n\varpi_\alpha$. Hence (iii).

(iv) follows from 3.10($*$) since $\pi_{\mu',\mu}$ commutes with $\mathcal{F}_\alpha$.

(v) follows from 3.10, 3.11 and the independence of the expansion of $fG_{\mu'}(b)$ on $\mu' \in \mu + \mathbb{N}\varpi$. □

Remark 1. One may give some credence to the no-gap hypothesis of 3.9. Suppose that $B(N(\mu))$ consists of just two strings S, S' with $\ell(S') = \ell(S) + 2$ (as in the second example of 3.9). Let s_{-n} (resp. $s'_{-(n+2)}$ be the lowest weight element of S (resp. S') and suppose that $fG_\mu(s_{-n})$ is a non-zero multiple of $G_\mu(s'_{-(n+2)})$. Now replace μ by

$\mu + m\varpi_\alpha : m = 1, 2, \ldots$, and let $s_{-(n+m)}$ and $s'_{-(n+m+2)}$ be the corresponding lowest weight elements in the lengthened strings (see (iii) above). By 3.12(∗) the expansion of $fG_{\mu+\varpi_\alpha}(s_{-n})$ still contains a non-zero multiple of $G_{\mu+\varpi_\alpha}(s'_{-(n+3)})$ and by 3.10 this forces $fG_{\mu+\varpi_\alpha}(s_{-(n+1)})$ to be a non-zero multiple of $G_{\mu+\varpi_\alpha}(s'_{-(n+4)})$. Repeating this argument it follows that the expansion of $fG_{\mu+m\varpi_\alpha}(s_{-(n+i)})$ contain a non-zero multiple of $G_{\mu+m\varpi_\alpha}(s'_{-(n+2+i)})$ for all m and all $i : 0 \leq i \leq m$.

There are three difficulties in deducing the no-gap hypothesis from this argument. First, it can break down even when there are just three strings. More seriously it only proves the no-gap hypothesis asymptotically. Finally it concerns f not e. This is trivially overcome if $\mathfrak{g}$ is finite dimensional by noting that highest weight modules are also lowest weight modules and using the Chevalley anti-automorphism. (Observe however that a given family of strings is not e, f interchange invariant; but one family gets mapped to a second one. This is again evidenced by the second example of 3.9.)

Remark 2. With respect to (v), the new α-strings obtained on passing from $\mu + m\varpi_\alpha$ to $\mu + (m+1)\varpi_\alpha$ cannot occur in an arbitrary fashion. Indeed by 3.11 and 3.12(∗) the global basis vectors they define must form a $U_q^{\mathbb{Z}}(\mathfrak{s}_\alpha)$ submodule.

3.14.

Fix $\alpha \in \pi$, $\mu \in P^+$ and take b, N, M as in 3.9. Then $B(M)$ consists of finitely many α-strings of which there is just one S of shortest length (by 3.5) and this contains b. Let $\mathcal{S}'$ be any union of α-strings of $B(M)$ excluding S.

Lemma ($\mathfrak{g}$ simply-laced). *Fix $n \in \mathbb{N}^+$. Suppose that $b', b'' := \tilde{e}_\alpha^n b' \in S$ and that*

$$e_\alpha^n G_\mu(b') \in K^* G_\mu(b'') + \sum_{b''' \in \mathcal{S}'} K G_\mu(b''').$$

Then

$$e_\alpha^i G_\mu(b') \in K^* G_\mu(\tilde{e}_\alpha^i b') + \sum_{b''' \in \mathcal{S}'} K G_\mu(b'''),$$

for all $i \leq n$.

Proof. Since $\mathfrak{g}$ is assumed simply-laced the positivity result [Lu2, Thm. 22.1.7] of Lusztig applies. This asserts that $e_\alpha G_\mu(c) : c \in B(\mu)$ is a linear combination of the $q^n G_\mu(c') : n \in \mathbb{Z},\ c' \in B(\mu)$ with non-negative integer coefficients. This means that no cancellations can occur. Thus if a term, say $G_\mu(b''')$, in a string different to one in $\mathcal{S}'$, appears in $e_\alpha^i G_\mu(b')$, then $G_\mu(\tilde{e}_\alpha^{n-i} b''')$ must occur in $e_\alpha^n G_\mu(b')$ because by 3.5(ii) it occurs in $e_\alpha^{n-i} G(b''')$. On the other hand $\tilde{e}_\alpha^{n-i} b'''$ belongs to the same α-string as b'''. Moreover it cannot be zero, since $\tilde{e}_\alpha^n b' = b'' \neq 0$ and because S is the shortest α-string in $B(M)$. This contradiction proves the lemma. □

4. The combinatorics of Demazure crystals

4.1.

Let π (resp. $\pi^\vee$) denote the set of simple roots (resp. coroots) of a given symmetrizable Kac–Moody algebra.

Recall that a normal crystal B is a set with maps $wt : B \to P, \varepsilon_\alpha, \varphi_\alpha : B \to \mathbb{Z}, \tilde{e}_\alpha, \tilde{f}_\alpha : B \to B \cup \{0\} : \alpha \in \pi$, satisfying

(1) $\varepsilon_\alpha(b) = \max\{n | \tilde{e}_\alpha^n b \neq 0\}, \varphi_\alpha(b) = \max\{n | \tilde{f}_\alpha^n b \neq 0\}, \ \forall\, \alpha \in \pi, b \in B$
(2) $\varphi_\alpha(b) - \varepsilon_\alpha(b) = \alpha^\vee(wtb), \ \forall\, \beta \in B$
(3) $b' = \tilde{e}_\alpha b \Longleftrightarrow b = \tilde{f}_\alpha b', \ \forall\, b, b' \in B$
(4) $\tilde{e}_\alpha b \neq 0 \Longrightarrow wt\tilde{e}_\alpha b = wtb + \alpha, \ \forall\, \alpha \in \pi, b \in B.$

For example a crystal basis $B(\lambda)$ of a simple integral module has the structure of a normal crystal with respect to the Kashiwara operators $\tilde{e}_\alpha, \tilde{f}_\alpha : \alpha \in \pi$. In an upper normal crystal, part two of (1) is replaced by $\varphi_\alpha(\tilde{e}_\alpha b) = \varphi_\alpha(b) + 1$ if $\tilde{e}_\alpha b \neq 0$.

4.2.

Let B be a normal crystal and define $\tilde{\mathcal{E}}, \tilde{\mathcal{F}}$ as in 2.3. A crystal is said to be of highest weight $\lambda \in P$ if there exists an element $b_\lambda \in B$ of weight λ such that $\tilde{\mathcal{E}} b_\lambda = 0$, $B = \tilde{\mathcal{F}} b_\lambda$. Assuming B normal forces $\lambda \in P^+$. Not unexpectedly $B(\lambda)$ is a highest weight crystal.

4.3.

It is possible to form the tensor product $B_1 \otimes B_2$ of two crystals B_1, B_2. It is $B_1 \times B_2$ as a set and satisfies in particular (writing $b_1 \otimes b_2$ for (b_1, b_2))

$$\tilde{e}_\alpha(b_1 \otimes b_2) = \begin{cases} \tilde{e}_\alpha b_1 \otimes b_2, & \text{if } \varphi_\alpha(b_1) \geq \varepsilon_\alpha(b_2) \\ b_1 \otimes \tilde{e}_\alpha b_2, & \text{otherwise.} \end{cases}$$

$$\tilde{f}_\alpha(b_1 \otimes b_2) = \begin{cases} \tilde{f}_\alpha b_1 \otimes b_2, & \text{if } \varphi_\alpha(b_1) > \varepsilon_\alpha(b_2) \\ b_1 \otimes \tilde{f}_\alpha b_2, & \text{otherwise.} \end{cases}$$

for all $\alpha \in \pi, \ b_1 \in B_1, b_2 \in B_2$.

In $B(\lambda) \otimes B(\mu)$, the element $b_\lambda \otimes b_\mu$ satisfies $\tilde{\mathcal{E}}(b_\lambda \otimes b_\mu) = 0$, by the above rules. It is true but not obvious that $\tilde{\mathcal{F}}(b_\lambda \otimes b_\mu)$ is a crystal and further is isomorphic to $B(\lambda + \mu)$. This property and normality characterizes the family $B(\lambda) : \lambda \in P^+$, up to isomorphism [J2, Thm. 6.4.21]. Then even more remarkably one has

$$B(\lambda) \otimes B(\mu) = \coprod_{\nu \in S_{\lambda,\mu}} B(\nu)$$

for some multisubset $S_{\lambda,\mu}$ of P^+. This follows from Kashiwara's deep work on the $q \to 0$ limit; but Littelmann has also given a path model construction of this (unique) family of normal crystals and a (difficult) combinatorial proof [L2] of the above decomposition. An exposition of this latter proof is given in [J5, Sect. 15] where it is noted to be valid even in the non-symmetrizable case.

Recall the order relation on P^+ defined through $\mu' \succ \mu \implies \mu' - \mu \in P^+$. Kashiwara introduced a highest weight (upper normal) crystal $B(\infty)$ generated over $\mathcal{F}$ by an element b_∞ of weight 0. Moreover [J2, 5.3.13] there exists a one element crystal $S_\lambda = \{s_\lambda\}$ and a strict crystal embedding $\psi_\lambda : B(\lambda) \to B(\infty) \otimes S_\lambda$ given by $\psi_\lambda(\tilde{f} b_\lambda) = \tilde{f}(b_\infty \otimes s_\lambda)$, $\forall\, \tilde{f} \in \tilde{\mathcal{F}}$. Whilst $\tilde{\mathcal{F}}$ acts injectively on $B(\infty)$, it cannot do so on $B(\lambda)$. Indeed eventually $\tilde{f}_\alpha(\tilde{f} b_\infty \otimes s_\lambda) = \tilde{f} b_\infty \otimes \tilde{f}_\alpha s_\lambda = 0$, which by the crystal rules exactly occurs when $\varphi_\alpha(\tilde{f} b_\infty) \leq \varepsilon_\alpha(s_\lambda) := -\alpha^\vee(\lambda)$. Suppose $\mu' \succ \mu$. Then $-\alpha^\vee(\mu') \leq -\alpha^\vee(\mu)$, for all $\alpha \in \pi$. It follows that $b \otimes s_{\mu'} \in \operatorname{Im} \psi_{\mu'}$ for all $b \in B(\mu)$ and so this construction embeds $B(\mu)$ as a subcrystal of $B(\mu')$, up to a shift of weights. This embedding commutes with $\tilde{e}_\alpha, \varepsilon_\alpha : \alpha \in \pi$ and with $\varphi_\alpha : \alpha \in \pi$, up to a shift imposed by 4.1(2). It does not quite (see 4.1(3)) commute with $\tilde{f}_\alpha : \alpha \in \pi$. Moreover $B(\infty)$ is just the direct limit of the $B(\mu) : \mu \in P^+$ with respect to the above order relation. (This is exactly analogous to the dual Verma module $\delta M(0)$ being a direct limit of the images of the $V(\mu)$ under the (unique) $U_q(\mathfrak{n}^+)$ module map sending u_μ to a fixed zero weight vector in $\delta M(0)$).

4.4.

For every $\nu \in S_{\lambda,\mu}$, the unique element of $B(\nu)$ in $B(\lambda) \otimes B(\mu)$ of weight ν takes the form $b_\lambda \otimes b$, for some $b \in B(\mu)_{\nu-\lambda}$ (this is easy!). We showed in [J4, Sect. 5] how to compute b in terms of the embedding $B(\lambda) \hookrightarrow B(\infty) \otimes S_\lambda$ and Kashiwara's involution $\star$ on $B(\infty)$. More precisely call $b_\lambda \otimes b \in B(\lambda) \otimes B(\mu)$ primitive if $\tilde{\mathcal{E}}(b_\lambda \otimes b) = 0$. Then every such primitive element $b_\lambda \otimes b$ satisfies $\psi_\mu(b) = b'^\star \otimes s_\mu$, for some $b' \otimes s_\lambda \in \operatorname{Im} \psi_\lambda$. More simply, although less precisely, $b_\lambda \otimes b$ is primitive if and only if $b \in B(\lambda)^\star \cap B(\mu)$. If $b \in B(\lambda)^\star$, but $b \notin B(\mu)$ we use the convention that $b = 0$ and in this sense the set of primitive elements $b_\lambda \otimes b \in B(\lambda) \otimes B(\mu)$ depends only on λ.

4.5.

Define $\tilde{\mathcal{E}}_\alpha, \tilde{\mathcal{F}}_\alpha : \alpha \in \pi$ as in 3.6. Obviously $\tilde{\mathcal{E}}_\alpha^2 = \tilde{\mathcal{E}}_\alpha$ and $\tilde{\mathcal{F}}_\alpha^2 = \tilde{\mathcal{F}}_\alpha$. As already noted by Kashiwara (see also [J4, 4.2]) the $\tilde{\mathcal{E}}_\alpha$ (resp. $\tilde{\mathcal{F}}_\alpha$) $: \alpha \in \pi$ as monoids further satisfy the Coxeter relations on elements of the (unique) family $B(\lambda) : \lambda \in P^+$. For example $\tilde{\mathcal{E}}_\alpha \tilde{\mathcal{E}}_\beta \tilde{\mathcal{E}}_\alpha = \tilde{\mathcal{E}}_\beta \tilde{\mathcal{E}}_\alpha \tilde{\mathcal{E}}_\beta$ if $\{\alpha, \beta\}$ is of type A_2. A simple proof of this fact using the Littelmann path model may be found in [J5, 16.15]. Let W denote the Weyl group (generated by the $s_\alpha : \alpha \in \pi$). Thus for each $w \in W$, the subset $\tilde{\mathcal{E}}_w$ of $\tilde{\mathcal{E}}$ (resp. $\tilde{\mathcal{F}}_w$ of $\tilde{\mathcal{F}}$) is defined (on the above family) by taking a reduced decomposition and setting $\tilde{\mathcal{E}}_e = \tilde{\mathcal{F}}_e = \{1\}$.

With respect to $b_\lambda \in B(\lambda)$, $w \in W$ a Demazure "crystal" $B_w(\lambda)$ is defined to be $\tilde{\mathcal{F}}_w b_\lambda$. It is a subset of $B(\lambda)$ and nearly a crystal. Indeed $B_w(\lambda)$ is $\tilde{\mathcal{E}}$ stable and admits what we call the string property, namely for all $b \in B_w(\lambda)$ and all $\alpha \in \pi$ such that $\varepsilon_\alpha(b) > 0$ one has $\tilde{\mathcal{F}}_\alpha b \subset B_w(\lambda)$. Remarkably this holds regardless of whether $s_\alpha w < w$; but in this case $B_w(\lambda)$ itself is $\tilde{\mathcal{F}}_\alpha$ stable. This string property is the basis of the Demazure character formula and as we shall see a key combinatorial component of the existence of Demazure flags.

4.6.

In [L3, Thm. 4], see also [LLM, 2.4], or in [J4, 2.11] it was shown for all $\lambda, \mu \in P^+$, $w \in W$, there exists a finite set I such that

$$b_\lambda \otimes B_w(\mu) = \coprod_{i \in I} B_{y_i}(\nu_i) \qquad (*)$$

for some $y_i \in W$, $\nu_i \in P^+$. More precisely, the $B_{y_i}(\nu_l)$ are subsets of $b_\lambda \otimes B_w(\mu)$ of the form $\tilde{\mathcal{F}}_{y_i}(b_\lambda \otimes b_i)$, where $\tilde{\mathcal{E}}(b_\lambda \otimes b_i) = 0$ (and so $\tilde{\mathcal{F}}(b_\lambda \otimes b_i)$ is the subcrystal $B(\nu_i)$ of $B(\lambda) \otimes B(\mu)$). In [J4, 4.4] we described the y_i explicitly. A crucial point in the proof of the above result is that setting $\tilde{\mathcal{F}}^\lambda_{w,b_i} = \{\tilde{f} \in \tilde{\mathcal{F}} | \tilde{f}(b_\lambda \otimes b_i) \subset b_\lambda \otimes B_w(\mu)\}$ one has

$$\tilde{\mathcal{F}}^\lambda_{w,b_i} \supset \tilde{\mathcal{F}}_{y_i} \text{ and } \tilde{\mathcal{F}}^\lambda_{w,b_i}(b_\lambda \otimes b_i) = \tilde{\mathcal{F}}_{y_i}(b_\lambda \otimes b_i). \qquad (**)$$

(This corrects slightly [J4, 4.4] and is the assertion actually proved there. See also [J5, Sect. 19 and in particular Remark 2 of 19.3].)
Moreover we note (and this will be important in varying μ) that y_i is independent of μ, depending only on b_i. This independence follows from [J4, 3.4].

In particular, suppose we fix i, write $y = y_i$, take a reduced decomposition $y = s_{\alpha_1} s_{\alpha_2} \cdots s_{\alpha_k}$ and set $\alpha = \alpha_j$, $z_j = s_{\alpha_{j+1}} s_{\alpha_{j+2}} \cdots s_{\alpha_k}$, $z_k = 1$. Then $\tilde{\mathcal{F}}_{z_j}(b_\lambda \otimes b_i) = b_\lambda \otimes \tilde{\mathcal{F}}_{z_j} b_i$. Furthermore for each $b \in \tilde{\mathcal{F}}_{z_j} b_i$, we have $\tilde{f}_\alpha(b_\lambda \otimes b) = b_\lambda \otimes \tilde{f}_\alpha b$ and so $\varepsilon_\alpha(b) \geq \alpha^\vee(\lambda)$, by 4.3. (As one might expect it is rather that one first proves this inequality and deduces $(*)$. This is achieved in [J4, 4.4]). The above inequality will play a key role in our analysis.

We remark that I is finite even under the convention of 4.4, since $B(\lambda)^\star \cap B_w(\infty)$ is finite [J4, last paragraph of 5.3]. If $B_w(\mu)$ is replaced by its limiting value $B(\mu)$, then I is still countable. This countable set can be identified with $\mathbb{N}^+$.

4.7.

In the notation of 4.6 we set $B_i^- = \tilde{\mathcal{F}}_{y_i} b_i$ (resp. $B_i^-(\mu) = \tilde{\mathcal{F}}_{y_i} b_i$) when b_i is viewed as an element of $B(\infty)$ (resp. $B(\mu)$). In view of [J4, 3.4] one has $B_i^-(\mu) = B(\mu) \cap B_i^-$. This means that we can omit μ by using our convention that $b = 0$ if $b \notin B(\mu)$. One has $b_\lambda \otimes B_i^- = B_{y_i}(\nu_i)$. Similarly we set $B_{j,i}^- = \tilde{\mathcal{F}}_{z_j} b_i$, with a corresponding convention. One has $b_\lambda \otimes B_{j,i}^- = \tilde{\mathcal{F}}_{z_j}(b_\lambda \otimes b_i) \cong B_{z_j}(\nu_i)$. These sets $B_i^-, B_{j,i}^-$ are neither $\tilde{\mathcal{E}}$ nor $\tilde{\mathcal{F}}$ stable. Yet let us say that $B \subset B_w(\mu)$ has the proper string-λ (resp. string-λ) property if for each $b \in B$ and all $\beta \subset \pi$ such that $\varepsilon_\beta(b) \geq \beta^\vee(\lambda)$ (resp. $\varepsilon_\beta(b) > \beta^\vee(\lambda)$) one has $\tilde{\mathcal{F}}_\beta b \subset B \cup \{0\}$.

Lemma. *Take n as above. For all $i \in I$, $j \in \{0, 1, 2, \ldots, k\}$ the set $B^-_{j,i}$ has the string-λ property.*

Proof. Suppose $b \in B^-_{j,i}$ satisfies $\varepsilon_\beta(b) > \beta^\vee(\lambda)$. Since $\varphi_\beta(b_\lambda) = \beta^\vee(\lambda)$ it follows from 4.3 and 4.1(1) that $\tilde{e}_\beta(b_\lambda \otimes b) = b_\lambda \otimes \tilde{e}_\beta b \neq 0$ and so $\varepsilon_\beta(b_\lambda \otimes b) > 0$. (This also follows from a rule for tensor product [J2, 5.2.4(2)] which we did not give). Since $b_\lambda \otimes B^-_{j,i} \cong B_{z_j}(\nu_i)$, it has the string property and so it follows that $\tilde{\mathcal{F}}_\beta(b_\lambda \otimes b) \subset B_{z_j}(\nu_i)$. Again since $\varepsilon_\beta(b) > \varphi_\beta(b_\lambda)$, the latter equals $b_\lambda \otimes \tilde{\mathcal{F}}_\beta b$ by 4.3 and so $\tilde{\mathcal{F}}_\beta b \subset B^-_{j,i}$, as required. □

Remark. If we take $\beta = \alpha$ with α defined as in 4.6 we have $\varepsilon_\alpha(b) \geq \varphi_\alpha(b_\lambda)$, for all $b \in B^-_{j,i}$. In some sense to be made precise later, we are mainly reduced to analyzing the case $\varepsilon_\alpha(b) = \varphi_\alpha(b_\lambda) = \alpha^\vee(\lambda)$. This results in a simplification in the proof of the existence of Demazure flags.

4.8.

Clearly B^-_i has the string-λ property; but not quite the proper string-λ property. To take care of this we set $\pi' = \{\alpha \in \pi \,|\, \alpha^\vee(\lambda) = 0\}$ and let $W_{\pi'}$ be the subgroup of W generated by the $s_\alpha : \alpha \in \pi'$. Since $\lambda \in P^+$, this is just $\mathrm{Stab}_W \lambda$ by say [J2, A.1.1(vii)]. Set

$$\tilde{\mathcal{F}}_{\hat{y}_i} = \varinjlim\{\tilde{\mathcal{F}}_x \tilde{\mathcal{F}}_{y_i} | x \in W_{\pi'}\}, \quad \hat{B}^-_i = \tilde{\mathcal{F}}_{\hat{y}_i} b_i.$$

Of course if $W_{\pi'}$ is a finite group and $w_{\pi'}$ is its unique longest element, then $\tilde{\mathcal{F}}_{\hat{y}_i} = \tilde{\mathcal{F}}_{w_{\pi'}} \tilde{\mathcal{F}}_{y_i}$. Then $\hat{y}_i$ can be taken to be the unique longest element in $W_{\pi'} y_i$. Observe that

$$\tilde{\mathcal{F}}_{\hat{y}_i}(b_\lambda \otimes b_i) = b_\lambda \otimes \tilde{\mathcal{F}}_{\hat{y}_i} b_i = b_\lambda \otimes \hat{B}^-_i \subset b_\lambda \otimes B(\mu).$$

Lemma. *For all $\beta \in \pi \setminus \pi'$, $b \in B^-_i$ with $\varepsilon_\beta(b) \geq \beta^\vee(\lambda)$ one has $\tilde{\mathcal{F}}_\beta b \in B^-_i$.*

Proof. Indeed $\tilde{f}^n_\beta(b_\lambda \otimes b) = b_\lambda \otimes \tilde{f}^n_\beta b$, for all $n \in \mathbb{N}$, by 4.3. Yet $b = \tilde{f} b_i$, for some $\tilde{f} \in \tilde{\mathcal{F}}$ which by 4.6(**) satisfies $\tilde{f}(b_\lambda \otimes b_i) = b_\lambda \otimes \tilde{f} b_i$. Set $\tilde{f}' = \tilde{f}^n_\beta \tilde{f}$. Then $\tilde{f}'(b_\lambda \otimes b_i) = b_\lambda \otimes \tilde{f}^n_\beta b$. Yet $\varepsilon_\beta(b) \geq \beta^\vee(\lambda) > 0$ and so $\tilde{\mathcal{F}}_\beta b \in B_w(\mu)$ by the string property of the latter. Hence $\tilde{\mathcal{F}}_\beta b \in B^-_i$, by 4.6(**). □

Remark. This is just the proper string-λ property off the orthogonal of λ in π.

4.9.

Recall the following identity for q-binomial coefficients

$$q^m \begin{bmatrix} n \\ m \end{bmatrix} + q^{m-n-1} \begin{bmatrix} n \\ m-1 \end{bmatrix} = \begin{bmatrix} n+1 \\ m \end{bmatrix}, \ \forall\, n \geq m \geq 0. \qquad (*)$$

We need the following result which is no doubt well known. For all $n, r \in \mathbb{N}$, let $M^{n,r}$ denote the matrix with entries

$$M^{n,r}_{s,m} = \begin{bmatrix} n+r-m \\ r-s \end{bmatrix} q^{(r-m)(n+s-m)} \ : \ s, m \in \{0, 1, 2, \ldots, r\}.$$

Lemma. *One has* det $M^{n,r}_{s,m} = 1$. *Moreover the* $s = m = 0$ *cofactor of* det $M^{n,r}_{s,m}$ *equals 1 as well.*

Proof. Since the assertion is trivial for $r = 0$, it is enough to show that det $M^{n,r}_{s,m} =$ det $M^{n,r-1}_{s,m}$.

The bottom row $M^{n,r}_{r,m}$ of $M^{n,r}$ is just $q^{(r-m)(n+r-m)}$. Starting at $m = 0$, multiply the $(m+1)^{\text{th}}$ column by $q^{2(r-m)+n+1}$, and subtract it from the mth column. Then the entries in the bottom row become zero except for the last. Moreover for $s < r$ the s, m entry becomes

$$q^{(r-m)(n+s-m)} \left(\begin{bmatrix} n+r-m \\ r-s \end{bmatrix} - q^{r,s} \begin{bmatrix} n+r-m-1 \\ r-s \end{bmatrix} \right)$$

$$= q^{(r-m)(n+s-m)+r-s-(n+r-m)} \begin{bmatrix} n+r-1-m \\ r-1-m \end{bmatrix}, \quad \textit{by } (*)$$

$$= M^{n,r-1}_{s,m}.$$

Since $M^{n,r}_{r,r} = 1$, the assertion results. The proof of the last part obtains by increasing r, s, m by 1. □

5. Demazure flags

5.1.

Fix $\mu \in P^+, w \in W$. By construction our first generator of u_μ of $V(\mu)$ satisfies $u_\mu = G(b_\mu)$. The crystal basis $B(\mu)$ has precisely one element $b_{w\mu}$ of weight $w\mu$ and we set $u_{w\mu} = G(b_{w\mu})$. By 3.5(i) it may be alternatively described as in [J2, 6.3.9].

5.2.

Recall the definition of the A modules $\mathcal{E}_\alpha, \mathcal{F}_\alpha : \alpha \in \pi$ given in 3.7. Obviously $\mathcal{E}_\alpha^2 = \mathcal{E}_\alpha$. Moreover they satisfy the Coxeter relations by virtue of the Verma relations on the $e_\alpha : \alpha \in \pi$. Thus $\mathcal{E}_y$ and similarly $\mathcal{F}_y$ is defined for all $y \in W$ by taking a reduced decomposition and the A module generated by the appropriate products. More precisely, $\mathcal{E}_y = A\mathcal{E}_{\alpha_1}\mathcal{E}_{\alpha_2} \cdots \mathcal{E}_{\alpha_n}$ given a reduced decomposition $y = s_{\alpha_1}s_{\alpha_2} \cdots s_{\alpha_n}$. If $y = 1$, we set $\mathcal{E}_y = \mathcal{F}_y = A$.

5.3.

The Demazure module defined over $\mathbb{Z}[q, q^{-1}]$ for the pair $\mu \in P^+, w \in W$ is by definition $V^{\mathbb{Z}}_w(\mu) := U^{\mathbb{Z}}_q(n^+)u_{w\mu}$. It is clearly also T stable. We also set $V_w(\mu) = U_q(\mathfrak{n}^+)u_{w\mu}$. Obviously $U_q(\mathfrak{n}^+)u_{w\mu} = KV^{\mathbb{Z}}_w(\mu)$. The following result is noted by Kashiwara in [K2, Remark 3.2.6] and also in [R, 3.5]. We give the proof for completion.

Proposition. *For all $\mu \in P^+$, $w \in W$ one has*

(i) $V_w^{\mathbb{Z}}(\mu) = \mathcal{F}_w u_\mu$.
(ii) $\mathcal{F}_w u_\mu = \bigoplus_{b \in B_w(\mu)} AG(b)$.

Proof. (ii) results from 3.8 by induction on the length $\ell(w)$ of w. Indeed it is enough to recall that if $w \in W \setminus \{e\}$, then there exists $\alpha \in \pi$ such that $\ell(s_\alpha w) = \ell(w) - 1$. Then $\mathcal{F}_w u_\mu = \mathcal{F}_\alpha(\mathcal{F}_{s_\alpha w} u_\mu)$ and one applies 3.8 with $N = \mathcal{F}_{s_\alpha w} u_\mu$ and $M = \mathcal{F}_\alpha N$ to obtain (ii).

Recall the Kac relation

$$e_\alpha^{(r)} f_\alpha^{(s)} = \sum_{j=0}^{\min(r,s)} f_\alpha^{(s-j)} \begin{bmatrix} t_\alpha; 2j - r - s \\ j \end{bmatrix} e_\alpha^{(r-j)}$$

where

$$\begin{bmatrix} t_\alpha;\ n \\ m \end{bmatrix} = \prod_{j=1}^{m} \left(\frac{t_\alpha q_\alpha^{(n-j+1)} - t_\alpha^{-1} q_\alpha^{-(n-j+1)}}{q_\alpha^j - q_\alpha^{-j}} \right)$$

with $q_\alpha = q^{\frac{1}{2}(\alpha,\alpha)}$. Evaluated on a weight vector of weight $\lambda \in P^+$, this latter expression becomes

$$\begin{bmatrix} \alpha^\vee(\lambda) + n \\ m \end{bmatrix}_{q_\alpha}$$

and so is an element of A. Again $[e_\alpha^{(r)}, f_\beta^{(s)}] = 0$ if $\alpha \neq \beta$. From these relations one deduces that $\mathcal{F}_w u_\mu$ is $U_q^{\mathbb{Z}}(\mathfrak{n}^+)$ stable. From the definition of $u_{w\mu}$ one checks that $u_{w\mu} \in \mathcal{F}_w u_\mu$. We conclude that $\mathcal{F}_w u_\mu \supset U_q^{\mathbb{Z}}(\mathfrak{n}^+) u_{w\mu}$.

Finally take $y \in W$ and $\alpha \in \pi$ such that $s_\alpha y > y$. Then $e_\alpha u_{y\lambda} = 0$, so exactly as above $\mathcal{F}_\alpha u_{y\lambda}$ is $\mathcal{E}_\alpha$ stable and contains $u_{s_\alpha y\lambda}$ and so $\mathcal{F}_\alpha u_{y\lambda} \supset \mathcal{E}_\alpha u_{s_\alpha y\lambda}$. Similarly $\mathcal{E}_\alpha u_{s_\alpha y\lambda}$ is $\mathcal{F}_\alpha$ stable and contains $u_{y\lambda}$ giving the opposite inclusion. Then by induction on $\ell(w)$ we obtain $\mathcal{F}_w u_\mu = \mathcal{E}_{w^{-1}} u_{w\mu} \subset U_q^{\mathbb{Z}}(\mathfrak{n}^+) u_{w\mu}$, proving (i). □

Remark 1. The Kac relation implies that $\mathcal{E}_\alpha \mathcal{F}_\alpha v = \mathcal{F}_\alpha \mathcal{E}_\alpha v$, for any weight vector v.

Remark 2. By contrast the inclusion $\tilde{\mathcal{E}} b_{w\mu} \subset B_w(\mu)$ can be strict. For example take the adjoint module (of highest weight ρ) in type A_2 with $w = s_\alpha s_\beta$ of length 2. Then $\tilde{e}_\alpha \tilde{e}_\beta b_{w\rho} = 0$, whilst $e_\alpha e_\beta G(b_{w\rho}) \neq 0$, also illustrating the triangularity in 3.8.

5.4.

As usual, for a semisimple T module M with weight A-submodules which are finitely generated and free, one may define the formal character of M through

$$\operatorname{ch} M = \bigoplus_{\nu \in P} (rk_A M_\nu) e^\nu.$$

Similarly for any crystal B one may define its formal character through

$$\text{ch } B = \bigoplus_{\nu \in P} (\# B_\nu) e^\nu.$$

From 5.3 we obtain

$$\text{ch } V_w^{\mathbb{Z}}(\mu) = \text{ch } B_w(\mu).$$

One may remark that using the string property of $B_w(\mu)$ the right-hand side is given by the Demazure operator Δ_w applied to e^μ. Specialization at $q = 1$, gives in particular Kashiwara's proof of the Demazure character formula for the Demazure module defined with respect to the Kostant $\mathbb{Z}$ form of $U(\mathfrak{n}^+)$, namely

$$\text{ch } V_w^{\mathbb{Z}}(\mu) = \Delta_w e^\mu.$$

5.5.

Define an order relation on $P(\pi)$ through $\mu \geq \nu$ if $\mu - \nu \in \mathbb{N}\pi$. Recall 4.3 and 4.6 and define a linear order on I so that $wt\ b_i < wt\ b_j \implies j < i$. We note that this order relation does not depend on μ. With respect to this ordering we obtain the

Lemma. *If* $V^{\mathbb{Z}}(\nu_j)_{\nu_i + n\alpha} \neq 0 : n \in \mathbb{N}^+, \alpha \in \pi,$ *then* $j < i$.

Proof. Indeed the hypothesis implies that $\lambda + wt\ b_i + n\alpha \leq \lambda + wt\ b_j$ and so $wt\ b_i < wt\ b_j$ forcing $j < i$. □

5.6.

By the associativity of the tensor product, the condition that $u_\lambda \otimes V_w(\mu)$ admits a Demazure flag reduces to the case λ fundamental. Although most of our analysis is setup to avoid making this reduction, we shall soon meet some clear advantages in so doing. Thus in 5.8–5.21 we shall assume $\alpha^\vee(\lambda) \leq 1$ for all $\alpha \in \pi$. Aside from ensuring that previous induction steps hold, this is not needed until 5.17.

5.7.

Recall 4.7 and set

$$X_i^- = \bigoplus_{b \in B_i^-} AG(b),\ Y_i^- = \bigoplus_{j<i} X_j^-,\ X_i = G(b_\lambda) \otimes X_i^-,\ Y_i = G(b_\lambda) \otimes Y_i^-.$$

It is also convenient to set $C_i^- = \bigcup_{j<i} B_j^-$, $\hat{C}_i^- = \bigcup_{j<i} \hat{B}_j^-$.

Let $\hat{X}_i^-, \hat{Y}_i^-, \hat{X}_i, \hat{Y}_i$ denote the corresponding A modules obtained by replacing B_i^- by $\hat{B}_i^-$. We shall eventually prove that Y_i^- is $U_q^{\mathbb{Z}}(\mathfrak{n}^+)$ stable, although this is not obvious for the moment. However $Y_i \subset G(b_\lambda) \otimes V_w^{\mathbb{Z}}(\mu)$ and the latter is $U_q^{\mathbb{Z}}(\mathfrak{n}^+)$ stable by 5.3. Recall the definitions of b_i, ν_i in 4.6.

Lemma. *For all $n \in \mathbb{N}^+, \alpha \in \pi$ one has*

$$e_\alpha^{(n)}(G(b_\lambda) \otimes G(b_i)) \subset Y_i.$$

Proof. By the above remark the left-hand side is an A linear combination of the $G(b_\lambda) \otimes G(b) : b \in B_w(\mu)$ having weight $n\alpha + \nu_i$. By 4.6(∗) it is enough to show that $b \notin B_j^-$, for $j \geq i$. If not, then $G(b_\lambda) \otimes G(b) \in V_{y_j}^{\mathbb{Z}}(\nu_j)_{n\alpha+\nu_i} \subset V^{\mathbb{Z}}(\nu_j)_{n\alpha+\nu_i}$, which by 5.5 forces $j < i$. □

5.8.

Fix $w \in W$. Recall the last paragraph of 4.6. Choose n so that the $b_\lambda \otimes b_i : i \in \{1, 2, \dots, n\}$ are the primitive elements of $b_\lambda \otimes V_w(\mu')$ with $\mu' \in P^+$ sufficiently large.

Set

$$Z_i = \sum_{j<i} U_q^{\mathbb{Z}}(\mathfrak{n}^-)(G(b_\lambda) \otimes G(b_j))$$

which is a submodule of $V^{\mathbb{Z}}(\lambda) \otimes V^{\mathbb{Z}}(\mu)$.

We shall prove inductively that Z_i is $U_q^{\mathbb{Z}}(\mathfrak{g})$ stable and

$$Z_i \cap (G(b_\lambda) \otimes V_w^{\mathbb{Z}}(\mu)) = Y_i.$$

Admitting this last result we obtain from 5.7 that

$$e_\alpha^{(n)}(G(b_\lambda) \otimes G(b_i)) \subset Z_i, \ \forall\, n \in \mathbb{N}, \alpha \in \pi. \qquad (*)$$

If $b_i \neq 0$ (recall 4.4) it follows that $G(b_\lambda) \otimes G(b_i)$ is a (non-zero) highest weight vector mod Z_i and so Z_{i+1} is a $U_q^{\mathbb{Z}}(\mathfrak{g})$ module with quotient isomorphic to $V^{\mathbb{Z}}(\nu_i)$. We shall eventually show (5.25) that $Z_1 \subset Z_2 \subset \cdots \subset Z_{n+1}$ is a Weyl flag for $U_q^{\mathbb{Z}}(\mathfrak{g})(G(b_\lambda) \otimes V_w^{\mathbb{Z}}(\mu))$. Replacing $V_w^{\mathbb{Z}}(\mu)$ by $V^{\mathbb{Z}}(\mu)$ similarly gives a Weyl flag for $V^{\mathbb{Z}}(\lambda) \otimes V^{\mathbb{Z}}(\mu)$.

5.9.

Our key result can be expressed as follows. Fix $w \in W$ and $n \in \mathbb{N}^+$ as in 5.8.

Theorem ($\mathfrak{g}$ simply-laced, $\alpha^\vee(\lambda) \leq 1, \ \forall\, \alpha \in \pi$). *For all $i \in \{1, 2, \dots, n\}$, and all $\mu \in P^+$, one has*

(i) $\mathcal{F}_{y_i}(G(b_\lambda) \otimes G(b_i)) + Z_i = X_i + Z_i$.
(ii) Z_{i+1} *is* $U_q^{\mathbb{Z}}(\mathfrak{g})$ *stable and* $Z_{i+1} \cap (G(b_\lambda) \otimes V_w^{\mathbb{Z}}(\mu)) = Y_{i+1}$.

Remark 1. Note that X_i, Y_i, Z_i depend on μ; but the indexing and the b_i, y_i do not.

Remark 2. Notice that (ii) implies that Y_{i+1} is $U_q^{\mathbb{Z}}(\mathfrak{n}^+)$ stable.

Remark 3. By the $(i-1)^{\text{th}}$ induction step of (ii) one has $Y_i \subset Z_i$, so the first assertion (in the ith induction step) of (ii) follows from 5.7 (see also 5.8(∗)). In view of 2.6 and the definition (4.8) of $\hat{B}_i^-$, this further implies $\hat{Y}_i \subset Z_i$.

5.10.

Let us first observe that 5.9 implies that $G(b_\lambda) \otimes V_w^{\mathbb{Z}}(\mu)$ admits a Demazure flag. Here we can ignore those $b_i \notin B_w(\mu)$. Clearly $X_i \cong Y_{i+1}/Y_i$ which by the Remark 2 of 5.9 has a $U_q^{\mathbb{Z}}(\mathfrak{n}^+)$ module structure. Since $Y_i \subset Z_i$, we obtain

$$\begin{aligned}(\mathcal{F}_{y_i}(G(b_\lambda) \otimes G(b_i)) + Z_i)/Z_i \\ &= (X_i + Y_i + Z_i)/Z_i, \text{ by (i)}, \\ &= (Y_{i+1} + Z_i)/Z_i, \\ &= Y_{i+1}/Y_i, \text{ by (ii)}.\end{aligned}$$

Since $G(b_\lambda) \otimes G(b_i)$ is a highest weight vector of highest weight ν_i, it follows from 5.3(i) that the left-hand side is an image of the Demazure module $V_{y_i}^{\mathbb{Z}}(\nu_i)$. Yet by 5.3(ii) and the definition of x_i we obtain ch Y_{i+1}/Y_i = ch $V_{y_i}^{\mathbb{Z}}(\nu_i)$. Now by 5.3(ii) $V_{y_i}^{\mathbb{Z}}(\nu_i)$ is a direct sum of its A weight submodules which are free of finite rank, whilst so is Y_{i+1}/Y_i by construction. Then equality of characters forces the surjection $V_{y_i}^{\mathbb{Z}}(\nu_i) \twoheadrightarrow Y_{i+1}/Y_i$ to be an isomorphism. Since $Y_{n+1} = G(b_\lambda) \otimes V_w^{\mathbb{Z}}(\mu)$ and $Y_1 = 0$, we obtain the

Corollary ($\mathfrak{g}$ simply-laced, $\alpha^\vee(\lambda) \leq 1$, $\forall\, \alpha \in \pi$). *For all $\lambda, \mu \in P^+, w \in W$, $G(b_\lambda) \otimes V_w^{\mathbb{Z}}(\mu)$ admits an increasing and exhaustive $U_q^{\mathbb{Z}}(\mathfrak{n}^+)$ filtration $0 = Y_1 \subset Y_2 \subset \cdots Y_{n+1}$ with quotients Y_{i+1}/Y_i isomorphic to the Demazure module $V_{y_i}^{\mathbb{Z}}(\nu_i)$ with generator $G(b_\lambda) \otimes G(b_i)$* mod $Y_i : i = 1, 2, \ldots, n$.

5.11.

The proof of 5.9 is by induction on $i \in I \cup \{0\}$, taking $b_0 = 0$ in (i). This is carried out in 5.11–5.20 below in which the hypotheses of 5.9 are assumed. The main idea is very simple. We consider the submodule N of $V_w^{\mathbb{Z}}(\nu)$ with global basis given by $B_{j,i}^- \cup C_i^-$ and show that it is $U_q^{\mathbb{Z}}(\mathfrak{n}^+)$ stable. For this, the considerations of 3.7–3.8 apply with $\alpha = \alpha_j$. Had 3.9(∗∗) been true then 4.7 would allow us to apply 3.8 and 4.8 to carry out the induction on j in the proposition below. This in turn gives the induction step i of $i+1$ of 5.9(i). Because 3.9(∗∗) fails we are forced into a third induction step (Prop. 5.12) on chain lengths which we are only able to carry out when $\alpha^\vee(\lambda) \leq 1$ and $\mathfrak{g}$ is simply-laced. The case $\alpha^\vee(\lambda) = 0$ is easy apart from a little sting in its tail from B_i^- not having the proper λ-string property (see 4.8). For the case $\alpha^\vee(\lambda) = 1$ we show using notably the compatibility of the global basis with the action of $U_q^{\mathbb{Z}}(\mathfrak{n}^-)$ that the third induction step holds asymptotically (Lemma 5.17). Here we need to pass to the field K and use a dimensionality estimate (Lemma 5.15). Finally we make use of Lusztig's positivity result and again the dimensionality estimate to deduce the general case (5.18). Then we recover the assertion over A (5.19) using 5.13. We conclude the induction process in 5.20.

Since $Z_1 = Y_1 = 0$ the assertion of 5.9 is trivial in this case. Assume now that 5.9 holds for $i-1$ and establish it for i. Set $y_i = y$ and recall the notation of 4.6. Set

$$X^-_{j,i} = \bigoplus_{b\in B^-_{j,i}} AG(b), \ X_{j,i} = G(b_\lambda) \otimes X^-_{j,i}.$$

Proposition. *For all* $j = \{0, 1, 2 \cdots, \ell(y)\}$, *all* $\mu \in P^+$ *one has*

(i)′ $\mathcal{F}_{z_j}(G(b_\lambda) \otimes G(b_i)) + Z_i = X_{j,i} + Z_i$.

Proof (first step). The proof begins by decreasing induction on j. If $j = \ell(y)$, then $z_j = e$ and the assertions are trivial. Assume the result for j and set $\alpha = \alpha_j, z = z_j$. Then $\mathcal{F}_{z_{j-1}} = \mathcal{F}_\alpha \mathcal{F}_z$. In what follows (ii) refers to 5.9(ii). Recall that we are assuming 5.9 holds for $i - 1$.

Set $N = X^-_{j,i} + Y^-_i$. By definition it admits a global basis. We claim that

$$U^{\mathbb{Z}}_q(\mathfrak{n}^+)N \subset N. \qquad (*)$$

Take $\beta \in \pi, m \in \mathbb{N}$. Then by 2.6

$$\begin{aligned} G(b_\lambda) \otimes e^{(m)}_\beta N &= e^{(m)}_\beta (G(b_\lambda) \otimes N), \\ &\subset e^{(m)}_\beta (X_{j,i} + Z_i), \quad \text{since } Y_i \subset Z_i \text{ by (ii)}, \\ &= e^{(m)}_\beta \mathcal{F}_{z_j}(G(b_\lambda) \otimes G(b_i)) + Z_i, \quad \text{by (i)}', \\ &\subset \mathcal{F}_{z_j}(G(b_\lambda) \otimes G(b_i)) + Z_i, \quad \text{by 5.3 and 5.8}, \qquad (*) \\ &\subset G(b_\lambda) \otimes N + Z_i, \quad \text{by (i)}'. \end{aligned}$$

Yet $N \subset V^{\mathbb{Z}}_w(\mu)$ and the latter is $U^{\mathbb{Z}}_q(\mathfrak{n}^+)$ stable by 5.3. Hence intersection with $G(b_\lambda) \otimes V^{\mathbb{Z}}_w(\mu)$ and use of (ii) gives the required claim.

In the next step we shall just use that $\mathcal{E}_\alpha N \subset N$ and eventually apply 3.5–3.12. Observe that since Y_i is $U^{\mathbb{Z}}_q(\mathfrak{n}^+)$ stable we have $U^{\mathbb{Z}}_q(\mathfrak{n}^+)Y^-_i \subset Y^-_i$ and so in particular $\mathcal{E}_\alpha Y^-_i \subset Y^-_i$. By 3.7 we conclude that

$$\text{both } B^-_{j,i} \cup C^-_i \text{ and } C^-_i \text{ are } \tilde{\mathcal{E}}_\alpha \text{ stable.}$$

5.12.

In order to continue the proof of (i)′ of 5.11 we shall need a further induction parameter. Consider $b \in B^-_{j,i}$ and recall that we are setting $\alpha = \alpha_j$. Then $\tilde{f}^n_\alpha b \in B^-_{j-1,i}$ and moreover the definition of α implies (see 4.6) that $\varepsilon_\alpha(b) \geq \alpha^\vee(\lambda)$. Then for the inclusion $\supset$ in (i)′ we must show that

$$G(b_\lambda) \otimes G(\tilde{f}^n_\alpha b) \in \mathcal{F}_\alpha \mathcal{F}_{z_j}(G(b_\lambda) \otimes G(b_i)) + Z_i, \quad \text{for all } n \in \mathbb{N}. \qquad (*)$$

By 4.7 and since $B^-_{j,i} \cup C^-_i$ is $\tilde{\mathcal{E}}_\alpha$ stable, every α-chain S meeting $B^-_{j,i}$ either lies entirely in $B^-_{j,i} \cup C^-_i$ or there is a single element $b \in S \cap B^-_{j,i}$ satisfying $\varepsilon_\alpha(b) = \alpha^\vee(\lambda)$.

Let $\mathcal{T}$ denote the union of all α-chains with the latter property and set $B^{\sim}_{j,i} = \mathcal{T} \cap B^{-}_{j,i}$. One has

$$B^{\sim}_{j,i} = \{b \in B^{-}_{j,i} | \varepsilon_\alpha(b) = \alpha^\vee(\lambda),\ \tilde{f}^n_\alpha b \notin B^{-}_{j,i},\ \forall\, n \in \mathbb{N}^+\}.$$

By the induction hypothesis it is enough to establish $(*)$ for $b \in B^{\sim}_{j,i}$.

Now suppose $b \in B^{\sim}_{j,i}$ and let $\ell(b)$ be the length of the α-string generated by b, that is $\ell(b) = \varepsilon(b) + \varphi(b)$. Our further (induction) hypothesis is that the required assertion has been proved for all such b belonging to an α-string of strictly greater length. However since we wish to vary μ (which also varies string length) it is more convenient to view b as an element of $B(\infty)$ and set $\gamma = -wt\ b$. Then we only need to consider $b' \in B^{-}_{j,i}$ satisfying $\varepsilon_\alpha(b') \geq \alpha^\vee(\lambda)$ and take $wt\ b' \in -\gamma + \mathbb{N}^+\alpha$ to recover these longer strings in $B_{-\gamma+\mathbb{Z}\alpha}$. Notice further that $wt\ b' + \gamma = wt\ b' - wt\ b$ does not depend on which $B(\mu) : \mu \in P^+$ we consider b, b' to belong (as long as they do belong!). Recall (4.6, 4.7) that $\varepsilon_\alpha(b') \geq \alpha^\vee(\lambda)$, for all $b' \in B^{-}_{j,i}$. Then with $b \in B^{\sim}_{j,i}$ fixed we set

$$B^{-}_{j,i,b} = \{b' \in B^{-}_{j,i} | wt\ b' - wt\ b \in \mathbb{N}^+\alpha\},$$

$$X^{-}_{j,i,b} = \bigoplus_{b' \in B^{-}_{j,i,b}} AG(b'),\ X_{j,i,b} = G(b_\lambda) \otimes X^{-}_{j,i,b}.$$

$$\hat{X}^{-}_{j,i,b} = \bigoplus_{b' \in \tilde{\mathcal{F}}_\alpha B^{-}_{j,i,b}} AG(b'),\ \hat{X}_{j,i,b} = G(b_\lambda) \otimes \hat{X}^{-}_{j,i,b}.$$

We show by induction on $\ell(b)$ the

Proposition. *For all $b \in B^{\sim}_{j,i}$ one has*

$$\sum_{n \in \mathbb{N}} A(G(b_\lambda) \otimes G(\tilde{f}^n_\alpha b)) \subset \mathcal{F}_\alpha(G(b_\lambda) \otimes G(b)) + Z_i + \mathcal{F}_\alpha X_{j,i,b}.$$

5.13.

The proof of the above proposition is given in several steps in 5.13–5.18 below. First we set up the general construction. For this we will need a little extra book-keeping due to repetitions in the reduced decompositions of y_i. Recall our fixed reduced decomposition $s_{\alpha_{j+1}} \cdots s_{\alpha_k}$ of z_j. If $\alpha \notin \{\alpha_{j+1}, \ldots, \alpha_k\}$ set $Z_{j,i} = 0$. Otherwise let j_α be the unique minimal element of $\{j+1, \ldots, k\}$ such that $\alpha = \alpha_{j_\alpha}$ and set $Z_{j,i} = \mathcal{F}_{z_{j_\alpha - 1}}(G(b_\lambda) \otimes G(b_i))$. When j_α is defined we can assume $b \notin B^{-}_{j_\alpha,i}$ since by definition of j_α every α-string in $B^{-}_{j_\alpha,i}$ is already complete. Again we can assume $b \notin \tilde{\mathcal{F}}_\alpha B^{-}_{j,i,b}$ through the (last) induction hypothesis. When j_α is not defined, we set $X_{j_\alpha,i} = 0$ and $B^{-}_{j_\alpha,i} = \emptyset$.

Let N_b be the smallest $\mathcal{E}_\alpha$ stable submodule of $V^{\mathbb{Z}}_w(\mu)$ containing $G(b)$ and admitting a global basis, so $B(N_b)$ is defined and is $\mathcal{E}_\alpha$ stable (3.7). Since $G(b) \in N$, it follows from 5.11(∗) that $N_b \subset N$. Set $M_b = \mathcal{F}_\alpha N_b$. By 3.8, M_b admits a global basis and $B(M_b) = \tilde{\mathcal{F}}_\alpha B(N_b)$. Furthermore $B(M_b)$ is a union of α-strings and there is just one S_b of minimal length $\ell(b)$ and this contains b. Let $\mathcal{S}_b$ denote the union of the remaining strings and set

$$M_{\mathcal{S}} = \bigoplus_{b' \in \mathcal{S}_b} AG(b')$$

which by 3.5 is a $U_q^{\mathbb{Z}}(\mathfrak{s}_\alpha)$ submodule of M. Set $\alpha^\vee(\lambda) = r$. (We do not need to suppose $\alpha^\vee(\lambda) \leq 1$ for the moment, except to ensure that earlier induction steps have been completed.) Recall that $b \in B^{\sim}_{j,i}$, so $\varepsilon_\alpha(b) = \alpha^\vee(\lambda) = r$.

Lemma. *One has*

$$G(b_\lambda)\otimes G(\tilde{f}_\alpha^n b) \in f_\alpha^{(n)}(G(b_\lambda)\otimes G(b)) + \sum_{s=1}^{r} A f_\alpha^{(n+s)}(G(b_\lambda)\otimes G(\tilde{e}_\alpha^s b)) + G(b_\lambda)\otimes M_{\mathcal{S}}, \tag{$*$}$$

for all $n \in \mathbb{N}$.

Proof. Recall that by 2.6

$$\Delta(f_\alpha) = x + y, \quad \text{where } x = f_\alpha \otimes 1,\ y = t_\alpha \otimes f_\alpha.$$

From now on we omit the α-subscript on $e_\alpha, f_\alpha, t_\alpha, q_\alpha, \tilde{e}_\alpha, \tilde{f}_\alpha$. One checks that $xy = q^2 yx$. Then by say [J2, 1.2.12(3)] one has

$$\begin{aligned}(x+y)^{(n)} &= \frac{1}{[n]_q^!} \sum_{m=0}^{n} \begin{bmatrix} n \\ m \end{bmatrix}_{q^{-1}} q^{-(n-m)m} x^{n-m} y^m \\ &= \sum_{m=0}^{n} q^{-(n-m)m} f^{(n-m)} t^m \otimes f^{(m)}.\end{aligned}$$

Setting $c_m = f^{(m)} G(b_\lambda)\otimes G(\tilde{f}^{n-m} b)$, we obtain from 3.5 that $\mod G(b_\lambda)\otimes M_{\mathcal{S}}$ one has

$$\begin{aligned} f^{(n+s)}(G(b_\lambda)\otimes G(\tilde{e}^s b)) &= \sum_{m=0}^{n+s} q^{-(n+s-m)m} (f^{(m)} t^{n+s-m} G(b_\lambda) \otimes f^{(n+s-m)} G(\tilde{e}^s b)) \\ &= \sum_{m=0}^{r} q^{(r-m)(n+s-m)} \begin{bmatrix} r+n-m \\ r-s \end{bmatrix} c_m, \end{aligned}$$

Thus the required assertion results from the two assertions in Lemma 4.9. □

5.14.

For the moment assume $\alpha^\vee(\lambda) = 0$. Of course this case should be rather easy and will serve as a warm-up.

Lemma. *Suppose $\alpha^\vee(\lambda) = 0$ in 5.12. Then its conclusion holds.*

Proof. Since $B(N_b) \subset B(N) = B^-_{j,i} \cup C^-_i$ and $\alpha^\vee(\lambda) = 0$, it follows from 4.8 and the definition of $B^-_{j,i,b}$ that $\mathcal{S}_b \subset \tilde{\mathcal{F}}_\alpha B^-_{j,i,b} \cup \hat{C}^-_i$. Then by Remark 3 of 5.9, and 5.12 applied to $b' \in B^-_{j,i,b} \cap B^\sim_{j,i}$ we obtain

$$G(b_\lambda) \otimes M_{\mathcal{S}} \subset \mathcal{F}_\alpha X_{j,i,b} + \hat{Y}_i \subset \mathcal{F}_\alpha X_{j,i,b} + Z_i.$$

Yet by 3.5 we have

$$\bigoplus_{m \in \mathbb{N}} AG(\tilde{f}^m_\alpha b) \subset \mathcal{F}_\alpha G(b) + M_{\mathcal{S}}.$$

Since $\alpha^\vee(\lambda) = 0$, we have $f_\alpha G(b_\lambda) = 0$, so by 2.6,

$$\begin{aligned} \mathcal{F}_\alpha(G(b_\lambda) \otimes G(b)) + Z_i + \mathcal{F}_\alpha X_{j,i,b} &\supset G(b_\lambda) \otimes (\mathcal{F}_\alpha G(b) + M_{\mathcal{S}}) \\ &\supset \bigoplus_{n \in \mathbb{N}} A(G(b_\lambda) \otimes G(\tilde{f}^n_\alpha b)), \end{aligned}$$

as required. □

5.15.

For the general case recall that $e_\alpha G(b) \in N_b \subset N$ and suppose $G(b')$ occurs in its decomposition. Then $b' \in B^-_{j,i} \cup C^-_i$, by 5.11(∗) and 3.7. Moreover either $b' \in S_b$ or $b' \in \mathcal{S}_b$. In the former case $b' = \tilde{e}_\alpha b \notin B^-_{j,i}$, because $\varepsilon_\alpha(\tilde{e}_\alpha b) = \varepsilon_\alpha(b) - 1 < \alpha^\vee(\lambda)$ and so $b' \in C^-_i$. (Notice that this implies that all terms in the summation on the right hand side of 5.13(∗) lie in Z_i.) In the latter case $\varepsilon_\alpha(b') > \varepsilon_\alpha(b) - 1 = \alpha^\vee(\lambda) - 1$ and so either $b \in C^-_i$ or $b \in B^-_{j,i,b}$. Thus in both cases

$$b' \in C^-_i \cup B^-_{j,i,b}.$$

Thus

$$e_\alpha G(b) \in Y^-_i + X^-_{j,i,b}. \tag{$*$}$$

Similarly (the whole of) $Y^-_i + X^-_{j,i,b}$ is $\mathcal{E}_\alpha$ stable. Hence $C^-_i \cup B^-_{j,i,b}$ is $\tilde{\mathcal{E}}_\alpha$ stable by 3.7. Moreover $Y_i + X_{j,i,b}$ is $\mathcal{E}_\alpha$ stable by 2.6. By the $(i-1)$ induction step of 5.9(ii) one has $Y_i \subset Z_i$. Then through Remark 1 of 5.3 and the above,

$$Z_i + \mathcal{F}_\alpha X_{j,i,b} \ \ is \ \ U^{\mathbb{Z}}_q(\mathfrak{s}_\alpha) \text{ stable}.$$

Now let $\mathcal{S}'_b$ denote the union of all α-strings in $\mathcal{S}_b$ each of which lie entirely in $\hat{C}^-_i \cup \tilde{\mathcal{F}}_\alpha B^-_{j,i,b}$ and set $M_{\mathcal{S}'} = \bigoplus_{b' \in \mathcal{S}'_b} AG(b')$. One has $G(b_\lambda) \otimes M_{\mathcal{S}'} \subset Z_i + \mathcal{F}_\alpha X_{j,i,b}$. Unfortunately because 3.9(∗∗) can fail there is no reason to suppose that $\mathcal{S}_b = \mathcal{S}'_b$. For example, see 5.21. Otherwise 5.12 would follow from 5.13.

To overcome the above difficulty we first replace A by the field $K = \mathbb{Q}(q)$ here and in 5.16–5.18 below. Here we do not have to care for divided powers.

Recall the definitions of j_α, $Z_{j,i}$, $X_{j_\alpha,i}$ given in 5.11 and 5.13.

We assume as part of the induction hypothesis that over K one has

$$(Z_i + Z_{j,i} + \mathcal{F}_\alpha X_{j,i,b}) \cap (G(b_\lambda) \otimes V_w(\mu)) = Y_i + X_{j_\alpha,i} + \hat{X}_{j,i,b}. \qquad (**)$$

With respect to the right-hand side we recall that $B^-_{j_\alpha,i}$ (when j_α is defined) consists of complete α-strings and those of length $> \ell(b)$ already lie in $\tilde{\mathcal{F}}_\alpha B^-_{j,i,b}$ (by definition of the latter).

Lemma. *Consider* $\mathcal{F}_\alpha(G(b_\lambda) \otimes G(b)) \bmod (Z_i + \mathcal{F}_\alpha X_{j,i,b})$. *Then either*

$$\dim_K \mathcal{F}_\alpha(G(b_\lambda) \otimes G(b)) = \dim_K \left(\sum_{n \in \mathbb{N}} K G(b_\lambda) \otimes G(\tilde{f}^n_\alpha b) \right)$$

or the left-hand side is zero.

Proof. In view of 2.6

$$e_\alpha(G(b_\lambda) \otimes G(b)) = G(b_\lambda) \otimes e_\alpha G(b) \subset Z_i + \mathcal{F}_\alpha X_{j,i,b},$$

by $(*)$. Thus $\mathcal{F}_\alpha(G(b_\lambda) \otimes G(b)) \mod (Z_i + \mathcal{F}_\alpha X_{j,i,b})$ is $\mathcal{E}_\alpha$ stable with $G(b_\lambda) \otimes G(b)$ its highest weight vector. We can assume that the latter is non-zero, otherwise the second conclusion of the lemma holds. Define $wt\ b$ when b is viewed as an element of $B(\mu)$. Then $\mathcal{F}_\alpha(G(b_\lambda) \otimes G(b)) \mod Z_i + \mathcal{F}_\alpha X_{j,i,b}$ has dimension equal to $\alpha^\vee(\lambda + wt\ b) + 1$, by $U_q(\mathfrak{sl}(2))$ theory, specifically [J2, 4.3]. On the other hand $\tilde{e}_\alpha(b_\lambda \otimes b) = \tilde{e}_\alpha b_\lambda \otimes b = 0$ and $\tilde{f}^n_\alpha(b_\lambda \otimes b) = b_\lambda \otimes \tilde{f}^n_\alpha b$, for all $n \in \mathbb{N}$, since $\varepsilon_\alpha(b) = \alpha^\vee(\lambda)$. Hence

$$\operatorname{card}(b_\lambda \otimes \tilde{\mathcal{F}}_\alpha b) = \alpha^\vee(\lambda + wt\ b) + 1$$

by crystal $\mathfrak{sl}(2)$ theory, specifically 3.5$(*)$. Hence the assertion. □

Remark 1. The equality of the lemma does not depend on which $B(\mu) : \mu \in P^+$ we consider b to belong (as long as it belongs!).

Remark 2. We show in 5.17$(*)$ that only the first assertion of the lemma can hold (when $\alpha^\vee(\lambda) = 1$ and under our hypotheses $b \in B^\sim_{j,i}$, $b \notin B^-_{j_\alpha,i}$, $b \notin \tilde{\mathcal{F}}_\alpha B^-_{j,i,b}$).

5.16.

Up until now $\mu \in P^+$ has been fixed. However we now take $\mu' = \mu + n\varpi_\alpha$, where $n+\mathbb{N}$ avoids the finite set F in the conclusion of 3.13. Specifically we take $n > \operatorname{Sup} F$, so $(n + \mathbb{N}) \cap F = \emptyset$. We claim that it will be enough to prove 5.12 for μ' to obtain it for μ. Indeed all computations involve just applying $\mathcal{F}_\alpha$ to the $G(b_\lambda) \otimes G_\mu(b)$. From the form of the coproduct (2.6) we have

$$f_\alpha(G(b_\lambda) \otimes G_\mu(b)) = f_\alpha G(b_\lambda) \otimes G_\mu(b) + q_\alpha^{\alpha^\vee(\lambda)}(G(b_\lambda) \otimes f_\alpha G_\mu(b).$$

Since λ is fixed and the decomposition of $f_\alpha G_\mu(b)$ does not depend on μ by 3.13(iv), the required claim follows. This observation will be used in various forms throughout.

5.17.

We now make use of the restriction $\alpha^\vee(\lambda) \leq 1$. (This could perhaps be avoided; but the calculation is already complicated enough.) By 5.14 we can assume $\alpha^\vee(\lambda) = 1$. By 4.8 we can replace $\hat{C}_i^-$, as used in the definition of $\mathcal{S}'_b$ in 5.15 above, by C_i^-.

Set

$$M'_b = M_{\mathcal{S}'} + \sum_{b' \in B(M_b) | \varepsilon_\alpha(b')=0} KG(b').$$

We first show that

$$(Z_i + \mathcal{F}_\alpha X_{j,i,b}) \cap (G(b_\lambda) \otimes M_b) = G(b_\lambda) \otimes M'_b. \qquad (*)$$

Indeed by 5.12 applied to $b' \in B^-_{j,i,b} \cap B^\sim_{j,i}$, the inclusion $\supset$ follows by the definition of $\mathcal{S}'_b$ and of M'_b. On the other hand by 5.15(∗∗), the left-hand side belongs to $G(b_\lambda) \otimes (M_b \cap (Y_i^- + X^-_{j_\alpha,i} + \hat{X}^-_{j,i,b}))$. Now since all terms in this latter intersection admit a global basis, it must be spanned by the $G(b') : b' \in B(M_b) \cap (C_i^- \cup B^-_{j_\alpha,i} \cup \tilde{\mathcal{F}}_\alpha B^-_{j,i,b})$. Since we have assumed $b \notin B^-_{j_\alpha,i} \cup \tilde{\mathcal{F}}_\alpha B^-_{j,i,b}$ and $b \in B^\sim_{j,i}$ (so that $b \notin C_i^-$) it follows that b cannot belong to this last intersection which is hence contained in $B(M'_b) \setminus \{b\}$. This gives the opposite inclusion in (∗).

Take μ' as in 5.16 and set $t = \ell(b) - 1$, with b viewed as element of $B(\mu')$ which we recall contains $B(\mu)$. Then S_b consists of the $t+2$ elements $\tilde{e}_\alpha b \in C_i^-$, $b \in B^-_{j,i}$, $\tilde{f}^s_\alpha b \in B^-_{j-1,i} : s = 1, 2, \dots, t$, whilst $\tilde{f}^s_\alpha b = 0$, for $s > t$. When we replace μ' by $\mu' + \varpi_\alpha$, then S_b gains one further element namely $\tilde{f}^{t+1}_\alpha b$ and similarly all other α-strings in $B(M_b)$ increase in length by 1, by 3.13(iii).

Set $\mu_m = \mu' + m\varpi_\alpha : m \in \mathbb{N}$. Fix t as above.

Lemma. *For all $m \in \mathbb{N}^+$, $s : 1 \leq s \leq m$ one has*

$$G(b_\lambda) \otimes G_{\mu_m}(\tilde{f}^{t+s}_\alpha b) \in \mathcal{F}_\alpha(G(b_\lambda) \otimes G_{\mu_m}(b)) + Z_i + \mathcal{F}_\alpha X_{j,i,b}.$$

Proof. Set

$$B(N_b)^+ = \{b' \in B(N_b) | \varepsilon(b') = 0\}.$$

It is clear that card $B(N_b)^+$ is the number of α-strings in $B(M_b)$. Indeed $B(N_b)^+$ is just the set of primitive elements (relative to α) of $B(M_b)$. By 5.15(∗) $B(N_b)^+ \subset C_i^- \cup B^-_{j,i,b}$. Observe that $\alpha^\vee(wt\ b' - wt\ b)$ is a positive even integer for all $b' \in B(N_b)^+$.

Consider the K vector space J spanned by the

$$f_\alpha^{t+\frac{1}{2}\alpha^\vee(wt\ b' - wt\ b)+1}(G(b_\lambda) \otimes G_{\mu_m}(b')) : b' \in B(N_b)^+. \qquad (**)$$

Since $f_\alpha^2 G(b_\lambda) = 0$ each $f^s_\alpha(G(b_\lambda) \otimes G_{\mu_m}(b')) : b' \in B(N_b)^+$, $s = t + \frac{1}{2}\alpha^\vee(wt\ b' - wt\ b) + 1$, has just two terms, namely $f_\alpha G(b_\lambda) \otimes f^{s-1}_\alpha G_{\mu_m}(b')$ and $G(b_\lambda) \otimes f^s_\alpha G_{\mu_m}(b')$. Moding out by the second term one checks from 3.5 by induction on length of strings that the resulting space J' has dimension exactly card

$B(N_b)^+$. This means that J' contains the $f_\alpha G(b_\lambda) \otimes G_{\mu_m}(b') : b' \in B(M_b)$ satisfying $wt\ b' - wt\ b = -t\alpha$. Consequently all such terms can be eliminated from $f_\alpha^{t+1}(G(b_\lambda) \otimes G_{\mu_m}(b))$ by subtracting appropriate linear combinations of elements of J. This is of course just linear algebra; but it is why we needed to replace A by the field K. This leaves us with a term of the form

$$x := \sum_{b' \in B(M_b) | wt\ b' - wt\ b = -(t+1)\alpha} c_{b'}(G(b_\lambda) \otimes G_{\mu_m}(b')) : c_{b'} \in K,$$

which lies in $\mathcal{F}_\alpha(G(b_\lambda) \otimes G(b)) + Z_i + \mathcal{F}_\alpha X_{j,i,b}$. By the reasoning in 5.16 the $c_{b'}$ do not depend (up to an overall multiplicative scalar) on $m \in \mathbb{N}$.

Now take $m = 0$. Then $b' := \tilde{f}_\alpha^{t+1} b = 0$, so that $B(M_b)$ may be replaced by $\mathcal{S}_b$ in the above sum. *We claim that it may further be replaced by* $\mathcal{S}'_b$. Otherwise there will be some shortest string $S'_b \subset \mathcal{S}_b$ with $S'_b \cap \mathcal{S}'_b = \emptyset$ for which $c_{b'} : b' \in S'_b$ is non-zero. Now consider the $e_\alpha^s x : s = 0, 1, 2, \dots, t+2$. These still belong to $\mathcal{F}_\alpha(G(b_\lambda) \otimes G(b)) + Z_i + \mathcal{F}_\alpha X_{j,i,b}$. Moreover notice that every $G(b''')$ that occurs in $e_\alpha^s x$ satisfies $\varepsilon_\alpha(b''') \geq \alpha^\vee(\lambda) + t + 2 - s$.

Take $b'' \in \mathcal{S}_b \setminus \mathcal{S}'_b$. Suppose $e_\alpha G(b'')$ has a term $G(b''')$ with $b''' \in \mathcal{S}'_b$. Then $b'' \in \mathcal{S}'_b$ by 3.5 and the hypothesis on $\mathcal{S}'_b$, so $G(b_\lambda) \otimes G(b'') \in Z_i + \mathcal{F}_\alpha X_{j,i,b}$. As noted in 5.15 the latter is $\mathcal{E}_\alpha$ stable. Then by 5.15(**) and the remark following it, recalling that b'' belongs to S'_b which has length $> \ell(b)$, we obtain $b''' \in C_i^- \cup \tilde{\mathcal{F}}_\alpha B_{j,i,b}^-$. If such a term occurs in $e_\alpha^s x : s \leq t+2$, then $\varepsilon_\alpha(b''') \geq \alpha^\vee(\lambda)$ by the previous paragraph and so by 4.8 again S'_b would belong to $\mathcal{S}'_b$. This contradiction excludes cancellations from strings in $\mathcal{S}_b$ strictly shorter than S'_b (necessarily lying in $\mathcal{S}'_b$). We conclude from 3.5 that each $e_\alpha^s x : s = 0, 1, 2, \dots, t+2$ has a non-zero coefficient of the unique term $G(b'')$ with $b'' \in S'_b$ of weight $(wt\ x) + s\alpha$. Consequently by (*) the $e_\alpha^s x : s = 0, 1, 2, \dots, t+2$ are linearly independent $\mod Z_i + \mathcal{F}_\alpha X_{j,i,b}$. They span a space of dimension $t+3$, whilst $t+2$ is just right hand side in Lemma 5.15. This contradicts its conclusion and proves our claim. The truth of the claim means that $x \in Z_i + \mathcal{F}_\alpha X_{j,i,b}$.

Now take $m = 1$. Since the $c_{b'}$ do not depend on m, we conclude from the above that

$$x = c_{b'}(G(b_\lambda) \otimes G_{\mu_m}(b')) \mod Z_i + \mathcal{F}_\alpha X_{j,i,b}$$

with $b' = \tilde{f}_\alpha^{t+1} b$ which is a non-zero element of $B(\mu_1)$. Finally $c_{b'} \neq 0$ by 5.13 and 3.5. This proves the lemma for the case $m = 1$. Taking $m = 2$, we must still have its conclusion for $s = 1$. For $s = 2$ we apply the above reasoning to μ_0 replaced by μ_1. This eventually gives the assertion for all $m \in \mathbb{N}^+$. □

Remark. Let N' be the smallest $\mathcal{E}_\alpha$ submodule containing $M_{\mathcal{S}'}$ and admitting a global basis. The above reasoning shows that $\mathcal{S}' \subset B(N') \subset \mathcal{S}' \cup B(N_b)^+$. In particular, $B(N')_{wt\ b} = (\mathcal{S}')_{wt\ b}$.

5.18.

We now make use of the assumption that $\mathfrak{g}$ is simply-laced so that 3.14 applies. Applying e_α to the left-hand side occurring in Lemma 5.17 we conclude that

$$G(b_\lambda) \otimes e_\alpha^s G_{\mu_m}(\tilde{f}_\alpha^{t+m} b) \in \mathcal{F}_\alpha(G(b_\lambda) \otimes G_{\mu_m}(b)) + Z_i + \mathcal{F}_\alpha X_{j,i,b}$$

$s = 0, 1, 2, \dots, t+m+1$. Moreover, since S_b is the shortest string in $B(M_b)$ it follows by 3.5 and 5.17(∗) that these elements are all linearly independent mod $Z_i + \mathcal{F}_\alpha X_{j,i,b}$. Yet $t + m + 2$ is just the value of the right-hand side of 5.15. Since $B(M'_b)_{wt\ b} = (S'_b)_{wt\ b}$ it follows by 5.17(∗) that

$$e_\alpha^{t+m} G_{\mu_m}(\tilde{f}_\alpha^{t+m} b) \in K^* G_{\mu_m}(b) + M_{\mathcal{S}'}.$$

Then by 3.14 we obtain for all $s : 0 \leq s \leq t+m$ that

$$e_\alpha^s G_{\mu_m}(\tilde{f}_\alpha^{t+m} b) \in K^* G_{\mu_m}(\tilde{f}_\alpha^{t+m-s} b) + M_{\mathcal{S}'}.$$

Since $m \in \mathbb{N}$ is arbitrary, we conclude that

$$G(b_\lambda) \otimes G_{\mu_m}(\tilde{f}_\alpha^s b) \in \mathcal{F}_\alpha(G(b_\lambda) \otimes G_{\mu_m}(b)) + Z_i + \mathcal{F}_\alpha X_{j,i,b} \qquad (*)$$

for all $s \in \mathbb{N}$. Moreover by 5.16, this assertion is valid for μ_m replaced by μ. This establishes the conclusion of 5.12 over K. □

Finally we establish 5.15(∗∗) at the next induction level in our induction on α-string length. Since we may restrict to weights of $C_i^- \cup B_{j,i}^-$, lying in some $\xi + \mathbb{Z}\alpha$, it is appropriate to take $c \in B_{j,i}^\sim$ with $\ell(c) = \ell(b) - 2$.

Recall the definition of j_α given in 5.15. Since $X_{j_\alpha,i} \subset G(b_\lambda) \otimes V_w(\mu)$, by definition it is immediate from the $(i-1)^{\text{th}}$ induction step in 5.9(ii) and 5.11(i)′ that

$$(Z_i + Z_{j,i}) \cap (G(b_\lambda) \otimes V_w(\mu)) = Y_i + X_{j_\alpha,i}.$$

Now consider $b' \in B_{j,i,c}^-$ and let S be the α-string it generates.

If $S \subset C_i^-$, then $\mathcal{F}_\alpha(G(b_\lambda) \otimes G(b'')) \subset \mathcal{F}_\alpha Y_i \subset Z_i$, for all $b'' \in S$.

If $S \not\subset C_i^-$, but $S \subset C_i^- \cup B_{j,i}^-$, then s_α must have occurred in our fixed reduced decomposition of z_j. Thus $\alpha \in \{\alpha_{j+1}, \cdots, \alpha_k\}$ in the notation of 4.6, say $\alpha = \alpha_{j_2}$. Since $\alpha = \alpha_j$, then $\alpha \neq \alpha_{j+1}$ and so $j_2 - 1 \geq j + 1$. Hence we obtain a reduced decomposition $z_j = z' s_\alpha z_{j_2}$, where $z' = s_{\alpha_{j+1}} \cdots s_{\alpha_{j_2-1}}$. Moreover we can assume $S \not\subset C_i^- \cup B_{j_2,i}^-$ and $S \subset C_i^- \cup B_{j_3,i}^-$, where $z_{j_3} = s_\alpha z_{j_2}$, that is $j_3 = j_2 - 1$. Observe that $j_3 \geq j_\alpha$. By 5.11(i)′ applied to (j_3, i), we must have

$$\mathcal{F}_\alpha(G(b_\lambda) \otimes G(b'')) \subset \mathcal{F}_\alpha(Z_{j_3,i} + Z_i) = Z_{j_3,i} + Z_i \subset Z_{j,i} + Z_i, \quad \text{for all } b'' \in S.$$

It remains to consider the case $b' \in B_{j,i}^\sim \cap B_{j,i,c}^-$. Then 5.12 applied to b', gives through the equality of dimension in Lemma 5.15 (see Remark 2 of 5.15) that

$$\sum_{n \in \mathbb{N}} K(G(b_\lambda) \otimes G(\tilde{f}_\alpha^n b')) = \mathcal{F}_\alpha(G(b_\lambda) \otimes G(b')) \mod Z_i + \mathcal{F}_\alpha X_{j,i,b'}. \qquad (**)$$

Now $\tilde{\mathcal{F}}_\alpha B_{j,i}^- \subset B_{j-1,i}^-$, so the left-hand side also belongs to $G(b_\lambda) \otimes V_w(\mu)$. Now $\ell(b') \geq \ell(b)$, so 5.15(∗∗) applies to b' through the induction hypothesis. Substitution from (∗∗) extends the validity of 5.15(∗∗) to include the term $\mathcal{F}_\alpha(G(b_\lambda) \otimes G(b'))$ occurring in $\mathcal{F}_\alpha X_{j,i,c}$. Combined with the previous cases treated above establishes 5.15(∗∗) with respect to c, as required. □

Remark 1. Let N' be the $\mathcal{E}_\alpha$ submodule of M_b defined in 5.17, Remark. Factoring out by this submodule the conclusion of 5.18 would follow if one could show that $e_\alpha^s G(\tilde{f}_\alpha^{t+m} b)$ is proportional to $G(\tilde{f}_\alpha^{t+m-s} b)$, for all $s = 1, 2, \dots, t+m-1$ (as a consequence of this being true for $s = t+m$). For example this would follow from the no-gap hypothesis of 3.9.

Remark 2. Even if we could refine 5.17 so that it applies to μ this would still not obviate the need for 3.14 because $\tilde{f}_\alpha b$ need not be zero in $B(\mu)$.

5.19.

To complete the proof of Proposition 5.12 we must show that 5.18(∗) is valid when K is replaced by A. By 5.13 and 3.5 the left-hand side lies in $\mathcal{F}_\alpha(G(b_\lambda) \otimes G(b)) + \mathcal{F}_\alpha(G(b_\lambda) \otimes G(\tilde{e}_\alpha b)) \mod (G(b_\lambda) \otimes M_{\mathcal{S}})$. Recalling that $\tilde{e}_\alpha b \in C_i^-$, this must also hold $\mod (G(b_\lambda) \otimes M_{\mathcal{S}'})$ for otherwise, passing back to K, it would contradict equality in 5.18(∗∗). □

5.20.

We now complete the proof of (i)′. In view of 4.6 and 5.12 we obtain

$$X_{j-1,i} + Z_i \subset \mathcal{F}_{z_{j-1}}(G(b_\lambda) \otimes G(b_i)) + Z_i.$$

Now by (ii) of Theorem 5.9 at the $(i-1)^{\text{th}}$ induction step, the sum in the left-hand side is direct and so $X_{j-1,i}$ embeds in $(\mathcal{F}_{z_{j-1}}(G(b_\lambda) \otimes G(b_i)) + Z_i)/Z_i$. The latter is just an image of $V_{z_{j-1}}^{\mathbb{Z}}(v_i)$ which has formal character ch $B_{z_{j-1}}(v_i)$ by 5.4. Yet $B_{z_{j-1}}(v_i) \cong b_\lambda \otimes \tilde{\mathcal{F}}_{z_{j-1}} b_i$ and the right-hand side is just the formal character of $X_{j-1,i}$. Since in addition $X_{j-1,i}$ has a basis formed from a subset of the basis of $V^{\mathbb{Z}}(\lambda) \otimes V^{\mathbb{Z}}(\mu)$, this forces equality throughout and proves (i)′. □

We now complete the proof (of the ith step) of Theorem 5.9. Part (i) results from 5.11(i)′ taking $j = 0$. Since $Y_i \subset Z_i$ by the induction hypothesis, part (i) then gives

$$Y_{i+1} \subset \mathcal{F}_{y_i}(G(b_\lambda) \otimes G(b_i)) + Z_i \subset Z_{i+1}$$

which gives the inclusion $\supset$ in (ii).

Recall 5.3 and consider $\subset$ in (ii). Since Y_{i+1} is the free A module with basis a subset of the basis of the free A module $G(b_\lambda) \otimes V_w^{\mathbb{Z}}(\mu)$ it suffices to show that the inclusion

$$KY_{i+1} \subset G(b_\lambda) \otimes V_w(\mu) \cap \sum_{j=1}^{i} U_q(\mathfrak{n}^-)(G(b_\lambda) \otimes G(b_j)) =: \hat{N} \qquad (*)$$

is an equality. This will be proved by showing that these KT modules have the same formal character. Here ch M will always be with respect to the ring over which the weight submodules of M are defined (and free). We write ch $M \geq$ ch M' if $rk\ M_\nu \geq rk\ M'_\nu$, for all $\nu \in P$.

Set

$$N = G(b_\lambda) \otimes V_w(\mu) \cap \sum_{j=1}^{i} A^+ \tilde{\mathcal{F}}(G(b_\lambda) \otimes G(b_j)).$$

Then $\hat{N} \supset KN \supset \bigcup_{s \in \mathbb{N}} q^{-s} N \supset \hat{N}$. Again N is an A^+ submodule of the free A^+ module $M := L(\lambda) \otimes L(\mu)$. Since the latter has finite rank A^+ weight submodules and A^+ is a principal ideal domain it follows that N is a free A^+ module.

Consider $S = \sum_{j=1}^{i} A^+ \tilde{\mathcal{F}}(G(b_\lambda) \otimes G(b_j))$. It is a free A^+ module by the argument above. Set $B = \bigcup_{j=1}^{i} \tilde{\mathcal{F}}(b_\lambda \otimes b_j)$ and $B' = \bigcup_{j=1}^{i} b_\lambda \otimes \tilde{\mathcal{F}}_{y_j} b_j$. Recall (4.4, 4.6) that $B = \coprod_{j=1}^{i} B(\nu_j)$ and $B' = B \cap b_\lambda \otimes B_w(\mu)$.

By 5.8, $G(b_\lambda) \otimes G(b_i) \mod Z_i$ is a highest weight vector of weight ν_i generating Z_{i+1}/Z_i. Hence ch S = ch $Z_{i+1} \leq \sum_{j=1}^{i}$ch $V(\nu_j)$ = ch B. On the other hand by [J2, 5.1.12(i)] the image of S in M/qM is $\mathbb{Q}B$. For each $b_1 \otimes b_2 \in B_\nu$ we may view $G(b_1) \in L(\lambda)$, $G(b_2) \in L(\mu)$ as representatives of b_1, b_2. By the above there exists $c(b_1, b_2) \in qM_\nu$ such that $G(b_1) \otimes G(b_2) + c(b_1, b_2) \in S_\nu$. Since $rk\ S_\nu < \infty$ and A^+ is a local ring, these elements form an A^+ basis of S_ν. Consequently N is contained in the A^+ module generated by the subset of those elements for which $b_1 \otimes b_2 \in B'$. Hence ch $\hat{N}$ = ch $N \leq$ ch B' = ch Y_{i+1} = ch Z_{i+1}, as required. □

5.21.

We give an example when $\mathcal{S}'_b \subsetneq \mathcal{S}_b$. Take $\pi = \{\alpha, \beta\}$ of type A_2 and set $\lambda = \varpi_\alpha + \varpi_b$. Take $w = s_\alpha s_\beta$. Then in the convention of 3.9, taking μ sufficiently large, we have $B(\lambda)^\star \cap B_w(\mu) = \{1, \tilde{f}_\alpha, \tilde{f}_\beta, \tilde{f}_\beta \tilde{f}_\alpha, \tilde{f}_\beta \tilde{f}_\alpha^2\}$. Recall (3.9) that $G(\tilde{f}_\beta \tilde{f}_\alpha^n) = f_\beta f_\alpha^{(n)}$, $\forall\, n \in \mathbb{N}$. Take $b = \tilde{f}_\beta \tilde{f}_\alpha^2$. Since $e_\alpha G(\tilde{f}_\beta \tilde{f}_\alpha^2) = G(\tilde{f}_\beta \tilde{f}_\alpha)$ and $e_\alpha G(\tilde{f}_\beta \tilde{f}_\alpha) = G(\tilde{f}_\beta)$, it follows that $\mathcal{S}'_b = \emptyset$, whilst $\mathcal{S}_b = \{\tilde{f}_\alpha^n \tilde{f}_\beta : n \in \mathbb{N}\}$. In this case one may check that the conclusion of 5.12 directly using the Serre relation $f_\alpha^{(2)} f_\beta - f_\alpha f_\beta f_\alpha + f_\beta f_\alpha^{(2)} = 0$.

Notice that $\tilde{f}_\alpha \tilde{f}_\beta \in B(\lambda)^\star \cap B(\mu)$. One can hope to rework the proof of 5.9 by including the primitive element $b_\lambda \otimes \tilde{f}_\alpha \tilde{f}_\beta$ in the induction step. This would force $\mathcal{S}'_b = \mathcal{S}_b$ in the above; but whether this equality can be made to hold in general is another question.

5.22.

We now relax the hypothesis that $\alpha^\vee(\lambda) \leq 1 : \alpha \in \pi$. Recall the definitions in 4.6(∗), 4.7 and 5.7. By 5.6 and 5.10 and induction on the order relation $\prec$, we obtain the

Theorem ($\mathfrak{g}$ is simply-laced). *For all $\lambda, \mu \in P^+$, $w \in W$, one has a $U_q^{\mathbb{Z}}(\mathfrak{n}^+)$ filtration $0 = Y_1 \subset Y_2 \subset \cdots \subset Y_{n+1} - G(b_\lambda) \otimes V_w^{\mathbb{Z}}(\mu)$, with Y_{i+1}/Y_i isomorphic to the Demazure module $V_{y_i}^{\mathbb{Z}}(\nu_i)$ with generator $G(b_\lambda) \otimes G(b_i)$ mod $Y_i : i = 1, 2, \ldots, n$.*

5.23.

In the above theorem we would like to make the stronger assertion that Y_{i+1}/Y_i identifies with

$$G(b_\lambda) \otimes \bigoplus_{b \in B_i^-(\mu)} AG(b).$$

This is exactly what results from 5.9 and 5.10 in the case when $\alpha^\vee(\lambda) \leq 1$, $\forall \alpha \in \pi$. To deduce this result by the induction argument used in 5.22, we just need to show that the image of $G(b_\lambda) \otimes G(b) : b \in B_i^-(\mu)$ in $Z_{i+1}/Z_i \cong V^{\mathbb{Z}}(\nu_i)$ is an element of the canonical basis of $V^{\mathbb{Z}}(\nu_i)$. This is an interesting question in its own right and would give an elegant interpretation of the purely combinatorial fact expressed by $\tilde{\mathcal{F}}_{y_i}(b_\lambda \otimes b_i) = b_\lambda \otimes \tilde{\mathcal{F}}_{y_i} b_i = b_\lambda \otimes B_i^-(\mu)$. An indication that this holds is obtained from 5.13. For example if we take $b = b_i$, it follows from 5.8(∗) that all terms in the right-hand side of 5.13(∗) except the first lie in Z_i. Hence

$$G(b_\lambda) \otimes G(\tilde{f}_\alpha^n b_i) = f_\alpha^{(n)}(G(b_\lambda) \otimes G(b_i)) \mod Z_i.$$

Yet by 5.8(∗) again, the image of $G(b_\lambda) \otimes G(b_i)$ in $Z_{i+1}/Z_i \cong V^{\mathbb{Z}}(\nu_i)$ can be taken to be $G(b_{\nu_i})$. Then $f_\alpha^{(n)}(G(b_\lambda) \otimes G(b_i))$ has image $G(\tilde{f}_\alpha^n b_{\nu_i})$, as required. This proves in particular that our question has a positive answer for $\mathfrak{sl}(2)$. Notice also that by induction on $\prec$, it is enough to resolve this question for λ fundamental to obtain the general case. Finally by 2.6 the presence of $G(b_\lambda)$ does not affect the action of the $e_\alpha^{(n)}$, that is $e_\alpha^{(n)}(G(b_\lambda) \otimes G(b)) = G(b_\lambda) \otimes e_\alpha^{(n)} G(b)$. In some sense this is dual to 3.12(∗).

A further advantage of the more precise result we seek is that the resulting Demazure flag on $G(b_\lambda) \otimes V_w^{\mathbb{Z}}(\mu)$ would only depend on how we lift the partial order on I induced by the order relation $\geq$ on P to a linear order (see 5.5). In this sense it would be canonical, improving thereby the Mathieu–Polo flags described implicitly by van der Kallen's criterion (see 1.1). In any case the multiplicity of the quotients being independent of the chosen flag was already immediate from the linear independence of the Demazure characters $\Delta_{y_i} e^{\nu_i}$.

5.24. Lemma. *For all $\lambda, \mu \in P^+$ one has*

$$U_q^{\mathbb{Z}}(\mathfrak{n}^-)(G(b_\lambda) \otimes V^{\mathbb{Z}}(\mu)) = V^{\mathbb{Z}}(\lambda) \otimes V^{\mathbb{Z}}(\mu).$$

Proof. Recall that $U_q^{\mathbb{Z}}(\mathfrak{n}^-)G(b_\lambda) = V^{\mathbb{Z}}(\lambda)$. Then the assertion follows from 2.6 by induction on weight submodules of the first factor with respect to the order relation $\leq$ on weights. □

5.25.

Recall 5.8 and set $V_\infty^{\mathbb{Z}}(\mu) = V^{\mathbb{Z}}(\mu)$, $\forall \mu \in P^+$.

Theorem (𝔤 simply-laced). *For all $\lambda, \mu \in P^+$, $w \in \{W, \infty\}$ one has an increasing exhaustive filtration $Z_1 \subset Z_2 \subset \cdots$, of $U_q^{\mathbb{Z}}(\mathfrak{g})(G(b_\lambda) \otimes V_w^{\mathbb{Z}}(\mu))$ by $U_q^{\mathbb{Z}}$ modules with $Z_{i+1}/Z_i \cong V^{\mathbb{Z}}(\nu_i)$, for all $i \in I$.*

Proof. By 5.22, $G(b_\lambda) \otimes G(b_i) \mod Z_i$ is a highest weight vector and so Z_{i+1}/Z_i is an image of $V^{\mathbb{Z}}(\nu_i)$. Now each weight submodule of $V^{\mathbb{Z}}(\nu_i)$ has an A basis (provided by the global or canonical basis) and so is free of finite rank. Since A is a domain any proper quotient of $V^{\mathbb{Z}}(\nu_i)$, even as just an A module, must admit at least one weight submodule of strictly smaller rank. Since rank is additive on exact sequences, any Z_{i+1}/Z_i being a proper image of $V^{\mathbb{Z}}(\nu_i)$ would imply that $U_q^{\mathbb{Z}}(\mathfrak{n}^-)(G(b_\lambda) \otimes V^{\mathbb{Z}}(\mu))$ has a weight submodule of strictly smaller rank than that given by the sum of the ranks of the corresponding weight submodules of the $V^{\mathbb{Z}}(\nu_i)$. Now the b_i are just those elements of $B(\lambda) \otimes B(\mu)$ for which $b_\lambda \otimes b_i$ is primitive so (see 4.3 and 4.4)

$$B(\lambda) \otimes B(\mu) = \coprod_{i \in I} B(\nu_i),$$

which implies

$$\operatorname{ch}(V^{\mathbb{Z}}(\lambda) \otimes V^{\mathbb{Z}}(\mu)) = \sum_{i \in I} \operatorname{ch} V^{\mathbb{Z}}(\nu_i).$$

Consequently this latter sum is just the rank (which is again finite) of the corresponding weight submodule of $V^{\mathbb{Z}}(\lambda) \otimes V^{\mathbb{Z}}(\mu)$. Combined with 5.24 this gives a contradiction proving the theorem. □

6. The PRV theorem

6.1.

Take $\lambda, \mu \in P^+, w \in W$. Parthasarathy–Ranga Rao–Varadarajan conjectured that $V(\lambda) \otimes V(\mu)$ admits an irreducible component with extreme weight $\lambda + w\mu$ (that is its highest weight ν was the unique dominant element in $W(\lambda + w\mu)$). Set $W_\lambda = \operatorname{Stab}_w \lambda$. Obviously ν does not depend on the choice of w in its double coset $W_\lambda w W_\mu$ and one would not expect the irreducible component to depend on this choice either. Kostant then refined this conjecture to determine $V(\nu)$ canonically from λ, μ, w by taking it to be the presumed unique component of this highest weight in $U(\mathfrak{g})(u_\lambda \otimes u_{w\mu})$. This refinement was proved independently by Kumar [Ku1] and Mathieu [M2]. Later Littelmann gave a crystal basis argument [L2] which combined with Kashiwara's globalization technique provided a further proof [J3]. It was a crucial element in the main result of [J3].

6.2.

The proofs of Mathieu and Kumar were based on the annihilator formula mentioned in 1.1 together with Frobenius splitting. That which derives from Littelmann though short became a little messy especially if multiplicities had to be taken into account. Here we note that the result is a very natural consequence of our main theorem and indeed holds in this more general simply-laced Kac–Moody setting. Indeed we show that the required $V(\nu)$ is generated from the "radical" of $G(b_\lambda) \otimes V_w^{\mathbb{Z}}(\mu)$. More precisely our Demazure flag of the latter admits $V_w^{\mathbb{Z}}(\nu)$ as its top component. Since the corresponding primitive elements $b^w \in B_\lambda \otimes B_w(\mu)$ are distinct (6.3) for distinct double cosets so are the resulting $V(\nu)$.

6.3.

By 5.3 it follows that

$$U_q^{\mathbb{Z}}(\mathfrak{n}^+)(G(b_\lambda) \otimes G(b_{w\mu})) = G(b_\lambda) \otimes V_w^{\mathbb{Z}}(\mu).$$

By 5.22, there exists a Demazure flag

$$0 = Y_1 \subsetneq Y_2 \subsetneq \cdots \subsetneq Y_{n+1} = G(b_\lambda) \otimes V_w^{\mathbb{Z}}(\mu).$$

Consequently $G(b_\lambda) \otimes G(b_{w\mu})$ admits a non-zero image in $Y_{n+1}/Y_n \cong V_{y_n}^{\mathbb{Z}}(\nu_n)$ and is a generator for the latter. In particular, $y_n\nu_n = \lambda + w\mu$. Moreover by 4.5 there exists a unique primitive element $b^w \in b_\lambda \otimes B_w(\mu)$ generating the copy of $B_{y_n}(\nu_n)$ corresponding to $V_{y_n}(\nu_n)$. This is just $b^w = b_\lambda \otimes b_n$ in our previous indexation. More precisely b_n is the (necessarily unique) element of $B(\lambda)^\star \cap B_w(\mu)$ of minimal weight with respect to $\leq$. (Thus we obtain b_n *automatically* whilst Littelmann had to do some fancy footwork [L2]). It remains to show that

Lemma. $b^{w'} = b^w \iff w' \in W_\lambda w W_\mu$.

Proof. Recall that there exist $y, y' \in W$ such that $b_\lambda \otimes b_{w\mu} \in \tilde{\mathcal{F}}_y b^w$, $b_\lambda \otimes b_{w'} \in \tilde{\mathcal{F}}_{y'} b^{w'}$. Moreover $b^w, b^{w'}$ are primitive elements of their respective highest weight crystals in the disjoint union 4.3(∗). Thus

$$b^{w'} = b^w \iff \tilde{\mathcal{E}}(b_\lambda \otimes b_{w'\mu}) \cap \tilde{\mathcal{E}}(b_\lambda \otimes b_{w\mu}) \neq \emptyset.$$

Now $\varphi_\alpha(b_\lambda) = \alpha^\vee(\lambda)$, whilst $\varepsilon_\alpha(w'\mu) = \max\{0, -\alpha^\vee(w'\mu)\}$. Consequently by the tensor product rule (4.3) we can replace w (resp. w') by its unique minimal length element w_0 (resp. w_0') in its $W_\lambda \setminus W/W_\mu$ double coset. This gives $\Longleftarrow$ of the lemma. For the opposite implication it suffices to show that w_0 is the unique minimal element (for the Bruhat order) such that $b^w \in b_\lambda \otimes B_{w_0}(\mu)$. This was shown in [J3, 2.7(ii)] using the Littelmann path model. Indeed b^w is given by a Bruhat sequence [J2, 6.4.2] which strictly ends in w_0, that is to say the LS partition of $wt\ b^w$ corresponding to b^w [J2, 6.4.2] has a non-zero coefficient of $w_0(wt\ b^w)$. (Regrettably we could not find anything easier. A more detailed version of the proof of this last assertion can be found in [J5, Sect. 17]. □

Index of Notation

Symbols occurring frequently are given below at the place they are first defined.

1.4. $\mathfrak{g}$.
2.2. q, K, ψ, A, A^+, A^-, $[n]_q$, $[n]_q^!$, $\left[\begin{smallmatrix} n \\ m \end{smallmatrix}\right]_q$, $U_q(\mathfrak{g})$, e_i, f_i, t_i, π, P, P^+, $\alpha^\vee$, q_i, $e_i^{(n)}$, $f_i^{(n)}$, $U_q^{\mathbb{Z}}(\mathfrak{n}^+)$, $U_q^{\mathbb{Z}}(\mathfrak{n}^-)$, $\mathcal{E}$, $\mathcal{F}$, T, $U_q^{\mathbb{Z}}(\mathfrak{g})$, $U_q(\mathfrak{s}_\alpha)$, $U_q^{\mathbb{Z}}(\mathfrak{s}_\alpha)$.
2.3. $V(\lambda)$, $\tilde{f}_\alpha$, $\tilde{e}_\alpha$, $\tilde{\mathcal{E}}$, $\tilde{\mathcal{F}}$.
2.4. u_λ, $L(\lambda)$, $B(\lambda)$.
2.5. $V^{\mathbb{Z}}(\lambda)$, $L^{\mathbb{Z}}(\lambda)$, $L(\lambda)^-$, $L^{\mathbb{Z}}(\lambda)^-$, $L(\infty)$, $L(\infty)^-$, E, G_λ.
2.6. Δ.
3.5. ε_α, φ_α, $\ell_\alpha(b)$, $I^r(B)$, $W^r(B)$, $B_{\xi+\mathbb{Z}\alpha}$, $V_{\xi+\mathbb{Z}\alpha}$, $V_{\xi+\mathbb{Z}\alpha}^{\mathbb{Z}}$.
3.6. $\mathcal{E}_\alpha$, $\mathcal{F}_\alpha$, $\tilde{\mathcal{E}}_\alpha$, $\tilde{\mathcal{F}}_\alpha$.
3.7. $W_s(V_{\xi+\mathbb{Z}\alpha}^{\mathbb{Z}})$, $B(N)$.
3.10. $\succ$.
3.11. ρ, ϖ_α.
4.1. wt.
4.2. b_λ.
4.4. $\star$.
4.5. W, $\tilde{\mathcal{E}}_w$, $\tilde{\mathcal{F}}_w$.
4.6. I, y_i, v_i, b_i, z_j.
4.7. B_i^-, $B_{j,i}^-$.
4.8. $\hat{B}_i^-$.
5.1. $b_{w\mu}$.
5.2. $\mathcal{E}_w$, $\mathcal{F}_w$.
5.3. $V_w^{\mathbb{Z}}(\mu)$.
5.4. $\mathrm{ch}\, M$, $\mathrm{ch}\, B$, Δ_w.
5.5. $\geq$.
5.7. X_i^-, Y_i^-, X_i, Y_i, C_i^-, $\hat{C}_i^-$, $\hat{X}_i^-$, $\hat{Y}_i^-$, $\hat{X}_i$, $\hat{Y}_i$.
5.8. Z_i.
5.11. $X_{j,i}^-$, $X_{j,i}$.
5.12. $B_{j,i}^\sim$, $B_{j,i,b}^-$, $X_{j,i,b}^-$, $X_{j,i,b}$, $\hat{X}_{j,i,b}^-$, $\hat{X}_{j,i,b}$.
5.13. N_b, M_b, $\mathcal{S}_b$, $M_\mathcal{S}$.
5.15. $\mathcal{S}_b'$, $M_{\mathcal{S}'}$, j_α, $Z_{j,i}$.
5.17. μ_m', $B(N_b)^+$.

Acknowledgments

I would like to thank Mrs. Annie Abraham for her patience in typing this manuscript.

Some corrections were made to this manuscript whilst I visited the Department of Mathematics of Sydney University during November 2003–January 2004, where the main results were presented in a workshop held during 12–14 November, 2003. I would like to thank the members for their hospitality. These results were also presented at a meeting in Luminy during 29 March–2 April, 2004. A few further corrections and clarifications were made in August 2004 and February 2005.

References

[B] P. Baumann, Canonical bases and the conjugation representation of a semisimple Lie group, *Pacific J. Math.* **206** (2002), 25–37.

[D] M. Demazure, Une nouvelle formule des caractères, *Bull. Sci. Math. (2)* **98** (1974), no. 3, 163–172.

[Do] S. Donkin, *A filtration for rational modules, Math. Z.* **177** (1981), 1–8.

[Do] S. Donkin, *Rational Representations of Algebraic Groups, LNM* 1140, Springer, Berlin 1985.

[G] W.A. de Graaf, Constructing canonical bases of quantized enveloping algebras, *Exper. Math.* **11** (2002), 161–170.

[J1] A. Joseph, On the Demazure character formula, *Ann. Sci. École Norm. Sup. (4)* **18** (1985), no. 3, 389–419.

[J2] A. Joseph, *Quantum Groups and Their Primitive Ideals*, Ergebnisse der Mathematik und ihrer Grenzgebiete (3) **29**, Springer-Verlag, Berlin, 1995.

[J3] A. Joseph, On a Harish-Chandra homomorphism, *C. R. Acad. Sci. Paris Sér. I Math.* no. 7 **324** (1997), 759–764.

[J4] A. Joseph, A decomposition theorem for Demazure crystals, *J. Algebra*, **265** (2003), 562–578.

[J5] A. Joseph, Lie algebras, their representations and crystals, Lecture Notes, Weizmann 2004, available from www.wisdom.weizmann.ac.il/~gorelik/agrt.htm.

[Ka] M. Kaneda, Based modules and good filtrations in algebraic groups, *Hiroshima Math. J.*, **28** (1998), 337–344.

[K1] M. Kashiwara, On crystal bases of the q-analogue of the enveloping algebra, *Duke Math. J.* no. 2, **63** (1991), 465–516.

[K2] M. Kashiwara, The crystal basis and Littelmann's modified Demazure formula, *Duke Math. J.* **71**, no. 3 (1993), 839–858.

[K3] M. Kashiwara, Global crystal bases of quantum groups, *Duke Math. J.*, **69** (1993), 455–485.

[Ku1] S. Kumar, Proof of the Parthasarathy–Ranga Rao–Varadarajan conjecture, *Invent. Math.*, no. 1 **93** (1988), 117–130.

[Ku2] S. Kumar, A refinement of the PRV conjecture, *Invent. Math.* no. 2, **97** (1989), 305–311.

[LLM] V. Lakshmibai, P. Littelmann and P. Magyar, *Standard monomial theory for Bott–Samelson varieties, Compos. Math.* **130** (2002), 293–318.

[L1] P. Littelmann, A Richardson–Littlewood formula for symmetrizable Kac–Moody algebras, *Invent. Math.* **116** (1994), 329–346.

[L2] P. Littelmann, Path and root operators in representation theory, *Ann. Math.* **142** (1995), 499–525.

[L3] P. Littelmann, The path model, the quantum Frobenius map and standard monomial theory, *Algebraic Groups and Their Representations* (Cambridge, 1997), 175–212, *NATO Adv. Sci. Inst. Ser. C Math. Phys. Sci.*, **517**, Kluwer Acad. Publ., Dordrecht, 1998.

[L4] P. Littelmann, Contracting modules and standard monomial theory for symmetrizable Kac–Moody algebras, *J. Amer. Math. Soc.* **11** (1998), 551–567.

[L5] P. Littelmann, Cones, crystals and patterns, *Trans. Groups* **3** (1998), 145–179.

[Lu1] G. Lusztig, Canonical bases arising from quantized enveloping algebras, *J.A.M.S.* **3** (1990), 447–498.

[Lu2] G. Lusztig, Introduction to Quantum Groups, *Prog. Math.*, **110**, Birkhäuser Boston, 1993.

[Lu3] G. Lusztig, Tight monomials in quantized enveloping algebras, in *Quantum deformations of algebras and their representations*, Israel Mathematical Conference Proceedings, **7** (1993), 117–132.

[M1] O. Mathieu, Filtrations of B-modules, *Duke Math. J.*, no. 2, **59** (1989), 421–442.

[M2] O. Mathieu, Construction d'un groupe de Kac–Moody et applications, *Compos. Math.*, no. 1, **69** (1989), 37–60.

[M3] O. Mathieu, Filtrations of G-modules, *Ann. Sci. École Norm. Sup. (4)*, **23** (1990), 625–644.

[M4] O. Mathieu, Positivity of some intersections in $K_0(G/B)$. Commutative algebra, homological algebra and representation theory (Catania/Genoa/Rome), (1998), *J. Pure Appl. Algebra,* no. 1–3 **152** (2000), 231–243.

[M5] O. Mathieu, Frobenius actions on the B-cohomology, *Adv. Ser. Math. Phys.*, Vol. 7, 1989, World Scientific, Singapore, 39–51.

[Pa] J. Paradowski, Filtrations of modules over quantum algebras, *Proc. Symp. Pure Math.* **56** (1994), 93–108

[P] P. Polo, Variétés de Schubert et excellentes filtrations. Orbites unipotentes et représentations, III,. *Astérisque No. 173–174* (1989), 10–11, 281–311.

[R] S. Ryom-Hansen, *A q-analogue of Kempf's vanishing theorem, Moscow Math. J.* **3** (2003), 173–187.

[V1] W. van der Kallen, Longest weight vectors and excellent filtrations, *Math. Z.*, no. 1, **201** (1989), 19–31.

[V2] W. van der Kallen, Lectures on Frobenius splittings and B modules, Tata Institute Lecture Notes, Springer, Berlin, 1993.

[W] J.-P. Wang, Sheaf cohomology on G/B and tensor products of Weyl modules, *J. Alg.* **77** (1982), 162–185.

Microlocalization of ind-sheaves

M. Kashiwara[1*], P. Schapira[2], F. Ivorra[3] and I. Waschkies[4†]

[1] Research Institute for Mathematical Sciences
Kyoto University
Kyoto 606–8502
Japan
masaki@kurims.kyoto-u.ac.jp

[2] Université Pierre et Marie Curie
Institut de Mathématiques
175 rue du Chevaleret
75013 Paris, France
schapira@math.jussieu.fr

[3] Université Pierre et Marie Curie
Institut de Mathématiques
175 rue du Chevaleret
75013 Paris, France
fivorra@math.jussieu.fr

[4] Université de Nice - Sophia Antipolis Laboratoire J.A.
Dieudonné Parc Valrose 06108
Nice, France
ingo@math.nice.fr

Summary. Let X be a C^∞-manifold and T^*X its cotangent bundle. We construct a microlocalization functor $\mu_X : \mathrm{D^b}(\mathrm{I}(\mathbb{K}_X)) \longrightarrow \mathrm{D^b}(\mathrm{I}(\mathbb{K}_{T^*X}))$, where $\mathrm{D^b}(\mathrm{I}(\mathbb{K}_X))$ denotes the bounded derived category of ind-sheaves of vector spaces on X over a field $\mathbb{K}$. This functor satisfies $\mathrm{R}\mathcal{H}om(\mu_X(\mathcal{F}), \mu_X(\mathcal{G})) \simeq \mu hom(\mathcal{F}, \mathcal{G})$ for any $\mathcal{F}, \mathcal{G} \in \mathrm{D^b}(\mathbb{K}_X)$, thus generalizing the classical theory of microlocalization. Then we discuss the functoriality of μ_X. The main result is the existence of a microlocal convolution morphism

$$\mu_{X\times Y}(\mathcal{K}_1) \overset{a}{\circ} \mu_{Y\times Z}(\mathcal{K}_2) \longrightarrow \mu_{X\times Z}(\mathcal{K}_1 \circ \mathcal{K}_2)$$

which is an isomorphism under suitable non-characteristic conditions on $\mathcal{K}_1$ and $\mathcal{K}_2$.

Key words: microlocalization, ind-sheaves.

Subject Classifications: Primary: 35A27; Secondary: 32C38

* The author M.K. is partially supported by Grant-in-Aid for Scientific Research (B1) 13440006, Japan Society for the Promotion of Science and the 21st century COE program "Formation of an International Center of Excellence in the Frontier of Mathematics and Fostering of Researchers in Future Generations."

† The author I.W. is partially supported by the same 21st century COE program.

0. Introduction

This paper is based on ideas of the authors M.K. and P.S. announced in [KS5] and developed in a preliminary manuscript of M.K.

The idea of microlocalization goes back to M. Sato [S] in 1969 who invented the functor of microlocalization of sheaves (along a smooth submanifold of a real manifold) in order to analyze the singularities of hyperfunction solutions of systems of differential equations in the cotangent bundle. This microlocalization procedure then allowed Sato, Kashiwara and Kawai [SKK] to define functorially the sheaf of rings of microdifferential operators on the cotangent bundle T^*X of a complex manifold X, a sheaf whose direct image is the sheaf of differential operators on X.

Then in the 1980s, M.K. and P.S. (See [KS2], [KS3]) developed a microlocal theory of sheaves on a C^∞-manifold X, based on the notion of microsupport (a conic involutive closed subset of the cotangent bundle to X) and introduced in particular the functor μhom. This is roughly speaking a functor that associates to a pair of sheaves on X the sheaf of microlocal morphisms between them.

On the other hand, the Riemann–Hilbert problem, solved by M.K., tells us that there is a one-to-one correspondence between the regular holonomic modules over the ring of differential operators and the perverse sheaves. The notion of regular holonomic modules over the ring of differential operators can be easily microlocalized to the notion of regular holonomic modules over the ring of microdifferential operators and it is a natural question to ask if there is a natural notion of microlocalization of perverse sheaves, or, more generally a functor μ of microlocalization for sheaves, the microsupport of a sheaf being the support of its microlocalization and the functor μhom being the internal *hom* applied to the microlocalization. This is indeed what we do in this paper.

As an application of the new functor μ, the author I.W. [W] has recently constructed the stack of microlocal perverse sheaves on the cotangent bundle, after M.K. [K] had constructed the stack of microdifferential modules.

The paper consists of two parts. The first is the technical heart of the paper. We define kernels on a C^∞-manifold X, attached to the data of a closed submanifold Z and a 1-form σ vanishing on Z. Then we study its functorial properties. These kernels can be seen as "general" microlocalization kernels, although their only role in this paper is to provide us with the tools for the proofs of the functorial properties of μ.

In the second part we introduce the functor μ, which is the integral transform with respect to the kernel K_{T^*X} on $T^*X \times T^*X$ associated with the fundamental 1-form. We discuss the functorial properties of μ, deduced from the corresponding properties of the kernels studied in the first part. We then show how some classical microlocal properties can be generalized to ind-sheaves. We give a comparison theorem between the micro-support of ind-sheaves $\mathcal{F}$ and the support of its microlocalization $\mu(\mathcal{F})$.

As an application, we prove that, on a complex manifold X, μhom induces a well-defined functor

$$\mu hom(\bullet, \mathcal{O}_X)\colon \mathrm{D}^{\mathrm{b}}(\mathbb{C}_X)^{\mathrm{op}} \to \mathrm{D}^{\mathrm{b}}(\mathcal{E}_X),$$

where $\mathcal{E}_X$ is the ring of microdifferential operators.

1. Microlocal kernels

Throughout this paper, $\mathbb{K}$ denotes a field.

1.1. Review on Ind-sheaves on manifolds

In this section we shall give a short overview on the theory of ind-sheaves of [KS1].

Let X be a locally compact topological space with finite cohomological dimension, $\mathrm{Mod}(\mathbb{K}_X)$ the category of sheaves of $\mathbb{K}$-vector spaces on X, and $\mathrm{Mod}^{\mathrm{c}}(\mathbb{K}_X)$ its full subcategory of sheaves with compact supports.

We denote by $\mathrm{I}(\mathbb{K}_X)$ the category of ind-sheaves, which is by definition the category of ind-objects of $\mathrm{Mod}^{\mathrm{c}}(\mathbb{K}_X)$. Then $\mathrm{I}(\mathbb{K}_X)$ is an abelian category, and its bounded derived category is denoted by $\mathrm{D}^{\mathrm{b}}(\mathrm{I}(\mathbb{K}_X))$.

There is a fully faithful exact functor

$$\iota_X \colon \mathrm{Mod}(\mathbb{K}_X) \to \mathrm{I}(\mathbb{K}_X) \quad \text{given by} \quad F \mapsto \text{“}\varinjlim_{U \subset\subset X}\text{”}\, F_U,$$

where the direct limit on the right is taken over the family of relatively compact open subsets U of X. In the sequel, we will regard $\mathrm{Mod}(\mathbb{K}_X)$ as a full subcategory of $\mathrm{I}(\mathbb{K}_X)$.

The functor ι_X admits an exact left adjoint functor

$$\alpha_X \colon \mathrm{I}(\mathbb{K}_X) \to \mathrm{Mod}(\mathbb{K}_X), \quad \text{“}\varinjlim_{i\in I}\text{”}\, F_i \mapsto \varinjlim_{i\in I} F_i .$$

Since ι_X is fully faithful, we have $\alpha_X \circ \iota_X \simeq \mathrm{Id}_{\mathrm{Mod}(\mathbb{K}_X)}$. The functor α_X admits an exact fully faithful left adjoint

$$\beta_X \colon \mathrm{Mod}(\mathbb{K}_X) \to \mathrm{I}(\mathbb{K}_X).$$

Since β_X is fully faithful, we get $\alpha_X \circ \beta_X \simeq \mathrm{Id}_{\mathrm{Mod}(\mathbb{K}_X)}$. The functor β_X is less easy to define than α_X and ι_X. However, for a locally closed subset $S \subset X$,

$$\widetilde{\mathbb{K}}_S := \beta_X(\mathbb{K}_S)$$

is described as follows. Let Z be a closed subset; then we have

$$\widetilde{\mathbb{K}}_Z \simeq \text{“}\varinjlim_{Z \subset W}\text{”}\, \mathbb{K}_{\overline{W}},$$

where W runs through the open subsets containing Z. If $U \subset X$ is an open subset, then

$$\widetilde{\mathbb{K}}_U \simeq \text{“}\varinjlim_{V \subset\subset U}\text{”}\, \mathbb{K}_V,$$

where V runs through the family of relatively compact open subsets of U. If $S \subset X$ is locally closed, then we can write $S = Z \cap U$ where U is open and Z is closed, and

$$\widetilde{\mathbb{K}}_S \simeq \widetilde{\mathbb{K}}_U \otimes \widetilde{\mathbb{K}}_W \simeq \text{“}\varinjlim_{V \subset\subset U,\, Z \subset W}\text{”}\, \mathbb{K}_{V \cap \overline{W}} .$$

Therefore $\mathbb{K}_{V\cap\overline{W}} \to \mathbb{K}_S$ induces a morphism $\widetilde{\mathbb{K}}_S \to \mathbb{K}_S$ which is not an isomorphism in general.

Note that if Z is closed and $S \subset Z$ is a locally closed subset, then

$$\mathbb{K}_S \otimes \widetilde{\mathbb{K}}_Z \simeq \mathbb{K}_S\,.$$

The machinery of Grothendieck's six operations is also applied to this context. We have the functors:

$$\begin{aligned}
f^{-1},\ f^{!} &: \mathrm{D}^{\mathrm{b}}(\mathrm{I}(\mathbb{K}_Y)) \to \mathrm{D}^{\mathrm{b}}(\mathrm{I}(\mathbb{K}_X)),\\
\mathrm{R}f_*,\ \mathrm{R}f_{!!} &: \mathrm{D}^{\mathrm{b}}(\mathrm{I}(\mathbb{K}_X)) \to \mathrm{D}^{\mathrm{b}}(\mathrm{I}(\mathbb{K}_Y)),\\
\mathrm{R}\mathcal{I}\mathcal{H}om &: \mathrm{D}^{\mathrm{b}}(\mathrm{I}(\mathbb{K}_X))^{\mathrm{op}} \times \mathrm{D}^{\mathrm{b}}(\mathrm{I}(\mathbb{K}_X)) \to \mathrm{D}^{+}(\mathrm{I}(\mathbb{K}_X)),\\
\otimes &: \mathrm{D}^{\mathrm{b}}(\mathrm{I}(\mathbb{K}_X)) \times \mathrm{D}^{\mathrm{b}}(\mathrm{I}(\mathbb{K}_X)) \to \mathrm{D}^{\mathrm{b}}(\mathrm{I}(\mathbb{K}_X)),
\end{aligned}$$

(here, $f\colon X \to Y$ is a continuous map) and we have the stack-theoretical hom

$$\mathrm{R}\mathcal{H}om\colon \mathrm{D}^{\mathrm{b}}(\mathrm{I}(\mathbb{K}_X))^{\mathrm{op}} \times \mathrm{D}^{\mathrm{b}}(\mathrm{I}(\mathbb{K}_X)) \to \mathrm{D}^{+}(\mathbb{K}_X).$$

Note that the functor $\mathrm{R}\mathcal{I}\mathcal{H}om$ sends $\mathrm{D}^{\mathrm{b}}(\mathbb{K}_X)^{\mathrm{op}} \times \mathrm{D}^{\mathrm{b}}(\mathrm{I}(\mathbb{K}_X))$ to $\mathrm{D}^{\mathrm{b}}(\mathrm{I}(\mathbb{K}_X))$ and $\mathrm{R}\mathcal{H}om$ sends $\mathrm{D}^{\mathrm{b}}(\mathbb{K}_X)^{\mathrm{op}} \times \mathrm{D}^{\mathrm{b}}(\mathrm{I}(\mathbb{K}_X))$ to $\mathrm{D}^{\mathrm{b}}(\mathbb{K}_X)$.

The inverse image functor f^{-1} is a left adjoint of the direct image functor $\mathrm{R}f_*$. The functor of direct image with proper support $\mathrm{R}f_{!!}$ has a right adjoint functor $f^{!}$. Most formulas of sheaves have their counterpart in the theory of ind-sheaves, but some formulas are new. We shall not repeat them here and refer to [KS1]. As an example we state the following propositions:

Proposition 1.1.1. *Consider a cartesian square*

$$\begin{array}{ccc}
X' & \xrightarrow{f'} & Y' \\
{\scriptstyle g'}\downarrow & & \downarrow{\scriptstyle g} \\
X & \xrightarrow[f]{} & Y.
\end{array}$$

Then we have canonical isomorphisms

$$\mathrm{R}f'_{!!}g'^{-1} \simeq g^{-1}\mathrm{R}f_{!!}, \qquad \mathrm{R}f'_{*}g'^{!} \simeq g^{!}\mathrm{R}f_{*}, \qquad \mathrm{R}f'_{!!}g'^{!} \simeq g^{!}\mathrm{R}f_{!!}.$$

Note that the last isomorphism has no counterpart in sheaf theory.

Proposition 1.1.2. *For a morphism $f\colon X \to Y$ and for $K \in \mathrm{D}^{\mathrm{b}}(\mathbb{K}_Y)$, $\mathcal{F} \in \mathrm{D}^{\mathrm{b}}(\mathrm{I}(\mathbb{K}_X))$, we have*

$$\mathrm{R}f_{!!}\,\mathrm{R}\mathcal{I}\mathcal{H}om(f^{-1}K, \mathcal{F}) \simeq \mathrm{R}\mathcal{I}\mathcal{H}om(K, \mathrm{R}f_{!!}\mathcal{F}) \ \textit{in}\ \mathrm{D}^{\mathrm{b}}(\mathrm{I}(\mathbb{K}_Y)),$$

$$\mathrm{R}f_{!}\,\mathrm{R}\mathcal{H}om(f^{-1}K, \mathcal{F}) \simeq \mathrm{R}\mathcal{H}om(K, \mathrm{R}f_{!!}\mathcal{F}) \quad \textit{in}\ \mathrm{D}^{\mathrm{b}}(\mathbb{K}_Y).$$

Remark 1.1.3. Let Z be a closed subset of X and let $i\colon Z \to X$, $j\colon X \setminus Z \to Z$ be the inclusion morphisms. Then for $\mathcal{F}, \mathcal{F}' \in \mathrm{D}^{\mathrm{b}}(\mathrm{I}(\mathbb{K}_X))$, we have

$$\begin{aligned} &\mathrm{R}j_{!!}j^{-1}\mathcal{F} \simeq \widetilde{\mathbb{K}}_{X\setminus Z} \otimes \mathcal{F}, && \mathrm{R}i_*i^{-1}\mathcal{F} \simeq \mathbb{K}_Z \otimes \mathcal{F}, \\ &\mathrm{R}j_*j^{-1}\mathcal{F} \simeq \mathrm{R}\mathcal{I}\mathcal{H}om(\widetilde{\mathbb{K}}_{X\setminus Z}, \mathcal{F}), && \mathrm{R}i_*i^{!}\mathcal{F} \simeq \mathrm{R}\mathcal{I}\mathcal{H}om(\mathbb{K}_Z, \mathcal{F}), \\ &\mathrm{R}j_*j^{-1}\,\mathrm{R}\mathcal{H}om(\mathcal{F}', \mathcal{F}) \simeq \mathrm{R}\mathcal{H}om(\widetilde{\mathbb{K}}_{X\setminus Z} \otimes \mathcal{F}', \mathcal{F}). \end{aligned} \tag{1.1}$$

Hence there are *not* distinguished triangles

$$\mathrm{R}j_{!!}j^{-1}\mathcal{F} \to \mathcal{F} \to \mathrm{R}i_*i^{-1}\mathcal{F} \xrightarrow{+1} \text{ nor } \mathrm{R}i_*i^{!}\mathcal{F} \to \mathcal{F} \to \mathrm{R}j_*j^{-1}\mathcal{F} \xrightarrow{+1},$$

and instead there are distinguished triangles

$$\begin{aligned} &\mathrm{R}j_{!!}j^{-1}\mathcal{F} \to \mathcal{F} \to \mathcal{F} \otimes \widetilde{\mathbb{K}}_Z \xrightarrow{+1} \text{ and} \\ &\mathrm{R}\mathcal{I}\mathcal{H}om(\widetilde{\mathbb{K}}_Z, \mathcal{F}) \to \mathcal{F} \to \mathrm{R}j_*j^{-1}\mathcal{F} \xrightarrow{+1}. \end{aligned} \tag{1.2}$$

The functor β satisfies the following properties:

$$\beta_X(F) \otimes \beta_X(G) \simeq \beta_X(F \otimes G) \text{ for } F, G \in \mathrm{D}^{\mathrm{b}}(\mathbb{K}_X). \tag{1.3 a}$$

For $f\colon X \to Y$ and $G \in \mathrm{D}^{\mathrm{b}}(\mathbb{K}_Y)$ and $\mathcal{G} \in \mathrm{D}^{\mathrm{b}}(\mathrm{I}(\mathbb{K}_X))$, we have

$$f^{-1}\beta_Y(G) \simeq \beta_X(f^{-1}G) \quad \text{and} \quad f^{!}(\mathcal{G} \otimes \beta_Y(G)) \simeq f^{!}\mathcal{G} \otimes \beta_X(f^{-1}G). \tag{1.3 b}$$

For $\mathcal{F} \in \mathrm{D}^{\mathrm{b}}(\mathrm{I}(\mathbb{K}_X))$ and $K, K' \in \mathrm{D}^{\mathrm{b}}(\mathbb{K}_X)$, we have

$$\begin{aligned} \mathrm{R}\mathcal{I}\mathcal{H}om(K, \mathcal{F}) \otimes \beta_X(K') &\simeq \mathrm{R}\mathcal{I}\mathcal{H}om\left(K, \mathcal{F} \otimes \beta_X(K')\right) \text{ in } \mathrm{D}^{\mathrm{b}}(\mathrm{I}(\mathbb{K}_X)), \\ \mathrm{R}\mathcal{H}om(K, \mathcal{F}) \otimes K' &\simeq \mathrm{R}\mathcal{H}om\left(K, \mathcal{F} \otimes \beta_X(K')\right) \text{ in } \mathrm{D}^{\mathrm{b}}(\mathbb{K}_X). \end{aligned} \tag{1.3 c}$$

In general β does not commute with direct image.

Lemma 1.1.4. *Consider a closed embedding $i\colon Z \hookrightarrow X$ and $F \in \mathrm{D}^{\mathrm{b}}(\mathbb{K}_Z)$. Then we have an isomorphism*

$$\beta_X(\mathrm{R}i_*F) \otimes \mathbb{K}_Z \simeq \mathrm{R}i_*\beta_Z(F).$$

Proof. We have

$$\beta_X(\mathrm{R}i_*F) \otimes \mathbb{K}_Z \simeq \mathrm{R}i_*i^{-1}\beta_X(\mathrm{R}i_*F) \simeq \mathrm{R}i_*\beta_Z(i^{-1}\,\mathrm{R}i_*F) \simeq \mathrm{R}i_*\beta_Z(F). \qquad \square$$

The following fact will be used frequently in the paper:

$$\begin{aligned} &\text{A morphism } u\colon \mathcal{F} \to \mathcal{G} \text{ in } \mathrm{D}^{\mathrm{b}}(\mathrm{I}(\mathbb{K}_X)) \text{ is an isomorphism if and} \\ &\text{only if } \mathcal{F} \otimes \widetilde{\mathbb{K}}_x \to \mathcal{G} \otimes \widetilde{\mathbb{K}}_x \text{ is an isomorphism for all } x \in X. \end{aligned} \tag{1.4}$$

We list the commutativity of various functors. Here, "○" means that the functors commute, and "×" that they do not.

	ι	α	β	$\varinjlim$
$\otimes$	○	○	○	○
f^{-1}	○	○	○	○
Rf_*	○	○	×	×
$Rf_{!!}$	×	○	×	○
$f^!$	○	×	×	○
$\varinjlim$	×	○	○	

In the table, $\varinjlim$ means filtrant inductive limits. For example, the commutativity of $Rf_{!!}$ and $\varinjlim$ should be understood as in Proposition 2.3.2 (i) below.

Notation 1.1.5. *For a continuous map $f\colon X \to Y$, we denote by $\omega_{X/Y}$ the topological dualizing sheaf $f^!\mathbb{K}_Y$, and $\omega_X = \omega_{X/\{\mathrm{pt}\}}$. If X and Y are manifolds, $\omega_{X/Y} \simeq \omega_X \otimes f^{-1}\omega_Y^{\otimes -1}$.*

For three manifolds X_i $(i = 1, 2, 3)$ and for kernels $K \in \mathrm{D^b}(\mathrm{I}(\mathbb{K}_{X_1\times X_2}))$ and $K' \in \mathrm{D^b}(\mathrm{I}(\mathbb{K}_{X_2\times X_3}))$, we define their convolution by

$$K \underset{X_2}{\circ} K' = \mathrm{R}p_{13!!}(p_{12}^{-1}K \otimes p_{23}^{-1}K'), \tag{1.5}$$

where p_{ij} is the projection from $X_1 \times X_2 \times X_3$ to $X_i \times X_j$. We sometimes denote it simply by $K \circ K'$ when there is no risk of confusion.

This product of kernels satisfies the associative law:

$$(K \circ K') \circ K'' \simeq K \circ (K' \circ K'')$$

for $K \in \mathrm{D^b}(\mathrm{I}(\mathbb{K}_{X_1\times X_2}))$, $K' \in \mathrm{D^b}(\mathrm{I}(\mathbb{K}_{X_2\times X_3}))$ and $K'' \in \mathrm{D^b}(\mathrm{I}(\mathbb{K}_{X_3\times X_4}))$. By taking $\{\mathrm{pt}\}$ as X_3 in (1.5), we obtain the integral transform functor:

$$K\circ\,\colon\ \mathrm{D^b}(\mathrm{I}(\mathbb{K}_{X_2})) \to \mathrm{D^b}(\mathrm{I}(\mathbb{K}_{X_1})).$$

The following lemma is frequently used in Section 2.

Lemma 1.1.6. *Let $f_k\colon X_k \to Y_k$ $(k=1,2,3)$ be morphisms and $\mathcal{K}_{ij} \in \mathrm{D^b}(\mathrm{I}(\mathbb{K}_{X_i\times X_j}))$ and $\mathcal{L}_{ij} \in \mathrm{D^b}(\mathrm{I}(\mathbb{K}_{Y_i\times Y_j}))$.*

(i) $\big((f_1 \times \mathrm{id}_{Y_2})^{-1}\mathcal{L}_{12}\big) \underset{Y_2}{\circ} \big((\mathrm{id}_{Y_2} \times f_3)^{-1}\mathcal{L}_{23}\big) \simeq (f_1 \times f_3)^{-1}(\mathcal{L}_{12} \underset{Y_2}{\circ} \mathcal{L}_{23})$ *in* $\mathrm{D^b}(\mathrm{I}(\mathbb{K}_{X_1\times X_3}))$,

(ii) $\big((f_1 \times \mathrm{id}_{X_2})_{!!}\mathcal{K}_{12}\big) \underset{Y_2}{\circ} \big((\mathrm{id}_{X_2} \times f_3)_{!!}\mathcal{K}_{23}\big) \simeq (f_1 \times f_3)_{!!}(\mathcal{K}_{12} \underset{X_2}{\circ} \mathcal{K}_{23})$ *in* $\mathrm{D^b}(\mathrm{I}(\mathbb{K}_{Y_1\times Y_3}))$,

(iii) $\big((\mathrm{id}_{Y_1} \times f_2)^{-1}\mathcal{L}_{12}\big) \underset{X_2}{\circ} \mathcal{K}_{23} \simeq \mathcal{L}_{12} \underset{Y_2}{\circ} \mathrm{R}(f_2 \times \mathrm{id}_{X_3})_{!!}\mathcal{K}_{23}$ *in* $\mathrm{D^b}(\mathrm{I}(\mathbb{K}_{Y_1\times X_3}))$.

1.2. Kernels attached to 1-forms

Let us denote by $\pi_X\colon T^*X \to X$ the cotangent bundle to X. For a closed submanifold Z of X, we denote by $T^*_Z X$ its conormal bundle. In particular, $T^*_X X$ is the zero section of T^*X. To a differentiable map $f\colon X \to Y$, we associate the diagram

$$T^*X \xleftarrow[f_d]{} T^*Y \underset{Y}{\times} X \xrightarrow[f_\pi]{} T^*Y.$$

Notation 1.2.1. *For a vector bundle $p\colon E \to X$, we denote by $\dot{E}$ the space E with the zero section removed, and by $\dot{p}$ the projection $\dot{E} \to X$. For example, we use the notations $\dot{\pi}_X\colon \dot{T}^*X \to X$, $\dot{T}^*_Z X$, and so on.*

Definition 1.2.2. A kernel data is a triple (X, Z, σ), where X is a manifold, Z is a closed submanifold of X and σ is a section of $T^*X \underset{X}{\times} Z \to Z$.

We set $\mathcal{T}(\sigma) = \sigma^{-1}(T^*_Z X)$ and $\mathcal{Z}(\sigma) = \sigma^{-1}(T^*_X X)$. We have therefore

$$\mathcal{Z}(\sigma) \subset \mathcal{T}(\sigma) \subset Z.$$

Each kernel data (X, Z, σ) defines a closed cone P_σ in $T_Z X \underset{X}{\times} \mathcal{T}(\sigma)$ by

$$P_\sigma = \{(x, v) \in T_Z X;\, x \in \mathcal{T}(\sigma) \text{ and } \langle v, \sigma(x)\rangle \geqslant 0\}.$$

Consider the deformation of the normal bundle to Z in X which will be denoted by $\widetilde{X}_Z$ or simply by $\widetilde{X}$ (see e.g., [KS2]). We have the following commutative diagram where the squares marked by $\square$ are cartesian:

$$\begin{array}{ccccccc}
 & & \{0\} & \hookrightarrow & \mathbb{R} & \hookleftarrow & \{t \in \mathbb{R};\, t > 0\} \\
 & & \uparrow & \square & \uparrow{\scriptstyle t} & \square & \uparrow \\
P_\sigma & \hookrightarrow & T_Z X & \overset{s}{\hookrightarrow} & \widetilde{X}_Z & \overset{j}{\hookleftarrow} & \Omega \\
 & & \downarrow{\scriptstyle \tau_Z} & & \downarrow{\scriptstyle p} & \swarrow{\scriptstyle \widetilde{p}} & \\
 & & Z & \overset{i}{\hookrightarrow} & X. & &
\end{array} \tag{1.6}$$

Here Ω is the open subset defined by $\Omega = \{t > 0\}$ for the natural smooth map $t\colon \widetilde{X}_Z \to \mathbb{R}$. The normal bundle $T_Z X$ is identified with the inverse image of $0 \in \mathbb{R}$ by t. With a local coordinate system $(x, z) = (x_1, \dots, x_n, z_1, \dots, z_m)$ of X such that Z is given by $x = 0$, $\widetilde{X}_Z$ has the coordinates $(t, \tilde{x}, z) = (t, \tilde{x}_1, \dots, \tilde{x}_n, z_1, \dots, z_m)$ and p is given by $p(t, \tilde{x}, z) = (t\tilde{x}, z)$.

Recall that the normal cone $C_Z(A)$ of a subset A of X is a closed cone of $T_Z X$ defined by

$$C_Z(A) = T_Z X \cap \overline{p^{-1}(A) \cap \Omega}. \tag{1.7}$$

Note that p is not smooth but the relative dualizing complex $\omega_{\widetilde{X}/X}$ is isomorphic to $\mathbb{K}_{\widetilde{X}}[1]$. In the sequel we will usually regard P_σ as a closed subset of $\widetilde{X}_Z$ by $P_\sigma \subset T_Z X \subset \widetilde{X}_Z$.

Definition 1.2.3. (i) Let (X, Z, σ) be a kernel data. We define the kernel $\mathcal{L}_\sigma(Z, X) \in \mathrm{D^b}(\mathrm{I}(\mathbb{K}_X))$ by

$$\mathcal{L}_\sigma(Z, X) = \mathrm{R}p_{!!}(\mathbb{K}_{\overline{\Omega}} \otimes \widetilde{\mathbb{K}}_{P_\sigma}) \otimes \beta_X(\mathrm{R}i_*\omega_{Z/X}^{\otimes -1}).$$

(ii) A morphism of kernel data $f\colon (X_1, Z_1, \sigma_1) \to (X_2, Z_2, \sigma_2)$ is a morphism of manifolds $f\colon X_1 \to X_2$ satisfying
 (i) $f(Z_1) \subset Z_2$,
 (ii) $\sigma_1 = f^*\sigma_2$.

Remark 1.2.4. Note that $\mathcal{L}_\sigma(Z, X)$ is supported on $\mathcal{T}(\sigma)$, i.e.,

$$\mathcal{L}_\sigma(Z, X) \xrightarrow{\sim} \mathcal{L}_\sigma(Z, X) \otimes \widetilde{\mathbb{K}}_{\mathcal{T}(\sigma)}.$$

This kernel behaves differently on $\mathcal{Z}(\sigma)$ and outside. We have

$$\mathcal{L}_\sigma(Z, X) \otimes \widetilde{\mathbb{K}}_{\mathcal{Z}(\sigma)} \simeq \mathbb{K}_Z \otimes \widetilde{\mathbb{K}}_{\mathcal{Z}(\sigma)}$$

and $\mathcal{L}_\sigma(Z, X)|_{X \setminus \mathcal{Z}(\sigma)}$ is concentrated in degree $-\operatorname{codim} Z$ (see Corollary 1.2.13).

In order to prove these facts, we shall start by the following vanishing lemma.

Lemma 1.2.5. *(i)* $\mathrm{R}p_{!!}(\mathbb{K}_\Omega \otimes \widetilde{\mathbb{K}}_{T_Z X}) \simeq 0$ *and* $\mathrm{R}p_{!!}(\mathbb{K}_{\overline{\Omega}} \otimes \widetilde{\mathbb{K}}_{T_Z X}) \simeq \mathrm{R}i_*\omega_{Z/X}$.
(ii) Regarding Z as the zero section of $T_Z X \subset \widetilde{X}_Z$, we have

$$\mathrm{R}p_{!!}\left(\mathbb{K}_{\overline{\Omega}} \otimes \widetilde{\mathbb{K}}_Z\right) \simeq \widetilde{\mathbb{K}}_Z.$$

(iii) $\left(\mathrm{R}p_{!!}(\mathbb{K}_{T_Z X} \otimes \widetilde{\mathbb{K}}_{P_\sigma})\right) \otimes \widetilde{\mathbb{K}}_{Z \setminus \mathcal{Z}(\sigma)} \simeq 0.$

Proof. (i) Since the problem is local, we may assume that X is affine endowed with a system of global coordinates (x, z) such that $Z = \{x = 0\}$, $\widetilde{X}_Z = (t, \widetilde{x}, z)$ and $p(t, \widetilde{x}, z) = (t\widetilde{x}, z)$. We then have for all integer j

$$\begin{aligned} \mathrm{R}^j p_{!!}\left(\mathbb{K}_\Omega \otimes \widetilde{\mathbb{K}}_{T_Z X}\right) &\simeq \mathrm{R}^j p_{!!}\left(\underset{R>0,\, \varepsilon>0}{\text{“}\varinjlim\text{”}}\ \mathbb{K}_{\{0<t\leqslant\varepsilon,\, |\widetilde{x}|<R\}}\right) \\ &\simeq \underset{R>0,\, \varepsilon>0}{\text{“}\varinjlim\text{”}}\ \mathrm{R}^j p_! \mathbb{K}_{\{0<t\leqslant\varepsilon,\, |\widetilde{x}|<R\}} \simeq 0, \end{aligned}$$

which implies the first statement. The last one follows from the distinguished triangle

$$\mathrm{R}p_{!!}(\mathbb{K}_\Omega \otimes \widetilde{\mathbb{K}}_{T_Z X}) \to \mathrm{R}p_{!!}(\mathbb{K}_{\overline{\Omega}} \otimes \widetilde{\mathbb{K}}_{T_Z X}) \to \mathrm{R}p_{!!}(\mathbb{K}_{T_Z X}) \xrightarrow{+1}$$

and $\mathrm{R}p_{!!}(\mathbb{K}_{T_Z X}) \simeq \mathrm{R}i_*\omega_{Z/X}$.

(ii) We have a chain of morphisms

$$\mathrm{R}p_{!!}\left(\mathbb{K}_{\overline{\Omega}} \otimes \widetilde{\mathbb{K}}_Z\right) \to \mathrm{R}p_{!!}\left(\mathbb{K}_{\overline{\Omega}} \otimes \mathbb{K}_Z\right) \simeq \mathrm{R}p_{!!}\,\mathbb{K}_Z \simeq \mathbb{K}_Z,$$

which allows us to prove the isomorphism locally on X. With the coordinate system as above, we get for all integer j

$$\mathrm{R}^j p_{!!}\left(\mathbb{K}_{\overline{\Omega}} \otimes \widetilde{\mathbb{K}}_Z\right) \simeq \mathrm{R}^j p_{!!}\left(\underset{\varepsilon>0}{\text{“}\varinjlim\text{”}}\, \mathbb{K}_{\{0\leqslant t\leqslant\varepsilon, |\tilde{x}|\leqslant\varepsilon\}}\right) \simeq \underset{\varepsilon>0}{\text{“}\varinjlim\text{”}}\, \mathrm{R}^j p_!\, \mathbb{K}_{\{0\leqslant t\leqslant\varepsilon, |\tilde{x}|\leqslant\varepsilon\}}$$

$$\simeq \begin{cases} \underset{\varepsilon>0}{\text{“}\varinjlim\text{”}}\, \mathbb{K}_{\{|x|\leqslant\varepsilon^2\}} \simeq \widetilde{\mathbb{K}}_Z & \text{if } j=0, \\ 0 & \text{if } j\neq 0. \end{cases}$$

(iii) For $z_0 \in \mathcal{T}(\sigma) \setminus \mathcal{Z}(\sigma)$, we have

$$\left(\mathrm{R}p_{!!}(\mathbb{K}_{T_ZX} \otimes \widetilde{\mathbb{K}}_{P_\sigma})\right) \otimes \widetilde{\mathbb{K}}_{z_0} \simeq \mathrm{R}p_{!!}\left(\mathbb{K}_{T_ZX} \otimes \widetilde{\mathbb{K}}_{P_\sigma \cap p^{-1}(z_0)}\right).$$

Set $\sigma(z_0) = \langle \xi_0, dx\rangle \neq 0$. Then we have

$$\mathbb{K}_{T_ZX} \otimes \widetilde{\mathbb{K}}_{P_\sigma\cap p^{-1}(z_0)} \simeq \underset{R>0,\varepsilon>0}{\text{“}\varinjlim\text{”}}\ \mathbb{K}_{\{t=0,\ -\varepsilon\leq\langle\xi_0,\tilde{x}\rangle,\ |\tilde{x}|<R\}},$$

and for all integer j

$$\left(\mathrm{R}^j p_{!!}(\mathbb{K}_{T_ZX} \otimes \widetilde{\mathbb{K}}_{P_\sigma})\right) \otimes \widetilde{\mathbb{K}}_{z_0} \simeq \widetilde{\mathbb{K}}_{z_0} \otimes \underset{R>0,\,\varepsilon>0}{\text{“}\varinjlim\text{”}}\ \mathrm{R}^j p_!\left(\mathbb{K}_{\{t=0,\ -\varepsilon\leq\langle\xi_0,\tilde{x}\rangle,\ |\tilde{x}|<R\}}\right) \simeq 0.$$ □

Lemma 1.2.6. *There is a natural morphism*

$$\mathcal{L}_\sigma(Z,X) \to \widetilde{\mathbb{K}}_{\mathcal{T}(\sigma)} \otimes \beta_X\left(\mathrm{R}i_*\omega_{Z/X}^{\otimes -1}\right).$$

Proof. Regard $\mathcal{T}(\sigma)$ as a subset of $\widetilde{X}_Z$ by $\mathcal{T}(\sigma) \subset Z \subset T_ZX \subset \widetilde{X}_Z$. Then we get a natural morphism

$$\mathcal{L}_\sigma(Z,X) \to \mathrm{R}p_{!!}\left(\mathbb{K}_{\overline{\Omega}} \otimes \widetilde{\mathbb{K}}_{\mathcal{T}(\sigma)}\right) \otimes \beta_X\left(\mathrm{R}i_*\omega_{Z/X}^{\otimes -1}\right).$$

Hence the desired morphism is obtained by Lemma 1.2.5 (ii). □

The following lemma provides a useful distinguished triangle to study some properties of the kernel $\mathcal{L}_\sigma(Z,X)$.

Lemma 1.2.7. *There is a natural distinguished triangle*

$$\mathrm{R}p_{!!}\left(\mathbb{K}_\Omega \otimes \widetilde{\mathbb{K}}_{P_\sigma}\right) \otimes \beta_X\left(\mathrm{R}i_*\omega_{Z/X}^{\otimes -1}\right) \to \mathcal{L}_\sigma(Z,X) \to \mathrm{R}p_{!!}\left(\mathbb{K}_{T_ZX} \otimes \widetilde{\mathbb{K}}_{P_\sigma}\right) \otimes \mathrm{R}i_*\omega_{Z/X}^{\otimes -1} \xrightarrow{+1}.$$

Proof. It is enough to apply the triangulated functor $\mathrm{R}p_{!!}(\,\cdot\,\otimes\widetilde{\mathbb{K}}_{P_\sigma})\otimes\beta_X(\mathrm{R}i_*\omega_{Z/X}^{\otimes-1})$ to the distinguished triangle

$$\mathbb{K}_\Omega\to\mathbb{K}_{\overline{\Omega}}\to\mathbb{K}_{T_Z^*X}\xrightarrow{+1}, \tag{1.8}$$

and to use $\mathbb{K}_Z\otimes\beta_X(\mathrm{R}i_*\omega_{Z/X}^{\otimes-1})\simeq\mathrm{R}i_*\omega_{Z/X}^{\otimes-1}$. □

Recall that $\mathcal{Z}(\sigma)$ is the set of zeroes of σ, i.e., $\mathcal{Z}(\sigma)=\sigma^{-1}(T_X^*X)\subset Z$.

Proposition 1.2.8. *We have*

$$\mathcal{L}_\sigma(Z,X)\otimes\widetilde{\mathbb{K}}_{\mathcal{Z}(\sigma)}\simeq\mathbb{K}_Z\otimes\widetilde{\mathbb{K}}_{\mathcal{Z}(\sigma)}\,.$$

In particular, if $\sigma=0$, then $\mathcal{L}_\sigma(Z,X)\simeq\mathbb{K}_Z$.

Proof. By the definition of $\mathcal{Z}(\sigma)$, the cone $P_\sigma\underset{Z}{\times}\mathcal{Z}(\sigma)$ coincides with $T_ZX\underset{Z}{\times}\mathcal{Z}(\sigma)$. Hence we have $\mathbb{K}_{\overline{\Omega}}\otimes\widetilde{\mathbb{K}}_{P\sigma}\otimes p^{-1}\widetilde{\mathbb{K}}_{\mathcal{Z}(\sigma)}\simeq\mathbb{K}_{\overline{\Omega}}\otimes p^{-1}\widetilde{\mathbb{K}}_{\mathcal{Z}(\sigma)}$, which implies that

$$\mathcal{L}_\sigma(Z,X)\otimes\widetilde{\mathbb{K}}_{\mathcal{Z}(\sigma)}\simeq\mathrm{R}p_{!!}(\mathbb{K}_{\overline{\Omega}}\otimes\widetilde{\mathbb{K}}_{T_ZX})\otimes\widetilde{\mathbb{K}}_{\mathcal{Z}(\sigma)}\otimes\beta_X(\mathrm{R}i_*\omega_{Z/X}^{\otimes-1}).$$

Hence the result follows from Lemma 1.2.5 (i). □

Proposition 1.2.9. *Let (X,Z,σ) be a kernel data, and set $X_0=X\setminus\mathcal{Z}(\sigma)$ and $Z_0=Z\setminus\mathcal{Z}(\sigma)$. Then there is a natural distinguished triangle*

$$\mathrm{R}j_{!!}\mathcal{L}_{\sigma_0}(Z_0,X_0)\to\mathcal{L}_\sigma(Z,X)\to\mathbb{K}_Z\otimes\widetilde{\mathbb{K}}_{\mathcal{Z}(\sigma)}\xrightarrow{+1},$$

where σ_0 is the restriction of σ to Z_0 and j denotes the open immersion $X_0\hookrightarrow X$.

Proof. We have the distinguished triangle

$$\mathcal{L}_\sigma(Z,X)\otimes\widetilde{\mathbb{K}}_{X_0}\to\mathcal{L}_\sigma(Z,X)\to\mathcal{L}_\sigma(Z,X)\otimes\widetilde{\mathbb{K}}_{\mathcal{Z}(\sigma)}\xrightarrow{+1}.$$

The first term is isomorphic to $\mathrm{R}j_{!!}\mathcal{L}_{\sigma_0}(Z_0,X_0)$, and the last term is isomorphic to $\mathbb{K}_Z\otimes\widetilde{\mathbb{K}}_{\mathcal{Z}(\sigma)}$ by Lemma 1.2.8. □

Corollary 1.2.10. *There are natural morphisms*

$$\mathbb{K}_Z\to\mathcal{L}_\sigma(Z,X)\to\widetilde{\mathbb{K}}_{\mathcal{T}(\sigma)}\otimes\beta_X\left(\mathrm{R}i_*\omega_{Z/X}^{\otimes-1}\right).$$

Proof. The first arrow is constructed as an immediate consequence of the preceding proposition and the obvious inclusion $P_\sigma\subset P_0=T_ZX$. The last arrow follows from Lemma 1.2.6. □

Proposition 1.2.11. *Assume the section σ never vanishes. Then*

$$\mathcal{L}_\sigma(Z, X) \simeq \mathrm{R}p_{!!}\left(\mathbb{K}_\Omega \otimes \widetilde{\mathbb{K}}_{P_\sigma}\right) \otimes \beta_X\left(\mathrm{R}i_*\omega_{Z/X}^{\otimes -1}\right)$$

$$\simeq \text{“}\varinjlim_U\text{''}\, \mathbb{K}_U \otimes \beta_X\left(\mathrm{R}i_*\omega_{Z/X}^{\otimes -1}\right) \otimes \widetilde{\mathbb{K}}_{\mathcal{T}(\sigma)},$$

where the inductive limit is taken over the family of open subsets U of X such that

$$P_\sigma \cap \mathrm{C}_Z(U) \subset Z.$$

Here, Z is regarded as the zero section of $T_Z X$.

Remark that the set of such U's is a filtrant ordered set by the inclusion order.

Proof. By Lemma 1.2.7 and Lemma 1.2.5 (iii), we have

$$\mathcal{L}_\sigma(Z, X) \simeq \mathrm{R}p_{!!}\left(\mathbb{K}_\Omega \otimes \widetilde{\mathbb{K}}_{P_\sigma}\right) \otimes \beta_X\left(\mathrm{R}i_*\omega_{Z/X}^{\otimes -1}\right).$$

Hence it is enough to show that

$$\mathrm{R}p_{!!}\left(\mathbb{K}_\Omega \otimes \widetilde{\mathbb{K}}_{P_\sigma}\right) \simeq \text{“}\varinjlim_U\text{''}\, \mathbb{K}_U \otimes \widetilde{\mathbb{K}}_{\mathcal{T}(\sigma)}.$$

Since we have $Z \cap U = \emptyset$ on a neighborhood of $\mathcal{T}(\sigma)$, $p^{-1}(U) \cap \Omega = p^{-1}(U) \cap \overline{\Omega}$ is a closed subset of Ω and we get the following chain of natural morphisms :

$$p^{-1}\mathbb{K}_U \simeq \mathbb{K}_{p^{-1}(U)} \to \mathbb{K}_{p^{-1}(U)\cap\Omega} \to \mathbb{K}_\Omega \to \mathbb{K}_\Omega \otimes \widetilde{\mathbb{K}}_{P_\sigma}.$$

Since $\overline{p^{-1}(U) \cap \Omega} \cap P_\sigma = \mathrm{C}_Z(U) \cap P_\sigma$ is contained in the zero section of $T_Z X$, $\mathrm{Supp}(p^{-1}\mathbb{K}_U \otimes \widetilde{\mathbb{K}}_{P_\sigma})$ is proper over Z. Hence we have a chain of morphisms

$$\mathbb{K}_U \to p_*(p^{-1}\mathbb{K}_U \otimes \widetilde{\mathbb{K}}_{P_\sigma}) \simeq p_{!!}(p^{-1}\mathbb{K}_U \otimes \widetilde{\mathbb{K}}_{P_\sigma}) \to p_{!!}\left(\mathbb{K}_\Omega \otimes \widetilde{\mathbb{K}}_{P_\sigma}\right),$$

which provides a natural morphism

$$\text{“}\varinjlim_U\text{''}\, \mathbb{K}_U \to \mathrm{R}p_{!!}\left(\mathbb{K}_\Omega \otimes \widetilde{\mathbb{K}}_{P_\sigma}\right).$$

By tensoring we get the morphism

$$\text{“}\varinjlim_U\text{''}\, \mathbb{K}_U \otimes \widetilde{\mathbb{K}}_{\mathcal{T}(\sigma)} \to \mathrm{R}p_{!!}\left(\mathbb{K}_\Omega \otimes \widetilde{\mathbb{K}}_{P_\sigma}\right). \tag{1.9}$$

We shall now show that this morphism is an isomorphism. It is enough to show that (1.9) is an isomorphism after tensoring by $\widetilde{\mathbb{K}}_{x_0}$ for any $x_0 \in \mathcal{T}(\sigma)$. Let us take local coordinate system (x, z) of X such that $Z = \{x = 0\}$. We may assume $x_0 = (0, 0)$, and we set $\sigma(x_0) = \langle \xi_0, dx \rangle$. We then have

$$\mathrm{R}p_{!!}\left(\mathbb{K}_\Omega\otimes\widetilde{\mathbb{K}}_{P_\sigma}\right)\otimes\widetilde{\mathbb{K}}_{x_0}\simeq\mathrm{R}p_{!!}\left(\mathbb{K}_\Omega\otimes\widetilde{\mathbb{K}}_{P_\sigma}\otimes\widetilde{\mathbb{K}}_{p^{-1}(x_0)}\right)$$
$$\simeq\mathrm{R}p_{!!}\left(\mathbb{K}_\Omega\otimes\widetilde{\mathbb{K}}_{P_\sigma\cap p^{-1}(x_0)}\right),$$

and

$$\mathbb{K}_\Omega\otimes\widetilde{\mathbb{K}}_{P_\sigma\cap p^{-1}(x_0)}\simeq\underset{V\subset\subset\widetilde{X}_Z,\ P_\sigma\cap p^{-1}(x_0)\subset V'}{\text{“}\varinjlim\text{”}}\mathbb{K}_{\Omega\cap V\cap\overline{V'}}$$
$$\simeq\widetilde{\mathbb{K}}_{x_0}\otimes\underset{R>0,\,\varepsilon_1>0,\,\varepsilon_2>0}{\text{“}\varinjlim\text{”}}\mathbb{K}_{A_{R,\varepsilon_1,\varepsilon_2}},$$

where we have set

$$A_{R,\varepsilon_1,\varepsilon_2}=\left\{(t,\widetilde{x},z)\in\widetilde{X}_Z\ ;\ 0<t\leqslant\varepsilon_1,\ -\varepsilon_2\leqslant\langle\xi_0,\widetilde{x}\rangle,|\widetilde{x}|<R\right\}.$$

Hence for all integer j, we have

$$\mathrm{R}^j p_{!!}\left(\mathbb{K}_\Omega\otimes\widetilde{\mathbb{K}}_{P_\sigma}\right)\otimes\widetilde{\mathbb{K}}_{x_0}\simeq\widetilde{\mathbb{K}}_{x_0}\otimes\underset{R>0,\,\varepsilon_1>0,\,\varepsilon_2>0}{\text{“}\varinjlim\text{”}}\mathrm{R}^j p_!\,\mathbb{K}_{A_{R,\varepsilon_1,\varepsilon_2}}.$$

We have

$$p^{-1}((x,z))\simeq\{t\in\mathbb{R};0<t\le\varepsilon_1,\ -\varepsilon_2\le\langle\xi_0,t^{-1}x\rangle,\ |t^{-1}x|<R\}$$
$$\simeq\{t\in\mathbb{R};\ R^{-1}|x|<t\le\varepsilon_1,\ -\varepsilon_2^{-1}\langle\xi_0,x\rangle\le t\},$$

and hence

$$\mathrm{R}p_!(\mathbb{K}_{A_{R,\varepsilon_1,\varepsilon_2}})\simeq\mathbb{K}_{\left\{R^{-1}|x|<-\varepsilon_2^{-1}\langle x,\xi_0\rangle\leqslant\varepsilon_1\right\}}.$$

Taking the limit we can use a cofinality argument to get

$$\mathrm{R}p_{!!}\left(\mathbb{K}_\Omega\otimes\widetilde{\mathbb{K}}_{P_\sigma}\right)\otimes\widetilde{\mathbb{K}}_{x_0}\simeq\widetilde{\mathbb{K}}_{x_0}\otimes\underset{\varepsilon>0}{\text{“}\varinjlim\text{”}}\mathbb{K}_{\{(x,z)\in X\ ;\ -\langle\xi_0,x\rangle>\varepsilon|x|\}}.$$

Then the theorem follows from the following easy sublemma. □

Sublemma 1.2.12. *(i) Let $U=\{(x,z)\in X;\varepsilon|x|<-\langle\xi_0,x\rangle\}$. Then $P_\sigma\cap C_Z(U)\subset Z$.*

(ii) Let $U\subset X$ be an open subset such that $P_\sigma\cap C_Z(U)\subset Z$. Then there exist $\varepsilon>0$ and $\delta>0$ such that

$$U\cap\{|(x,z)|\leqslant\delta\}\subset\{(x,z)\in X;-\langle x,\xi_0\rangle>\varepsilon|x|\}.$$

Corollary 1.2.13. *Let (X,Z,σ) be a kernel data. Assume that X is endowed with a local coordinate system (x,z) such that $Z=\{x=0\}$ and σ is a nowhere vanishing section. Then, writing $\sigma(z)=\langle\sigma_1(z),dx\rangle+\langle\sigma_2(z),dz\rangle$, we have*

$$\mathcal{L}_\sigma(Z,X)\simeq\widetilde{\mathbb{K}}_{\{x=0,\,\sigma_2(z)=0\}}\otimes\underset{\varepsilon>0}{\text{“}\varinjlim\text{”}}\mathbb{K}_{\left\{(x,z);-\langle\sigma_1(z),x\rangle>\varepsilon|x|\right\}}[\operatorname{codim}Z].$$

Remark 1.2.14. (i) We have

$$\alpha_X\big(\mathcal{L}_\sigma(Z,X)\big) \simeq \mathbb{K}_{\mathcal{Z}(\sigma)}\,.$$

(ii) Let (X, Z, σ_1) and (X, Z, σ_2) be kernel data, and let W be a closed subset of Z such that $\sigma_1(x) = \sigma_2(x)$ for all $x \in W$. Since $P_{\sigma_1} \cap \tau_Z^{-1}W = P_{\sigma_2} \cap \tau_Z^{-1}W$, we have

$$\mathcal{L}_{\sigma_1}(X,Z) \otimes \widetilde{\mathbb{K}}_W \simeq \mathcal{L}_{\sigma_2}(X,Z) \otimes \widetilde{\mathbb{K}}_W\,.$$

1.3. Functorial Properties

In this subsection, we will investigate the behavior of microlocal kernels $\mathcal{L}_\sigma(Z,X)$ under inverse and proper direct images, and under convolution.

Let $f : (X_1, Z_1, \sigma_1) \to (X_2, Z_2, \sigma_2)$ be morphism of kernel data. We have the diagrams of manifolds

$$\begin{array}{ccccc} T^*_{Z_1}X_1 & \xleftarrow{f_d} & T^*_{Z_2}X_2 \underset{Z_2}{\times} Z_1 & \xrightarrow{f_\pi} & T^*_{Z_2}X_2 \\ \uparrow{\scriptstyle \sigma_1} & & \uparrow & & \uparrow{\scriptstyle \sigma_2} \\ \mathcal{T}(\sigma_1) & \hookleftarrow & \mathcal{T}(\sigma_2) \underset{Z_2}{\times} Z_1 & \longrightarrow & \mathcal{T}(\sigma_2) \end{array} \qquad \text{and} \qquad \begin{array}{ccccc} \widetilde{X}_1 & \xrightarrow{\tilde{f}} & \widetilde{X}_2 & \xrightarrow{t} & \mathbb{R} \\ {\scriptstyle p_1}\downarrow & & \downarrow{\scriptstyle p_2} & & \\ X_1 & \xrightarrow{f} & X_2 & & \end{array}$$

where $\widetilde{X}_k = \widetilde{X_k}_{Z_k}$ $(k = 1, 2)$. We denote by $i_k : Z_k \hookrightarrow X_k$ the inclusion map. We have

$$P_{\sigma_1} \underset{X_2}{\times} \mathcal{T}(\sigma_2) = \tilde{f}^{-1}(P_{\sigma_2}). \tag{1.10}$$

Proposition 1.3.1. *Let $f : (X_1, Z_1, \sigma_1) \to (X_2, Z_2, \sigma_2)$ be a morphism of kernel data. Assume that $Z_1 = f^{-1}(Z_2)$ and the morphism $f : X_1 \to X_2$ is clean with respect to Z_2 (i.e., $(T_{Z_1}X_1)_x \to (T_{Z_2}X_2)_{f(x)}$ is injective for any $x \in Z_1$). Then there exists a natural morphism*

$$f^{-1}\mathcal{L}_{\sigma_2}(Z_2, X_2) \to \mathcal{L}_{\sigma_1}(Z_1, X_1) \otimes \beta_{X_1}(\mathrm{R}i_{1*}\omega_{Z_1/Z_2}) \otimes \omega^{\otimes -1}_{X_1/X_2} \otimes \widetilde{\mathbb{K}}_{f^{-1}\mathcal{T}(\sigma_2)}\,.$$

Proof. Since f is clean, $\widetilde{X}_1 \to \widetilde{X}_2 \underset{X_2}{\times} X_1$ is a closed embedding and there is a morphism of functors $f^{-1}\mathrm{R}p_{2!!} \to \mathrm{R}p_{1!!}\tilde{f}^{-1}$ which induces a natural morphism

$$\begin{aligned} f^{-1}\mathcal{L}_{\sigma_2}(Z_2, X_2) &\simeq f^{-1}\mathrm{R}p_{2!!}\left(\mathbb{K}_{\overline{\Omega}_2} \otimes \widetilde{\mathbb{K}}_{P_{\sigma_2}}\right) \otimes f^{-1}\beta_{X_2}(\mathrm{R}i_{2*}\omega^{\otimes -1}_{Z_2/X_2}) \\ &\to \mathrm{R}p_{1!!}\tilde{f}^{-1}\left(\mathbb{K}_{\overline{\Omega}_2} \otimes \widetilde{\mathbb{K}}_{P_{\sigma_2}}\right) \otimes f^{-1}\beta_{X_2}(\mathrm{R}i_{2*}\omega^{\otimes -1}_{Z_2/X_2}) \qquad (1.11) \\ &\simeq \mathrm{R}p_{1!!}\left(\mathbb{K}_{\overline{\Omega}_1} \otimes \widetilde{\mathbb{K}}_{\tilde{f}^{-1}(P_{\sigma_2})}\right) \otimes \beta_{X_1}(f^{-1}\mathrm{R}i_{2*}\omega^{\otimes -1}_{Z_2/X_2}). \end{aligned}$$

By (1.10), we have a morphism

$$f^{-1}\mathcal{L}_{\sigma_2}(Z_2, X_2) \to \mathrm{R}p_{1!!}\left(\mathbb{K}_{\overline{\Omega}_1} \otimes \widetilde{\mathbb{K}}_{P_{\sigma_1}}\right) \otimes f^{-1}(\widetilde{\mathbb{K}}_{\mathcal{T}(\sigma_2)})$$
$$\otimes \beta_{X_1}(f^{-1}\,\mathrm{R}i_{2*}\omega_{Z_2/X_2}^{\otimes -1}). \tag{1.12}$$

Hence, to get the desired morphism, it is enough to remark that

$$f^{-1}\,\mathrm{R}i_{2*}\omega_{Z_2/X_2}^{\otimes -1} \simeq \mathrm{R}i_{1*}\left(\omega_{Z_1/X_1}^{\otimes -1} \otimes \omega_{Z_1/Z_2} \otimes i_1^{-1}\omega_{X_1/X_2}^{\otimes -1}\right)$$
$$\simeq \mathrm{R}i_{1*}\omega_{Z_1/X_1}^{\otimes -1} \otimes \mathrm{R}i_{1*}\omega_{Z_1/Z_2} \otimes \omega_{X_1/X_2}^{\otimes -1}. \qquad \square$$

By adjunction, we obtain:

Corollary 1.3.2. *Under the hypothesis of the Proposition 1.3.1, we have a natural morphism*

$$\mathcal{L}_{\sigma_2}(Z_2, X_2) \to \mathrm{R}f_*\left(\mathcal{L}_{\sigma_1}(Z_1, X_1) \otimes \beta_{X_1}(\mathrm{R}i_{1*}\omega_{Z_1/Z_2}) \otimes \omega_{X_1/X_2}^{\otimes -1} \otimes f^{-1}\,\widetilde{\mathbb{K}}_{\mathcal{T}(\sigma_2)}\right).$$

Proposition 1.3.3. *Let $f\colon (X_1, Z_1, \sigma_1) \to (X_2, Z_2, \sigma_2)$ be a morphism of kernel data. Assume that $f^{-1}(Z_2) = Z_1$ and f is transversal to Z_2. Then we have a natural isomorphism*

$$f^{-1}\mathcal{L}_{\sigma_2}(Z_2, X_2) \xrightarrow{\sim} \mathcal{L}_{\sigma_1}(Z_1, X_1).$$

Proof. Indeed if f is transversal, $\widetilde{X}_1 \to \widetilde{X}_2 \underset{X_2}{\times} X_1$ is an isomorphism and $Z_1 \cap f^{-1}(\mathcal{T}(\sigma_2)) = \mathcal{T}(\sigma_1)$, which implies that the morphism (1.11) as well as (1.12) is an isomorphism. We have furthermore $\omega_{Z_1/Z_2} \simeq i_1^{-1}\omega_{X_1/X_2}$. $\square$

Proposition 1.3.4. *Let $f\colon (X_1, Z_1, \sigma_1) \to (X_2, Z_2, \sigma_2)$ be a morphism of kernel data. Then there is a natural morphism*

$$\mathrm{R}f_{!!}\Big(\mathcal{L}_{\sigma_1}(Z_1, X_1) \otimes \beta_{X_1}(\mathrm{R}i_{1*}\omega_{Z_1/Z_2})\Big) \to \mathcal{L}_{\sigma_2}(Z_2, X_2).$$

Proof. The left-hand side is isomorphic to

$$\mathrm{R}f_{!!}\Big(\mathrm{R}p_{1!!}(\mathbb{K}_{\overline{\Omega}_1} \otimes \widetilde{\mathbb{K}}_{P_{\sigma_1}}) \otimes \beta_{X_1}(\mathrm{R}i_{1*}\omega_{Z_1/X_1}^{\otimes -1}) \otimes \beta_{X_1}(\mathrm{R}i_{1*}\omega_{Z_1/Z_2})\Big)$$
$$\simeq \mathrm{R}f_{!!}\Big(\mathrm{R}p_{1!!}(\mathbb{K}_{\overline{\Omega}_1} \otimes \widetilde{\mathbb{K}}_{P_{\sigma_1}}) \otimes \omega_{X_1/X_2} \otimes \beta_{X_1}(f^{-1}\,\mathrm{R}i_{2*}\omega_{Z_2/X_2}^{\otimes -1})\Big)$$
$$\simeq \mathrm{R}f_{!!}\,\mathrm{R}p_{1!!}\Big((\mathbb{K}_{\overline{\Omega}_1} \otimes \widetilde{\mathbb{K}}_{P_{\sigma_1}}) \otimes p_1^{-1}\omega_{X_1/X_2}\Big) \otimes \beta_{X_2}(\mathrm{R}i_{2*}\omega_{Z_2/X_2}^{\otimes -1}) \tag{1.13}$$
$$\simeq \mathrm{R}p_{2!!}\,\mathrm{R}\widetilde{f}_{!!}\Big(\widetilde{f}^{-1}\,\mathbb{K}_{\overline{\Omega}_2} \otimes \widetilde{\mathbb{K}}_{P_{\sigma_1}} \otimes p_1^{-1}\omega_{X_1/X_2}\Big) \otimes \beta_{X_2}(\mathrm{R}i_{2*}\omega_{Z_2/X_2}^{\otimes -1})$$
$$\simeq \mathrm{R}p_{2!!}\Big(\mathbb{K}_{\overline{\Omega}_2} \otimes \mathrm{R}\widetilde{f}_{!!}(\widetilde{\mathbb{K}}_{P_{\sigma_1}} \otimes \omega_{\widetilde{X}_1/\widetilde{X}_2})\Big) \otimes \beta_{X_2}(\mathrm{R}i_{2*}\omega_{Z_2/X_2}^{\otimes -1}).$$

Hence, it is enough to construct a morphism

$$\mathrm{R}\widetilde{f}_{!!}\left(\widetilde{\mathbb{K}}_{P_{\sigma_1}} \otimes \omega_{\widetilde{X}_1/\widetilde{X}_2}\right) \to \widetilde{\mathbb{K}}_{P_{\sigma_2}}\,. \tag{1.14}$$

By adjunction it is enough to construct a morphism $\widetilde{\mathbb{K}}_{P_{\sigma_1}} \otimes \omega_{\widetilde{X}_1/\widetilde{X}_2} \to \widetilde{f}^{!}\,\widetilde{\mathbb{K}}_{P_{\sigma_2}}$. However by (1.10), we have

$$\widetilde{\mathbb{K}}_{P_{\sigma_1}} \otimes \omega_{\widetilde{X}_1/\widetilde{X}_2} \to \widetilde{\mathbb{K}}_{P_{\sigma_1} \underset{X_2}{\times} \mathcal{T}(\sigma_2)} \otimes \omega_{\widetilde{X}_1/\widetilde{X}_2} \simeq \widetilde{f}^{-1}\,\widetilde{\mathbb{K}}_{P_{\sigma_2}} \otimes \omega_{\widetilde{X}_1/\widetilde{X}_2} \simeq \widetilde{f}^{!}\,\widetilde{\mathbb{K}}_{P_{\sigma_2}},$$

where the last isomorphism follows from (1.3 a). □

Corollary 1.3.5. *Let $f\colon (X_1, Z_1, \sigma_1) \to (X_2, Z_2, \sigma_2)$ be a morphism, and assume that f is smooth and induces an isomorphism from Z_1 to Z_2. Then we have a natural isomorphism*

$$\mathrm{R}f_{!!}\mathcal{L}_{\sigma_1}(Z_1, X_1) \xrightarrow{\sim} \mathcal{L}_{\sigma_2}(Z_2, X_2).$$

Proof. By the assumption, we have $\mathcal{T}(\sigma_2) \underset{Z_2}{\times} Z_1 = \mathcal{T}(\sigma_1)$. By (1.13), it is enough to prove that (1.14) is an isomorphism. Since $P_{\sigma_1} = \widetilde{f}^{-1}(P_{\sigma_2})$, we have

$$\mathrm{R}\widetilde{f}_{!!}\left(\widetilde{\mathbb{K}}_{P_{\sigma_1}} \otimes \omega_{\widetilde{X}_1/\widetilde{X}_2}\right) \simeq \widetilde{\mathbb{K}}_{P_{\sigma_2}} \otimes \mathrm{R}\widetilde{f}_{!!}\left(\widetilde{\mathbb{K}}_{T_{Z_1}X_1} \otimes \omega_{\widetilde{X}_1/\widetilde{X}_2}\right).$$

Hence we have reduced the problem to

$$\mathrm{R}\widetilde{f}_{!!}\left(\widetilde{\mathbb{K}}_{T_{Z_1}X_1} \otimes \omega_{\widetilde{X}_1/\widetilde{X}_2}\right) \simeq \widetilde{\mathbb{K}}_{T_{Z_2}X_2}\,.$$

Since f is smooth, we can take local coordinate systems (x, z) on X_2 and (x, y, z) on X_1 such that $Z_2 = \{x = 0\}$, $Z_1 = \{x = 0,\ y = 0\}$ and f is given by the projection. We then take a coordinate system $(t, \tilde{x}, z)$ on $\widetilde{X_2}$ and $(t, \tilde{x}, \tilde{y}, z)$ on $\widetilde{X_1}$. The associated morphism $\widetilde{f}\colon \widetilde{X_1} \to \widetilde{X_2}$ is given by $(t, \tilde{x}, \tilde{y}, z) \to (t, \tilde{x}, z)$. Then we can check easily $\mathrm{R}\widetilde{f}_{!!}(\widetilde{\mathbb{K}}_{T_{Z_1}X_1} \otimes \omega_{\widetilde{X}_1/\widetilde{X}_2}) \simeq \mathrm{R}\widetilde{f}_{!!}(\widetilde{\mathbb{K}}_{\{t=0\}} \otimes \omega_{\widetilde{X}_1/\widetilde{X}_2}) \simeq \widetilde{\mathbb{K}}_{\{t=0\}}$. □

Lemma 1.3.6. *Let (X, Z, σ) be a kernel data on X, and let $f\colon X \to Y$ be a smooth morphism which induces a closed embedding $Z \hookrightarrow Y$. Assume that $\sigma(x) \notin T^*_{f(x)}Y$ for any $x \in \mathcal{T}(\sigma)$. Then we have*

$$\mathrm{R}f_{!!}\mathcal{L}_{\sigma}(Z, X) \simeq 0.$$

Proof. For any $x_0 \in \mathcal{T}(\sigma)$, take a local coordinate system $(y, z) = (y_1, \dots, y_n, z_1, \dots, z_m)$ of Y in a neighborhood of $f(x_0)$ such that $f(Z)$ is given by $y = 0$. Then we can take a local coordinate system (t, x, y, z) of X in a neighborhood of x_0 such that Z is given by $\{t = 0,\ x = 0,\ y = 0\}$, and $\sigma(x_0) = -dt(x_0)$. Then we have

$$\mathcal{L}_{\sigma}(Z, X) \otimes \widetilde{\mathbb{K}}_{x_0} \simeq \left(\underset{\delta>0,\, \varepsilon>0}{\text{“}\varinjlim\text{”}}\ \mathbb{K}_{F_{\delta,\varepsilon}}\right) \otimes \beta_X(\mathrm{R}i_*\omega_{Z/X}^{\otimes -1}) \otimes \widetilde{\mathbb{K}}_{x_0},$$

where

$$F_{\delta,\varepsilon} = \{(t, x, y, z); \delta \geq t > \varepsilon(|x| + |y|)\}.$$

Hence, $\big(\mathrm{R}f_{!!}(\mathcal{L}_\sigma(Z, X)\big) \otimes \widetilde{\mathbb{K}}_{f(x_0)} \simeq \mathrm{R}f_{!!}(\mathcal{L}_\sigma(Z, X) \otimes \widetilde{\mathbb{K}}_{x_0}) \simeq 0$ follows from

$$\mathrm{R}^j f_!(\mathbb{K}_{F_{\delta,\varepsilon}}) \simeq 0 \quad \text{for any } j \in \mathbb{Z}. \qquad \square$$

Proposition 1.3.7. *Let $f\colon (X_1, Z_1, \sigma_1) \to (X_2, Z_2, \sigma_2)$ be a morphism of kernel data, and assume that f is a closed immersion which induces an isomorphism $Z_1 \xrightarrow{\sim} Z_2$. Then there is a natural isomorphism*

$$\mathcal{L}_{\sigma_1}(Z_1, X_1) \xrightarrow{\sim} f^!\mathcal{L}_{\sigma_2}(Z_2, X_2).$$

Proof. Since f is a closed immersion, we get the commutative diagrams

$$\begin{array}{ccccccc}
Z_1 & \xhookrightarrow{i_1} & X_1 & \xleftarrow{p_1} & \widetilde{X_1} & \xhookleftarrow{j_1} & \Omega_1 \\
\sim\downarrow & \square & \downarrow f & & \downarrow \widetilde{f} & \square & \downarrow \\
Z_2 & \xhookrightarrow[i_2]{} & X_2 & \xleftarrow[p_2]{} & \widetilde{X}_2 & \xhookleftarrow[j_2]{} & \Omega_2
\end{array}
\quad \text{and} \quad
\begin{array}{ccc}
T_{Z_1}X_1 & \xhookrightarrow{s_1} & \widetilde{X_1} \\
T_Z f \downarrow & \square & \downarrow \widetilde{f} \\
T_{Z_2}X_2 & \xhookrightarrow[s_2]{} & \widetilde{X}_2\,,
\end{array}$$

in which the squares marked by $\square$ are cartesian. Recall the adjunction isomorphism $f^!\,\mathrm{R}f_{!!} \simeq \mathrm{id}$. Hence it is enough to construct an isomorphism

$$\mathrm{R}f_{!!}\mathcal{L}_{\sigma_1}(Z_1, X_2) \xrightarrow{\sim} \mathrm{R}f_{!!}f^!\mathcal{L}_{\sigma_2}(Z_2, X_2).$$

Next recall that

$$\mathrm{R}f_{!!}f^!\mathcal{L}_{\sigma_2}(Z_2, X_2) \simeq \mathrm{R}\mathcal{I}\mathcal{H}om\left(\mathbb{K}_{X_1}, \mathcal{L}_{\sigma_2}(Z_2, X_2)\right).$$

Therefore we may write:

$$\begin{aligned}
\mathrm{R}f_{!!}f^!\mathcal{L}_{\sigma_2}(Z_2, X_2) &\simeq \mathrm{R}\mathcal{I}\mathcal{H}om\Big(\mathbb{K}_{X_1}, \mathrm{R}p_{2!!}\Big(\mathbb{K}_{\overline{\Omega_2}} \otimes \beta_{\widetilde{X}_2}\big(\mathbb{K}_{P_{\sigma_2}} \otimes p_2^{-1}\,\mathrm{R}i_{2*}\omega^{\otimes -1}_{Z_2/X_2}\big)\Big)\Big) \\
&\simeq \mathrm{R}p_{2!!}\,\mathrm{R}\mathcal{I}\mathcal{H}om\Big(p_2{}^{-1}\,\mathbb{K}_{X_1}, \mathbb{K}_{\overline{\Omega_2}} \otimes \beta_{\widetilde{X_2}}\big(\mathbb{K}_{P_{\sigma_2}} \otimes p_2^{-1}\,\mathrm{R}i_{2*}\omega^{\otimes -1}_{Z_2/X_2}\big)\Big) \\
&\simeq \mathrm{R}p_{2!!}\Big(\mathrm{R}\mathcal{I}\mathcal{H}om\,(p_2^{-1}\,\mathbb{K}_{X_1}, \mathbb{K}_{\overline{\Omega_2}}) \otimes \beta_{\widetilde{X_2}}(\mathbb{K}_{P_{\sigma_2}} \otimes p_2^{-1}\,\mathrm{R}i_{2*}\omega^{\otimes -1}_{Z_2/X_2})\Big).
\end{aligned}$$

On the other hand, $P_{\sigma_1} = \widetilde{f}^{-1}P_{\sigma_2}$ implies that

$$\begin{aligned}
\mathrm{R}f_{!!}\mathcal{L}_{\sigma_1}(Z_1, X_1) &\simeq \mathrm{R}f_{!!}\,\mathrm{R}p_{1!!}\Big(\mathbb{K}_{\overline{\Omega_1}} \otimes \beta_{\widetilde{X_1}}\big(\mathbb{K}_{P_{\sigma_1}} \otimes p_1^{-1}\,\mathrm{R}i_{1*}\omega^{\otimes -1}_{Z_1/X_1}\big)\Big) \\
&\simeq \mathrm{R}p_{2!!}\,\mathrm{R}\widetilde{f}_{!!}\Big(\mathbb{K}_{\widetilde{f}^{-1}(\overline{\Omega_2})} \otimes \beta_{\widetilde{X_1}}\big(\widetilde{f}^{-1}\,\mathbb{K}_{P_{\sigma_1}} \otimes \widetilde{f}^{-1}p_2^{-1}\,\mathrm{R}s_{2*}\omega^{\otimes -1}_{Z_2/X_2} \otimes p_1{}^{-1}\omega_{X_1/X_2}\big)\Big) \\
&\simeq \mathrm{R}p_{2!!}\,\mathrm{R}\widetilde{f}_{!!}\Big(\widetilde{f}^{-1}\Big(\mathbb{K}_{\overline{\Omega_2}} \otimes \beta_{\widetilde{X_2}}\big(\mathbb{K}_{P_{\sigma_2}} \otimes p_2^{-1}\,\mathrm{R}i_{2*}\omega^{\otimes -1}_{Z_2/X_2}\big)\Big) \otimes \omega_{\widetilde{X}_1/\widetilde{X}_2}\Big) \\
&\simeq \mathrm{R}p_{2!!}\Big(\mathbb{K}_{\overline{\Omega_2}} \otimes \beta_{\widetilde{X_2}}\big(\mathbb{K}_{P_{\sigma_2}} \otimes p_2^{-1}\,\mathrm{R}i_{2*}\omega^{\otimes -1}_{Z_2/X_2}\big) \otimes \mathrm{R}\widetilde{f}_{!!}\omega_{\widetilde{X}_1/\widetilde{X}_2}\Big),
\end{aligned}$$

and it is enough to show that

$$\mathrm{R}\mathcal{IH}om(p_2^{-1}\mathbb{K}_{X_1}, \mathbb{K}_{\overline{\Omega_2}}) \simeq \mathbb{K}_{\overline{\Omega_2}} \otimes \mathrm{R}\widetilde{f}_{!!}\omega_{\widetilde{X}_1/\widetilde{X}_2}.$$

However we have the natural chain of isomorphisms

$$\begin{aligned}\mathrm{R}\mathcal{IH}om(p_2^{-1}\mathbb{K}_{X_1}, \mathbb{K}_{\overline{\Omega_2}}) &\simeq \mathrm{R}\mathcal{IH}om(p_2^{-1}\mathbb{K}_{X_1}, \mathrm{R}j_{2*}\mathbb{K}_{\Omega_2})\\ &\simeq \mathrm{R}j_{2*}\mathrm{R}\mathcal{IH}om(j_2{}^{-1}p_2^{-1}\mathbb{K}_{X_1}, \mathbb{K}_{\Omega_2})\\ &\simeq \mathrm{R}j_{2*}\mathrm{R}\mathcal{IH}om(\mathbb{K}_{\Omega_1}, \mathbb{K}_{\Omega_2}).\end{aligned}$$

On the other hand, we have, as an object of $\mathrm{D^b}(\mathrm{I}(\mathbb{K}_{\Omega_2}))$,

$$\mathrm{R}\mathcal{IH}om(\mathbb{K}_{\Omega_1}, \mathbb{K}_{\Omega_2}) \simeq j_2^{-1}\,\mathrm{R}\widetilde{f}_{*}\omega_{\widetilde{X}_1/\widetilde{X}_2},$$

and hence

$$\begin{aligned}\mathrm{R}\mathcal{IH}om(p_2^{-1}\mathbb{K}_{X_1}, \mathbb{K}_{\overline{\Omega_2}}) &\simeq \mathrm{R}j_{2*}j_2^{-1}\,\mathrm{R}\widetilde{f}_{*}\omega_{\widetilde{X}_1/\widetilde{X}_2}\\ &\simeq \mathrm{R}j_{2*}\mathbb{K}_{\Omega_1}\otimes\mathrm{R}\widetilde{f}_{*}\omega_{\widetilde{X}_1/\widetilde{X}_2} \simeq \mathbb{K}_{\overline{\Omega_1}} \otimes \mathrm{R}\widetilde{f}_{!!}\omega_{\widetilde{X}_1/\widetilde{X}_2}.\end{aligned}$$ □

Proposition 1.3.8. *Let (X, Z_1, σ_1) and (X, Z_2, σ_2) be kernel data on the same base manifold X. Assume that Z_1, Z_2 are transversal submanifolds. Then there is a natural morphism*

$$\mathcal{L}_{\sigma_1}(Z_1, X) \otimes \mathcal{L}_{\sigma_2}(Z_2, X) \to \mathcal{L}_{\sigma_1+\sigma_2}(Z_1 \cap Z_2, X) \otimes \widetilde{\mathbb{K}}_{\mathcal{T}(\sigma_1)\cap\mathcal{T}(\sigma_2)}.$$

Proof. Set $Z = Z_1 \cap Z_2$, $\sigma = \sigma_1 + \sigma_2$ and $N = \mathcal{T}(\sigma_1) \cap \mathcal{T}(\sigma_2) \subset \mathcal{T}(\sigma) \subset Z$.

(i) Assume first that $\sigma_1(x)$ and $\sigma_2(x)$ are linearly independent vectors of T^*X for every $x \in Z$. Then we have

$$\mathcal{L}_{\sigma_k}(Z_k, X) \otimes \widetilde{\mathbb{K}}_N \simeq \underset{\overrightarrow{U_k}}{\text{“lim”}}\, \mathbb{K}_{U_k} \otimes \widetilde{\mathbb{K}}_N \otimes \beta_X\left(\mathrm{R}i_{k*}\omega_{Z_k/X}^{\otimes -1}\right),$$

where the inductive limits is taken over the family of open subsets U_k of X such that $\mathrm{C}_{Z_k}(U_k) \cap P_{\sigma_k} \subset Z_k$. For such open subsets U_1, U_2, we have

$$\mathrm{C}_Z(U_1 \cap U_2) \cap \left(P_\sigma \underset{Z}{\times} N\right) \subset Z,$$

since $P_\sigma \underset{Z}{\times} N \subset P_{\sigma_1} \cup P_{\sigma_2}$. Hence we get a natural morphism

$$\begin{aligned}&\mathcal{L}_{\sigma_1}(Z_1, X) \otimes \mathcal{L}_{\sigma_2}(Z_2, X) \otimes \widetilde{\mathbb{K}}_N\\ &\quad\simeq \left(\underset{\overrightarrow{U_1}}{\text{“lim”}}\, \mathbb{K}_{U_1} \otimes \beta\left(\mathrm{R}i_{1*}\omega_{Z_1/X}^{\otimes -1}\right)\right)\\ &\qquad\otimes \left(\underset{\overrightarrow{U_2}}{\text{“lim”}}\, \mathbb{K}_{U_2} \otimes \beta\left(\mathrm{R}i_{2*}\omega_{Z_2/X}^{\otimes -1}\right)\right) \otimes \widetilde{\mathbb{K}}_N\\ &\quad\to \left(\underset{\overrightarrow{U}}{\text{“lim”}}\, \mathbb{K}_U\right) \otimes \beta\left(\mathrm{R}i_{1*}\omega_{Z_1/X}^{\otimes -1}\right) \otimes \beta\left(\mathrm{R}i_{2*}\omega_{Z_2/X}^{\otimes -1}\right) \otimes \widetilde{\mathbb{K}}_N,\end{aligned}$$

where U ranges over the family of open subsets of X such that $C_Z(U) \cap \left(P_\sigma \underset{Z}{\times} N\right) \subset Z$.
Since Z_1 and Z_2 are transversal submanifolds of X, we have

$$\omega_{Z/X}^{\otimes -1} \simeq \left(\omega_{Z_1/X}^{\otimes -1}|_Z\right) \otimes \left(\omega_{Z_2/X}^{\otimes -1}|_Z\right).$$

Hence we obtain

$$\text{“}\varinjlim_U\text{''}\, \mathbb{K}_U \otimes \beta_X\left(\mathrm{R}i_{1*}\omega_{Z_1/X}^{\otimes -1}\right) \otimes \beta_X\left(\mathrm{R}i_{2*}\omega_{Z_2/X}^{\otimes -1}\right) \otimes \widetilde{\mathbb{K}}_N \simeq \mathcal{L}_\sigma(Z, X) \otimes \widetilde{\mathbb{K}}_N,$$

which provides the desired morphism.

(ii) Consider the general case. We set $\mathbb{A}_X^n = X \times \mathbb{R}^n$ for $n = 1, 2$. We use coordinates (x, t_1, t_2) on $\mathbb{A}_X^2$. We regard the manifold $\mathbb{A}_{Z_k}^1$ as a submanifold of $\mathbb{A}_X^2$ by

$$\mathbb{A}_{Z_k}^1 := \{(x, t_1, t_2)\,;\; x \in Z_k,\ t_k = 0\},$$

and $\mathbb{A}_X^1$ as the submanifold $\{t_2 = 0\}$ of $\mathbb{A}_X^2$. We identify Z with

$$\mathbb{A}_{Z_1}^1 \cap \mathbb{A}_{Z_2}^1 = \{(x, t_1, t_2)\,;\; x \in Z,\ t_1 = t_2 = 0\}.$$

Thus we obtain the following commutative diagrams

$$\begin{array}{ccccccc}
 & & X & \xhookrightarrow{i} & \mathbb{A}_X^1 & \xhookrightarrow{i'} & \mathbb{A}_X^2 \\
 & & \uparrow & & \uparrow & \mathrm{tr} & \uparrow{\scriptstyle j_1} \\
Z & \hookrightarrow & Z_1 & \xrightarrow{\sim} & Z_1 & \hookrightarrow & \mathbb{A}_{Z_1}^1
\end{array}
\quad \text{and} \quad
\begin{array}{ccccccc}
 & & X & \xhookrightarrow{i} & \mathbb{A}_X^1 & \xhookrightarrow{i'} & \mathbb{A}_X^2 \\
 & & \uparrow & \mathrm{tr} & \uparrow & & \uparrow{\scriptstyle j_2} \\
Z & \hookrightarrow & Z_2 & \hookrightarrow & \mathbb{A}_{Z_2}^1 & \xrightarrow{\sim} & \mathbb{A}_{Z_2}^1
\end{array}$$

where $j_1(z_1, t) = (z_1, 0, t)$ and $j_2(z_2, t) = (z_2, t, 0)$. Note that the squares marked with tr are transversal. Define the sections

$$\begin{aligned}
\tilde{\sigma}_1 &= \sigma_1 + dt_1 : \mathbb{A}_{Z_1}^1 \to T^*\mathbb{A}_X^2, \\
\tilde{\sigma}_2 &= \sigma_2 + dt_2 : \mathbb{A}_{Z_2}^1 \to T^*\mathbb{A}_X^2, \\
\tilde{\sigma} &= \sigma_1 + \sigma_2 + dt_1 + dt_2 : Z \to T^*\mathbb{A}_X^2.
\end{aligned}$$

Clearly $\tilde{\sigma}_1$ and $\tilde{\sigma}_2$ are linearly independent at each point, and the result in the first part gives a morphism

$$\mathcal{L}_{\tilde{\sigma}_1}\left(\mathbb{A}_{Z_1}^1, \mathbb{A}_X^2\right) \otimes \mathcal{L}_{\tilde{\sigma}_2}\left(\mathbb{A}_{Z_2}^1, \mathbb{A}_X^2\right) \to \mathcal{L}_{\tilde{\sigma}}(Z, \mathbb{A}_X^2) \otimes \widetilde{\mathbb{K}}_N.$$

We then deduce morphisms with the help of Propositions 1.3.3 and 1.3.7

$$\begin{aligned}\mathcal{L}_{\sigma_1}(Z_1,X)\otimes\mathcal{L}_{\sigma_2}(Z_2,X) &\simeq i^!\mathcal{L}_{\tilde{\sigma}_1}\left(Z_1,\mathbb{A}^1_X\right)\otimes i^{-1}\mathcal{L}_{\sigma_2}\left(\mathbb{A}^1_{Z_2},\mathbb{A}^1_X\right)\\ &\to i^!\left(\mathcal{L}_{\tilde{\sigma}_1}\left(Z_1,\mathbb{A}^1_X\right)\otimes\mathcal{L}_{\sigma_2}\left(\mathbb{A}^1_{Z_2},\mathbb{A}^1_X\right)\right)\\ &\simeq i^!\left(i'^{-1}\mathcal{L}_{\tilde{\sigma}_1}\left(\mathbb{A}^1_{Z_1},\mathbb{A}^2_X\right)\otimes i'^!\mathcal{L}_{\tilde{\sigma}_2}\left(\mathbb{A}^1_{Z_2},\mathbb{A}^2_X\right)\right)\\ &\to i^!i'^!\left(\mathcal{L}_{\tilde{\sigma}_1}\left(\mathbb{A}^1_{Z_1},\mathbb{A}^2_X\right)\otimes\mathcal{L}_{\tilde{\sigma}_2}\left(\mathbb{A}^1_{Z_2},\mathbb{A}^2_X\right)\right)\\ &\to i^!i'^!\left(\mathcal{L}_{\tilde{\sigma}}(Z,\mathbb{A}^2_X)\otimes\widetilde{\mathbb{K}}_N\right)\simeq\mathcal{L}_{\sigma}(Z,X)\otimes\widetilde{\mathbb{K}}_N,\end{aligned}$$

which completes the proof. □

Remark 1.3.9. Although we do not give proofs, the following two facts hold.

(i) If σ_1 and σ_2 are linearly independent, the two morphisms constructed in the parts (i) and (ii) of the proof of Proposition 1.3.8 coincide.
(ii) If (X, Z_3, σ_3) is a third kernel data such that (Z_1, Z_2), (Z_1, Z_3) and (Z_2, Z_3) are transversal in X and that $(Z_1 \cap Z_3, Z_2 \cap Z_3)$ is transversal in Z_3, then the following diagram is commutative where $N = \mathcal{T}(\sigma_1) \cap \mathcal{T}(\sigma_2) \cap \mathcal{T}(\sigma_3)$:

$$\begin{array}{ccc}\begin{array}{c}\mathcal{L}_{\sigma_1}(Z_1,X)\otimes\mathcal{L}_{\sigma_2}(Z_2,X)\\ \otimes\mathcal{L}_{\sigma_3}(Z_3,X)\end{array} & \longrightarrow & \begin{array}{c}\mathcal{L}_{\sigma_1+\sigma_2}(Z_1\cap Z_2,X)\\ \otimes\mathcal{L}_{\sigma_3}(Z_3,X)\otimes\widetilde{\mathbb{K}}_N\end{array}\\ \downarrow & & \downarrow\\ \begin{array}{c}\mathcal{L}_{\sigma_1}(Z_1,X)\otimes\mathcal{L}_{\sigma_2+\sigma_3}\\ \times(Z_2\cap Z_3,X)\otimes\widetilde{\mathbb{K}}_N\end{array} & \longrightarrow & \begin{array}{c}\mathcal{L}_{\sigma_1+\sigma_2+\sigma_3}(Z_1\cap Z_2\cap Z_3,X)\\ \otimes\widetilde{\mathbb{K}}_N,\end{array}\end{array}$$

i.e., the composition morphisms are associative.

Lemma 1.3.10. *Let (X, Z_1, σ_1), (X, Z_2, σ_2) be kernel data on X and assume that Z_1, Z_2 are transversal submanifolds of X and that σ_1 and σ_2 never vanish. Let $f: X \to Y$ be a smooth morphism which induces a closed embedding $Z_1 \cap Z_2 \hookrightarrow Y$. Assume the following condition:*

$$\left(\mathbb{R}_{\geq 0}\sigma_1(x)+\mathbb{R}_{\geq 0}\sigma_2(x)\right)\cap T^*_{f(x)}Y=\{0\}\ \textit{for every}\ x\in\mathcal{T}(\sigma_1)\cap\mathcal{T}(\sigma_2).$$

Here $T^*_{f(x)}Y$ is regarded as a subspace of T^*_xX by f_d. Then we have

$$\mathrm{R}f_{!!}\left(\mathcal{L}_{\sigma_1}(Z_1,X)\otimes\mathcal{L}_{\sigma_2}(Z_2,X)\right)\simeq 0.$$

Proof. Let us show that

$$\mathrm{R}f_{!!}\left(\mathcal{L}_{\sigma_1}(Z_1,X)\otimes\mathcal{L}_{\sigma_2}(Z_2,X)\otimes\widetilde{\mathbb{K}}_{x_0}\right)\simeq 0$$

for any $x_0 \in \mathcal{T}(\sigma_1) \cap \mathcal{T}(\sigma_2)$. We first reduce the proof to the case where X is of relative dimension one over Y. Assume the assertion to be true for relative one-dimensional morphisms. Set $E = T_{x_0}(f^{-1}f(x_0))$. Then by the assumption, E satisfies $\left(\mathbb{R}_{\geq 0}\sigma_1(x_0) + \mathbb{R}_{\geq 0}\sigma_2(x_0)\right) \cap E^{\perp} = \{0\}$. Hence there exists a line $\ell \subset E$

such that $\left(\mathbb{R}_{\geq 0}\sigma_1(x_0) + \mathbb{R}_{\geq 0}\sigma_2(x_0)\right) \cap \ell^{\perp} = \{0\}$. Decompose f into the composition of smooth morphisms $X \xrightarrow{g} Y' \xrightarrow{h} Y$ on a neighborhood of x_0 such that g and h are smooth and $T_{x_0}(g^{-1}g(x_0)) = \ell$. Then g satisfies the conditions in the lemma. Hence applying to g the relative one-dimensional morphism case, we obtain $\mathrm{R}g_{!!}\left(\mathcal{L}_{\sigma_1}(Z_1, X) \otimes \mathcal{L}_{\sigma_2}(Z_2, X) \otimes \widetilde{\mathbb{K}}_{x_0}\right) \simeq 0$, which implies the desired result.

Now assume that f has relative dimension one. Since $\sigma_k(x_0) \notin T^*_{f(x_0)}Y$, the map $Z_k \to Y$ is a (local) embedding, and $T_{x_0}Z_k = f_*^{-1}\left(T_{f(x_0)}Z'_k\right) \cap \sigma_k(x_0)^{-1}(0)$, where $Z'_k := f(Z_k) \subset Y$. Then Z'_1 and Z'_2 are transversal, and $f(Z_1 \cap Z_2)$ is a hypersurface of $Z'_1 \cap Z'_2$ since

$$\begin{aligned}\operatorname{codim}_Y(f(Z_1 \cap Z_2)) &= \operatorname{codim}_X(Z_1 \cap Z_2) - 1 = \operatorname{codim}_X(Z_1) + \operatorname{codim}_X(Z_2) - 1\\ &= \operatorname{codim}_Y(Z'_1) + \operatorname{codim}_Y(Z'_2) + 1 = \operatorname{codim}_Y(Z'_1 \cap Z'_2) + 1.\end{aligned}$$

Since $T_{x_0}(Z_1 \cap Z_2) = f_*^{-1}\left(T_{f(x_0)}(Z'_1 \cap Z'_2)\right) \cap \sigma_1(x_0)^{-1}(0) \cap \sigma_2(x_0)^{-1}(0)$, the vectors $\sigma_1(x_0)$ and $\sigma_2(x_0)$ are linearly independent. By multiplying by a positive constant, we may therefore assume that

$$\sigma_1(x_0) - \sigma_2(x_0) \in T^*_{f(x_0)}Y \setminus \{0\}.$$

Take a local coordinate system (t, y_1, y_2, z) of Y such that

$$Z'_k = \{y_k = 0\} \text{ and } \sigma_2(x_0) - \sigma_1(x_0) = dt.$$

Then take a local coordinate system (x, t, y_1, y_2, z) of X such that $\sigma_1(x_0) = -dx$ (and hence $\sigma_2(x_0) = dt - dx$), and $Z_1 = \{y_1 = 0,\ x = 0\}$ and f is given by forgetting x. Set $Z_2 = \{y_2 = 0,\ x = \varphi(t, y_1, z)\}$. Then replacing $\varphi(t, y_1, z)$ with t, we may assume from the beginning that

$$Z_2 = \{y_2 = 0,\ x = t\},\ Z_1 \cap Z_2 = \{y_1 = 0,\ y_2 = 0,\ x = t = 0\}.$$

Then we have

$$\begin{aligned}\mathcal{L}_{\sigma_1}(Z_1, X) \otimes \mathcal{L}_{\sigma_2}(Z_2, X) \otimes \widetilde{\mathbb{K}}_{x_0} \simeq \ & \text{“}\varinjlim_{\delta>0,\, \varepsilon>0}\text{”} \left(\mathbb{K}_{U^1_{\delta,\varepsilon}} \otimes \mathbb{K}_{U^2_{\delta,\varepsilon}}\right)\\ & \otimes \beta_X\left(\mathrm{R}i_{1*}\omega^{\otimes -1}_{Z_1/X} \otimes \mathrm{R}i_{2*}\omega^{\otimes -1}_{Z_2/X}\right) \otimes \widetilde{\mathbb{K}}_{x_0},\end{aligned}$$

where the open sets $U^k_{\delta,\varepsilon}$ are given by

$$U^1_{\delta,\varepsilon} = \{\varepsilon|y_1| < x \leq \delta\} \quad \text{and} \quad U^2_{\delta,\varepsilon} = \{\varepsilon|y_2| < x - t \leq \delta\}.$$

Hence we have

$$U^1_{\delta,\varepsilon} \cap U^2_{\delta,\varepsilon} = \{\max(\varepsilon|y_1|,\ \varepsilon|y_2| + t) < x \leq \min(\delta,\ \delta + t)\}.$$

Then the result follows from

$$\mathrm{R}f_!(\mathbb{K}_{U^1_{\delta,\varepsilon} \cap U^2_{\delta,\varepsilon}}) \simeq 0. \qquad \square$$

Proposition 1.3.11. *Let (X, Z_1, σ_1), (X, Z_2, σ_2) be kernel data on X and (Y, Z, σ) a kernel data on Y. Assume that Z_1, Z_2 are transversal submanifolds of X. Let $f\colon X \to Y$ be a smooth morphism which induces an isomorphism $Z_1 \cap Z_2 \xrightarrow{\sim} Z$. Let N be a closed subset of $\mathcal{T}(\sigma_1) \cap \mathcal{T}(\sigma_2)$ satisfying the following conditions:*

(i) $\mathcal{Z}(\sigma_1) \cap \mathcal{Z}(\sigma_2) \subset N$,
(ii) $f^*\sigma(x) = \sigma_1(x) + \sigma_2(x)$ for every $x \in N$,
(iii) $\sigma_1(x) \notin T^*_{f(x)}Y$ for any $x \in N \setminus \big(\mathcal{Z}(\sigma_1) \cup \mathcal{Z}(\sigma_2)\big)$,
(iv) $\big(\mathbb{R}_{\geq 0}\sigma_1(x) + \mathbb{R}_{\geq 0}\sigma_2(x)\big) \cap T^*_{f(x)}Y = \{0\}$ for every $x \in \big(\mathcal{T}(\sigma_1) \cap \mathcal{T}(\sigma_2)\big) \setminus N$,
(v) the morphism $Z_k \to Y$ is smooth at each point of $\mathcal{Z}(\sigma_k)$ for $k = 1, 2$.

Then there is a natural isomorphism

$$\mathrm{R}f_{!!}\big(\mathcal{L}_{\sigma_1}(Z_1, X) \otimes \mathcal{L}_{\sigma_2}(Z_2, X)\big) \xrightarrow{\sim} \mathcal{L}_\sigma(Z, Y) \otimes \widetilde{\mathbb{K}}_{f(N)}\,.$$

Proof. The morphism is obtained as the composition

$$\begin{aligned}\mathrm{R}f_{!!}\big(\mathcal{L}_{\sigma_1}(Z_1, X) \otimes \mathcal{L}_{\sigma_2}(Z_2, X)\big) &\to \mathrm{R}f_{!!}\big(\mathcal{L}_{\sigma_1+\sigma_2}(Z_1 \cap Z_2, X) \otimes \widetilde{\mathbb{K}}_N\big)\\ &\simeq \mathrm{R}f_{!!}\big(\mathcal{L}_{f^*\sigma}(Z_1 \cap Z_2, X) \otimes \widetilde{\mathbb{K}}_N\big)\\ &\to \mathcal{L}_\sigma(Z, Y) \otimes \widetilde{\mathbb{K}}_{f(N)}\,.\end{aligned}$$

In order to see that it is an isomorphism, it is enough to prove the isomorphism

$$\mathrm{R}f_{!!}\big(\mathcal{L}_{\sigma_1}(Z_1, X) \otimes \mathcal{L}_{\sigma_2}(Z_2, X) \otimes \widetilde{\mathbb{K}}_{x_0}\big) \xrightarrow{\sim} \mathcal{L}_\sigma(Z, Y) \otimes \widetilde{\mathbb{K}}_{f(N)} \otimes \widetilde{\mathbb{K}}_{f(x_0)}$$

for any $x_0 \in \mathcal{T}(\sigma_1) \cap \mathcal{T}(\sigma_2)$.

(a) Assume first that $\sigma_1(x_0) = \sigma_2(x_0) = 0$. Then, (i) implies $x_0 \in N$, and we have $\sigma(f(x_0)) = 0$ by (ii). Hence Proposition 1.2.8 implies

$$\begin{aligned}\mathrm{R}f_{!!}\big(\mathcal{L}_{\sigma_1}(Z_1, X) \otimes \mathcal{L}_{\sigma_2}(Z_2, X) \otimes \widetilde{\mathbb{K}}_{x_0}\big) &\simeq \mathrm{R}f_{!!}\big(\mathbb{K}_{Z_1} \otimes \mathbb{K}_{Z_2} \otimes \widetilde{\mathbb{K}}_{x_0}\big)\\ &\simeq \mathbb{K}_Z \otimes \widetilde{\mathbb{K}}_{f(x_0)} \simeq \mathcal{L}_\sigma(Z, Y) \otimes \widetilde{\mathbb{K}}_{f(N)} \otimes \widetilde{\mathbb{K}}_{f(x_0)}\,.\end{aligned}$$

(b) Assume $\sigma_1(x_0) = 0$ and $\sigma_2(x_0) \neq 0$. Then we have

$$\begin{aligned}\mathrm{R}f_{!!}\big(\mathcal{L}_{\sigma_1}(Z_1, X) \otimes \mathcal{L}_{\sigma_2}(Z_2, X) \otimes \widetilde{\mathbb{K}}_{x_0}\big) &\simeq \mathrm{R}f_{!!}\big(\mathbb{K}_{Z_1} \otimes \mathcal{L}_{\sigma_2}(Z_2, X) \otimes \widetilde{\mathbb{K}}_{x_0}\big)\\ &\simeq \mathrm{R}f_{!!}i_{1!!}i_1^{-1}\mathcal{L}_{\sigma_2}(Z_2, X) \otimes \widetilde{\mathbb{K}}_{f(x_0)},\end{aligned}$$

where $i_1\colon Z_1 \to X$ is the inclusion. Proposition 1.3.3 implies $i_1^{-1}\mathcal{L}_{\sigma_2}(Z_2, X) \simeq \mathcal{L}_{\sigma_2}(Z_1 \cap Z_2, Z_1)$. Note that $Z_1 \to Y$ is smooth at x_0 by the assumption (v). If $x_0 \in N$, then Corollary 1.3.5, along with by the hypothesis (ii), implies $\mathrm{R}f_{!!}i_{1!!}\mathcal{L}_{\sigma_2}(Z_1 \cap Z_2, Z_1) \simeq \mathcal{L}_\sigma(Z, Y)$. Assume $x \in \big(\mathcal{T}(\sigma_1) \cap \mathcal{T}(\sigma_2)\big) \setminus N$. Then (iv) implies that $\sigma_2(x_0) \notin T^*{}_{f}(x_0)Y$, and hence Lemma 1.3.6 implies $\mathrm{R}f_{!!}i_{1!!}\mathcal{L}_{\sigma_2}(Z_1 \cap Z_2, Z_1) \simeq 0$.

(c) Therefore we may assume that $\sigma_1(x_0) \neq 0$ and $\sigma_2(x_0) \neq 0$. If $x_0 \notin N$, then the result follows from (iv) and Lemma 1.3.10. We may assume therefore $x_0 \in N$. Similar to the proof of Lemma 1.3.10, we first reduce the proof to the case where X is of relative dimension one over Y. Assume the theorem to be true in the relative one-dimensional morphism case. Set $E = T_{x_0}(f^{-1}f(x_0))$. Let us choose a line $\ell \subset E$ such that $\sigma_1(x_0)|_\ell \neq 0$, and then decompose f into $X \xrightarrow{g} Y' \xrightarrow{h} Y$ on a neighborhood of x_0 such that g and h are smooth, and $T_{x_0}(g^{-1}g(x_0)) = \ell$. Then g satisfies the conditions (i)–(iv), and applying the relative dimension one case to g, we obtain

$$\mathrm{R}f_{!!}\left(\mathcal{L}_{\sigma_1}(Z_1, X) \otimes \mathcal{L}_{\sigma_2}(Z_2, X)\right) \simeq \mathrm{R}h_{!!}\mathcal{L}_{h^*\sigma}\left(g(Z_1 \cap Z_2), Y'\right) \simeq \mathcal{L}_\sigma(Z, Y),$$

where the last isomorphism is deduced from Corollary 1.3.5.

Hence we may assume that the relative dimension of X over Y is one. By the assumption (iii), $Z_k \to Y$ is a (local) embedding and $T_{x_0}Z_k = f_*^{-1}(T_{f(x_0)}Z'_k) \cap \sigma_k(x_0)^{-1}(0)$ where $Z'_k := f(Z_k)$. Then Z'_1 and Z'_2 are transversal submanifolds of Y and Z is a one-codimensional submanifold of $Z' := Z'_1 \cap Z'_2$. We have

$$\sigma(f(x_0)) \notin T^*_{Z'}Y.$$

Indeed, we have

$$\begin{aligned} T_{x_0}(Z_1 \cap Z_2) &= f_*^{-1}(T_{f(x_0)}Z') \cap \sigma_1(x_0)^{-1}(0) \cap \sigma_2(x_0)^{-1}(0) \\ &= f_*^{-1}\Big(T_{f(x_0)}Z' \cap \sigma(f(x_0))^{-1}(0)\Big) \cap \sigma_1(x_0)^{-1}(0), \end{aligned}$$

which implies $T_{f(x_0)}Z = T_{f(x_0)}Z' \cap \sigma(f(x_0))^{-1}(0) \neq T_{f(x_0)}Z'$.

Hence we can take local coordinates $(t, y_1, y_2, z) \in \mathbb{R} \times \mathbb{R}^{m_1} \times \mathbb{R}^{m_2} \times \mathbb{R}^n$ of Y such that $\sigma(f(x_0)) = -dt(f(x_0))$ and $Z'_k = \{y_k = 0\}$ $(k = 1, 2)$. Then we can choose a system of coordinates (x, t, y_1, y_2, z) on X such that f is given by forgetting x, $\sigma_1(x_0) = -dx(x_0)$ by (iii) (and hence $\sigma_2(x_0) = dx(x_0) - dt(x_0)$) and that $Z_1 = \{y_1 = 0,\ x = 0\}$. Set $Z_2 = \{y_2 = 0,\ x = \varphi(t, y_1, z)\}$. Replacing $\varphi(t, y_1, z)$ with t, we may assume from the beginning that

$$Z_2 = \{y_2 = 0,\ x = t\} \text{ and } Z = \{y_1 = 0,\ y_2 = 0,\ t = 0\}.$$

We then have using Corollary 1.2.13

$$\mathcal{L}_{\sigma_1}(Z_1, X) \otimes \widetilde{\mathbb{K}}_{x_0} \simeq \widetilde{\mathbb{K}}_{x_0} \otimes \text{“}\varinjlim_{\varepsilon>0}\text{”}\, \mathbb{K}_{U^1_\varepsilon} \otimes \beta_X\left(\mathrm{R}i_{1*}\omega^{\otimes -1}_{Z_1/X}\right),$$

$$\mathcal{L}_{\sigma_2}(Z_2, X) \otimes \widetilde{\mathbb{K}}_{x_0} \simeq \widetilde{\mathbb{K}}_{x_0} \otimes \text{“}\varinjlim_{\varepsilon>0}\text{”}\, \mathbb{K}_{U^2_\varepsilon} \otimes \beta_X\left(\mathrm{R}i_{2*}\omega^{\otimes -1}_{Z_2/X}\right),$$

where the open sets U_ε^k are given by

$$U_\varepsilon^1 = \{\varepsilon|y_1| < x\} \quad \text{and} \quad U_\varepsilon^2 = \{\varepsilon|y_2| < t - x\}.$$

We may therefore write

$$\begin{aligned}
&\mathcal{L}_{\sigma_1}(Z_1, X) \otimes \mathcal{L}_{\sigma_2}(Z_2, X) \otimes \widetilde{\mathbb{K}}_{x_0} \\
&\qquad \simeq \widetilde{\mathbb{K}}_{x_0} \otimes \text{“}\varinjlim_{\varepsilon>0}\text{”}\, \mathbb{K}_{U_\varepsilon^1 \cap U_\varepsilon^2} \otimes \beta_X\big(\mathrm{R}i_{1*}\omega_{Z_1/X}^{\otimes -1}\big) \otimes \beta_X\big(\mathrm{R}i_{2*}\omega_{Z_2/X}^{\otimes -1}\big) \\
&\qquad \simeq \widetilde{\mathbb{K}}_{x_0} \otimes \text{“}\varinjlim_{\varepsilon>0}\text{”}\, \mathbb{K}_{U_\varepsilon^1 \cap U_\varepsilon^2} \otimes \beta_X\big(f^{-1}\,\mathrm{R}i_*\omega_{Z/Y}^{\otimes -1}\big) \otimes \omega_{X/Y}^{\otimes -1}.
\end{aligned}$$

Since the relative dimension of X over Y is one, we have $\omega_{X/Y}^{\otimes -1} \otimes \widetilde{\mathbb{K}}_{x_0} \simeq \widetilde{\mathbb{K}}_{x_0}[1]$, and we deduce an isomorphism

$$\begin{aligned}
&\mathrm{R}f_{!!}\Big(\mathcal{L}_{\sigma_1}(Z_1, X) \otimes \mathcal{L}_{\sigma_2}(Z_2, X) \otimes \widetilde{\mathbb{K}}_{x_0}\Big) \\
&\qquad \simeq \mathrm{R}f_{!!}\left(\widetilde{\mathbb{K}}_{x_0} \otimes \text{“}\varinjlim_{\varepsilon>0}\text{”}\, \mathbb{K}_{U_\varepsilon^1 \cap U_\varepsilon^2} \otimes \omega_{X/Y}\right) \otimes \beta_Y\, \mathrm{R}i_*\omega_{Z/Y}^{\otimes -1} \\
&\qquad \simeq \mathrm{R}f_{!!}\left(\text{“}\varinjlim_{\varepsilon>0}\text{”}\, \mathbb{K}_{U_\varepsilon^1 \cap U_\varepsilon^2}\right)[1] \otimes \widetilde{\mathbb{K}}_{f(x_0)} \otimes \beta_Y\, \mathrm{R}i_*\omega_{Z/Y}^{\otimes -1}.
\end{aligned}$$

Since $U_\varepsilon^1 \cap U_\varepsilon^2 = \{\varepsilon|y_1| < x < t - \varepsilon|y_2|\}$, we have

$$\mathrm{R}f_!\big(\mathbb{K}_{U_\varepsilon^1 \cap U_\varepsilon^2}\big) \simeq \mathbb{K}_{\{\varepsilon(|y_1|+|y_2|)<t\}}[-1].$$

Hence we finally deduce that

$$\begin{aligned}
\mathrm{R}f_{!!}\Big(\mathcal{L}_{\sigma_1}(Z_1, X) \otimes \mathcal{L}_{\sigma_2}(Z_2, X) \otimes \widetilde{\mathbb{K}}_{x_0}\Big) &\simeq \left(\text{“}\varinjlim_{\varepsilon>0}\text{”}\, \mathbb{K}_{\{\varepsilon(|y_1|+|y_2|)<t\}}\right) \\
&\quad \otimes \beta_Y\, \mathrm{R}i_*\omega_{Z/Y}^{\otimes -1} \otimes \widetilde{\mathbb{K}}_{f(x_0)} \simeq \mathcal{L}_\sigma(Z, Y) \otimes \widetilde{\mathbb{K}}_{f(x_0)}. \qquad \square
\end{aligned}$$

Proposition 1.3.12. *Let (X_1, X_2, X_3) be a triplet of manifolds and $(X_i \times X_j, Z_{ij}, \sigma_{ij})$ a kernel data for $1 \le i < j \le 3$. Assume that $Z_{12} \times X_3$ and $X_1 \times Z_{23}$ are transversal in $X_1 \times X_2 \times X_3$ and that the projections $p_{ij}\colon X_1 \times X_2 \times X_3 \to X_i \times X_j$ induce an isomorphism $Z_{12} \underset{X_2}{\times} Z_{23} \xrightarrow{\sim} Z_{13}$. Let us denote by $p_2\colon X_1 \times X_2 \times X_3 \to X_2$ the second projection and by $p_{2*}\colon T^*(X_1 \times X_2 \times X_3) \to T^*X_2$ the induced projection. Let $N \subset \mathcal{T}(\sigma_{12}) \underset{X_2}{\times} \mathcal{T}(\sigma_{23})$ be a closed subset satisfying the following conditions:*

(i) $\mathcal{Z}(\sigma_{12}) \underset{X_2}{\times} \mathcal{Z}(\sigma_{23}) \subset N$,

(ii) $p_{13}^*\sigma_{13}(x) = p_{12}^*\sigma_{12}(x) + p_{23}^*\sigma_{23}(x)$ *for every* $x \in N$,

(iii) $p_{2*}\sigma_{12}(x) \notin T^*_{X_2}X_2$ *for any* $x \in N \setminus \big(\mathcal{Z}(\sigma_{12}) \times X_3 \cup X_1 \times \mathcal{Z}(\sigma_{23})\big)$,

(iv) $\mathbb{R}_{\geq 0}p_{2*}\sigma_{12}(x) \neq \mathbb{R}_{\leq 0}p_{2*}\sigma_{23}(x)$ *for every* $x \in \big(\mathcal{T}(\sigma_{12}) \underset{X_2}{\times} \mathcal{T}(\sigma_{23})\big) \setminus N$,

(v) the morphism $Z_{12} \to X_1$ *is smooth at each point of* $\mathcal{Z}(\sigma_{12})$ *and the morphism* $Z_{23} \to X_3$ *is smooth at each point of* $\mathcal{Z}(\sigma_{23})$.

Then we have an isomorphism

$$\mathcal{L}_{\sigma_{12}}(Z_{12}, X_1 \times X_2) \circ \mathcal{L}_{\sigma_{23}}(Z_{23}, X_2 \times X_3) \xrightarrow{\sim} \mathcal{L}_{\sigma_{13}}(Z_{13}, X_1 \times X_3) \otimes \widetilde{\mathbb{K}}_{f(N)}\,.$$

Proof. By Proposition 1.3.3, we have

$$p_{12}{}^{-1}\mathcal{L}_{\sigma_{12}}(Z_{12}, X_1 \times X_2) \simeq \mathcal{L}_{p_{12}^*\sigma_{12}}(Z_{12} \times X_3, X_1 \times X_2 \times X_3),$$

$$p_{23}{}^{-1}\mathcal{L}_{\sigma_{23}}(Z_{23}, X_2 \times X_3) \simeq \mathcal{L}_{p_{23}^*\sigma_{23}}(X_1 \times Z_{23}, X_1 \times X_2 \times X_3),$$

and Proposition 1.3.11 implies

$$\mathrm{R}p_{13!!}\Big(\mathcal{L}_{p_{12}^*\sigma_{12}}(Z_{12}\times X_3, X_1\times X_2\times X_3) \otimes \mathcal{L}_{p_{23}^*\sigma_{23}}(X_1\times Z_{23}, X_1\times X_2 \times X_3)\Big)$$
$$\simeq \mathcal{L}_{\sigma_{13}}(Z_{13}, X_1 \times X_3) \otimes \widetilde{\mathbb{K}}_{f(N)}\,. \qquad \square$$

2. Microlocalization of ind-sheaves

2.1. The kernel $\mathrm{K}_{\mathfrak{X}}$ of ind-microlocalization

We shall construct the kernel of microlocalization by the methods of the preceding section using the fundamental 1-form ω_X of T^*X. Since the construction uses only a 1-form, we shall discuss it on homogeneous symplectic manifolds. A *homogeneous symplectic manifold* is a manifold $\mathfrak{X}$ of even dimension endowed with a 1-form $\omega_{\mathfrak{X}}$ such that $(d\,\omega_{\mathfrak{X}})^{\dim \mathfrak{X}/2}$ never vanishes. It is a classical result that there locally exists a coordinate system $(x_1, \dots, x_n; \xi_1, \dots, \xi_n)$ where $\omega_{\mathfrak{X}}$ does not vanish and

$$\omega_{\mathfrak{X}} = \sum_{i=1}^{n} \xi_i dx_i. \tag{2.1}$$

Let $p_i\colon \mathfrak{X} \times \mathfrak{X} \to \mathfrak{X}$ $(i = 1, 2)$ be the projection and let $\Delta_{\mathfrak{X}}$ denote the diagonal of $\mathfrak{X} \times \mathfrak{X}$. Then $\sigma_{\mathfrak{X}} = p_1^*\,\omega_{\mathfrak{X}} - p_2^*\,\omega_{\mathfrak{X}}$ gives a section of $T^*_{\Delta_{\mathfrak{X}}}(\mathfrak{X} \times \mathfrak{X}) \to \Delta_{\mathfrak{X}}$.

Definition 2.1.1. The microlocalization kernel is the kernel defined on $\mathfrak{X} \times \mathfrak{X}$ by

$$\mathrm{K}_{\mathfrak{X}} = \mathcal{L}_{\sigma_{\mathfrak{X}}}(\Delta_{\mathfrak{X}}, \mathfrak{X} \times \mathfrak{X}) \in \mathrm{D^b}(\mathrm{I}(\mathbb{K}_{\mathfrak{X}\times\mathfrak{X}})).$$

Lemma 2.1.2. *There is a natural morphism*

$$\varepsilon_{\mathfrak{X}}\colon \mathbb{K}_{\Delta_{\mathfrak{X}}} \to \mathrm{K}_{\mathfrak{X}}$$

such that the compositions

$$\mathrm{K}_{\mathfrak{X}} \simeq \mathrm{K}_{\mathfrak{X}} \circ \mathbb{K}_{\Delta_{\mathfrak{X}}} \xrightarrow{\mathrm{K}_{\mathfrak{X}} \circ \varepsilon_{\mathfrak{X}}} \mathrm{K}_{\mathfrak{X}} \circ \mathrm{K}_{\mathfrak{X}},$$

$$\mathrm{K}_{\mathfrak{X}} \simeq \mathbb{K}_{\Delta_{\mathfrak{X}}} \circ \mathrm{K}_{\mathfrak{X}} \xrightarrow{\varepsilon_{\mathfrak{X}} \circ \mathrm{K}_{\mathfrak{X}}} \mathrm{K}_{\mathfrak{X}} \circ \mathrm{K}_{\mathfrak{X}}$$

are isomorphisms, and these two isomorphisms coincide.

Proof. We have constructed the morphism $\varepsilon_{\mathfrak{X}}$ in Corollary 1.2.10. The second statement easily follows from Proposition 1.3.12. The last statement follows from Lemma 2.1.3 below. □

Lemma 2.1.3. *Let $F : \mathcal{C} \to \mathcal{C}$ be a functor and $\alpha\colon \mathrm{id}_{\mathcal{C}} \to F$ a morphism of functors. Assume that for any object $X \in \mathrm{Ob}(\mathcal{C})$ the morphisms*

$$\alpha_{F(X)}\colon F(X) \to F(F(X)) \qquad\qquad F(\alpha_X)\colon F(X) \to F(F(X))$$

are isomorphisms. Then

(i) for any two objects $X, Y \in \mathrm{Ob}(\mathcal{C})$, the composition with α_X defines a bijection

$$\mathrm{Hom}_{\mathcal{C}}(F(X), F(Y)) \xrightarrow{\sim} \mathrm{Hom}_{\mathcal{C}}(X, F(Y)),$$

(ii) $\alpha_{F(X)} = F(\alpha_X)$ for any $X \in \mathrm{Ob}(\mathcal{C})$.

Lemma 2.1.4. *For two homogeneous symplectic manifolds $\mathfrak{X}$ and $\mathfrak{Y}$, we have*

$$\mathrm{K}_{\mathfrak{X}\times\mathfrak{Y}} \circ (\mathrm{K}_{\mathfrak{X}} \boxtimes \mathrm{K}_{\mathfrak{Y}}) \simeq \mathrm{K}_{\mathfrak{X}\times\mathfrak{Y}} \quad \textit{and} \quad \mathrm{K}_{\mathfrak{X}\times\mathfrak{Y}} \circ \mathrm{K}_{\mathfrak{X}} \simeq \mathrm{K}_{\mathfrak{X}\times\mathfrak{Y}}.$$

Proof. The last isomorphism is obtained by applying Proposition 1.3.12 to $(\mathfrak{X} \times \mathfrak{Y} \times \mathfrak{Y}, \mathfrak{X}, \mathfrak{X})$, and the first isomorphism follows from the second since

$$\mathrm{K}_{\mathfrak{X}\times\mathfrak{Y}} \circ (\mathrm{K}_{\mathfrak{X}} \boxtimes \mathrm{K}_{\mathfrak{Y}}) \simeq (\mathrm{K}_{\mathfrak{X}\times\mathfrak{Y}} \circ \mathrm{K}_{\mathfrak{X}}) \circ \mathrm{K}_{\mathfrak{Y}}.$$ □

Now let X be a manifold and set $\mathfrak{X} := T^*X$. Then $\mathfrak{X}$ has a canonical structure of a homogeneous symplectic manifold. The microlocalization functor is defined by:

$$\mu_X\colon \mathrm{D}^{\mathrm{b}}(\mathrm{I}(\mathbb{K}_X)) \to \mathrm{D}^{\mathrm{b}}(\mathrm{I}(\mathbb{K}_{\mathfrak{X}})); \quad \mathcal{F} \mapsto \mu_X\mathcal{F} := \mathrm{K}_{\mathfrak{X}} \circ \pi_X^{-1}\mathcal{F}.$$

The microlocalization functor μ_X may also be obtained as an integral transform associated with a kernel $\mathrm{L}_X \in \mathrm{D}^{\mathrm{b}}(\mathrm{I}(\mathbb{K}_{T^*X\times X}))$ which is often easier to manipulate than $\mathrm{K}_{\mathfrak{X}}$.

Definition 2.1.5. The kernel $\mathrm{L}_X \in \mathrm{D}^{\mathrm{b}}(\mathrm{I}(\mathbb{K}_{\mathbb{K}_{T^*X\times X}}))$ is given by

$$\mathrm{L}_X = \mathcal{L}_{\sigma_X}\left(T^*X \underset{X}{\times} X, T^*X \times X\right),$$

where σ_X is induced by ω_X on the first factor and -id on the second factor.

Remark 2.1.6. Let $(x; \xi)$ be a local coordinate system on $\mathfrak{X} = T^*X$ and let $(x, \xi; \eta, y)$ denote the associated coordinates on $T^*\mathfrak{X}$. Then σ_X is defined by

$$\sigma_X(x; \xi) = ((x, \xi; \xi, 0), (x; -\xi)) \in T^*\mathfrak{X} \times T^*X.$$

Therefore $\mathcal{T}(\sigma_X) = T^*X \underset{X}{\times} X$.

Proposition 2.1.7. *Let* $\mathcal{F} \in \mathrm{D}^{\mathrm{b}}(\mathrm{I}(\mathbb{K}_X))$. *There is a canonical isomorphism*

$$\mu_X\mathcal{F} \simeq L_X \circ \mathcal{F}.$$

Proof. Consider the following diagram

$$\begin{array}{ccccc} & & T^*X\times T^*X & & \\ & \overset{p_1}{\swarrow} & \downarrow{\scriptstyle q} & \overset{p_2}{\searrow} & \\ & & T^*X\times X & & T^*X \\ & \overset{p'_1}{\swarrow} & & \overset{p'_2}{\searrow} & \downarrow{\scriptstyle \pi_X} \\ T^*X & & & & X. \end{array} \tag{2.2}$$

Since q satisfies the assumptions of Corollary 1.3.5, we have the isomorphism $\mathrm{R}q_{!!}\,\mathrm{K}_{\mathfrak{X}} \simeq \mathrm{L}_X$, which implies

$$\begin{aligned} \mathrm{L}_X \circ \mathcal{F} &\simeq \mathrm{R}p'_{1!!}\left(\mathrm{R}q_{!!}\,\mathrm{K}_{\mathfrak{X}} \otimes p'^{-1}_2\mathcal{F}\right) \simeq \mathrm{R}p'_{1!!}\,\mathrm{R}q_{!!}\left(\mathrm{K}_{\mathfrak{X}} \otimes q^{-1}p'^{-1}_2\mathcal{F}\right) \\ &\simeq \mathrm{R}p_{1!!}\left(\mathrm{K}_{\mathfrak{X}} \otimes p_2^{-1}\pi_X^{-1}\mathcal{F}\right) \\ &\simeq \mathrm{K}_{\mathfrak{X}} \circ \pi_X^{-1}\mathcal{F} \simeq \mu_X\mathcal{F}. \end{aligned}$$

□

The next lemma immediately follows from Lemma 2.1.2.

Lemma 2.1.8. *For* $\mathcal{F} \in \mathrm{D}^{\mathrm{b}}(\mathrm{I}(\mathbb{K}_X))$, *we have*

$$\mathrm{K}_{T^*X} \circ \mu_X\mathcal{F} \simeq \mu_X\mathcal{F}.$$

Example 2.1.9. Let $Z \subset X$ be a closed submanifold. Then

$$\mu_X(\mathbb{K}_Z) \simeq \mathcal{L}_{\omega_X}\left(T^*X \underset{X}{\times} Z, T^*X\right).$$

Indeed, noting that $\mathbb{K}_Z \simeq \mathcal{L}_0(Z, X)$, it is enough to apply Proposition 1.3.12 to the triplet (T^*X, X, pt) with $N = T^*X \underset{X}{\times} Z$.

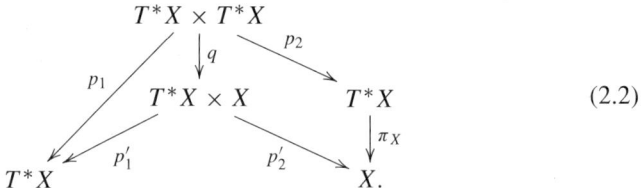

Note that the support of $\mu_X(\mathbb{K}_Z)$ is $T^*_Z X$. Let us take a local coordinate system (x, z) on X such that $Z = \{x = 0\}$. Let $(x, z; \xi, \zeta)$ be the corresponding coordinates on T^*X. Then on $\dot{T}^*X$, we have

$$\mu_X(\mathbb{K}_Z) \simeq \text{“}\varinjlim_{\varepsilon>0}\text{”}\, \mathbb{K}_{\{-\langle\xi,x\rangle>\varepsilon|x|\}} \otimes \widetilde{\mathbb{K}}_{\{x=0,\,\zeta=0\}}[\operatorname{codim} Z].$$

Note that

$$\mu_X(\widetilde{\mathbb{K}}_Z) \simeq \widetilde{\mathbb{K}}_{T^*_X X \underset{X}{\times} Z}\,. \tag{2.3}$$

Lemma 2.1.10. *Let $\mathcal{F} \in \mathrm{D}^{\mathrm{b}}(\mathrm{I}(\mathbb{K}_{T^*X}))$. Then*

$$(\mathrm{K}_{T^*X} \circ \mathcal{F}) \otimes \widetilde{\mathbb{K}}_{T^*_X X} \simeq \mathcal{F} \otimes \widetilde{\mathbb{K}}_{T^*_X X},$$

In particular if $\mathcal{F} \in \mathrm{D}^{\mathrm{b}}(\mathrm{I}(\mathbb{K}_X))$, then

$$\mu_X\mathcal{F} \otimes \widetilde{\mathbb{K}}_{T^*_X X} \simeq \pi_X^{-1}\mathcal{F} \otimes \widetilde{\mathbb{K}}_{T^*_X X}\,.$$

Proof. With the notations in (2.2),we have an isomorphism by Proposition 1.2.8:

$$\mathrm{K}_{T^*X} \otimes p_1^{-1}\widetilde{\mathbb{K}}_{T^*_X X} \simeq \mathbb{K}_{\Delta_{T^*X}} \otimes p_1^{-1}\widetilde{\mathbb{K}}_{T^*_X X}\,.$$

Therefore we have for $\mathcal{F} \in \mathrm{D}^{\mathrm{b}}(\mathrm{I}(\mathbb{K}_{T^*X}))$

$$\begin{aligned}
(\mathrm{K}_{T^*X} \circ \mathcal{F}) \otimes \widetilde{\mathbb{K}}_{T^*_X X} &\simeq \mathrm{R}p_{1!!}\left(\mathrm{K}_{T^*X} \otimes p_2^{-1}\mathcal{F}\right) \otimes \widetilde{\mathbb{K}}_{T^*_X X} \\
&= \mathrm{R}p_{1!!}\left(\mathrm{K}_{T^*X} \otimes p_1^{-1}\widetilde{\mathbb{K}}_{T^*_X X} \otimes p_2^{-1}\mathcal{F}\right) \\
&\simeq \mathrm{R}p_{1!!}\left(\mathbb{K}_{\Delta_{T^*X}} \otimes p_1^{-1}\widetilde{\mathbb{K}}_{T^*_X X} \otimes p_2^{-1}\mathcal{F}\right) \\
&= \mathrm{R}p_{1!!}\left(\mathbb{K}_{\Delta_{T^*X}} \otimes p_2^{-1}\mathcal{F}\right) \otimes \widetilde{\mathbb{K}}_{T^*_X X} \\
&\simeq \mathcal{F} \otimes \widetilde{\mathbb{K}}_{T^*_X X}\,.
\end{aligned}$$

□

Remark 2.1.11. The ind-sheaf $\mu_X\mathcal{F}$ is conical in the sense that it is equivariant with respect to the $\mathbb{R}_{>0}$-action on T^*X. We will not develop here the theory of conic ind-sheaves but simply give some consequences sufficient for our purpose. Let $\dot{T}^*X$ be the cotangent bundle with its zero section removed, and S^*X the associated sphere bundle. Let $\gamma : \dot{T}^*X \to S^*X$ be the natural projection and $\mathcal{F} \in \mathrm{D}^{\mathrm{b}}(\mathrm{I}(\mathbb{K}_X))$. Then we have the following isomorphism:

$$\mu_X\mathcal{F}|_{\dot{T}^*X} \simeq \gamma^{-1}\mathrm{R}\gamma_*\mu_X\mathcal{F}|_{\dot{T}^*X}.$$

Indeed, the kernel L_X satisfies a similar property.

Lemma 2.1.12. *Let X be a real manifold and $\pi_E : E \to X$ a real vector bundle over X. Denote by SE the spherical bundle associated with E and by*

$$j : \dot{E} \hookrightarrow E \qquad p : \dot{E} \to \mathrm{S}E$$

the natural morphisms. Assume that $\mathcal{F} \in \mathrm{D^b}(\mathrm{I}(\mathbb{K}_E))$ satisfies $j^{-1}\mathcal{F} \simeq p^{-1}\mathcal{G}$ for some $\mathcal{G} \in \mathrm{D^b}(\mathrm{I}(\mathbb{K}_{SE}))$. Then

(i) $\mathrm{R}\pi_{E*}\,\mathrm{R}j_{!!}j^{-1}\mathcal{F} \simeq 0$,

(ii) $\mathrm{R}\pi_{E*}(\mathcal{F}) \xrightarrow{\sim} \mathrm{R}\pi_{E*}(\widetilde{\mathbb{K}}_X \otimes \mathcal{F})$, *where X is associated to the zero section of E,*

(iii) there is a natural distinguished triangle

$$\mathrm{R}\dot{\pi}_{E!!}j^{-1}\mathcal{F} \to \mathrm{R}\pi_{E!!}\mathcal{F} \to \mathrm{R}\pi_{E*}\mathcal{F} \xrightarrow{+1} .$$

Proof. (a) Let E_X denote the real blow up of E along X identified with the zero section, *i.e.*, $E_X = \big(\dot{E} \times \mathbb{R}_{\geq 0}\big)/\mathbb{R}_{>0}$, hence $E_X = \dot{E} \sqcup SE$ as a set. We have the following commutative diagram:

$$\begin{array}{ccccc} \dot{E} & \overset{i}{\hookrightarrow} & E_X & \xrightarrow{q} & SE \\ & {\scriptstyle j}\searrow & \downarrow{\scriptstyle \pi_{E_X}} & & \downarrow{\scriptstyle \pi_{SE}} \\ & & E & \xrightarrow[\pi_E]{} & X \end{array}$$

where π_{E_X} and π_{SE} are proper.

(b) We shall first show

$$\mathrm{R}q_*\,\mathrm{R}i_{!!}j^{-1}\mathcal{F} \simeq 0.$$

Since q is locally trivial with fiber $R_{\geq 0}$, we have $q^!\mathcal{G} \simeq q^{-1}\mathcal{G} \otimes q^!\,\mathbb{K}_{SE} \simeq q^{-1}\mathcal{G} \otimes \mathbb{K}_{i(\dot{E})}[1]$. Therefore we have

$$\begin{aligned} \mathrm{R}q_*(\mathbb{K}_{i(\dot{E})} \otimes q^{-1}\mathcal{G}) &\simeq \mathrm{R}q_*\,\mathrm{R}\mathcal{IH}om\left(\mathbb{K}_{E_X}[1], q^!\mathcal{G}\right) \\ &\simeq \mathrm{R}\mathcal{IH}om\left(\mathrm{R}q_{!!}\,\mathbb{K}_{E_X}[1], \mathcal{G}\right) \simeq 0 \end{aligned}$$

since $\mathrm{R}q_{!!}\,\mathbb{K}_{E_X} = 0$. On the other hand, we have

$$\begin{aligned} \mathrm{R}q_*\left((\mathbb{K}_{i(\dot{E})} / \widetilde{\mathbb{K}}_{i(\dot{E})}) \otimes q^{-1}\mathcal{G}\right) &\simeq \mathrm{R}q_{!!}\left((\mathbb{K}_{i(\dot{E})} / \widetilde{\mathbb{K}}_{i(\dot{E})}) \otimes q^{-1}\mathcal{G}\right) \\ &\simeq \mathrm{R}q_{!!}\left((\mathbb{K}_{i(\dot{E})} / \widetilde{\mathbb{K}}_{i(\dot{E})})\right) \otimes \mathcal{G} \simeq 0. \end{aligned}$$

Hence the desired result follows from the distinguished triangle:

$$\begin{aligned} \mathrm{R}q_*(\widetilde{\mathbb{K}}_{i(\dot{E})} \otimes q^{-1}\mathcal{G}) &\to \mathrm{R}q_*(\mathbb{K}_{i(\dot{E})} \otimes q^{-1}\mathcal{G}) \\ &\to \mathrm{R}q_*\left((\mathbb{K}_{i(\dot{E})} / \widetilde{\mathbb{K}}_{i(\dot{E})}) \otimes q^{-1}\mathcal{G}\right) \xrightarrow{+1}, \end{aligned}$$

in which the first term is isomorphic to $\mathrm{R}q_*\,\mathrm{R}i_{!!}j^{-1}\mathcal{F}$.

(i) We have a chain of isomorphisms

$$\mathrm{R}\pi_{E*}\,\mathrm{R}j_{!!}j^{-1}\mathcal{F} \simeq \mathrm{R}\pi_{E*}\,\mathrm{R}\pi_{E_X\,!!}\,\mathrm{R}i_{!!}j^{-1}\mathcal{F}$$

$$\simeq \mathrm{R}\pi_{E*}\,\mathrm{R}\pi_{E_X*}\,\mathrm{R}i_{!!}j^{-1}\mathcal{F} \simeq \mathrm{R}\pi_{SE*}\,\mathrm{R}q_*\,\mathrm{R}i_{!!}j^{-1}\mathcal{F},$$

which vanishes by (b).

(ii) Applying the functor $\mathrm{R}\pi_{E*}(\bullet\otimes\mathcal{F})$ to the distinguished triangle

$$\widetilde{\mathbb{K}}_{\dot{E}} \to \mathbb{K}_E \to \widetilde{\mathbb{K}}_X \xrightarrow{+1}, \tag{2.4}$$

we obtain the distinguished triangle

$$\mathrm{R}\pi_{E*}(\widetilde{\mathbb{K}}_{\dot{E}}\otimes\mathcal{F}) \to \mathrm{R}\pi_{E*}\mathcal{F} \to \mathrm{R}\pi_{E*}(\widetilde{\mathbb{K}}_X\otimes\mathcal{F}) \xrightarrow{+1},$$

in which the first term vanishes by (i).

(iii) Applying the functor $\mathrm{R}\pi_{E\,!!}(\bullet\otimes\mathcal{F})$ to the distinguished triangle (2.4), we obtain the distinguished triangle

$$\mathrm{R}\pi_{E\,!!}(\widetilde{\mathbb{K}}_{\dot{E}}\otimes\mathcal{F}) \to \mathrm{R}\pi_{E\,!!}\mathcal{F} \to \mathrm{R}\pi_{E\,!!}(\widetilde{\mathbb{K}}_X\otimes\mathcal{F}) \xrightarrow{+1},$$

in which the first term is isomorphic to $\mathrm{R}\dot{\pi}_{E\,!!}j^{-1}\mathcal{F}$ and the last term is isomorphic to $\mathrm{R}\pi_{E*}\mathcal{F}$ by (ii). □

Proposition 2.1.13. *Let* $\mathcal{F}\in \mathrm{D^b}(\mathrm{I}(\mathbb{K}_X))$. *Then*

(i) $\mathrm{R}\pi_{X*}\mu_X\mathcal{F}\simeq\mathcal{F}$,

(ii) $\mathrm{R}\pi_{X\,!!}\mu_X\mathcal{F}\simeq\widetilde{\mathbb{K}}_{\Delta_X}\circ\mathcal{F}$,

(iii) $\mathrm{R}\dot{\pi}_{X\,!!}\left(\mu_X\mathcal{F}|_{\dot{T}^*X}\right)\simeq\left(\mathbb{K}_{X\times X\setminus\Delta_X}\otimes\widetilde{\mathbb{K}}_{\Delta_X}\right)\circ\mathcal{F}$,

(iv) there is a natural distinguished triangle

$$\mathrm{R}\dot{\pi}_{X\,!!}\left(\mu_X\mathcal{F}|_{\dot{T}^*X}\right)\to\mathrm{R}\pi_{X\,!!}\mu_X\mathcal{F}\to\mathcal{F}\xrightarrow{+1}.$$

Proof. (i) By Lemma 2.1.12 (ii), we have

$$\mathrm{R}\pi_{X*}\mu_X\mathcal{F}\simeq\mathrm{R}\pi_{X*}\left(\mu_X\mathcal{F}\otimes\widetilde{\mathbb{K}}_{T^*_XX}\right)\simeq\mathrm{R}\pi_{X!!}\left(\pi_X^{-1}\mathcal{F}\otimes\widetilde{\mathbb{K}}_{T^*_XX}\right)$$

$$\simeq\mathcal{F}\otimes\mathrm{R}\pi_{X!!}\,\widetilde{\mathbb{K}}_{T^*_XX}\simeq\mathcal{F},$$

where the second isomorphism follows from Lemma 2.1.10.

(ii) and (iii) Let us denote by $p\colon T^*X\times X\to X\times X$ the canonical morphism. Then we have isomorphisms:

$$\mathrm{R}\pi_{X!!}\,\mu_X\mathcal{F}\simeq(\mathrm{R}p_{!!}L_X)\circ\mathcal{F},$$

$$\mathrm{R}\dot{\pi}_{X!!}(\mu_X\mathcal{F}|_{\dot{T}^*X})\simeq\left(\mathrm{R}p_{!!}(L_X\otimes\widetilde{\mathbb{K}}_{\dot{T}^*X\times X})\right)\circ\mathcal{F}.$$

Hence, it is enough to show the isomorphism

$$\mathrm{R}p_{!!}L_X \simeq \widetilde{\mathbb{K}}_{\Delta_X}, \tag{2.5}$$

$$\mathrm{R}p_{!!}(L_X \otimes \widetilde{\mathbb{K}}_{\dot{T}^*X\times X}) \simeq \mathbb{K}_{X\times X\setminus\Delta_X} \otimes \widetilde{\mathbb{K}}_{\Delta_X}\,. \tag{2.6}$$

The natural morphism given in Corollary 1.2.10

$$\mathrm{L}_X \to \widetilde{\mathbb{K}}_{T^*X\underset{X}{\times}X} \otimes \beta_{T^*X\times X}\left(\omega^{\otimes -1}_{T^*X\underset{X}{\times}X/T^*X\times X}\right) = p^!\,\widetilde{\mathbb{K}}_{\Delta_X}$$

provides a morphism $\mathrm{R}p_{!!}L_X \to \widetilde{\mathbb{K}}_{\Delta_X}$.

We shall first show (2.6). Take a local coordinate system $x = (x_1, \dots, x_n)$ on X and let $((x;\xi), x')$ be the associated local coordinates on $T^*X \times X$. We have

$$\begin{aligned}
\mathrm{L}_X \otimes \widetilde{\mathbb{K}}_{\dot{T}^*X\times X} &\simeq \text{“}\varinjlim_{\varepsilon>0}\text{”}\, \mathbb{K}_{\{((x;\xi),x')\,;\,\langle\xi,x'-x\rangle>\varepsilon|x'-x|\}} \otimes \widetilde{\mathbb{K}}_{\dot{T}^*X\underset{X}{\times}X} \\
&\qquad \otimes \beta\left(\mathrm{R}i_*\omega^{\otimes -1}_{T^*X\underset{X}{\times}X/T^*X\times X}\right) \\
&\simeq \text{“}\varinjlim_{\varepsilon>0}\text{”}\, \mathbb{K}_{\{((x;\xi),x')\,;\,\langle\xi,x'-x\rangle>\varepsilon|x'-x|\}} \otimes p^{-1}\,\widetilde{\mathbb{K}}_{\Delta_X}[n].
\end{aligned}$$

Hence

$$\begin{aligned}
&\mathrm{R}p_{!!}\left(\mathrm{L}_X \otimes \widetilde{\mathbb{K}}_{\dot{T}^*X\times X}\right) \\
&\quad \simeq \mathrm{R}p_{!!}\left(\text{“}\varinjlim_{\varepsilon>0,\,R>0}\text{”}\ \mathbb{K}_{\{((x;\xi),x')\,;\,\langle\xi,x'-x\rangle>\varepsilon|x'-x|,\ |\xi|<R\}}\right) \otimes \widetilde{\mathbb{K}}_{\Delta_X}[n].
\end{aligned}$$

For $0 < \varepsilon < R$, we have

$$\mathrm{R}^k p_!\left(\mathbb{K}_{\{((x;\xi),x')\,;\,\langle\xi,x'-x\rangle>\varepsilon|x'-x|,\ |\xi|<R\}}\right) \simeq \begin{cases} \mathbb{K}_{\{0<|x'-x|<\varepsilon^{-1}R\}} & \text{if } k = n. \\ 0 & \text{if } k \neq n. \end{cases}$$

Hence we have shown that

$$\mathrm{R}p_{!!}\left(\mathrm{L}_X \otimes \widetilde{\mathbb{K}}_{\dot{T}^*X\times X}\right) \simeq \mathbb{K}_{X\times X\setminus\Delta_X}[-n] \otimes \widetilde{\mathbb{K}}_{\Delta_X}[n] \simeq \mathbb{K}_{X\times X\setminus\Delta_X} \otimes \widetilde{\mathbb{K}}_{\Delta_X},$$

which proves (2.6). In the morphism of distinguished triangles

$$\begin{array}{ccccccc}
\mathrm{R}p_{!!}\left(\mathrm{L}_X \otimes \widetilde{\mathbb{K}}_{\dot{T}^*X\times X}\right) & \longrightarrow & \mathrm{R}p_{!!}\left(\mathrm{L}_X\right) & \longrightarrow & \mathrm{R}p_{!!}\left(\mathrm{L}_X \otimes \widetilde{\mathbb{K}}_{T^*_XX\times X}\right) & \xrightarrow{+1} & \\
\downarrow{\scriptstyle\sim} & & \downarrow & & \downarrow & & \\
\mathbb{K}_{X\times X\setminus\Delta_X} \otimes \widetilde{\mathbb{K}}_{\Delta_X} & \longrightarrow & \widetilde{\mathbb{K}}_{\Delta_X} & \longrightarrow & \mathbb{K}_{\Delta_X} & \xrightarrow{+1} & ,
\end{array}$$

the left vertical arrow is an isomorphism by (2.6) and the right vertical arrow is an isomorphism since

$$\mathrm{R}p_{!!}\left(\mathrm{L}_X \otimes \widetilde{\mathbb{K}}_{T^*_X X\times X}\right) \simeq \mathrm{R}p_{!!}\left(\mathbb{K}_{T^*X\underset{X}{\times}X} \otimes \widetilde{\mathbb{K}}_{T^*_X X\times X}\right)$$

$$\simeq \mathbb{K}_{\Delta_X} \otimes \mathrm{R}p_{!!}(\widetilde{\mathbb{K}}_{T^*_X X\times X}) \simeq \mathbb{K}_{\Delta_X} .$$

Hence we obtain (2.5).

(iv) follows immediately from Lemma 2.1.12 and (i). □

Proposition 2.1.14. *For* $\mathcal{F} \in \mathrm{D^b}(\mathrm{I}(\mathbb{K}_X))$ *and* $\mathcal{G} \in \mathrm{D^b}(\mathrm{I}(\mathbb{K}_Y))$*, we have an isomorphism*

$$\mu_{X\times Y}(\mathcal{F} \boxtimes \mathcal{G}) \simeq \mathrm{K}_{T^*(X\times Y)} \circ (\mu_X \mathcal{F} \boxtimes \mu_Y \mathcal{G}) .$$

Proof. This follows immediately from Lemma 2.1.4. □

2.2. The link with μhom and classical microlocalization

Proposition 2.2.1. *Let* $\sigma \in \Gamma(X, \Omega^1_X)$ *and* $\mathcal{F}, \mathcal{G} \in \mathrm{D^b}(\mathbb{K}_X)$*. Then we have an isomorphism*

$$\sigma^{-1}\mu hom(\mathcal{F}, \mathcal{G}) \simeq \mathrm{R}\mathcal{H}om\left(\mathcal{F}, \mathcal{L}_{\widetilde{\sigma}}(\Delta_X, X \times X) \circ \mathcal{G}\right),$$

where $\widetilde{\sigma} = q_1^*\sigma - q_2^*\sigma$ *and* $q_i \colon X \times X \to X$ *is the* i*-th projection* $(i = 1, 2)$.

Proof. By definition we have

$$\mu hom(\mathcal{F}, \mathcal{G}) \simeq \nu hom(\mathcal{F}, \mathcal{G})^\wedge,$$

where νhom is the specialization of the functor $\mathrm{R}\mathcal{H}om$ (see below), and $(\cdot)^\wedge$ is the Fourier–Sato transform. Setting

$$P' = \left\{((x;\xi),(x;v)) \in T^*X \underset{X}{\times} TX;\ \langle \xi, v\rangle \leqslant 0\right\},$$

the Fourier–Sato transform is the integral transform with kernel $\mathbb{K}_{P'}$. Consider the following commutative diagram

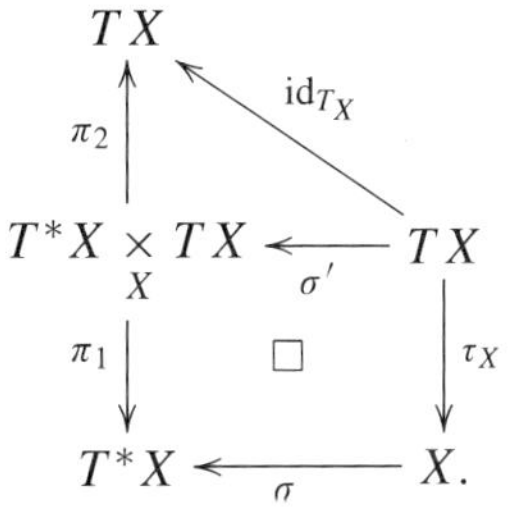

Then $\mu hom(\mathcal{F}, \mathcal{G}) \simeq \nu hom(\mathcal{F}, \mathcal{G})^\wedge \simeq \mathrm{R}\pi_{1!}\left(\pi_2^{-1}\, \nu hom(\mathcal{F}, \mathcal{G}) \otimes \mathbb{K}_{P'}\right)$. Hence

$$\begin{aligned}\sigma^{-1}\, \mu hom(\mathcal{F}, \mathcal{G}) &\simeq \sigma^{-1}\, \mathrm{R}\pi_{1!}\left(\pi_2^{-1}\, \nu hom(\mathcal{F}, \mathcal{G}) \otimes \mathbb{K}_{P'}\right)\\ &\simeq \mathrm{R}\tau_{X!}\sigma'^{-1}\left(\pi_2^{-1}\, \nu hom(\mathcal{F}, \mathcal{G}) \otimes \mathbb{K}_{P'}\right)\\ &\simeq \mathrm{R}\tau_{X!}\left(\nu hom(\mathcal{F}, \mathcal{G}) \otimes \mathbb{K}_{P'_\sigma}\right),\end{aligned}$$

where we have set $P'_\sigma = \sigma'^{-1}(P') = \{(x, v) \in TX;\ \langle \sigma(x), v\rangle \leqslant 0\}$. Consider the normal deformation of Δ_X in $X \times X$, visualized by the diagram:

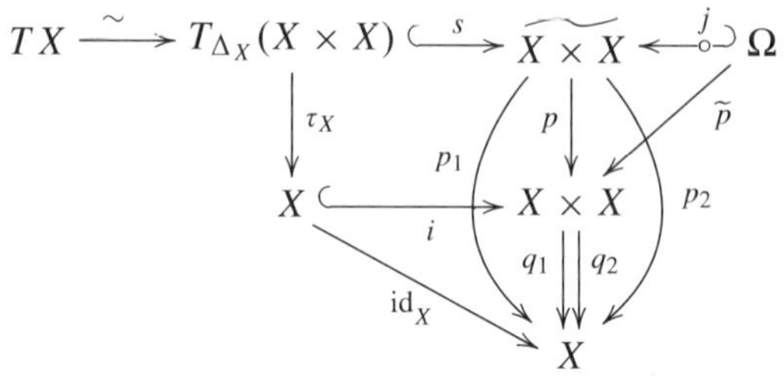

Then $\nu hom(\mathcal{F}, \mathcal{G})$ is by definition $s^{-1}\, \mathrm{R}j_*\widetilde{p}^{-1}\, \mathrm{R}\mathcal{H}om(q_2^{-1}\mathcal{F}, q_1^!\mathcal{G})$. Since $\widetilde{p}$ is smooth we have

$$\begin{aligned}\widetilde{p}^{-1}\, \mathrm{R}\mathcal{H}om(q_2^{-1}\mathcal{F}, q_1^!\mathcal{G}) &\simeq \mathrm{R}\mathcal{H}om(\widetilde{p}^{-1}q_2^{-1}\mathcal{F}, \widetilde{p}^{-1}q_1^!\mathcal{G})\\ &\simeq \mathrm{R}\mathcal{H}om(j^{-1}p_2^{-1}\mathcal{F}, j^{-1}p^{-1}q_1^!\mathcal{G}).\end{aligned}$$

Hence we have

$$\begin{aligned}\nu hom(\mathcal{F}, \mathcal{G}) &\simeq s^{-1}\, \mathrm{R}j_*\, \mathrm{R}\mathcal{H}om(j^{-1}p_2^{-1}\mathcal{F}, j^{-1}p^{-1}q_1^!\mathcal{G})\\ &\simeq s^{-1}\, \mathrm{R}\mathcal{H}om(p_2^{-1}\mathcal{F}, \mathrm{R}j_*j^{-1}p^{-1}q_1^!\mathcal{G})\\ &\simeq s^{-1}\, \mathrm{R}\mathcal{H}om(p_2^{-1}\mathcal{F}, \mathrm{R}j_*j^{-1}p_1^{-1}\mathcal{G}) \otimes \tau_X^{-1}\omega_X.\end{aligned}$$

Since p_1 is smooth, we have the estimate

$$\begin{aligned}\mathrm{SS}(p_1^{-1}\mathcal{G}) \cap \mathrm{SS}\,\mathbb{K}_\Omega &\subset (p_1)_\pi^{-1}(T^*X) \cap \left(T^*_{T_{\Delta_X}(X\times X)}\widetilde{X \times X} \cup T^*_{\widetilde{X\times X}}\widetilde{X \times X}\right)\\ &\subset T^*_{\widetilde{X\times X}}\widetilde{X \times X},\end{aligned}$$

which implies

$$\begin{aligned}\mathrm{R}j_*j^{-1}p_1^{-1}\mathcal{G} \simeq \mathrm{R}\mathcal{H}om(\mathbb{K}_\Omega, p_1^{-1}\mathcal{G}) &\simeq \mathrm{R}\mathcal{H}om(\mathbb{K}_\Omega, \mathbb{K}_X) \otimes p_1^{-1}\mathcal{G}\\ &\simeq \mathbb{K}_{\overline{\Omega}} \otimes p_1^{-1}\mathcal{G}.\end{aligned}$$

Applying this result we obtain

$$\begin{aligned}\nu hom(\mathcal{F},\mathcal{G}) &\simeq s^{-1}\,\mathrm{R}\mathcal{H}om(p_2^{-1}\mathcal{F}, p_1^{-1}\mathcal{G}\otimes\mathbb{K}_{\overline{\Omega}})\otimes\tau_X^{-1}\omega_X\\ &\simeq s^{-1}\,\mathrm{R}\mathcal{H}om(p_2^{-1}\mathcal{F}, p_1^{-1}\mathcal{G}\otimes\mathbb{K}_{\overline{\Omega}})\otimes\tau_X^{-1}\omega_{\Delta_X/X\times X}^{\otimes-1},\end{aligned}$$

and finally

$$\begin{aligned}\sigma^{-1}\,\mu hom(\mathcal{F},\mathcal{G}) &\simeq \mathrm{R}\tau_{X!}\Big(s^{-1}\,\mathrm{R}\mathcal{H}om\left(p_2^{-1}\mathcal{F}, p_1^{-1}\mathcal{G}\otimes\mathbb{K}_{\overline{\Omega}}\right)\\ &\qquad\otimes\tau_X^{-1}\omega_{\Delta_X/X\times X}^{\otimes-1}\otimes\mathbb{K}_{P'_\sigma}\Big)\\ &\simeq \mathrm{R}p_{2!}\,\mathrm{R}s_!\Big(s^{-1}\,\mathrm{R}\mathcal{H}om\left(p_2^{-1}\mathcal{F}, p_1^{-1}\mathcal{G}\otimes\mathbb{K}_{\overline{\Omega}}\right)\\ &\qquad\otimes\tau_X^{-1}\omega_{\Delta_X/X\times X}^{\otimes-1}\otimes\mathbb{K}_{P'_\sigma}\Big)\\ &\simeq \mathrm{R}p_{2!}\Big(\mathrm{R}\mathcal{H}om\left(p_2^{-1}\mathcal{F}, p_1^{-1}\mathcal{G}\otimes\mathbb{K}_{\overline{\Omega}}\right)\\ &\qquad\otimes\mathbb{K}_{P'_\sigma}\otimes p^{-1}\,\mathrm{R}i_*\omega_{\Delta_X/X\times X}^{\otimes-1})\Big).\end{aligned}$$

Note that this intermediate result is obtained by means of classical sheaf theory. However, formulas in the derived category of ind-sheaves allow us to continue the calculations. Using the properties (1.3 c) of the functor β and Proposition 1.1.2, we have

$$\begin{aligned}\sigma^{-1}\,\mu hom(\mathcal{F},\mathcal{G}) &\simeq \mathrm{R}p_{2!}\,\mathrm{R}\mathcal{H}om\Big(p_2^{-1}\mathcal{F}, p_1^{-1}\mathcal{G}\otimes\mathbb{K}_{\overline{\Omega}}\\ &\qquad\otimes\beta_{\widetilde{X\times X}}(\mathbb{K}_{P'_\sigma}\otimes p^{-1}\,\mathrm{R}i_*\omega_{\Delta_X/X\times X}^{\otimes-1})\Big)\\ &\simeq \mathrm{R}\mathcal{H}om\Big(\mathcal{F}, \mathrm{R}p_{2!!}(p_1^{-1}\mathcal{G}\otimes\mathbb{K}_{\overline{\Omega}}\otimes\widetilde{\mathbb{K}}_{P'_\sigma}\\ &\qquad\otimes p^{-1}\beta_{X\times X}(\mathrm{R}i_*\omega_{\Delta_X/X\times X}^{\otimes-1}))\Big).\end{aligned}$$

We have furthermore

$$\begin{aligned}&\mathrm{R}p_{2!!}(p_1^{-1}\mathcal{G}\otimes\mathbb{K}_{\overline{\Omega}}\otimes\widetilde{\mathbb{K}}_{P'_\sigma}\otimes p^{-1}\beta_{X\times X}(\mathrm{R}i_*\omega_{\Delta_X/X\times X}^{\otimes-1}))\\ &\quad\simeq \mathrm{R}q_{2!!}\,\mathrm{R}p_{!!}\left(p^{-1}q_1^{-1}\mathcal{G}\otimes\mathbb{K}_{\overline{\Omega}}\otimes\widetilde{\mathbb{K}}_{P'_\sigma}\otimes p^{-1}\beta_{X\times X}(\mathrm{R}i_*\omega_{\Delta_X/X\times X}^{\otimes-1})\right)\\ &\quad\sim \mathrm{R}q_{2!!}\left(q_1^{-1}\mathcal{G}\otimes\mathrm{R}p_{!!}(\mathbb{K}_{\overline{\Omega}}\otimes\widetilde{\mathbb{K}}_{P'_\sigma})\otimes\beta_{X\times X}(\mathrm{R}i_*\omega_{\Delta_X/X\times X}^{\otimes-1})\right)\end{aligned}$$

$$\simeq \mathrm{R}q_{1!!}\Big(q_2^{-1}\mathcal{G}\otimes \mathrm{R}p_{!!}(\mathbb{K}_{\overline{\Omega}}\otimes\widetilde{\mathbb{K}}_{P_{\widetilde{\sigma}}})\otimes\beta_{X\times X}(\mathrm{R}i_*\omega^{\otimes -1}_{\Delta_X/X\times X})\Big)$$

$$\simeq \mathrm{R}q_{1!!}\Big(q_2^{-1}\mathcal{G}\otimes\mathcal{L}_{\widetilde{\sigma}}(\Delta_X, X\times X)\Big)\simeq\mathcal{L}_{\widetilde{\sigma}}(\Delta_X, X\times X)\circ\mathcal{G}. \qquad \square$$

Corollary 2.2.2. *Let $\mathcal{F},\mathcal{G}\in\mathrm{D}^{\mathrm{b}}(\mathbb{K}_X)$. Then we have an isomorphism*

$$\mu hom(\mathcal{F},\mathcal{G})\simeq \mathrm{R}\mathcal{H}om(\pi_X^{-1}\mathcal{F},\mu_X\mathcal{G})\simeq\mathrm{R}\mathcal{H}om(\mu_X\mathcal{F},\mu_X\mathcal{G}).$$

Proof. Consider the fundamental 1-form $\omega_X\in\Gamma(T^*X,\Omega^1_{T^*X})$ of the cotangent bundle of X. Then we have

$$\mu hom(\mathcal{F},\mathcal{G})\simeq\omega_X^{-1}\ \mu hom(\pi_X^{-1}\mathcal{F},\pi_X^{-1}\mathcal{G})$$

and by Proposition 2.2.1 we get a natural isomorphism

$$\mu hom(\mathcal{F},\mathcal{G})=\mathrm{R}\mathcal{H}om(\pi_X^{-1}\mathcal{F},\mathrm{K}_{T^*X}\circ\pi_X^{-1}\mathcal{G})\simeq\mathrm{R}\mathcal{H}om(\pi_X^{-1}\mathcal{F},\mu_X\mathcal{G})$$

The last isomorphism is a consequence of Lemma 2.1.3 and Lemma 2.1.2. $\square$

Proposition 2.2.3. *Let $\mathcal{F}\in\mathrm{D}^{\mathrm{b}}(\mathbb{K}_X)$ and let Z be a closed submanifold of X. Denote by i the closed immersion $i\colon T^*X\underset{X}{\times}Z\hookrightarrow T^*X$. Then we have a natural isomorphism*

$$\mu_Z(\mathcal{F})\simeq\alpha_{T^*X\underset{X}{\times}Z}(i^!\mu_X\mathcal{F})|_{T^*_ZX})\simeq\mathrm{R}\mathcal{H}om(\mathbb{K}_{T^*X\underset{X}{\times}Z},\mu_X\mathcal{F})|_{T^*_ZX}.$$

Here $\mu_Z(\mathcal{F})$ denotes the classical functor of Sato's microlocalization

See [KS2], Chapter IV for definitions and a detailed study for μ_Z. We only remark here that $\mu_Z(\mathcal{F})\simeq\mu hom(\mathbb{K}_Z,\mathcal{F})|_{T^*_ZX}$.

Proof. We have by Corollary 2.2.2

$$\mu_Z(\mathcal{F})\simeq\mathrm{R}\mathcal{H}om(\pi_X^{-1}\mathbb{K}_Z,\mu_X(\mathcal{F}))|_{T^*_ZX}\simeq\mathrm{R}\mathcal{H}om\big(\mathrm{R}i_{!!}\mathbb{K}_{T^*X\underset{X}{\times}Z},\mu_X(\mathcal{F})\big)|_{T^*_ZX}$$

$$\simeq\mathrm{R}\mathcal{H}om\big(\mathbb{K}_{T^*X\underset{X}{\times}Z},i^!\mu_X(\mathcal{F})\big)|_{T^*_ZX}\simeq\big(\alpha_{T^*X\underset{X}{\times}Z}i^!\mu_X(\mathcal{F})\big)|_{T^*_ZX}. \qquad\square$$

2.3. Review on the microsupport of ind-sheaves

In this section we shall give a short overview on the results of [KS4] on the microsupport of ind-sheaves.

The microsupport $\mathrm{SS}(\mathcal{F})$ of an object $\mathcal{F}\in\mathrm{D}^{\mathrm{b}}(\mathbb{K}_X)$ is a closed involutive cone in the cotangent bundle T^*X which describes the codirections in which the cohomology of $\mathcal{F}$ does not propagate (See [KS2], [KS3]). The corresponding notions for ind-sheaves are more intricate.

Let $\mathcal{C}$ be an abelian category, and consider the functor

$$\mathrm{J}\colon\mathrm{D}^{\mathrm{b}}(\mathrm{Ind}(\mathcal{C}))\to\mathrm{D}^{\mathrm{b}}(\mathcal{C})^\wedge\qquad\text{given by}\qquad\mathcal{F}\mapsto\mathrm{Hom}_{\mathrm{D}^{\mathrm{b}}(\mathrm{Ind}(\mathcal{C}))}(\,\cdot\,,\mathcal{F}).$$

Here $\mathrm{D}^{\mathrm{b}}(\mathcal{C})^\wedge$ is the category of contravariant functors from $\mathrm{D}^{\mathrm{b}}(\mathcal{C})$ to the category of sets. Then it can be shown that J factors through $\mathrm{Ind}(\mathrm{D}^{\mathrm{b}}(\mathcal{C}))$. Note that J is conservative, which is a consequence of the commutative diagram

$$\begin{array}{ccc} \mathrm{D}^{\mathrm{b}}(\mathrm{Ind}(\mathcal{C})) & \xrightarrow{\ \mathrm{J}\ } & \mathrm{Ind}(\mathrm{D}^{\mathrm{b}}(\mathcal{C})) \\ & {\scriptstyle \mathrm{H}^k}\searrow \quad \swarrow{\scriptstyle \mathrm{IH}^k} & \\ & \mathrm{Ind}(\mathcal{C}) & \end{array}$$

Finally assume

$$\mathcal{C} \text{ has enough injectives and finite homological dimension.} \tag{2.7}$$

Recall that in this case $\varphi\colon \mathcal{F} \to \mathcal{G}$ is an isomorphism in $\mathrm{Ind}(\mathrm{D}^{\mathrm{b}}(\mathcal{C}))$ if and only if $\mathrm{IH}^k(\varphi)$ is an isomorphism for all k. Then we easily get the following result.

Lemma 2.3.1. *Assume* (2.7). *Let* $\mathcal{F} \in \mathrm{D}^{\mathrm{b}}(\mathrm{Ind}(\mathcal{C}))$ *and let* $\{\mathcal{F}_i \to \mathcal{F}\}_{i\in I}$ *be a filtrant inductive system of morphisms in* $\mathrm{D}^{\mathrm{b}}(\mathrm{Ind}(\mathcal{C}))$. *Then* $\text{“}\varinjlim\text{”}\, \mathrm{J}(\mathcal{F}_i) \xrightarrow{\sim} \mathrm{J}(\mathcal{F})$ *if and only if* $\text{“}\varinjlim_{i\in I}\text{”}\, \mathrm{H}^k(\mathcal{F}_i) \xrightarrow{\sim} \mathrm{H}^k(\mathcal{F})$.

In particular if $\text{“}\varinjlim\text{”}\, \mathrm{J}(\mathcal{F}_i) \xrightarrow{\sim} \mathrm{J}(\mathcal{F})$, then we have $\text{“}\varinjlim\text{”}\, \mathrm{J}(\tau^{\leqslant n}\mathcal{F}_i) \xrightarrow{\sim} \mathrm{J}(\tau^{\leqslant n}\mathcal{F})$ for all k.

We shall apply the results above to the case of ind-sheaves, by taking $\mathrm{Mod}^{\mathrm{c}}(\mathbb{K}_X)$ as $\mathcal{C}$. For a C^∞-manifold X, let

$$J_X\colon\ \mathrm{D}^{\mathrm{b}}(\mathrm{I}(\mathbb{K}_X)) \to \big(\mathrm{D}^{\mathrm{b}}(\mathrm{Mod}^{\mathrm{c}}(\mathbb{K}_X))\big)^\wedge$$

be the canonical functor.

Proposition 2.3.2. *Let* $f\colon X \to Y$ *be a continuous map. Let* $\{\mathcal{F}_i \to \mathcal{F}\}_{i\in I}$ *be a filtrant inductive system of morphisms in* $\mathrm{D}^{\mathrm{b}}(\mathrm{I}(\mathbb{K}_X))$ *and* $\{\mathcal{G}_j \to \mathcal{G}\}_{j\in J}$ *a filtrant inductive system in* $\mathrm{D}^{\mathrm{b}}(\mathrm{I}(\mathbb{K}_Y))$ *such that*

$$J_X(\mathcal{F}) \simeq \text{“}\varinjlim_{i\in I}\text{”}\, J_X(\mathcal{F}_i) \qquad \text{and} \qquad J_Y(\mathcal{G}) \simeq \text{“}\varinjlim_{j\in J}\text{”}\, J_Y(\mathcal{G}_j).$$

Then

(i)

$$J_Y(\mathrm{R}f_{!!}\mathcal{F}) \simeq \text{“}\varinjlim_{i\in I}\text{”}\, J_Y(\mathrm{R}f_{!!}\mathcal{F}_i),$$

(ii) For $\mathcal{K} \in \mathrm{D}^{\mathrm{b}}(\mathrm{I}(\mathbb{K}_X))$, *we have*

$$J_X(\mathcal{K} \otimes \mathcal{F}) = \text{“}\varinjlim_{i\in I}\text{”}\, J_X(\mathcal{K} \otimes \mathcal{F}_i),$$

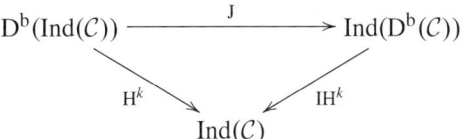

(iii)

$$J_X(f^{-1}\mathcal{G}) \simeq \text{“}\varinjlim_{j\in J}\text{”}\, J_X(f^{-1}\mathcal{G}_j) \quad \textit{and} \quad J_X(f^!\mathcal{G}) \simeq \text{“}\varinjlim_{j\in J}\text{”}\, J_X(f^!\mathcal{G}_j)$$

(iv)

$$J_{T^*X}(\mu_X\mathcal{F}) \simeq \text{“}\varinjlim_{i\in I}\text{”}\, J_{T^*X}(\mu_X\mathcal{F}_i).$$

Proof. By Lemma 2.3.1, we can reduce the situation by dévissage to usual ind-sheaves, where the formulas are obvious. □

Definition 2.3.3. (i) Let $\mathcal{F} \in \mathrm{D}^{\mathrm{b}}(\mathrm{I}(\mathbb{K}_X))$. The micro-support of $\mathcal{F}$, denoted $\mathrm{SS}(\mathcal{F})$, is the closed conic subset of T^*X whose complementary is the set of points $p \in T^*X$ such that there exist a conic open neighborhood U of p in T^*X, an open neighborhood W of $\pi_X(p)$ and a small filtrant inductive system $\{\mathcal{F}_i\}_{i\in I}$ of objects $\mathcal{F}_i \in \mathrm{D}^{\mathrm{b}}(\mathrm{Mod}^{\mathrm{c}}(\mathbb{K}_X))$ such that $\mathrm{SS}(\mathcal{F}_i) \cap U = \emptyset$ and

$$J_X(\mathcal{F} \otimes \mathbb{K}_W) \simeq \text{“}\varinjlim_{i\in I}\text{”}\, \mathcal{F}_i \otimes \mathbb{K}_W.$$

(ii) For $\mathcal{F} \in \mathrm{D}^{\mathrm{b}}(\mathrm{I}(\mathbb{K}_X))$, one sets $\mathrm{SS}_0(\mathcal{F}) = \mathrm{Supp}(\mu_X\mathcal{F})$.

Remark 2.3.4. The micro-support defined above coincides with the classical definition for objects of $\mathrm{D}^{\mathrm{b}}(\mathbb{K}_X)$; it satisfies the triangular inequality (in a distinguished triangle, the micro-support of an object is contained in the union of the micro-supports of the two others), and we have

$$\mathrm{Supp}(\mathcal{F}) = \mathrm{SS}(\mathcal{F}) \cap T^*_XX, \quad \mathrm{SS}(\alpha_X(\mathcal{F})) \subset \mathrm{SS}(\mathcal{F}) \quad \text{for } \mathcal{F} \in \mathrm{D}^{\mathrm{b}}(\mathrm{I}(\mathbb{K}_X)).$$

In general, it is no longer an involutive subset of T^*X.

Proposition 2.3.5. *Let* $\mathcal{F} \in \mathrm{D}^{\mathrm{b}}(\mathrm{I}(\mathbb{K}_X))$*. Then*

$$\mathrm{SS}_0(\mathcal{F}) \subset \mathrm{SS}(\mathcal{F}).$$

If $\mathcal{F} \in \mathrm{D}^{\mathrm{b}}(\mathbb{K}_X)$*, then*

$$\mathrm{SS}_0(\mathcal{F}) = \mathrm{SS}(\mathcal{F}).$$

Proof. The result for sheaves is actually an obvious consequence of Corollary 2.2.2 since

$$\mathrm{SS}(\mathcal{F}) = \mathrm{Supp}(\mu hom(\mathcal{F}, \mathcal{F})) = \mathrm{Supp}(\mathrm{R}\mathcal{H}om(\mu_X\mathcal{F}, \mu_X\mathcal{F})) = \mathrm{Supp}(\mu_X\mathcal{F}).$$

Now assume that $\mathcal{F} \in \mathrm{D}^{\mathrm{b}}(\mathrm{I}(\mathbb{K}_X))$ and $p \notin \mathrm{SS}\,\mathcal{F}$. Consider a filtrant inductive system $\mathcal{F}_i$ in $\mathrm{D}^{\mathrm{b}}(\mathrm{Mod}^c(\mathbb{K}_X))$ and an open neighborhood W of $\pi_X(p)$, a neighborhood $U \subset {\pi_{T^*X}}^{-1}(W)$ of p such that

$$J_X(\mathcal{F} \otimes \mathbb{K}_W) \simeq \text{“}\varinjlim_{i}\text{”}(\mathcal{F}_i \otimes \mathbb{K}_W)$$

and $SS(\mathcal{F}_i) \cap \overline{U} = \emptyset$. We have by Proposition 2.3.2

$$J_X\big(\mu_X(\mathcal{F} \otimes \mathbb{K}_W)\big) \simeq \text{“}\varinjlim_{i}\text{”}\, J_X\big(\mu_X(\mathcal{F}_i \otimes \mathbb{K}_W)\big),$$

and we get $\mu_X\mathcal{F}|_U \simeq 0$ since $\mathrm{Supp}(\mu_X\mathcal{F}_i) = \mathrm{SS}\,(F_i)$. □

Example 2.3.6. For a closed submanifold Z of X, we have

$$\mathrm{SS}_0(\mathbb{K}_Z) = \mathrm{SS}(\mathbb{K}_Z) = T^*_Z X \quad \text{and}$$
$$\mathrm{SS}_0(\widetilde{\mathbb{K}}_Z) = T^*_X X \underset{X}{\times} Z, \quad \mathrm{SS}(\widetilde{\mathbb{K}}_Z) = T^*_Z X.$$

Lemma 2.3.7. *Let Ω be an open subset of $\dot{T}^*X$ and let $\mathcal{F} \in \mathrm{D^b}(\mathbb{K}_\Omega)$, $\mathcal{G} \in \mathrm{D^b}(\mathrm{I}(\mathbb{K}_\Omega))$. Assume that $\mathcal{F}$ is cohomologically constructible (see* [KS2, Definition 3.4.1]*). Assume further that*

$$\omega_X^{-1}\big(\mathrm{SS}(\mathcal{F})\big) \cap \mathrm{Supp}(\mathcal{G}) = \emptyset,$$

where ω_X is considered as a map $T^*X \to T^*(T^*X)$. Then we have an isomorphism

$$\mathrm{R}\mathcal{H}om(\mathcal{F}, \mathbb{K}_\Omega) \otimes (\mathrm{K}_\Omega \circ \mathcal{G}) \xrightarrow{\ \sim\ } \mathrm{R}\mathcal{I}\mathcal{H}om(\mathcal{F}, \mathrm{K}_\Omega \circ \mathcal{G}) \quad \text{in } \mathrm{D^b}(\mathrm{I}(\mathbb{K}_\Omega)).$$

Proof. By shrinking Ω, we may assume from the beginning that $\omega_X^{-1}\big(\mathrm{SS}(\mathcal{F})\big) = \emptyset$.

(i) Assume first that $\mathcal{G} \in \mathrm{D^b}(\mathbb{K}_\Omega)$. For $p = (x_0, \xi_0) \in \Omega$, we shall prove that

$$\mathrm{R}\mathcal{H}om(\mathcal{F}, \mathbb{K}_\Omega) \otimes (\mathrm{K}_\Omega \circ \mathcal{G}) \otimes \widetilde{\mathbb{K}}_p \xrightarrow{\ \sim\ } \mathrm{R}\mathcal{I}\mathcal{H}om(\mathcal{F}, \mathrm{K}_\Omega \circ \mathcal{G}) \otimes \widetilde{\mathbb{K}}_p\,.$$

Since $p \notin T^*_X X$, we have

$$\begin{aligned}(\mathrm{K}_\Omega \circ \mathcal{G}) \otimes \widetilde{\mathbb{K}}_p &\simeq (\mathrm{K}_\Omega \otimes \widetilde{\mathbb{K}}_{(p,p)}) \circ \mathcal{G} \\ &\simeq \text{“}\varinjlim_{\rho>0}\text{”}\, \mathbb{K}_{K_\rho} \otimes \left(\left(\text{“}\varinjlim_{\delta>0,\varepsilon>0}\text{”}\ \mathbb{K}_{F_{\delta,\varepsilon}}\right)[-n] \circ \mathcal{G}\right),\end{aligned} \tag{2.8}$$

where

$$K_\rho = \{(x,\xi);\, |x - x_0| \le \rho,\ |\xi - \xi_0| \le \rho\}$$

and

$$F_{\delta,\varepsilon} = \big\{\delta \geqslant \langle \xi_0, x' - x\rangle > \varepsilon(|x' - x| + |\xi' - \xi|)\big\}.$$

Let $p_1 \colon T^*\Omega \times T^*\Omega \to T^*\Omega$ be the first projection. For sufficiently small ε, δ and ρ, $\pi_{\mathfrak{X}}^{-1} K_\rho \cap p_1\big(\mathrm{SS}(\mathbb{K}_{F_{\delta,\varepsilon}})\big)$ is contained in a sufficiently small neighborhood of $\omega_X(p)$, and hence so is $\pi_{\mathfrak{X}}^{-1} K_\rho \cap \mathrm{SS}(\mathbb{K}_{F_{\delta,\varepsilon}} \circ \mathcal{G})$. Thus we obtain by assumption

$$\pi_{\mathfrak{X}}^{-1} K_\rho \cap \mathrm{SS}\,\mathcal{F} \cap \mathrm{SS}(\mathbb{K}_{F_{\delta,\varepsilon}} \circ \mathcal{G}) \subset T^*_{\mathfrak{X}}\mathfrak{X}.$$

Then by [KS2, Corollary 6.4.3], we have an isomorphism

$$\mathbb{K}_{K_\rho} \otimes \mathrm{R}\mathcal{H}om(\mathcal{F}, \mathbb{K}_\Omega) \otimes (\mathbb{K}_{F_{\delta,\varepsilon}} \circ \mathcal{G}) \xrightarrow{\sim} \mathbb{K}_{K_\rho} \otimes \mathrm{R}\mathcal{H}om(\mathcal{F}, \mathbb{K}_{F_{\delta,\varepsilon}} \circ \mathcal{G})$$

in $\mathrm{D^b}(\mathbb{K}_\Omega)$. Therefore we have

$$\begin{aligned} J_\Omega\,(\mathrm{R}\mathcal{H}om(\mathcal{F}, \mathbb{K}_\Omega) &\otimes (\mathrm{K}_\Omega \circ \mathcal{G}) \otimes \widetilde{\mathbb{K}}_p) \\ &\simeq \underset{\delta>0,\,\varepsilon>0,\,\rho>0}{\text{“}\varinjlim\text{”}} J_\Omega\left(\mathbb{K}_{K_\rho} \otimes \mathrm{R}\mathcal{H}om(\mathcal{F}, \mathbb{K}_\Omega) \otimes (\mathbb{K}_{F_{\delta,\varepsilon}} \circ \mathcal{G})[-n]\right) \\ &\simeq \underset{\delta>0,\,\varepsilon>0,\,\rho>0}{\text{“}\varinjlim\text{”}} J_\Omega\Big(\mathbb{K}_{K_\rho} \otimes \mathrm{R}\mathcal{H}om\big(\mathcal{F}, \mathbb{K}_{F_{\delta,\varepsilon}} \circ \mathcal{G}[-n]\big)\Big) \\ &\simeq J_\Omega\left(\mathrm{R}\mathcal{I}\mathcal{H}om(\mathcal{F}, \mathrm{K}_\Omega \circ \mathcal{G}) \otimes \widetilde{\mathbb{K}}_p\right), \end{aligned}$$

and the lemma is proved when $\mathcal{G} \in \mathrm{D^b}(\mathbb{K}_\Omega)$.

In the general case, taking a filtrant inductive system $\mathcal{G}_k$ in $\mathrm{D^b}(\mathbb{K}_\Omega)$ such that $J_\Omega(\mathcal{G}) \simeq \text{“}\varinjlim\text{”}\, \mathcal{G}_k$. we have

$$\begin{aligned} J_\Omega\,(\mathrm{R}\mathcal{H}om(\mathcal{F}, \mathbb{K}_\Omega) \otimes (\mathrm{K}_\Omega \circ \mathcal{G})) &\simeq \underset{k}{\text{“}\varinjlim\text{”}}\, J_\Omega\,(\mathrm{R}\mathcal{H}om(\mathcal{F}, \mathbb{K}_\Omega) \otimes (\mathrm{K}_\Omega \circ \mathcal{G}_k)) \\ &\simeq \underset{k}{\text{“}\varinjlim\text{”}}\, J_\Omega\,(\mathrm{R}\mathcal{H}om(\mathcal{F}, K_\Omega \circ \mathcal{G}_k)) \simeq J_\Omega\Big(\mathrm{R}\mathcal{H}om(\mathcal{F}, \mathrm{K}_\Omega \circ \mathcal{G})\Big), \end{aligned}$$

which completes the proof. □

We prove now in the framework of ind-sheaves a well-known result for sheaves.

Proposition 2.3.8. *Let $\mathcal{F} \in \mathrm{D^b}(\mathbb{K}_X)$ and $\mathcal{G} \in \mathrm{D^b}(\mathrm{I}(\mathbb{K}_X))$. Assume that $\mathcal{F}$ is cohomologically constructible. Assume further the non-characteristic condition*

$$\mathrm{SS}(\mathcal{F}) \cap \mathrm{SS}_0(\mathcal{G}) \subset T^*_X X.$$

Then we have an isomorphism

$$\mathrm{R}\mathcal{H}om(\mathcal{F}, \mathbb{K}_X) \otimes \mathcal{G} \xrightarrow{\sim} \mathrm{R}\mathcal{I}\mathcal{H}om(\mathcal{F}, \mathcal{G}).$$

Proof. Since $\omega_X^{-1}\,\mathrm{SS}(\pi_X^{-1}\mathcal{F}) = \mathrm{SS}\,\mathcal{F}$, the non-characteristic condition may be rewritten as

$$\omega_X^{-1}\,\mathrm{SS}(\pi_X^{-1}\mathcal{F}) \cap \mathrm{Supp}\,\mu_X\mathcal{G} \cap \dot{T}^*X = \emptyset,$$

and Lemma 2.3.7 assures that

$$\begin{aligned} \left(\pi_X^{-1}\,\mathrm{R}\mathcal{H}om(\mathcal{F}, \mathbb{K}_X) \otimes \mu_X\mathcal{G}\right)|_{\dot{T}^*X} &\simeq \left(\mathrm{R}\mathcal{H}om(\pi_X^{-1}\mathcal{F}, \mathbb{K}_{T^*X}) \otimes \mu_X\mathcal{G}\right)|_{\dot{T}^*X} \\ &\simeq \mathrm{R}\mathcal{I}\mathcal{H}om(\pi_X^{-1}\mathcal{F}, \mu_X\mathcal{G})|_{\dot{T}^*X}. \end{aligned}$$

Applying the functor $\mathrm{R}\dot{\pi}_{X!!}$, we obtain

$$\mathrm{R}\mathcal{H}om(\mathcal{F}, \mathbb{K}_X) \otimes \mathrm{R}\dot{\pi}_{X!!}\left(\mu_X \mathcal{G}|_{\dot{T}^*X}\right) \simeq \mathrm{R}\mathcal{IH}om\left(\mathcal{F}, \mathrm{R}\dot{\pi}_{X!!}\left(\mu_X \mathcal{G}|_{\dot{T}^*X}\right)\right).$$

Now Proposition 2.1.13 gives the following morphism of distinguished triangles where $\mathcal{F}^* = \mathrm{R}\mathcal{H}om(\mathcal{F}, \mathbb{K}_X)$:

$$\begin{array}{ccccccc}
\begin{array}{c}\mathcal{F}^* \otimes \mathrm{R}\dot{\pi}_{X!!} \\ \left(\mu_X \mathcal{G}|_{\dot{T}^*X}\right)\end{array} & \longrightarrow & \mathcal{F}^* \otimes (\widetilde{\mathbb{K}}_{\Delta_X} \circ \mathcal{G}) & \longrightarrow & \mathcal{F}^* \otimes \mathcal{G} & \xrightarrow{+1} & \\
\downarrow\wr & & \downarrow & & \downarrow & & \\
\begin{array}{c}\mathrm{R}\mathcal{IH}om \\ \left(\mathcal{F}, \mathrm{R}\dot{\pi}_{X!!}(\mu_X \mathcal{G}|_{\dot{T}^*X})\right)\end{array} & \longrightarrow & \mathrm{R}\mathcal{IH}om(\mathcal{F}, \widetilde{\mathbb{K}}_{\Delta_X} \circ \mathcal{G}) & \longrightarrow & \mathrm{R}\mathcal{IH}om(\mathcal{F}, \mathcal{G}) & \xrightarrow{+1} & .
\end{array}$$

The middle vertical arrow is an isomorphism by the following lemma, and hence the right arrow is an isomorphism. □

Lemma 2.3.9. *Let $\mathcal{F} \in \mathrm{D}^{\mathrm{b}}(\mathbb{K}_X)$ and $\mathcal{G} \in \mathrm{D}^{\mathrm{b}}(\mathrm{I}(\mathbb{K}_X))$. Assume that $\mathcal{F}$ is cohomologically constructible. Then we have an isomorphism*

$$\mathrm{R}\mathcal{H}om(\mathcal{F}, \mathbb{K}_X) \otimes \left(\widetilde{\mathbb{K}}_{\Delta_X} \circ \mathcal{G}\right) \xrightarrow{\sim} \mathrm{R}\mathcal{IH}om(\mathcal{F}, \widetilde{\mathbb{K}}_{\Delta_X} \circ \mathcal{G}).$$

Proof. Let $p_k \colon X \times X \to X$ be the k-th projection ($k = 1, 2$). Then we have

$$p_1^{-1}\mathrm{R}\mathcal{H}om(\mathcal{F}, \mathbb{K}_X) \otimes p_2^{-1}\mathcal{G} \xrightarrow{\sim} \mathrm{R}\mathcal{H}om(p_1^{-1}\mathcal{F}, p_2^{-1}\mathcal{G}) \quad \text{for any } \mathcal{G} \in \mathrm{D}^{\mathrm{b}}(\mathrm{I}(\mathbb{K}_X)).$$

Hence we have

$$\begin{aligned}
\mathrm{R}\mathcal{H}om(\mathcal{F}, \mathbb{K}_X) \otimes \left(\widetilde{\mathbb{K}}_{\Delta_X} \circ \mathcal{G}\right) &\simeq \mathrm{R}p_{1!!}\left(p_1^{-1}\mathrm{R}\mathcal{H}om(\mathcal{F}, \mathbb{K}_X) \otimes p_2^{-1}\mathcal{G} \otimes \widetilde{\mathbb{K}}_{\Delta_X}\right) \\
&\simeq \mathrm{R}p_{1!!}\left(\mathrm{R}\mathcal{IH}om(p_1^{-1}\mathcal{F}, p_2^{-1}\mathcal{G}) \otimes \widetilde{\mathbb{K}}_{\Delta_X}\right) \\
&\simeq \mathrm{R}p_{1!!}\,\mathrm{R}\mathcal{IH}om(p_1^{-1}\mathcal{F}, p_2^{-1}\mathcal{G} \otimes \widetilde{\mathbb{K}}_{\Delta_X}) \\
&\simeq \mathrm{R}\mathcal{IH}om\Big(\mathcal{F}, \mathrm{R}p_{1!!}\big(p_2^{-1}\mathcal{G} \otimes \widetilde{\mathbb{K}}_{\Delta_X}\big)\Big) \\
&\simeq \mathrm{R}\mathcal{IH}om(\mathcal{F}, \widetilde{\mathbb{K}}_{\Delta_X} \circ \mathcal{G}).
\end{aligned}$$

□

Corollary 2.3.10. *Assume that $i \colon Z \hookrightarrow X$ is a closed immersion and $\mathcal{F} \in \mathrm{D}^{\mathrm{b}}(\mathrm{I}(\mathbb{K}_X))$ satisfies the condition*

$$\mathrm{SS}_0(\mathcal{F}) \cap T^*_Z X \subset T^*_X X.$$

Then we have an isomorphism

$$i^{-1}\mathcal{F} \otimes \omega_{Z/X} \xrightarrow{\sim} i^{!}\mathcal{F}.$$

Proof. We have $i^{-1}\mathcal{F} \otimes \omega_{Z/X} \simeq i^{-1}\mathcal{F} \otimes i^{-1}\mathrm{R}\mathcal{IH}om(\mathbb{K}_Z, \mathbb{K}_X) \simeq i^{-1}\mathrm{R}\mathcal{IH}om(\mathbb{K}_Z, \mathcal{F}) \simeq i^{!}\mathcal{F}$. □

Lemma 2.3.11. *Let $\Omega \subset \dot{T}^*X$ be an open subset and $\mathcal{K} \in \mathrm{D}^{\mathrm{b}}(\mathrm{I}(\mathbb{K}_{Y\times\Omega}))$. Assume that*

$$\mathrm{SS}(\mathcal{K})^a \cap \big(T^*Y \times \omega_X(\Omega)\big) = \emptyset,$$

where a denotes the antipodal map. Then

$$(\mathcal{K} \circ \mathrm{K}_{T^*X})|_{Y\times\Omega} = 0.$$

Proof. We can easily reduce to the case where $\mathcal{K} \in \mathrm{D}^{\mathrm{b}}(\mathbb{K}_{Y\times\Omega})$. In this case, let us prove that

$$(\mathcal{K} \circ \mathrm{K}_{T^*X}) \otimes \widetilde{\mathbb{K}}_p \simeq 0 \quad \text{for } p \in Y \times \Omega.$$

We may assume that X, Y are affine and $p = (y_0, x_0; \xi_0)$. We have

$$\mathrm{K}_{T^*X} \otimes \widetilde{\mathbb{K}}_{(x_0,\xi_0)} \simeq \text{“}\varinjlim_{\delta>0,\varepsilon>0}\text{”}\ \mathbb{K}_{F_{\delta,\varepsilon}}[2\dim X],$$

where we have set $F_{\delta,\varepsilon} = \{\delta \geqslant \langle \xi_0, x' - x\rangle > \varepsilon(|x' - x| + |\xi' - \xi|)\}$. Hence it is enough to show that there exists a neighborhood U of p such that

$$(\mathcal{K} \circ \mathbb{K}_{F_{\delta,\varepsilon}})|_U \simeq 0$$

for $0 < \delta \ll \varepsilon \ll 1$. Let p_{ij} be the (i, j)-th projection from $Y \times \Omega \times \Omega$ to $Y \times \Omega$ or $\Omega \times \Omega$. Then we have

$$\mathcal{K} \circ \mathbb{K}_{F_{\delta,\varepsilon}} \simeq \mathrm{R}p_{13!}(p_{12}^{-1}\mathcal{K} \otimes p_{23}^{-1}\mathbb{K}_{F_{\delta,\varepsilon}}).$$

For $\mathrm{SS}(F_{\delta,\varepsilon})$ contained in a sufficiently small neighborhood of $(\omega_X(p), -\omega_X(p))$, $\mathrm{SS}\big(p_{12}^{-1}\mathcal{K} \otimes p_{23}^{-1}\mathbb{K}_{F_{\delta,\varepsilon}}\big)$ does not intersect $T^*Y \times \{-\langle\xi_0, dx\rangle\} \times T^*\Omega$. Since the map $Y \times \mathrm{Supp}(\mathbb{K}_{F_{\delta,\varepsilon}}) \to Y \times \mathbb{R} \times T^*X$ induced by $\langle \xi_0, x\rangle$ is proper, Proposition 5.4.17 in [KS2] implies that $\big(\mathcal{K} \circ \mathbb{K}_{F_{\delta,\varepsilon}}\big)|_U \simeq 0$. □

Proposition 2.3.12. *Let $\mathcal{K} \in \mathrm{D}^{\mathrm{b}}(\mathrm{I}(\mathbb{K}_{Y\times X}))$ be a kernel and $\mathcal{F} \in \mathrm{D}^{\mathrm{b}}(\mathrm{I}(\mathbb{K}_X))$. Assume that*

$$\mathrm{SS}(\mathcal{K})^a \cap \big(T^*Y \times \mathrm{SS}_0(\mathcal{F})\big) \subset T^*Y \times T^*_XX.$$

Then we have an isomorphism

$$\mathcal{K} \circ \widetilde{\mathbb{K}}_{\Delta_X} \circ \mathcal{F} \xrightarrow{\ \sim\ } \mathcal{K} \circ \mathcal{F}.$$

Proof. It is enough to show that $\mathcal{K} \circ \big(\mathrm{Ker}(\widetilde{\mathbb{K}}_{\Delta_X} \to \mathbb{K}_{\Delta_X})\big) \circ \mathcal{F} \simeq 0$.

Let $p\colon Y \times T^*X \to Y \times X$ be the projection. We have

$$\mathrm{SS}\left(p^{-1}\mathcal{K}\right) \subset \Big\{((y;\eta), (x, \xi; \xi', 0));\ ((y;\eta), (x;\xi')) \in \mathrm{SS}(\mathcal{K})\Big\}.$$

Hence

$$\begin{aligned}\mathrm{SS}\left(\pi_X^{-1}\mathcal{K}\right) &\cap \Big(T^*Y \times \omega_X(\mathrm{SS}_0(\mathcal{F}) \setminus T^*_XX)\Big) \\ &\subset \Big\{((y;\eta), (x, \xi; \xi, 0));\ ((y;\eta), (x;\xi)) \in \mathrm{SS}(\mathcal{K}),\ (x,\xi) \in \mathrm{SS}_0(\mathcal{F}) \setminus T^*_XX\Big\}\end{aligned}$$

is empty by assumption. Therefore Lemma 2.3.11 assures that

$$\operatorname{Supp}(p^{-1}\mathcal{K} \circ \mathrm{K}_{T^*X}) \cap \big(Y \times \mathrm{SS}_0(\mathcal{F})\big) \subset Y \times T^*_X X.$$

Let $p_1 \colon Y \times T^*X \to Y$ and $p_2 \colon Y \times T^*X \to T^*X$ be the projections. Then

$$\begin{aligned} p^{-1}\mathcal{K} \circ \big(\mu_X\mathcal{F} \otimes \widetilde{\mathbb{K}}_{\dot{T}^*X}\big) &\simeq p^{-1}\mathcal{K} \circ \mathrm{K}_{T^*X} \circ \big(\mu_X\mathcal{F} \otimes \widetilde{\mathbb{K}}_{\dot{T}^*X}\big) \\ &\simeq \mathrm{R}p_{1!}\Big((p^{-1}\mathcal{K} \circ \mathrm{K}_{T^*X}) \otimes p_2^{-1}(\mu_X\mathcal{F} \otimes \widetilde{\mathbb{K}}_{\dot{T}^*X})\Big) \simeq 0. \end{aligned}$$

This proves the proposition since $p^{-1}\mathcal{K} \circ (\mu_X\mathcal{F} \otimes \widetilde{\mathbb{K}}_{\dot{T}^*X}) \simeq \mathcal{K} \circ \mathrm{R}\pi_{X!!}(\mu_X\mathcal{F} \otimes \widetilde{\mathbb{K}}_{\dot{T}^*X})$ by Lemma 1.1.6 (iii), and $\mathrm{R}\pi_{X!!}(\mu_X\mathcal{F} \otimes \widetilde{\mathbb{K}}_{\dot{T}^*X}) \simeq \mathrm{Ker}(\widetilde{\mathbb{K}}_{\Delta_X} \to \mathbb{K}_{\Delta_X}) \circ \mathcal{F}$ by Proposition 2.1.13 (iii). □

2.4. Functorial properties of microlocalization

To study the functorial behavior of the functor μ_X, it is convenient to introduce various transfer kernels. They will be used exclusively inside the proofs in order to keep notations as simple as possible. In the sequel, we frequently use Lemma 1.1.6 without mentioning it.

Let $f \colon X \to Y$ be a morphism of manifolds. Let us recall the commutative diagram:

$$\begin{array}{ccccc} T^*X & \xleftarrow{f_d} & T^*Y \underset{Y}{\times} X & \xrightarrow{f_\pi} & T^*Y \\ {\scriptstyle \pi_X}\downarrow & & \downarrow & & \\ X & \xrightarrow{f} & Y. & & \end{array}$$

We have $f_d^*\omega_X = f_\pi^*\omega_Y$. Consider the maps

$$\begin{aligned} \Big(T^*Y \underset{Y}{\times} X\Big) \times X &\xrightarrow{f_d \times \mathrm{id}_X} T^*X \times X, \\ \Big(T^*Y \underset{Y}{\times} X\Big) \times Y &\xrightarrow{f_\pi \times \mathrm{id}_Y} T^*Y \times Y, \\ T^*Y \times X &\xrightarrow{\mathrm{id}_{T^*Y} \times f} T^*Y \times Y. \end{aligned}$$

They define morphisms

$$\begin{aligned} \Gamma\Big(T^*X \underset{X}{\times} X, \Omega^1_{T^*X \times X}\Big) &\to \Gamma\Big(T^*Y \underset{Y}{\times} X, \Omega^1_{(T^*Y \underset{Y}{\times} X) \times X}\Big), \\ \Gamma\Big(T^*Y, \Omega^1_{T^*Y \times Y}\Big) &\to \Gamma\Big(T^*Y \underset{Y}{\times} X, \Omega^1_{(T^*Y \underset{Y}{\times} X) \times Y}\Big), \\ \Gamma\Big(T^*Y, \Omega^1_{T^*Y \times Y}\Big) &\to \Gamma\Big(T^*Y \underset{Y}{\times} X, \Omega^1_{T^*Y \times X}\Big). \end{aligned}$$

We denote by $\sigma_{Y\leftarrow X}$, $\sigma_{X\to Y}$ and $\sigma_{X|Y}$ the images of the section σ_X, σ_Y and σ_Y (defined in 2.1.5), respectively. We set

$$\mathrm{L}_{Y\leftarrow X} = \mathcal{L}_{\sigma_{Y\leftarrow X}}\left(\left(T^*Y \underset{Y}{\times} X\right) \underset{X}{\times} X, \left(T^*Y \underset{Y}{\times} X\right) \times X\right),$$

$$\mathrm{L}_{X\to Y} = \mathcal{L}_{\sigma_{X\to Y}}\left(\left(T^*Y \underset{Y}{\times} X\right) \underset{Y}{\times} Y, (T^*Y \underset{Y}{\times} X) \times Y\right),$$

$$\mathrm{L}_{X|Y} = \mathcal{L}_{\sigma_{X|Y}}\left(T^*Y \underset{Y}{\times} X, T^*Y \times X\right).$$

Note that if $f = \mathrm{id}_X \colon X \to X$, then these three kernels coincide and are isomorphic to L_X.

Lemma 2.4.1. *Let $f\colon X \to Y$ be a morphism of manifolds. There are natural isomorphisms*

(i) $\mathrm{L}_X \simeq \mathrm{R}\,(\mathrm{id}_{T^*X} \times \pi_X)_{!!}\, \mathrm{K}_{T^*X}$,
(ii) $(f_d \times \mathrm{id}_X)^{-1}\, \mathrm{L}_X \simeq \mathrm{L}_{Y\leftarrow X}$,
(iii) $\mathrm{L}_{X|Y} \simeq (\mathrm{id}_{T^*Y} \times f)^{-1}\, \mathrm{L}_Y$,
(iv) $\mathrm{K}_{T^*Y} \underset{T^*Y}{\circ} \mathrm{L}_{X|Y} \simeq \mathrm{L}_{X|Y}$,
(v) $\mathrm{R}(f_\pi \times \mathrm{id}_X)_{!!}\, \mathrm{L}_{Y\leftarrow X} \to \mathrm{K}_{T^*Y} \underset{T^*Y}{\circ} \mathrm{R}(f_\pi \times \mathrm{id}_X)_{!!}\, \mathrm{L}_{Y\leftarrow X} \xrightarrow{\sim} \mathrm{L}_{X|Y}$,
(vi) $\mathrm{R}(f_\pi \times \mathrm{id}_X)_{!!}\, \mathrm{L}_{Y\leftarrow X} \xrightarrow{\sim} \mathrm{L}_{X|Y}$ *if f is smooth,*
(vii) $(f_\pi \times \mathrm{id}_Y)^{-1}\, \mathrm{L}_Y \simeq \mathrm{L}_{X\to Y}$.
(viii) Moreover, there is a morphism $\mathrm{R}(\mathrm{id}_{T^*Y\underset{Y}{\times}X} \times f)_{!!}\, \mathrm{L}_{Y\leftarrow X} \to \mathrm{L}_{X\to Y}$ *which is an isomorphism if f is smooth.*

The results easily follow from the first part of the paper.

Theorem 2.4.2 (proper direct image). *Let $f\colon X \to Y$ be a morphism of manifolds and $\mathcal{F} \in \mathrm{D}^{\mathrm{b}}(\mathrm{I}(\mathbb{K}_X))$. Then*

(i) we have a natural morphism and a natural isomorphism

$$\mathrm{R}f_{\pi\,!!}{f_d}^{-1}\mu_X\mathcal{F} \to \mathrm{K}_{T^*Y} \circ \mathrm{R}f_{\pi\,!!}{f_d}^{-1}\mu_X\mathcal{F} \xrightarrow{\sim} \mu_Y\,(\mathrm{R}f_{!!}\mathcal{F})\,;$$

(ii) if f is smooth we get an isomorphism

$$\mathrm{R}f_{\pi\,!!}{f_d}^{-1}\mu_X\mathcal{F} \xrightarrow{\sim} \mu_Y\,(\mathrm{R}f_{!!}\mathcal{F})\,.$$

Proof. We have ${f_d}^{-1}\mu_X\mathcal{F} \simeq \mathrm{L}_{Y\leftarrow X} \circ \mathcal{F}$ by Lemma 2.4.1 (ii), and a natural morphism by Lemma 2.4.1 (v),

$$\mathrm{R}(f_\pi \times \mathrm{id}_X)_{!!}\, \mathrm{L}_{Y\leftarrow X} \to \mathrm{K}_{T^*Y} \underset{T^*Y}{\circ} \mathrm{R}(f_\pi \times \mathrm{id}_X)_{!!}\, \mathrm{L}_{Y\leftarrow X} \xrightarrow{\sim} \mathrm{L}_{X|Y}\,.$$

However $(\mathrm{R}(f_\pi \times \mathrm{id}_X)_{!!} L_{Y \leftarrow X}) \circ \mathcal{F} \simeq \mathrm{R} f_{\pi\,!!} {f_d}^{-1} \mu_X \mathcal{F}$ and $L_{X|Y} \circ \mathcal{F} \simeq \mu_Y (\mathrm{R} f_{!!} \mathcal{F})$. Hence we get natural morphisms

$$\mathrm{R} f_{\pi\,!!} {f_d}^{-1} \mu_X \mathcal{F} \to \mathrm{K}_{T^*Y} \circ \mathrm{R} f_{\pi\,!!} {f_d}^{-1} \mu_X \mathcal{F} \xrightarrow{\sim} \mu_Y (\mathrm{R} f_{!!} \mathcal{F}),$$

which are isomorphisms if f is smooth by Lemma 2.4.1 (vi). □

Proposition 2.4.3 (inverse image). *Let $f \colon X \to Y$ be a morphism of manifolds and $\mathcal{G} \in \mathrm{D}^{\mathrm{b}}(\mathrm{I}(\mathbb{K}_Y))$. Then*

(i) we have a natural morphism

$${f_d}^{-1} \mu_X (f^{-1} \mathcal{G}) \to {f_\pi}^{-1} \mu_Y \mathcal{G},$$

which is an isomorphism if f is smooth,

(ii) we have a natural morphism

$$\mu_X (f^{-1} \mathcal{G}) \to \mathrm{R} f_{d*} {f_\pi}^{-1} \mu_Y \mathcal{G}.$$

Proof. We have

$${f_d}^{-1} \mu_X \left(f^{-1} \mathcal{G}\right) \simeq \mathrm{L}_{Y \leftarrow X} \circ f^{-1} \mathcal{G} \qquad \text{and} \qquad {f_\pi}^{-1} \mu_Y \mathcal{G} \simeq \mathrm{L}_{X \to Y} \circ \mathcal{G}.$$

Since $\mathrm{L}_{Y \leftarrow X} \circ f^{-1} \mathcal{G} \simeq \left(\mathrm{R}\,(\mathrm{id}_{T^*Y \underset{Y}{\times} X} \times f)_{!!}\, \mathrm{L}_{Y \leftarrow X}\right) \circ \mathcal{G}$, we deduce a morphism by Lemma 2.4.1 (viii):

$${f_d}^{-1} \mu_X \left(f^{-1} \mathcal{G}\right) \simeq \left(\mathrm{R}\,(\mathrm{id}_{T^*Y \underset{Y}{\times} X} \times f)_{!!}\, \mathrm{L}_{Y \leftarrow X}\right) \circ \mathcal{G} \to \mathrm{L}_{X \leftarrow Y} \circ \mathcal{G} \simeq {f_\pi}^{-1} \mu_Y \mathcal{G},$$

which is an isomorphism whenever f is smooth. By adjunction we get then the inverse image morphism $\mu_X (f^{-1} \mathcal{G}) \to \mathrm{R} f_{d*} {f_\pi}^{-1} \mu_Y \mathcal{G}$. □

Theorem 2.4.4 (embedding case). *Let $f \colon X \hookrightarrow Y$ be a closed embedding. Then the following statements hold: for $\mathcal{G} \in \mathrm{D}^{\mathrm{b}}(\mathrm{I}(\mathbb{K}_Y))$:*

(i) *we have a natural morphism*

$$\mathrm{R} f_{d!!} {f_\pi}^{-1} \mu_Y (\mathcal{G}) \to \mu_X (f^{-1} \mathcal{G}),$$

(ii) *if X is non-characteristic for $\mathcal{G}$* (i.e., $\mathrm{SS}_0(\mathcal{G}) \cap T^*_X Y \subset T^*_Y Y$), *then the morphism in* (i) *is an isomorphism and* $\mathrm{SS}_0(f^{-1} \mathcal{G}) \subset f_d {f_\pi}^{-1} \mathrm{SS}_0(\mathcal{G})$.

Proof. (i) Consider the following diagrams:

$$\begin{array}{ccc} T^*X & \xleftarrow{f_d} & T^*Y \underset{Y}{\times} X \\ & & \downarrow{\scriptstyle f_\pi} \\ & & T^*Y \end{array} \quad \text{and} \quad \begin{array}{ccccccc} X & \xleftarrow{p} & T^*Y \underset{Y}{\times} X & \xleftarrow{p_1} & \left(T^*Y \underset{Y}{\times} X\right) \times X & \xrightarrow{p_2} & X \\ & & \| & & \downarrow{\scriptstyle f'} & & \downarrow{\scriptstyle f} \\ & & T^*Y \underset{Y}{\times} X & \xleftarrow{p_1'} & \left(T^*Y \underset{Y}{\times} X\right) \times Y & \xrightarrow{p_2'} & Y. \end{array}$$

We have

$$f_d{}^{-1}\mu_X\left(f^{-1}\mathcal{G}\right) \simeq \mathrm{L}_{Y\leftarrow X}\circ f^{-1}\mathcal{G} \qquad \text{and} \qquad f_\pi{}^{-1}\mu_Y\mathcal{G} \simeq \mathrm{L}_{X\to Y}\circ\mathcal{G}.$$

Since f is a closed immersion, f_d is smooth and we get

$$f_d{}^!\mu_X\left(f^{-1}\mathcal{G}\right) \simeq (\mathrm{L}_{Y\leftarrow X}\circ f^{-1}\mathcal{G})\otimes\omega_{T^*Y\underset{Y}{\times}X/T^*X}.$$

Since the cotangent bundles are canonically orientable, we have

$$\omega_{T^*Y\underset{Y}{\times}X/T^*X} \simeq p^{-1}\omega_{X/Y}[2(\dim Y-\dim X)] \simeq p^{-1}\omega_{X/Y}^{\otimes -1},$$

where $p\colon T^*Y\underset{Y}{\times}X\to X$ is the projection. Hence we get

$$f_d{}^!\mu_X\left(f^{-1}\mathcal{G}\right) \simeq (\mathrm{L}_{Y\leftarrow X}\circ f^{-1}\mathcal{G})\otimes p^{-1}\omega_{X/Y}^{\otimes -1}.$$

Now since f' is a closed immersion, $\mathrm{L}_{Y\leftarrow X}\simeq f'^{\,!}\mathrm{L}_{X\to Y}$ using Proposition 1.3.7, which induces a morphism

$$\begin{aligned}f'^{-1}\mathrm{L}_{X\to Y} &\to \mathrm{L}_{Y\leftarrow X}\otimes\omega^{\otimes -1}_{X\times(T^*Y\underset{Y}{\times}X)/Y\times(T^*Y\underset{Y}{\times}X)}\\ &\simeq \mathrm{L}_{Y\leftarrow X}\otimes p_2^{-1}\omega_{X/Y}^{\otimes -1}\simeq \mathrm{L}_{Y\leftarrow X}\otimes p_1^{-1}p^{-1}\omega_{X/Y}^{\otimes -1}.\end{aligned}$$

Then the preceding morphism together with the adjunction morphism $\mathrm{id}\to \mathrm{R}f'_*f'^{-1}\simeq \mathrm{R}f'_{!!}f'^{-1}$ provides a morphism

$$\begin{aligned}f_\pi{}^{-1}\mu_Y\mathcal{G} &\simeq \mathrm{L}_{X\to Y}\circ\mathcal{G} = \mathrm{R}p'_{1!!}(\mathrm{L}_{X\to Y}\otimes p'^{-1}_2\mathcal{G})\\ &\simeq \mathrm{R}p'_{1!!}\mathrm{R}f'_{!!}f'^{-1}(\mathrm{L}_{X\to Y}\otimes p'^{-1}_2\mathcal{G})\\ &\to \mathrm{R}p_{1!!}(\mathrm{L}_{Y\leftarrow X}\otimes p_1^{-1}p^{-1}\omega_{X/Y}^{\otimes -1}\otimes p_2^{-1}f^{-1}\mathcal{G})\\ &\simeq (\mathrm{L}_{Y\leftarrow X}\circ f^{-1}\mathcal{G})\otimes p^{-1}\omega_{X/Y}^{\otimes -1}.\end{aligned}$$

Finally we obtain a morphism

$$\begin{aligned}f_\pi{}^{-1}\mu_Y\mathcal{G} &\to (\mathrm{L}_{Y\leftarrow X}\circ f^{-1}\mathcal{G})\otimes p^{-1}\omega_{X/Y}^{\otimes -1}\\ &\simeq f_d{}^{-1}\mu_X\left(f^{-1}\mathcal{G}\right)\otimes p^{-1}\omega_{X/Y}^{\otimes -1}\simeq f_d{}^!\mu_X\left(f^{-1}\mathcal{G}\right),\end{aligned}$$

and by adjunction the desired morphism

$$\mathrm{R}f_{d\,!!}f_\pi{}^{-1}\mu_Y(\mathcal{G})\to\mu_X(f^{-1}\mathcal{G}).$$

(ii) Assume now that X is non-characteristic for $\mathcal{G}$. By induction we may assume that X is a hypersurface in Y. For $p\in T^*X$, let us show that $\mathrm{R}f_{d\,!!}f_\pi{}^{-1}\mu_Y(\mathcal{G})\otimes\widetilde{\mathbb{K}}_p \xrightarrow{\sim} \mu_X(f^{-1}\mathcal{G})\otimes\widetilde{\mathbb{K}}_p$.

Assume first that $p \in T^*_X X$. Since X is non characteristic for $\mathcal{G}$ we get

$$\begin{aligned}
\mathrm{R}f_{d!!} f_\pi^{-1} \mu_Y \mathcal{G} \otimes \widetilde{\mathbb{K}}_p &\simeq \mathrm{R}f_{d!!}\big(f_\pi^{-1} \mu_Y \mathcal{G} \otimes \widetilde{\mathbb{K}}_{T^*_X Y}\big) \otimes \widetilde{\mathbb{K}}_p \\
&\simeq \mathrm{R}f_{d!!}\big(f_\pi^{-1}\big(\mu_Y \mathcal{G} \otimes \widetilde{\mathbb{K}}_{T^*_Y Y}\big)\big) \otimes \widetilde{\mathbb{K}}_p \\
&\simeq \mathrm{R}f_{d!!}\big(f_\pi^{-1}(\pi_Y^{-1} \mathcal{G} \otimes \widetilde{\mathbb{K}}_{T^*_Y Y})\big) \otimes \widetilde{\mathbb{K}}_p \\
&\simeq \mathrm{R}f_{d!!}\big(f_d^{-1} \pi_X^{-1} f^{-1} \mathcal{G} \otimes \widetilde{\mathbb{K}}_{T^*_Y Y \times_Y X}\big) \otimes \widetilde{\mathbb{K}}_p \\
&\simeq \pi_X^{-1} f^{-1} \mathcal{G} \otimes \mathrm{R}f_{d!!} \widetilde{\mathbb{K}}_{T^*_Y Y \times_Y X} \otimes \widetilde{\mathbb{K}}_p \simeq \pi_X^{-1} f^{-1} \mathcal{G} \otimes \widetilde{\mathbb{K}}_p \\
&\simeq \mu_X f^{-1} \mathcal{G} \otimes \widetilde{\mathbb{K}}_p .
\end{aligned}$$

Assume now that $p \notin T^*_X X$. Consider the following diagram:

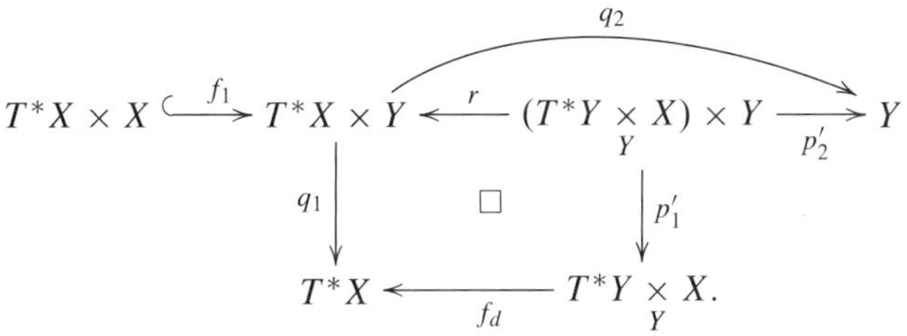

$$\begin{array}{ccccccc}
T^*X \times X & \xhookrightarrow{f_1} & T^*X \times Y & \xleftarrow{r} & (T^*Y \times_Y X) \times Y & \xrightarrow{p_2'} & Y \\
 & & \downarrow{\scriptstyle q_1} & \square & \downarrow{\scriptstyle p_1'} & & \\
 & & T^*X & \xleftarrow[f_d]{} & T^*Y \times_Y X . & &
\end{array}$$

(with $q_2 \colon T^*X \times Y \to Y$ drawn as an arc over the top row.)

Note that

$$\mathrm{R}f_{d!!} f_\pi^{-1} \mu_Y \mathcal{G} \simeq (\mathrm{R}r_{!!}\, \mathrm{L}_{X \to Y}) \circ \mathcal{G} \quad \text{and} \quad \mu_X f^{-1} \mathcal{G} \simeq \mathrm{L}_X \circ f^{-1} \mathcal{G} \simeq (\mathrm{R}f_{1!!}\, \mathrm{L}_X) \circ \mathcal{G}.$$

Hence we have to prove that

$$\big(\mathrm{R}r_{!!}\, \mathrm{L}_{X \to Y} \otimes \widetilde{\mathbb{K}}_p\big) \circ \mathcal{G} \simeq \big(\mathrm{R}f_{1!!}\, \mathrm{L}_X \otimes \widetilde{\mathbb{K}}_p\big) \circ \mathcal{G}.$$

Here we identify $p \in T^*X$ with $\big(p, f(\pi_X(p))\big) \in T^*X \times Y$. Take a local coordinate system $(t, x) = (t, x_1, \dots, x_n)$ of Y such that X is given by $t = 0$ and denote by (t, x, τ, ξ) and (x, ξ) the associated coordinates on T^*Y and T^*X, respectively. Set $p = (0, \xi_0)$. Let $((x, \tau, \xi), (t', x'))$ be the coordinates of $(T^*Y \times_Y X) \times Y$. Then $r((x, \tau, \xi), (t', x')) = ((x, \xi), (t', x'))$. We have

$$\begin{aligned}
&\mathrm{R}r_{!!}\, \mathrm{L}_{X \to Y} \otimes \widetilde{\mathbb{K}}_p \\
&\simeq \mathrm{R}r_{!!} \left(\underset{\varepsilon>0,\, R>0}{\text{“}\varinjlim\text{”}}\ \mathbb{K}_{\{\tau t' + \langle \xi_0, x'-x\rangle > \varepsilon(|t'|+|x'-x|),\, |\tau|<R\}}[\dim Y] \right) \otimes \widetilde{\mathbb{K}}_p .
\end{aligned}$$

Since the fiber of $\{\tau t' + \langle \xi_0, x' - x\rangle > \varepsilon(|t'| + |x' - x|),\ |\tau| < R\}$ over $((x, \xi), t', x')$ is a non-empty open interval if $R|t'| + \langle \xi_0, x' - x\rangle > \varepsilon(|t'| + |x' - x|)$, and empty otherwise, we obtain

$$\mathrm{R}r_{!!}\,\mathrm{L}_{X\to Y}\otimes\widetilde{\mathbb{K}}_p\simeq\left(\underset{\varepsilon>0,\,R>0}{\text{“}\varinjlim\text{”}}\ \mathbb{K}_{\{R|t'|+\langle\xi_0,x'-x\rangle>\varepsilon(|t'|+|x'-x|)\}}[\dim Y-1]\right)\otimes\widetilde{\mathbb{K}}_p\,.$$

Therefore

$$\left(\mathrm{R}r_{!!}\,\mathrm{L}_{X\to Y}\otimes\widetilde{\mathbb{K}}_p\right)\circ\mathcal{G}\simeq\left(\underset{\varepsilon>0,\,R>0}{\text{“}\varinjlim\text{”}}\ \mathbb{K}_{\{R|t'|+\langle\xi_0,x'-x\rangle>\varepsilon|x'-x|\}}[\dim X]\otimes\widetilde{\mathbb{K}}_p\right)\circ\mathcal{G}.$$

On the other hand we have

$$\begin{aligned}\left(\mathrm{R}f_{1!!}\,\mathrm{L}_X\otimes\widetilde{\mathbb{K}}_p\right)\circ\mathcal{G}&\simeq\left(\mathrm{R}f_{1!!}\big(\underset{\varepsilon>0}{\text{“}\varinjlim\text{”}}\,\mathbb{K}_{\{\langle\xi_0,x'-x>\varepsilon|x'-x|\}}[\dim X]\big)\otimes\widetilde{\mathbb{K}}_p\right)\circ\mathcal{G}\\&\simeq\left(\underset{\varepsilon>0}{\text{“}\varinjlim\text{”}}\,\mathbb{K}_{\{\langle\xi_0,x'-x\rangle>\varepsilon|x'-x|,\,t'=0\}}[\dim X]\otimes\widetilde{\mathbb{K}}_p\right)\circ\mathcal{G}.\end{aligned}$$

Hence it is enough to show that

$$\left(\underset{\varepsilon>0,\,R>0}{\text{“}\varinjlim\text{”}}\ \mathbb{K}_{\{R|t'|+\langle\xi_0,x'-x\rangle>\varepsilon|x'-x|,\,0<t'\}}\otimes\widetilde{\mathbb{K}}_p\right)\circ\mathcal{G}\simeq 0.$$

Let us set $U_{\varepsilon,\delta,R}=\{Rt'+\langle\xi_0,x'-x\rangle>|x-x|,\ 0<t'\le\delta\}$. For ε, δ sufficiently small and R sufficiently large, $\mathrm{SS}(\mathbb{K}_{U_{\varepsilon,\delta,R}})$ is contained in a sufficiently small neighborhood of $-Rdt'+\langle\xi_0,d(x-x')\rangle$ on a neighborhood of p. Hence we obtain

$$\mathrm{SS}(\mathbb{K}_{U_{\varepsilon,\delta,R}})^a\cap T^*(T^*X)\times\mathrm{SS}_0(\mathcal{G})\subset T^*(T^*X)\times T^*_YX$$

on a neighborhood of p for $R\gg0$.

Therefore Proposition 2.3.12 implies that

$$\left(\mathbb{K}_{U_{\varepsilon,\delta,R}}\circ\widetilde{\mathbb{K}}_{\Delta_Y}\right)\circ\mathcal{G}\simeq\mathbb{K}_{U_{\varepsilon,\delta,R}}\circ\mathcal{G}\quad\text{on a neighborhood of } p \text{ for } R\gg0.$$

Hence we have reduced the problem to

$$\left(\underset{\varepsilon>0,\,\delta>0\,R>0}{\text{“}\varinjlim\text{”}}\ \mathbb{K}_{U_{\varepsilon,\delta,R}}\otimes\widetilde{\mathbb{K}}_p\right)\circ\widetilde{\mathbb{K}}_{\Delta_Y}\simeq0.$$

Consider the projection on the first and third factors

$$h\colon T^*X\times Y\times Y\to T^*X\times Y\qquad \textit{i.e.}, ((x;\xi),(t',x'),(t'',x''))\mapsto((x;\xi),(t'',x'')).$$

Then

$$\left(\underset{\varepsilon>0,\,\delta>0\,R>0}{\text{“}\varinjlim\text{”}}\ \mathbb{K}_{U_{\varepsilon,\delta,R}}\otimes\widetilde{\mathbb{K}}_p\right)\circ\widetilde{\mathbb{K}}_{\Delta_Y}\simeq\mathrm{R}h_{!!}\left(\underset{\varepsilon>0,\,\delta>0\,R>0}{\text{“}\varinjlim\text{”}}\ \mathbb{K}_{V_{\varepsilon,\delta,R}}\right)\otimes\widetilde{\mathbb{K}}_p,$$

where $V_{\varepsilon,\delta,R}=\{Rt'+\langle\xi_0,x'-x\rangle>\varepsilon|x'-x|,\ 0<t'\le\delta,\ |x'-x''|\le\delta,\ |t'-t''|\le\delta\}$. This vanishes by the following lemma. □

Sublemma 2.4.5. *Let $(t, t', x, y) = (t, t', x_1, \dots, x_n, y_1, \dots, y_n)$ be the coordinates of $\mathbb{R} \times \mathbb{R} \times \mathbb{R}^n \times \mathbb{R}^n$, and let $h\colon \mathbb{R} \times \mathbb{R} \times \mathbb{R}^n \times \mathbb{R}^n \to \mathbb{R} \times \mathbb{R}^n$ be the projection, $h(t, t', x, y) = (t', y)$. For $\xi \in \mathbb{R}^n \setminus \{0\}$ and $\delta > 0$, set $V_\delta = \{(t, t', x, y);\ t + \langle \xi_0, x\rangle > |x|,\ |x - y| \le \delta, 0 < t \le \delta,\ |t - t'| \le \delta\}$. Then*

$$\mathrm{Supp}\big(\mathrm{R}h_!\ \mathbb{K}_{V_\delta}\big) \not\ni 0.$$

Proof. Let us decompose h into $\mathbb{R} \times \mathbb{R} \times \mathbb{R}^n \times \mathbb{R}^n \xrightarrow{h_1} \mathbb{R} \times \mathbb{R}^n \times \mathbb{R}^n \xrightarrow{h_2} \mathbb{R} \times \mathbb{R}^n$, where $h_1(t, t', x, y) = (t', x, y)$ and $h_2(t', x, y) = (t', y)$. When $|x - y| \le \delta$, the fiber $V_\delta \cap h_1^{-1}(t', x, y)$ is $\{t;\ \max(0,\ |x| - \langle \xi_0, x\rangle) < t \le \min(\delta, t' + \delta),\ t' - \delta \le t\}$. Hence, setting

$$W_\delta = \{(t', x, y);\ \max(0,\ |x| - \langle \xi_0, x\rangle) < t' - \delta \le \min(\delta, t' + \delta),\ |x - y| \le \delta\},$$

we have $\mathrm{R}h_{1!}\ \mathbb{K}_{V_\delta} \simeq \mathbb{K}_{W_\delta}$. Since $\mathrm{Supp}(\mathbb{K}_{W_\delta}) \subset \{(t', x, y);\ \delta \le t'\}$, we obtain

$$\mathrm{Supp}(\mathrm{R}h_!\ \mathbb{K}_{V_\delta}) \subset \{(t', y);\ \delta \le t'\}.$$

□

2.5. Microlocal convolution of kernels

Let X, Y and Z be manifolds, and let p_{ij} be the (i, j)-th projection from $T^*X \times T^*Y \times T^*Z$. As usual, denote by $a\colon T^*X \to T^*X$ the antipodal map. Then define the antipodal projection p_{12}^a by

$$p_{12}^a\colon T^*X \times T^*Y \times T^*Z \xrightarrow{p_{12}} T^*X \times T^*Y \xrightarrow{\mathrm{id} \times a} T^*X \times T^*Y.$$

For $\mathcal{F} \in \mathrm{D^b}(\mathrm{I}(\mathbb{K}_{T^*X \times T^*Y}))$ and $\mathcal{G} \in \mathrm{D^b}(\mathrm{I}(\mathbb{K}_{T^*Y \times T^*Z}))$, we set

$$\mathcal{F} \overset{a}{\circ} \mathcal{G} = \mathrm{R}p_{13!!}\big(p_{12}^{a-1}\mathcal{F} \otimes p_{23}^{-1}\mathcal{G}\big).$$

In an analogous way, for $S_1 \subset T^*X \times T^*Y$ and $S_2 \subset T^*Y \times T^*Z$, we set

$$S_1 \underset{T^*Y}{\overset{a}{\times}} S_2 = p_{12}^{a-1}(S_1) \cap p_{23}^{-1}(S_2) \subset T^*X \times T^*Y \times T^*Z.$$

Now we are ready to state the main theorem:

Theorem 2.5.1 (Microlocal convolution of kernels). *Let $\mathcal{K}_1 \in \mathrm{D^b}(\mathrm{I}(\mathbb{K}_{X \times Y}))$ and $\mathcal{K}_2 \in \mathrm{D^b}(\mathrm{I}(\mathbb{K}_{Y \times Z}))$.*

(i) There is a natural morphism

$$\mu_{X \times Y}\mathcal{K}_1 \overset{a}{\circ} \mu_{Y \times Z}\mathcal{K}_2 \to \mu_{X \times Z}(\mathcal{K}_1 \circ \mathcal{K}_2). \tag{2.9}$$

(ii) Assume the non-characteristic condition

$$\mathrm{SS}_0(\mathcal{K}_1) \underset{T^*Y}{\overset{a}{\times}} \mathrm{SS}_0(\mathcal{K}_2) \cap (T_X^*X \times T^*Y \times T_Z^*Z) \subset T_X^*X \times T_Y^*Y \times T_Z^*Z. \tag{2.10}$$

Then (2.9) *is an isomorphism outside*

$$\overline{p_{13}\big(\mathrm{SS}_0(\mathcal{K}_1) \underset{T^*Y}{\overset{a}{\times}} \mathrm{SS}_0(\mathcal{K}_2) \cap T^*X \times T^*_Y Y \times T^*Z\big)}.$$

Proof. (a) We shall first construct the morphism. Consider the manifolds $\mathcal{X}_1 = X \times Y$, $\mathcal{X}_2 = Y \times Z$ and $\mathcal{X} = \mathcal{X}_1 \times \mathcal{X}_2 = X \times Y \times Y \times Z$ together with the diagonal embedding

$$\mathcal{Y} := X \times Y \times Z \overset{j}{\hookrightarrow} \mathcal{X}.$$

Denote by $\mathcal{Z} = X \times Z$, and let $q_{13} \colon \mathcal{Y} \to \mathcal{Z}$ be the projection. The map

$$T^*\mathcal{Y} \hookrightarrow \mathcal{Y} \underset{\mathcal{X}}{\times} T^*\mathcal{X} \qquad \text{given by} \qquad (x, y, z; \xi, \eta, \zeta) \mapsto (x, y, y, z; \xi, -\eta, \eta, \zeta)$$

defines the cartesian square in the following commutative diagram:

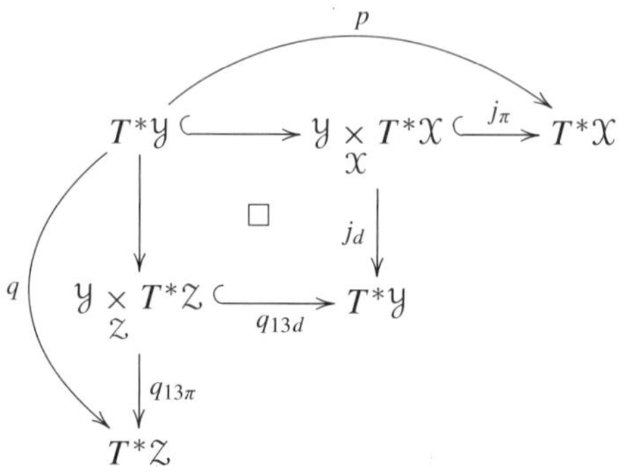

By Proposition 2.1.14, we have an isomorphism

$$\mathrm{K}_{T^*\mathcal{X}} \circ (\mu_{\mathcal{X}_1}\mathcal{K}_1 \boxtimes \mu_{\mathcal{X}_2}\mathcal{K}_2) \simeq \mu_{\mathcal{X}}(\mathcal{K}_1 \boxtimes \mathcal{K}_2).$$

By Theorem 2.4.4 we have a morphism

$$\mathrm{R}j_{d!!}j_\pi^{-1}\mu_{\mathcal{X}}(\mathcal{K}_1 \boxtimes \mathcal{K}_2) \to \mu_{\mathcal{Y}}(j^{-1}(\mathcal{K}_1 \boxtimes \mathcal{K}_2)). \tag{2.11}$$

Since q_{13} is smooth we also have an isomorphism by Theorem 2.4.2 (ii):

$$\mathrm{R}q_{13\pi!!}q_{13d}^{-1}\mu_{\mathcal{Y}}(j^{-1}(\mathcal{K}_1 \boxtimes \mathcal{K}_2)) \xrightarrow{\sim} \mu_{\mathcal{Z}}(\mathrm{R}q_{13!!}j^{-1}(\mathcal{K}_1 \boxtimes \mathcal{K}_2)) \simeq \mu_{\mathcal{Z}}(\mathcal{K}_1 \circ \mathcal{K}_2).$$

Hence we get a morphism

$$\mathrm{R}q_{!!}p^{-1}\Big(\mathrm{K}_{T^*\mathcal{X}} \circ (\mu_{\mathcal{X}_1}\mathcal{K}_1 \boxtimes \mu_{\mathcal{X}_2}\mathcal{K}_2)\Big) \to \mu_{\mathcal{Z}}(\mathcal{K}_1 \circ \mathcal{K}_2). \tag{2.12}$$

Hence we obtain

$$\begin{aligned}\mu_{\mathcal{X}_1}\mathcal{K}_1 \overset{a}{\circ} \mu_{\mathcal{X}_2}\mathcal{K}_2 &\simeq \mathrm{R}q_{!!}p^{-1}(\mu_{\mathcal{X}_1}\mathcal{K}_1 \boxtimes \mu_{\mathcal{X}_2}\mathcal{K}_2) \\ &\to \mathrm{R}q_{!!}p^{-1}\Big(\mathrm{K}_{T^*\mathcal{X}} \circ (\mu_{\mathcal{X}_1}\mathcal{K}_1 \boxtimes \mu_{\mathcal{X}_2}\mathcal{K}_2)\Big) \to \mu_{\mathcal{Z}}(\mathcal{K}_1 \circ \mathcal{K}_2).\end{aligned}$$

(b) By Theorem 2.4.4, (2.11) is an isomorphism under the non-characteristic hypothesis, and hence (2.12) is also an isomorphism under the same hypothesis.

Therefore in order to show (ii), it is enough to show that

$$\begin{gathered}\mu_{\mathcal{X}_1}\mathcal{K}_1 \overset{a}{\circ} \mu_{\mathcal{X}_2}\mathcal{K}_2 \simeq \mathrm{R}q_{!!}p^{-1}\Big(\mathrm{K}_{T^*\mathcal{X}} \circ (\mu_{\mathcal{X}_1}\mathcal{K}_1 \boxtimes \mu_{\mathcal{X}_2}\mathcal{K}_2)\Big) \\ \text{outside } p_{13}\big(\mathrm{SS}_0(\mathcal{K}_1) \underset{T^*Y}{\overset{a}{\times}} \mathrm{SS}_0(\mathcal{K}_2) \cap T^*X \times T^*_Y Y \times T^*Z\big).\end{gathered} \tag{2.13}$$

First note that

$$\begin{aligned}\mu_{\mathcal{X}_1}\mathcal{K}_1 \overset{a}{\circ} \mu_{\mathcal{X}_2}\mathcal{K}_2 &\simeq (\mathrm{K}_{T^*\mathcal{X}_1} \circ \mu_{\mathcal{X}_1}\mathcal{K}_1) \overset{a}{\circ} (\mathrm{K}_{T^*\mathcal{X}_1} \circ \mu_{\mathcal{X}_2}\mathcal{K}_2) \\ &\simeq \big(\mathrm{K}_{T^*\mathcal{X}_1} \underset{T^*Y}{\circ} \mathrm{K}_{T^*\mathcal{X}_2}\big) \circ \big(\mu_{\mathcal{X}_1}\mathcal{K}_1 \boxtimes \mu_{\mathcal{X}_2}\mathcal{K}_2\big).\end{aligned}$$

Consider the diagram

$$\begin{array}{ccc} & T^*\mathcal{Y} \times T^*\mathcal{X} & \\ {}^{q'=(q,\mathrm{id})}\swarrow & & \searrow^{p'=(p,\mathrm{id})} \\ T^*\mathcal{Z} \times T^*\mathcal{X} & & T^*\mathcal{X} \times T^*\mathcal{X} \end{array}$$

Then we have

$$\mathrm{R}q_{!!}p^{-1}\Big(\mathrm{K}_{T^*\mathcal{X}} \circ (\mu_{\mathcal{X}_1}\mathcal{K}_1 \boxtimes \mu_{\mathcal{X}_2}\mathcal{K}_2)\Big) \simeq \big(\mathrm{R}q'_{!!}p'^{-1}\,\mathrm{K}_{T^*\mathcal{X}}\big) \circ (\mu_{\mathcal{X}_1}\mathcal{K}_1 \boxtimes \mu_{\mathcal{X}_2}\mathcal{K}_2).$$

Using Proposition 1.3.3 and Corollary 1.3.5, we have

$$\mathrm{R}q'_{!!}p'^{-1}\,\mathrm{K}_{T^*\mathcal{X}} \simeq \mathcal{L}_\sigma\big(T^*\mathcal{Y}, T^*\mathcal{Z} \times T^*\mathcal{X}\big),$$

where $T^*\mathcal{Y}$ is embedded into $T^*\mathcal{Z} \times T^*\mathcal{X}$ by (q, p) and the section σ is given by

$$\sigma = (\omega_X, \omega_Z, -\omega_X, -\omega_Y, -\omega_Y, -\omega_Z).$$

In order to see (2.13) under the non-characteristic hypothesis, it is enough to show that

$$\begin{gathered}\mathrm{K}_{T^*\mathcal{X}_1} \underset{T^*Y}{\circ} \mathrm{K}_{T^*\mathcal{X}_2} \to \mathcal{L}_\sigma\big(T^*\mathcal{Z} \underset{T^*\mathcal{Z}}{\times} T^*\mathcal{Y}, T^*\mathcal{Z} \times T^*\mathcal{X}\big) \text{ is an isomorphism on } T^*\mathcal{Z} \times \\ \big(T^*X \times \dot{T}^*(Y \times Y) \times T^*Z\big) \subset T^*\mathcal{Z} \times T^*\mathcal{X}.\end{gathered} \tag{2.14}$$

However it is a consequence of Proposition 1.3.12 (note that (iii) and (v) in the proposition fail on $T^*X \times T^*_Y Y \times T^*Z$). □

2.6. A vanishing theorem for microlocal holomorphic functions

Theorem 2.6.1. *Let X be a complex manifold of dimension n. Then, $\mu_X(\mathcal{O}_X)|_{\dot{T}^*X}$ is concentrated in degree $-n$.*

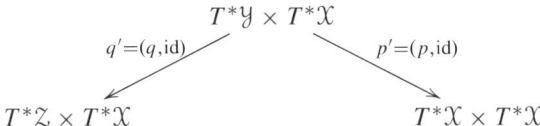

Proof. We may assume $X = \mathbb{C}^n$. Let $q_1 : T^*X \times X \to T^*X$ and $q_2 : T^*X \times X \to X$ be the projections. Let $p = (x_0, \xi_0) \in \dot{T}^*X$. Then, we have

$$\mu_X(\mathcal{O}_X) \otimes \widetilde{\mathbb{C}}_p \simeq \widetilde{\mathbb{C}}_p \otimes \mathrm{R}q_{1!!}\left(\underset{\varepsilon,\delta>0}{\text{“}\varinjlim\text{”}}(\mathbb{C}_{F_{\delta,\varepsilon}} \otimes q_2^{-1}\mathcal{O}_X)\right)[2n],$$

where $F_{\delta,\varepsilon} = \big\{\big((x,\xi),x'\big); \delta \geqslant \langle \xi_0, x'-x\rangle > \varepsilon|x'-x|\big\}$. Hence it is enough to show that

$$\mathrm{R}q_{1!}(\mathbb{C}_{F_{\delta,\varepsilon}} \otimes q_2^{-1}\mathcal{O}_X)$$

is concentrated in degree n. We have

$$\mathrm{R}q_{1!}(\mathbb{C}_{F_{\delta,\varepsilon}} \otimes q_2^{-1}\mathcal{O}_X)_{(x_1,\xi_1)} \simeq \mathrm{R}\Gamma_c\left(\{x' \in X; \delta \geqslant \langle \xi_0, x'-x_1\rangle > \varepsilon|x'-x_1|\}, \mathcal{O}_X\right).$$

The cohomology with compact support of $\mathcal{O}_X$ on the difference of two convex open subsets is concentrated in degree n. □

Now $H^{-n}\big(\mu_X(\mathcal{O}_X)|_{\dot{T}^*X}\big)$ has a structure of $\mathcal{E}_X|_{\dot{T}^*X}$-module, i.e., there exists a canonical ring homomorphism $\mathcal{E}_X|_{\dot{T}^*X} \to \mathcal{E}\mathrm{nd}\big(H^{-n}(\mu_X(\mathcal{O}_X)|_{\dot{T}^*X})\big)$.

Indeed, let $p_k : X \times X \to X$ be the k-th projection, and $\mathcal{O}^{(0,n)}_{X\times X} := \mathcal{O}_{X\times X} \otimes_{p_2^{-1}\mathcal{O}_X} p_2^{-1}\mathcal{O}^{(n)}_X$. We have morphisms $\mathrm{R}p_{1!}(\mathcal{O}^{(0,n)}_{X\times X}[n] \otimes p_2^{-1}\mathcal{O}_X) \to \mathrm{R}p_{1!}(\mathcal{O}^{(0,n)}_{X\times X}[n]) \to \mathcal{O}_X$ which induce $\mathcal{O}^{(0,n)}_{X\times X}[n] \to \mathrm{R}\mathcal{H}om(p_2^{-1}\mathcal{O}_X, p_1^!\mathcal{O}_X)$. Thus we obtain

$$\begin{aligned}\mathcal{E}_X \to \mu_{\Delta_X}(\mathcal{O}^{(0,n)}_{X\times X}[n]) \to \mu_{\Delta_X}\big(\mathrm{R}\mathcal{H}om(p_2^{-1}\mathcal{O}_X, p_1^!\mathcal{O}_X)\big)\\ \simeq \mu hom(\mathcal{O}_X, \mathcal{O}_X) \simeq \mathrm{R}\mathcal{H}om\big(\mu_X(\mathcal{O}_X), \mu_X(\mathcal{O}_X)\big).\end{aligned}$$

Hence, Theorem 2.6.1 implies that $\mu_X(\mathcal{O}_X)|_{\dot{T}^*X}$ belongs to

$$\mathrm{D^b}\big(\mathrm{Mod}\big(\mathcal{E}_X|_{\dot{T}^*X}, \mathrm{I}(\mathbb{C}_{\dot{T}^*X})\big)\big),$$

the derived category of the abelian category $\mathrm{Mod}\big(\mathcal{E}_X|_{\dot{T}^*X}, \mathrm{I}(\mathbb{C}_{\dot{T}^*X})\big)$ of ind-sheaves $\mathcal{F}$ on $\dot{T}^*X$ endowed with a ring homomorphism $\mathcal{E}_X|_{\dot{T}^*X} \to \mathcal{E}nd\,(\mathcal{F})$. This implies the following theorem.

Theorem 2.6.2. *Let X be a complex manifold. Then* $\mathcal{F} \mapsto \mu hom(\mathcal{F}, \mathcal{O}_X)|_{\dot{T}^*X}$ *is a well defined functor from* $\mathrm{D^b}(\mathbb{C}_X)$ *to* $\mathrm{D^b}(\mathcal{E}_X|_{\dot{T}^*X})$.

Acknowledgments

The authors would like to thank A. D'Agnolo for many helpful comments.

References

[K] M. Kashiwara, Quantization of contact manifold, *Publ. of Research Institute for Mathematical Sciences*, **32** no.1 (1996) 1–7.

[KS1] M. Kashiwara and P. Schapira, Ind-Sheaves, Astérisque, Vol. 271, *Soc. Math. France*, 2001.

[KS2] ——, *Sheaves on Manifolds*, Springer, 1990.

[KS3] ——, Microlocal Study of Sheaves, Astérisque, Vol. 128, *Soc. Math. France*, 1985.

[KS4] ——, Microlocal Study of Ind-Sheaves I: Micro-Support and Regularity, Astérisque, Vol. 284, *Soc. Math. France*, 2003.

[KS5] ——, *Ind-sheaves, distributions, and microlocalization*, Sém Ec. Polytechnique, May 18, 1999, arXiv:math.AG/9906200.

[S] M. Sato, Hyperfunctions and partial differential equations, in *Proc. Int. Conf. on Functional Analysis and Related Topics*, Tokyo Univ. Press, 1969.

[SKK] M. Sato, T. Kawai, and M. Kashiwara, Hyperfunctions and pseudo-differential equations, in Komatsu (ed.), Hyperfunctions and pseudo-differential equations, *Proceedings Katata* 1971, Lecture Notes in Mathematics, Springer, Vol. 287, 1973, pp. 265–529.

[W] I. Waschkies, The stack of microlocal perverse sheaves, *Bull. Soc. Math. France*, **132** no.3, (2004) pp 397–462.

Endoscopic decomposition of certain depth zero representations

David Kazhdan[1] and Yakov Varshavsky[2*]

Institute of Mathematics
The Hebrew University of Jerusalem
Givat-Ram
Jerusalem 91904
Israel
[1]kazhdan@math.huji.ac.il
[2]vyakov@math.huji.ac.il

Dedicated to A. Joseph on his 60th birthday

Summary. We construct an endoscopic decomposition for local L-packets associated to irreducible cuspidal Deligne–Lusztig representations. Moreover, the obtained decomposition is compatible with inner twistings.

Key words: Endoscopy, Deligne–Lusztig representations

Subject Classifications: Primary: 22E50; Secondary: 22E35

0. Introduction

Let E be a local non-archimedean field, $\Gamma \supset W \supset I$ the absolute Galois, the Weil and the inertia groups of E, respectively. Let G be a reductive group over E, and let ${}^L G = \widehat{G} \rtimes W$ be the complex Langlands dual group of G. Denote by $\mathcal{D}(G(E))$ the space of invariant generalized functions on $G(E)$, that is, the space of $\operatorname{Int} G$-invariant linear functionals on the space of locally constant compactly supported measures on $G(E)$.

Every admissible homomorphism $\lambda : W \to {}^L G$ (see [Ko1, §10]) gives rise to a finite group $S_\lambda := \pi_0(Z_{\widehat{G}}(\lambda)/Z(\widehat{G})^\Gamma)$, where $Z_{\widehat{G}}(\lambda)$ is the centralizer of $\lambda(W)$ in $\widehat{G}$. To every conjugacy class κ of S_λ, Langlands [La1] associated an endoscopic subspace $\mathcal{D}_{\kappa,\lambda}(G(E)) \subset \mathcal{D}(G(E))$. For simplicity, we will restrict ourselves to the elliptic case, where $\lambda(W)$ does not lie in any proper Levi subgroup of ${}^L G$.

* Both authors were supported by The Israel Science Foundation (Grants No. 38/01 and 241/03).

Langlands conjectured that every elliptic λ corresponds to a finite set Π_λ, called an L-packet, of cuspidal irreducible representations of $G(E)$. Moreover, the subspace $\mathcal{D}_\lambda(G(E)) \subset \mathcal{D}(G(E))$, generated by characters $\{\chi(\pi)\}_{\pi\in\Pi_\lambda}$, should have an endoscopic decomposition. More precisely, it is expected ([La1, IV, 2]) that there exists a basis $\{a_\pi\}_{\pi\in\Pi_\lambda}$ of the space of central functions on S_λ such that $\chi_{\kappa,\lambda} := \sum_{\pi\in\Pi_\lambda} a_\pi(\kappa)\chi(\pi)$ belongs to $\mathcal{D}_{\kappa,\lambda}(G)$ for every conjugacy class κ of S_λ.

The goal of this paper is to construct the endoscopic decomposition of $\mathcal{D}_\lambda(G(E))$ for tamely ramified λ's such that $Z_{\widehat{G}}(\lambda(I))$ is a maximal torus. In this case, G splits over an unramified extension of E, and λ factors through ${}^LT \hookrightarrow {}^LG$ for an elliptic unramified maximal torus T of G.

Each $\kappa \in S_\lambda = \widehat{T}^\Gamma/Z(\widehat{G})^\Gamma$ gives rise to an elliptic endoscopic triple $\mathcal{E}_{\kappa,\lambda}$ for G, while characters of S_λ are in bijection with conjugacy classes of embeddings $T \hookrightarrow G$, stably conjugate to the inclusion. By the local Langlands correspondence for tori ([La2]), a homomorphism $\lambda : W \to {}^LT$ defines a tamely ramified homomorphism $\theta : T(E) \to \mathbb{C}^\times$. Therefore each character a of S_λ gives rise to an irreducible cuspidal representation $\pi_{a,\lambda}$ of $G(E)$ (denoted by $\pi_{a,\theta}$ in Notation 2.1.3).

Our main theorem asserts that if the residual characteristic of E is sufficiently large, then each $\chi_{\kappa,\lambda} := \sum_a a(\kappa)\chi(\pi_{a,\lambda})$ is $\mathcal{E}_{\kappa,\lambda}$-stable (see Definition 1.6.6). More generally (see Corollary 1.6.12 (b)), for each inner form G' of G, we denote by $\chi'_{\kappa,\lambda}$ the corresponding generalized function on $G'(E)$, and our main theorem asserts that $\chi_{\kappa,\lambda}$ and $\chi'_{\kappa,\lambda}$ are "$\mathcal{E}_{\kappa,\lambda}$-equivalent" (see Definition 1.6.10 for a more precise term). Although in this work we show this result only for local fields of characteristic zero, the case of local fields of positive characteristic follows by approximation (see [KV2]).

Our argument goes as follows. First, we prove the equivalence of the restrictions of $\chi_{\kappa,\lambda}$ and $\chi'_{\kappa,\lambda}$ to the subsets of topologically unipotent elements of $G(E)$ and $G'(E)$. If the residual characteristic of E is sufficiently large, topologically unipotent elements of $G(E)$ and $G'(E)$ can be identified with topologically nilpotent elements of the Lie algebras $\mathcal{G}(E)$ and $\mathcal{G}'(E)$, respectively. Thus we are reduced to an analogous assertion about generalized functions on Lie algebras. Now the equivalence follows from a combination of a Springer hypothesis (Theorem A.1), which describes the trace of a Deligne–Lusztig representation in terms of Fourier transform of an orbit, and a generalization of a theorem of Waldspurger [Wa2] to inner forms, which asserts that up to a sign, Fourier transform preserves the equivalence.

To prove the result in general, we use the topological Jordan decomposition ([Ka2]). We would like to stress that in order to prove just the stability of $\chi_{\kappa,\lambda}$ one still needs a generalization of [Wa2] to inner forms.

This paper is organized as follows.

In the first section we give basic definitions and constructions of a rather general nature. In particular, most of the section is essentially a theory of endoscopy, which was developed by Langlands, Shelstad and Kottwitz. In order to incorporate both the case of algebraic groups and Lie algebras we work in a more general context of algebraic varieties equipped with an action of G^{ad}.

More precisely, in Subsection 1.1 we recall basic properties, results and constructions concerning inner twistings and stable conjugacy. In Subsections 1.2 and 1.3 we give basic definitions and properties of dual groups and of endoscopic triples. Then in Subsection 1.4 we prove that certain subsets of the group $Z(\widehat{G^{\mathrm{ad}}})^{\Gamma}$ are actually subgroups. Unfortunately, this result is proved case-by-case. In Subsection 1.5 we specialize previous results to the case of endoscopic triples over local non-archimedean fields.

In Subsection 1.6 we define the notions of stability and equivalence of generalized functions, while in Subsection 1.7 we write down explicitly the condition for stability and equivalence for generalized functions coming from invariant locally L^1 functions. Note that the notion of equivalence is much more subtle than that of stability. In particular, it depends not just on an endoscopic triple but also on a triple $(a, a'; [b])$, consisting of compatible embeddings of maximal tori into G, G' and the endoscopic group.

We finish the section by Subsection 1.8 in which we study basic properties of certain equivariant maps from reductive groups to their Lie algebras, which we call quasi-logarithms. We use these maps to identify topologically unipotent elements of the group with topologically nilpotent elements of the Lie algebra.

The second section is devoted to the formulation and the proof of the main theorem. More precisely, in Subsection 2.1 we give two equivalent formulations of our main result. In Subsection 2.2 we prove the equivalence of the restrictions of $\chi_{\kappa,\lambda}$ and $\chi'_{\kappa,\lambda}$ to topologically unipotent elements.

In Subsection 2.3 we rewrite character $\chi(\pi_{a,\theta})$ of $G(E)$ in terms of restrictions to topologically unipotent elements of corresponding characters of centralizers $G_{\delta}(E)$. For this we use the topological Jordan decomposition. In the next Subsection 2.4 we compare endoscopic triples for the group G and for its centralizers $G_{\delta}(E)$. Finally, in Subsections 2.5–2.7 we carry out the proof itself.

We finish the paper by two appendices of independent interest, crucially used in Subsection 2.2. In Appendix A we prove the Springer hypothesis. In the case of large characteristic this result was proved by the first author in [Ka1]. For the proof in general, we use Lusztig's interpretation of a trace of the Deligne–Lusztig representation in terms of character sheaves [Lu] and results of Springer [Sp] on the Fourier transform.

In Appendix B we prove a generalization of both the theorem of Waldspurger [Wa2] and that of Kazhdan–Polishchuk [KP, Thm. 2.7.1] (see also Remark B.1.3). Our strategy is very similar to those of [Wa2] and [KP]. More precisely, using stationary phase principle and the results of Weil [We], we construct in Subsection B.2 certain measures whose Fourier transform can be explicitly calculated. Then in Subsection B.3 we extend our data over a local field to a corresponding data over a number field. Finally, in Section 1.4 we deduce our result from a simple form of the trace formula.

For the convenience of the reader, we also include a list of main terms and symbols, indexed by the page number in which they first appear.

This work is an expanded version of the announcement [KV1]. In the process of writing, we have learned that DeBacker and Reeder obtained similar results (see [DBR]). After the work was completed, it was pointed out to us that our scheme of the argument is similar to the one used by Mœglin–Waldspurger in [MW].

We would like to thank the referee for his numerous valuable remarks.

Notation and conventions

For a finite abelian group A, we denote by A^D the group of complex characters of A.

For an algebraic group G, we denote by G^0, $Z(G)$, G^{ad}, G^{der} and $W(G)$ the connected component of the identity of G, the center of G, the adjoint group of G, the derived group of G, and the Weyl group of G, respectively. Starting from 1.1.9, G will always be assumed to be reductive and connected, in which case we denote by G^{sc} the simply connected covering of G^{der}.

We denote by $\mathcal{G}$, $\mathcal{H}$, $\mathcal{T}$ and $\mathcal{L}$ Lie algebras of algebraic groups G, H, T and L, respectively.

Let an algebraic group G act on an algebraic variety X. For each $x \in X$, we denote by G_x and $\mathcal{G}_x$ the stabilizers of x in G and $\mathcal{G}$, respectively. Explicitly, $\mathcal{G}_x$ is the kernel of the differential at $g = 1$ of the morphism $G \to X$ $(g \mapsto g(x))$. (Note that x and therefore also G_x and $\mathcal{G}_x$ will have different meanings starting from 1.8.5.)

Each algebraic group acts on itself by inner automorphisms and by the adjoint action on its Lie algebra. For each $g \in G$ we denote by $\mathrm{Int}\, g$ and $\mathrm{Ad}\, g$ the corresponding elements in $\mathrm{Int}\, G \subset \mathrm{Aut}(G)$ and $\mathrm{Ad}\, G \subset \mathrm{Aut}(\mathcal{G})$, respectively.

For a field E, we denote by $\overline{E}$ a fixed algebraic closure of E, and by E^{sep} the maximal separable extension of E in $\overline{E}$. Γ will always be the absolute Galois group of E. When Γ acts on a set X we will write ${}^{\sigma}x$ instead of $\sigma(x)$.

For a reductive algebraic group G over E, we denote by $\mathrm{rk}_E(G)$ the rank of G over E, and put $e(G) := (-1)^{\mathrm{rk}_E(G^{\mathrm{ad}})}$. We also set $e'(G) := e(G)e(G^*)$, where G^* is the quasi-split inner form of G. Then $e'(G)$ coincides with the sign defined by Kottwitz ([Ko5]).

Starting from Subsection 1.5, E will be a local non-archimedean field with ring of integers $\mathcal{O}$, maximal ideal $\mathfrak{m}$, and residue field $\mathbb{F}_q$ of characteristic p. We denote by E^{nr} the maximal unramified extension of E in $\overline{E}$. Starting from Subsection 2.2, we will assume that the characteristic of E is zero.

1. Basic definitions and constructions

1.1. Stable conjugacy

In this subsection we recall basic definitions and constructions concerning inner forms and stable conjugacy.

Let G be an algebraic group over a field E. Starting from 1.1.9, we will assume that G is reductive and connected.

Let X be an algebraic variety over E (that is, a reduced scheme locally of finite type over E) equipped with an action of G^{ad} (that is, with an action of G, trivial on $Z(G)$).

Our basic examples will be $X = G$ and $X = \mathcal{G}$ with the natural action of G^{ad}.

1.1.1. Inner twistings (a) Let G' be an algebraic group over E. Recall that an *inner twisting* $\varphi : G \to G'$ is an isomorphism $\varphi : G_{E^{\rm sep}} \overset{\sim}{\to} G'_{E^{\rm sep}}$ such that for each $\sigma \in \Gamma$ the automorphism $c_\sigma := \varphi^{-1\sigma}\varphi \in \mathrm{Aut}(G)$ is inner. In this case, $\{c_\sigma\}_\sigma$ form a cocycle of Γ in $\mathrm{Int}\, G = G^{\rm ad}$, and we denote by $\mathrm{inv}(\varphi) = \mathrm{inv}(G, G') \in H^1(E, G^{\rm ad})$ the corresponding cohomology class.

(b) Two inner twistings are called *isomorphic* if they differ by an inner automorphism. Then the map $(\varphi : G \to G') \mapsto \mathrm{inv}(G, G')$ gives a bijection between the set of isomorphism classes of inner twistings of G and $H^1(E, G^{\rm ad})$.

(c) Each inner twisting $\varphi : G \to G'$ gives rise to a twisting $\varphi_X : X \to X'$, where X' is an algebraic variety over E equipped with an action of G', and φ_X is a $G_{E^{\rm sep}} \cong G'_{E^{\rm sep}}$-equivariant isomorphism $X_{E^{\rm sep}} \overset{\sim}{\to} X'_{E^{\rm sep}}$. Explicitly, X' is a twist of X by the image of the cocycle $\{c_\sigma\}_\sigma \subset G^{\rm ad}$ in $\mathrm{Aut}(X)$. In particular, for each $\sigma \in \Gamma$ we have ${}^\sigma\varphi_X = \varphi \circ c_\sigma$.

By construction, for each $x \in X$ and $g \in G$, we have $\varphi_X(g(x)) = \varphi(g)(\varphi_X(x))$.

(d) An inner twisting φ is called *trivial*, if $\mathrm{inv}(\varphi) = 1$. Explicitly, φ is trivial if and only if there exists $g \in G(E^{\rm sep})$ such that $\varphi \circ \mathrm{Int}\, g$ induces an isomorphism $G \overset{\sim}{\to} G'$ over E. In particular, the identity map $\mathrm{Id}_G : G \to G$ is a trivial inner twisting.

Definition 1.1.2. (a) Two points $x, x' \in X(E)$ are called *conjugate* if there exists $g \in G(E)$ such that $x' = g(x)$.

(b) Let $\varphi : G \to G'$ be an inner twisting, and $\varphi_X : X \to X'$ the corresponding twisting. Elements $x \in X(E)$ and $x' \in X'(E)$ are called $E^{\rm sep}$*-conjugate* if there exists $g \in G(E^{\rm sep})$ such that $x' = \varphi_X(g(x))$.

(c) When G and G_x (and hence also G' and $G'_{x'}$) are connected reductive groups, $E^{\rm sep}$-conjugate x and x' are also called *stably conjugate*.

Remark 1.1.3. All of our examples will satisfy assumption (c). In this case, our notion of stable conjugacy generalizes the standard one (see [Ko3]).

1.1.4. Cohomological invariants Let $x \in X(E)$ and $x' \in X'(E)$ be $E^{\rm sep}$-conjugate elements. Denote by $G_{x,x'}$ the set of $g \in G(E^{\rm sep})$ such that $x' = \varphi_X(g(x))$.

(a) Assume that $\varphi = \mathrm{Id}_G$. Then for each $g \in G_{x,x'}$, the map $\sigma \mapsto g^{-1\sigma}g$ defines a cocycle of Γ in G_x. Moreover, the corresponding cohomology class $\mathrm{inv}(x, x') \in H^1(E, G_x)$ is independent of g. Furthermore, the correspondence $x' \mapsto \mathrm{inv}(x, x')$ gives a bijection between the set of conjugacy classes of $x' \in X(E)$ stably conjugate to x and $\mathrm{Ker}\,[H^1(E, G_x) \to H^1(E, G)]$ (compare [Ko2, 4.1]).

(b) Let φ be general. Then for each $g \in G_{x,x'}$, the map $\sigma \mapsto g^{-1}(\varphi^{-1\sigma}\varphi)^\sigma g$ defines a cocycle of Γ in $G_x/Z(G) = (G^{\rm ad})_x \subset \mathrm{Int}\, G$. Moreover, the corresponding cohomology class $\overline{\mathrm{inv}}(x, x') \in H^1(E, (G^{\rm ad})_x)$ is independent of g. Furthermore, the correspondence $x' \mapsto \overline{\mathrm{inv}}(x, x')$ gives a surjection from the set of conjugacy classes of $x' \in X'(E)$ stably conjugate to x to the preimage of $\mathrm{inv}(G, G') \in H^1(E, G^{\rm ad})$ in $H^1(E, (G^{\rm ad})_x)$.

When $\varphi = \mathrm{Id}_G$, then $\overline{\mathrm{inv}}(x, x')$ is the image of $\mathrm{inv}(x, x')$ under the natural projection $H^1(E, G_x) \to H^1(E, (G^{\rm ad})_x)$.

(c) For each $g \in G_{x,x'}$, the map $h \mapsto \varphi(ghg^{-1})$ defines an inner twisting $G_x \to G'_{x'}$. Moreover, the corresponding invariant $\mathrm{inv}(G_x, G'_{x'}) \in H^1(E, (G_x)^{\mathrm{ad}})$ is just the image of $\overline{\mathrm{inv}}(x, x')$. In particular, $G'_{x'}$ is canonically identified with G_x, if G_x is abelian.

(d) Assume that G_x is abelian, and let $y \in X(E)$ and $y' \in X'(E)$ be E^{sep}-conjugates of x and x'. Then the identification $G'_{y'} = G'_{x'} = G_x = G_y$ identifies $\overline{\mathrm{inv}}(y, y')$ with the product of $\overline{\mathrm{inv}}(x, x')$ and the images of $\mathrm{inv}(y, x)$ and $\mathrm{inv}(x', y')$. Moreover, if $\varphi = \mathrm{Id}_G$, the same identification identifies $\mathrm{inv}(y, y')$ with the product $\mathrm{inv}(y, x)\,\mathrm{inv}(x, x')\mathrm{inv}(x', y')$.

1.1.5. Generalization (Compare [LS, (3.4)]). Let $\varphi : G \to G'$ be an inner twisting, $X_1, \dots, X_k$ a k-tuple of algebraic varieties over E equipped with an action of G^{ad}, and $X'_1, \dots, X'_k$ the corresponding inner twistings. Let $x_i \in X_i(E)$ and $x'_i \in X'_i(E)$ be E^{sep}-conjugate for each $i = 1, \dots, k$.

Choose representatives $\widetilde{c}_\sigma \in G(E^{\mathrm{sep}})$ of $c_\sigma = \varphi^{-1\sigma}\varphi \in G^{\mathrm{ad}}(E^{\mathrm{sep}})$ for all $\sigma \in \Gamma$ and choose elements $g_i \in G(E^{\mathrm{sep}})$ such that $x'_i = \varphi_{X_i}(g_i(x_i))$ for $i = 1, \dots, k$. Then the map $\sigma \mapsto [(g_i^{-1}\widetilde{c}_\sigma{}^\sigma g_i)_i] \in G^k/Z(G)$ gives a cocycle of Γ in $(\prod_i G_{x_i})/Z(G)$, independent of the choice of $\widetilde{c}_\sigma$'s, and the corresponding cohomology class

$$\overline{\mathrm{inv}}((x_1, x'_1); \dots; (x_k, x'_k)) \in H^1\left(E, \left(\prod_i G_{x_i}\right)\Big/ Z(G)\right)$$

of $[(g_i^{-1}\widetilde{c}_\sigma{}^\sigma g_i)_i]$ is independent of the g_i's.

Note that $\overline{\mathrm{inv}}((x_1, x'_1); \dots; (x_k, x'_k))$ lifts both

$$(\overline{\mathrm{inv}}(x_i, x'_i)_i) \in \prod_i H^1(E, G_{x_i}/Z(G))$$

and $\Delta(\mathrm{inv}(G, G')) \in H^1(E, (G^k)/Z(G))$. (Here $\Delta : G^{\mathrm{ad}} \to (G^k)/Z(G)$ is the diagonal embedding.)

The following result follows immediately from definitions.

Lemma 1.1.6. *(a)* $\overline{\mathrm{inv}}((x_1, x'_1); \dots; (x_k, x'_k))$ *depends only on the conjugacy classes of* x_i*'s and* x'_i*'s.*

(b) If $\varphi = \mathrm{Id}_G$*, then* $\overline{\mathrm{inv}}((x_1, x'_1); \dots; (x_k, x'_k))$ *is the image of* $((\mathrm{inv}(x_i, x'_i)_i) \in H^1(E, \prod_i G_{x_i})$.

(c) The canonical projection $(\prod_{i=1}^k G_{x_i})/Z(G) \to (\prod_{i=1}^{k-1} G_{x_i})/Z(G)$ *maps* $\overline{\mathrm{inv}}((x_1, x'_1); \dots; (x_k, x'_k))$ *to* $\overline{\mathrm{inv}}((x_1, x'_1); \dots; (x_{k-1}, x'_{k-1}))$.

(d) The diagonal map $G_x/Z(G) \hookrightarrow (G_x)^2/Z(G)$ *maps*

$$\overline{\mathrm{inv}}(x, x') \in H^1(E, G_x/Z(G))$$

to $\overline{\mathrm{inv}}((x, x'); (x, x')) \in H^1(E, (G_x)^2/Z(G))$.

(e) *Assume that each* G_{x_i} *is abelian,* $y_i \in X_i(E)$ *is an* E^{sep}*-conjugate of* x_i*, and* $y_i' \in X_i'(E)$ *is an* E^{sep}*-conjugate of* x_i'*. Then identifications* $G_{x_i} = G_{y_i} = G_{x_i'}$ *identify* $\overline{\mathrm{inv}}((y_1, y_1'); \ldots; (y_k, y_k'))$ *with the product of* $\overline{\mathrm{inv}}((x_1, x_1'); \ldots; (x_k, x_k'))$ *and the images of* $((\mathrm{inv}(y_i, x_i)_i) \in H^1(E, \prod_i G_{y_i})$ *and* $((\mathrm{inv}(x_i', y_i')_i) \in H^1(E, \prod_i G_{x_i'})$.

(f) *Let* $\varphi' : G' \to G''$ *be another inner twisting,* $\varphi'_{X'} : X' \to X''$ *the corresponding twisting of* X'*, and* $x_i'' \in X''(E)$ *a stable conjugate of* x_i *and* x_i' *for each* $i = 1, \ldots, k$*. If each* G_{x_i} *is abelian, then identifications* $G_{x_i} = G_{x_i'}$ *identify* $\overline{\mathrm{inv}}((x_1, x_1''); \ldots; (x_k, x_k''))\overline{\mathrm{inv}}((x_1, x_1'); \ldots; (x_k, x_k'))^{-1}$ *with* $\overline{\mathrm{inv}}((x_1', x_1''); \ldots; (x_k', x_k''))$.

Notation 1.1.7. (a) For each $x \in X(E)$, we denote by $[x] \subset X(E)$ the E^{sep}-conjugate class of x and by $a_x : G_x \hookrightarrow G$ the corresponding inclusion map.

(b) When G is reductive and connected, we denote by X^{sr} the set of $x \in X$ such that $G_x \subset G$ is a maximal torus and $\mathcal{G}_x \subset \mathcal{G}$ is a Lie algebra of G_x. We will call elements of X^{sr} *strongly regular*.

Remark 1.1.8. The condition on $\mathcal{G}_x$ holds automatically if the characteristic of E is zero.

From now on we will assume that G is reductive and connected, and T is a torus over E of the same absolute rank as G.

1.1.9. Embedding of tori (a) There exists an affine variety $\underline{\mathrm{Emb}}(T, G)$ over E equipped with an action of G^{ad} such that for every extension E'/E the set $\underline{\mathrm{Emb}}(T, G)(E')$ classifies embeddings $T_{E'} \hookrightarrow G_{E'}$, and G acts by conjugation.

To show the assertion, note that both G and T split over E^{sep}; therefore there exists an embedding $\iota : T_{E^{\mathrm{sep}}} \hookrightarrow G_{E^{\mathrm{sep}}}$. Consider the affine variety $\underline{\mathrm{Emb}}_\iota(T, G) := G \times_{\mathrm{Norm}_G(\iota(T))} \mathrm{Aut}(T)$ over E^{sep}. Then the map $[g, \alpha] \mapsto (\mathrm{Int}\, g) \circ \iota \circ \alpha$ defines a $G(E')$-equivariant bijection $\psi_\iota : \underline{\mathrm{Emb}}_\iota(T, G)(E') \xrightarrow{\sim} \underline{\mathrm{Emb}}(T, G)(E')$ for every extension E' of E^{sep} (compare (the proof of) Lemma 1.1.10 (a) below).

For every two embeddings $\iota_1, \iota_2 : T_{E^{\mathrm{sep}}} \hookrightarrow G_{E^{\mathrm{sep}}}$, there exists a unique isomorphism $\underline{\mathrm{Emb}}_{\iota_1}(T, G) \xrightarrow{\sim} \underline{\mathrm{Emb}}_{\iota_2}(T, G)$, compatible with the ψ_{ι_j}'s, and we define $\underline{\mathrm{Emb}}(T, G)$ to be the inverse limit of the $\underline{\mathrm{Emb}}_\iota(T, G)$'s. Finally, since $\underline{\mathrm{Emb}}_\iota(T, G)$ is a disjoint union of affine varieties, it descends to E.

(b) For each $a \in \underline{\mathrm{Emb}}(T, G)$, the stabilizer $G_a = a(T)$ is a maximal torus of G, which we will identify with T. It follows that $\underline{\mathrm{Emb}}(T, G)^{\mathrm{sr}} = \underline{\mathrm{Emb}}(T, G)$. Also if $\varphi : G \to G'$ is an inner twisting, then the corresponding inner twisting $\underline{\mathrm{Emb}}(T, G)'$ of $\underline{\mathrm{Emb}}(T, G)$ is naturally isomorphic to $\underline{\mathrm{Emb}}(T, G')$. In particular, we can speak about stable conjugacy of embeddings $a : T \hookrightarrow G$ and $a' : T \hookrightarrow G'$.

(c) If $x \in X^{\mathrm{sr}}(E)$ and $x' \in X'^{\mathrm{sr}}(E)$ are stably conjugate, then $a_x : G_x \hookrightarrow G$ and $a_{x'} : G_x \cong G'_{x'} \hookrightarrow G'$ are stably conjugate. Moreover, $\overline{\mathrm{inv}}(a_x, a_{x'}) = \overline{\mathrm{inv}}(x, x')$, (and $\mathrm{inv}(a_x, a_{x'}) = \mathrm{inv}(x, x')$ when $\varphi = \mathrm{Id}_G$.)

Conversely, for every stably conjugate embeddings $a : T \hookrightarrow G$ and $a' : T \hookrightarrow G'$ and each $t \in T(E)$, elements $a(t) \in G(E)$ and $a'(t) \in G'(E)$ are E^{sep}-conjugate.

Lemma 1.1.10. *(a) Every conjugacy class $[a]$ of embeddings $T_{\overline{E}} \hookrightarrow G_{\overline{E}}$ contains an E^{sep}-rational embedding a.*
(b) If G is quasi-split, then every Γ-invariant conjugacy class $[a]$ of embeddings $T_{\overline{E}} \hookrightarrow G_{\overline{E}}$ contains an E-rational embedding $a : T \hookrightarrow G$.
(c) Let $\varphi : G \to G'$ be an inner twisting such that G' is quasi-split. Then for every embedding $a : T \hookrightarrow G$ there exists an embedding $a' : T \hookrightarrow G'$ stably conjugate to a.

Proof. (a) Let $S \subset G$ be a maximal torus over E. Then there exists $a \in [a]$ such that $a(T_{\overline{E}}) = S_{\overline{E}}$. Since both T and S split over E^{sep}, we get that a is E^{sep}-rational.

(b) When E is perfect, the assertion was shown in [Ko3, Cor. 2.2]. In general, the proof is similar: Since there is a Γ-equivariant bijection between maximal tori of G and those of G^{sc}, we can assume that G is semisimple and simply connected. Next fix an E^{sep}-rational $a' \in [a]$, which exists by (a). Since every homogeneous space for a connected group over a finite field has a rational point, the assertion holds in this case. Thus we may assume that E is infinite; therefore there exists $t \in T(E)$ such that $a'(t) \in G(E^{\mathrm{sep}})$ is strongly regular.

The conjugacy class of $a'(t) \in G(E^{\mathrm{sep}})$ is E^{sep}-rational and Γ-invariant. Thus it is E-rational. By the theorem of Steinberg [St, Thm 1.7] (when E is perfect) and Borel and Springer [BS, 8.6] (in the general case) there exists $g \in G(E^{\mathrm{sep}})$ such that $ga'(t)g^{-1} \in G(E)$. Then $ga'(t)g^{-1} \in G^{\mathrm{sr}}(E)$, hence $a := ga'g^{-1} : T \hookrightarrow G$ is E-rational.

(c) Since φ is an inner twisting, the conjugacy class of $\varphi \circ a : T_{\overline{E}} \hookrightarrow G'_{\overline{E}}$ is Γ-invariant. Hence by (b), $[\varphi \circ a]$ contains an E-rational element a', which by definition is stably conjugate to a. □

Corollary 1.1.11. *Let $\varphi : G \to G'$ be an inner twisting, and $\varphi_X : X \to X'$ the corresponding twisting. If G' is quasi-split, then for every $x \in X^{\mathrm{sr}}(E)$ there exists $x' \in X'^{\mathrm{sr}}(E)$ stably conjugate to x'.*

Proof. Denote by $\iota : G/G_x \hookrightarrow X$ the canonical G-equivariant embedding $[g] \mapsto g(x)$, and by $\iota' : (G/G_x)' \hookrightarrow X'$ the twisted map. By Lemma 1.1.10 (c), there exists an embedding $a'_x : G_x \hookrightarrow G'$ stably conjugate to $a_x : G_x \hookrightarrow G$. Moreover, $(G/G_x)'$ is G'-equivariantly isomorphic to $G'/a'_x(G_x)$. It follows that the image of $[1] \in G'/a'_x(G_x)(E)$ under ι' is stably conjugate to x. □

Definition 1.1.12. By a *quasi-isogeny* we call a homomorphism $\pi : \widetilde{G} \to G$ such that $\pi(Z(\widetilde{G})) \subset Z(G)$ and the induced homomorphism $\pi^{\mathrm{ad}} : \widetilde{G}^{\mathrm{ad}} \to G^{\mathrm{ad}}$ is an isomorphism.

1.1.13. Quasi-isogenies Let $\pi : \widetilde{G} \to G$ be a quasi-isogeny.

(a) Each inner twisting $\varphi : G \to G'$ gives rise to an inner twisting $\widetilde{\varphi} : \widetilde{G} \to \widetilde{G}'$ such that $\mathrm{inv}(\widetilde{G}, \widetilde{G}') = \mathrm{inv}(G, G')$ (in $H^1(E, \widetilde{G}^{\mathrm{ad}}) = H^1(E, G^{\mathrm{ad}})$).

(b) X is equipped with an action of $\widetilde{G}$ trivial on $Z(\widetilde{G})$.

(c) There is a π- and Γ-equivariant bijection between embeddings of maximal tori $a : T \hookrightarrow G$ and the corresponding embeddings $\widetilde{a} : \widetilde{T} \hookrightarrow \widetilde{G}$. Indeed, given a, put

$\widetilde{T} = T \times_G \widetilde{G} := \{t \in T, \widetilde{g} \in \widetilde{G} \mid a(t) = \pi(\widetilde{g})\}$ and $\widetilde{a}(t, \widetilde{g}) := \widetilde{g}$. Conversely, given $\widetilde{a}$, define a be the embedding $\pi(\widetilde{a}(\widetilde{T})) \hookrightarrow G$. In particular, a_1 and a_2 are stably conjugate if and only if $\widetilde{a_1}$ and $\widetilde{a_2}$ are such.

We will call $\widetilde{a}$ *the lift of* a and $[\widetilde{a}]$ *the lift of* $[a]$.

(d) For each $i = 1, \dots, k$, let $a_i : T_i \hookrightarrow G$ and $a'_i : T_i \hookrightarrow G'$ be stable conjugate embeddings of maximal tori, and let $\widetilde{a}_i : \widetilde{T}_i \hookrightarrow \widetilde{G}$ and $\widetilde{a}'_i : \widetilde{T}_i \hookrightarrow \widetilde{G}'$ be the lifts of the a_i's and the a'_i's, respectively. Then $\overline{\mathrm{inv}}((a_1, a'_1); \dots; (a_k, a'_k)) \in H^1(E, (\prod_i T_i)/Z(G))$ is the image of

$$\overline{\mathrm{inv}}((\widetilde{a}_1, \widetilde{a}'_1); \dots; (\widetilde{a}_k, \widetilde{a}'_k)) \in H^1\left(E, \left(\prod_i \widetilde{T}_i\right) \Big/ Z(\widetilde{G})\right)$$

under the canonical map $(\prod_i \widetilde{T}_i)/Z(\widetilde{G})) \to (\prod_i T_i)/Z(G))$.

1.2. Preliminaries on dual groups

In this subsection, we will recall basic properties of Langlands dual groups. More specifically, we will study properties of triples $(G, H, [\eta])$ from 1.2.3. Constructions from this subsection will be later used in the case when H is an endoscopic group for G.

Notation 1.2.1. For each connected reductive group G over a field E, we denote by $\widehat{G}$ (or $\{G\}^\frown$) the complex connected Langlands dual group, and by $\rho_G : \Gamma \to \mathrm{Out}(\widehat{G})$ the corresponding Galois action.

1.2.2. Basic properties of dual groups (a) The map $G \mapsto (\widehat{G}, \rho_G)$ defines a surjection from the set of isomorphism classes of connected reductive groups over E to that of pairs consisting of a connected complex reductive group $\widehat{G}$, and a continuous homomorphism $\rho : \Gamma \to \mathrm{Out}(\widehat{G})$. Moreover, $(\widehat{G_1}, \rho_{G_1}) \cong (\widehat{G_2}, \rho_{G_2})$ if and only if G_2 is an inner twist of G_1. In particular, each pair $(\widehat{G}, \rho_G)$ comes from a unique quasi-split group G over E.

(b) Let T be a torus over E of the same absolute rank as G. Then there exists a canonical (hence Γ-equivariant) bijection $[a] \mapsto \widehat{[a]}$ between conjugacy classes of embeddings $T_{\overline{E}} \hookrightarrow G_{\overline{E}}$ and conjugacy classes of embeddings $\widehat{T} \hookrightarrow \widehat{G}$. In particular, $[a]$ is Γ-invariant if and only if $\widehat{[a]}$ is such.

(c) For each embeddings of maximal tori $T_{\overline{E}} \hookrightarrow G_{\overline{E}}$ and $\widehat{T} \hookrightarrow \widehat{G}$ related as in (b), the set of roots (resp. coroots) of $(G_{\overline{E}}, T_{\overline{E}})$ is canonically identified with the set of coroots (resp. roots) of $(\widehat{G}, \widehat{T})$. In particular, the Weyl group $W(\widehat{G})$ is canonically identified with $W(G)$.

(d) Each quasi-isogeny $\pi : G_1 \to G_2$ gives rise to a conjugacy class $[\widehat{\pi}]$ of quasi-isogenies $\widehat{G_2} \to \widehat{G_1}$. In particular, it induces a homomorphism $Z_{[\widehat{\pi}]} : Z(\widehat{G_2}) \to Z(\widehat{G_1})$.

1.2.3. Triple For the rest of this subsection, we fix a triple $(G, H, [\eta])$, consisting of a connected reductive group G over a field E, a quasi-split reductive group H over E of the same absolute rank as G, and a Γ-invariant $\widehat{G}$-conjugacy class $[\eta]$ of embeddings $\widehat{H} \hookrightarrow \widehat{G}$.

1.2.4. Properties of the triple $(G, H, [\eta])$ (a) Every stable conjugacy class $[b]$ of embeddings of maximal tori $T \hookrightarrow H$ defines a Γ-invariant conjugacy class $\widehat{[b]}$ of embeddings $\widehat{T} \hookrightarrow \widehat{H}$ (by 1.2.2), hence a Γ-invariant conjugacy class $\widehat{[b]}_G := [\eta] \circ \widehat{[b]}$ of embeddings $\widehat{T} \hookrightarrow \widehat{G}$, and thus a Γ-invariant conjugacy class $[b]_G$ of embeddings $T_{\overline{E}} \hookrightarrow G_{\overline{E}}$.

(b) There are exist canonical (Γ-equivariant) embeddings $Z(\widehat{G}) \hookrightarrow Z(\widehat{H})$, $Z(G) \hookrightarrow Z(H)$ and $W(H) \hookrightarrow W(G)$.

To see it, fix a maximal torus $\widehat{T} \subset \widehat{H}$ and an embedding $\eta : \widehat{H} \hookrightarrow \widehat{G}$ from $[\eta]$. Then $\widehat{T}$ is a maximal torus of $\widehat{G}$, hence the set of roots (therefore also of coroots) of $(\widehat{H}, \widehat{T})$ is naturally a subset of that of $(\widehat{G}, \widehat{T})$. It follows that $W(H) = W(\widehat{H})$ is naturally a subgroup of $W(\widehat{G}) = W(G)$. Also by 1.2.2 (c), the set of roots of (H, T) is naturally a subset of that of (G, T). Since $Z(\widehat{G}) \subset \widehat{T}$ (resp. $Z(G) \subset T$) is the intersection of kernels of all roots of $(\widehat{G}, \widehat{T})$ (resp. of (G, T)), and similarly for $Z(\widehat{H})$ (resp. $Z(H)$), we get an embedding $Z(\widehat{G}) \hookrightarrow Z(\widehat{H})$ (resp. $Z(G) \hookrightarrow Z(H)$).

(c) $[\eta]$ naturally gives rise to a conjugacy class $[\overline{\eta}]$ of embeddings $\widehat{H/Z(G)} \hookrightarrow \widehat{G^{\mathrm{ad}}}$. Namely, each $\eta : \widehat{H} \hookrightarrow \widehat{G}$ from $[\eta]$ has a unique lift $\overline{\eta} : \widehat{H/Z(G)} \hookrightarrow \widehat{G^{\mathrm{ad}}}$, and we denote by $[\overline{\eta}]$ the corresponding conjugacy class.

By (b), $[\eta]$ thus induces a homomorphism $Z_{[\overline{\eta}]} : Z(\widehat{G^{\mathrm{ad}}})^{\Gamma} \to \pi_0(Z(\widehat{H/Z(G)})^{\Gamma})$.

1.2.5. Example Any embedding $a : T \hookrightarrow G$ of a maximal torus gives rise to a triple 1.2.3 with $H = T$ and $[\eta] = [\widehat{a}]$. In particular, it gives rise to a Γ-equivariant embedding $Z(\widehat{G}) \hookrightarrow \widehat{T}$, hence to a homomorphism $\pi_0(Z(\widehat{G})^{\Gamma}) \to \pi_0(\widehat{T}^{\Gamma})$. We will denote both of these maps by $Z_{[\widehat{a}]}$.

1.2.6. Construction of an exact sequence For each maximal torus $T \subset H$, we consider the exact sequence

$$0 \to \widehat{T} \xrightarrow{\mu_T} \widehat{T^2/Z(G)} \xrightarrow{\nu_T} \widehat{T/Z(G)} \to 0, \tag{1.2.1}$$

dual to the exact sequence $0 \to T/Z(G) \xrightarrow{\widehat{\nu}_T} T^2/Z(G) \xrightarrow{\widehat{\mu}_T} T \to 0$, where $\widehat{\nu}_T$ is the diagonal morphism and $\widehat{\mu}_T([t_1, t_2]) := t_1/t_2$.

Since a center of a reductive group equals to the intersection of kernels of all roots, we conclude that $\mu_T(Z(\widehat{H})) \subset Z(\widehat{H^2/Z(G)})$, $\nu_T(Z(\widehat{H^2/Z(G)})) \subset Z(\widehat{H/Z(G)})$, and the induced sequence

$$0 \to Z(\widehat{H}) \xrightarrow{\mu_H} Z(\widehat{H^2/Z(G)}) \xrightarrow{\nu_H} Z(\widehat{H/Z(G)}) \tag{1.2.2}$$

is Γ-equivariant and exact. Furthermore, since over $\overline{E}$ all maximal tori of H are conjugate, the sequence (1.2.2) is independent of the choice of T.

Observe that the composition of μ_H with the projection $Z(\widehat{H^2/Z(G)}) \to Z(\widehat{H^2}) = Z(\widehat{H})^2$ is the map $z \mapsto (z, z^{-1})$.

Lemma 1.2.7. *For every $i, j \in \{1, 2, 3\}$, let $\mu_{i,j} : Z(\widehat{H}) \hookrightarrow Z(\widehat{H^3/Z(G)})$ be the composition of μ_H and the embedding $Z(\widehat{H^2/Z(G)}) \hookrightarrow Z(\widehat{H^3/Z(G)})$ corresponding to the projection $H^3/Z(G) \to H^2/Z(G)$ to the i-th and the j-th factors. Then $\mu_{1,3} = \mu_{1,2}\mu_{2,3}$.*

Proof. Consider the projection $\lambda_{i,j} : T^3/Z(G) \to T$ given by the rule $\lambda_{i,j}([t_1, t_2, t_3]) = t_j/t_i$. Since each $\mu_{i,j}$ is the restriction of $\lambda_{i,j} : \widehat{T} \hookrightarrow \widehat{T^3/Z(G)})$ to $Z(\widehat{H})$, the equality $\mu_{1,3} = \mu_{1,2}\mu_{2,3}$ follows from the equality $\lambda_{1,3} = \lambda_{1,2}\lambda_{2,3}$. □

1.2.8. Construction of two homomorphisms (a) Consider a pair $[b_i]$ of stable conjugacy classes of embeddings of maximal tori $T_i \hookrightarrow H$. Then the $[b_i]$'s give rise to a stable conjugacy class $[b_1, b_2]$ of embeddings $(T_1 \times T_2)/Z(G) \hookrightarrow H^2/Z(G)$, hence to a Γ-invariant embedding

$$\iota([b_1], [b_2]) := Z_{\widehat{[b_1, b_2]}} \circ \mu_H : Z(\widehat{H}) \hookrightarrow \{(T_1 \times T_2)/Z(G)\}\widehat{\ }. \tag{1.2.3}$$

In its turn, $\iota([b_1], [b_2])$ induces a homomorphism

$$\kappa([b_1], [b_2]) : \pi_0(Z(\widehat{H})^\Gamma) \to \pi_0((\{(T_1 \times T_2)/Z(G)\}\widehat{\ })^\Gamma). \tag{1.2.4}$$

(b) Assume that in the notation of (a), we have $T_1 = T_2 = T$ and $[b_1]_G = [b_2]_G$. Then $\iota([b_1]_G, [b_2]_G) = \iota([b_1], [b_2])|_{Z(\widehat{G})}$ factors through $\mu_T : \widehat{T} \hookrightarrow \widehat{T^2/Z(G)}$, hence the image of $\iota([b_1]_G, [b_2]_G)$ lies in $\operatorname{Ker} \nu_T$. Thus the composition $\nu_T \circ \iota([b_1], [b_2]) : Z(\widehat{H}) \to \widehat{T/Z(G)}$ factors through $Z(\widehat{H})/Z(\widehat{G})$ and induces a homomorphism

$$\kappa\left(\frac{[b_1]}{[b_2]}\right) : \pi_0(Z(\widehat{H})^\Gamma/Z(\widehat{G})^\Gamma) \to \pi_0(\widehat{T/Z(G)}^\Gamma). \tag{1.2.5}$$

(c) For every $i = 1, 2$, denote by $[\overline{b}_i]$ the stable conjugacy class of embeddings $T/Z(G) \hookrightarrow H/Z(G)$ induced by $[b_i]$. Then the composition of the projection $Z(\widehat{H/Z(G)}) \to Z(\widehat{H})$ and $\nu_T \circ \iota([b_1], [b_2]) : Z(\widehat{H}) \to \widehat{T/Z(G)}$ equals the quotient $Z_{\widehat{[\overline{b}_1]}}/Z_{\widehat{[\overline{b}_2]}} : Z(\widehat{H/Z(G)}) \to \widehat{T/Z(G)}$.

This gives the following description of the map $\kappa\left(\frac{[b_1]}{[b_2]}\right)$. For each representative $\widetilde{s} \in Z(\widehat{H/Z(G)})$ of $\overline{s} \in \pi_0(Z(\widehat{H})^\Gamma/Z(\widehat{G})^\Gamma)$, the quotient $Z_{\widehat{[\overline{b}_1]}}(\widetilde{s})/Z_{\widehat{[\overline{b}_2]}}(\widetilde{s}) \in \widehat{T/Z(G)}$ is Γ-invariant, and $\kappa\left(\frac{[b_1]}{[b_2]}\right)(\overline{s})$ equals the class of $Z_{\widehat{[\overline{b}_1]}}(\widetilde{s})/Z_{\widehat{[\overline{b}_2]}}(\widetilde{s})$. (Note that $\widetilde{s} \in Z(\widehat{H/Z(G)})$ is not always Γ-invariant).

Lemma 1.2.9. *(a) Let $[b_1]$ and $[b_2]$ be as in 1.2.8, and let $\check{s} \in \pi_0(Z(\widehat{H})^\Gamma)$ be a representative of $\overline{s} \in \pi_0(Z(\widehat{H})^\Gamma/Z(\widehat{G})^\Gamma)$. Then the quotient $Z_{\widehat{[b_1]}}(\check{s})/Z_{\widehat{[b_2]}}(\check{s}) \in \pi_0(\widehat{T}^\Gamma)$ equals the image of $\kappa\left(\frac{[b_1]}{[b_2]}\right)(\overline{s})$.*

(b) Let $[b_1], [b_2]$ and $[b_3]$ be stable conjugacy classes of embeddings of maximal tori $T \hookrightarrow H$ such that $[b_1]_G = [b_2]_G = [b_3]_G$. Then we have $\kappa\left(\frac{[b_1]}{[b_3]}\right) = \kappa\left(\frac{[b_1]}{[b_2]}\right)\kappa\left(\frac{[b_2]}{[b_3]}\right)$.

Proof. Both assertions follow from the description of $\kappa\left(\frac{[b_i]}{[b_j]}\right)$, given in 1.2.8 (c). □

1.3. Endoscopic triples: basic properties

Let G be a connected reductive group over a field E. In this subsection we give basic constructions and properties of endoscopic triples (compare [Ko1, §7]).

Definition 1.3.1. (a) A triple $\mathcal{E} = (H, [\eta], \overline{s})$ consisting of

- a quasi-split reductive group H over E of the same absolute rank as G;
- a Γ-invariant $\widehat{G}$-conjugacy class $[\eta]$ of embeddings $\widehat{H} \hookrightarrow \widehat{G}$;
- an element $\overline{s} \in \pi_0(Z(\widehat{H})^\Gamma/Z(\widehat{G})^\Gamma)$, where the Γ-equivariant embedding $Z(\widehat{G}) \hookrightarrow Z(\widehat{H})$ is induced by $[\eta]$ (see 1.2.4 (b)),

is called an *endoscopic triple* for G if for each generic representative $s \in Z(\widehat{H})^\Gamma/Z(\widehat{G})^\Gamma$ of $\overline{s}$ we have $\eta(\widehat{H}) = \widehat{G}_{\eta(s)}$ for all $\eta \in [\eta]$. Such a representative s will be called *$\mathcal{E}$-compatible*.

(b) An endoscopic triple $\mathcal{E}$ for G is called *elliptic* if the group $Z(\widehat{H})^\Gamma/Z(\widehat{G})^\Gamma$ is finite.

(c) An *isomorphism* from an endoscopic triple $\mathcal{E}_1 = (H_1, [\eta_1], \overline{s}_1)$ to $\mathcal{E}_2 = (H_2, [\eta_2], \overline{s}_2)$ is an isomorphism $\alpha : H_1 \overset{\sim}{\to} H_2$ such that the corresponding element $\widehat{[\alpha]}$ of $\mathrm{Isom}(\widehat{H_2}, \widehat{H_1})/\mathrm{Int}(\widehat{H_2})$ satisfies $[\eta_1] \circ \widehat{[\alpha]} = [\eta_2]$ and $\overline{s}_1 = \widehat{[\alpha]}(\overline{s}_2)$.

(d) We denote by $\mathrm{Aut}(\mathcal{E})$ the group of automorphisms of $\mathcal{E}$ and by $\Lambda(\mathcal{E})$ the quotient $\mathrm{Aut}(\mathcal{E})/H^{\mathrm{ad}}(E)$.

(e) An endoscopic triple $\mathcal{E}$ is called *split* (resp. *unramified*) if H is a split (resp. unramified) group over E.

Remark 1.3.2. (a) When E is a local non-archimedean field, our notion of an endoscopic triple is equivalent to the standard one ([Ko1, 7.4]). Indeed, let (H, η, s) be an endoscopic triple in the sense of [Ko1, 7.4]. Since $Z(\widehat{H})^\Gamma/Z(\widehat{G})^\Gamma$ is a subgroup of finite index in $[Z(\widehat{H})/Z(\widehat{G})]^\Gamma$, condition [Ko1, 7.4.3] asserts that the image of $s \in Z(\widehat{H})$ in $Z(\widehat{H})/Z(\widehat{G})$ belongs to $Z(\widehat{H})^\Gamma/Z(\widehat{G})^\Gamma$, thus defining an element $\overline{s} \in \pi_0(Z(\widehat{H})^\Gamma/Z(\widehat{G})^\Gamma)$. Moreover, the map $(H, \eta, s) \mapsto (H, [\eta], \overline{s})$ defines an equivalence of categories between endoscopic triples in the sense of [Ko1, 7.4] and those in our sense.

If (H, η_1, s_1) and (H, η_2, s_2) are two endoscopic triples in the sense of [Ko1, 7.4] such that $[\eta_1] = [\eta_2]$ and $\overline{s_1} = \overline{s_2}$, then they are canonically isomorphic. Therefore we have chosen to deviate from the Kottwitz notation and to identify these objects.

(b) When E is a number field, then our notion of an endoscopic triple slightly differs from the standard one ([Ko1, 7.4]). However in this paper we will only consider the case $G = G^{\mathrm{sc}}$ (see Appendix B). In this case, $Z(\widehat{G}) = 1$, hence both notions coincide.

Remark 1.3.3. Since an inner twisting $\varphi : G \to G'$ identifies the dual groups of G and G', every endoscopic triple for G defines the one for G'.

Remark 1.3.4. In the notation of Definition 1.3.1 (c), the map $[\alpha] \mapsto \widehat{[\alpha]}$ identifies $\mathrm{Isom}(\mathcal{E}_1, \mathcal{E}_2)/H_1^{\mathrm{ad}}(E)$ with the set of all Γ-invariant $\widehat{[\alpha]} \in \mathrm{Isom}(\widehat{H_2}, \widehat{H_1})/\mathrm{Int}(\widehat{H_2})$ such that $[\eta_1] \circ \widehat{[\alpha]} = [\eta_2]$ and $\overline{s}_1 = \widehat{[\alpha]}(\overline{s}_2)$.

In particular, for each endoscopic triple $\mathcal{E} = (H, [\eta], \overline{s})$, the map $[\alpha] \mapsto \widehat{[\alpha]}$ identifies $\Lambda(\mathcal{E})$ with the set of $g \in \mathrm{Out}(\widehat{H})^{\Gamma}$ such that $g(\overline{s}) = \overline{s}$ and $[\eta] \circ g = [\eta]$.

1.3.5. Homomorphisms, corresponding to endoscopic triples To every endoscopic triple $\mathcal{E} = (H, [\eta], \overline{s})$ for G, we associate a homomorphism $\pi_{\mathcal{E}} : \Lambda(\mathcal{E}) \to \pi_0(Z(\widehat{H/Z(G)})^{\Gamma})$. Moreover, the image of $\pi_{\mathcal{E}}$ lies in the image of the canonical map $Z_{[\overline{\eta}]} : Z(\widehat{G^{\mathrm{ad}}})^{\Gamma} \to \pi_0(Z(\widehat{H/Z(G)})^{\Gamma})$ from 1.2.4 (c).

To define $\pi_{\mathcal{E}}$, fix $\eta \in [\eta]$, and identify $\widehat{H}$ with its image $\eta(\widehat{H}) \subset \widehat{G}$. Then η lifts to an embedding $\widehat{H/Z(G)} \hookrightarrow \widehat{G^{\mathrm{ad}}} = \widehat{G}^{\mathrm{sc}}$. For every $\alpha \in \Lambda(\mathcal{E})$, choose $g \in \mathrm{Norm}_{\widehat{G}^{\mathrm{ad}}}(\widehat{H})$ inducing $\widehat{\alpha} \in \mathrm{Out}(\widehat{H})$. Then g normalizes $\widehat{H/Z(G)} \subset \widehat{G}^{\mathrm{sc}}$, hence it induces an element $\widetilde{\widetilde{\alpha}} \in \mathrm{Out}(\widehat{H/Z(G)})$. Moreover, $\widetilde{\widetilde{\alpha}}$ is Γ-invariant and independent of the choices of η and g.

Furthermore, $\widetilde{\widetilde{\alpha}}$ is trivial on the kernel $\mathrm{Ker}[\widehat{H/Z(G)} \to \widehat{H}/Z(\widehat{G})]$. Therefore for every $\widetilde{t} \in Z(\widehat{H/Z(G)})$, the quotient $\widetilde{\widetilde{\alpha}}(\widetilde{t})/\widetilde{t} \in Z(\widehat{H/Z(G)})$ depends only on the image $t \in Z(\widehat{H})/Z(\widehat{G})$ of $\widetilde{t}$. Hence the map $t \mapsto \widetilde{\widetilde{\alpha}}(\widetilde{t})/\widetilde{t}$ defines a Γ-equivariant homomorphism $Z(\widehat{H})/Z(\widehat{G}) \to Z(\widehat{H/Z(G)})$, which in turn induces a homomorphism $\widetilde{\alpha} : \pi_0(Z(\widehat{H})^{\Gamma}/Z(\widehat{G})^{\Gamma}) \to \pi_0(Z(\widehat{H/Z(G)})^{\Gamma})$.

We define $\pi_{\mathcal{E}}$ by the formula $\pi_{\mathcal{E}}(\alpha) := \widetilde{\alpha}(\overline{s}) \in \pi_0(Z(\widehat{H/Z(G)})^{\Gamma})$. Since $\widehat{\alpha}(\overline{s}) = \overline{s}$, we see that $\pi_{\mathcal{E}}$ is a homomorphism, and the image of $\pi_{\mathcal{E}}$ lies in the kernel $\mathrm{Ker}[\pi_0(Z(\widehat{H/Z(G)})^{\Gamma}) \to \pi_0(Z(\widehat{H})^{\Gamma}/Z(\widehat{G})^{\Gamma})]$. As the last group coincides with the image of $Z_{[\overline{\eta}]} : Z(\widehat{G^{\mathrm{ad}}})^{\Gamma} \to \pi_0(Z(\widehat{H/Z(G)})^{\Gamma})$ (see [Ko1, Cor. 2.3]), we get the assertion.

Lemma 1.3.6. *For every embedding of maximal torus $b : T \hookrightarrow H$ the corresponding map $Z_{\widehat{[b]}} : \pi_0(Z(\widehat{H/Z(G)})^{\Gamma})] \to \pi_0(\widehat{T/Z(G)}^{\Gamma})$ maps $\pi_{\mathcal{E}}(\alpha)$ to $\kappa\left(\frac{\alpha([b])}{[b]}\right)(\overline{s})$.*

Proof. The assertion follows immediately from 1.2.8 (c). □

Notation 1.3.7. Let $\mathcal{E} = (H, [\eta], \overline{s})$ be an endoscopic triple for G.

(a) Denote by $\Pi_{\mathcal{E}}$ the map sending a stable conjugacy class $[b]$ of embeddings of maximal tori $T \hookrightarrow H$ to a pair consisting of an E-rational conjugacy class $[b]_G$ (see 1.2.4 (c)) of embeddings $T_{\overline{E}} \hookrightarrow G_{\overline{E}}$ and an element $\overline{\kappa}_{[b]} := Z_{\widehat{[b]}}(\overline{s}) \in \pi_0(\widehat{T}^{\Gamma}/Z(\widehat{G})^{\Gamma})$.

(b) Denote by $Z(\mathcal{E}) \subset Z(\widehat{G^{\mathrm{ad}}})^{\Gamma}$ the preimage of $\mathrm{Im}\,\pi_{\mathcal{E}} \subset \pi_0(Z(\widehat{H/Z(G)})^{\Gamma})$ under $Z_{[\overline{\eta}]}$.

(c) For every pair $([a], \overline{\kappa}) \in \mathrm{Im}\,\Pi_{\mathcal{E}}$, we denote by $S_{([a],\overline{\kappa})} \subset \pi_0(\widehat{T/Z(G)}^{\Gamma})$ the subgroup consisting of elements $\kappa\left(\frac{[b_1]}{[b_2]}\right)(\overline{s})$, where b_1 and b_2 run through the set of embeddings $b : T \hookrightarrow H$ such that $\Pi_{\mathcal{E}}([b]) = ([a], \overline{\kappa})$. We denote by $Z(\mathcal{E}, [a], \overline{\kappa}) \subset Z(\widehat{G^{\mathrm{ad}}})^{\Gamma}$ the preimage of $S_{([a],\overline{\kappa})}$ under the map $Z_{[\widehat{a}]} : Z(\widehat{G^{\mathrm{ad}}})^{\Gamma} \to \pi_0(\widehat{T/Z(G)}^{\Gamma})$.

(d) We call embeddings of maximal tori $a : T \hookrightarrow G$ and $b : T \hookrightarrow H$ *compatible* if $a \in [b]_G$.

1.3.8. Endoscopic triples, corresponding to pairs (a) Following Langlands ([La1, II, 4]), we associate an endoscopic triple $\mathcal{E} = \mathcal{E}_{([a],\kappa)}$ for G to each pair $([a], \kappa)$, consisting of a stable conjugacy class of an embedding $a : T \hookrightarrow G$ of maximal torus and an element $\kappa \in \widehat{T}^{\Gamma}/Z(\widehat{G})^{\Gamma}$. Moreover, $\mathcal{E}_{([a],\kappa)}$ is elliptic if $a(T)$ is an elliptic torus of G.

For the convenience of the reader, we will recall this important construction. Choose an element $\widehat{a} : \widehat{T} \hookrightarrow \widehat{G}$ of $\widehat{[a]}$, and identify $\widehat{T}$ with $\widehat{a}(\widehat{T})$. Choose a representative $\widetilde{\kappa} \in \widehat{T}^{\Gamma}$ of κ, put $s := \widehat{a}(\widetilde{\kappa}) \in \widehat{G}/Z(\widehat{G})$, $\widehat{H} := \widehat{G}_s^0$, and let $[\eta]$ be the conjugacy class of the inclusion $\widehat{H} \hookrightarrow \widehat{G}$. Then the image of $\rho_T : \Gamma \to \mathrm{Out}(\widehat{T})$ lies in $\mathrm{Norm}_{\mathrm{Aut}(\widehat{G})}(\widehat{T})/\widehat{T}$, and $\rho_G : \Gamma \to \mathrm{Out}(\widehat{G})$ is the composition of $\rho_T : \Gamma \to \mathrm{Norm}_{\mathrm{Aut}(\widehat{G})}(\widehat{T})/\widehat{T}$ with the canonical homomorphism $\mathrm{Norm}_{\mathrm{Aut}(\widehat{G})}(\widehat{T})/\widehat{T} \hookrightarrow \mathrm{Aut}(\widehat{G}) \to \mathrm{Out}(\widehat{G})$.

As $\widetilde{\kappa}$ belongs to $\widehat{T}^{\Gamma}$, the image of ρ_T lies in $\mathrm{Norm}_{\mathrm{Aut}(\widehat{G})}(\widehat{T})_s/\widehat{T}$, where $\mathrm{Norm}_{\mathrm{Aut}(\widehat{G})}(\widehat{T})_s$ is the stabilizer of s in $\mathrm{Norm}_{\mathrm{Aut}(\widehat{G})}(\widehat{T})$. Since $\mathrm{Norm}_{\mathrm{Aut}(\widehat{G})}(\widehat{T})_s \subset \mathrm{Norm}_{\mathrm{Aut}(\widehat{G})}(\widehat{H})$, we can compose $\rho_T : \Gamma \to \mathrm{Norm}_{\mathrm{Aut}(\widehat{G})}(\widehat{T})_s/\widehat{T}$ with the canonical homomorphism $\mathrm{Norm}_{\mathrm{Aut}(\widehat{G})}(\widehat{T})_s/\widehat{T} \to \mathrm{Norm}_{\mathrm{Aut}(\widehat{G})}(\widehat{H})/\widehat{H} \subset \mathrm{Out}(\widehat{H})$. We denote the composition $\Gamma \to \mathrm{Out}(\widehat{H})$ by ρ, and let H be the unique quasi-split group over E corresponding to the pair $(\widehat{H}, \rho)$ (see 1.2.2 (a)).

By construction, $[\eta]$ is Γ-equivariant, and s belongs to $Z(\widehat{H})^{\Gamma}$. Denote by $\overline{s} \in \pi_0(Z(\widehat{H})^{\Gamma}/Z(\widehat{G})^{\Gamma})$ the class of s. Then $(H, [\eta], \overline{s})$ is an endoscopic triple for G, independent of the choices of $\widehat{a}$ and $\widetilde{\kappa}$. We denote this endoscopic triple by $\mathcal{E}_{([a],\kappa)}$.

(b) In the notation of (a), we have $([a], \overline{\kappa}) \in \mathrm{Im}\, \Pi_{\mathcal{E}}$.

To find $[b] \in \Pi_{\mathcal{E}}^{-1}([a], \overline{\kappa})$, we choose $\widehat{a}$ and $\widetilde{\kappa}$ as in (a) and identify $\widehat{T}$ with $\widehat{a}(\widehat{T})$. By construction, there exists $\eta \in [\eta]$ such that $\widehat{T} \subset \eta(\widehat{H})$ and $\eta(s) = \widetilde{\kappa}$. Let $\widehat{[b]}$ be the conjugacy class of the inclusion $\widehat{T} \hookrightarrow \eta(\widehat{H}) \cong \widehat{H}$. Then $\widehat{[b]}$ is Γ-equivariant; hence it gives rise to a stable conjugacy class $[b]$ of embeddings $T \hookrightarrow H$ (see Lemma 1.1.10 (b)), which belongs to $\Pi_{\mathcal{E}}^{-1}([a], \overline{\kappa})$.

Lemma 1.3.9. *Let $\mathcal{E} = (H, [\eta], \overline{s})$ be an endoscopic triple for G, $b : T \hookrightarrow H$ and $a : T \hookrightarrow G$ embeddings of maximal tori, $\overline{\kappa}$ an element of $\pi_0(\widehat{T}^{\Gamma}/Z(\widehat{G})^{\Gamma})$ such that $\Pi_{\mathcal{E}}([b]) = ([a], \overline{\kappa})$.*

(a) There exists a representative $\kappa \in \widehat{T}^{\Gamma}/Z(\widehat{G})^{\Gamma}$ of $\overline{\kappa}$ such that $\mathcal{E} \cong \mathcal{E}_{([a],\kappa)}$.

(b) The group $Z(\mathcal{E})$ is contained in $Z(\mathcal{E}, [a], \overline{\kappa})$. Moreover, if $b(T) \subset H$ is elliptic, then we have $Z(\mathcal{E}, [a], \overline{\kappa}) = Z(\mathcal{E})$. In particular, $Z(\mathcal{E}, [a], \overline{\kappa}) = Z(\mathcal{E})$ if $a(T) \subset G$ is elliptic.

Proof. (a) Let $s \in Z(\widehat{H})^\Gamma/Z(\widehat{G})^\Gamma$ be an $\mathcal{E}$-compatible representative of $\overline{s}$, and let $\kappa \in \widehat{T}^\Gamma/Z(\widehat{G})^\Gamma$ be the image of s under the embedding $Z(\widehat{H})^\Gamma/Z(\widehat{G})^\Gamma \hookrightarrow \widehat{T}^\Gamma/Z(\widehat{G})^\Gamma$, induced by $[\widehat{b}]$. Then κ is a representative of $\overline{\kappa}$, and we claim that $\mathcal{E}_{([a],\kappa)} \cong \mathcal{E}$. To show it, choose an embedding $\widehat{b} : \widehat{T} \hookrightarrow \widehat{H}$ such that $\widehat{b} \in [\widehat{b}]$. Then for each $\eta \in [\eta]$, the composition $\widehat{a} := \eta \circ \widehat{b} : \widehat{T} \hookrightarrow \widehat{G}$ belongs to $\widehat{[a]}$ and satisfies $\eta(\widehat{H}) = \widehat{G}^0_{\widehat{a}(\kappa)}$. Finally, since the conjugacy class of the embedding $\widehat{b} : \widehat{T} \hookrightarrow \widehat{H}$ is Γ-invariant, the homomorphism $\rho_H : \Gamma \to \mathrm{Out}(\widehat{H})$ is induced by $\rho_T : \Gamma \to \mathrm{Out}(\widehat{T})$, as in 1.3.8 (a).

(b) Since for each $\alpha \in \Lambda(\mathcal{E})$, the stable conjugacy class $\alpha([b])$ belongs to $\Pi_{\mathcal{E}}^{-1}([a],\overline{\kappa})$, the inclusion $Z(\mathcal{E}) \subset Z(\mathcal{E},[a],\overline{\kappa})$ follows from Lemma 1.3.6.

Assume now that $b(T) \subset H$ is elliptic. Then for the second assertion, it will suffice to check that for each $[b'] \in \Pi_{\mathcal{E}}^{-1}([a],\overline{\kappa})$, there exists $\alpha \in \Lambda(\mathcal{E})$ such that $[b'] = \alpha([b])$.

Embed $\widehat{H}$ into $\widehat{G}$ by means of an element of $[\eta]$, choose embeddings $\widehat{b} : \widehat{T} \hookrightarrow \widehat{H} \subset \widehat{G}$ and $\widehat{b'} : \widehat{T} \hookrightarrow \widehat{H} \subset \widehat{G}$ from $\widehat{[b]}$ and $\widehat{[b']}$, respectively, and identify $\widehat{T}$ with its image $\widehat{b}(\widehat{T}) \subset \widehat{H}$. Replacing $\widehat{b'}$ by its $\widehat{H}$-conjugate, we may assume that $\widehat{b'}(\widehat{T}) = \widehat{T}$.

Choose a representative $s \in Z(\widehat{H})^\Gamma$ of $\overline{s}$ such that $\widehat{H} = \widehat{G}^0_s$. Since $\Pi_{\mathcal{E}}([b']) = \Pi_{\mathcal{E}}([b])$, there exists an element $g \in \widehat{G}$ such that $\widehat{b'} = g\widehat{b}g^{-1}$ and $(g^{-1}sg)s^{-1}$ belongs to $(\widehat{T}^\Gamma)^0 Z(\widehat{G})^\Gamma = (Z(\widehat{H})^\Gamma)^0 Z(\widehat{G})^\Gamma \subset Z(\widehat{H})$. Therefore $g^{-1}sg \in Z(\widehat{H})$, hence $g\widehat{H}g^{-1} = \widehat{H}$. Let $\overline{g} \in \mathrm{Out}(\widehat{H})$ be the class of g. Since the conjugacy classes of $\widehat{b} : \widehat{T} \hookrightarrow \widehat{H}$ and $\widehat{b'} : \widehat{T} \hookrightarrow \widehat{H}$ are Γ-invariant, we get from the equality $\widehat{b'} = g\widehat{b}g^{-1}$ that $\overline{g}$ is Γ-invariant. In other words, there exists $\alpha \in \Lambda(\mathcal{E})$ such that $\widehat{\alpha} \in \mathrm{Out}(\widehat{H})$ equals $\overline{g}$. Then $[b'] = \alpha([b])$, as claimed. □

Lemma 1.3.10. *Let $\mathcal{E} = (H,[\eta],\overline{s})$ be an endoscopic triple for G, and let $\pi : \widetilde{G} \to G$ be a quasi-isogeny.*

(a) There exists a unique pair consisting of an endoscopic triple $\widetilde{\mathcal{E}} = (\widetilde{H},[\widetilde{\eta}],\overline{\widetilde{s}})$ for $\widetilde{G}$ and a stable conjugacy class of quasi-isogenies $\pi' : \widetilde{H} \to H$ such that $[\widetilde{\eta}] \circ [\widehat{\pi}'] = [\widehat{\pi}] \circ [\eta]$ and $Z_{[\widehat{\pi}']}$ maps $\overline{s}$ to $\overline{\widetilde{s}}$.

Furthermore, the endoscopic triple $\widetilde{\mathcal{E}}$ satisfies the following properties.

(b) For an embedding of maximal torus $a : T \hookrightarrow G$ and an element $\kappa \in \widehat{T}^\Gamma/Z(\widehat{G})^\Gamma$, denote by $\widetilde{a} : \widetilde{T} \hookrightarrow \widetilde{G}$ the lift of a, and by $\widetilde{\kappa} \in \widehat{\widetilde{T}}^\Gamma/Z(\widehat{\widetilde{G}})^\Gamma$ the image of κ. If $\mathcal{E} \cong \mathcal{E}_{([a],\kappa)}$, then $\widetilde{\mathcal{E}} \cong \mathcal{E}_{([\widetilde{a}],\widetilde{\kappa})}$.

(c) We have $\Lambda(\mathcal{E}) = \Lambda(\widetilde{\mathcal{E}})$ and $Z(\widetilde{\mathcal{E}}) = Z(\mathcal{E})$.

(d) In the notation of (b), the map sending an embedding of a maximal torus $b : T \hookrightarrow H$ to its lift $\widetilde{b} : \widetilde{T} \hookrightarrow \widetilde{H}$ induces a bijection between $\Pi_{\mathcal{E}}^{-1}([a],\overline{\kappa})$ and $\Pi_{\widetilde{\mathcal{E}}}^{-1}([\widetilde{a}],\overline{\widetilde{\kappa}})$. Moreover, for each two embeddings b_1 and b_2, we have $\kappa\left(\frac{[b_1]}{[b_2]}\right)(\overline{s}) = \kappa\left(\frac{[\widetilde{b}_1]}{[\widetilde{b}_2]}\right)(\overline{\widetilde{s}})$.

Proof. (a) Choose $\eta \in [\eta]$, a quasi-isogeny $\widehat{\pi} : \widehat{G} \to \widehat{\widetilde{G}}$ corresponding to π, and an $\mathcal{E}$-compatible representative $s \in \widehat{H}^\Gamma/Z(\widehat{G})^\Gamma$ of $\overline{s}$. Identify $\widehat{H}$ with $\eta(\widehat{H}) \subset \widehat{G}$ and put $\widetilde{s} := \widehat{\pi}(s) \in \widehat{\widetilde{G}}/Z(\widehat{\widetilde{G}})$.

Set $\widehat{\widetilde{H}} := (\widehat{\widetilde{G}})^0_{\widetilde{s}}$. Then $\widehat{\pi}$ induces a quasi-isogeny $\widehat{\pi}' : \widehat{H} \to \widehat{\widetilde{H}}$, and we have a canonical isomorphism $\widehat{\widetilde{H}} \xrightarrow{\sim} \widehat{H} \times_{Z(\widehat{G})} Z(\widehat{\widetilde{G}})$. Since both homomorphisms $\rho_H : \Gamma \to$

$\mathrm{Out}(\widehat{H})$ and $\rho_{\widetilde{G}} : \Gamma \to \mathrm{Aut}(Z(\widehat{\widetilde{G}}))$ induce the natural Galois action on $Z(\widehat{G})$, their product defines a homomorphism $\rho : \Gamma \to \mathrm{Out}(\widehat{\widetilde{H}})$.

Denote by $\widetilde{H}$ the quasi-split group over E, corresponding to the pair $(\widehat{\widetilde{H}}, \rho)$ (see 1.2.2 (a)), by $\overline{\widetilde{s}} \in \pi_0(Z(\widehat{\widetilde{H}})^{\Gamma}/Z(\widehat{\widetilde{G}})^{\Gamma})$ the class of $\widetilde{s}$, by $[\widetilde{\eta}]$ the conjugacy class of the inclusion $\widehat{\widetilde{H}} \hookrightarrow \widehat{\widetilde{G}}$, and by $[\pi']$ the conjugacy class of quasi-isogenies $\widetilde{H} \to H$ corresponding to $\widehat{\pi}'$. Then $\widetilde{\mathcal{E}} := (\widetilde{H}, [\widetilde{\eta}], \overline{\widetilde{s}})$ is an endoscopic triple for $\widetilde{G}$, and the pair $(\widetilde{\mathcal{E}}, [\pi'])$ satisfies the required properties. The proof of uniqueness is similar.

(b) follows immediately from the description of $\widetilde{\mathcal{E}}$ in (a).

(c) For every $\alpha \in \Lambda(\mathcal{E})$, the corresponding element $\widehat{\alpha} \in \mathrm{Out}(\widehat{H})$ is Γ-invariant and is induced by an element of $G^{\mathrm{ad}} = \widetilde{G}^{\mathrm{ad}}$. Therefore α induces a unique element $\widetilde{\alpha} \in \Lambda(\widetilde{\mathcal{E}})$ and vice versa. Also we have an equality $\pi_{\mathcal{E}} = \pi_{\widetilde{\mathcal{E}}}$, implying that $Z(\mathcal{E}) = Z(\widetilde{\mathcal{E}})$.

(d) is clear. □

1.4. Endoscopic triples: further properties

Let $\mathcal{E} = (H, [\eta], \overline{s})$ be an elliptic endoscopic triple for G, and $b : T \hookrightarrow H$ an embedding of a maximal torus. Put $([a], \overline{\kappa}) := \Pi_{\mathcal{E}}([b])$, and denote by $S_{[b]}$ the subset $\left\{\kappa\left(\frac{[b_1]}{[b]}\right)(\overline{s})\right\}_{[b_1]\in\Pi_{\mathcal{E}}^{-1}([a],\overline{\kappa})}$ of $\pi_0(\widehat{T/Z(G)}^{\Gamma})$.

The primary goal of this subsection is to prove the following result.

Proposition 1.4.1. *The subset $S_{[b]} \subset \pi_0(\widehat{T/Z(G)}^{\Gamma})$ is a subgroup.*

This proposition has the following corollary.

Corollary 1.4.2. *For each $z \in Z(\mathcal{E}, [a], \overline{\kappa})$, there exists $[b_1] \in \Pi_{\mathcal{E}}^{-1}([a], \overline{\kappa})$ such that $\kappa\left(\frac{[b_1]}{[b]}\right)(\overline{s}) = Z_{[\widehat{a}]}(z)$.*

Proof of the corollary. By definition, $Z_{[\widehat{a}]}(Z(\mathcal{E}, [a], \overline{\kappa})) = S_{([a],\overline{\kappa})}$, while $S_{([a],\overline{\kappa})}$ is the group generated by $S_{[b]}$. Since $S_{[b]}$ itself is a group, we get that $S_{([a],\overline{\kappa})} = S_{[b]}$, implying the assertion. □

To prove the proposition, we will first show several results of independent interest, while the proof itself will be carried out in 1.4.9.

As $\mathcal{E}$ is elliptic, we denote $\overline{s} \in \pi_0(Z(\widehat{H})^{\Gamma}/Z(\widehat{G})^{\Gamma}) = Z(\widehat{H})^{\Gamma}/Z(\widehat{G})^{\Gamma}$ simply by s. Note that $\pi_0(\widehat{G}_s)$ is a Γ-invariant subgroup of $\mathrm{Out}(\widehat{H})$.

Lemma 1.4.3. *Choose $\eta \in [\eta]$ and identify $\widehat{H}$ with $\eta(\widehat{H}) \subset \widehat{G}$.*

(a) There exists a natural isomorphism $\Lambda(\mathcal{E}) \cong \pi_0(\widehat{G}_s)^{\Gamma}$, and a Γ-equivariant injection $\iota : \pi_0(\widehat{G}_s) \hookrightarrow Z(\widehat{G}^{\mathrm{sc}})$. Moreover, ι induces an isomorphism $\Lambda(\mathcal{E}) \xrightarrow{\sim} Z(\mathcal{E})$.

(b) If G is split, then the image of $\rho_H : \Gamma \to \mathrm{Out}(\widehat{H})$ lies in $\pi_0(\widehat{G}_s)$, and $\Lambda(\mathcal{E})$ is canonically isomorphic to $\pi_0(\widehat{G}_s)$.

Proof. For the proof we can replace G by G^{sc} and $\mathcal{E}$ by the corresponding endoscopic triple; thus we can assume that $G = G^{\mathrm{sc}}$.

(a) Using the fact that $\mathcal{E}$ is elliptic, the first assertion follows from Remark 1.3.4. Next for each $g \in \widehat{G}_s$, choose a representative $\widetilde{g} \in \widehat{G}^{\mathrm{sc}}$ of g and a representative $\widetilde{s} \in \widehat{G}^{\mathrm{sc}}$ of s. Then the element $(\widetilde{g}\widetilde{s}\widetilde{g}^{-1})\widetilde{s}^{-1} \in Z(\widehat{G}^{\mathrm{sc}})$ does not depend on the choices, and the map $g \mapsto (\widetilde{g}\widetilde{s}\widetilde{g}^{-1})\widetilde{s}^{-1}$ defines a homomorphism $\widetilde{\iota}: \widehat{G}_s \to Z(\widehat{G}^{\mathrm{sc}})$. Moreover, $g \in \operatorname{Ker}\widetilde{\iota}$ if and only if $\widetilde{g} \in (\widehat{G}^{\mathrm{sc}})_{\widetilde{s}}$. By [St, §8], the last group is connected; therefore $\operatorname{Ker}\widetilde{\iota} = \widehat{G}_s^0$. Hence $\widetilde{\iota}$ induces an embedding $\iota : \pi_0(\widehat{G}_s) \hookrightarrow Z(\widehat{G}^{\mathrm{sc}})$, which is clearly Γ-equivariant.

Since $\mathcal{E}$ is elliptic, the group $Z(\widehat{H/Z(G)})^{\Gamma}$ is finite. As $\pi_{\mathcal{E}}$ is the composition of $\iota|_{\Lambda(\mathcal{E})}$ with the embedding $Z(\widehat{G}^{\mathrm{sc}})^{\Gamma} \hookrightarrow Z(\widehat{H/Z(G)})^{\Gamma} = \pi_0(Z(\widehat{H/Z(G)})^{\Gamma})$, the last assertion follows.

(b) The first assertion follows from the definition. Since embedding $\iota : \pi_0(\widehat{G}_s) \hookrightarrow Z(\widehat{G}^{\mathrm{sc}})$ is Γ-equivariant and Γ acts trivially on $Z(\widehat{G}^{\mathrm{sc}})$, we conclude from (a) that that $\Lambda(\mathcal{E}) \cong \pi_0(\widehat{G}_s)^{\Gamma} = \pi_0(\widehat{G}_s)$, as claimed. □

1.4.4. Action of $Z(\widehat{G}^{\mathrm{sc}})$ on the extended Dynkin diagram $\widetilde{D}_{\widehat{G}}$ of $\widehat{G}$ Let $\widehat{T}^{\mathrm{ad}}$ and $\widehat{T}^{\mathrm{sc}}$ be the abstract Cartan subgroups of $\widehat{G}^{\mathrm{ad}}$ and $\widehat{G}^{\mathrm{sc}}$, respectively, and let $C \subset X_*(\widehat{T}^{\mathrm{ad}}) \otimes \mathbb{R}$ be the fundamental alcove. For every $\mu \in X_*(\widehat{T}^{\mathrm{ad}})$ there exists a unique element $w_\mu \in W^{\mathrm{aff}}$ of the affine Weyl group of $\widehat{G}$ such that $w_\mu(C + \mu) = C$. Then the map $c_\mu : x \mapsto w_\mu(x + \mu)$ defines an affine automorphism of C, and hence an automorphism of $\widetilde{D}_{\widehat{G}}$. Moreover, the map $c : \mu \mapsto c_\mu$ is a homomorphism, and $X_*(\widehat{T}^{\mathrm{sc}}) \subset \operatorname{Ker} c$. Thus c induces an action of $Z(\widehat{G}^{\mathrm{sc}}) = X_*(\widehat{T}^{\mathrm{ad}})/X_*(\widehat{T}^{\mathrm{sc}})$ on C, hence on $\widetilde{D}_{\widehat{G}}$.

Lemma 1.4.5. *Let G be a split simple group. Then there exists a bijection $[\alpha] \mapsto \mathcal{E}_{[\alpha]} = (H_{[\alpha]}, s_{[\alpha]}, [\eta]_{[\alpha]})$ between the set of $Z(\widehat{G}^{\mathrm{sc}})$-orbits of vertices of $\widetilde{D}_{\widehat{G}}$ and the set of isomorphism classes of split elliptic endoscopic triples for G.*

Moreover, for every vertex $\alpha \in \widetilde{D}_{\widehat{G}}$, the stabilizer $Z(\widehat{G}^{\mathrm{sc}})_\alpha$ is canonically isomorphic to $\Lambda(\mathcal{E}_{[\alpha]})$, and the order $\operatorname{ord}(s_{[\alpha]})$ *is equal to the coefficient of α in the reduced linear dependence $\sum_{\alpha \in \widetilde{D}_{\widehat{G}}} n_\alpha \alpha = 0$.*

Proof. The set of isomorphisms of split endoscopic triples for G is in bijection with the set of conjugacy classes of semisimple elements $s \in \widehat{G}^{\mathrm{ad}}$ such that $\widehat{G}_s^0$ is semisimple, hence with the set of $W(\widehat{G})$-orbits of elements $s \in \widehat{T}^{\mathrm{ad}}$ such that $[X^*(\widehat{T}^{\mathrm{ad}}) : X^*(\widehat{T}^{\mathrm{ad}})_s] < \infty$.

Note that $X_*(\widehat{T}^{\mathrm{ad}}) \otimes \mathbb{R}/X_*(\widehat{T}^{\mathrm{ad}})(= \operatorname{Hom}(X_*(\widehat{T}^{\mathrm{ad}}), \mathbb{R}/\mathbb{Z}))$ is naturally isomorphic to $\widehat{T}(\mathbb{C})^1 = \operatorname{Hom}(X_*(\widehat{T}^{\mathrm{ad}}), S^1)$. This isomorphism induces a bijection between $W(\widehat{G})$-orbits on $\widehat{T}(\mathbb{C})^1$ and $(X_*(\widehat{T}^{\mathrm{ad}}) \rtimes W(\widehat{G}))$-orbits on $X_*(\widehat{T}^{\mathrm{ad}}) \otimes \mathbb{R}$. Since C is a fundamental domain for the action of $W^{\mathrm{aff}} = X_*(\widehat{T}^{\mathrm{sc}}) \rtimes W(\widehat{G})$, the latter set coincides with the set of $Z(\widehat{G}^{\mathrm{sc}})$-orbits on C.

For each $s \in \widehat{T}(\mathbb{C})^1$ and each representative $\widetilde{s} \in C \subset X_*(\widehat{T}^{\mathrm{ad}}) \otimes \mathbb{R}$ of s, the set of roots α of $\widehat{G}$ such that $\alpha(s) = 1$ are in bijection with the set of affine roots β of $\widehat{G}$

such that $\beta(\widetilde{s}) = 0$. Therefore $[X^*(\widehat{T}^{\mathrm{ad}}) : X^*(\widehat{T}^{\mathrm{ad}})_s] < \infty$ if and only if $\widetilde{s}$ is a vertex C, that is, a vertex of $\widetilde{D}_{\widehat{G}}$. The last assertion is clear. □

The proof of the following result is done case-by-case.

Claim 1.4.6. *Let G be an absolutely simple group over E such that $(G^*)^{\mathrm{sc}}$ is not isomorphic to SL_n. For every embedding $a : T \hookrightarrow G$ of a maximal torus, and $\overline{\kappa} \in \pi_0(\widehat{T}^{\Gamma}/Z(\widehat{G})^{\Gamma})$ such that $([a], \overline{\kappa}) \in \mathrm{Im}\,\Pi_{\mathcal{E}}$, we have*

$$[Z_{\widehat{[a]}}(Z(\mathcal{E}, [a], \overline{\kappa})) : Z_{\widehat{[a]}}(Z(\mathcal{E}))] \leq 2.$$

Proof. Replacing G by G^*, we can assume that G is quasi-split. Replacing G by G^{sc} and $\mathcal{E}$ by the corresponding endoscopic triple, we can assume that $G = G^{\mathrm{sc}}$.

Assume that our assertion is false, that is, $[Z_{\widehat{[a]}}(Z(\mathcal{E}, [a], \overline{\kappa})) : Z_{\widehat{[a]}}(Z(\mathcal{E}))] > 2$. Since $Z(\mathcal{E}, [a], \overline{\kappa})$ is a subgroup of $Z(\widehat{G}^{\mathrm{sc}})^{\Gamma}$, we conclude that $|Z(\widehat{G}^{\mathrm{sc}})^{\Gamma}| > 2$. Therefore by the classification of simple algebraic groups, we get that G (hence also $\widehat{G}$) is of type A, D or E_6. Moreover, since the group $\mathrm{Out}(\widehat{G})$ acts faithfully on $Z(\widehat{G}^{\mathrm{sc}})$, we see case-by-case that the assumption $|Z(\widehat{G}^{\mathrm{sc}})^{\Gamma}| > 2$ implies that G is split.

By our assumption, G is not of type A; therefore $|Z(\widehat{G}^{\mathrm{sc}})| \leq 4$. Since by our assumption $[Z(\widehat{G}^{\mathrm{sc}}) : Z(\mathcal{E})] > 2$, we get that $Z(\mathcal{E}) = 1$, hence $\Lambda(\mathcal{E}) = 1$ (see Lemma 1.4.3 (a)). It follows from Lemma 1.4.3 (b) that $\mathrm{Im}\,\rho_H = 1$, thus $\mathcal{E}$ is split. Therefore by Lemma 1.4.5, $\mathcal{E}$ corresponds to a $Z(\widehat{G}^{\mathrm{sc}})$-orbit $[\alpha] \subset \widetilde{D}_{\widehat{G}}$. Moreover, since $\Lambda(\mathcal{E}) = 1$, we get that $|[\alpha]| = |Z(\widehat{G}^{\mathrm{sc}})| > 2$.

Recall that $Z_{\widehat{[a]}}(Z(\mathcal{E}, [a], \overline{\kappa}))$ consists of images of $s = s_{[\alpha]}$ under certain homomorphisms $\kappa\left(\frac{[b_1]}{[b_2]}\right) : Z(\widehat{H}) \to \pi_0(\widehat{T/Z(G)}^{\Gamma})$. Therefore every $z \in Z_{\widehat{[a]}}(Z(\mathcal{E}, [a], \overline{\kappa}))$ satisfies $z^{\mathrm{ord}(s_{[\alpha]})} = 1$. Since $Z_{\widehat{[a]}}(Z(\mathcal{E}, [a], \overline{\kappa})) \neq \{1\}$, we get $(\mathrm{ord}(s_{[\alpha]}), |Z(\widehat{G}^{\mathrm{sc}})|) \neq 1$. In particular, the orbit $[\alpha]$ is non-special, that is, it consists of non-special vertices.

Now it is easy to get a contradiction. Indeed, in the case of D_n, there are no non-special $\mathrm{Aut}(\widetilde{D}_{\widehat{G}})$-orbits of cardinality greater than two, while in the case of E_6, there is only one such orbit. However in this case we have $\mathrm{ord}(s_{[\alpha]}) = 2$ and $|Z(\widehat{G}^{\mathrm{sc}})| = 3$, contradicting the assumption that $(\mathrm{ord}(s_{[\alpha]}), |Z(\widehat{G}^{\mathrm{sc}})|) \neq 1$. □

1.4.7. Restriction of scalars (a) Let E' be a finite separable extension of E, $\Gamma' = \mathrm{Gal}(\overline{E}/E')$, G' a reductive group over E', and $G = R_{E'/E}G'$. Then $(\widehat{G}, \rho_G)$ has the following description. First, $\widehat{G} = \prod_{\sigma \in \mathrm{Hom}_E(E', \overline{E})} \widehat{{}^{\sigma}G'}$, where ${}^{\sigma}G'$ is a group over $\sigma(E')$ induced from G' by σ. Every $\tau \in \Gamma$ induces a canonical element of $\rho_{G'}(\tau, \sigma) \in \mathrm{Isom}(\widehat{{}^{\sigma}G'}, \widehat{{}^{\tau\sigma}G'})/\mathrm{Int}(\widehat{{}^{\sigma}G'})$, which coincides with $\rho_{{}^{\sigma}G'}(\tau)$ if $\tau \in \mathrm{Gal}(\overline{E}/\sigma(E'))$. Then for every $\tau \in \Gamma$ we have $\rho_G(\tau) = \prod_{\sigma} \rho_{G'}(\tau, \sigma)$.

(b) There is a canonical isomorphism $Z(\widehat{G})^{\Gamma} \xrightarrow{\sim} Z(\widehat{G'})^{\Gamma'}$.

(c) Every endoscopic triple $\mathcal{E}' = (H', [\eta'], \overline{s}')$ for G' gives rise to an endoscopic triple $\mathcal{E} = (H, [\eta], \overline{s})$ for $G = R_{E'/E}G'$, denoted by $R_{E'/E}\mathcal{E}'$, defined as follows. $H = R_{E'/E}H'$, $[\eta]$ is a product $\prod_{\sigma}[\eta'_{\sigma}]$, where $[\eta'_{\sigma}]$ is the conjugacy class of embeddings $\widehat{{}^{\sigma}H'} \hookrightarrow \widehat{{}^{\sigma}G'}$ induced by $[\eta']$, and $\overline{s}$ is the preimage of $\overline{s'}$ under the canonical isomorphism $Z(\widehat{H})^{\Gamma}/Z(\widehat{G})^{\Gamma} \xrightarrow{\sim} Z(\widehat{H'})^{\Gamma'}/Z(\widehat{G'})^{\Gamma'}$.

Lemma 1.4.8. *(a) For every embedding of a maximal torus $a : T \hookrightarrow G = R_{E'/E}G'$, there exists an embedding $a' : T' \hookrightarrow G'$ of a maximal torus such that $T = R_{E'/E}T'$ and $a = R_{E'/E}a'$. Moreover, the map $[a'] \mapsto [R_{E'/E}a']$ induces a bijection between the sets of stable conjugacy classes of embeddings $T' \hookrightarrow G'$ and $T \hookrightarrow G$.*

(b) For every endoscopic triple $\mathcal{E} = (H, [\eta], \overline{s})$ for G, there exists a unique endoscopic triple $\mathcal{E}' = (H', [\eta'], \overline{s}')$ for G' such that $R_{E'/E}\mathcal{E}' \cong \mathcal{E}$.

(c) In the above notation, for each $([a'], \overline{\kappa}) \in \operatorname{Im} \Pi_{\mathcal{E}}$ the map $b' \mapsto b := R_{E'/E}b'$ induces a bijection between $\Pi_{\mathcal{E}'}^{-1}([a'], \overline{\kappa})$ and $\Pi_{\mathcal{E}}^{-1}([R_{E'/E}a'], \overline{\kappa})$. Moreover, for each two embeddings $b'_1, b'_2 : T' \hookrightarrow H'$, the isomorphism $Z(\widehat{H/Z(G)})^{\Gamma} \xrightarrow{\sim} Z(\widehat{H'/Z(G')})^{\Gamma'}$ maps $\kappa\left(\frac{[b_1]}{[b_2]}\right)(\overline{s})$ to $\kappa\left(\frac{[b'_1]}{[b'_2]}\right)(\overline{s}')$.

Proof. (a) Note first that G' is a direct factor of $G_{E'}$, and denote by T' the image of the composition map $T_{E'} \overset{a}{\hookrightarrow} G_{E'} \to G'$. Then $R_{E'/E}T' \cong T$, and the embedding $a' : T' \hookrightarrow G'$ satisfies $R_{E'/E}a' = a$. The second assertion follows from the first.

(b) Choose $a : T \hookrightarrow G$ and $\kappa \in \widehat{T}^{\Gamma}/Z(\widehat{G})^{\Gamma}$ such that $\mathcal{E} \cong \mathcal{E}_{([a],\kappa)}$. Let $a' : T' \hookrightarrow G'$ be the embedding as in (a), and let $\kappa' \in \widehat{T'}^{\Gamma'}/Z(\widehat{G'})^{\Gamma'}$ be the image of κ. Then $\mathcal{E}' := \mathcal{E}_{([a'],\kappa')}$ satisfies $R_{E'/E}\mathcal{E}' \cong \mathcal{E}$. The uniqueness is clear.

(c) follows immediately from (a). □

1.4.9. *Proof of Proposition* 1.4.1. The proof will be carried out in two steps. First we will treat the case $G = \mathrm{SL}_n$, and then reduce the general case to it.

Step 1: The case $G = \mathrm{SL}_n$. In this case, we will describe all the objects involved explicitly.

First, there exists a divisor $m|n$, a cyclic Galois extension $K \subset \overline{E}$ of E of degree m such that H is isomorphic to $(R_{K/E}\,\mathrm{GL}_{\frac{n}{m}})^1 = \{g \in R_{K/E}\,\mathrm{GL}_{\frac{n}{m}} \mid N_{K/E}(\det g) = 1\}$. Next $s \in Z(\widehat{H}) \cong (\mathbb{C}^{\times})^{\mathrm{Gal}(K/E)}/\mathbb{C}^{\times}$ is a class $[(\mu(\sigma))_{\sigma \in \mathrm{Gal}(K/E)}]$ for a certain isomorphism $\mu : \mathrm{Gal}(K/E) \xrightarrow{\sim} \{z \in \mathbb{C}^{\times} \mid z^m = 1\}$.

Moreover, if we embed $H \hookrightarrow \mathrm{SL}_n$ by means of any E-linear isomorphism $K \xrightarrow{\sim} E^m$, then for every embedding of maximal torus $b : T \hookrightarrow H$, we get that $[b]_G$ is the stable conjugacy class of the composition $T \overset{b}{\hookrightarrow} H \hookrightarrow \mathrm{SL}_n$.

Every maximal torus $T \subset H \subset SL_n$ is of the form $(\prod_{i=1}^{l} R_{K_i/E}\mathbb{G}_m)^1$, where $(\prod_{i=1}^{l} R_{K_i/E}\mathbb{G}_m)^1 = \{t_i \in \prod_{i=1}^{l} R_{K_i/E}\mathbb{G}_m \mid \prod_i N_{K_i/E}(t_i) = 1\}$, for certain finite extensions K_i/K with $\sum_i [K_i : K] = \frac{n}{m}$. Denote by b the inclusion $T \hookrightarrow H$. Then embeddings $b' : T \hookrightarrow H$ such that $[b'] = [b]$ (resp. $[b']_G = [b]_G$) are in canonical bijection with K-linear (resp. E-linear) algebra embeddings $\oplus_{i=1}^{l} K_i \hookrightarrow \mathrm{Mat}_{\frac{n}{m}}(K)$.

In particular, we get a bijection $[\imath] \mapsto b_{[\imath]}$ from the set of l-tuples $\overline{\imath} = (\imath_1, \ldots, \imath_l)$ of E-algebra embeddings $\imath_i : K \hookrightarrow K_i$ to that of stable conjugacy classes $[b']$ of embeddings $T \hookrightarrow H$ such that $[b']_G = [b]_G$. Therefore both sets are principal homogeneous spaces for the action of the group $\mathrm{Gal}(K/E)^l$.

The dual torus $\widehat{T}$ is $[\prod_{i=1}^{l}(\mathbb{C}^\times)^{\mathrm{Hom}_E(K_i,\overline{E})}]/\mathbb{C}^\times$, and $(\widehat{T}^\Gamma)^0$ is the image of the diagonal map $[(\mathbb{C}^\times)^l]/\mathbb{C}^\times \hookrightarrow \widehat{T}$. Also $\widehat{T/Z(G)}$ consists of elements $\{c_{i,\sigma_i}\}_{i,\sigma_i}$ of $\prod_i(\mathbb{C}^\times)^{\mathrm{Hom}_E(K_i,\overline{E})}$ such that $\prod_i\prod_{\sigma_i} c_{i,\sigma_i} = 1$.

For every stable conjugacy class $[b_{\overline{\iota}}]$ of embeddings $T \hookrightarrow H$, the corresponding embedding $Z_{[\widehat{b_{\overline{\iota}}}]} : Z(\widehat{H}) \hookrightarrow \widehat{T}$ sends $s \in Z(\widehat{H})$ to an element $[(c_{i,\sigma_i})_{i,\sigma_i}]$, given by the rule $c_{i,\sigma_i} := \mu(\sigma_i \circ \iota_i)$ (here $\sigma_i \circ \iota_i \in \mathrm{Hom}_E(K,\overline{E}) = \mathrm{Gal}(K/E)$). When $\overline{\iota}$ is replaced by $\overline{\iota}\circ\overline{\tau}$ for certain $\overline{\tau} = (\tau_1,\ldots,\tau_l)$, then each c_{i,σ_i} is multiplied by $\mu(\tau_i)$. It follows that $\kappa\left(\frac{[b_{\overline{\iota}\circ\overline{\tau}}]}{[b_{\overline{\iota}}]}\right)(s)$ is the class $[(\mu(\tau_i))_{i,\sigma_i}] \in \widehat{T/Z(G)}^\Gamma$. In particular, the image of each $\kappa\left(\frac{[b_{\overline{\iota}\circ\overline{\tau}}]}{[b_{\overline{\iota}}]}\right)(s)$ in $\pi_0(\widehat{T}^\Gamma)$ is trivial. Thus (see Lemma 1.2.9) we get that $\Pi_{\mathcal{E}}([b_{\overline{\iota}}])$ is independent of $\overline{\iota}$ (hence $\Pi_{\mathcal{E}}([b_{\overline{\iota}}]) = ([a],\overline{\kappa})$ for every $\overline{\iota}$). As a result, the subset $S_{[b]}$ consists of classes of elements $[(\mu(\tau_i))_{i,\sigma_i}]$ where $\overline{\tau}$ runs through $\mathrm{Gal}(K/E)^l$. Hence $S_{[b]}$ is a group, as claimed.

Step 2: The general case. It follows from Lemma 1.3.10 (d) that the subset $S_{[b]}$ will not change if we replace G by G^{sc}, $\mathcal{E}$ by the corresponding endoscopic triple for G^{sc} (see Lemma 1.3.10 (a)) and b by its lifting $b^{\mathrm{sc}} : T^{\mathrm{sc}} \hookrightarrow G^{\mathrm{sc}}$. Thus we are reduced to the case when G is semisimple and simply connected.

Then G is the product of its simple factors $G = \prod_i G_i$, and there exist embeddings of maximal tori $b_i : T_i \hookrightarrow G_i$ such that $T = \prod_i T_i$ and $b = \prod_i b_i$. Then $S_{[b]}$ decomposes as a product $\prod_i S_{[b_i]} \subset \prod_i \pi_0(\widehat{T_i/Z(G_i)}^\Gamma) = \pi_0(\widehat{T/Z(G)}^\Gamma)$. Thus it will suffice to show that each $S_{[b_i]}$ is a subgroup, thus reducing us to the case when G is simple and simply connected.

There exists a finite separable extension E' of E and an absolutely simple simply connected algebraic group G' over E' such that $G \cong R_{E'/E}G'$. Using Lemma 1.4.8, the subset $S_{[b]}$ will not change if we replace E by E', G by G', $\mathcal{E}$ by $\mathcal{E}'$ and $[b]$ by $[b']$. Thus we can assume that G is absolutely simple. Replacing G by G^*, we can assume that G is quasi-split, not isomorphic to SL_n.

For each embedding $b_1 : T \hookrightarrow H$ such that $\Pi_{\mathcal{E}}([b_1]) = \Pi_{\mathcal{E}}([b])$ and each $\alpha \in \Lambda(\mathcal{E})$ we have $\Pi_{\mathcal{E}}(\alpha([b_1])) = \Pi_{\mathcal{E}}([b])$. By Lemmas 1.2.9 and 1.3.6 we have $\kappa\left(\frac{\alpha([b_1])}{[b]}\right)(\overline{s}) = \kappa\left(\frac{[b_1]}{[b]}\right)(\overline{s})Z_{[\widehat{a}]}(\pi_{\mathcal{E}}(\alpha))$. In other words, $S_{[b]}$ is invariant under the multiplication by elements from $Z_{[\widehat{a}]}(Z(\mathcal{E}))$.

On the other hand, by definition, $S_{[b]} \subset Z_{[\widehat{a}]}(Z(\mathcal{E},[a],\overline{\kappa}))$. Since by Claim 1.4.6, we have $[Z_{[\widehat{a}]}(Z(\mathcal{E},[a],\overline{\kappa})) : Z_{[\widehat{a}]}(Z(\mathcal{E}))] \leq 2$, we conclude that $S_{[b]}$ is equal either to $Z_{[\widehat{a}]}(Z(\mathcal{E},[a],\overline{\kappa}))$ or to $Z_{[\widehat{a}]}(Z(\mathcal{E}))$. Hence it is a group, as claimed. □

1.5. The case of local fields

In this subsection we will apply the results from 1.1–1.3 to the case of endoscopic triples for a reductive group G over a local non-archimedean field E.

1.5.1. Tate–Nakayama duality For every torus T over E, Tate–Nakayama duality provides us with a functorial isomorphism $\mathcal{D}_T : H^1(E,T) \xrightarrow{\sim} \pi_0(\widehat{T}^\Gamma)^D$ of finite abelian groups.

Kottwitz showed that for every connected reductive group G over E its cohomology group $H^1(E, G)$ has a unique structure of a finite abelian group such that for every maximal torus $T \subset G$ the natural map $H^1(E, T) \to H^1(E, G)$ is a group homomorphism (see [Ko2, Thm 1.2]). Moreover, there exists a group isomorphism $\mathcal{D}_G : H^1(E, G) \xrightarrow{\sim} \pi_0(Z(\widehat{G})^\Gamma)^D$ such that for every maximal torus T of G, the embedding $T \hookrightarrow G$ induces a commutative diagram:

$$\begin{array}{ccc} H^1(E,T) & \longrightarrow & H^1(E,G) \\ {\scriptstyle \mathcal{D}_T}\downarrow & & {\scriptstyle \mathcal{D}_G}\downarrow \\ \pi_0(\widehat{T}^\Gamma)^D & \longrightarrow & \pi_0(Z(\widehat{G})^\Gamma)^D. \end{array}$$

In particular, we have a canonical surjection

$$\widehat{T}^\Gamma / Z(\widehat{G})^\Gamma \to \mathrm{Coker}[\pi_0(Z(\widehat{G})^\Gamma) \to \pi_0(\widehat{T}^\Gamma)] \xrightarrow{\sim} (\mathrm{Ker}\,[H^1(E,T) \to H^1(E,G)])^D.$$

Remark 1.5.2. Borovoi [Bo] showed that for every reductive group G over E there is a functorial group isomorphism $H^1(E, G) \xrightarrow{\sim} (\pi_1(G)_\Gamma)_{\mathrm{tor}}$, where $(\cdot)_{\mathrm{tor}}$ means for torsion. In particular, for every homomorphism of reductive groups $f : G_1 \to G_2$, the induced map $H^1(E, G_1) \to H^1(E, G_2)$ is a group homomorphism as well. Now the existence of the Kottwitz isomorphism $\mathcal{D}_G$ follows from the Γ-equivariant isomorphism $\pi_1(G) \xrightarrow{\sim} X^*(Z(\widehat{G}))$.

Lemma 1.5.3. *Let $a : T \hookrightarrow G$ be an embedding of a maximal elliptic torus. Then for every inner twisting $\varphi : G \to G'$, there exists an embedding $a' : T \hookrightarrow G'$ stably conjugate to a.*

Proof. By assumption, $T/Z(G)$ is anisotropic, therefore $\widehat{T/Z(G)}^\Gamma$ is finite. Hence the canonical map $\pi_0(Z(\widehat{G^{\mathrm{ad}}})^\Gamma) = Z(\widehat{G^{\mathrm{ad}}})^\Gamma \hookrightarrow \widehat{T/Z(G)}^\Gamma = \pi_0(\widehat{T/Z(G)}^\Gamma)$ is injective. By duality, the canonical map $H^1(E, T/Z(G)) \to H^1(E, G^{\mathrm{ad}})$ is surjective (see 1.5.1). This implies the assertion (see 1.1.4 (b)). □

Lemma 1.5.4. *Assume that we are in the situation of 1.1.5 with $k = 2$. Let $\mathcal{E} = (H, [\eta], \overline{s})$ be an endoscopic triple for G, $b_1 : T_1 \hookrightarrow H$ and $b_2 : T_2 \hookrightarrow H$ are embeddings of maximal tori compatible with a_1 and a_2, respectively. Then for every $z \in \pi_0(Z(\widehat{G})^\Gamma)$ its image $\widetilde{z} := \kappa([b_1], [b_2])(z) \in \pi_0(([(T_1 \times T_2)/Z(G)]\widehat{\ }\,)^\Gamma)$ (see 1.2.8) satisfies $\langle \overline{\mathrm{inv}}((a_1, a_1'); (a_2, a_2'), \widetilde{z}\rangle = 1$.*

Proof. Recall that $\widetilde{z}$ is the image of $\mu_G(z) \in \pi_0(Z(\widehat{G^2/Z(G)})^\Gamma)$ under the map $Z_{\widehat{[a_1,a_2]}} : \pi_0(Z(\widehat{G^2/Z(G)})^\Gamma) \to \pi_0(([(T_1 \times T_2)/Z(G)]\widehat{\ }\,)^\Gamma)$. Therefore by the commutative diagram of 1.5.1, we have

$$\langle \overline{\mathrm{inv}}((a_1, a_1'); (a_2, a_2')), \widetilde{z}\rangle = \langle \Delta(\mathrm{inv}(G, G')), \mu_G(z)\rangle.$$

Moreover, the latter expression equals $\langle \mathrm{inv}(G, G'), \nu_G(\mu_G(z))\rangle = \langle \mathrm{inv}(G, G'), 0\rangle = 1$, as claimed. □

Notation 1.5.5. Let $\mathcal{E} = (H, [\eta], \overline{s})$ be an endoscopic triple for G, $\varphi : G \to G'$ an inner twisting, $(a_i, a'_i, [b_i])$ two triples consisting of stably conjugate embeddings of maximal tori $a_i : T_i \hookrightarrow G$ and $a'_i : T_i \hookrightarrow G'$, and stably conjugate classes $[b_i]$ of embeddings of maximal tori $T_i \hookrightarrow H$, compatible with a_i. To these data one associates elements $\overline{\mathrm{inv}}((a_1, a'_1); (a_2, a'_2)) \in H^1(E, (T_1 \times T_2)/Z(G))$ (see 1.1.5) and $\kappa([b_1], [b_2])(\check{s}) \in \pi_0((\{(T_1 \times T_2)/Z(G)\}^\frown)^\Gamma)$ for every representative $\check{s} \in \pi_0(Z(\widehat{H})^\Gamma)$ of $\overline{s}$ (see 1.2.8). By Lemma 1.5.4, the pairing

$$\left\langle \frac{a_1, a'_1; [b_1]}{a_2, a'_2; [b_2]} \right\rangle = \left\langle \frac{a_1, a'_1; [b_1]}{a_2, a'_2; [b_2]} \right\rangle_{\mathcal{E}} := \langle \overline{\mathrm{inv}}((a_1, a'_1); (a_2, a'_2)), \kappa([b_1], [b_2])(\check{s}) \rangle \in \mathbb{C}^\times$$

is independent of the choice of $\check{s}$.

Remark 1.5.6. This invariant is essentially the term Δ_1 of Langlands–Shelstad ([LS, (3.4)]).

Lemma 1.5.7. *(a) For any three triples* $(a_i, a'_i, [b_i])$, $(i = 1, 2, 3)$, *we have*

$$\left\langle \frac{a_1, a'_1; [b_1]}{a_3, a'_3; [b_3]} \right\rangle = \left\langle \frac{a_1, a'_1; [b_1]}{a_2, a'_2; [b_2]} \right\rangle \left\langle \frac{a_2, a'_2; [b_2]}{a_3, a'_3; [b_3]} \right\rangle.$$

(b) Assume that $T_1 = T_2 = T$, $a_1 = a_2 = a$ *and* $a'_1 = a'_2 = a'$. *Then* $\left\langle \frac{a, a'; [b_1]}{a, a'; [b_2]} \right\rangle = \left\langle \overline{\mathrm{inv}}(a, a'), \kappa\left(\frac{[b_1]}{[b_2]}\right)(\overline{s}) \right\rangle$. *If, moreover,* $\kappa\left(\frac{[b_1]}{[b_2]}\right)(\overline{s}) = Z_{\widehat{[a]}}(z)$ *for some* $z \in Z(\widehat{G^{\mathrm{ad}}})^\Gamma$, *then* $\left\langle \frac{a, a'; [b_1]}{a, a'; [b_2]} \right\rangle = \langle \mathrm{inv}(G, G'), z \rangle$.

(c) Assume that $T_1 = T_2 = T$ *and* $b_1 = b_2 = b$. *Then*

$$\left\langle \frac{a_1, a'_1; [b]}{a_2, a'_2; [b]} \right\rangle = \langle \mathrm{inv}(a_1, a_2), \overline{\kappa}_{[b]} \rangle \langle \mathrm{inv}(a'_1, a'_2), \overline{\kappa}_{[b]} \rangle^{-1}.$$

(d) Assume that $\varphi = \mathrm{Id}_G$. *Then* $\left\langle \frac{a_1, a'_1; [b_1]}{a_2, a'_2; [b_2]} \right\rangle = \langle \mathrm{inv}(a_1, a'_1), \overline{\kappa}_{[b_1]} \rangle \langle \mathrm{inv}(a_2, a'_2), \ \overline{\kappa}_{[b_2]} \rangle^{-1}$.

(e) Let $\pi : \widetilde{G} \to G$ *be a quasi-isogeny,* $\widetilde{\mathcal{E}}$ *the endoscopic triple for* $\widetilde{G}$ *induced from* $\mathcal{E}$, $\widetilde{\varphi} : \widetilde{G} \to \widetilde{G}'$ *the inner twisting induced from* φ, *and* $\widetilde{a}_i$, $\widetilde{a}_i{}'$ *and* $[\widetilde{b}_i]$ *the lifts of* a_i, a'_i *and* $[b_i]$, *respectively. Then* $\left\langle \frac{\widetilde{a}_1, \widetilde{a}_1{}'; [\widetilde{b}_1]}{\widetilde{a}_2, \widetilde{a}_2{}'; [\widetilde{b}_2]} \right\rangle_{\widetilde{\mathcal{E}}} = \left\langle \frac{a_1, a'_1; [b_1]}{a_2, a'_2; [b_2]} \right\rangle_{\mathcal{E}}$.

(f) Let $\varphi' : G' \to G''$ *be another inner twisting, and for each* $i = 1, 2$, *let* $a''_i : T \hookrightarrow G''$ *be a stable conjugate of* a_i *and* a'_i. *Then*

$$\left\langle \frac{a_1, a''_1; [b_1]}{a_2, a''_2; [b_2]} \right\rangle = \left\langle \frac{a_1, a'_1; [b_1]}{a_2, a'_2; [b_2]} \right\rangle \left\langle \frac{a'_1, a''_1; [b_1]}{a'_2, a''_2; [b_2]} \right\rangle.$$

Proof. All assertions follow from the functoriality of the Tate–Nakayama duality 1.5.1.

(a) By Lemma 1.1.6 (c), $\left\langle \frac{a_i, a'_i; [b_i]}{a_j, a'_j; [b_j]} \right\rangle$ equals the pairing of $\overline{\mathrm{inv}}((a_1, a'_1); (a_2, a'_2); (a_3, a'_3))$ with the image of $\kappa([b_i], [b_j])(\check{s})$ in $\pi_0((\{(\prod_{i=1}^3 T_i)/Z(G)\}^\frown)^\Gamma)$. Thus the assertion follows from Lemma 1.2.7.

(b) The first assertion follows from Lemma 1.1.6 (d), while the second follows from the fact that the image of $\overline{\text{inv}}(a, a')$ in $H^1(E, G^{\text{ad}})$ equals $\text{inv}(G, G')$.

(c) Since $\kappa([b], [b])$ equals $\mu_T \circ Z_{\widehat{[b]}}$, we get that

$$\left\langle \frac{a_1, a_1'; [b]}{a_2, a_2'; [b]} \right\rangle = \left\langle \widehat{\mu_T}(\overline{\text{inv}}((a_1, a_1'); (a_2, a_2')), Z_{\widehat{[b]}}(\check{s}) \right\rangle.$$

Using Lemma 1.1.6 (d) and (e), we conclude that

$$\widehat{\mu_T}(\overline{\text{inv}}((a_1, a_1'); (a_2, a_2'))) = \text{inv}(a_1, a_2)\, \text{inv}(a_1', a_2')^{-1},$$

implying the assertion.

(d) By Lemma 1.1.6 (b), $\overline{\text{inv}}((a_1, a_1'); (a_2, a_2'))$ is the image of

$$(\text{inv}(a_1, a_1'), \text{inv}(a_2, a_2')).$$

Since the image of $\kappa([b_1], [b_2])(\check{s})$ in $\pi_0(\widehat{T_1}^{\Gamma}/Z(\widehat{G})^{\Gamma}) \times \pi_0(\widehat{T_2}^{1}/Z(\widehat{G})^{\Gamma})$ equals $(\overline{\kappa}_{[b_1]}, \overline{\kappa}_{[b_2]}^{-1})$, the assertion follows.

(e) By 1.1.13 (d), $\overline{\text{inv}}((a_1, a_1'); (a_2, a_2')) \in H^1(E, (T_1 \times T_2)/Z(G))$ is the image of $\overline{\text{inv}}((\widetilde{a_1}, \widetilde{a_1}'); (\widetilde{a_2}, \widetilde{a_2}')) \in H^1(E, (\widetilde{T_1} \times \widetilde{T_2})/Z(\widetilde{G}))$. Choose a representative $\check{s} \in \pi_0(Z(\widehat{H})^{\Gamma})$ of $\overline{s}$, and let $\widetilde{s} \in \pi_0(Z(\widehat{\widetilde{H}})^{\Gamma})$ be the image of $\check{s}$. Then $\kappa([\widetilde{b_1}], [\widetilde{b_2}])(\widetilde{s})$ is the image of $\kappa([b_1], [b_2])(\check{s})$, and the assertion follows.

(f) Follows from Lemma 1.1.6 (f). □

Definition 1.5.8. Let $\mathcal{E}$ be an endoscopic triple for G, and $([a], \overline{\kappa})$ a pair belonging to $\text{Im}\,\Pi_{\mathcal{E}}$. An inner twisting $\varphi : G \to G'$ is called *$\mathcal{E}$-admissible* (resp. *$(\mathcal{E}, [a], \overline{\kappa})$-admissible*) if the corresponding class $\text{inv}(G, G') \in H^1(E, G^{\text{ad}}) \cong (\widehat{Z(G^{\text{ad}})^{\Gamma}})^D$ is orthogonal to $Z(\mathcal{E}) \subset Z(\widehat{G^{\text{ad}}})^{\Gamma}$ (resp. orthogonal to $Z(\mathcal{E}, [a], \overline{\kappa}) \subset Z(\widehat{G^{\text{ad}}})^{\Gamma}$).

Lemma 1.5.9. *(a) If φ is $(\mathcal{E}, [a], \overline{\kappa})$-admissible, then φ is $\mathcal{E}$-admissible. The converse is true if $a(T) \subset G$ is elliptic.*

(b) For every $i = 1, 2$, the function $[b_i] \mapsto \left\langle \frac{a_1, a_1'; [b_1]}{a_2, a_2'; [b_2]} \right\rangle$ is constant on the fiber $\Pi_{\mathcal{E}}^{-1}([a_i], \overline{\kappa})$ if and only if φ is $(\mathcal{E}, [a_i], \overline{\kappa})$-admissible.

Proof. (a) The assertion is a translation of Lemma 1.3.9 (b).

(b) We will show the assertion for $i = 1$, while the case $i = 2$ is similar. For every $[b_1], [b_1'] \in \Pi_{\mathcal{E}}^{-1}([a_1], \overline{\kappa})$, the quotient $\left\langle \frac{a_1, a_1'; [b_1']}{a_2, a_2'; [b_2]} \right\rangle / \left\langle \frac{a_1, a_1'; [b_1]}{a_2, a_2'; [b_2]} \right\rangle$ equals the pairing $\left\langle \overline{\text{inv}}(a_1, a_1'), \kappa\left(\frac{[b_1]}{[b_1']}\right)(\overline{s}) \right\rangle$ (use Lemma 1.5.7 (a) and (b)). Since by definition, elements $\kappa\left(\frac{[b_1]}{[b_1']}\right)(\overline{s})$ run through $Z_{\widehat{[a]}}(Z(\mathcal{E}, [a_1], \overline{\kappa}))$, our assertion follows from the last assertion of Lemma 1.5.7 (b). □

Notation 1.5.10. (a) Assume that in Notation 1.5.5, φ is $(\mathcal{E}, [a_1], \overline{\kappa}_{[b_1]})$-admissible (resp. $(\mathcal{E}, [a_2], \overline{\kappa}_{[b_2]})$-admissible). Then we will denote $\left\langle \frac{a_1, a_1'; [b_1]}{a_2, a_2'; [b_2]} \right\rangle$ by $\left\langle \frac{a_1, a_1'; \overline{\kappa}_{[b_1]}}{a_2, a_2'; [b_2]} \right\rangle$ (resp. $\left\langle \frac{a_1, a_1'; [b_1]}{a_2, a_2'; \overline{\kappa}_{[b_2]}} \right\rangle$). This notion is well defined by Lemma 1.5.9 (b).

(b) If in addition, $\varphi_X : X \to X'$ is an inner twisting induced by φ, $a_1 = a_x$ and $a'_1 = a_{x'}$ (see 1.1.1 (c) and 1.1.9 (c)), then we will denote $\left\langle \frac{a_1, a'_1; \overline{\kappa}[b_1]}{a_2, a'_2; [b_2]} \right\rangle$ simply by $\left\langle \frac{x, x'; \overline{\kappa}[b_1]}{a_2, a'_2; [b_2]} \right\rangle$.

1.6. Definitions of stability and equivalence

In this subsection we will define the notions of stability and equivalence of invariant generalized functions.

1.6.1. Set up (a) Let G be a connected reductive group over a local non-archimedean field E, ω_G a non-zero translation invariant differential form on G of the top degree, and $dg := |\omega_G|$ the corresponding Haar measure on $G(E)$. Let $\omega_{\mathcal{G}}$ a non-zero translation invariant differential form on $\mathcal{G}$ of the top degree such that the identification $T_1(G) = \mathcal{G} = T_0(\mathcal{G})$ identifies $\omega_G|_{g=1}$ with $\omega_{\mathcal{G}}|_{x=0}$.

We also assume that $\mathcal{G}$ contain regular semisimple elements, which hold automatically if the characteristic of E is different from two.

(b) Let (X, ω_X) be a pair consisting of a smooth algebraic variety X over E, equipped with a G^{ad}-action, and a nowhere vanishing G-invariant top degree differential form ω_X on X, and let $dx := |\omega_X|$ be the corresponding measure on $X(E)$. We assume that $X^{\mathrm{sr}} \subset X$ is Zariski dense. This condition automatically implies that $X^{\mathrm{sr}} \subset X$ is open (see Lemma 1.7.6 (a), below). In order to avoid dealing with algebraic spaces in Subsection 1.7, we assume that X is quasi-projective.

The results from this and the next subsections will later be used in two particular cases: $(X, \omega_X) = (G, \omega_G)$ and $(X, \omega_X) = (\mathcal{G}, \omega_{\mathcal{G}})$ with the actions of $\operatorname{Int} G = G^{\mathrm{ad}}$ and $\operatorname{Ad} G = G^{\mathrm{ad}}$, respectively.

(c) Let $\varphi : G \to G'$ be an inner twisting. The inner twist X' of X is smooth, and the differential form $\omega_{X'} := (\varphi_X^{-1})^*(\omega_X)$ on X' is E-rational and G'^{ad}-invariant.

We also denote by G^* the quasi-split inner twist of G, and by X^* the corresponding twist of X.

(d) Let $\mathcal{E} = (H, [\eta], \overline{s})$ be an endoscopic triple for G.

Notation 1.6.2. (a) let $C_c^\infty(X(E))$ be the space of locally constant functions on $X(E)$ with compact support, and let $\mathcal{S}(X(E))$ be the space of locally constant measures on $X(E)$ with compact support, that is, measures of the form $\phi = f dx$, where $f \in C_c^\infty(X(E))$.

Denote by $\mathcal{D}(X(E)) = (\mathcal{S}(X(E))^*)^{G(E)}$ the space of $G(E)$-invariant linear functionals on $\mathcal{S}(X(E))$, which we call *(invariant) generalized functions*.

(b) Let $U \subset X(E)$ be an open and closed subset. For each $\phi = f dx \in \mathcal{S}(X(E))$, put $\phi|_U := (f|_U)dx \in \mathcal{S}(X(E))$. Moreover, if U is $G(E)$-invariant, then for each $F \in \mathcal{D}(X(E))$, the generalized function $F|_U$ given by the formula $F|_U(\phi) := F(\phi|_U)$ belongs to $\mathcal{D}(X(E))$.

(c) For every smooth morphism $\pi : X_1 \to X_2$, the integration along fibers $\pi_!$ maps $\mathcal{S}(X_1(E))$ to $\mathcal{S}(X_2(E))$. Moreover, if π is G-equivariant, then the dual of $\pi_!$ induces a map $\pi^* : \mathcal{D}(X_2(E)) \to \mathcal{D}(X_1(E))$.

Remark 1.6.3. The map $\phi \mapsto \frac{\phi}{dx}$ identifies $\mathcal{S}(X(E))$ with $C_c^\infty(X(E))$, hence the space $\mathcal{D}(X(E))$ with the space of invariant distributions on $X(E)$. Below we list several reasons why $\mathcal{S}(X(E))$ and $\mathcal{D}(X(E))$ are more convenient to work with.

(i) The space $C_c^\infty(X(E))$ is not functorial with respect to non-proper maps.
(ii) Characters of admissible representations of $G(E)$ belong to $\mathcal{D}(X(E))$.
(iii) Orbital integrals behave better (see Remark 1.6.5 below).

Notation 1.6.4. For each $x \in X^{\mathrm{sr}}(E)$ and $\overline{\xi} \in \pi_0(\widehat{G_x}^\Gamma/Z(\widehat{G})^\Gamma)$,

(i) fix an invariant measure dg_x on $G_x(E)$ such that the total measure of the maximal compact subgroup of $G_x(E)$ is 1, and define orbital integral $O_x \in \mathcal{D}(X(E))$ by the formula

$$O_x(\phi) := \int_{G(E)/G_x(E)} \left(\frac{\phi}{dx}\right)(g(x))\frac{dg}{dg_x}$$

for every $\phi \in \mathcal{S}(X(E))$;

(ii) denote by $O_x^{\overline{\xi}} \in \mathcal{D}(X(E))$ the sum $\sum_{x'} \langle \mathrm{inv}(x, x'), \overline{\xi} \rangle O_{x'}$, taken over a set of representatives $x' \in X(E)$ of conjugacy classes stably conjugate to x;

(iii) when $\overline{\xi} = 1$, we will write SO_x instead of $O_x^{\overline{\xi}}$. More generally, for each $x^* \in (X^*)^{\mathrm{sr}}(E)$ we define $SO_{x^*} \in \mathcal{D}(X(E))$ be zero unless there exists a stable conjugate $x \in X(E)$ of x^*, in which case $SO_{x^*} := SO_x$ (compare Corollary 1.1.11).

Remark 1.6.5. If (X, ω_X) is either (G, ω_G) or $(\mathcal{G}, \omega_{\mathcal{G}})$, then measure dx is induced by dg, and orbital integrals O_x are independent of a choice of dg.

Definition 1.6.6. (i) A measure $\phi \in \mathcal{S}(X(E))$ is called *$\mathcal{E}$-unstable* if $O_x^{\overline{\xi}}(\phi) = 0$ for all $x \in X^{\mathrm{sr}}(E)$ and $\overline{\xi} \in \pi_0(\widehat{G_x}^\Gamma/Z(\widehat{G})^\Gamma)$ such that $([a_x], \overline{\xi}) \in \mathrm{Im}\,\Pi_{\mathcal{E}}$.

(ii) A generalized function $F \in \mathcal{D}(X(E))$ is called *$\mathcal{E}$-stable* if $F(\phi) = 0$ for all $\mathcal{E}$-unstable $\phi \in \mathcal{S}(X(E))$.

Remark 1.6.7. Denote by $\mathcal{D}^0(X(E)) \subset \mathcal{D}(X(E))$ the closure of the linear span of $\{O_x\}_{x \in X^{\mathrm{sr}}(E)}$. Our notion of $\mathcal{E}$-stability (and of $(a, a'; [b])$-equivalence below) seems to be "correct" only for generalized functions belonging to $\mathcal{D}^0(X(E))$.

However, all generalized functions considered in this paper belong to $\mathcal{D}^0(X(E))$. Indeed, if (X, ω_X) is either (G, ω_G) or $(\mathcal{G}, \omega_{\mathcal{G}})$, it follows from the results of Harish-Chandra [HC2, Theorem 3.1] (at least when the characteristic of E is zero) that $\mathcal{D}^0(X(E)) = \mathcal{D}(X(E))$ (see also Remarks 1.7.2 and 1.7.12 below).

Notation 1.6.8. Fix a triple $(a, a'; [b])$, consisting of stably conjugate embeddings of maximal tori $a : T \hookrightarrow G, a' : T \hookrightarrow G'$, and a stable conjugacy class $[b]$ of embeddings $T \hookrightarrow H$, compatible with a and a'. For every $\phi' \in \mathcal{S}(X'(E)), x \in X^{\mathrm{sr}}(E)$ and embedding $c : G_x \hookrightarrow H$ compatible with $a_x : G_x \hookrightarrow G$, we define

$$(O_x^{[c]})_{(a,a';[b])} := \sum_{x'} \left\langle \frac{a_x, a_{x'}; [c]}{a, a'; [b]} \right\rangle O_{x'} \in \mathcal{D}(X'(E)),$$

where the sum is taken over a set of representatives $x' \in X'(E)$ of conjugacy classes stably conjugate to x.

Remark 1.6.9. If $\varphi = \mathrm{Id}_G$ (and $a' = a$), then $(O_x^{[c]})_{(a,a;[b])} = O_x^{\overline{\kappa}[c]}$ (by Lemma 1.5.7 (d)). In general, $(O_x^{[c]})_{(a,a';[b])}$ vanishes unless there exists a stable conjugate $x' \in X'(E)$ of x, in which case, $(O_x^{[c]})_{(a,a';[b])} = \left\langle \frac{a_x, a_{x'};[c]}{a,a';[b]} \right\rangle O_{x'}^{\overline{\kappa}[c]}$ (by Lemma 1.5.7 (a) and (c)).

Definition 1.6.10. Let $(a, a'; [b])$ be as in Notation 1.6.8. By Lemma 1.1.10 (b), we can choose a stably conjugate embedding $a^* : T \hookrightarrow G^*$ of a and a'.

(a) Measures $\phi \in \mathcal{S}(X(E))$ and $\phi' \in \mathcal{S}(X'(E))$ are called $(a, a'; [b])$-*indistinguishable*, if for each $x^* \in (X^*)^{\mathrm{sr}}(E)$ and each embedding $c : G^*_{x^*} \hookrightarrow H$ compatible with $a_{x^*} : G^*_{x^*} \hookrightarrow G^*$, we have

$$(O_{x^*}^{[c]})_{(a^*,a;[b])}(\phi) = (O_{x^*}^{[c]})_{(a^*,a';[b])}(\phi'). \tag{1.6.1}$$

(b) Generalized functions $F \in \mathcal{D}(X(E))$ and $F' \in \mathcal{D}(X'(E))$ are called $(a, a'; [b])$-*equivalent* if $F(\phi) = F'(\phi')$ for every pair of $(a, a'; [b])$-indistinguishable measures $\phi \in \mathcal{S}(X(E))$ and $\phi' \in \mathcal{S}(X'(E))$.

Lemma 1.6.11. *Measures $\phi \in \mathcal{S}(X(E))$ and $\phi' \in \mathcal{S}(X'(E))$ are $(a, a'; [b])$-indistinguishable if and only if the following conditions are satisfied:*

(i) For each $x \in X^{\mathrm{sr}}(E)$ and $\overline{\xi} \in \pi_0(\widehat{G_x}^{\Gamma}/Z(\widehat{G})^{\Gamma})$ such that $([a_x], \overline{\xi}) \in \mathrm{Im}\,\Pi_{\mathcal{E}}$ and x does not have a stable conjugate element in $X'(E)$, we have $O_x^{\overline{\xi}}(\phi) = 0$.

(ii) Condition (i) holds if x, X, G, ϕ are replaced by x', X', G', ϕ'.

(iii) For every stable conjugate $x \in X^{\mathrm{sr}}(E)$ and $x' \in X'^{\mathrm{sr}}(E)$ and every $\overline{\xi} \in \pi_0(\widehat{G_x}^{\Gamma}/Z(\widehat{G})^{\Gamma})$ such that $([a_x], \overline{\xi}) \in \mathrm{Im}\,\Pi_{\mathcal{E}}$ we have

$(iii)'$ $O_x^{\overline{\xi}}(\phi) = O_{x'}^{\overline{\xi}}(\phi') = 0$, if φ is not $(\mathcal{E}, [a_x], \overline{\xi})$-admissible, and

$(iii)''$ $O_x^{\overline{\xi}}(\phi) = \left\langle \frac{x,x';\overline{\xi}}{a,a';[b]} \right\rangle O_{x'}^{\overline{\xi}}(\phi')$, if φ is $(\mathcal{E}, [a_x], \overline{\xi})$-admissible.

Proof. Fix $x^* \in (X^*)^{\mathrm{sr}}(E)$ and $\overline{\xi} \in \pi_0(\widehat{G^*_{x^*}}^{\Gamma}/Z(\widehat{G})^{\Gamma})$ such that $([a_{x^*}], \overline{\xi}) \in \mathrm{Im}\,\Pi_{\mathcal{E}}$. Using Remark 1.6.9, we see that equalities (1.6.1) for all $[c] \in \Pi_{\mathcal{E}}^{-1}([a_{x^*}], \overline{\xi})$ are equivalent to equalities:

- $0 = 0$, if there are no stable conjugates of x^* either in $X(E)$ or in $X'(E)$;

- $O_x^{\overline{\xi}}(\phi) = 0$, if there exists a stable conjugate $x \in X(E)$ of x^* but there is no such conjugate in $X'(E)$;

- $O_{x'}^{\overline{\xi}}(\phi') = 0$, if there exists a stable conjugate $x' \in X'(E)$ of x^* but there is no such conjugate in $X(E)$;

- $O_x^{\overline{\xi}}(\phi) = \left\langle \frac{a_x, a_{x'};[c]}{a,a';[b]} \right\rangle O_{x'}^{\overline{\xi}}(\phi')$ for all $[c] \in \Pi_{\mathcal{E}}^{-1}([a_x], \overline{\xi})$, if there exist stable conjugates $x \in X(E)$ and $x' \in X'(E)$ of x^* (use Lemma 1.5.7 (f)).

Moreover, by Lemma 1.5.9 (b), the last equalities are equivalent to the equalities $(iii)'$ and $(iii)''$. Now the assertion follows from Corollary 1.1.11. □

Corollary 1.6.12. *(a) The notion of $(a, a'; [b])$-equivalence is independent of the choice of a^*.*

(b) *Every two* $(a, a'; [b])$*-equivalent generalized functions* F *and* F' *are* $\mathcal{E}$*-stable.*

(c) *Assume that* φ *is not* $\mathcal{E}$*-admissible. Then every* $\mathcal{E}$*-stable* $F \in \mathcal{D}(X(E))$ *and* $F' \in \mathcal{D}(X'(E))$ *are* $(a, a'; [b])$*-equivalent.*

(d) *Assume that* $a(T)$ *is elliptic. If* $F \in \mathcal{D}(X(E))$ *and* $F' \in \mathcal{D}(X'(E))$ *are* $(a, a'; [b])$*-equivalent, then they are* $(a, a'; [b'])$*-equivalent for all* $b' : T \hookrightarrow H$ *such that* $\Pi_{\mathcal{E}}([b']) = \Pi_{\mathcal{E}}([b])$*.*

Proof. All assertions follow almost immediately from Lemma 1.6.11.

(a) is clear.

(b) By duality, we have to check that every $\mathcal{E}$-unstable measures $\phi \in \mathcal{S}(X(E))$ and $\phi' \in \mathcal{S}(X'(E))$ are $(a, a'; [b])$-indistinguishable, which is clear.

(c) By duality, we have to check that every $(a, a'; [b])$-indistinguishable $\phi \in \mathcal{S}(X(E))$ and $\phi' \in \mathcal{S}(X'(E))$ are $\mathcal{E}$-unstable. Hence the assertion follows from the first assertion of Lemma 1.5.9 (a).

(d) When φ is not $\mathcal{E}$-admissible, the assertion was proved in (c). When φ is $\mathcal{E}$-admissible, the assertion follows from Lemma 1.5.9. □

Corollary 1.6.13. *Let* $\pi : \widetilde{G} \to G$ *be a quasi-isogeny, and let* $\widetilde{\varphi} : \widetilde{G} \to \widetilde{G}', \widetilde{\mathcal{E}}$, $(\widetilde{a}, \widetilde{a}'; [\widetilde{b}])$ *be the corresponding objects for* $\widetilde{G}$*. Generalized functions* $F \in \mathcal{D}(X(E))$ *and* $F' \in \mathcal{D}(X'(E))$ *are* $(\widetilde{a}, \widetilde{a}'; [\widetilde{b}])$*-equivalent if and only if they are* $(a, a'; [b])$*-equivalent.*

Proof. By duality, we have to show that measures $\phi \in \mathcal{S}(X(E))$ and $\phi' \in \mathcal{S}(X'(E))$ are $(a, a'; [b])$-indistinguishable if and only if they are $(\widetilde{a}, \widetilde{a}'; [\widetilde{b}])$-indistinguishable. It follows from Lemma 1.3.10, that for each $x \in X^{\mathrm{sr}}(E)$ and $\overline{\overline{\xi}} \in \pi_0(\widehat{\widetilde{G}_x}^{\Gamma} / Z(\widehat{\widetilde{G}})^{\Gamma})$ such that $([\widetilde{a}_x], \overline{\overline{\xi}}) \in \operatorname{Im} \Pi_{\widetilde{\mathcal{E}}}$, there exists $\overline{\xi} \in \pi_0(\widehat{G_x}^{\Gamma} / Z(\widehat{G})^{\Gamma})$ such that $(a_x, \overline{\xi}) \in \operatorname{Im} \Pi_{\mathcal{E}}$ and $\overline{\overline{\xi}}$ is the image of $\overline{\xi}$. Moreover, we have $O_x^{\overline{\overline{\xi}}} = c O_x^{\overline{\xi}}$, where $c \in \mathbb{C}^{\times}$ is such that measure $\frac{d\widetilde{g}}{d\widetilde{g}_x}$ on $(\widetilde{G}/\widetilde{G}_x)(E) = (G/G_x)(E)$ equals $c \frac{dg}{dg_x}$. Therefore the assertion follows from Lemmas 1.6.11, 1.5.7 (e), and 1.3.10 (c), (d). □

Definition 1.6.14. Let $a : T \hookrightarrow G$ and $a' : T \hookrightarrow G'$ be stable conjugate embeddings of maximal elliptic tori; let κ be an element of $\widehat{T}^{\Gamma} / Z(\widehat{G})^{\Gamma}$ such that $([a], \kappa) \in \operatorname{Im} \Pi_{\mathcal{E}}$. We say that $F \in \mathcal{D}(X(E))$ and $F' \in \mathcal{D}(X'(E))$ are $(a, a'; \kappa)$*-equivalent* if they are $(a, a'; [b])$-equivalent for some or, equivalently, for all $[b] \in \Pi_{\mathcal{E}}^{-1}([a], \kappa)$ (see Corollary 1.6.12 (d)).

Lemma 1.6.15. (a) *Let* $\pi : X_1 \to X_2$ *be a smooth* G*-equivariant morphism. For every* $\mathcal{E}$*-unstable* $\phi \in \mathcal{S}(X_1(E))$*, its push-forward* $\pi_!(\phi) \in \mathcal{S}(X_2(E))$ *is* $\mathcal{E}$*-unstable.*

(b) *Let* $\varphi : G \to G'$ *be an inner twisting, and* $\pi' : X_1' \to X_2'$ *the corresponding inner twisting of* π*. For every* $(a, a'; [b])$*-indistinguishable* $\phi \in \mathcal{S}(X_1(E))$ *and* $\phi' \in \mathcal{S}(X_1'(E))$*, their push-forwards* $\pi_!(\phi) \in \mathcal{S}(X_1(E))$ *and* $\pi_!'(\phi') \in \mathcal{S}(X_1'(E))$ *are* $(a, a'; [b])$*-indistinguishable.*

Proof. Recall that ω_{X_1} and ω_{X_2} are global nowhere vanishing sections of sheaves of top degree differential forms $\Omega_{X_1}^{\dim X_1}$ and $\Omega_{X_2}^{\dim X_2}$, respectively. Therefore

$\omega_{X_1} \otimes \pi^*(\omega_{X_2})^{-1}$ is a global nowhere vanishing section of $\Omega_{X_1}^{\dim X_1} \otimes \pi^*(\Omega_{X_2}^{\dim X_2})^{-1}$, which induces a measure, denoted by $dy := \frac{dx_1}{dx_2}$ on all fibers of $\pi(E) : X_1(E) \to X_2(E)$.

For every $x \in X_2^{\mathrm{sr}}(E)$ and $y \in \pi(E)^{-1}(x)$, we have $y \in X_1^{\mathrm{sr}}(E)$ and $G_x = G_y$. Moreover, for all $\overline{\xi} \in \pi_0(\widehat{G_x}^{\Gamma}/Z(\widehat{G})^{\Gamma})$, we have $O_x^{\overline{\xi}}(\pi_!(\phi)) = \int_{\pi(E)^{-1}(x)} O_y^{\overline{\xi}}(\phi)dy$. From this the assertion follows. □

Lemma 1.6.15 has the following corollary.

Corollary 1.6.16. *(a) Let $\pi : X_1 \to X_2$ be a smooth G-equivariant morphism. For every $\mathcal{E}$-stable $F \in \mathcal{D}(X_2(E))$, its pullback $\pi^*(F) \in \mathcal{D}(X_1(E))$ is $\mathcal{E}$-stable.*
(b) Let $\varphi : G \to G'$ be an inner twisting, and $\pi' : X_1' \to X_2'$ the inner twist of π. For every $(a, a'; [b])$-equivalent $F \in \mathcal{D}(X_2(E))$ and $F' \in \mathcal{D}(X_2'(E))$, their pullbacks $\pi^*(F) \in \mathcal{D}(X_1(E))$ and $\pi'^*(F') \in \mathcal{D}(X_1'(E))$ are $(a, a'; [b])$-equivalent.

1.7. Locally L^1 functions

The goal of this subsection is to write down explicitly the condition for $\mathcal{E}$-stability and $(a, a'; [b])$-equivalence of generalized functions coming from invariant locally L^1 functions.

Notation 1.7.1. (a) Denote by $L^1_{\mathrm{loc}}(X(E))$ the space of $G(E)$-invariant locally L^1 functions on $X(E)$, whose restriction to its open subset $X^{\mathrm{sr}}(E)$ is locally constant.

(b) We have a canonical embedding $L^1_{\mathrm{loc}}(X(E)) \hookrightarrow \mathcal{D}(X(E))$, which sends each $F \in L^1_{\mathrm{loc}}(X(E))$ to a generalized function $\phi \mapsto \int_{X(E)} F\phi$. For simplicity of notation, we identify functions from $L^1_{\mathrm{loc}}(X(E))$ with the corresponding generalized functions from $\mathcal{D}(X(E))$.

Remark 1.7.2. For every $F \in L^1_{\mathrm{loc}}(X(E)))$, the corresponding generalized function is contained in $\mathcal{D}^0(X(E))$.

Notation 1.7.3. For a $G(E)$-invariant function $F : X(E) \to \mathbb{C}$ and a pair $x \in X^{\mathrm{sr}}(E)$ and $\overline{\xi} \in \pi_0(\widehat{G_x}^{\Gamma}/Z(\widehat{G})^{\Gamma})$, we put $F(x, \overline{\xi}) := \sum_{x'} \langle \mathrm{inv}(x, x'), \overline{\xi}\rangle^{-1} F(x')$, where $x' \in X(E)$ runs over a set of representatives of $G(E)\backslash[x] \subset G(E)\backslash X(E)$.

Proposition 1.7.4. *In the notation of Definition 1.6.10,*

(a) $F \in L^1_{\mathrm{loc}}(X(E)) \subset \mathcal{D}(X(E))$ is $\mathcal{E}$-stable if and only if for every pair $x \in X^{\mathrm{sr}}(E)$ and $\overline{\xi} \in \pi_0(\widehat{G_x}^{\Gamma}/Z(\widehat{G})^{\Gamma})$ such that $([a_x], \overline{\xi}) \notin \mathrm{Im}\,\Pi_{\mathcal{E}}$, we have $F(x, \overline{\xi}) = 0$.
(b) $F \in L^1_{\mathrm{loc}}(X(E))$ and $F' \in L^1_{\mathrm{loc}}(X'(E))$ are $(a, a'; [b])$-equivalent if and only if the following two conditions are satisfied:
(i) F and F' are $\mathcal{E}$-stable;

(ii) for all stably conjugate $x \in X^{\mathrm{sr}}(E)$ *and* $x' \in X'^{\mathrm{sr}}(E)$ *and all* $\overline{\xi} \in \pi_0(\widehat{G_x}^{\Gamma}/Z(\widehat{G})^{\Gamma})$ *such that* $([a_x], \overline{\xi}) \in \operatorname{Im}\Pi_{\mathcal{E}}$ *and* φ *is* $(\mathcal{E}, [a_x], \overline{\xi})$*-admissible, we have*

$$F'(x', \overline{\xi}) = \left\langle \frac{x, x'; \overline{\xi}}{a, a'; [b]} \right\rangle F(x, \overline{\xi}).$$

After certain preparations, the proof of the proposition will be carried out in 1.7.14.

Notation 1.7.5. Denote by $\underline{Tor}$ the variety of all maximal tori in G.

Lemma 1.7.6. *(a) The subset* $X^{\mathrm{sr}} \subset X$ *is open, and there exists a smooth morphism* $\pi : X^{\mathrm{sr}} \to \underline{Tor}$ *such that* $\pi(x) = G_x$ *for each* $x \in X^{\mathrm{sr}}$.

(b) There exists a geometric quotient $Y = G\backslash X^{\mathrm{sr}}$*. Moreover, the canonical projection* $f : X^{\mathrm{sr}} \to Y$ *is smooth, the restriction of* f *to each fiber of* π *is étale, and the induced map* $f(E) : X^{\mathrm{sr}}(E) \to Y(E)$ *is a (locally) trivial fibration.*

Proof. (a) Denote by X^{reg} the set of $x \in X$ such that $\dim \mathcal{G}_x = \mathrm{rk}_{\overline{E}}(G)$. Then X^{reg} contains X^{sr}, and therefore X^{reg} is Zariski dense in X. Our first step will be to show that X^{reg} is open in X, and the map $x \mapsto \mathcal{G}_x$ gives an algebraic morphism $\overline{\pi}$ from X^{reg} to the Grassmannian $\mathrm{Gr}_{\mathcal{G},\mathrm{rk}_{\overline{E}}(G)}$, classifying linear subspaces of $\mathcal{G}$ of dimension $\mathrm{rk}_{\overline{E}}(G)$.

Observe that the action $\mu : G \times X \to X$ induces a map $T(\mu) : T(G) \times T(X) \to T(X)$ of tangent bundles. The restriction of $T(\mu)$ to $\mathcal{G} \times X$, where $\mathcal{G} = T_1(G) \subset T(G)$, and $X \subset T(X)$ is the zero section, is a map of vector bundles $f : \mathcal{G} \times X \to T(X)$ such that for every $x \in X$, the kernel of $f_x : \mathcal{G} \to T_x(X)$ is $\mathcal{G}_x$. In other words, X^{reg} can be described as the set of $x \in X$ such that $\mathrm{rk}\, f_x = \dim \mathcal{G} - \mathrm{rk}_{\overline{E}}(G)$. Since X^{reg} is dense in X, we get that $\mathrm{rk}\, f_x \leq \dim \mathcal{G} - \mathrm{rk}_{\overline{E}}(G)$ for each $x \in X$, and $X^{\mathrm{reg}} \subset X$ is open. Moreover, the restriction $\mathrm{Ker}\, f|_{X^{\mathrm{reg}}}$ is a vector subbundle of $\mathcal{G} \times X^{\mathrm{reg}}$; therefore it gives rise to a morphism $\overline{\pi} : X^{\mathrm{reg}} \to \mathrm{Gr}_{\mathcal{G},\mathrm{rk}_{\overline{E}}(G)}$ such that $\overline{\pi}(x) = \mathcal{G}_x$.

Next consider a subset X^{rss} of X^{reg} consisting of points x such that $\mathcal{G}_x \subset \mathcal{G}$ is a Cartan subalgebra of $\mathcal{G}$ (hence $G_x^0 \subset G$ is a maximal torus). Since we assumed that $\mathcal{G}^{\mathrm{rss}} \neq \emptyset$, every Cartan subalgebra of $\mathcal{G}$ has a non-zero intersection with $\mathcal{G}^{\mathrm{rss}}$. Hence X^{rss} equals the set of $x \in X^{\mathrm{reg}}$ such that $\mathcal{G}_x \cap \mathcal{G}^{\mathrm{rss}} \neq \emptyset$. Since $\mathcal{G}^{\mathrm{rss}}$ is open in $\mathcal{G}$, we conclude that X^{rss} is open in X^{reg}, hence in X.

Note that the map $T \mapsto \mathcal{T} = \mathrm{Lie}\, T$ identifies $\underline{Tor}$ with the variety of Cartan subalgebras of $\mathcal{G}$, which is a locally closed subvariety of $\mathrm{Gr}_{\mathcal{G},\mathrm{rk}_{\overline{E}}(G)}$. Therefore the restriction of $\overline{\pi}|_{X^{\mathrm{rss}}}$ can be viewed as a morphism $\pi : X^{\mathrm{rss}} \to \underline{Tor}$ such that $\pi(x) = G_x^0$ for each $x \in X^{\mathrm{rss}}$.

We claim that π is smooth. Since both X^{rss} and $\underline{Tor}$ are smooth, we only have to check that the differential $d\pi_x : T_x(X) \to T_{\pi(x)}(\underline{Tor})$ is surjective for each $x \in X^{\mathrm{rss}}$. Put $T := \pi(x)$. Then $G_x^0 = T$, $G_x \subset \mathrm{Norm}_G(T)$, $G(x) \cong G/G_x$ and $\underline{Tor} \cong G/\mathrm{Norm}_G(T)$. Hence $\pi|_{G(x)}$ is étale, thus $d\pi_x|_{T_x(G(x))}$ is surjective, which implies the surjectivity of $d\pi_x$.

It remains to show that X^{sr} is open in X^{rss}. Fix $T \in \underline{Tor}$, and put $W = \mathrm{Norm}_G(T)/T$. Then W acts on $\pi^{-1}(T)$, and $Z_T := \pi^{-1}(T) \cap X^{\mathrm{sr}}$ consists of points

$x \in \pi^{-1}(T)$ such that $w(x) \neq x$ for all $w \neq 1$. Hence $Z_T \subset \pi^{-1}(T)$ is open, and W acts freely on Z_T. In particular, Z_T is smooth.

Consider the natural map $\iota : (G/T) \times_W Z_T \to X^{\mathrm{rss}} : [g, x] \mapsto g(x)$. This is a map between smooth spaces, which induces an isomorphism between tangent spaces, hence ι is étale. Since ι induces a bijection between $(G/T) \times_W Z_T$ and $X^{\mathrm{sr}} \subset X^{\mathrm{rss}}$, we get that $X^{\mathrm{sr}} \cong (G/T) \times_W Z_T$ is open in X^{rss}.

(b) Since X is quasi-projective, Z_T is quasi-projective as well. Hence a quasi-projective scheme $Y := W\backslash Z_T$ is a geometric quotient $G\backslash X^{\mathrm{sr}} = G\backslash[(G/T) \times_W Z_T]$. Moreover, the projection $f : X^{\mathrm{sr}} \to Y$ is smooth, and $f|_{Z_T}$ is étale.

To show the last assertion, choose $x \in X^{\mathrm{sr}}(E)$ and put $T := G_x$. Since the projection $f|_{Z_T}$ is étale, there exist open neighborhoods $U \subset Z_T(E)$ of x and $V \subset Y(E)$ of $f(x)$ such that f induces a homeomorphism $U \overset{\sim}{\to} V$. Then the map $(G/T)(E) \times U \to X^{\mathrm{sr}}(E)$ sending $([g], u)$ to $g(u)$ induces a $G(E)$-equivariant isomorphism between $(G/T)(E) \times U \cong (G/T)(E) \times V$ and $f(E)^{-1}(V)$. □

Construction 1.7.7. (a) For every $T \in \underline{Tor}\,(E)$ and an open and compact subset $U \subset \pi^{-1}(T)(E)$, there exists a measure $\phi_U \in \mathcal{S}(X(E))$ such that $O_x(\phi_U) = 1$ for each $x \in G(E)(U)$, and $O_x(\phi_U) = 0$ otherwise.

Explicitly, for each open and compact subgroup $K \subset G(E)$, the measure $\phi_U := \frac{|dt|(K \cap T(E))}{|dg|(K)} \chi_{K(U)} dx$, where $\chi_{K(U)}$ is the characteristic function of $K(U) \subset X^{\mathrm{sr}}(E)$ and dt is an invariant measure on $T(E)$ such that the total measure of the maximal compact subgroup of $T(E)$ is 1, satisfies the required properties.

(b) For every stable conjugates $x \in X^{\mathrm{sr}}(E)$ and $x' \in X'^{\mathrm{sr}}(E)$, there exists a natural isomorphism $\varphi_{x,x'}$ between $\pi^{-1}(G_x) \subset X^{\mathrm{sr}}$ and $\pi'^{-1}(G'_{x'}) \subset X'^{\mathrm{sr}}$.

Explicitly, choose $g \in G(E^{\mathrm{sep}})$ such that $x' = \varphi_X(g(x))$. Then the map $X_{E^{\mathrm{sep}}} \to X'_{E^{\mathrm{sep}}} : y \mapsto \varphi_X(g(y))$ maps $\pi^{-1}(G_x)$ to $\pi'^{-1}(G'_{x'})$, and the corresponding morphism $\varphi_{x,x'} : \pi^{-1}(G_x) \to \pi'^{-1}(G'_{x'})$ is E-rational, independent of g and $\varphi_{x,x'}(x) = x'$.

Corollary 1.7.8. *(a) Given $x \in X^{\mathrm{sr}}(E)$ and $\overline{\xi} \in \pi_0(\widehat{G_x}^{\Gamma}/Z(\widehat{G})^{\Gamma})$ such that $([a_x], \overline{\xi}) \notin \operatorname{Im} \Pi_{\mathcal{E}}$, there exists an $\mathcal{E}$-unstable measure $\phi \in \mathcal{S}(X(E))$ such that $O_x^{\overline{\xi}}(\phi) \neq 0$ and $O_x^{\overline{\xi}'}(\phi) = 0$ for each $\overline{\xi}' \neq \overline{\xi}$.*

(b) Let $x \in X^{\mathrm{sr}}(E)$ and $\overline{\xi} \in \pi_0(\widehat{G_x}^{\Gamma}/Z(\widehat{G})^{\Gamma})$ be such that $([a_x], \overline{\xi}) \in \operatorname{Im} \Pi_{\mathcal{E}}$, φ is $(\mathcal{E}, [a_x], \overline{\xi})$-admissible and there exists a stable conjugate $x' \in X'(E)$ of x. Then there exist $(a, a'; [b])$-indistinguishable measures $\phi \in \mathcal{S}(X(E))$ and $\phi' \in \mathcal{S}(X'(E))$ such that $O_x^{\overline{\xi}}(\phi) \neq 0$, $O_{x'}^{\overline{\xi}}(\phi') \neq 0$ and $O_x^{\overline{\xi}'}(\phi) = O_{x'}^{\overline{\xi}'}(\phi') = 0$ for each $\overline{\xi}' \neq \overline{\xi}$.

Proof. (a) Let $x_1 = x, \dots, x_n \in X(E)$ be a set of representatives of conjugacy classes stably conjugate to x. Choose an open neighborhood $U \subset \pi^{-1}(G_x)$ of x, and for every $i = 1, \dots, n$ put $U_i := \varphi_{x,x_i}(U) \subset \pi^{-1}(G_{x_i})(E)$, and let $\phi_{U_i} \in \mathcal{S}(X(E))$ be as in Construction 1.7.7 (a). Then measure $\phi := \sum_i \langle \mathrm{inv}(x, x_i), \overline{\xi} \rangle^{-1} \phi_{U_i}$ satisfies the required property.

(b) Now choose a set of representatives $x'_1, \dots, x'_n \in X'(E)$ of conjugacy classes stably conjugate to x, and put $U'_i := \varphi_{x,x'_i}(U) \subset \pi'^{-1}(G'_{x_i})(E)$. Then measures $\phi :=$

$\sum_i \langle \text{inv}(x, x_i), \overline{\xi} \rangle^{-1} \phi_{U_i}$ and $\phi' := \sum_i \left\langle \frac{x, x'_i; \overline{\xi}}{a, a'; [b]} \right\rangle^{-1} \phi_{U_i}$ satisfy the required property (use Lemma 1.6.11). □

Lemma 1.7.9. *(a) Let $F \in \mathcal{D}(X(E))$ be of the form $F = \sum_{\overline{\xi}} c_{x,\overline{\xi}} O_x^{\overline{\xi}}$, where $x \in X^{\text{sr}}(E)$, $c_{x,\overline{\xi}} \in \mathbb{C}$ and $\overline{\xi}$ runs over $\pi_0(\widehat{G_x}^{\Gamma}/Z(\widehat{G})^{\Gamma})$. Then F is $\mathcal{E}$-stable if and only if $c_{x,\overline{\xi}} = 0$ for each $\overline{\xi}$ with $([a_x], \overline{\xi}) \notin \text{Im}\,\Pi_{\mathcal{E}}$.*

(b) Let $F \in \mathcal{D}(X(E))$ and let $F' \in \mathcal{D}(X'(E))$ be of the form $F = \sum_{\overline{\xi}} c_{x,\overline{\xi}} O_x^{\overline{\xi}}$ and $F' = \sum_{\overline{\xi}} c_{x',\overline{\xi}} O_{x'}^{\overline{\xi}}$ for some stable conjugate $x \in X^{\text{sr}}(E)$ and $x' \in X'^{\text{sr}}(E)$. Then F and F' are $(a, a'; [b])$-equivalent if and only if they satisfy the following two conditions:

(i) F and F' are $\mathcal{E}$-stable;

(ii) for each $\overline{\xi} \in \pi_0(\widehat{G_x}^{\Gamma}/Z(\widehat{G})^{\Gamma}) = \pi_0(\widehat{G'_{x'}}^{\Gamma}/Z(\widehat{G'})^{\Gamma})$ such that $([a_x], \overline{\xi}) \in \text{Im}\,\Pi_{\mathcal{E}}$ and φ is $(\mathcal{E}, [a_x], \overline{\xi})$-admissible, we have $c_{x',\overline{\xi}} = \left\langle \frac{x, x'; \overline{\xi}}{a, a'; [b]} \right\rangle c_{x,\overline{\xi}}$.

Proof. (a) The "if" assertion is clear. The "only if" assertion follows from the equality $F(\phi) = 0$ applied to measure ϕ from Corollary 1.7.8 (a).

(b) Assume that F and F' satisfy assertions (i) and (ii). Then it follows from Lemma 1.6.11 and assertion (a), that for every $(a, a'; [b])$-indistinguishable $\phi \in \mathcal{S}(X(E))$ and $\phi' \in \mathcal{S}(X'(E))$ we have $F(\phi) = F'(\phi')$. Conversely, assume that F and F' are $(a, a'; [b])$-equivalent. Then condition (i) was proved in Corollary 1.6.12 (b) and condition (ii) follows from the equality $F(\phi) = F'(\phi')$ applied to measures ϕ and ϕ' from Corollary 1.7.8 (b). □

The following result is clear.

Lemma 1.7.10. *Let $f : Z \to Y$ be a morphism of smooth algebraic varieties over E such that the induced map $f(E) : Z(E) \to Y(E)$ is a locally trivial fibration. Fix a measure μ on $Y(E)$, and let $U_1 \supset U_2 \supset \dots$ be a basis of open and compact neighborhoods of $y \in Y(E)$. Then for every locally constant function F on $Z(E)$ and every $\phi \in \mathcal{S}(Z(E))$, the sequence $\frac{1}{\mu(U_i)} F|_{f^{-1}(U_i)}(\phi)$ stabilizes.*

Notation 1.7.11. For each $x \in X^{\text{sr}}(E)$ and $F \in L^1_{\text{loc}}(X(E))$, denote by $F_x \in \mathcal{D}(X(E))$ the generalized function $\phi \mapsto SO_x(F\phi)$. For each $x^* \in (X^*)^{\text{sr}}(E)$, $F \in L^1_{\text{loc}}(X(E))$ and $F' \in L^1_{\text{loc}}(X'(E))$, we denote by $F_{x^*} \in \mathcal{D}(X(E))$ and $F'_{x^*} \in \mathcal{D}(X'(E))$ the generalized functions $\phi \mapsto SO_{x^*}(F\phi)$ and $\phi' \mapsto SO_{x^*}(F'\phi')$, respectively (see Notation 1.6.4 (iii)).

Remark 1.7.12. Clearly, F_x and F_{x^*} belong to $\mathcal{D}^0(X(E))$.

Claim 1.7.13. *(a) $F \in L^1_{\text{loc}}(X(E))$ is $\mathcal{E}$-stable if and only if $F_x \in \mathcal{D}(X(E))$ is $\mathcal{E}$-stable for all $x \in X^{\text{sr}}(E)$.*

(b) $F \in L^1_{\text{loc}}(X(E))$ and $F' \in L^1_{\text{loc}}(X'(E))$ are $(a, a'; [b])$-equivalent if and only if $F_{x^*} \in \mathcal{D}(X(E))$ and $F'_{x^*} \in \mathcal{D}(X'(E))$ are $(a, a'; [b])$-equivalent for all $x^* \in (G^*)^{\text{sr}}(E)$.

Proof. (a) Since X^{sr} is Zariski dense in X, the complement $X(E) \smallsetminus X^{\mathrm{sr}}(E)$ is nowhere dense. As $F \in L^1_{\mathrm{loc}}(X(E))$, we get that F is $\mathcal{E}$-stable if and only if the restriction $F|_{X^{\mathrm{sr}}(E)}$ is $\mathcal{E}$-stable.

Consider the map $f : X^{\mathrm{sr}} \to Y$ from Lemma 1.7.6. Then for every $x \in X^{\mathrm{sr}}(E)$ and $\phi \in \mathcal{S}(X(E))$, the value $F_x(\phi)$ is the limit of the stabilizing sequence $\frac{1}{\mu(U_i)} F|_{f^{-1}(U_i)}(\phi)$, where U_i is any basis of open and compact neighborhoods of $f(x) \in Y(E)$ (use Lemma 1.7.10). From this the assertion follows.

Indeed, if $F|_{X^{\mathrm{sr}}(E)}$ is $\mathcal{E}$-stable, then each $F|_{f^{-1}(U_i)}$ is $\mathcal{E}$-stable. In particular, for every $\mathcal{E}$-unstable ϕ, we have $F|_{f^{-1}(U_i)}(\phi) = 0$ for each i, hence $F_x(\phi) = 0$. Conversely, assume that each F_x is $\mathcal{E}$-stable, and pick an $\mathcal{E}$-unstable ϕ. Then there exists an open disjoint covering $\{U_\alpha\}_\alpha$ of $f(\mathrm{Supp}\,\phi)$ such that each $F|_{f^{-1}(U_\alpha)}(\phi) = 0$ for each α, hence $F(\phi) = \sum_\alpha F|_{f^{-1}(U_\alpha)}(\phi) = 0$. This shows that F is $\mathcal{E}$-stable.

(b) As in (a), F and F' are $(a, a'; [b])$-equivalent if and only if $F|_{X^{\mathrm{sr}}(E)}$ and $F'|_{X'^{\mathrm{sr}}(E)}$ are $(a, a'; [b])$-equivalent. Next since G acts trivially on Y, we get identifications $Y' = Y^* = Y$, and the projection $f : X^{\mathrm{sr}} \to Y$ induces maps $f' : X'^{\mathrm{sr}} \to Y$ and $f^* : (X^*)^{\mathrm{sr}} \to Y$. Moreover, by Corollary 1.1.11, both $\mathrm{Im}\, f(E)$ and $\mathrm{Im}\, f'(E)$ are contained in $\mathrm{Im}\, f^*(E)$. Now the assertion follows from Lemma 1.7.10 by exactly the same arguments as (a). □

1.7.14. *Proof of Proposition* 1.7.4. (a) For each $x \in X^{\mathrm{sr}}(E)$, we denote by N_x the cardinality of $\pi_0(\widehat{G_x}^{\Gamma}/Z(\widehat{G})^{\Gamma})$. By Claim 1.7.13 (a), F is $\mathcal{E}$-stable if and only if each F_x is stable. Since $F_x = \frac{1}{N_x}\sum_{\overline{\xi}} F(x, \xi) O_x^{\overline{\xi}}$, the assertion then is just a reformulation of Lemma 1.7.9 (a).

(b) The proof of (b) is similar. By Claim 1.7.13 (b), $F \in L^1_{\mathrm{loc}}(X(E))$ and $F' \in L^1_{\mathrm{loc}}(X'(E))$ are $(a, a'; [b])$-equivalent if and only if each $F_{x^*} \in \mathcal{D}(X(E))$ and $F'_{x^*} \in \mathcal{D}(X'(E))$ are $(a, a'; [b])$-equivalent. By definition this means that each F_x and $F'_{x'}$ are $\mathcal{E}$-stable, and for every stable conjugate $x \in X^{\mathrm{sr}}(E)$ and $x' \in X'^{\mathrm{sr}}(E)$, F_x and $F'_{x'}$ are $(a, a'; [b])$-equivalent. But by Claim 1.7.13 (a) and Lemma 1.7.9 (b), these conditions are equivalent to conditions (i) and (ii), respectively. □

Corollary 1.7.15. *(a) Let $\pi : X_1 \to X_2$ be a smooth G-equivariant morphism of varieties as in 1.6.1 (b) and $U \subset X_1(E)$ an open $G(E)$-invariant subset. For every $F \in L^1_{\mathrm{loc}}(X_2(E))$ such that $\pi^*(F)|_U$ is $\mathcal{E}$-stable, the restriction $F|_{\pi(U)}$ is $\mathcal{E}$-stable.*

(b) Let $\pi' : X'_1 \to X'_2$ be the inner twisting of π and $U' \subset X'_1(E)$ an open $G'(E)$-invariant subset. For every $F' \in L^1_{\mathrm{loc}}(X'_2(E))$ such that $\pi^*(F)|_U$ and $\pi'^*(F')|_{U'}$ are $(a, a'; [b])$-equivalent, the restrictions $F|_{\pi(U)}$ and $F'|_{\pi'(U')}$ are $(a, a'; [b])$-equivalent.

Proof. (a) Since π is smooth and G-equivariant, the subset $\pi(U) \subset X_2(E)$ is open and $G(E)$-invariant. Thus $F|_{\pi(U)}$ is defined and belongs to $L^1_{\mathrm{loc}}(X_2(E))$. Let $(x, \overline{\xi})$ be as in Proposition 1.7.4 (a). If $[x] \cap \pi(U) = \emptyset$, then $F|_{\pi(U)}$ vanishes on $[x]$, hence $F|_{\pi(U)}(x, \overline{\xi}) = 0$. Assume now that $[x] \cap \pi(U) \neq \emptyset$. Replacing x by a stable conjugate, we can assume that $x = \pi(x_1)$ for some $x_1 \in X_1(E)$. Then $x_1 \in X_1^{\mathrm{sr}}(E)$, $G_{x_1} = G_x$, $a_{x_1} = a_x$, and π induces a $G(E)$- and Γ-equivariant isomorphism

$[x_1] \xrightarrow{\sim} [x]$. Hence $F|_{\pi(U)}(x, \overline{\xi}) = \pi^*(F)|_U(x_1, \overline{\xi}) = 0$. Thus the assertion follows from Proposition 1.7.4 (a).

(b) Follows from Proposition 1.7.4 (b) by exactly the same arguments as (a). □

1.8. Quasi-logarithm maps

Starting from 1.8.5, E will be a local non-archimedean field, $\mathcal{O}$ the ring of integers of E, $\mathfrak{m}$ the maximal ideal of $\mathcal{O}$, $\mathbb{F}_q$ the residue field of E, p the characteristic of $\mathbb{F}_q$, and G a reductive group over E split over E^{nr}.

Definition 1.8.1. Let G be an algebraic group over a field k. By a *quasi-logarithm* we call a G^{ad}-equivariant algebraic morphism $\Phi : G \to \mathcal{G}$ such that $\Phi(1) = 0$ and $d\Phi_1 : \mathcal{G} = T_1(G) \to \mathcal{G}$ is the identity map.

Example 1.8.2. Let $\rho : G \to \operatorname{Aut} V$ be a representation such that the corresponding G-invariant pairing $\langle a, b\rangle_\rho := \operatorname{Tr}(\rho(a)\rho(b))$ on $\mathcal{G}$ is non-degenerate. Denote by $\operatorname{pr}_\rho : \operatorname{End} V \to \mathcal{G}$ be the projection given by the rule $\operatorname{Tr}(\operatorname{pr}_\rho(A)\rho(b)) = \operatorname{Tr}(A\rho(b))$ for each $b \in \mathcal{G}$. Then the map $\Phi_\rho : g \mapsto \operatorname{pr}_\rho(\rho(g) - \operatorname{Id}_V)$ is a quasi-logarithm $G \to \mathcal{G}$.

Lemma 1.8.3. *Let $\Phi : G \to \mathcal{G}$ be a quasi-logarithm map.*

(a) For every Borel subgroup B of G, we have $\Phi(B) \subset \operatorname{Lie} B$;
(b) If a Cartan subgroup of G is a maximal torus, then Φ induces a quasi-logarithm map $G_{\mathrm{red}} := G/R_u(G) \to \mathcal{G}_{\mathrm{red}}$

Proof. Let $T \subset B$ be a maximal torus, and $C := \operatorname{Cent}_G(T)$ the corresponding Cartan subgroup.

(a) Since Φ is G^{ad}-equivariant, $\Phi(C)$ is contained in the set of fixed points of $\operatorname{Ad} T$ in $\mathcal{G}$, that is, $\Phi(C) \subset \operatorname{Lie} C$. Therefore

$$\Phi(\operatorname{Int} B(C)) = \operatorname{Ad} B(\Phi(T)) \subset \operatorname{Ad} B(\operatorname{Lie} C) \subset \operatorname{Lie} B.$$

Since C is a Cartan subgroup of B, $\operatorname{Int} B(C) \subset B$ is Zariski dense, hence $\Phi(B)$ is contained in $\operatorname{Lie} B$.

(b) We have to show that for each $g \in G$ and $u \in R_u(G)$, we have

$$\Phi(gu) - \Phi(g) \in \operatorname{Lie} R_u(G). \tag{1.8.1}$$

Since we assumed that T is a Cartan subgroup, $\operatorname{Int} G(T)$ is Zariski dense in G. Therefore it is enough to check the equality (1.8.1) only for $g \in \operatorname{Int} G(T)$, hence (since Φ is G^{ad}-equivariant and $R_u(G)$ is normal in G), only for $g = t \in T$. Consider subgroup $H = TR_u(G) \subset G$. Then T is a Cartan subgroup of H, hence $\operatorname{Int} H(T)$ is Zariski dense in H. Since $H = T \ltimes R_u(G)$, it will therefore suffice to check (1.8.1) under the additional assumption that $tu = vtv^{-1}$ for certain $v \in R_u(G)$. In this case,

$$\Phi(tu) - \Phi(t) = (\operatorname{Ad} v - \operatorname{Id})(\Phi(t)) \in (\operatorname{Ad} v - \operatorname{Id})(\mathcal{T}) \subset \operatorname{Lie} R_u(G),$$

as claimed. □

Remark 1.8.4. If we do not assume that the Cartan subgroup of G is a maximal torus, then the assertion (b) of the lemma is obviously false. For example, it is false for abelian groups.

From now on, G is a reductive group over a local non-archimedean field E, which is split over E^{nr}.

1.8.5. Bruhat–Tits building (a) Denote by $\mathcal{B}(G)$ the (non-reduced) Bruhat–Tits building of G. For every point $x \in \mathcal{B}(G)$, we denote by $G_x \subset G(E)$ (resp. $\mathcal{G}_x \subset \mathcal{G}$) the corresponding parahoric subgroup (resp. subalgebra), and let $G_{x^+} \subset G_x$ (resp. $\mathcal{G}_{x^+} \subset \mathcal{G}_x$) be the pro-unipotent (resp. pro-nilpotent) radical of G_x (resp. of $\mathcal{G}_x$) (compare [MP1]).

(b) For every $x \in \mathcal{B}(G)$, denote by $\underline{G}_x$ the canonical smooth connected group scheme over $\mathcal{O}$ whose generic fiber is G and $\underline{G}_x(\mathcal{O}) = G_x$, and let $\overline{G}_x$ be the special fiber of $\underline{G}_x$. Then $\overline{G}_x$ is a connected group over $\mathbb{F}_q$, whose Cartan subgroup is a maximal torus. (Here we use the assumption that G splits over E^{nr}).

(c) For every $x \in \mathcal{B}(G)$, denote by L_x the quotient $(\overline{G}_x)_{\mathrm{red}} = \overline{G}_x/R_u(\overline{G}_x)$. We have canonical identifications $L_x(\mathbb{F}_q) = G_x/G_{x^+}$ and $\mathcal{L}_x := \mathrm{Lie}\, L_x = \mathcal{G}_x/\mathcal{G}_{x^+}$. For every $g \in G_x$ and $a \in \mathcal{G}_x$, we put $\overline{g} := gG_{x^+} \in L_x(\mathbb{F}_q)$ and $\overline{a} := a + \mathcal{G}_{x^+} \in \mathcal{L}_x(\mathbb{F}_q)$.

(d) If $G = T$ is a torus, then the group scheme $\underline{T}_x$ is independent of $x \in \mathcal{B}(T)$, and coincides with the canonical $\mathcal{O}$-structure $T_{\mathcal{O}}$ of T. We denote by $\overline{T}$ the special fiber of $T_{\mathcal{O}}$, and will write $T(\mathcal{O})$ instead of $T_{\mathcal{O}}(\mathcal{O})$.

Notation 1.8.6. (a) We call an invariant pairing $\langle\cdot,\cdot\rangle$ on $\mathcal{G}$ *non-degenerate at* $x \in \mathcal{B}(G)$ if it is non-degenerate over E and the dual lattice

$$(\mathcal{G}_x)^{\perp} := \{x \in \mathcal{G}(E) \mid \langle x, y\rangle \in \mathfrak{m} \text{ for each } y \in \mathcal{G}_x\}$$

equals $\mathcal{G}_{x^+}$.

(b) We call a quasi-logarithm $\Phi : G \to \mathcal{G}$ *defined at* $x \in \mathcal{B}(G)$ if Φ extends to the morphism $\Phi_x : \underline{G}_x \to \mathcal{G}_x$ of schemes over $\mathcal{O}$.

Lemma 1.8.7. *(a) If an invariant pairing $\langle\cdot,\cdot\rangle$ on $\mathcal{G}$ is non-degenerate at x for some $x \in \mathcal{B}(G)$, then it is non-degenerate at x for all $x \in \mathcal{B}(G)$. In this case, $\langle\cdot,\cdot\rangle$ defines an invariant non-degenerate pairing $\langle\cdot,\cdot\rangle_x$ on $\mathcal{L}_x$ for all $x \in \mathcal{B}(G)$.*

(b) If a quasi-logarithm $\Phi : G \to \mathcal{G}$ over E is defined at x for some $x \in \mathcal{B}(G)$, then it is defined at x for all $x \in \mathcal{B}(G)$. In this case, Φ gives rise to a quasi-logarithm map $\overline{\Phi}_x : L_x \to \mathcal{L}_x$ for all $x \in \mathcal{B}(G)$.

Proof. (a) Assume that $(\mathcal{G}_x)^{\perp} = \mathcal{G}_{x^+}$. Since $\mathfrak{m}\mathcal{G}_x \subset \mathcal{G}_{x^+}$, we get that $\langle a, b\rangle \in \mathcal{O}$ for every $a, b \in \mathcal{G}_x$, and the pairing on $\mathcal{L}_x$ given by the formula $\langle \overline{a}, \overline{b}\rangle_x := \overline{\langle a, b\rangle} \in \mathbb{F}_q$ for every $a, b \in \mathcal{G}_x$ is well defined, L_x-invariant and non-degenerate.

It remains to show that for every $x, y \in \mathcal{B}(G)$, the equalities $(\mathcal{G}_x)^{\perp} = \mathcal{G}_{x^+}$ and $(\mathcal{G}_y)^{\perp} = \mathcal{G}_{y^+}$ are equivalent. For this we can extend scalars to E^{nr}. Also we can assume that $\mathcal{G}_y$ is an Iwahori subalgebra of $\mathcal{G}_x$.

Assume first that $(\mathcal{G}_x)^{\perp} = \mathcal{G}_{x^+}$. Since $\mathcal{G}_{x^+} \subset \mathcal{G}_y \subset \mathcal{G}_x$, we get that $\mathcal{G}_{x^+} \subset (\mathcal{G}_y)^{\perp} \subset \mathcal{G}_x$. Moreover, $(\mathcal{G}_y)^{\perp}/\mathcal{G}_{x^+} \subset \mathcal{L}_x$ is the orthogonal complement of the Borel subalgebra $\mathcal{G}_y/\mathcal{G}_{x+} \subset \mathcal{L}_x$ with respect to the non-degenerate pairing $\langle\cdot,\cdot\rangle_x$. Hence $(\mathcal{G}_y)^{\perp}/\mathcal{G}_{x+}$ is the nilpotent radical of $\mathcal{G}_y/\mathcal{G}_{x+}$, thus $(\mathcal{G}_y)^{\perp} = \mathcal{G}_{y+}$. Conversely, assume that $(\mathcal{G}_y)^{\perp} = \mathcal{G}_{y+}$. Since $\mathcal{G}_x = \cup_{g\in G_x} \operatorname{Ad} g(\mathcal{G}_y)$ and $\mathcal{G}_{x+} = \cap_{g\in G_x} \operatorname{Ad} g(\mathcal{G}_{y+})$, we get that $\mathcal{G}_{x+} = (\mathcal{G}_x)^{\perp}$, as claimed.

(b) The strategy will be similar to that of (a). Assume that Φ extends to a morphism $\Phi_x : \underline{G}_x \to \mathcal{G}_x$. Then the special fiber of Φ_x is a quasi-logarithm $\overline{G}_x \to \overline{\mathcal{G}}_x$. Since $L_x = (\overline{G}_x)_{\text{red}}$, the existence $\overline{\Phi}_x$ follows from Lemma 1.8.3 (b) and the observation of 1.8.5 (b).

It remains to show that for every $x, y \in \mathcal{B}(G)$, the existence of Φ_x is equivalent to that of Φ_y. Notice first that the existence of Φ_x is equivalent to the fact that $\Phi(\underline{G}_x(\mathcal{O}_{E^{\text{nr}}})) \subset \mathcal{G}_x \otimes_{\mathcal{O}} \mathcal{O}_{E^{\text{nr}}}$ (see [BT, Prop. 1.7.6]). Thus we can extend scalars to E^{nr}, and we are required to check that the inclusions $\Phi(G_x) \subset \mathcal{G}_x$ and $\Phi(G_y) \subset \mathcal{G}_y$ are equivalent. Also we can assume that G_y is an Iwahori subgroup of G_x.

Assume that $\Phi(G_x) \subset \mathcal{G}_x$. As we have shown, Φ induces a quasi-logarithm $\overline{\Phi}_x : L_x \to \mathcal{L}_x$. By Lemma 1.8.3 (a), we get $\overline{\Phi}_x(G_y/G_{x+}) \subset \mathcal{G}_y/\mathcal{G}_{x+}$, thus $\Phi(G_y) \subset \mathcal{G}_y$. Conversely, assume that $\Phi(G_y) \subset \mathcal{G}_y$. As Φ is G^{ad} equivariant, the inclusion $\Phi(G_x) \subset \mathcal{G}_x$ follows from equalities $G_x = \cup_{g\in G_x} gG_yg^{-1}$ and $\mathcal{G}_x = \cup_{g\in G_x} \operatorname{Ad} g(\mathcal{G}_y)$. □

Lemma 1.8.7 allows us to give the following definition.

Definition 1.8.8. (a) We call an invariant pairing $\langle\cdot,\cdot\rangle$ on $\mathcal{G}$ *non-degenerate over* $\mathcal{O}$ if it is non-degenerate at x for some (or, equivalently, for all) $x \in \mathcal{B}(G)$.

(b) We call a quasi-logarithm $\Phi : G \to \mathcal{G}$ *defined over* $\mathcal{O}$ if it is defined at x for some (or, equivalently, for all) $x \in \mathcal{B}(G)$.

Lemma 1.8.9. *(a) Let $\varphi : G \to G'$ be an inner twisting defined over E^{nr}. Every non-degenerate over $\mathcal{O}$ invariant pairing $\langle\cdot,\cdot\rangle$ on $\mathcal{G}$ gives rise to the corresponding pairing $\langle\cdot,\cdot\rangle'$ on $\mathcal{G}'$. Every quasi-logarithm $\Phi : G \to \mathcal{G}$ defined over $\mathcal{O}$ gives rise to the corresponding quasi-logarithm $\Phi' : G' \to \mathcal{G}'$.*

(b) If the pairing $\langle\cdot,\cdot\rangle_\rho$ on $\mathcal{G}$ corresponding to a representation $\rho : G \to \operatorname{Aut}(V)$ is non-degenerate over $\mathcal{O}$, then the corresponding quasi-logarithm $\Phi_\rho : G \to \mathcal{G}$ is defined over $\mathcal{O}$.

Proof. (a) Recall that φ induces isomorphisms $G_{E^{\text{nr}}} \overset{\sim}{\to} G'_{E^{\text{nr}}}$ and $\mathcal{G}_{E^{\text{nr}}} \overset{\sim}{\to} \mathcal{G}'_{E^{\text{nr}}}$. Hence Φ and $\langle\cdot,\cdot\rangle$ give rise to a quasi-logarithm $\Phi' : G'_{E^{\text{nr}}} \to \mathcal{G}'_{E^{\text{nr}}}$ and a pairing $\langle\cdot,\cdot\rangle' : \mathcal{G}_{E^{\text{nr}}} \times \mathcal{G}_{E^{\text{nr}}} \to E^{\text{nr}}$, respectively. Furthermore, since the twisting φ is inner, while Φ and $\langle\cdot,\cdot\rangle$ are G^{ad}-equivariant, the quasi-logarithm Φ' and the inner twisting $\langle\cdot,\cdot\rangle'$ are defined over E. Finally, to show that Φ' and $\langle\cdot,\cdot\rangle'$ are defined over $\mathcal{O}$, we can extend scalars to E^{nr}. Then the assertion follows from the corresponding assertion for Φ and $\langle\cdot,\cdot\rangle$.

(b) For the proof we can replace E by its finite unramified extension so that G is split over E. Fix a hyperspecial vertex $x \in \mathcal{B}(G)$; we have to show that $\Phi_\rho(G_x) \subset \mathcal{G}_x$. As $(\mathcal{G}_x)^{\perp} = \mathcal{G}_{x+}$, it is enough to show that $\operatorname{Tr}(\rho(g)\rho(a)) \in \mathfrak{m}$ for every $g \in G_x$ and $a \in \mathcal{G}_{x+} = \mathfrak{m}\mathcal{G}_x$. Choose any $\rho(G_x)$-invariant $\mathcal{O}$-lattice $V_x \subset V$. Then V_x is $\rho(\mathcal{G}_x)$-invariant, hence $\rho(g)\rho(a)(V_x) \subset \mathfrak{m}V_x$. Thus $\operatorname{Tr}(\rho(g)\rho(a)) \in \mathfrak{m}$, as claimed. □

Definition 1.8.10. We say that the group G over E *satisfies property* (vg) if G^{sc} admits a quasi-logarithm map $G^{\mathrm{sc}} \to \mathcal{G}^{\mathrm{sc}}$ defined over $\mathcal{O}$, $\mathcal{G}^{\mathrm{sc}}$ admits an invariant pairing non-degenerate over $\mathcal{O}$, and p does not divide the order of $Z(G^{\mathrm{sc}})$.

Remark 1.8.11. By Lemma 1.8.9 (a), for every inner twisting $\varphi : G \to G'$, the group G satisfies property (vg) if and only if G' satisfies property (vg).

Lemma 1.8.12. *Write* $(G^*)^{\mathrm{sc}}$ *in the form* $\prod_i R_{E_i/E} H_i$, *where each* H_i *is a quasi-split absolutely simple over a finite unramified extension* E_i *of* E. *Then* G *satisfies property* (vg), *if the following conditions are satisfied:*

(i) p *is good for each* H_i *in the sense of [SS, I, §4];*
(ii) p *does not divide the order of each* $Z(H_i)$*;*
(iii) p *does not divide* $[E_i[H_i] : E_i]$, *where* $E_i[H_i]$ *is the splitting field of* H_i.

Proof. Assume that p satisfies assumptions (i)–(iii) of the lemma. By Lemma 1.8.9 (a), we may replace G by $(G^*)^{\mathrm{sc}}$ hence by each $R_{E_i/E} H_i$, thus assuming that G is quasi-split simple and simply connected. By Lemma 1.8.9 (b), it will suffice to construct a representation ρ of G such that the corresponding pairing $\langle \cdot, \cdot \rangle_\rho$ on $\mathcal{G}$ is non-degenerate over $\mathcal{O}$. We will construct such a ρ in three steps.

Assume first that $G = H_i$ is split. In this case, take ρ to be the standard representation if G is classical, and the adjoint representation if G is exceptional. Then assumptions (i) and (ii) imply (as in [SS, I, Lemma. 5.3]) that the pairing $\langle \cdot, \cdot \rangle_\rho$ is non-degenerate at every hyperspecial vertex of $\mathcal{B}(G)$, hence non-degenerate over $\mathcal{O}$.

Next we assume that $G = H_i$ is absolutely simple, set $E' := E[G]$, and put $G' := G_{E'}$. Then by the claims proven above, there exists a representation $\rho' : G' \to \mathrm{Aut}_{E'}(V)$ such that the pairing $\langle \cdot, \cdot \rangle_{\rho'}$ on $\mathcal{G}'$ is non-degenerate over $\mathcal{O}_{E'}$. Take ρ to be the restriction of $R_{E'/E}\rho' : R_{E'/E} G' \to \mathrm{Aut}_E(V)$ to G. Extending scalars to E', we get $\rho_{E'} \cong (\rho')^{[E':E]}$, hence $\langle \cdot, \cdot \rangle_{\rho_{E'}} = [E' : E] \langle \cdot, \cdot \rangle_{\rho'}$. Therefore by assumption (iii), $\langle \cdot, \cdot \rangle_{\rho_{E'}}$ is non-degenerate over $\mathcal{O}_{E'}$, and thus $\langle \cdot, \cdot \rangle_\rho$ is non-degenerate over $\mathcal{O}$.

In the general case, choose a representation $\rho' : H_i \to \mathrm{Aut}_{E_i}(V)$ such that $\langle \cdot, \cdot \rangle_{\rho'}$ is non-degenerate over $\mathcal{O}_{E_i}$. Then the representation $\rho := R_{E_i/E}\rho' : G \to \mathrm{Aut}_E(V)$ satisfies the required property. □

Remark 1.8.13. The designation (vg) was chosen to indicate the fact that it is closely related to the notion of a very good prime. (Recall that p is called very good for H_i if it satisfies properties (i) and (ii) of Lemma 1.8.12.)

Notation 1.8.14. (a) For an algebraic group H, we denote by $\mathcal{U}(H) \subset H$ and $\mathcal{N}(\mathcal{H}) \subset \mathcal{H}$ the subvarieties of unipotent elements of H and of nilpotent elements of $\mathcal{H}$, respectively.

(b) For every $x \in \mathcal{B}(G)$, we denote by $G_{x,\mathrm{tu}} \subset G_x$ and $\mathcal{G}_{x,\mathrm{tn}} \subset \mathcal{G}_x$ the preimages of $\mathcal{U}(L_x)(\mathbb{F}_q) \subset L_x(\mathbb{F}_q)$ and $\mathcal{N}(\mathcal{L}_x)(\mathbb{F}_q) \subset \mathcal{L}_x$, respectively.

(c) Put $G(E)_{\mathrm{tu}} := \cup_{x \in \mathcal{B}(G)} G_{x,\mathrm{tu}}$ and $\mathcal{G}(E)_{\mathrm{tn}} := \cup_{x \in \mathcal{B}(G)} \mathcal{G}_{x,\mathrm{tn}}$.

Lemma 1.8.15. *For every* $x \in \mathcal{B}(G)$,

(a) $G_{x,\mathrm{tu}} = \cup_y G_{y+}$ *and* $\mathcal{G}_{x,\mathrm{tn}} = \cup_y \mathcal{G}_{y+}$, *where* y *runs over the union of alcoves in* $\mathcal{B}(G)$, *whose closures contain* x.

(b) $G_{x,\mathrm{tu}} = G_x \cap G(E)_{\mathrm{tu}} \subset G(E)$ *and* $\mathcal{G}_{x,\mathrm{tn}} = \mathcal{G}_x \cap \mathcal{G}(E)_{\mathrm{tn}} \subset \mathcal{G}(E)$.

Proof. (a) is clear.

(b) The first assertion follows from the equality $G_{x,\mathrm{tu}} = \{g \in G_x \mid g^{p^n} \underset{n\to\infty}{\longrightarrow} 1\}$. For the second equality, we have to show that for each $z \in \mathcal{B}(G)$, there exists y as in (a), such that $\mathcal{G}_x \cap \mathcal{G}_{z+} \subset \mathcal{G}_{y+}$. But every y, lying in the segment $[x, z] \subset \mathcal{B}(G)$, satisfies this property. □

Proposition 1.8.16. *Let* $\Phi : G \to \mathcal{G}$ *be a quasi-logarithm defined over* $\mathcal{O}$.

(a) For every $x \in \mathcal{B}(G)$, Φ *induces measure preserving analytic isomorphisms* $G_{x,\mathrm{tu}} \xrightarrow{\sim} \mathcal{G}_{x,\mathrm{tu}}$ *and* $G_{x+} \xrightarrow{\sim} \mathcal{G}_{x+}$ *(with respect to measures* $|\omega_G|$ *and* $|\omega_{\mathcal{G}}|$ *chosen in 1.6.1).*

(b) Φ *induces a measure preserving analytic isomorphism* $G(E)_{\mathrm{tu}} \xrightarrow{\sim} \mathcal{G}(E)_{\mathrm{tn}}$.

(c) Let ${}^0G \subset G$ *be the biggest open subset* $U \subset G$ *such that* $\Phi|_U$ *is étale. Then* ${}^0G(E)$ *contains* $G(E)_{\mathrm{tu}}$.

Proof. (a) For the proof, one can replace E by a finite unramified extension, so we can assume that G splits over E. By Lemma 1.8.7 (b), we have $\Phi(G_{x+}) \subset \mathcal{G}_{x+}$ for each $x \in \mathcal{B}(G)$; therefore by Lemma 1.8.15 (a), $\Phi(G_{x,\mathrm{tu}}) \subset \mathcal{G}_{x,\mathrm{tn}}$. Since $|\omega_G|(G_{x+}) = |\omega_{\mathcal{G}}|(\mathcal{G}_{x+})$, the second assertion follows from the first.

Let us first show the assertion for a hyperspecial vertex $x \in \mathcal{B}(G)$. By Lemma 1.8.7 (b), Φ extends to the morphism $\Phi_x : \underline{G_x} \to \mathcal{G}_x$ of schemes over $\mathcal{O}$, whose special fiber is a quasi-logarithm map $\overline{\Phi}_x : L_x \to \mathcal{L}_x$. By [BR, 9.1, 9.2, 9.3.3 and 6.3] (compare [BR, 9.4]), $\overline{\Phi}_x$ induces an isomorphism $\mathcal{U}(L_x) \xrightarrow{\sim} \mathcal{N}(\mathcal{L}_x)$. Moreover, there exists an open affine neighborhood $V \subset L_x$ of $\mathcal{U}(L_x)$ such that $\overline{\Phi}_x|_V$ is étale (see [BR, Thm 6.2 and 9.1]). Therefore by Hensel's lemma, Φ_x induces an analytic isomorphism $G_{x,\mathrm{tu}} \xrightarrow{\sim} \mathcal{G}_{x,\mathrm{tn}}$.

Since Φ_x is an algebraic morphism over $\mathcal{O}$, we get that $\Phi(gG_{x,r}) = \Phi(g) + \mathcal{G}_{x,r}$ for each $g \in G_{x,\mathrm{tu}}$ and $r \in \mathbb{N}$. But $|\omega_G|(gG_{x,r}) = |\omega_{\mathcal{G}}|(\Phi(g) + \mathcal{G}_{x,r})$, and $\{gG_{x,r}\}_{g,r}$ form a basis of open neighborhoods of $G_{x,\mathrm{tu}}$. Hence the analytic isomorphism $G_{x,\mathrm{tu}} \xrightarrow{\sim} \mathcal{G}_{x,\mathrm{tn}}$ is measure-preserving.

It remains to show that for every $x, y \in \mathcal{B}(G)$, the assertions for x and y are equivalent. Moreover, we can assume that G_y is an Iwahori subgroup of G_x (compare the proof of Lemma 1.8.7). Then $G_{y,\mathrm{tu}} = G_{x,\mathrm{tu}} \cap G_y$ and $\mathcal{G}_{y,\mathrm{tn}} = \mathcal{G}_{x,\mathrm{tn}} \cap \mathcal{G}_y$ (see Lemma 1.8.15), so the assertion for x implies that for y. The opposite direction follows from equalities $G_{x,\mathrm{tu}} = \cup_{g\in G_x} gG_{y,\mathrm{tu}}g^{-1}$ and $\mathcal{G}_{x,\mathrm{tn}} = \cup_{g\in G_x} \mathrm{Ad}\, g(\mathcal{G}_{y,\mathrm{tn}})$.

(b) By (a), we get that $\Phi(G(E)_{\mathrm{tu}}) = \mathcal{G}(E)_{\mathrm{tn}}$, and that the induced map $G(E)_{\mathrm{tu}} \to \mathcal{G}(E)_{\mathrm{tn}}$ is open. Thus we have to check that the restriction of Φ to $G(E)_{\mathrm{tu}}$ is one-to-one.

Assume that $g_1, g_2 \in G(E)_{\mathrm{tu}}$ satisfy $\Phi(g_1) = \Phi(g_2)$. Choose $x, y \in \mathcal{B}(G)$ such that $g_1 \in G_{x+}$ and $g_2 \in G_{y+}$ (use Lemma 1.8.15). By (a), Φ induces a measure-preserving embedding $G_{x+} \cap G_{y+} \hookrightarrow \mathcal{G}_{x+} \cap \mathcal{G}_{y+}$. Since measures of both sides are equal, the last embedding is surjective. But $\Phi(g_1) = \Phi(g_2)$ belongs to $\mathcal{G}_{x+} \cap \mathcal{G}_{y+}$, hence there exists $g_3 \in G_{x+} \cap G_{y+}$ such that $\Phi(g_1) = \Phi(g_3) = \Phi(g_2)$. Since $\Phi|_{G_{x+}}$ and $\Phi|_{G_{y+}}$ are injective, we get that $g_1 = g_3 = g_2$, as claimed.

(c) We have to check that for each $g \in G(E)_{\mathrm{tu}}$, the differential $d\Phi_g : T_g(G) \to T_{\Phi(g)}(\mathcal{G})$ is an isomorphism. But this follows from (b). □

We finish this subsection with a result, which we will need later.

Lemma 1.8.17. *(a) Let $\pi : \widetilde{G} \to G$ be an isogeny (that is, a finite surjective quasi-isogeny) of order prime to p. Then $\pi(\widetilde{G}(E)_{\mathrm{tu}}) = G(E)_{\mathrm{tu}}$.*
(b) Let $\pi : \widetilde{G} \to G$ be a surjective quasi-isogeny such that $S = \ker \pi$ is a torus split over E^{nr}. Then for every $x \in \mathcal{B}(\widetilde{G})$, we have $\pi(\widetilde{G}_x) = G_{\pi(x)}$.

Proof. (a) Since π is an isogeny, it identifies $\mathcal{B}(\widetilde{G})$ with $\mathcal{B}(G)$. Thus we have to check that for each $x \in \mathcal{B}(\widetilde{G}) = \mathcal{B}(G)$, we have $\pi(\widetilde{G}_{x,\mathrm{tu}}) = G_{x,\mathrm{tu}}$. Since the order of π is prime to p, the corresponding map $\pi_x : \underline{\widetilde{G}}_x \to \underline{G}_x$ of group schemes over $\mathcal{O}$ is étale. Since the special fiber $\overline{\pi}_x : \overline{\widetilde{G}}_x \to \overline{G}_x$ induces an isomorphism $\mathcal{U}(\overline{\widetilde{G}}_x) \xrightarrow{\sim} \mathcal{U}(\overline{G}_x)$, the assertion follows from Hensel's lemma.

(b) The homomorphism π gives rise to an exact sequence $1 \to S_{\mathcal{O}} \to \underline{\widetilde{G}}_x \xrightarrow{\pi_x} \underline{G}_{\pi(x)} \to 1$ of group schemes over $\mathcal{O}$. In particular, π_x is smooth. Passing to special fibers, we get an exact sequence $1 \to \overline{S} \to \overline{\widetilde{G}}_x \xrightarrow{\overline{\pi}_x} \overline{G}_{\pi(x)} \to 1$ of groups over $\mathbb{F}_q$. Since $\overline{S}$ is connected, we get that $H^1(\mathbb{F}_q, \overline{S}) = 1$, hence the map $\overline{\pi}_x(\mathbb{F}_q) = \pi_x(\mathbb{F}_q)$ is surjective. Therefore the surjectivity of $\pi_x(\mathcal{O})$ follows from Hensel's lemma. □

Corollary 1.8.18. *Let $\iota : G^{\mathrm{sc}} \to G$ be a canonical map. Then for every $x \in \mathcal{B}(G)$ and an unramified maximal torus $T \subset G$, we have $G_x \subset \iota(G^{\mathrm{sc}})(E) \cdot T(\mathcal{O})$.*

Proof. Assume first that $G^{\mathrm{der}} = G^{\mathrm{sc}}$. Denote by q the projection $G \to G^{\mathrm{ab}}$. Then we have to check that $q(G_x) \subset q(T(\mathcal{O}))$. Since $q(G_x) \subset G^{\mathrm{ab}}(\mathcal{O})$, the assertion follows from part (b) of the lemma applied to the morphism $q|_T : T \to G^{\mathrm{ab}}$.

For a general G, there exists a surjective quasi-isogeny $\pi : \widetilde{G} \to G$ such that $\widetilde{G}^{\mathrm{der}} = \widetilde{G}^{\mathrm{sc}} (= G^{\mathrm{sc}})$, and $\mathrm{Ker}\,\pi$ is an induced torus splitting over E^{nr} (see [MS, Prop. 3.1]). Then for every $\widetilde{x} \in \mathcal{B}(\widetilde{G})$ such that $\pi(\widetilde{x}) = x$ we have $\pi(\widetilde{G}_{\widetilde{x}}) = G_x$, so the assertion for G_x and T follows from that for $\widetilde{G}_{\widetilde{x}}$ and $\pi^{-1}(T) \subset \widetilde{G}$. □

2. Endoscopic decomposition

2.1. Main Theorem

In this subsection we will give two equivalent formulations of the main result of the paper.

2.1.1. Deligne–Lusztig representations Let L be a connected reductive group over $\mathbb{F}_q$, $\overline{a} : \overline{T} \hookrightarrow L$ an embedding of a maximal torus of L, and $\overline{\theta} : \overline{T}(\mathbb{F}_q) \to \mathbb{C}^\times$ a character. To this data Deligne and Lusztig [DL] associate a virtual representation $R^{\theta}_{\overline{a}(\overline{T})}$ of $L(\mathbb{F}_q)$. Moreover, if the torus $\overline{a}(\overline{T}) \subset L$ is elliptic and the character $\overline{\theta}$ is non-singular, then $\rho_{\overline{a},\overline{\theta}} := (-1)^{\mathrm{rk}_{\mathbb{F}_q}(L) - \mathrm{rk}_{\mathbb{F}_q}(\overline{T})} R^{\theta}_{\overline{a}(\overline{T})} (= e(L) R^{\theta}_{\overline{a}(\overline{T})})$ is a cuspidal representation (see [DL, Proposition 7.4 and Theorem 8.3]). In particular, $\rho_{\overline{a},\overline{\theta}}$ is a genuine representation and not a virtual one. Moreover, $\rho_{\overline{a},\overline{\theta}}$ is irreducible if $\overline{\theta}$ is in general position.

2.1.2. Recall that there is an equivalence of categories $T \mapsto \overline{T}$ between tori over E splitting over E^{nr} and tori over $\mathbb{F}_q$. Moreover, every such T has a canonical $\mathcal{O}$-structure. We denote by $T(\mathcal{O})^+$ and $\mathcal{T}(\mathcal{O})^+$ the kernels of the reduction maps $\mathrm{Ker}[T(\mathcal{O}) \to \overline{T}(\mathbb{F}_q)]$ and $\mathrm{Ker}[\mathcal{T}(\mathcal{O}) \to \overline{\mathcal{T}}(\mathbb{F}_q)]$, respectively.

Notation 2.1.3. (a) Let G be a reductive group over E, T a torus over E splitting over E^{nr}, and $a : T \hookrightarrow G$ an embedding of a maximal elliptic torus of G.

For every vertex x of $\mathcal{B}(T)$, $a(x)$ is a vertex of $\mathcal{B}(G)$. Moreover, since T is elliptic, we have $\mathcal{B}(T) = \mathcal{B}(Z(G)^0)$. Thus the $Z(G)^0(E)$-orbit of $a(x)$, hence also the parahoric subgroup $G_{a(x)}$ does not depend on x. Therefore we can denote $G_{a(x)}$ by G_a, and similarly for $\mathcal{G}_{a(x)}$, $G_{a(x)^+}$, $\mathcal{G}_{a(x)^+}$, $L_{a(x)}$ and $\mathcal{L}_{a(x)}$. We also set $\widetilde{G_a} := Z(G)(E)G_a$.

An embedding $a : T \hookrightarrow G$ induces an embedding $\overline{a} : \overline{T} \hookrightarrow L_a$ of a maximal elliptic torus of L_a.

(b) Let $\theta : T(E) \to \mathbb{C}^\times$ be a non-singular character (that is, θ is not orthogonal to any coroot of (G, T)), trivial on $T(\mathcal{O})^+$. Denote by $\overline{\theta} : \overline{T}(\mathbb{F}_q) \to \mathbb{C}^\times$ the character of $\overline{T}(\mathbb{F}_q)$ defined by θ. Then there exists a unique representation $\rho_{a,\theta}$ of $\widetilde{G_a}$, whose central character is the restriction of θ, extending the inflation to G_a of the Deligne–Lusztig representation $\rho_{\overline{a},\overline{\theta}}$ of $L_a(\mathbb{F}_q)$. We denote by $\pi_{a,\theta}$ the induced representation $\mathrm{Ind}_{\widetilde{G_a}}^{G(E)} \rho_{a,\theta}$ of $G(E)$. Since for each irreducible factor $\rho' \subset \rho_{a,\theta}$, the induced representation $\mathrm{Ind}_{\widetilde{G_a}}^{G(E)} \rho'$ is cuspidal and irreducible (see [MP2, Prop. 6.6]), we get that $\pi_{a,\theta}$ is a semisimple cuspidal representation of finite length, which is irreducible, if θ is in general position.

Definition 2.1.4. (a) Let $a : T \hookrightarrow G$ be an embedding of a maximal torus split over E^{nr}. We say that an element $\overline{t} \in \overline{T}(\mathbb{F}_q)$ is *a-strongly regular* if $\overline{t}$ is not fixed by a non-trivial element of the Weyl group $W(G, a(T)) \subset \mathrm{Aut}(T_{E^{\mathrm{nr}}}) = \mathrm{Aut}(\overline{T}_{\overline{\mathbb{F}_q}})$.

(b) Let $a^{\mathrm{sc}} : T^{\mathrm{sc}} \hookrightarrow G^{\mathrm{sc}}$ be the lift of a. We say that G *satisfies property* $(\upsilon g)_a$ if G satisfies property (υg) (see Definition 1.8.10) and there exists an a^{sc}*-strongly regular* element $\overline{t} \in \overline{T}^{\mathrm{sc}}(\mathbb{F}_q)$.

Notation 2.1.5. To each $\kappa \in \widehat{T}^{\Gamma}/Z(\widehat{G})^{\Gamma}$, an embedding $a_0 : T \hookrightarrow G$, and a character θ of $T(E)$ as in Notation 2.1.3, we associate an invariant generalized function

$$\chi_{a_0,\kappa,\theta} := e(G) \sum_{a}{}' \langle \mathrm{inv}(a_0, a), \kappa \rangle \, \chi(\pi_{a,\theta}) \in \mathcal{D}(G(E)).$$

Here a runs over a set of representatives of conjugacy classes of embeddings $T \hookrightarrow G$ which are stably conjugate to a_0, and $\chi(\pi_{a,\theta})$ denotes the character of $\pi_{a,\theta}$.

Now we are ready to formulate our main result of the paper.

Theorem 2.1.6. *Let (a_0, κ, θ) be as in Notation 2.1.5. Assume that the characteristic of E is zero and G satisfies property $(vg)_{a_0}$. Then*

(a) The generalized function $\chi_{a_0,\kappa,\theta}$ is $\mathcal{E}_{([a_0],\kappa)}$-stable.
(b) For each inner twisting $\varphi : G \to G'$ and each embedding $a'_0 : T \hookrightarrow G'$, stably conjugate to a_0, the generalized functions $\chi_{a_0,\kappa,\theta}$ on $G(E)$ and $\chi_{a'_0,\kappa,\theta}$ on $G'(E)$ are $(a_0, a'_0; \kappa)$-equivalent.

Remark 2.1.7. By Corollary 1.6.12 (b), assertion (a) is a particular case of (b). Moreover, if φ is not $\mathcal{E}_{([a_0],\kappa)}$-admissible, then assertions (a) and (b) are equivalent (by Corollary 1.6.12 (c)).

Notation 2.1.8. (a) To each $a : T \hookrightarrow G$ and $\theta : T(E) \to \mathbb{C}^\times$ as in Notation 2.1.3 we associate a function $t_{a,\theta}$ on $G(E)$, supported on $\widetilde{G_a}$ and equal there to Tr $\rho_{a,\theta}$.

(b) Since $t_{a,\theta}$ is cuspidal, it follows from [HC1, Lem. 23] that for every $\gamma \in G^{\mathrm{sr}}(E)$ and every compact open subgroup $K \subset G(E)$, the sum $\sum_{g\in D_b} t_{a,\theta}(g\gamma g^{-1})$, where $D_b := \widetilde{G_a}\backslash\widetilde{G_a}bK$, does not vanish only for finitely many $b \in \widetilde{G_a}\backslash G(E)/K$. Therefore the sum

$$F_{a,\theta}(\gamma) := \sum_{b\in\widetilde{G_a}\backslash G(E)/K}\left[\sum_{g\in D_b} t_{a,\theta}(g\gamma g^{-1})\right]$$

stabilizes, and the resulting value is independent of K.

Explicitly, $F_{a,\theta}(\gamma) = \sum_{g\in\widetilde{G_a}\backslash\Omega} t_{a,\theta}(g\gamma g^{-1})$ for each sufficiently large compact modulo center subset $\Omega = \widetilde{G_a}\Omega K \subset G(E)$. In particular, $F_{a,\theta}$ is a locally constant invariant function on $G^{\mathrm{sr}}(E)$.

(c) For every $\kappa \in \widehat{T}^\Gamma/Z(\widehat{G})^\Gamma$, put

$$F_{a_0,\kappa,\theta} := e(G)\sum_a \langle \mathrm{inv}(a_0, a), \kappa\rangle\, F_{a,\theta},$$

where a runs over a set of representatives of conjugacy classes of embeddings $T \hookrightarrow G$ which are stably conjugate to a_0.

Lemma 2.1.9. *Assume that the characteristic of E is zero. Then for each a and θ as in Notation 2.1.3, $F_{a,\theta}$ belongs to $L^1_{\mathrm{loc}}(G(E))$, and the corresponding generalized function is equal to $\chi(\pi_{a,\theta})$.*

Proof. Since $\pi_{a,\theta}$ is cuspidal, the assertion is a combination of the theorem of Harish-Chandra ([HC1, Theorem 16]) and a formula for characters of induced representations. □

For the next result, we will use Notation 1.7.3.

Theorem 2.1.10. *Under the assumptions of Theorem 2.1.6, let* $\gamma \in G^{\mathrm{sr}}(E)$ *and* $\overline{\xi} \in \pi_0(\widehat{G_\gamma}^{\Gamma}/Z(\widehat{G})^{\Gamma})$ *be such that* $F_{a_0,\kappa,\theta}(\gamma, \overline{\xi}) \neq 0$. *Then*

(i) $([a_\gamma], \overline{\xi}) \in \operatorname{Im}\Pi_{\mathcal{E}}$;
(ii) *if* $\varphi : G \to G'$ *is* $(\mathcal{E}, [a_\gamma], \overline{\xi})$*-admissible, then for every stable conjugate* $\gamma' \in G'(E)$ *of* γ *we have* $F'_{a'_0,\kappa,\theta}(\gamma', \overline{\xi}) = \left\langle \frac{\gamma,\gamma';\overline{\xi}}{a,a';\kappa} \right\rangle F_{a_0,\kappa,\theta}(\gamma, \overline{\xi})$.

Lemma 2.1.11. *Theorem 2.1.10 is equivalent to Theorem 2.1.6.*

Proof. The equivalence follows from Lemma 2.1.9 and Proposition 1.7.4. More precisely, Proposition 1.7.4 (a) implies the equivalence between Theorem 2.1.6 (a) and Theorem 2.1.10 (i), while Proposition 1.7.4 (b) implies the equivalence between Theorem 2.1.6 (b) and a combination of Theorem 2.1.6 (a) and Theorem 2.1.10 (ii). □

Remark 2.1.12. If the characteristic of E is positive, then it is not known that $\chi(\pi_{a,\theta})$ belongs to $L^1_{\mathrm{loc}}(G(E))$. However the restriction $\chi(\pi_{a,\theta})|_{G^{\mathrm{sr}}(E)}$ belongs to $L^1_{\mathrm{loc}}(G^{\mathrm{sr}}(E))$; therefore Proposition 1.7.4 implies that Theorem 2.1.10 for E is equivalent to an analog of Theorem 2.1.6 for restrictions $\chi_{a_0,\kappa,\theta}|_{G^{\mathrm{sr}}(E)}$ and $\chi_{a'_0,\kappa,\theta}|_{G'^{\mathrm{sr}}(E)}$.

Moreover, Theorem 2.1.10 for local fields of positive characteristic follows from that for local fields of characteristic zero by approximation arguments of [Ka3] and [De] (see [KV2]).

2.2. Stability of the restriction to $G(E)_{\mathrm{tu}}$

Starting from this subsection we will assume that the characteristic of E is zero. In this subsection we will strongly use definitions and results from Subsection 1.8.

2.2.1. Assumptions Assume that G admits a quasi-logarithm map $\Phi : G \to \mathcal{G}$ defined over $\mathcal{O}$, $\mathcal{G}$ admits an invariant pairing $\langle\cdot,\cdot\rangle$ non-degenerate over $\mathcal{O}$, and there exists $t \in \mathcal{T}(\mathcal{O})$, whose reduction $\overline{t} \in \overline{\mathcal{T}}(\mathbb{F}_q)$ is a_0-strongly regular (see Definition 2.1.4 (a)).

Notation 2.2.2. (a) For every generalized function $F \in \mathcal{D}(G(E))$, denote by F_{tu} the restriction of $F|_{G(E)_{\mathrm{tu}}}$ (see Notation 1.6.2 and Notation 1.8.14). Since $G(E)_{\mathrm{tu}} \subset {}^0G(E)$ (see Proposition 1.8.16 (c)), we can consider F_{tu} as an element either of $\mathcal{D}(G(E))$ or of $\mathcal{D}({}^0G(E))$.

(b) Denote by ${}^0\Phi : {}^0G \to \mathcal{G}$ the restriction of Φ to ${}^0G$.

The goal of this subsection is to prove the following particular case of Theorem 2.1.6.

Theorem 2.2.3. *Let* (a_0, κ, θ) *be as in Notation 2.1.5. Under the assumptions of 2.2.1, the generalized functions* $(\chi_{a_0,\kappa,\theta})_{\mathrm{tu}}$ *and* $(\chi_{a'_0,\kappa,\theta})_{\mathrm{tu}}$ *are* $(a_0, a'_0; \kappa)$*-equivalent. In particular, each* $(\chi_{a_0,\kappa,\theta})_{\mathrm{tu}}$ *is* $\mathcal{E}_{([a_0],\kappa)}$*-stable.*

Theorem 2.2.3 will be deduced in 2.2.13 from the corresponding statement about generalized functions on Lie algebras.

Notation 2.2.4. For every $a : T \hookrightarrow G$ as in Notation 2.1.3, we denote by $\overline{\Omega}_{a,t} \subset \mathcal{L}_a(\mathbb{F}_q)$ the $\operatorname{Ad} L_a(\mathbb{F}_q)$-orbit of $\overline{a}(\overline{t})$, by $\Omega_{a,t} \subset \mathcal{G}_a \subset \mathcal{G}(E)$ the preimage of $\overline{\Omega}_{a,t}$, and let $\delta_{a,t}$ and $\overline{\delta}_{a,t}$ be the characteristic functions of $\Omega_{a,t}$ and $\overline{\Omega}_{a,t}$, respectively.

Lemma 2.2.5. *(a) For each $y \in \Omega_{a,t}$, its stabilizer $G_y \subset G$ is G_a-conjugate to $a(T)$;* *(b) for each $y \in \Omega_{a,t}$ and $g \in G(E)$ such that $\operatorname{Ad} g(y) \in \Omega_{a,t}$, we have $g \in \widetilde{G_a}$.*

Proof. (a) Since $\overline{t} \in \overline{\mathcal{T}}(\mathbb{F}_q)$ is a_0-strongly regular, we see that $a(t) \in a(\mathcal{T}(E)) \subset \mathcal{G}(E)$ is strongly regular, hence $G_{a(t)} = a(T)$.

First, we will show that for every $y \in a(t) + \mathcal{G}_{a+}$, we have $y \in \mathcal{G}^{\mathrm{sr}}(E)$, and G_y is G_{a+}-conjugate to $a(T)$. By [DB, Lemma. 2.2.2], it will suffice to prove that $y \in \mathcal{G}(E)$ is G-regular, and G_y splits over E^{nr}. For this we can replace E by an unramified extension, so we may assume that T splits over E. Under this assumption we will show that y is $G(E)$-conjugate to an element of $a(t + \mathcal{T}(\mathcal{O})^+)$.

Choose an Iwahori subgroup $I \subset G_a$, containing $a(T(\mathcal{O}))$, and let $\mathcal{I}$, I^+ and $\mathcal{I}^+$ be the corresponding Iwahori subalgebra, the pro-unipotent radical of I and the pro-nilpotent radical of $\mathcal{I}$, respectively. Since $\alpha(a(t)) \in \mathcal{O}^\times$ for each root α of $(G, a(T))$, it follows from direct calculations that every element of $a(t) + \mathcal{I}^+$ is I^+-conjugate to an element of $a(t + \mathcal{T}(\mathcal{O})^+)$. But $y \in a(t) + \mathcal{G}_{a+} \subset a(t) + \mathcal{I}^+$, and therefore we get the assertion in this case.

For an arbitrary $y \in \Omega_{a,t}$, there exists $h \in G_a$ such that $\operatorname{Ad} h(y) \in a(t) + \mathcal{G}_{a+}$. So the general case follows from the previous one.

(b) Replacing y and $\operatorname{Ad} g(y)$ by their G_a-conjugates, we can assume that $y \in a(t) + \mathcal{G}_{a+}$ and $\operatorname{Ad} g(y) \in a(t) + \mathcal{G}_{a+}$. Then by the claim shown in (a), one can further replace y and $\operatorname{Ad} g(y)$ by their G_{a+}-conjugates, so that both G_y and $G_{\operatorname{Ad} g(y)} = gG_y g^{-1}$ equal $a(T)$. Thus $g \in \operatorname{Norm}_G(a(T))$.

Since $\overline{\operatorname{Ad} g(y)} = \overline{a}(\overline{t}) = \overline{y}$ is not fixed by a non-trivial element of the Weyl group $W(G, a(T))$, we get that $g \in a(T)(E)$. By Corollary 2.2.7 (a) below, g therefore belongs to $\widetilde{G_a}$, as claimed. □

Lemma 2.2.6. *Let T be an unramified torus over E, and $S \subset T$ a maximal split subtorus. Then $T(E) = T(\mathcal{O})S(E)$.*

Proof. By a very particular case of Lemma 1.8.17 (b), the projection $T(\mathcal{O}) \to (T/S)(\mathcal{O})$ is surjective, and therefore we have to check that $(T/S)(E) = (T/S)(\mathcal{O})$. Since T/S is anisotropic over E, the group $(T/S)(E)$ is compact. Hence $(T/S)(E)$ is contained in $(T/S)(E) \cap (T/S)(\mathcal{O}_{E^{\mathrm{sep}}}) = (T/S)(\mathcal{O})$, as claimed. □

Corollary 2.2.7. *(a) $\widetilde{G_a} = a(T)(E)G_a$;* *(b) G_a is the unique maximal compact subgroup of $\widetilde{G_a}$.*

Proof. (a) Since $Z(G) \subset a(T)$, we get the inclusion $\widetilde{G_a} \subset a(T)(E)G_a$. It remains to show that $a(T)(E)$ is contained in $\widetilde{G_a}$. Let $S \subset T$ be the maximal split subtorus. Since $a(T) \subset G$ is elliptic, we get $a(S) \subset Z(G)$. Now the assertion follows from the inclusion $a(T(\mathcal{O})) \subset G_a$ and Lemma 2.2.6.

(b) Assume that $g \in \widetilde{G_a}$ belongs to a compact subgroup. Choose $g_a \in G_a$ and $z \in Z(G)(E) \subset a(T)(E)$ such that $g = g_a z$. Since G_a is compact and the sequence

$\{g^n\}_n = \{g_a^n z^n\}_n \subset \widetilde{G_a}$ has a convergent subsequence, the sequence $\{z^n\}_n \subset a(T)(E)$ has a convergent subsequence. Hence z is contained in $a(T)(\mathcal{O}) \subset G_a$, thus $g \in G_a$. □

Notation 2.2.8. It follows from Lemma 2.2.5 (b) that for every $x \in \mathcal{G}(E)$ there exists at most one coset $g \in \widetilde{G_a}\backslash G(E)$ such that $\delta_{a,t}(\operatorname{Ad} g(x)) \neq 0$. Therefore

$$\Delta_{a,t}(x) := \sum_{g \in \widetilde{G_a}\backslash G(E)} \delta_{a,t}(\operatorname{Ad} g(x))$$

is the characteristic function of an open and closed subset $\operatorname{Ad} G(E)(\Omega_{t,a}) \subset \mathcal{G}(E)$. In particular, $\Delta_{a,t}$ lies in $L^1_{\mathrm{loc}}(\mathcal{G}(E)) \subset \mathcal{D}(\mathcal{G}(E))$. Similar to Notation 2.1.5, we define elements $\Delta_{a_0,\kappa,t} := e(G)\sum_a \langle \operatorname{inv}(a_0,a), \ \kappa\rangle\, \Delta_{a,t} \in L^1_{\mathrm{loc}}(\mathcal{G}(E))$ and $\Delta'_{a'_0,\kappa,t} = e(G')\sum_{a'} \left\langle \operatorname{inv}(a'_0,a'),\kappa\right\rangle \Delta_{a',t} \in L^1_{\mathrm{loc}}(\mathcal{G}'(E))$.

Lemma 2.2.9. *$e(G)\Delta_{a_0,\kappa,t}$ is $(a_0,a'_0;\kappa)$-equivalent to $e(G')\Delta'_{a'_0,\kappa,t}$.*

Proof. By Proposition 1.7.4, we have to show that for each $x_0 \in \mathcal{G}^{\mathrm{sr}}(E)$ and $\overline{\xi} \in \pi_0(\widehat{G_{x_0}}^{\Gamma}/Z(\widehat{G})^{\Gamma})$ such that $\Delta_{a_0,\kappa,t}(x_0,\overline{\xi}) \neq 0$, we have

(i) $([a_{x_0}],\overline{\xi}) \in \operatorname{Im}\Pi_{\mathcal{E}}$;

(ii) if $\varphi : G \to G'$ is $(\mathcal{E},[a_{x_0}],\overline{\xi})$-admissible, then for every stable conjugate $x'_0 \in \mathcal{G}'(E)$ of x_0, we have $e(G')\Delta'_{a'_0,\kappa,\theta}(x'_0,\overline{\xi}) = \left\langle \frac{x_0,x'_0;\overline{\xi}}{a_0,a'_0;\kappa}\right\rangle e(G)\Delta_{a_0,\kappa,\theta}(x_0,\overline{\xi})$.

Recall that

$$e(G)\Delta_{a_0,\kappa,t}(x_0,\overline{\xi}) = \sum_x\sum_a \left\langle \operatorname{inv}(x_0,x),\overline{\xi}\right\rangle^{-1} \langle \operatorname{inv}(a_0,a),\kappa\rangle\, \Delta_{a,t}(x),$$

where x runs over a set of representatives of $G(E)\backslash[x_0] \subset G(E)\backslash\mathcal{G}(E)$. We identify T with $a_0(T) \subset G$ and $\mathcal{T}$ with $a_0(\mathcal{T}) \subset \mathcal{G}$. By Lemma 2.2.5 (a), the support of each $\Delta_{a,t}$ consists of elements, stably conjugate to $\mathcal{T}(E)$. Hence replacing x_0 by a stable conjugate, we can assume that $x_0 \in \mathcal{T}(E)$. Then $G_{x_0} = T$, $a_{x_0} = a_0$, and thus $\overline{\xi}$ is an element of $\widehat{T}^{\Gamma}/Z(\widehat{G})^{\Gamma}$.

For a stable conjugate x of x_0, we have $\Delta_{a,t}(x) = 1$ if and only if $\operatorname{inv}(x_0,x) = \operatorname{inv}(a_0,a)$. Therefore $e(G)\Delta_{a_0,\kappa,t}(x_0,\overline{\xi}) = \sum_a \left\langle \operatorname{inv}(a_0,a),\kappa\overline{\xi}^{-1}\right\rangle$. Since the latter sum is non-zero, we get that $\overline{\xi} = \kappa$, and $e(G)\Delta_{a_0,\kappa,t}(x_0,\overline{\xi}) = |\widehat{T}^{\Gamma}/Z(\widehat{G})^{\Gamma}|$. Since $\mathcal{E} = \mathcal{E}_{([a_0],\kappa)}$, we get that $([a_{x_0}],\overline{\xi}) = ([a_0],\kappa) \in \operatorname{Im}\Pi_{\mathcal{E}}$, showing the assertion (i).

To show (ii), we can replace x'_0 by a stably conjugate $a'_0(a_0^{-1}(x_0)) \in a'_0(\mathcal{T}(E)) \subset \mathcal{G}'(E)$. Then the same arguments show that $e(G')\Delta'_{a'_0,\kappa,t}(x'_0,\overline{\xi}) = |\widehat{T}^{\Gamma}/Z(\widehat{G})^{\Gamma}|$. Now the required assertion $\left\langle \frac{x_0,x'_0;\overline{\xi}}{a_0,a'_0,\kappa}\right\rangle = 1$ follows from equalities $a_{x_0} = a_0$, $a_{x'_0} = a'_0$ and $\overline{\xi} = \kappa$. □

2.2.10. Fourier transform Fix an additive character $\psi : E \to \mathbb{C}^\times$ such that $\psi|_{\mathcal{O}}$ is non-trivial, but $\psi|_{\mathfrak{m}}$ is trivial.

(a) The pairing $\langle\cdot,\cdot\rangle$ and the measure $dx = |\omega_{\mathcal{G}}|$ on $\mathcal{G}(E)$ give rise to the Fourier transform $\mathcal{F} = \mathcal{F}(\psi, \langle\cdot,\cdot\rangle, dx)$ on $C_c^\infty(\mathcal{G}(E))$. Then $\mathcal{F}$ induces Fourier transforms on $\mathcal{S}(\mathcal{G}(E))$ and $\mathcal{D}(\mathcal{G}(E))$ given by the formulas $\mathcal{F}(f dx) := \mathcal{F}(f)dx$ for each $f \in C_c^\infty(\mathcal{G}(E))$ and $\mathcal{F}(F)(\phi) := F(\mathcal{F}(\phi))$ for each $\phi \in \mathcal{S}(\mathcal{G}(E))$ and $F \in \mathcal{D}(\mathcal{G}(E))$.

(b) For each $f_1, f_2 \in C_c^\infty(\mathcal{G}(E))$ we have $\int_{\mathcal{G}(E)} f_1\mathcal{F}(f_2)dx = \int_{\mathcal{G}(E)} \mathcal{F}(f_1)f_2dx$. Therefore the embedding $C_c^\infty(\mathcal{G}(E)) \hookrightarrow \mathcal{D}(\mathcal{G}(E))$ commutes with the Fourier transform.

(c) For each parahoric subalgebra $\mathcal{G}_a \subset \mathcal{G}(E)$, we denote by $\overline{\mathcal{F}} = \overline{\mathcal{F}}(\overline{\psi}, \langle\cdot,\cdot\rangle_a, \mu)$ the Fourier transform on $\mathcal{L}_a(\mathbb{F}_q)$, where the character $\overline{\psi} : \mathbb{F}_q \to \mathbb{C}^\times$ is induced by ψ, pairing $\langle\cdot,\cdot\rangle_a$ is induced by $\langle\cdot,\cdot\rangle$ (see Lemma 1.8.7 (a)) and $\mu(l) = 1$ for each $l \in L_a(\mathbb{F}_q)$.

Lemma 2.2.11. *Denote by $\mathcal{I}^+$ the pro-nilpotent radical of an Iwahori subalgebra of $\mathcal{G}$. Then for each $u \in G(E)_{\mathrm{tu}}$, we have*

$$t_{a,\theta}(u) = \mathcal{F}(\delta_{a,t})(\Phi(u))|\omega_{\mathcal{G}}|(\mathcal{I}^+)^{-1}.$$

Proof. First, we claim that $\mathcal{F}(\delta_{a,t})(x) = 0$ for each $x \in \mathcal{G}(E) \smallsetminus \mathcal{G}_a$, and $\mathcal{F}(\delta_{a,t})(x) = \overline{\mathcal{F}}(\overline{\delta}_{a,t})(\overline{x})|\omega_{\mathcal{G}}|(\mathcal{G}_{a^+})$ for each $x \in \mathcal{G}(E) \smallsetminus \mathcal{G}_a$. Indeed, since $\delta_{a,t}$ vanishes outside of $\mathcal{G}_a$, we have an equality

$$\mathcal{F}(\delta_{a,t})(x) = \int_{\mathcal{G}(E)} \psi(\langle x, y\rangle)\delta_{a,t}(y)dy = \int_{\mathcal{G}_a} \psi(\langle x, y\rangle)\delta_{a,t}(y)dy$$

for each $x \in \mathcal{G}(E)$. Since $\delta_{a,t}(y + y') = \delta_{a,t}(y)$ for each $y' \in \mathcal{G}_{a^+}$, we conclude that $\mathcal{F}(\delta_{a,t})(x)$ equals

$$\sum_z \psi(\langle x, z\rangle)\delta_{a,t}(z) \int_{\mathcal{G}_{a^+}} \psi(\langle x, y\rangle)dy,$$

where z runs over a set of representatives of $\mathcal{G}_a/\mathcal{G}_{a^+}$ in $\mathcal{G}_a$.

The assumptions on ψ and $\langle\cdot,\cdot\rangle$ imply that $\mathcal{G}_a$ is the orthogonal complement of $\mathcal{G}_{a^+}$ with respect to the pairing $(x, y) \mapsto \psi(\langle x, y\rangle)$. Therefore $\mathcal{F}(\delta_{a,t})(x) = 0$ for each $x \notin \mathcal{G}_a$, and

$$\mathcal{F}(\delta_{a,t})(x) = \sum_{\overline{z}\in\mathcal{L}_a(\mathbb{F}_q)} \psi(\langle\overline{x}, \overline{z}\rangle)\overline{\delta}_{a,t}(\overline{z})|\omega_{\mathcal{G}}|(\mathcal{G}_{a^+}) = \overline{\mathcal{F}}(\overline{\delta}_{a,t})(\overline{x})|\omega_{\mathcal{G}}|(\mathcal{G}_{a^+})$$

for each $x \in \mathcal{G}_a$.

Now we are ready to prove the lemma. Assume first that $u \in G_{a,\mathrm{tu}}$. It follows from Lemma 1.8.9 (b) that $\Phi(u) \in \mathcal{G}_a$ and Φ induces a quasi-logarithm $\overline{\Phi}_a : L_a \to \mathcal{L}_a$ satisfying $\overline{\Phi(u)} = \overline{\Phi}_a(\overline{u})$. Therefore $\mathcal{F}(\delta_{a,t})(\Phi(u))$ equals $\overline{\mathcal{F}}(\overline{\delta}_{a,t})(\overline{\Phi}_a(\overline{u}))|\omega_{\mathcal{G}}|(\mathcal{G}_{a^+})$. It now follows from a combination of Theorem A.1 (see Appendix A) and the equality $|\omega_{\mathcal{G}}|(\mathcal{G}_{a^+}) = q^{-\frac{1}{2}\dim(L_a/\overline{T})}|\omega_{\mathcal{G}}|(\mathcal{I}^+)$ that $\mathcal{F}(\delta_{a,t})(\Phi(u))$ equals

$$|\omega_{\mathcal{G}}|(\mathcal{I}^+)\operatorname{Tr}\rho_{\overline{a},\overline{\theta}}(\overline{u}) = |\omega_{\mathcal{G}}|(\mathcal{I}^+)t_{a,\theta}(u).$$

Finally, assume that $u \in G(E)_{\mathrm{tu}} \smallsetminus G_{a,\mathrm{tu}}$. In this case, $t_{a,\theta}(u) = 0$. On the other hand, by Proposition 1.8.16, Φ induces bijections $G(E)_{\mathrm{tu}} \xrightarrow{\sim} \mathcal{G}(E)_{\mathrm{tn}}$ and $G_{a,\mathrm{tu}} \xrightarrow{\sim} \mathcal{G}_{a,\mathrm{tn}}$; therefore $\Phi(u) \in \mathcal{G}(E)_{\mathrm{tn}} \smallsetminus \mathcal{G}_{a,\mathrm{tn}}$. Using the equality $\mathcal{G}_{a,\mathrm{tn}} = \mathcal{G}(E)_{\mathrm{tn}} \cap \mathcal{G}_a$ from Lemma 1.8.15 (b), we conclude that $\Phi(u) \notin \mathcal{G}_a$, hence $\mathcal{F}(\Delta_{a,t})(\Phi(u)) = 0$. This completes the proof of the lemma. □

Corollary 2.2.12. *For each a, we have* $\chi(\pi_{a,\theta})_{\mathrm{tu}} = {}^0\Phi^*(\mathcal{F}(\Delta_{a,t}))_{\mathrm{tu}}|\omega_{\mathcal{G}}|(\mathcal{I}^+)^{-1}$.

Proof. Note first that ${}^0\Phi$ is étale, hence smooth, therefore the pullback ${}^0\Phi^*(\mathcal{F}(\Delta_{a,t}))$ is defined. Consider generalized functions $t_{a,\theta} \in C_c^\infty(G(E)) \subset \mathcal{D}(G(E))$ and $\delta_{a,t} \in C_c^\infty(\mathcal{G}(E)) \subset \mathcal{D}(\mathcal{G}(E))$. In light of 2.2.10 (b), Lemma 2.2.11 implies the equality of generalized functions

$$(t_{a,\theta})_{\mathrm{tu}} = {}^0\Phi^*(\mathcal{F}(\delta_{a,t}))_{\mathrm{tu}}|\omega_{\mathcal{G}}|(\mathcal{I}^+)^{-1}. \tag{2.2.1}$$

Since $\pi_{a,\theta} = \mathrm{Ind}_{\widetilde{G_a}}^{G(E)} \rho_{a,\theta}$ is admissible, it follows from the formula for characters of induced representations that

$$\chi(\pi_{a,\theta}) = \sum_{g \in \widetilde{G_a}\backslash G(E)} (\mathrm{Int}\, g)^*(t_{a,\theta}). \tag{2.2.2}$$

Explicitly, $\chi(\pi_{a,\theta})(\phi) = \sum_{g \in \widetilde{G_a}\backslash G(E)} (\mathrm{Int}\, g)^*(t_{a,\theta})(\phi)$ for each $\phi \in \mathcal{S}(G(E))$, where only finitely many terms in the sum are non-zero. On the other hand, by the very definition of $\Delta_{a,t}$, we have

$$\Delta_{a,t} = \sum_{g \in \widetilde{G_a}\backslash G(E)} (\mathrm{Ad}\, g)^*(\delta_{a,t}). \tag{2.2.3}$$

Since $\langle\cdot,\cdot\rangle$ and Φ are G-equivariant, we get the equality

$${}^0\Phi^*(\mathcal{F}((\mathrm{Ad}\, g)^*\delta_{a,t})) = {}^0\Phi^*(\mathrm{Ad}\, g)^*(\mathcal{F}(\delta_{a,t})) = (\mathrm{Int}\, g)^{*0}\Phi^*(\mathcal{F}(\delta_{a,t})). \tag{2.2.4}$$

Now our corollary is an immediate consequence of equalities (2.2.1)–(2.2.4). □

2.2.13. *Proof of Theorem* 2.2.3. Let $\Phi' : G' \to \mathcal{G}'$ and $\langle\cdot,\cdot\rangle'$ be the quasi-logarithm map and the pairing on $\mathcal{G}'$ induced by Φ and $\langle\cdot,\cdot\rangle$, respectively (see Lemma 1.8.9 (a)). We denote by $\mathcal{F} = \mathcal{F}(\psi, \langle\cdot,\cdot\rangle', |\omega_{\mathcal{G}'}|)$ the corresponding Fourier transform $\mathcal{G}'(E)$ and by $\mathcal{I}'^+$ the pro-nilpotent radical of an Iwahori subalgebra of $\mathcal{G}'(E)$.

By Lemma 2.2.9, $e(G)\Delta_{a_0,\kappa,t}$ is $(a_0, a_0'; \kappa)$-equivalent to $e(G')\Delta'_{a_0',\kappa,t}$. Using the equality $e(G)e(G') = e'(G)e'(G')$, it follows from Theorem B.1.2 (see Appendix B) that $\mathcal{F}(\Delta_{a_0,\kappa,t})$ is $(a_0, a_0'; \kappa)$-equivalent to $\mathcal{F}(\Delta'_{a_0',\kappa,t})$. Hence by Corollary 1.6.16, the pullback ${}^0\Phi^*(\mathcal{F}(\Delta_{a_0,\kappa,t}))$ is $(a_0, a_0'; \kappa)$-equivalent to ${}^0\Phi'^*(\mathcal{F}(\Delta'_{a_0',\kappa,t}))$. By Corollary 2.2.12, we thus get that $|\omega_{\mathcal{G}}|(\mathcal{I}^+)(\chi_{a_0,\kappa,\theta})_{\mathrm{tu}}$ and $|\omega_{\mathcal{G}'}|(\mathcal{I}'^+)(\chi_{a_0',\kappa,\theta})_{\mathrm{tu}}$ are $(a_0, a_0'; \kappa)$-equivalent. Since $|\omega_{\mathcal{G}}|(\mathcal{I}^+) = |\omega_{\mathcal{G}'}|(\mathcal{I}'^+)$ (see, for example, [Ko1, p. 632]), the assertion follows. □

2.3. Reduction formula

In this subsection we will assume that $G^{\text{der}} = G^{\text{sc}}$. Our goal is to rewrite character $\chi(\pi_{a,\theta})$ in terms of restrictions to topologically unipotent elements of the corresponding characters of the centralizers $G_\delta(E)$.

Lemma 2.3.1. *Assume that $G^{\text{der}} = G^{\text{sc}}$. Then*

(a) For each semisimple element $\delta \in G$, the centralizer G_δ is connected.
(b) The stabilizer in $G(E)$ of each $x \in \mathcal{B}(G)$ is G_x.

Proof. (a) was shown in [St, Cor. 8.5] when G is semisimple, and in [Ko3, pp. 788–789] in the general case.

(b) When G is semisimple, the result was proved in [BT, Prop. 4.6.32]. For a general G, we can replace E by an unramified extension so that G is split over E. Choose a split maximal torus $T \subset G$ such that $x \in \mathcal{B}(T)$. Since $\text{Stab}_{G(E)}(x)$ is compact, we see as in Corollary 1.8.18 that $\text{Stab}_{G(E)}(x)$ is contained in $G^{\text{der}}(E)T(\mathcal{O})$. Since $T(\mathcal{O}) \subset \text{Stab}_{G(E)}(x)$, we get that $\text{Stab}_{G(E)}(x)$ is contained in $\text{Stab}_{G^{\text{der}}(E)}(x)T(\mathcal{O})$, hence (use [BT, Prop. 4.6.32]) in $(G^{\text{der}})_x T(\mathcal{O}) = G_x$, as claimed. □

Notation 2.3.2. (a) We will call an element $\gamma \in G(E)$ *compact* if it generates a relatively compact subgroup of $G(E)$.

(b) We will call an element $\gamma \in G(E)$ *topologically unipotent* if the sequence $\{\gamma^{p^n}\}_n$ converges to 1.

Corollary 2.3.3. *(a) The set of compact elements of $G(E)$ is $\cup_{x \in \mathcal{B}(G)} G_x$.*
(b) The set of topologically unipotent elements of $G(E)$ is $G(E)_{\text{tu}}$ (see Notation 1.8.14).

Proof. As each G_x is compact and every compact element of $G(E)$ stabilizes a point of $\mathcal{B}(G)$, (a) follows from Lemma 2.3.1 (b). Since every topologically unipotent element is compact, (b) follows from (a) and the fact that $G_{x,\text{tu}}$ is a set of all topologically unipotent elements of G_x. □

The following result is a straightforward generalization of [Ka2, Lemma 2, p. 226].

Lemma 2.3.4. *For every compact element $\gamma \in G(E)$, there exists a unique decomposition $\gamma = \delta u = u\delta$ such that δ is of finite order prime to p, and u is topologically unipotent. In particular, this decomposition is compatible with conjugation and field extensions.*

Notation 2.3.5. The decomposition $\gamma = \delta u$ from Lemma 2.3.4 is called the *topological Jordan decomposition* of γ.

Remark 2.3.6. If $\delta \in G(E)$ is an element of finite order prime to p, then δ is automatically semisimple.

The goal of this subsection is to prove the following result.

Proposition 2.3.7. *For every embedding $a : T \hookrightarrow G$ and a compact element $\gamma \in G(E)$ with topological Jordan decomposition $\gamma = \delta u$, we have the following formula*

$$e(G)F_{a,\theta}(\gamma) = e(G_\delta)\sum_b \theta(b^{-1}(\delta))F_{b,\theta}(u). \qquad (2.3.1)$$

Here b runs over a set of representatives of conjugacy classes of embeddings $T \hookrightarrow G_\delta$, whose composition with the inclusion $G_\delta \hookrightarrow G$ is conjugate to a.

First we need to prove two preliminary results.

Lemma 2.3.8. *(a) For each vertex x of $\mathcal{B}(G)$, we have $e(L_x) = e(G)$.*

(b) Let $\delta \in G(E)$ be an element of finite order, prime to p. Then the centralizer G_δ splits over E^{nr}, and the building $\mathcal{B}(G_\delta)$ is canonically identified with the set of invariants $\mathcal{B}(G)^\delta \subset \mathcal{B}(G)$.

(c) For each $x \in \mathcal{B}(G_\delta) \subset \mathcal{B}(G)$, the parahoric subgroup $(G_\delta)_x$ is a subgroup of finite index in $(G_x)_\delta$, and the canonical map $(G_x)_\delta \hookrightarrow G_x \to L_x(\mathbb{F}_q)$ induces isomorphisms $(G_x)_\delta/(G_\delta)_{x^+} \xrightarrow{\sim} (L_x)_{\overline{\delta}}(\mathbb{F}_q)$ and $(G_\delta)_x/(G_\delta)_{x^+} \xrightarrow{\sim} (L_x)^0_{\overline{\delta}}(\mathbb{F}_q)$.

(d) Let $\delta, \delta' \in G_x$ be two elements of finite orders prime to p. Then δ and δ' are G_x-conjugate if and only if their reductions $\overline{\delta}, \overline{\delta}' \in L_x(\mathbb{F}_q)$ are $L_x(\mathbb{F}_q)$-conjugate.

Proof. (a) For every $x \in \mathcal{B}(G)$, the maximal split torus of L_x is the reduction of that of G, therefore $\mathrm{rk}_{\mathbb{F}_q}(L_x) = \mathrm{rk}_E(G)$. If, moreover, x is a vertex, then $\mathrm{rk}_{\mathbb{F}_q}(Z(L_x)^0) = \mathrm{rk}_E(Z(G)^0)$, hence $e(L_x) = e(G)$.

(b) The second assertion is shown in [PY]. For the first, recall that $G_{E^{\mathrm{nr}}}$ splits, hence there exists a split maximal torus $T \subset G_{E^{\mathrm{nr}}}$. Choose $g \in G(\overline{E})$ such that $g\delta g^{-1} \in T(\overline{E})$. Then $g\delta g^{-1}$ is of finite order, prime to p, therefore it follows from Hensel's lemma that $g\delta g^{-1} \in T(\mathcal{O}_{E^{\mathrm{nr}}}) \subset T(E^{\mathrm{nr}})$. Hence g gives rise to a cocycle $\sigma \mapsto g^{-1\sigma}g \in G_\delta(\overline{E})$ over E^{nr}. Since $H^1(E^{\mathrm{nr}}, G_\delta) = 0$, there exists $h \in G_\delta(\overline{E})$ such that $h^{-1\sigma}h = g^{-1\sigma}g$ for each $\sigma \in \mathrm{Gal}(\overline{E}/E^{\mathrm{nr}})$. It follows that $gh^{-1} \in G(E^{\mathrm{nr}})$, and $(gh^{-1})\delta(gh^{-1})^{-1} = g\delta g^{-1} \in T(E^{\mathrm{nr}})$. Therefore $(gh^{-1})^{-1}T(gh^{-1})$ is a split maximal torus of $(G_\delta)_{E^{\mathrm{nr}}}$.

(c) As $(G_x)_\delta \subset G_\delta(E)$ is the stabilizer of $x \in \mathcal{B}(G_\delta)$, it is compact. Therefore $(G_\delta)_x$ is a subgroup of finite index in $(G_x)_\delta$, thus the corresponding group scheme $\underline{G_\delta}_x$ over $\mathcal{O}$ is the connected component of $(\underline{G}_x)_\delta$. In particular, $(\underline{G}_x)_\delta$ is smooth over $\mathcal{O}$. Therefore by Hensel's lemma, the reduction map $G_x \to \overline{G}_x(\mathbb{F}_q)$ surjects $(G_x)_\delta = (\underline{G}_x)_\delta(\mathcal{O})$ onto $(\overline{G}_x)_\delta(\mathbb{F}_q) = (\underline{G}_x)_\delta(\mathbb{F}_q)$.

As L_x is the quotient $\overline{G}_x/R_u(\overline{G}_x)$, we see that $(L_x)_{\overline{\delta}}$ is the quotient $(\overline{G}_x)_\delta/R_u(\overline{G}_x)_\delta$. Since $R_u(\overline{G}_x)_\delta$ is a connected group over $\mathbb{F}_q$, Lang's theorem implies that the projection $\overline{G}_x(\mathbb{F}_q) \to L_x(\mathbb{F}_q)$ surjects $(\overline{G}_x)_{\overline{\delta}}(\mathbb{F}_q)$ onto $(L_x)_{\overline{\delta}}(\mathbb{F}_q)$. Therefore the projection $G_x \to L_x(\mathbb{F}_q)$ induces a surjection $(G_x)_\delta \to (L_x)_{\overline{\delta}}(\mathbb{F}_q)$, whose kernel is $(G_x)_\delta \cap G_{x^+} = (G_\delta)_{x^+}$. This shows the first isomorphism, while the proof of the second one is similar but easier.

(d) The "only if" assertion is clear. Assume now that $\overline{\delta}$ and $\overline{\delta}'$ are $L_x(\mathbb{F}_q)$-conjugate. Let us first show that δ and δ' are $\underline{G}_x(\mathcal{O}_{E^{\mathrm{nr}}})$-conjugate. For this we can replace E by an unramified extension, so that G, G_δ and $G_{\delta'}$ are split over E (use (b)). Since δ lies in G_x, we get that x belongs to $\mathcal{B}(G)^\delta = \mathcal{B}(G_\delta)$ (use (b)). Therefore there exists a split maximal torus $T \subset G_\delta \subset G$ such that $x \in \mathcal{B}(T)$. Similarly, there exists a split maximal torus $T' \subset G_{\delta'} \subset G$ such that $x \in \mathcal{B}(T')$. By a property of buildings, there exists $g \in G_x$ such that $gTg^{-1} = T'$. Replacing δ' by $g^{-1}\delta' g$, we may assume that $\delta, \delta' \in T(E)$.

Next we observe that the projection $\mathrm{Norm}_{G_x}(T) \to \mathrm{Norm}_{L_x(\mathbb{F}_q)}(\overline{T})$ is surjective. Indeed, for each $\overline{g} \in \mathrm{Norm}_{L_x(\mathbb{F}_q)}(\overline{T}) \subset L_x(\mathbb{F}_q)$, choose a representative $g \in G_x$. Then $\overline{gTg^{-1}} = \overline{T}$, hence by [DB, Lemma 2.2.2], there exists $h \in G_{x^+}$ such that $h(gTg^{-1})h^{-1} = T$. In other words, $hg \in \mathrm{Norm}_{G_x}(T)$ is a preimage of $\overline{g}$.

By the assumption, $\overline{\delta}, \overline{\delta'} \in \overline{T}(\mathbb{F}_q)$ are conjugate in $L_x(\mathbb{F}_q)$, therefore they are conjugate in $\mathrm{Norm}_{L_x(\mathbb{F}_q)}(\overline{T})$. Hence there exists $g \in \mathrm{Norm}_{G_x}(T)$ such that $\overline{g^{-1}\delta' g} = \overline{\delta}$. But the projection $T(\mathcal{O}) \to \overline{T}(\mathbb{F}_q)$ defines a bijection between elements of $T(\mathcal{O})$ of finite order prime to p and elements of $\overline{T}(\mathbb{F}_q)$. Hence $g^{-1}\delta' g = \delta$, implying that δ and δ' are conjugate by an element of $\underline{G}_x(\mathcal{O}_{E^{\mathrm{nr}}})$.

To show that δ and δ' are conjugate by G_x, consider the closed subscheme Z (resp. Z') of $\underline{G}_x$ (resp. L_x) consisting of elements g (resp. $\overline{g}$) such that $g\delta g^{-1} = \delta'$ (resp. $\overline{g}\overline{\delta}\overline{g}^{-1} = \overline{\delta}'$). By the assumption, $Z'(\mathbb{F}_q) \neq \emptyset$, and we have to show that $Z(\mathcal{O}) \neq \emptyset$. By the assertion shown above, $Z(\mathcal{O}_{E^{\mathrm{nr}}}) \neq \emptyset$. Thus Z and $(\underline{G}_x)_\delta$ are isomorphic over $\mathcal{O}_{E^{\mathrm{nr}}}$. In particular, Z is smooth over $\mathcal{O}$, thus by Hensel's lemma it suffices to show that the projection $Z(\mathbb{F}_q) \to Z'(\mathbb{F}_q)$ is surjective.

Denote by $\overline{Z} \subset \overline{G}_x$ the special fiber of Z. Since all fibers of the projection $\overline{Z} \to Z'$ are principal homogeneous spaces for the connected group $R_u(\overline{G}_x)_\delta$, the surjectivity of the projection $Z(\mathbb{F}_q) = \overline{Z}(\mathbb{F}_q) \to Z'(\mathbb{F}_q)$ follows from Lang's theorem. □

Lemma 2.3.9. *For every $\gamma \in G_a$ with topological Jordan decomposition $\gamma = \delta u$, we have an equality*

$$e(G)t_{a,\theta}(\gamma) = e(G_\delta)\sum_b \sum_{h\in(G_\delta)_b\backslash(G_a)_\delta} \theta(b^{-1}(\delta))t_{b,\theta}(huh^{-1}), \tag{2.3.2}$$

where b runs over a set of representatives of conjugacy classes of embeddings $T \hookrightarrow G_\delta$, which are G_a-conjugate to $a : T \hookrightarrow G$.

Proof. We start from the following claim.

Claim 2.3.10. *The correspondence $b \mapsto \overline{b}$ induces a bijection between the set of conjugacy classes of embeddings $T \hookrightarrow G_\delta$ which are G_a-conjugate to $a : T \hookrightarrow G$ and the set of conjugacy classes of embeddings $\overline{b} : \overline{T} \hookrightarrow (L_a)_{\overline{\delta}}$, which are $L_a(\mathbb{F}_q)$-conjugate to $\overline{a} : \overline{T} \hookrightarrow L_a$.*

Proof. For simplicity of notation, we identify T with $a(T)$ and $\overline{T}$ with $\overline{a}(\overline{T})$. Then the maps $b \mapsto t := b^{-1}(\delta)$ and $\overline{b} \mapsto \overline{t} := \overline{b}^{-1}(\overline{\delta})$ identify our sets with the sets of elements $t \in T(E)$, which are G_a-conjugate to δ, and elements $\overline{t} \in \overline{T}(\mathbb{F}_q)$, which are

$L_a(\mathbb{F}_q)$-conjugate to $\overline{\delta}$, respectively. Since the reduction map $t \mapsto \overline{t}$ induces a bijection between elements of $T(E)$ of finite order prime to p and elements of $\overline{T}(\mathbb{F}_q)$, we get the injectivity. The surjectivity follows from Lemma 2.3.8 (d). $\square$

Now we are ready to prove the lemma. By Lemma 2.3.8 and Claim 2.3.10, the right-hand side of (2.3.2) equals

$$e((L_a)^0_{\overline{\delta}}) \sum_{\overline{b}} \left[\sum_{\overline{h} \in (L_a)^0_{\overline{\delta}}(\mathbb{F}_q) \backslash (L_a)_{\overline{\delta}}(\mathbb{F}_q)} \overline{\theta}(\overline{b}^{-1}(\overline{\delta})) \operatorname{Tr} \rho_{\overline{b},\overline{\theta}}(\overline{h}\overline{u}\overline{h}^{-1}) \right], \tag{2.3.3}$$

where $\overline{b}$ runs over a set of representatives of $(L_a)_{\overline{\delta}}(\mathbb{F}_q)$-conjugacy classes of embeddings $\overline{T} \hookrightarrow (L_a)_{\overline{\delta}}$, which are $L_a(\mathbb{F}_q)$-conjugate to $\overline{a}$.

Next note that (2.3.3) can be rewritten as

$$e((L_a)^0_{\overline{\delta}}) \sum_{\overline{b}:\overline{T} \hookrightarrow (L_a)^0_{\overline{\delta}}} \theta(\overline{b}^{-1}(\overline{\delta})) \operatorname{Tr} \rho_{\overline{b},\overline{\theta}}(\overline{u}), \tag{2.3.4}$$

where $\overline{b}$ runs over a set of representatives of conjugacy classes of embeddings, $L_a(\mathbb{F}_q)$-conjugate to $\overline{a}$. By the formula of Deligne–Lusztig [DL, Theorem 4.2], (2.3.4) equals $e(L_a) \operatorname{Tr} \rho_{\overline{a},\overline{\theta}}(\overline{\gamma})$. Hence it is equal to $e(G)t_{a,\theta}(\gamma)$, as claimed. $\square$

2.3.11. *Proof of Proposition* 2.3.7. Notice first that the map $b \mapsto b^{-1}(\delta)$ embeds the set of conjugacy classes of embeddings $b : T \hookrightarrow G_\delta$, conjugate to $a : T \hookrightarrow G$, into the finite set $\{t \in T(E) \mid t^{\operatorname{ord}\delta} = 1\}$. Therefore the sum in (2.3.1) is finite.

Fix a set of representatives $J \subset G(E)$ of double classes $\widetilde{G_a} \backslash G(E)/G_\delta(E)$. For every $h \in J$, put $\gamma_h = h\gamma h^{-1}$, and let $\gamma_h = \delta_h u_h$ be the topological Jordan decomposition of γ_h.

By [HC1, Lemma 23] (compare Notation 2.1.8 (b)), for each sufficiently large compact modulo center $\widetilde{G_a}$-bi-invariant subset $\Omega \subset G(E)$, we have

$$F_{a,\theta}(\gamma) = e(G) \sum_{g \in \widetilde{G_a} \backslash \Omega} t_{a,\theta}(g\gamma g^{-1}).$$

Since $G(E)$ decomposes as a disjoint union $\sqcup_{h \in J} \widetilde{G_a} h G_\delta(E) = \sqcup_{h \in J} \widetilde{G_a} G_{\delta_h}(E) h$, we have a finite decomposition $\widetilde{G_a} \backslash \Omega = \sqcup_{h \in J} \widetilde{G_a} \backslash [\widetilde{G_a} G_{\delta_h}(E) h \cap \Omega]$. Therefore

$$F_{a,\theta}(\gamma) = e(G) \sum_{h \in J} \sum_{g \in \widetilde{G_a} \backslash [\widetilde{G_a} G_{\delta_h}(E) h \cap \Omega]} t_{a,\theta}(g\gamma g^{-1}).$$

Using the identifications

$$\widetilde{G_a} \backslash [\widetilde{G_a} G_{\delta_h}(E) h \cap \Omega] = (\widetilde{G_a})_{\delta_h} \backslash [G_{\delta_h}(E) h \cap \Omega] = (\widetilde{G_a})_{\delta_h} \backslash [G_{\delta_h}(E) \cap \Omega h^{-1}] h,$$

we get that $F_{a,\theta}(\gamma)$ equals

$$\sum_{h\in J} e(G)\left[\sum_{g\in(\widetilde{G_a})_{\delta_h}\backslash[G_{\delta_h}(E)\cap\Omega h^{-1}]} t_{a,\theta}(g\gamma_h g^{-1})\right]. \tag{2.3.5}$$

Using Corollary 2.2.7, we see that for each embedding $b : T \hookrightarrow G_{\delta_h}$ which is G_a-conjugate to a, the group $\widetilde{(G_{\delta_h})_b}$ is contained in $(\widetilde{G_a})_{\delta_h}$ and we have a natural isomorphism $(G_{\delta_h})_b\backslash(G_a)_{\delta_h} \cong \widetilde{(G_{\delta_h})_b}\backslash(\widetilde{G_a})_{\delta_h}$. Then by Lemma 2.3.9 applied to $\gamma_h = \delta_h u_h$, the contribution of each $h \in J$ to (2.3.5) equals

$$e(G_{\delta_h}) \sum_{b:T\hookrightarrow G_{\delta_h}} \theta(b^{-1}(\delta_h)) \sum_{g\in\widetilde{(G_{\delta_h})_b}\backslash[G_{\delta_h}(E)\cap\Omega h^{-1}]} t_{b,\theta}(g u_h g^{-1}), \tag{2.3.6}$$

where b runs over the a set of representatives of conjugacy classes of embeddings, which are G_a-conjugate to $a : T \hookrightarrow G$.

Conjugating by h^{-1}, we can rewrite (2.3.6) in the form

$$e(G_\delta) \sum_{b:T\hookrightarrow G_\delta} \theta(b^{-1}(\delta)) \sum_{g\in\widetilde{(G_\delta)_b}\backslash[G_\delta(E)\cap h^{-1}\Omega]} t_{b,\theta}(gug^{-1}), \tag{2.3.7}$$

where b runs over a set of representatives of conjugacy classes of embeddings such that $a = gbg^{-1}$ for some $g \in \widetilde{G_a}hG_\delta(E)$. In particular, b has a non-trivial contribution to (2.3.7) only for a unique $h \in J$, which we denote by h_b. It follows that $F_{a,\theta}(\gamma)$ equals

$$e(G_\delta) \sum_{b:T\hookrightarrow G_\delta} \theta(b^{-1}(\delta)) \sum_{g\in\widetilde{(G_\delta)_b}\backslash[G_\delta(E)\cap h_b^{-1}\Omega]} t_{b,\theta}(gug^{-1}), \tag{2.3.8}$$

where b runs over a (finite) set of representatives of conjugacy classes to embeddings, which are $G(E)$-conjugate to a.

Replacing b's by their $G_\delta(E)$-conjugates, we can assume that $a = gbg^{-1}$ for some $g \in \widetilde{G_a}h_b$. Then $h_b(G_\delta)_b h_b^{-1} \subset G_a$, hence $h_b\widetilde{(G_\delta)_b}h_b^{-1} \subset \widetilde{G_a}$ (use Corollary 2.2.7). It follows that the subset $G_\delta(E) \cap h_b^{-1}\Omega \subset G_\delta(E)$ is compact modulo center, $\widetilde{(G_\delta)_b}$-invariant from the left, and $h_b\widetilde{(G_\delta)_b}h_b^{-1}$-invariant from the right. Since the number of b's is finite, it follows from [HC1, Lemma 23] that for each sufficiently large $\Omega \subset G(E)$, the contribution of each b to (2.3.8) equals $\theta(b^{-1}(\delta))F_{b,\theta}(u)$. This completes the proof.

□

2.4. Endoscopy for G and G_δ

Let G be a connected reductive group over E such that $G^{\text{der}} = G^{\text{sc}}$, $\delta \in G(E)$ a semisimple element, and $\iota : G_\delta \hookrightarrow G$ the canonical embedding. In this subsection we will compare endoscopic triples for G and for G_δ.

2.4.1. (a) Similar to 1.2.4 (b), there exists a natural embedding $Z(\widehat{G}) \hookrightarrow Z(\widehat{G_\delta})$. Indeed, every maximal torus T of G_δ is a maximal torus of G, and the set of roots of (G_δ, T) equals the set of those roots of (G, T) which vanish on δ. Hence the set of coroots (hence also of roots) of $(\widehat{G_\delta}, \widehat{T})$ is naturally a subset of those of $(\widehat{G}, \widehat{T})$. Thus $Z(\widehat{G})$ is naturally a subgroup of $Z(\widehat{G_\delta})$.

(b) Fix an embedding $a_\delta : T \hookrightarrow G_\delta$ of a maximal torus and $\kappa \in \widehat{T}^\Gamma/Z(\widehat{G})$. Set $a := \iota \circ a_\delta : T \hookrightarrow G$, let $\overline{\kappa} \in \pi_0(\widehat{T}^\Gamma/Z(\widehat{G})^\Gamma)$ and $\overline{\kappa}' \in \pi_0(\widehat{T}^\Gamma/Z(\widehat{G_\delta})^\Gamma)$ be the classes of κ, and put $\mathcal{E} := \mathcal{E}_{([a],\kappa)} = (H, [\eta], \overline{s})$ and $\mathcal{E}' := \mathcal{E}_{([a_\delta],\kappa)} = (H', [\eta'], \overline{s}')$.

Also we fix embeddings $c' : T \hookrightarrow H'$ and $c : T \hookrightarrow H$ such that $\Pi_{\mathcal{E}}([c]) = ([a], \overline{\kappa})$ and $\Pi_{\mathcal{E}'}([c']) = ([a_\delta], \overline{\kappa}')$ (see 1.3.8 (b)).

Lemma 2.4.2. *(a) There is a natural Γ-equivariant embedding $Z(\widehat{H}) \hookrightarrow Z(\widehat{H}')$ mapping $Z(\widehat{G})$ into $Z(\widehat{G_\delta})$ and an embedding $W(H') \hookrightarrow W(H)$, both of which depend on c and c'. The induced map*

$$\pi_0(Z(\widehat{H})^\Gamma/Z(\widehat{G})^\Gamma) \to \pi_0(Z(\widehat{H}')^\Gamma/Z(\widehat{G_\delta})^\Gamma)$$

sends $\overline{s}$ to $\overline{s}'$.

(b) There is a natural map $[b'] \mapsto [b]$, depending on c and c', from the set of stable conjugacy classes of embeddings of maximal tori $S \hookrightarrow H'$ to those of embeddings $S \hookrightarrow H$.

(c) In the notation of (b), we have $[b]_G = \iota \circ [b']_{G_\delta}$, and $Z_{\widehat{[b]}} : Z(\widehat{H}) \hookrightarrow \widehat{S}$ is the restriction of $Z_{\widehat{[b']}} : Z(\widehat{H}') \hookrightarrow \widehat{S}$ (see (a)). In particular, $\overline{\kappa}_{[b']} \in \pi_0(\widehat{S}^\Gamma/Z(\widehat{G_\delta})^\Gamma)$ is the image of $\overline{\kappa}_{[b]} \in \pi_0(\widehat{S}^\Gamma/Z(\widehat{G})^\Gamma)$.

(d) Let $[b_i']$ be two stable conjugacy classes of embeddings of maximal tori $T_i \hookrightarrow H'$, and let $[b_i]$ be the corresponding stable conjugacy classes of embeddings $T_i \hookrightarrow H$. Then the following diagram is commutative:

$$\begin{array}{ccc} Z(\widehat{H}) & \xrightarrow{\iota([b_1],[b_2])} & \{(T_1 \times T_2)/Z(G)\}\widehat{} \\ \downarrow & & \uparrow \\ Z(\widehat{H}') & \xrightarrow{\iota([b_1'],[b_2'])} & \{(T_1 \times T_2)/Z(G_\delta)\}\widehat{}. \end{array}$$

(Here the left vertical map was defined in (a), the right one is induced by the inclusion $Z(G) \hookrightarrow Z(G_\delta)$, and the horizontal maps are the homomorphisms (1.2.3) from 1.2.8.)

Proof. (a) Embed T into G, G_δ, H and H' by a, a_δ, c and c', respectively. Then the set of roots of $(\widehat{G_\delta}, \widehat{T})$ (resp. of $(\widehat{H}, \widehat{T})$, resp. of $(\widehat{H}', \widehat{T})$) is the set of those roots $\hat{\alpha}$ of $(\widehat{G}, \widehat{T})$ such that $\alpha(\delta) = 1$ (resp. $\hat{\alpha}(\kappa) = 1$, resp. $\alpha(\delta) = 1$ and $\hat{\alpha}(\kappa) = 1$). In particular, the set of roots of $(\widehat{H}', \widehat{T})$ is canonically a subset of those of $(\widehat{H}, \widehat{T})$. This gives us the required embeddings $W(H') \hookrightarrow W(H)$ and $Z(\widehat{H}) \hookrightarrow Z(\widehat{H}')$. The last assertion follows from the fact that both $\overline{s}$ and $\overline{s}'$ are the classes of κ.

(b) Choose an embedding $b' : S \hookrightarrow H'$ from $[b']$, and identify S with $b'(S) \subset H'$ and T with $c'(T) \subset H'$. Choose $g \in H'(\overline{E})$ such that $gSg^{-1} = T$. Then $\operatorname{Int} g$ defines an isomorphism $S_{\overline{E}} \xrightarrow{\sim} T_{\overline{E}}$. Let $b : S_{\overline{E}} \hookrightarrow H$ be the composition $c \circ \operatorname{Int} g$. We claim that the $H(\overline{E})$-conjugacy class $[b]$ of b is Γ-invariant and independent of the choices of g and b'.

If $g' \in H'(\overline{E})$ is another element such that $g'Sg'^{-1} = T$, then $g^{-1}g' \in \operatorname{Norm}_{H'}(S)$, and $b' := c \circ \operatorname{Int} g'$ equals $b \circ \operatorname{Int}(g^{-1}g')$. But $\operatorname{Int}(g^{-1}g') : S_{\overline{E}} \to S_{\overline{E}}$ is induced by an element of $W(H') \subset W(H)$, therefore b' is conjugate to b. Thus $[b]$ is independent of the choice of g. For each $\sigma \in \Gamma$, we have ${}^{\sigma}b = c \circ \operatorname{Int}({}^{\sigma}g)$ and ${}^{\sigma}gS({}^{\sigma}g)^{-1} = T$. Hence from the above, ${}^{\sigma}b$ is conjugate to b. Finally, if b' is replaced by $\operatorname{Int} h \circ b'$ and g by gh^{-1} for some $h \in H'(\overline{E})$, then the resulting isomorphism $S_{\overline{E}} \xrightarrow{\sim} T_{\overline{E}}$ (and, therefore, b) do not change.

(c) Since the assertion is over $\overline{E}$, we can identify $S_{\overline{E}} \xrightarrow{\sim} T_{\overline{E}}$, as in (b), and thus replace $[b']$ by $[c']$ and $[b]$ by $[c]$. Now the first assertion follows from the fact that $\iota \circ [c']_{G_\delta} = \iota \circ [a_\delta] = [a] = [c]_G$, while the second was the definition of the embedding $Z(\widehat{H}) \hookrightarrow Z(\widehat{H}')$.

(d) Mimicking the proof of (a) and (b), we see that there exists a Γ-equivariant embedding $Z(\widehat{H^2/Z(G)}) \hookrightarrow Z(\widehat{(H')^2/Z(G)})$ characterized by the following property: For each pair $[c'_i]$ of stable conjugacy classes of embeddings of maximal tori $S_i \hookrightarrow H'$ with the corresponding stable conjugacy classes $[c_i]$ of embeddings of maximal tori $S_i \hookrightarrow H$, the embedding $Z_{\widehat{[c_1,c_2]}} : Z(\widehat{H^2/Z(G)}) \hookrightarrow \{(S_1 \times S_2)/Z(G)\}^{\wedge}$ is the restriction of $Z_{\widehat{[c'_1,c'_2]}} : Z(\widehat{(H')^2/Z(G)}) \hookrightarrow \{(S_1 \times S_2)/Z(G)\}^{\wedge}$. Then our diagram extends to the following diagram:

$$\begin{array}{ccccc}
Z(\widehat{H}) & \xrightarrow{\mu_H} & Z(\widehat{H^2/Z(G)}) & \xrightarrow{Z_{\widehat{[b_1,b_2]}}} & \{(T_1 \times T_2)/Z(G)\}^{\wedge} \\
\downarrow & & \downarrow & & \| \\
Z(\widehat{H'}) & \xrightarrow{\mu_{H'}} & Z(\widehat{(H')^2/Z(G)}) & \xrightarrow{Z_{\widehat{[b'_1,b'_2]}}} & \{(T_1 \times T_2)/Z(G)\}^{\wedge} \\
\| & & \uparrow & & \uparrow \\
Z(\widehat{H'}) & \xrightarrow{\mu_{H'}} & Z(\widehat{(H')^2/Z(G_\delta)}) & \xrightarrow{Z_{\widehat{[b'_1,b'_2]}}} & \{(T_1 \times T_2)/Z(G_\delta)\}^{\wedge}.
\end{array}$$

It remains to show that each inner square of the diagram is commutative. The commutativity of the top right square follows from the characterization of the embedding $Z(\widehat{H^2/Z(G)}) \hookrightarrow Z(\widehat{(H')^2/Z(G)})$. The commutativity of the top left square follows from the characterization of the vertical maps and the fact that both μ_H and $\mu_{H'}$ are restrictions of $\mu_T : \widehat{T} \hookrightarrow \widehat{T^2/Z(G)}$. The commutativity of the two bottom squares is clear. □

Corollary 2.4.3. *Let $[b'_1]$ and $[b'_2]$ be stable conjugacy classes of embeddings of maximal tori $S \hookrightarrow H'$ such that $[b'_1]_{G_\delta} = [b'_2]_{G_\delta}$, and let $[b_1]$ and $[b_2]$ be the corresponding stable conjugacy classes of embeddings $S \hookrightarrow H$.*

Then $[b_1]_G = [b_2]_G$*, and the image of* $\kappa\left(\frac{[b'_1]}{[b'_2]}\right)(\overline{s}') \in \pi_0(\widehat{S/Z(G_\delta)}^{\Gamma})$ *in* $\pi_0(\widehat{S/Z(G)}^{\Gamma})$ *equals* $\kappa\left(\frac{[b_1]}{[b_2]}\right)(\overline{s})$.

Proof. The first assertion follows from Lemma 2.4.2 (c). For the second, choose a representative $\check{s} \in \pi_0(Z(\widehat{H})^{\Gamma})$ of $\overline{s}$, and let $\check{s}' \in \pi_0(Z(\widehat{H}')^{\Gamma})$ be the image of $\check{s}$. Then $\kappa(\frac{[b_1]}{[b_2]})(\overline{s}) = \nu_S(\iota([b_1],[b_2])(\check{s}))$ and $\kappa(\frac{[b'_1]}{[b'_2]})(\overline{s}') = \nu_S(\iota([b'_1],[b'_2])(\check{s}'))$. So the assertion follows from Lemma 2.4.2 (d). □

2.4.4. Let $\varphi : G \to G'$ be an inner twisting such that $\delta' := \varphi(\delta)$ belongs to $G'(E)$. For every $i = 1, 2$, let $(a_\delta)_i : T_i \hookrightarrow G_\delta$ and $(a'_\delta)_i : T_i \hookrightarrow G'_{\delta'}$ be stably conjugate embeddings of maximal tori, and let $b'_i : T_i \hookrightarrow H'$ be an embedding of a maximal torus compatible with $(a_\delta)_i$.

Let $\iota' : G'_{\delta'} \hookrightarrow G'$ be the natural embedding, and for each $i = 1, 2$, set $a_i := \iota \circ (a_\delta)_i : T_i \hookrightarrow G$ and $a'_i := \iota' \circ (a'_\delta)_i : T_i \hookrightarrow G'$, and denote by $[b_i]$ the stable conjugacy class of embeddings of maximal tori $T_i \hookrightarrow H'$ corresponding to $[b'_i]$ (and compatible with a_i by Lemma 2.4.2 (b)).

Lemma 2.4.5. *(a) The image of* $\overline{\mathrm{inv}}((a_1, a'_1); (a_2, a'_2)) \in H^1(E, (T_1 \times T_2)/Z(G))$ *in* $H^1(E, (T_1 \times T_2)/Z(G_\delta))$ *equals* $\overline{\mathrm{inv}}(((a_\delta)_1, (a'_\delta)_1); ((a_\delta)_2, (a'_\delta)_2))$.

(b) We have an equality $\left\langle \frac{a_1, a'_1; [b_1]}{a_2, a'_2; [b_2]} \right\rangle_{\mathcal{E}} = \left\langle \frac{(a_\delta)_1, (a'_\delta)_1; [b'_1]}{(a_\delta)_2, (a'_\delta)_2; [b'_2]} \right\rangle_{\mathcal{E}'}$.

Proof. (a) Follows immediately from the definition of the invariant.

(b) Let $\check{s}$ and $\check{s}'$ be as in the proof of Corollary 2.4.3. Then by Lemma 2.4.2 (a) and (d), we obtain that $\kappa([b_1],[b_2])(\check{s}) \in \pi_0((\{(T_1 \times T_2)/Z(G)\}^{\frown})^{\Gamma})$ is the image of $\kappa([b'_1],[b'_2])(\check{s}') \in \pi_0((\{(T_1 \times T_2)/Z(G_\delta)\}^{\frown})^{\Gamma})$. Now the assertion follows from (a) and the functoriality of the Tate–Nakayama duality. □

2.4.6. Let $d_\delta : S \hookrightarrow G_\delta$ be an embedding of a maximal torus, $\overline{\xi}$ an element of $\pi_0(\widehat{S}^{\Gamma}/Z(\widehat{G})^{\Gamma})$, $d := \iota \circ d_\delta : S \hookrightarrow G$, $\overline{\xi}_\delta \in \pi_0(\widehat{S}^{\Gamma}/Z(\widehat{G_\delta})^{\Gamma})$ the class of $\overline{\xi}$, and $\varphi : G \to G'$ an $(\mathcal{E}, [d], \overline{\xi})$-admissible inner twisting.

Lemma 2.4.7. *Assume that there exists* $[b'] \in \Pi^{-1}_{\mathcal{E}'}([d_\delta], \overline{\xi}_\delta)$ *such that the corresponding stable conjugacy class* $[b]$ *of embeddings* $S \hookrightarrow H$ *satisfies* $\Pi_{\mathcal{E}}([b]) = ([d], \overline{\xi})$.

If there exists a stably conjugate embedding $d' : S \hookrightarrow G'$ *of* d*, then there exists a stable conjugate* d' *of* d *for which* $G'_{d'(d^{-1}(\delta))}$ *is an* $(\mathcal{E}', [d_\delta], \overline{\xi}_\delta)$*-admissible inner form of* G_δ.

Proof. For brevity, we will denote $Z(\mathcal{E}, [d], \overline{\xi}) \subset Z(\widehat{G^{\mathrm{ad}}})^{\Gamma}$ by Z and $Z(\mathcal{E}', [d_\delta], \overline{\xi}_\delta) \subset Z(\widehat{G^{\mathrm{ad}}_\delta})^{\Gamma}$ by Z_δ. The embedding $d_\delta : S \hookrightarrow G_\delta \subset G$ induces homomorphisms

$$H^1(E, (G_\delta)^{\mathrm{ad}}) \xleftarrow{g} H^1(E, S/Z(G)) \xrightarrow{f} H^1(E, G^{\mathrm{ad}}),$$

hence dual homomorphisms $Z(\widehat{(G_\delta)^{\mathrm{ad}}}^{\Gamma}) \xrightarrow{g^D} \pi_0(\widehat{S/Z(G)}^{\Gamma}) \xleftarrow{f^D} Z(\widehat{G^{\mathrm{ad}}})^{\Gamma}$.

First we claim that the assertion of the lemma is equivalent to the inclusion

$$g^D(Z_\delta) \cap \operatorname{Im} f^D \subset f^D(Z). \tag{2.4.1}$$

Indeed, put $x := \operatorname{inv}(G, G')$. By our assumptions, $x \in Z^\perp \cap \operatorname{Im} f$, and the lemma is equivalent to the assertion that there exists $y \in f^{-1}(x)$ such that $g(y) \in (Z_\delta)^\perp$. Equivalently, we have to show that $f^{-1}(x) \cap g^{-1}((Z_\delta)^\perp) \neq \emptyset$. Since $g^{-1}((Z_\delta)^\perp) = [g^D(Z_\delta)]^\perp$, we have to check that x belongs to $f([g^D(Z_\delta)]^\perp) = [(f^D)^{-1}(g^D(Z_\delta))]^\perp$. In other words, the lemma asserts that $\operatorname{Im} f \cap Z^\perp \subset [(f^D)^{-1}(g^D(Z_\delta))]^\perp$, or by duality that $(f^D)^{-1}(g^D(Z_\delta)) \subset Z + \operatorname{Ker} f^D$. But the last inclusion is equivalent to (2.4.1).

To show (2.4.1), take any element $y \in g^D(Z_\delta) \cap \operatorname{Im} f^D \subset \pi_0(\widehat{S/Z(G)}^\Gamma)$. Note that g^D factors through $Z_{\widehat{[d_\delta]}} : Z(\widehat{(G_\delta)^{\mathrm{ad}}})^\Gamma \to \pi_0(\widehat{S/Z(G_\delta)}^\Gamma)$. Therefore it follows from Corollary 1.4.2 that there exists $[b'_1] \in \Pi_{\mathcal{E}'}^{-1}([d_\delta], \overline{\xi}_\delta)$ such that y is the image of $\kappa\left(\frac{[b'_1]}{[b']}\right)(\overline{s}') \in \pi_0(\widehat{S/Z(G_\delta)}^\Gamma)$. Let $[b_1]$ be the stable conjugacy class of embeddings $S \hookrightarrow H$ corresponding to $[b'_1]$. Then $[b_1]_G = [b]_G$ and $y = \kappa\left(\frac{[b_1]}{[b]}\right)(\overline{s})$ (by Corollary 2.4.3). Since $y \in \operatorname{Im} f^D$, the image of $\kappa\left(\frac{[b_1]}{[b]}\right)(\overline{s})$ in $\pi_0(\widehat{S}^\Gamma/Z(\widehat{G})^\Gamma)$ is trivial. Since this image equals $\overline{\kappa}_{[b_1]}/\overline{\kappa}_{[b]}$ (use Lemma 1.2.9 (a)), we conclude that $\Pi_{\mathcal{E}}([b_1]) = \Pi_{\mathcal{E}}([b]) = ([d], \overline{\xi})$. Therefore $y = \kappa\left(\frac{[b_1]}{[b]}\right)(\overline{s})$ belongs to $f^D(Z)$, as claimed. □

2.5. Preparation for the proof of the Main Theorem

Lemma 2.5.1. *Let $\pi : \widetilde{G} \to G$ be a quasi-isogeny such that $\widetilde{G}$ splits over E^{nr}, $\widetilde{a}_0 : \widetilde{T} \hookrightarrow \widetilde{G}$ the lift of $a_0 : T \hookrightarrow G$, and $\widetilde{\theta}$ the composition $\widetilde{T}(E) \to T(E) \xrightarrow{\theta} \mathbb{C}^\times$.*

For every stable conjugate a of a_0, the representation $\pi_{a,\theta} \circ \pi$ of $\widetilde{G}(E)$ is isomorphic to the direct sum $\sum_{\widetilde{a}} \pi_{\widetilde{a},\widetilde{\theta}}$, taken over the set of all conjugacy classes of embeddings $\widetilde{a} : \widetilde{T} \hookrightarrow \widetilde{G}$ such that $\pi \circ \widetilde{a} : T \hookrightarrow G$ is conjugate to a.

Proof. Observe first that for each quasi-isogeny $\overline{\pi} : \widetilde{L} \to L_a$, the representation $\rho_{\overline{a},\overline{\theta}} \circ \overline{\pi}$ of $\widetilde{L}(\mathbb{F}_q)$ is isomorphic to the Deligne–Lusztig representation $\rho_{\widetilde{\overline{a}},\widetilde{\overline{\theta}}}$, where $\widetilde{\overline{a}} : \widetilde{\overline{T}} \hookrightarrow \widetilde{L}$ is the lift of $\overline{a}$, and $\widetilde{\overline{\theta}}$ is the composition $\widetilde{\overline{T}}(\mathbb{F}_q) \to \overline{T}(\mathbb{F}_q) \xrightarrow{\overline{\theta}} \mathbb{C}^\times$.

For each $\widetilde{a}$ as in the lemma, denote by $\pi_{\widetilde{a}} : \widetilde{G_{\widetilde{a}}} \to \widetilde{G_a}$ the restriction of π. Then by the above observation, the representation $\rho_{a,\theta} \circ \pi_{\widetilde{a}}$ is isomorphic to $\rho_{\widetilde{a},\widetilde{\theta}}$. It follows that each $\pi_{\widetilde{a},\widetilde{\theta}}$ is a subrepresentation of $\pi_{a,\theta} \circ \pi$.

Since conjugacy classes of the $\widetilde{a}$'s are naturally identified with the double coset $\pi(\widetilde{G}(E))\backslash G(E)/a(T(E))$, while the set of irreducible factors of $\pi_{a,\theta} \circ \pi$ is naturally identified with $\pi(\widetilde{G}(E))\backslash G(E)/\widetilde{G_a}$, it remains to check that these two double cosets coincide.

By Corollary 2.2.7, we have $\widetilde{G_a} = a(T(E))G_a$; therefore it will suffice to show that $G_a \subset \pi(\widetilde{G}(E))\, a(T(\mathcal{O}))$. But this inclusion follows from Corollary 1.8.18. □

Corollary 2.5.2. *In the notation of Lemma 2.5.1, let $\widetilde{\kappa} \in \widehat{\widetilde{T}}^{\Gamma}/Z(\widehat{\widetilde{G}})^{\Gamma}$ be the image of $\kappa \in \widehat{T}^{\Gamma}/Z(\widehat{G})^{\Gamma}$. Then $\pi^*(\chi_{a_0,\kappa,\theta}) = \chi_{\widetilde{a}_0,\widetilde{\kappa},\widetilde{\theta}}$.*

Proof. Since each $\widetilde{a}$ as in the lemma satisfies $\langle \mathrm{inv}(\widetilde{a}_0, \widetilde{a}), \widetilde{\kappa}\rangle = \langle \mathrm{inv}(a_0, a), \kappa\rangle$, the assertion follows from the lemma. □

Lemma 2.5.3. *It will suffice to prove Theorem 2.1.6 under the assumption that the derived group of G is simply connected.*

Proof. Let G be an arbitrary group satisfying the assumptions of Theorem 2.1.6. Since G splits over E^{nr}, there exists a surjective quasi-isogeny $\pi : \widetilde{G} \to G$ such that $\widetilde{G}^{\mathrm{der}} = \widetilde{G}^{\mathrm{sc}}$, and $\operatorname{Ker}\pi$ is an induced torus splitting over E^{nr} (use [MS, Prop. 3.1]). (Such a quasi-isogeny Kottwitz calls a z-extension.) Let $\widetilde{a}_0 : \widetilde{T} \hookrightarrow \widetilde{G}$ be the lift of a_0. Then $\widetilde{T}$ splits over E^{nr}, hence $\widetilde{G}$ satisfies all the assumptions of Theorem 2.1.6.

Let $\widetilde{\kappa} \in \widehat{\widetilde{T}}^{\Gamma}/Z(\widehat{\widetilde{G}})^{\Gamma}$ be the image of κ, $\widetilde{\theta}$ the composition $\widetilde{T}(E) \to T(E) \xrightarrow{\theta} \mathbb{C}^{\times}$, $\pi' : \widetilde{G}' \to G'$ the inner twist of π, induced by φ, and $\widetilde{a}'_0 : \widetilde{T} \hookrightarrow \widetilde{G}'$ the lift of $a'_0 : T \hookrightarrow G'$. By the assumption, generalized functions $\chi_{\widetilde{a}_0,\widetilde{\kappa},\widetilde{\theta}}$ and $\chi_{\widetilde{a}'_0,\widetilde{\kappa},\widetilde{\theta}}$ are $(\widetilde{a}_0, \widetilde{a}'_0; \widetilde{\kappa})$-equivalent. Since $H^1(E, \operatorname{Ker}\pi) = 0$, we get that $\pi(\widetilde{G}(E)) = G(E)$ and $\pi'(\widetilde{G}'(E)) = G'(E)$. Therefore it follows from Corollaries 2.5.2 and 1.7.15 that generalized functions $\chi_{a_0,\kappa,\theta}$ and $\chi_{a'_0,\kappa,\theta}$ are $(\widetilde{a}_0, \widetilde{a}'_0; \widetilde{\kappa})$-equivalent. Thus by Corollary 1.6.13, they are $(a_0, a'_0; \kappa)$-equivalent, as claimed. □

From now on we will assume that $G^{\mathrm{der}} = G^{\mathrm{sc}}$.

Lemma 2.5.4. *Let G and a_0 be as in Theorem 2.1.6, $\delta \in G(E)$ an element of finite order prime to p, and $b_0 : T \hookrightarrow G_\delta$ an embedding stably conjugate to a_0. Then*

(a) the conclusion of Theorem 2.2.3 holds for G_δ and b_0,
(b) the set of topologically unipotent elements of $G_\delta(E)$ equals $G_\delta(E)_{\mathrm{tu}}$.

Proof. (a) First, it follows from Lemma 2.3.8 (b) that G_δ splits over E^{nr}. Consider the canonical isogeny $\pi : G^{\mathrm{sc}} \times Z(G)^0 \to G$. It induces the isogeny $\pi_\delta : (G^{\mathrm{sc}})_\delta \times Z(G)^0 \to G_\delta$, where $(G^{\mathrm{sc}})_\delta := \{g \in G^{\mathrm{sc}} \mid \operatorname{Int}\delta(g) = g\}$ is connected by [St, Thm. 8.1].

Since G satisfies property (vg), the order of $Z(G^{\mathrm{sc}})$ is prime to p. Therefore π and hence π_δ are of order prime to p. As in the proof of Lemma 2.5.3, we see (using Lemma 1.8.17 (a) and Corollaries 2.5.2, 1.7.15 and 1.6.13) that we can replace G_δ by $(G^{\mathrm{sc}})_\delta \times Z(G)^0$ and b_0 by its lift. Thus it will suffice to show that $(G^{\mathrm{sc}})_\delta$ and the lift $b_0^{\mathrm{sc}} : T^{\mathrm{sc}} \hookrightarrow (G^{\mathrm{sc}})_\delta$ of b_0 satisfy the assumptions of 2.2.1.

Since G satisfies property (vg), G^{sc} admits a quasi-logarithm $\Phi : G^{\mathrm{sc}} \to \mathcal{G}^{\mathrm{sc}}$ defined over $\mathcal{O}$, and $\mathcal{G}^{\mathrm{sc}}$ admits a non-degenerate over $\mathcal{O}$ invariant pairing $\langle\cdot,\cdot\rangle$. As $\mathrm{Lie}(G^{\mathrm{sc}})_\delta = (\mathcal{G}^{\mathrm{sc}})_\delta$, the quasi-logarithm Φ induces a quasi-logarithm Φ_δ for $(G^{\mathrm{sc}})_\delta$, and $\langle\cdot,\cdot\rangle$ induces an invariant pairing $\langle\cdot,\cdot\rangle_\delta$ on $\mathrm{Lie}(G^{\mathrm{sc}})_\delta$. Furthermore, as δ is of finite order prime to p, we get that Φ_δ is defined over $\mathcal{O}$ and $\langle\cdot,\cdot\rangle_\delta$ is non-degenerate over $\mathcal{O}$. Finally, since $b_0^{\mathrm{sc}} : T^{\mathrm{sc}} \hookrightarrow (G^{\mathrm{sc}})_\delta$ is stably conjugate to $a_0^{\mathrm{sc}} : T^{\mathrm{sc}} \hookrightarrow G^{\mathrm{sc}}$ and since

G satisfies property $(vg)_{a_0}$, there exists a b_0^{sc}-strongly regular element of $\overline{T}^{\mathrm{sc}}(\mathbb{F}_q)$, implying the last assumption of 2.2.1.

(b) The proof is a generalization of Corollary 2.3.3. Each topologically unipotent element $u \in G_\delta(E)$ stabilizes some point $x \in \mathcal{B}(G_\delta) \subset \mathcal{B}(G)$. Hence u belongs to $G_\delta(E) \cap G_{x,\mathrm{tu}}$ (by Lemma 2.3.1 (b)). Therefore $\overline{u} \in L_x(\mathbb{F}_q)$ belongs to $(L_x)_{\overline{\delta}} \cap \mathcal{U}(L_x)$ (by Lemma 2.3.8 (b)). By Lemma 2.5.5 below, $\overline{u}$ belongs to $(L_x)^0_{\overline{\delta}} \cap \mathcal{U}(L_x) = \mathcal{U}((L_x)^0_{\overline{\delta}})$, hence u belongs to $(G_\delta)_{x,\mathrm{tu}} \subset G_\delta(E)_{\mathrm{tu}}$, as claimed.

□

Lemma 2.5.5. *Let L be a connected reductive group, and $s \in L$ a semisimple element. Then $\mathcal{U}(L)_s = \mathcal{U}(L) \cap L_s$ is contained in L_s^0.*

Proof. Recall that the canonical homomorphism $\iota : L^{\mathrm{sc}} \to L$ induces an L-equivariant isomorphism $\mathcal{U}(L^{\mathrm{sc}}) \xrightarrow{\sim} \mathcal{U}(L)$, hence an isomorphism $\mathcal{U}(L^{\mathrm{sc}})_s \xrightarrow{\sim} \mathcal{U}(L)_s$. Therefore $\mathcal{U}(L)_s = \iota(\mathcal{U}(L^{\mathrm{sc}})_s)$ is contained in $\iota((L^{\mathrm{sc}})_s)$. Since $(L^{\mathrm{sc}})_s$ is connected (by [St, Theorem. 8.1]), the latter group is contained in L_s^0, as claimed. □

Notation 2.5.6. (a) For a compact element $\gamma_0 \in G^{\mathrm{sr}}(E)$ with a topological Jordan decomposition $\gamma_0 = \delta_0 u_0$, we say that $t \in T(E)$ is (G, a_0, γ_0)-*relevant* if there exists an embedding $b_0 : T \hookrightarrow G_{\delta_0} \subset G$ stably conjugate to a_0 such that $b_0(t) = \delta_0$.

(b) Assume that $t \in T(E)$ is (G, a_0, γ_0)-relevant. Since $b_0(T) \subset G_{\delta_0}$ is elliptic, Kottwitz's theorem (see 1.5.1) implies that $H^1(E, T) \to H^1(E, G_{\delta_0})$ is surjective (compare the proof of Lemma 1.5.3). Hence for each $\delta \in G(E)$ stably conjugate to δ_0 there exists an embedding $b_{t,\delta} : T \hookrightarrow G_\delta \subset G$ stably conjugate to a_0 such that $b_{t,\delta}(t) = \delta$. Furthermore, $b_{t,\delta}$ is unique up to a stable conjugacy, and the endoscopic triple $\mathcal{E}_t := \mathcal{E}_{([b_{t,\delta}],\kappa)} = (H_t, [\eta_t], \overline{s}_t)$ of G_{δ_0} is independent of δ.

Lemma 2.5.7. *For every compact element $\gamma_0 \in G^{\mathrm{sr}}(E)$ with topological Jordan decomposition $\gamma_0 = \delta_0 u_0$ and each $\overline{\xi} \in \pi_0(\widehat{G_{\gamma_0}}^{\Gamma}/Z(\widehat{G})^{\Gamma})$, we have*

$$F_{a_0,\kappa,\theta}(\gamma_0, \overline{\xi}) = \sum_t \theta(t) \sum_\delta I_{t,\delta}, \tag{2.5.1}$$

where

(i) t runs over the set of (G, a_0, γ_0)-relevant elements of $T(E)$;
(ii) δ runs over a set of representatives of the conjugacy classes in $G(E)$ stably conjugate to δ_0, for which there exists a stably conjugate γ of γ_0 with topological Jordan decomposition $\gamma = \delta u$;
(iii) $I_{t,\delta}$ equals

$$\langle \mathrm{inv}(a_0, b_{t,\delta}), \kappa \rangle \langle \mathrm{inv}(\gamma_0, \gamma), \overline{\xi} \rangle^{-1} F_{b_{t,\delta},\kappa,\theta}(u, \overline{\xi}) \tag{2.5.2}$$

for every $\gamma = \delta u$ as in (ii).

Proof. Recall that

$$F_{a_0,\kappa,\theta}(\gamma_0, \overline{\xi}) = e(G) \sum_a \sum_\gamma \langle \mathrm{inv}(a_0, a), \kappa \rangle \langle \mathrm{inv}(\gamma_0, \gamma), \overline{\xi} \rangle^{-1} F_{a,\theta}(\gamma),$$

where $a : T \hookrightarrow G$ and $\gamma \in G(E)$ run over sets of representatives of conjugacy classes within the stable conjugacy classes of a_0 and γ_0, respectively.

Using Proposition 2.3.7, we see that $F_{a_0,\kappa,\theta}(\gamma_0, \overline{\xi})$ equals the triple sum

$$\sum_{\gamma=\delta u}\sum_{a} \langle \mathrm{inv}(a_0, a), \kappa\rangle \left\langle \mathrm{inv}(\gamma_0, \gamma), \overline{\xi}\right\rangle^{-1} e(G_\delta) \sum_b \theta(b^{-1}(\delta)) F_{b,\theta}(u), \tag{2.5.3}$$

where γ and a are as above, and b runs over conjugacy classes of embeddings $T \hookrightarrow G_\delta$, which are conjugate to $a : T \hookrightarrow G$.

Then (2.5.3) can be rewritten in the form

$$\sum_{\gamma=\delta u} e(G_\delta) \sum_b \langle \mathrm{inv}(a_0, b), \kappa\rangle \left\langle \mathrm{inv}(\gamma_0, \gamma), \overline{\xi}\right\rangle^{-1} \theta(b^{-1}(\delta)) F_{b,\theta}(u), \tag{2.5.4}$$

where b runs over the set of conjugacy classes of embeddings $T \hookrightarrow G_\delta$, whose composition with the inclusion $G_\delta \hookrightarrow G$ is stably conjugate to a_0.

Furthermore, each $t := b^{-1}(\delta) \in T(E)$, appearing in the sum, is (G, a_0, γ_0)-relevant, and the contribution of each such t to (2.5.4) is

$$\theta(t) \sum_{\gamma=\delta u} \left\langle \mathrm{inv}(a_0, b_{t,\delta}), \kappa\right\rangle \left\langle \mathrm{inv}(\gamma_0, \gamma), \overline{\xi}\right\rangle^{-1} e(G_\delta) \left[\sum_b \left\langle \mathrm{inv}(b_{t,\delta}, b), \kappa\right\rangle F_{b,\theta}(u)\right], \tag{2.5.5}$$

where b runs over a set of representatives of conjugacy classes of embeddings $T \hookrightarrow G_\delta$ stably conjugate to $b_{t,\delta}$. But (2.5.5) coincides with the sum $\theta(t)\sum_\delta I_{t,\delta}$ as in the lemma. □

2.5.8. We fix $\gamma_0 \in G^{\mathrm{sr}}(E)$ and $\overline{\xi} \in \pi_0(\widehat{G_{\gamma_0}}^{\Gamma}/Z(\widehat{G})^{\Gamma})$ such that $F_{a_0,\kappa,\theta}(\gamma_0, \overline{\xi}) \neq 0$, and we are going to show that $(\gamma_0, \overline{\xi})$ satisfies the conditions (i), (ii) of Theorem 2.1.10.

2.6. Proof of Theorem 2.1.10 (i)

2.6.1. Since the support of each $t_{a,\theta}$ is contained in $\widetilde{G_a}$, the assumption on γ_0 implies that there exists $z \in Z(G)(E)$ such that $z\gamma_0$ is compact. But $F_{a_0,\kappa,\theta}(z\gamma_0, \overline{\xi}) = \theta(z)$ $F_{a_0,\kappa,\theta}(\gamma_0, \overline{\xi})$ for each $z \in Z(G)(E)$; therefore the assertions of Theorem 2.1.10 for γ_0 are equivalent to those for $z\gamma_0$. Hence we can and will assume that γ_0 is compact. In particular, Lemma 2.5.7 holds for γ_0.

Every stably conjugate γ of γ_0 is compact as well, and we denote by $\gamma_0 = \delta_0 u_0$ and $\gamma = \delta u$ their topological Jordan decompositions. We also let $\overline{\xi}_{\delta_0}$ be the image of $\overline{\xi}$ in $\pi_0(\widehat{G_{\gamma_0}}^{\Gamma}/Z(\widehat{G_{\delta_0}})^{\Gamma})$.

Notation 2.6.2. We will say that stable conjugates $\delta_1, \delta_2 \in G(E)$ of δ_0 are $(\gamma_0, \overline{\xi})$-*equivalent* if there exist stable conjugate γ_1 and γ_2 of γ_0 with topological Jordan decompositions $\gamma_i = \delta_i u_i$, and G_{δ_2} is an $(\mathcal{E}_t, [a_{\gamma_0}], \overline{\xi}_{\delta_0})$-admissible inner form of G_{δ_1}. In this case we will write $\delta_1 \sim_{(\gamma_0, \overline{\xi})} \delta_2$.

2.6.3. Fix t which has a non-zero contribution to (2.5.1). Since the set of conjugacy classes of δ in Lemma 2.5.7 (ii) decomposes as a union of $(\gamma_0, \overline{\xi})$-equivalent classes, we can replace γ_0 by a stably conjugate element, so that $\sum_{\delta \sim_{(\gamma_0,\overline{\xi})} \delta_0} I_{t,\delta} \neq 0$. Further replacing γ_0, we can moreover assume that $I_{t,\delta_0} \neq 0$, thus $F_{b_{t,\delta_0},\kappa,\theta}(u_0, \overline{\xi}_{\delta_0}) \neq 0$.

We also fix embeddings of maximal tori $c : T \hookrightarrow H$ and $c' : T \hookrightarrow H_t$ such that $\Pi_{\mathcal{E}}([c]) = ([a_0], \overline{\kappa})$ and $\Pi_{\mathcal{E}_t}([c']) = ([b_{t,\delta_0}], \overline{\kappa})$ (see 2.4.1 (b)). This enables us to apply the results of Subsection 2.4.

2.6.4. Since $G^{\mathrm{der}} = G^{\mathrm{sc}}$, we get that $u_0 \in G_{\delta_0}(E)_{\mathrm{tu}}$ and the conclusion of Theorem 2.2.3 holds for G_{δ_0} and b_{t,δ_0} (see Lemma 2.5.4). Therefore as in Lemma 2.1.11, there exists an embedding $b' : G_{\gamma_0} = (G_{\delta_0})_{u_0} \hookrightarrow H_t$ such that $\Pi_{\mathcal{E}_t}([b']) = ([a_{\gamma_0}], \overline{\xi}_{\delta_0})$, where $a_{\gamma_0} : G_{\gamma_0} \hookrightarrow G_{\delta_0}$ is the natural inclusion.

Let $[b]$ be the stable conjugacy class of embeddings $G_{\gamma_0} \hookrightarrow H$ corresponding to $[b']$ (see Lemma 2.4.2 (b)), and put $\overline{\xi}_{[b']} := \overline{\kappa}_{[b]} \in \pi_0(\widehat{G_{\gamma_0}}^{\Gamma}/Z(\widehat{G})^{\Gamma})$.

To prove the assertion (i) of Theorem 2.1.10, it will suffice to show the existence of $[b'] \in \Pi_{\mathcal{E}_t}^{-1}([a_{\gamma_0}], \overline{\xi}_{\delta_0})$ such that $\overline{\xi}_{[b']} = \overline{\xi}$.

2.6.5. For each $[b'] \in \Pi_{\mathcal{E}_t}^{-1}([a_{\gamma_0}], \overline{\xi}_{\delta_0})$, we denote by

$$z_{[b']} \in \pi_0(\widehat{G_{\gamma_0}}^{\Gamma}/Z(\widehat{G})^{\Gamma} Z(\mathcal{E}_t, [a_{u_0}], \overline{\xi}_{\delta_0}))$$

the image of the quotient $\overline{\xi}_{[b']}/\overline{\xi} \in \pi_0(\widehat{G_{\gamma_0}}^{\Gamma}/Z(\widehat{G})^{\Gamma})$, where $Z(\mathcal{E}_t, [a_{u_0}], \overline{\xi}_{\delta_0}) \subset Z(\widehat{(G_{\delta_0})^{\mathrm{ad}}})^{\Gamma}$ is mapped into $\widehat{G_{\gamma_0}}^{\Gamma}$ via the homomorphism

$$Z(\widehat{(G_{\delta_0})^{\mathrm{ad}}}) \to \widehat{G_{\gamma_0}/Z(G_{\delta_0})} \to \widehat{G_{\gamma_0}}.$$

We claim that $z_{[b']}$ does not depend on $[b']$. Indeed, for each $[b'_1], [b'_2] \in \Pi_{\mathcal{E}_t}^{-1}([a_{\gamma_0}], \overline{\xi}_{\delta_0})$, the quotient $\overline{\xi}_{[b'_1]}/\overline{\xi}_{[b'_2]}$ is the image of

$$\kappa\begin{pmatrix}[b'_1]\\ [b'_2]\end{pmatrix}(\overline{s}_t) \in \pi_0(\widehat{G_{\gamma_0}/Z(G_{\delta_0})}^{\Gamma})$$

(by Corollary 2.4.3 and Lemma 1.2.9 (a)). Thus $\overline{\xi}_{[b'_1]}/\overline{\xi}_{[b'_2]}$ belongs to the image of $Z(\mathcal{E}_t, [a_{u_0}], \overline{\xi}_{\delta_0})$. It follows that $z := z_{[b']}$ is independent of $[b']$, as claimed.

2.6.6. Our next goal is to show that $z = 1$. By definition, z belongs to the image

$$\mathrm{Im}\left[\pi_0(Z(\widehat{G_{\delta_0}})^{\Gamma}/Z(\widehat{G})^{\Gamma}) \to \pi_0(\widehat{G_{\gamma_0}}^{\Gamma}/Z(\widehat{G})^{\Gamma} Z(\mathcal{E}_t, [a_{u_0}], \overline{\xi}_{\delta_0}))\right]. \tag{2.6.1}$$

Denote by $V \subset \mathrm{Ker}[H^1(E, G_{\delta_0}) \to H^1(E, G)]$ the intersection of the image of $\mathrm{Ker}[H^1(E, G_{\gamma_0}) \to H^1(E, G)]$ and the preimage of

$$Z(\mathcal{E}_t, [a_{u_0}], \overline{\xi}_{\delta_0})^{\perp} \subset H^1(E, (G_{\delta_0})^{\mathrm{ad}}).$$

Then for a stable conjugate $\delta \in G(E)$ of δ_0, we have $\delta \sim_{(\gamma_0,\overline{\xi})} \delta_0$ if and only if the invariant $\mathrm{inv}(\delta_0, \delta) \in \mathrm{Ker}[H^1(E, G_{\delta_0}) \to H^1(E, G)]$ lies in V.

By Kottwitz's theorem (1.5.1), the dual group V^D of V is naturally identified with the group (2.6.1). In particular, z belongs to V^D. Therefore for each $\delta \sim_{(\gamma_0,\overline{\xi})} \delta_0$ one can form a pairing $\langle \mathrm{inv}(\delta_0, \delta), z\rangle$. Explicitly,

$$\langle \mathrm{inv}(\delta_0, \delta), z\rangle = \langle \mathrm{inv}(\gamma_0, \gamma), \widetilde{z}\rangle \tag{2.6.2}$$

for every stably conjugate γ of γ_0 with topological Jordan decomposition $\gamma = \delta u$, and every representative $\widetilde{z} \in \pi_0(\widehat{G_{\gamma_0}}^{\Gamma}/Z(\widehat{G})^{\Gamma})$ of z.

Claim 2.6.7. *For each $\delta \sim_{(\gamma_0,\overline{\xi})} \delta_0$, we have $I_{t,\delta} = \langle \mathrm{inv}(\delta_0, \delta), z\rangle\, I_{t,\delta_0}$.*

Proof. Let $\gamma \in G(E)$ be a stable conjugate of γ_0 with topological Jordan decomposition $\gamma = \delta u$. Then $u \in G_\delta(E)$ is a stable conjugate of $u_0 \in G_{\delta_0}(E)$. Since by assumption $F_{b_{t,\delta_0},\kappa,\theta}(u_0, \overline{\xi}_{\delta_0}) \neq 0$ and $u_0 \in G_{\delta_0}(E)$ is topologically unipotent, we conclude from Lemma 2.5.4 (as in Lemma 2.1.11) that

$$F_{b_{t,\delta},\kappa,\theta}(u, \overline{\xi}_{\delta_0}) = \left\langle \frac{u_0, u; \overline{\xi}_{\delta_0}}{b_{t,\delta_0}, b_{t,\delta}; \kappa} \right\rangle_{\mathcal{E}_t} F_{b_{t,\delta_0},\kappa,\theta}(u_0, \overline{\xi}_{\delta_0}).$$

Hence the quotient $I_{t,\delta}/I_{t,\delta_0}$ equals

$$\left\langle \frac{u_0, u; \overline{\xi}_{\delta_0}}{b_{t,\delta_0}, b_{t,\delta}; \kappa} \right\rangle_{\mathcal{E}_t} \big\langle \mathrm{inv}(b_{t,\delta_0}, b_{t,\delta}), \kappa\big\rangle \big\langle \mathrm{inv}(\gamma_0, \gamma), \overline{\xi}\big\rangle^{-1}.$$

Thus our claim is equivalent to the equality

$$\left\langle \frac{u_0, u; \overline{\xi}_{\delta_0}}{b_{t,\delta_0}, b_{t,\delta}; \kappa} \right\rangle_{\mathcal{E}_t} = \langle \mathrm{inv}(\delta_0, \delta), z\rangle \big\langle \mathrm{inv}(\gamma_0, \gamma), \overline{\xi}\big\rangle \big\langle \mathrm{inv}(b_{t,\delta_0}, b_{t,\delta}), \kappa\big\rangle^{-1}. \tag{2.6.3}$$

The left-hand side of (2.6.3) equals $\left\langle \frac{a_{u_0}, a_u; [b']}{b_{t,\delta_0}, b_{t,\delta}; [c']} \right\rangle_{\mathcal{E}_t}$ for each $[b'] \in \Pi_{\mathcal{E}_t}^{-1}([a_{\gamma_0}], \overline{\xi}_{\delta_0})$. By Lemma 2.4.5 (b) and Lemma 1.5.7 (d), it therefore equals

$$\left\langle \frac{a_{\gamma_0}, a_\gamma; [b]}{b_{t,\delta_0}, b_{t,\delta}; [c]} \right\rangle_{\mathcal{E}} = \big\langle \mathrm{inv}(\gamma_0, \gamma), \overline{\kappa}_{[b]}\big\rangle \big\langle \mathrm{inv}(b_{t,\delta_0}, b_{t,\delta}), \overline{\kappa}_{[c]}\big\rangle^{-1}.$$

But $\overline{\kappa}_{[c]} = \kappa$, $\overline{\kappa}_{[b]} = (\overline{\xi}_{[b']}/\overline{\xi})\overline{\xi}$ and $\overline{\xi}_{[b']}/\overline{\xi}$ is a representative of z. Therefore equality (2.6.3) follows from (2.6.2). □

2.6.8. By Claim 2.6.7, the sum $\sum_{\delta \sim_{(\gamma_0,\overline{\xi})} \delta_0} I_{t,\delta_0}$ equals $I_{t,\delta_0}(\sum_{v\in V} \langle v, z\rangle)$. It follows that $\sum_{v\in V} \langle v, z\rangle \neq 0$, hence $z = 1$.

Choose now an arbitrary $[b'] \in \Pi_{\mathcal{E}_t}^{-1}([a_{\gamma_0}], \overline{\xi}_{\delta_0})$. Since $z = 1$, the quotient $\overline{\xi}_{[b']}/\overline{\xi} \in \pi_0(\widehat{G_{\gamma_0}}^{\Gamma}/Z(\widehat{G})^{\Gamma})$ lies in the image of $Z(\mathcal{E}_t, [a_{u_0}], \overline{\xi}_{\delta_0})$. Hence by Corollary 1.4.2, there exists $[b'_1] \in \Pi_{\mathcal{E}_t}^{-1}([a_{\gamma_0}], \overline{\xi}_{\delta_0})$ such that $\overline{\xi}_{[b']}/\overline{\xi}$ equals the image of $\kappa\left(\frac{[b']}{[b'_1]}\right)(\overline{s}_t)$. Since by Corollary 2.4.3 and Lemma 1.2.9 (a), the image of $\kappa\left(\frac{[b']}{[b'_1]}\right)(\overline{s}_t)$ in $\pi_0(\widehat{G_{\gamma_0}}^{\Gamma}/Z(\widehat{G})^{\Gamma})$ equals $\overline{\xi}_{[b']}/\overline{\xi}_{[b'_1]}$, we get that $\overline{\xi}_{[b'_1]} = \overline{\xi}$, completing the proof of (i).

2.7. Proof of Theorem 2.1.10 (ii)

2.7.1. Let $\gamma_0' \in G'(E)$ be a stable conjugate of γ_0. Since γ_0 is compact, so is γ_0', and we denote by $\gamma_0' = \delta_0' u_0'$ its topological Jordan decomposition. By Lemma 2.5.7, we can write $F_{a_0',\kappa,\theta}(\gamma_0', \overline{\xi})$ in the form

$$F_{a_0',\kappa,\theta}(\gamma_0', \overline{\xi}) = \sum_{t'} \theta(t') \sum_{\delta'} I_{t',\delta'},$$

where t', δ' and $I_{t',\delta'}$ have the same meaning as in Lemma 2.5.7.

First, we claim that an element $t \in T(E)$ is (G, a_0, γ_0)-relevant if and only if it is (G', a_0', γ_0')-relevant. Indeed, assume that t is (G, a_0, γ_0)-relevant, and let $b_0 : T \hookrightarrow G_{\delta_0} \subset G$ be the corresponding embedding. Since $T/Z(G)$ is anisotropic, Kottwitz's theorem implies that the map $H^1(E, T/Z(G)) \to H^1(E, G_{\delta_0}/Z(G))$ is surjective (compare the proof of Lemma 1.5.3). Hence there exists a stable conjugate $b_0' : T \hookrightarrow G'_{\delta_0'} \subset G'$ of b_0 such that $b_0'(t) = \delta_0'$, thus t is (G', a_0', γ_0')-relevant.

Therefore it will suffice to show that for each (G, a_0, γ_0)-relevant t, we have

$$\sum_{\delta'} I_{t,\delta'} = \left\langle \frac{\gamma_0, \gamma_0'; \overline{\xi}}{a_0, a_0', \kappa} \right\rangle_{\mathcal{E}} \sum_{\delta} I_{t,\delta}. \tag{2.7.1}$$

2.7.2. Fix (G, a_0, γ_0)-relevant t. Generalizing Notation 2.6.2, we will say that stable conjugates $\delta \in G(E)$ and $\delta' \in G'(E)$ of δ_0 are $(\gamma_0, \overline{\xi})$-*equivalent* (and we will write $\delta \sim_{(\gamma_0,\overline{\xi})} \delta'$) if there exist stable conjugates $\gamma \in G(E)$ and $\gamma' \in G'(E)$ of γ_0 with topological Jordan decompositions $\gamma = \delta u$ and $\gamma' = \delta' u'$ such that $G'_{\delta'}$ is an $(\mathcal{E}_t, [a_{u_0}], \overline{\xi}_{\delta_0})$-admissible inner form of G_δ.

Assume that $\varphi : G \to G'$ is $(\mathcal{E}, [a_{\gamma_0}], \overline{\xi})$-admissible. We claim that for every stable conjugate $\gamma \in G(E)$ of γ_0 with topological Jordan decomposition $\gamma = \delta u$, there exists a stable conjugate $\gamma' \in G'(E)$ of γ_0' with topological Jordan decomposition $\gamma' = \delta' u'$ such that $\delta \sim_{(\gamma_0,\overline{\xi})} \delta'$.

Indeed, we have shown in Subsection 2.6 that there exists $[b'] \in \Pi_{\mathcal{E}_t}^{-1}([a_{\gamma_0}], \overline{\xi}_{\delta_0})$ such that the corresponding stable conjugacy class $[b]$ of embeddings $G_{\gamma_0} \hookrightarrow H$ satisfies $\Pi_{\mathcal{E}}([b]) = ([a_{\gamma_0}], \overline{\xi})$. Since the inclusion $a_\gamma : G_\gamma \hookrightarrow G$ has a stable conjugate $a_{\gamma_0'} : G_\gamma \cong G'_{\gamma_0'} \hookrightarrow G'$, the assertion follows from Lemma 2.4.7.

Therefore equality (2.7.1) follows from the following generalization of Claim 2.6.7.

Claim 2.7.3. *For each* $\delta' \sim_{(\gamma_0,\overline{\xi})} \delta$, *we have* $I_{t,\delta'} = \left\langle \frac{\gamma_0,\gamma_0';\overline{\xi}}{a_0,a_0';\kappa} \right\rangle_{\mathcal{E}} I_{t,\delta}$.

Proof. The proof is very similar to that of Claim 2.6.7. By Theorem 2.2.3 for the inner twisting $G_\delta \to G'_{\delta'}$ (use Lemma 2.5.4), we have

$$F_{b_{t,\delta'},\kappa,\theta}(u', \overline{\xi}_{\delta_0}) = \left\langle \frac{u, u'; \overline{\xi}_{\delta_0}}{b_{t,\delta}, b_{t,\delta'}; \kappa} \right\rangle_{\mathcal{E}_t} F_{b_{t,\delta},\kappa,\theta}(u, \overline{\xi}_{\delta_0}).$$

Hence the quotient $I_{t,\delta'}/I_{t,\delta}$ equals

$$\left\langle \frac{u, u'; \overline{\xi}_{\delta_0}}{b_{t,\delta}, b_{t,\delta'}; \kappa} \right\rangle_{\mathcal{E}_t} \langle \operatorname{inv}(a'_0, b_{t,\delta'}), \kappa \rangle \langle \operatorname{inv}(a_0, b_{t,\delta}), \kappa \rangle^{-1} \langle \operatorname{inv}(\gamma'_0, \gamma'), \overline{\xi} \rangle^{-1} \langle \operatorname{inv}(\gamma_0, \gamma), \overline{\xi} \rangle.$$

Thus we have to check that $\left\langle \frac{u,u';\overline{\xi}_{\delta_0}}{b_{t,\delta},b_{t,\delta'};\kappa} \right\rangle_{\mathcal{E}_t} = \left\langle \frac{a_\gamma, a_{\gamma'}; [b']}{b_{t,\delta}, b_{t,\delta'}; [c']} \right\rangle_{\mathcal{E}_t}$ equals

$$\left\langle \frac{\gamma_0, \gamma'_0; \overline{\xi}}{a_0, a'_0; \kappa} \right\rangle_{\mathcal{E}} \langle \operatorname{inv}(a'_0, b_{t,\delta'}), \kappa \rangle^{-1} \langle \operatorname{inv}(a_0, b_{t,\delta}), \kappa \rangle \langle \operatorname{inv}(\gamma'_0, \gamma'), \overline{\xi} \rangle \langle \operatorname{inv}(\gamma_0, \gamma), \overline{\xi} \rangle^{-1}.$$

Since the latter expression equals $\left\langle \frac{a_\gamma, a_{\gamma'}; [b]}{b_{t,\delta}, b_{t,\delta'}; [c]} \right\rangle_{\mathcal{E}}$ (use Lemma 1.5.7 (a), (c)), the assertion follows from Lemma 2.4.5 (b). □

This completes the proof of Theorems 2.1.10 and 2.1.6.

Appendix A. Springer Hypothesis

The goal of this appendix is to prove the following result, conjectured by Springer and playing a crucial role in Subsection 2.2. In the case of large characteristic this result was first proved in [Ka1].

Theorem A.1. *Let L be a reductive group over a finite field $\mathbb{F}_q$, $\Phi : L \to \mathcal{L}$ a quasi-logarithm (see Definition 1.8.1), $\langle \cdot, \cdot \rangle$ a non-degenerate invariant pairing on $\mathcal{L}$, $T \subset L$ a maximal torus, θ a character of $T(\mathbb{F}_q)$, ψ a character of $\mathbb{F}_q$, and t an element of $\mathcal{T}(\mathbb{F}_q) \cap \mathcal{L}^{\mathrm{sr}}(\mathbb{F}_q)$.*

Denote by δ_t the characteristic function of the $\operatorname{Ad} L(\mathbb{F}_q)$*-orbit of t, by $\mathcal{F}(\delta_t)$ its Fourier transform, and by $(-1)^{\mathrm{rk}_{\mathbb{F}_q}(L)-\mathrm{rk}_{\mathbb{F}_q}(T)}\rho_{T,\theta}$ the Deligne–Lusztig [DL] virtual representation of $L(\mathbb{F}_q)$ corresponding to T and θ.*

For every unipotent $u \in L(\mathbb{F}_q)$, we have

$$\operatorname{Tr} \rho_{T,\theta}(u) = q^{-\frac{1}{2}\dim(L/T)} \mathcal{F}(\delta_t)(\Phi(u)).$$

Notation A.2. For each Weil sheaf $\mathcal{A}$ over a variety X over a finite field $\mathbb{F}_q$, we denote by $\mathrm{Func}(\mathcal{A})$ the corresponding function on $X(\mathbb{F}_q)$.

Proof. The theorem is a consequence of the results of Lusztig ([Lu]) and Springer ([Sp]), and seems to be well known to experts. Set $\mathcal{U} := \mathcal{U}(L) \subset L$ and $\mathcal{N} := \mathcal{N}(\mathcal{L}) \subset \mathcal{L}$. To carry out the proof, we will construct a Weil sheaf $\mathcal{A}$ on $\mathcal{L} \times \mathcal{T}$ such that the restrictions $\mathcal{A}_t$ to $\mathcal{L} \times \{t\} \cong \mathcal{L}$ are perverse for all $t \in \mathcal{T}(\mathbb{F}_q)$ and satisfy the following properties:

(i) The restriction $\mathcal{A}_t|_{\mathcal{N}}$ is independent of t.

(ii) If $t \in \mathcal{L}^{\mathrm{sr}}(\mathbb{F}_q)$, then $\mathrm{Func}(\mathcal{A}_t) = q^{-\frac{1}{2}\dim(L/T)}\mathcal{F}(\delta_t)$.

(iii) $\mathrm{Func}(\Phi^*(\mathcal{A}_0))|_{\mathcal{U}} = \operatorname{Tr} \rho_{T,\theta}|_{\mathcal{U}}$.

The existence of such an $\mathcal{A}$ implies the Theorem. Indeed, fix $u \in \mathcal{U}(\mathbb{F}_q)$. By (iii), $\operatorname{Tr} \rho_{T,\theta}(u)$ equals $\mathrm{Func}(\mathcal{A}_0)(\Phi(u))$. Since $\Phi(\mathcal{U}) \subset \mathcal{N}$ (see [BR, 9.1, 9.2]), we have $\Phi(u) \in \mathcal{N}$. Therefore the assertion follows from (i) and (ii).

A. 3. Construction of $\mathcal{A}$

Let T' be the abstract Cartan subgroup of L, and $W \subset \mathrm{Aut}(T')$ the Weyl group of L (see [DL, 1.1]). Denote by $\widetilde{\mathcal{L}}$ the Springer resolution of $\mathcal{L}$ classifying pairs $(\mathcal{B}, x)$, where $\mathcal{B} \subset \mathcal{L}$ is a Borel subalgebra and x is an element of $\mathcal{B}$. Consider the diagram

$$\mathcal{L} \times \mathcal{T}' \overset{\pi\times\mathrm{Id}}{\longleftarrow} \widetilde{\mathcal{L}} \times \mathcal{T}' \overset{\alpha\times\mathrm{Id}}{\longrightarrow} \mathcal{T}' \times \mathcal{T}' \overset{\langle\cdot,\cdot\rangle'}{\longrightarrow} \mathbb{A}^1,$$

where π and α send $(\mathcal{B}, x)$ to x and $\overline{x} \in \mathcal{B}/[\mathcal{B}, \mathcal{B}] = \mathcal{T}'$, respectively, and $\langle\cdot,\cdot\rangle'$ is the form on $\mathcal{T}'$ induced by $\langle\cdot,\cdot\rangle$. Put

$$\mathcal{A}' := (\pi \times \mathrm{Id})_!(\alpha \times \mathrm{Id})^*\langle\cdot,\cdot\rangle'^*(\mathcal{L}_\psi)[\dim L],$$

where $\mathcal{L}_\psi$ is the Artin–Schreier local system on $\mathbb{A}^1$ corresponding to ψ.

To construct $\mathcal{A}$, we will show first that for every $w \in W$, there exists a canonical isomorphism $(\mathrm{Id} \times w)^*\mathcal{A}' \overset{\sim}{\to} \mathcal{A}'$. Denote by upper index $(\cdot)^{\mathrm{sr}}$ the restriction to (the preimage of) $\mathcal{L}^{\mathrm{sr}}$. First we will show that $(\mathrm{Id} \times w)^*\mathcal{A}^{\mathrm{sr}}$ is canonically isomorphic to $\mathcal{A}^{\mathrm{sr}}$. As $\pi^{\mathrm{sr}} : \widetilde{\mathcal{L}}^{\mathrm{sr}} \to \mathcal{L}^{\mathrm{sr}}$ is an unramified Galois covering with the Galois group W, the functor $\pi^{\mathrm{sr}}_!$ is isomorphic to $\pi^{\mathrm{sr}}_! \circ w^*$. Since α^{sr} is W-equivariant, and $\langle\cdot,\cdot\rangle' = \langle\cdot,\cdot\rangle' \circ (w \times w)$, we have a canonical isomorphism of functors

$$(\mathrm{Id} \times w)^*(\pi^{\mathrm{sr}} \times \mathrm{Id})_!(\alpha^{\mathrm{sr}} \times \mathrm{Id})^*\langle\cdot,\cdot\rangle'^* \cong (\pi^{\mathrm{sr}} \times \mathrm{Id})_!(\mathrm{Id} \times w)^*(\alpha^{\mathrm{sr}} \times \mathrm{Id})^*\langle\cdot,\cdot\rangle'^* \cong$$

$$(\pi^{\mathrm{sr}} \times \mathrm{Id})_!(w \times w)^*(\alpha^{\mathrm{sr}} \times \mathrm{Id})^*\langle\cdot,\cdot\rangle'^* \cong (\pi^{\mathrm{sr}} \times \mathrm{Id})_!(\alpha^{\mathrm{sr}} \times \mathrm{Id})^*\langle\cdot,\cdot\rangle'^*,$$

implying the isomorphism $(\mathrm{Id} \times w)^*\mathcal{A}'^{\mathrm{sr}} \overset{\sim}{\to} \mathcal{A}'^{\mathrm{sr}}$. Since α and $\langle\cdot,\cdot\rangle'$ are smooth morphisms, while π is small, we see that $\mathcal{A}'[\dim T]$ is a semisimple perverse sheaf, which is the intermediate extension of $\mathcal{A}'^{\mathrm{sr}}[\dim T]$. Thus the constructed above isomorphism $(\mathrm{Id} \times w)^*\mathcal{A}'^{\mathrm{sr}} \overset{\sim}{\to} \mathcal{A}'^{\mathrm{sr}}$ uniquely extends to an isomorphism $(\mathrm{Id} \times w)^*\mathcal{A}' \overset{\sim}{\to} \mathcal{A}'$.

Denote by $\mathrm{Fr} : T' \to T'$ the geometric Frobenius morphism corresponding to the $\mathbb{F}_q$-structure of T', and choose an isomorphism between T and T' over $\overline{\mathbb{F}_q}$. Then there exists $w \in W$ such that the $\mathbb{F}_q$-structure of T corresponds to the morphism $\mathrm{Fr}_w := w \circ \mathrm{Fr} : T' \to T'$. Denote by $\mathcal{A}$ the Weil sheaf $\mathcal{A}$ on $\mathcal{L} \times \mathcal{T}$, which is isomorphic to $\mathcal{A}'$ over $\overline{\mathbb{F}_q}$, and the Weil structure corresponds to the composition $\mathrm{Fr}^*(\mathrm{Id} \times w)^*\mathcal{A}' \overset{\sim}{\to} \mathrm{Fr}^*\mathcal{A}' \overset{\sim}{\to} \mathcal{A}'$, where the first isomorphism was constructed above, and the second one comes from the Weil structure of $\mathcal{A}'$.

It remains to show that $\mathcal{A}$ satisfies properties (i)–(iii).

A. 4. Proof of properties (i)–(iii)

(i) Put $\widetilde{\mathcal{N}} := \pi^{-1}(\mathcal{N}) \subset \widetilde{\mathcal{L}}$. Then $\alpha(\widetilde{\mathcal{N}}) = 0$, hence $(\alpha \times \mathrm{Id})^*\langle\cdot,\cdot\rangle'^*(\mathcal{L}_\psi)|_{\widetilde{\mathcal{N}}\times\mathcal{T}'} \cong \overline{\mathbb{Q}_l}$. This implies the assertion for $\mathcal{A}'$. To show the assertion for $\mathcal{A}$, notice that $\widetilde{\mathcal{N}}$ is smooth of dimension $\dim(L/T)$, and the projection $\widetilde{\mathcal{N}} \to \mathcal{N}$ is semi-small. It follows that $\mathcal{A}'|_{\mathcal{N}\times\mathcal{T}}$ is a semisimple perverse sheaf. Therefore for each $t \in \mathcal{T}(\mathbb{F}_q)$ the restriction map

$$\mathrm{Hom}(\mathrm{Fr}_w^*\,\mathcal{A}'|_{\mathcal{N}\times\mathcal{T}}, \mathcal{A}'|_{\mathcal{N}\times\mathcal{T}}) \to \mathrm{Hom}(\mathrm{Fr}_w^*\,\mathcal{A}'|_{\mathcal{N}\times\{t\}}, \mathcal{A}'|_{\mathcal{N}\times\{t\}})$$

is an isomorphism. Thus the Weil structure of $\mathcal{A}_t|_{\mathcal{N}}$ is independent of t as well.

(ii) Denote by IC_t the constant perverse sheaf $\overline{\mathbb{Q}_l}(\frac{\dim(L/T)}{2})[\dim(L/T)]$ on the orbit $\mathrm{Ad}\, L(t) \subset \mathcal{L}$, and let $\mathcal{F}(\mathrm{IC}_t)$ be the Fourier–Deligne transform of IC_t. As $\dim(L/T)$ is even, we get that $\mathrm{Func}(\mathcal{F}(\mathrm{IC}_t)) = q^{-\frac{\dim(L/T)}{2}} \mathcal{F}(\delta_t)$. Therefore the assertion follows from well-known equality $\mathcal{F}(\mathrm{IC}_t) = \mathcal{A}_t$ (see for example [Sp]).

(iii) By [DL, Thm. 4.2], $\mathrm{Tr}\,\rho_{T,\theta}|_{\mathcal{U}(\mathbb{F}_q)}$ does not depend on θ, hence we can assume that $\theta = 1$. It was proved by Lusztig (see [Lu, Theorem. 1.14 (a) and Proposition. 8.15] and compare [BP]) that there exists a perverse sheaf $\mathcal{K}_T$ on L such that $\mathrm{Tr}\,\rho_{T,1} = \mathrm{Func}(\mathcal{K}_T)$. More precisely, in [Lu, Theorem. 1.14 (a)] Lusztig showed the corresponding result for general character sheaves if q is sufficiently large, while by [Lu, Prop. 8.15] the restriction on q is unnecessary in our situation.

Thus it will suffice to check that $\mathcal{K}_T|_{\mathcal{U}} = \Phi^*(\mathcal{A}_0)|_{\mathcal{U}}$. The description of $\mathcal{K}_T$ is very similar to that of $\mathcal{A}$. Let $\pi_L : \widetilde{L} \to L$ be the Springer resolution, put $\widetilde{\mathcal{U}} := \pi_L^{-1}(\mathcal{U})$, and we denote by $\pi_L^{\mathrm{sr}} : \widetilde{L}^{\mathrm{sr}} \to L^{\mathrm{sr}}$ the restriction of π to the preimage of L^{sr}. Then the semisimple perverse sheaf $\mathcal{K}_T$ is equal to $(\pi_L)_!(\overline{\mathbb{Q}_l})[\dim L]$ over $\overline{\mathbb{F}_q}$, while the Weil structure of $\mathcal{K}_T$ is induced (as in A.3) by isomorphism of functors $\mathrm{Fr}_w^*(\pi_L^{\mathrm{sr}})_! \xrightarrow{\sim} (\pi_L^{\mathrm{sr}})_!$.

By [BR, Theorem. 6.2 and 9.1], there exists a L^{ad}-invariant open affine neighborhood $V \supset \mathcal{U}$ in L such that $\Phi|_V : V \to \mathcal{L}$ is étale. We claim that $\mathcal{K}_T|_V$ is isomorphic to $\Phi^*(\mathcal{A}_0)|_V = (\Phi|_V)^*(\mathcal{A}_0)$. As both $\mathcal{K}_T|_V$ and $(\Phi|_V)^*(\mathcal{A}_0)$ are semisimple perverse sheaves, which are immediate extensions of their restrictions to $V \cap \Phi^{-1}(\mathcal{L}^{\mathrm{sr}})$, it will suffice to show that $\Phi^*(\mathcal{A}_0)|_{\Phi^{-1}(\mathcal{L}^{\mathrm{sr}})} \cong \mathcal{K}_T|_{\Phi^{-1}(\mathcal{L}^{\mathrm{sr}})}$.

Note that $\Phi^{-1}(\mathcal{L}^{\mathrm{sr}}) \subset L^{\mathrm{sr}}$ and that Φ gives rise to the commutative diagram

$$\begin{array}{ccc} \widetilde{L} & \xrightarrow{\widetilde{\Phi}} & \widetilde{\mathcal{L}} \\ {\scriptstyle \pi_L}\downarrow & & \downarrow{\scriptstyle \pi} \\ L & \xrightarrow{\Phi} & \mathcal{L} \end{array}$$

(use Lemma 1.8.3 (a)), whose the restriction to $\mathcal{L}^{\mathrm{sr}}$ is Cartesian and W-equivariant. Therefore the required isomorphism $\Phi^*(\mathcal{A}_0)|_{\Phi^{-1}(\mathcal{L}^{\mathrm{sr}})} \xrightarrow{\sim} \mathcal{K}_T|_{\Phi^{-1}(\mathcal{L}^{\mathrm{sr}})}$ follows from the proper base change theorem.

Appendix B. $(a, a'; [b])$-equivalence and Fourier transform

B.1. Formulation of the result

B.1.1. Let G be a reductive group over a local field E of characteristic zero, $\mathcal{E} = (H, [\eta], \overline{s})$ an endoscopic triple for G, and $\varphi : G \to G'$ an inner twisting. Fix a triple $(a, a'; [b])$, where $a : T \hookrightarrow G$ and $a' : T \hookrightarrow G'$ are stably conjugate embeddings of maximal tori, and $[b]$ is a stable conjugacy class of embeddings of maximal tori $T \hookrightarrow H$, compatible with a and a'.

Fix a non-trivial character $\psi : E \to \mathbb{C}^\times$, a non-degenerate G-invariant pairing $\langle \cdot, \cdot \rangle$ on $\mathcal{G}$, and a non-zero translation invariant top degree differential form ω_G on

G. Denote by $\langle \cdot, \cdot \rangle'$ the G'-invariant pairing on $\mathcal{G}'$, induced $\langle \cdot, \cdot \rangle$, and let $dx = |\omega_{\mathcal{G}}|$ and $dx' = |\omega_{\mathcal{G}'}|$ be the invariant measures on $\mathcal{G}(E)$ and $\mathcal{G}'(E)$ induced by ω_G. These data define Fourier transforms $F \mapsto \mathcal{F}(F)$ on $\mathcal{G}(E)$ and $F' \mapsto \mathcal{F}(F')$ on $\mathcal{G}'(E)$ (see 2.2.10).

The following result generalizes both the theorem of Waldspurger [Wa2] (who treated the case $\phi' = 0$) and of Kazhdan–Polishchuk [KP, Thm. 2.7.1] (where the stable case is treated).

Theorem B.1.2. *Generalized functions $F \in \mathcal{D}(\mathcal{G}(E))$ and $F' \in \mathcal{D}(\mathcal{G}'(E))$ are $(a, a'; [b])$-equivalent if and only if $e'(G)\mathcal{F}(F)$ and $e'(G')\mathcal{F}(F')$ are $(a, a'; [b]))$-equivalent.*

Remark B.1.3. When this work was already written, we learned that our Theorem B.1.2 seems to follow from the recent work of Chaudouard [Ch].

By duality, Theorem B.1.2 follows from the following result.

Theorem B.1.4. *Measures $\phi \in \mathcal{S}(\mathcal{G}(E))$ and $\phi' \in \mathcal{S}(\mathcal{G}'(E))$ are $(a, a'; [b])$-indistinguishable if and only if $e'(G)\mathcal{F}(\phi)$ and $e'(G')\mathcal{F}(\phi')$ are $(a, a'; [b])$-indistinguishable.*

For the proof, we will combine arguments from [Wa1, Wa2] with those from [KP].

Lemma B.1.5. *(a) The validity of Theorem B.1.4 is independent of the choice of ω_G, ψ and $\langle ., . \rangle$.*

(b) It will suffice to show Theorem B.1.4 under the assumption that $G = G^{\mathrm{sc}}$.

(c) It will suffice to show Theorem B.1.4 under the assumption that $\mathcal{E}$ is elliptic. (Note that this is the only case used in this paper).

Proof. The proof follows by essentially the same arguments as [Wa2, II].

(a) Another choice of the data results in replacing the Fourier transform $\mathcal{F}$ by $B^* \circ \mathcal{F}$ for a certain linear automorphism B of $\mathcal{G}$ commuting with $\operatorname{Ad} G$. Thus the assertion follows from Lemma 1.6.15.

(b) Let $\widetilde{\mathcal{E}}$ be the endoscopic triple for G^{sc} induced by $\mathcal{E}$, and let $(\widetilde{a}, \widetilde{a}'; [\widetilde{b}])$ be the lift of $(a, a'; [b])$ (see Lemma 1.3.10). Fix a pair of $(a, a'; [b])$-indistinguishable measures $\phi \in \mathcal{S}(\mathcal{G}(E))$ and $\phi' \in \mathcal{S}(\mathcal{G}'(E))$.

Denote by $\mathcal{Z}$ the Lie algebra of $Z(G) = Z(G')$. Then $\mathcal{G} = \mathcal{G}^{\mathrm{sc}} \oplus \mathcal{Z}$ and $\mathcal{G}' = \mathcal{G}'^{\mathrm{sc}} \oplus \mathcal{Z}$, hence there exist measures $h_i \in \mathcal{S}(\mathcal{Z}(E))$, $f_i \in \mathcal{S}(\mathcal{G}^{\mathrm{sc}}(E))$ and $f_i' \in \mathcal{S}(\mathcal{G}'^{\mathrm{sc}}(E))$ such that the h_i's are linearly independent, $\phi = \sum_i f_i \times h_i$ and $\phi' = \sum_i f_i' \times h_i$.

For each $x \in (\mathcal{G}^{\mathrm{sc}})^{\mathrm{sr}}(E)$, $z \in \mathcal{Z}(E)$ and $\overline{\kappa} \in \pi_0(\widehat{G_{x+z}}^{\Gamma}/Z(\widehat{G}^{\Gamma}))$, we have $O^{\overline{\kappa}}_{x+z}(\phi) = \sum_i O^{\overline{\overline{\kappa}}}_x(f_i) O_z(h_i)$, where $\overline{\overline{\kappa}} \in \pi_0(\widehat{G^{\mathrm{sc}}_x}^{\Gamma})$ is the image of $\overline{\kappa}$, and similarly for ϕ'. Since the h_i's are linearly independent, it follows from Lemmas 1.6.11 and 1.3.10 that f_i and f_i' are $(\widetilde{a}, \widetilde{a}'; [\widetilde{b}])$-indistinguishable for each i (compare Corollary 1.6.13 and its proof).

Let $\langle ., . \rangle$ be a direct sum of pairings on $\mathcal{G}^{\mathrm{sc}}$ and $\mathcal{Z}$. Then $\mathcal{F}(\phi) = \sum_i \mathcal{F}(f_i) \times \mathcal{F}(h_i)$ and $\mathcal{F}(\phi') = \sum_i \mathcal{F}(f_i') \times \mathcal{F}(h_i)$. By our assumptions, $e'(G)\mathcal{F}(f_i)$ and $e'(G')\mathcal{F}(f_i')$

are $(\widetilde{a}, \widetilde{a}'; [\widetilde{b}])$-indistinguishable for each i, therefore $e'(G)\mathcal{F}(\phi)$ and $e'(G')\mathcal{F}(\phi')$ are $(a, a'; [b])$-indistinguishable, as claimed.

(c) The assertion follows from the arguments of [Wa2, II. 3]. Since we do not use this result in the main body of the paper, we omit the details. □

From now on, we assume that $\mathcal{E}$ is elliptic, G is semisimple and simply connected, and $\langle ., . \rangle$ is the Killing form.

Notation B.1.6. (a) Consider the natural map $[y] \mapsto [y]_G$ from the set of stable conjugacy classes of elements of $\mathcal{H}^{\mathrm{sr}}(E)$ to the set of E-rational conjugacy classes in $\mathcal{G}^{\mathrm{sr}}(\overline{E})$, defined as follows. For each $y \in \mathcal{H}^{\mathrm{sr}}(E)$, denote by b_y the inclusion $H_y \hookrightarrow H$. Each embedding $a : (H_y)_{\overline{E}} \hookrightarrow G_{\overline{E}}$ from $[b_y]_G$ defines an embedding $da : (\mathcal{H}_y)_{\overline{E}} \hookrightarrow \mathcal{G}_{\overline{E}}$, and we denote by $[y]_G$ be the conjugacy class of $da(y) \in \mathcal{G}(\overline{E})$.

(b) We say that $y \in \mathcal{H}^{\mathrm{sr}}(E)$ and $x \in \mathcal{G}^{\mathrm{sr}}(E)$ are *compatible* if $x \in [y]_G$.

(c) For each $x \in \mathcal{G}^{\mathrm{sr}}(E)$, the map $b \mapsto y := db(x)$ defines a bijection between embeddings of maximal tori $b : G_x \hookrightarrow H$ compatible with $a_x : G_x \hookrightarrow G$ and elements $y \in \mathcal{H}^{\mathrm{sr}}(E)$ compatible with x. Let $y \mapsto b_y$ be the inverse map.

We will write $\overline{\kappa}_{[y]} \in \pi_0(\widehat{G_x}^{\Gamma})$ instead of $\overline{\kappa}_{[b_y]}$, $O_x^{[y]}$ instead of $O_x^{\overline{\kappa}_{[y]}}$, and $\left\langle \frac{x,x';[y]}{a,a';[b]} \right\rangle$, where $x' \in \mathcal{G}'(E)$ is a stable conjugate of x, instead of $\left\langle \frac{a_x,a_{x'};[b_y]}{a,a';[b]} \right\rangle$.

B.2. Local calculations

The primary goal of this subsection is to construct $(a, a'; [b])$-indistinguishable measures whose Fourier transforms are $(a, a'; [b])$-indistinguishable in some region. We mostly follow [KP].

B.2.1. (a) For every $t \in \mathcal{G}^{\mathrm{sr}}(E)$, fix a top degree form $\omega_{\mathcal{G}_t} \neq 0$ on the vector space $\mathcal{G}_t$ and identify it with the corresponding top degree translation invariant differential form. Then $\omega_{\mathcal{G}_t}$ defines a G_t-invariant top degree form $\nu = \nu_t := \omega_{\mathcal{G}} \otimes (\omega_{\mathcal{G}_t})^{-1}$ on $\mathcal{G}/\mathcal{G}_t$, which uniquely extends to a non-zero top degree G-invariant form on G/G_t, which we will also denote by ν.

(b) We denote by $d\overline{g}$ and du the measures $|\nu|$ on $(G/G_t)(E)$ and $|\omega_{\mathcal{G}_t}|$ on $\mathcal{G}_t(E)$, respectively.

(c) Consider the map $\Pi : (G/G_t) \times \mathcal{G}_t^{\mathrm{sr}} \to \mathcal{G}^{\mathrm{sr}}$ given by the rule $\Pi(\overline{g}, x) = \mathrm{Ad}\,\overline{g}(x)$. Then Π is étale, and we have an equality $\Pi^*(\omega_{\mathcal{G}}) = \nu \wedge \omega_{\mathcal{G}_t}$.

B.2.2. Preliminaries on quadratic forms over local fields (a) To every non-degenerate quadratic form q on an E-vector space V one associates a rank $\mathrm{rk}\, q = \dim V$, a determinant $\det q \in (\det V)^{\otimes(-2)}$ and a Hasse–Witt invariant $e(q) \in \{-1, 1\}$. Any trivialization $\det V \xrightarrow{\sim} E$ associates to $\det q$ an element of $E^{\times}$. Moreover, the class of $\det q$ in $E^{\times}/(E^{\times})^2$ is independent on the trivialization.

To each isomorphism class (q, ψ), where ψ is a non-trivial additive character of E, Weil [We] associated an 8th root of unity $\gamma(q, \psi)$. For each non-zero top degree form ν on V, we set $c(q, \nu, \psi) := \gamma(q, \psi)|\det(q)/\nu^2|^{-1/2}$.

(b) Weil proved that for every non-degenerate quadratic forms q and q' satisfying $\operatorname{rk} q = \operatorname{rk} q'$ and $\det q \equiv \det q' \mod (E^\times)^2$, we have $\gamma(q, \psi)/\gamma(q, \psi) = e(q)e(q')$.

(c) To each $t \in \mathcal{G}^{\mathrm{sr}}(E)$ and $y, z \in \mathcal{G}_t^{\mathrm{sr}}(E)$ we associate a non-degenerate quadratic form $q = q_{y,z} : \overline{x} \mapsto \langle y, (\operatorname{ad}\overline{x})^2(z)\rangle$ on $V := \mathcal{G}/\mathcal{G}_t$. Then the form ν on V, chosen in B.2.1 (a) gives rise to an invariant $c(q_{y,z}, \nu, \psi)$.

(d) Let $t' \in \mathcal{G}'^{\mathrm{sr}}(E)$ be a stable conjugate of t, ν' the form on $G'/G'_{t'}$ induced by ν, and $\varphi_{t,t'}$ the canonical isomorphism $\mathcal{G}_t \xrightarrow{\sim} \mathcal{G}'_{t'}$ from Construction 1.7.7 (b). For every $y, z \in \mathcal{G}_t^{\mathrm{sr}}$, we set $y' = \varphi_{t,t'}(y)$ and $z' = \varphi_{t,t'}(z)$.

It follows from results of [KP] that $c(q_{y,z}, \nu, \psi)/c(q_{y',z'}, \nu', \psi) = e'(G)e'(G')$. In particular, $c(q_{y,z}, \nu, \psi) = c(q_{y',z'}, \nu', \psi)$ if $G' = G$. Indeed, since $\det q_{y,z}/\nu^2 = \det q_{y',z'}/\nu'^2$ and $\operatorname{rk} q = \operatorname{rk} q'$, the assertion is a combination of the result of Weil (see (b)) and [KP, Lemma. 2.7.5 and the remark following it].

Notation B.2.3. Fix $t \in \mathcal{G}^{\mathrm{sr}}(E)$ and sufficiently small open compact subgroups $K \subset G(E)$ and $U \subset \mathcal{G}_t(E)$ such that $t+U \subset \mathcal{G}_t^{\mathrm{sr}}(E)$ and $K \cap \mathrm{Norm}_{G(E)}(\mathcal{G}_t) = K \cap G_t(E)$.

For each $a \in E^\times$, put $S_a := \operatorname{Ad} K(at + U) \subset \mathcal{G}(E)$, and we denote by $\chi_a = \chi_{a,t,K,U}$ be the characteristic function of S_a.

B.2.4. Stationary phase principle For each $u \in \mathcal{G}^{\mathrm{sr}}(E)$ and $x \in \mathcal{G}(E)$, we define function $f_{x,u} : (G/G_u)(E) \to E$ by the rule $f_{x,u}(\overline{g}) := \langle x, \operatorname{Ad}\overline{g}(u)\rangle$. Then $\overline{g} \in (G/G_u)(E)$ is a critical point for $f_{x,u}$ if and only if $x \in \operatorname{Ad}\overline{g}(\mathcal{G}_u) = \mathcal{G}_{\operatorname{Ad}\overline{g}(u)}$. In this case, the corresponding quadratic form on $T_{\overline{g}}(G/G_u) = \mathcal{G}/\mathcal{G}_u$ is $q_{\operatorname{Ad}\overline{g}^{-1}(x),u}$.

By the stationary phase principle (see [KP, Lem 2.5.1]), for each compact subset $C \subset G^{\mathrm{sr}}(E)$ there exists $N_0 = N_0(t, K, U, c) \in \mathbb{N}$ such that for each $x \in C$, $u \in t+U$ and $a \in E^\times$ with $|a| > N_0$, the integral $\int_{K/K\cap G_t(E)} \psi(a\langle x, \operatorname{Ad}\overline{g}(u)\rangle)d\overline{g}$ equals

$$c(q_{\operatorname{Ad}\overline{k}(x),u}, \nu, \psi)\psi(a\langle x, u\rangle)|a|^{-\frac{1}{2}\dim \mathcal{G}/\mathcal{G}_t}, \tag{B.2.1}$$

if there exists (a unique) element $\overline{k} \in K/K \cap G_t(E)$ such that $\operatorname{Ad}\overline{k}(x) \in \mathcal{G}_t^{\mathrm{sr}}(E)$, and vanishes otherwise.

Indeed, the map $f_{x,u}|_{K/K\cap G_t(E)}$ has a unique non-degenerate critical point $\overline{g} = \overline{k}^{-1}$ in the former case and has no critical points in the latter one. Thus the assertion for $a \in (E^\times)^2$ follows from [KP, Lem 2.5.1]. Since the quotient $E^\times/(E^\times)^2$ is finite, the general case now follows from the previous one applied to the compact set $\sqcup_b bC$, where $b \in E^\times$ runs over a set of representatives of $E^\times/(E^\times)^2$.

Lemma B.2.5. *Let t, K and U be as in Notation B.2.3, and let $[x] \subset \mathcal{G}^{\mathrm{sr}}(E)$ be a stable conjugacy class. Then there exists $N = N(t, K, U, [x]) \in \mathbb{N}$ such that for each $x \in [x]$ and $a \in E^\times$ with $|a| > N$, the Fourier transform $\mathcal{F}(\chi_a)(x)$ equals*

$$c(q_{\operatorname{Ad}\overline{k}(x),t}, \nu, \psi)\psi(a\langle x, t\rangle)|a|^{\frac{1}{2}\dim \mathcal{G}/\mathcal{G}_t}\int_U \psi(\langle x, u\rangle)du, \tag{B.2.2}$$

if there exists (a unique) element $\overline{k} \in K/K \cap G_t(E)$ such that $\operatorname{Ad}\overline{k}(x) \in \mathcal{G}_t^{\mathrm{sr}}(E)$, and vanishes otherwise.

Proof. First we claim that there exists a compact set $C_0 \subset \mathcal{G}(E)$ containing the support of $\mathcal{F}(\chi_a)$ for all $a \in E \smallsetminus \mathcal{O}$. To show this we will find an $\mathcal{O}$-lattice $L \subset \mathcal{G}$ such that $S_a + L = S_a$ for all $a \in E \smallsetminus \mathcal{O}$. Then the dual lattice $C_0 := \{x \in \mathcal{G} \mid \psi(\langle x, L\rangle) = 1\}$ satisfies the required property.

Recall that the map $(\overline{k}, u) \mapsto \operatorname{Ad} k(t + u)$ gives an analytic isomorphism F : $(K/K \cap G_t(E)) \times U \overset{\sim}{\to} S_1$, and we denote by $\pi : S_1 \to U$ the composition $\mathrm{pr}_2 \circ F^{-1}$. Since S_1 is compact, π has a bounded derivative. Therefore there exists a lattice $L \subset \mathcal{G}(E)$ such that for each $b \in E$ and $x \in S_1$, we have $\pi(x + bL) \subset \pi(x) + bU$. Shrinking L if necessary, we can moreover assume that $S_1 + L = S_1$.

Fix $a \in E \smallsetminus \mathcal{O}$. Then $a^{-1}S_a = \operatorname{Ad} K(t + a^{-1}U) \subset S_1$ and, moreover, $a^{-1}S_a$ is the preimage of $a^{-1}U \subset U$ under π. Therefore $a^{-1}S_a + a^{-1}L \subset S_1 + L = S_1$ and $\pi(a^{-1}S_a + a^{-1}L) \subset \pi(a^{-1}S_a) + a^{-1}U = a^{-1}U$. Hence $a^{-1}S_a + a^{-1}L \subset \pi^{-1}(a^{-1}U) = a^{-1}S_a$, thus $S_a + L = S_a$, as claimed.

As the intersection $\mathcal{G}_t^{\mathrm{sr}} \cap [x]$ is finite, the set $\operatorname{Ad} K(\mathcal{G}_t^{\mathrm{sr}} \cap [x])$ and therefore also $C := C_0 \cup \operatorname{Ad} K(\mathcal{G}_t^{\mathrm{sr}} \cap [x])$ is compact. Since $[x] \subset \mathcal{G}(E)$ is closed, the intersection $C \cap [x]$ is compact as well. Take any $N \geq N_0(t, K, U, C \cap [x])$ (see B.2.4) such that quadratic forms $q_{x,u}$ and $q_{x,t}$ are isomorphic for each element x of the finite set $\mathcal{G}_t(E) \cap [x]$ and each $u \in t + a^{-1}U$ with $|a| > N$. We claim that this N satisfies the required properties.

Indeed, the Fourier transform $\mathcal{F}(\chi_a)(x) = \int_{S_a} \psi(\langle x, y\rangle)dy$ equals

$$|a|^{\dim \mathcal{G}} \int_{a^{-1}S_a} \psi(a\langle x, y\rangle)dy = |a|^{\dim \mathcal{G}} \int_{t+a^{-1}U} du \int_{K/K\cap G_t(E)} \psi(x\langle a, \operatorname{Ad}\overline{g}(u)\rangle)d\overline{g}$$

(by B.2.1 (c) and our assumptions in Notation B.2.3). Therefore the assertion for $x \in [x] \cap C$ follows from B.2.4. Finally, if $x \in [x] \smallsetminus C$, then $x \notin \operatorname{Ad} K(\mathcal{G}_t^{\mathrm{sr}})$ and $\mathcal{F}(\chi_a)(x) = 0$ (by B.4), implying the assertion in the remaining case. □

Corollary B.2.6. *For each triple (y, x, x'), where $x \in \mathcal{G}^{\mathrm{sr}}(E)$ and $x' \in \mathcal{G}'^{\mathrm{sr}}(E)$ are stably conjugate, and $y \in \mathcal{H}^{\mathrm{sr}}(E)$ is compatible with x, there exist measures $\phi \in \mathcal{S}(\mathcal{G}(E))$ and $\phi' \in \mathcal{S}(\mathcal{G}'(E))$ satisfying the following properties:*

(i) ϕ and ϕ' are supported on elements stably conjugate to $\mathcal{G}_x^{\mathrm{sr}}(E)$.

(ii) $O_u^{[y]}(\phi) = \left\langle \frac{x,x';[y]}{a,a';[b]} \right\rangle O_{u'}^{[y]}(\phi')$ for each $u \in \mathcal{G}_x^{\mathrm{sr}}(E)$ and $u' = \phi_{x,x'}(u) \in \mathcal{G}'^{\mathrm{sr}}_{x'}(E)$.

(iii) $O_u^{\overline{\xi}}(\phi) = O_{u'}^{\overline{\xi}}(\phi') = 0$ for each $u \in \mathcal{G}_x^{\mathrm{sr}}(E)$, $u' = \phi_{x,x'}(u)$ and $\overline{\xi} \neq \overline{\kappa}_{[y]}$.

(iv) $O_x^{\overline{\xi}}(\mathcal{F}(\phi)) = O_{x'}^{\overline{\xi}}(\mathcal{F}(\phi')) = 0$ for each $\overline{\xi} \neq \overline{\kappa}_{[y]}$.

(v) $e'(G)O_x^{[y]}(\mathcal{F}(\phi)) = e'(G')\left\langle \frac{x,x';[y]}{a,a';[b]} \right\rangle O_{x'}^{[y]}(\mathcal{F}(\phi')) \neq 0$.

Proof. (Compare Corollary 1.7.8). Let $x_1 = x, x_2, \ldots, x_l$ be all elements of $\mathcal{G}_x(E) \cap [x]$. Pick $t \in \mathcal{G}_x^{\mathrm{sr}}(E)$ such that all $\langle x_j, t\rangle$ are distinct, and put $t' := \phi_{x,x'}(t) \in \mathcal{G}'^{\mathrm{sr}}_{x'}(E)$. Choose sufficiently small subgroups $K \subset G(E)$, $K' \subset G'(E)$ and $U \subset \mathcal{G}_t(E)$ satisfying the assumptions of Notation B.2.3 and such that $\psi(\langle x_j, u\rangle) = 1$ for each $u \in U$ and each $j = 1, \ldots, l$.

Let $t_1 = t, t_2, \dots, t_k \in \mathcal{G}(E)$ and $t'_1 = t', t'_2, \dots, t'_k \in \mathcal{G}'(E)$ be sets of representatives of conjugacy classes stably conjugate to t and t'. For each $i = 1, \dots, k$, put $U_i := \varphi_{t,t_i}(U) \subset \mathcal{G}_{t_i}(E)$ and $U'_i := \varphi_{t,t'_i}(U) \subset \mathcal{G}'_{t'_i}(E)$.

For each $i = 1, \dots, k$, let ν_i and ν'_i be the forms on G/G_{t_i} and $G'/G'_{t'_i}$ respectively, induced by ν. For each $a \in E \smallsetminus \mathcal{O}$, put $\phi_i = \frac{1}{|\nu_i|(K/K\cap G_{t_i}(E))}\chi_{a,t_i,U_i,K}dx$ and $\phi'_i = \frac{1}{|\nu'_i|(K'/K'\cap G'_{t'_i}(E))}\chi_{a,t'_i,U'_i,K'}dx'$. Define ϕ and ϕ' by the formulas $\phi := \sum_i \langle \text{inv}(t, t_i), \overline{\kappa}_{[y]}\rangle^{-1} \phi_i$ and $\phi' := \sum_i \left\langle \frac{a_t, a_{t'_i}; [b_y]}{a, a'; [b]} \right\rangle^{-1} \phi'_i$. Then ϕ and ϕ' clearly satisfy properties (i)–(iii) for each $a \in E \smallsetminus \mathcal{O}$, so it remains to show the existence of a for which properties (iv) and (v) are satisfied.

Let N be the maximum of the $N(t_i, U_i, K, [x])$'s and the $N(t'_i, U'_i, K', [x'])$'s. Then Lemma B.2.5 implies that for each $\overline{\xi} \in \pi_0(\widehat{G_x}^{\Gamma})$ and $a \in E^{\times}$ with $|a| > N$, we have

$$O_x^{\overline{\xi}}(\mathcal{F}(\phi_1)) = |a|^{\frac{1}{2}\dim \mathcal{G}/\mathcal{G}_t} |\omega_{\mathcal{G}_t}|(U) \sum_{j=1}^{l} \langle \text{inv}(x, x_j), \overline{\xi}\rangle c(q_{x_j,t}, \nu, \psi)\psi(a\langle x_j, t\rangle). \tag{B.2.3}$$

Moreover, for each $i = 1, \dots, n$, observation B.2.2 (d) and the equality $\varphi^*_{t,t_i}(\omega_{\mathcal{G}_{t_i}}) = \omega_{\mathcal{G}_t}$ imply that

$$O_x^{\overline{\xi}}(\mathcal{F}(\phi_i)) = \langle \text{inv}(t, t_i), \overline{\xi}\rangle O_x^{\overline{\xi}}(\mathcal{F}(\phi_1)). \tag{B.2.4}$$

Then (B.2.3) and (B.2.4) together with similar formulas for the ϕ'_i's imply the equalities from (iv) and (v) and that $O_x^{[y]}(\mathcal{F}(\phi)) = k O_x^{[y]}(\mathcal{F}(\phi_1))$.

It remains to show the existence of $a \in E^{\times}$ with $|a| > N$ such that $O_x^{[y]}(\mathcal{F}(\phi_1)) \neq 0$. Since all $\langle x_j, t\rangle$ are distinct, the functions $a \mapsto \psi(a\langle x_j, t\rangle)$ are linearly independent. The assertion now follows from (B.2.3). □

Later we will need the following result.

Lemma B.2.7. *Let G be an unramified semisimple simply connected group over E, $K \subset G(E)$ a hyperspecial subgroup, $\mathcal{K} \subset \mathcal{G}(E)$ the corresponding subalgebra, $1_{\mathcal{K}}$ the characteristic function of $\mathcal{K}$, and $\mathcal{E} = (H, [\eta], \overline{s})$ an endoscopic triple for G.*

(a) If H is ramified, then $O_x^{[y]}(1_{\mathcal{K}}dx) = 0$ for all compatible elements $y \in \mathcal{H}^{\text{sr}}(E)$ and $x \in \mathcal{G}^{\text{sr}}(E)$.

(b) If H is unramified, then there exists an open neighborhood of zero $\Omega \subset \mathcal{H}(E)$ such that $O_x^{[y]}(1_{\mathcal{K}}dx) \neq 0$ for all compatible elements $y \in \Omega \cap \mathcal{H}^{\text{sr}}(E)$ and $x \in \mathcal{G}^{\text{sr}}(E)$.

(c) If $x \in \mathcal{K}$ has a regular reduction modulo $\mathfrak{m}$, then every stably conjugate $x' \in \mathcal{K}$ of x is K-conjugate. In particular, $O_x^{\overline{\xi}}(1_{\mathcal{K}}dx) \neq 0$ for each $\overline{\xi} \in \pi_0(\widehat{G_x}^{\Gamma})$.

Proof. Since in the notation of [Wa1, Wa2], $O_x^{[y]}(1_{\mathcal{K}}dx)$ is a non-zero multiple of $J^{G,H}(y, 1_{\mathcal{K}})$, the assertions follow from [Wa1, 7.2 and 7.4] and [Wa2, III, Prop.] (compare [Ko2, Prop. 7.1 and 7.5]). □

B.3. Global results

For the proof of Theorem B.1.4 we will use global methods. In this subsection we will recall necessary notation and results.

B.3.1. (a) Let $\underline{E}$ be a number field, which we will always assume to be totally imaginary, $\underline{\Gamma}$ the absolute Galois group of $\underline{E}$, and $\mathbb{A}$ the ring of adèles of $\underline{E}$. We denote by V, V_∞ and V_f the set of all places, all infinite places and all finite places of $\underline{E}$, respectively. For each $v \in V_f$, we have a natural conjugacy class of embeddings $\underline{\Gamma}_v \hookrightarrow \underline{\Gamma}$. For every object $\underline{S}$ over $\underline{E}$ and each $v \in V$, we will denote by $\underline{S}_v$ the corresponding object over $\underline{E}_v$.

(b) For every reductive group $\underline{G}$ over $\underline{E}$, consider a sequence

$$H^1(\underline{E}, \underline{G}) \to \bigoplus_{v \in V_f} H^1(\underline{E}_v, \underline{G}) \xrightarrow{*} \pi_0(Z(\widehat{\underline{G}})^{\underline{\Gamma}})^D, \tag{B.3.1}$$

where the restriction of $*$ to $H^1(\underline{E}_v, \underline{G})$ is the composition of the isomorphism $\mathcal{D}_G : H^1(\underline{E}_v, \underline{G}) \xrightarrow{\sim} \pi_0(Z(\widehat{\underline{G}})^{\underline{\Gamma}_v})^D$ from 1.5.1 and the projection $\pi_0(Z(\widehat{\underline{G}})^{\underline{\Gamma}_v})^D \to \pi_0(Z(\widehat{\underline{G}})^{\underline{\Gamma}})^D$. Kottwitz proved (see [Ko2, Prop 2.6]) that the sequence (B.3.1) is exact.

(c) For each $c \in H^1(\underline{E}, \underline{G})$, $\kappa \in \pi_0(Z(\widehat{\underline{G}})^{\underline{\Gamma}})$ and $v \in V_f$, we denote by $c_v \in H^1(\underline{E}_v, \underline{G})$ and $\kappa_v \in \pi_0(Z(\widehat{\underline{G}})^{\underline{\Gamma}_v})$ the images of c and κ, respectively. Then Kottwitz's theorem from (b) asserts that $\prod_{v \in V_f} \langle c_v, \kappa_v \rangle = 1$.

Lemma B.3.2. *Let $\underline{G}$ be a reductive group over $\underline{E}$, and $u \in V_f$.*

(a) If u is inert in the splitting field $\underline{E}[\underline{G}^*]$ of the quasi-split inner form $\underline{G}^*$ of $\underline{G}$, then the diagonal map $H^1(\underline{E}, \underline{G}) \to \bigoplus_{v \neq u} H^1(\underline{E}_v, \underline{G})$ is surjective.

Assume in addition that $\underline{G}$ is either semisimple and simply connected or adjoint. Then

(b) The map from (a) is an isomorphism.

(c) Let $\underline{T}$ be a maximal torus of $\underline{G}$, and let $c \in H^1(\underline{E}, \underline{G})$ be such that $c_v \in H^1(\underline{E}_v, \underline{G})$ belongs to $\mathrm{Im}[H^1(\underline{E}_v, \underline{T}) \to H^1(\underline{E}_v, \underline{G})]$ for each $v \neq u$. Then $c \in \mathrm{Im}[H^1(\underline{E}, \underline{T}) \to H^1(\underline{E}, \underline{G})]$ in each of the following cases:

(i) u is inert in the splitting field $\underline{E}[\underline{T}]$ of $\underline{T}$;

(ii) u inert in $\underline{E}[\underline{G}^*]$, and $\underline{T}_u \subset \underline{G}_u$ is elliptic.

Proof. (a), (b) By assumption, we have $Z(\widehat{\underline{G}})^{\underline{\Gamma}} = Z(\widehat{\underline{G}})^{\underline{\Gamma}_u}$, so assertion (a) follows from the exactness of (B.3.1) while assertion (b) follows from the Hasse principle.

(c) Consider commutative diagram

$$\begin{array}{ccc} H^1(\underline{E},\underline{T}) & \xrightarrow{A} & \bigoplus_{v\neq u} H^1(\underline{E}_v,\underline{T}) \\ \downarrow & & D\downarrow \\ H^1(\underline{E},\underline{G}) & \xrightarrow{B} & \bigoplus_{v\neq u} H^1(\underline{E}_v,\underline{G}). \end{array}$$

In both cases, u is inert in $\underline{E}[\underline{G}^*]$, hence B is injective (by (b)). Since by our assumption, $B(c) = (c_v)_{v\neq u}$ belongs to Im D, it will suffice to show that A is surjective. In the case (i), the surjectivity of A follows from (a). In the case (ii), the canonical map $\pi_0(\widehat{\underline{T}}^{\underline{\Gamma}}) = \widehat{\underline{T}}^{\underline{\Gamma}}/Z(\widehat{\underline{G}})^{\underline{\Gamma}} \hookrightarrow \widehat{\underline{T}}^{\underline{\Gamma}_v}/Z(\widehat{\underline{G}})^{\underline{\Gamma}_v} = \pi_0(\widehat{\underline{T}}^{\underline{\Gamma}_u})$ is injective. Therefore surjectivity of A follows from the exactness of (B.3.1). □

From now on, $\underline{G}$ is a semisimple and simply connected group over $\underline{E}$.

B.3.3. (a) Let $\mathcal{E}' = (H', [\eta'], \overline{s}')$ be an endoscopic triple for $\underline{G}$. For each $v \in V_f$, $\mathcal{E}'$ gives rise to an (isomorphism class of an) endoscopic triple $\mathcal{E}'_v = (H'_v, [\eta'_v], \overline{s}'_v)$ for $\underline{G}_v$.

In particular, if $\mathcal{E}' \cong \mathcal{E}_{([\underline{a}],\underline{\kappa})}$ for a certain pair $(\underline{a}, \underline{\kappa})$ consisting of an embedding of a maximal torus $\underline{a} : \underline{T} \hookrightarrow \underline{G}$ and $\underline{\kappa} \in \widehat{T}^{\underline{\Gamma}}$, then $\mathcal{E}'_v \cong \mathcal{E}_{([\underline{a}_v],\underline{\kappa}_v)}$.

(b) Let $\mathcal{E}'_i = (H'_i, [\eta'_i], \overline{s}'_i)$, $i = 1, 2$, be a pair of endoscopic triples for $\underline{G}$. Assume that there exists $v \in V_f$, inert in both splitting fields $\underline{E}[H'_1]$ and $\underline{E}[H'_2]$, such that $(\mathcal{E}'_1)_v \cong (\mathcal{E}'_2)_v$. Then $\mathcal{E}'_1 \cong \mathcal{E}'_2$.

Indeed, the image of $\rho_{H'_i} : \underline{\Gamma} \to \mathrm{Out}(\widehat{H'_i})$ coincides with that of $\rho_{(H'_i)_v} : \underline{\Gamma}_v \to \mathrm{Out}(\widehat{H'_i})$ for each $i = 1, 2$, and therefore by Remark 1.3.4, the map

$$\mathrm{Isom}(\mathcal{E}'_1, \mathcal{E}'_2)/(H'_1)^{\mathrm{ad}}(\underline{E}) \to \mathrm{Isom}((\mathcal{E}'_1)_v, (\mathcal{E}'_2)_v)/(H'_1)^{\mathrm{ad}}(\underline{E}_v)$$

is bijective.

(c) Let $\mathcal{E}' = (H', [\eta'], \overline{s}')$ be an endoscopic triple for G, $\underline{\varphi} : \underline{G} \to \underline{G}'$ an inner twisting, and $(\underline{a}'_i, \underline{a}_i, [\underline{b}_i])$ be two triples consisting of stably conjugate embeddings of maximal tori $\underline{a}_i : \underline{T}_i \hookrightarrow \underline{G}$ and $\underline{a}'_i : \underline{T}_i \hookrightarrow \underline{G}'$, and stably conjugate classes $[\underline{b}_i]$ of embeddings of maximal tori $\underline{T}_i \hookrightarrow H'$, compatible with $\underline{a}_i$. Then we have the following product formula

$$\prod_{v\in V_f} \left\langle \frac{(\underline{a}_1)_v, (\underline{a}'_1)_v; [\underline{b}_1]_v}{(\underline{a}_2)_v, (\underline{a}'_2)_v; [\underline{b}_2]_v} \right\rangle_{\mathcal{E}'_v} = 1.$$

Indeed, consider elements $c := \overline{\mathrm{inv}}((\underline{a}_1, \underline{a}'_1); (\underline{a}_2, \underline{a}'_2)) \in H^1(\underline{E}, (\underline{T}_1 \times \underline{T}_2)/Z(\underline{G}))$ (see 1.1.5) and $\kappa := \kappa([b_1], [b_2])(\overline{s}') \in \pi_0((\widehat{\{(\underline{T}_1 \times \underline{T}_2)/Z(\underline{G})\}})^{\underline{\Gamma}}$ (see 1.2.8). Since $\left\langle \frac{(\underline{a}_1)_v, (\underline{a}'_1)_v; [\underline{b}_1]_v}{(\underline{a}_2)_v, (\underline{a}'_2)_v; [\underline{b}_2]_v} \right\rangle_{\mathcal{E}'_v} = \langle c_v, \kappa_v \rangle$ for each $v \in V_f$, the product formula follows from Kottwitz's theorem (see B.3.1 (c)).

B.3.4. Denote by dg and $dx = \prod_v dx_v$ the Tamagawa measures on $\underline{G}(\mathbb{A})$ and $\underline{\mathcal{G}}(\mathbb{A})$, respectively (defined by a non-zero translation invariant top degree differential form $\omega_{\underline{G}}$ on $\underline{G}$).

For each $v \in V_\infty$, denote by $\mathcal{S}(\underline{G}(\underline{E}_v))$ the space of measures on $\underline{G}(\underline{E}_v)$ of the form $f_v dx_v$, where f_v is a Schwartz function. Put $\mathcal{S}(\underline{\mathcal{G}}(\mathbb{A})) := \otimes'_{v \in V} \mathcal{S}(\underline{\mathcal{G}}(\underline{E}_v))$, and fix a non-trivial character $\underline{\psi} = \prod_v \psi_v : \mathbb{A}/\underline{E} \to \mathbb{C}^\times$. Then $\underline{\psi}$ gives rise to a Fourier transform $\mathcal{F} : \mathcal{S}(\underline{\mathcal{G}}(\mathbb{A})) \to \mathcal{S}(\underline{\mathcal{G}}(\mathbb{A}))$ such that $\mathcal{F}(\otimes_v \phi_v) = \otimes_v \mathcal{F}(\phi_v)$, where $\mathcal{F}(\phi_v)$ is the Fourier transform of ϕ_v corresponding to a measure dx_v and a character ψ_v.

For each $\underline{x} \in \underline{\mathcal{G}}^{\mathrm{sr}}(\underline{E})$, $\underline{\kappa} \in \widehat{\underline{G}_{\underline{x}}}^{\Gamma}$, and $\phi = \otimes_v \phi_v \in \mathcal{S}(\underline{\mathcal{G}}(\mathbb{A}))$ put $O^{\underline{\kappa}}_{\underline{x}}(\phi) =: \prod_{v \in V} O^{\underline{\kappa}_v}_{\underline{x}_v}(\phi_v)$, where $O^{\underline{\kappa}_v}_{\underline{x}_v} := O_{\underline{x}_v}$ for each $x \in V_\infty$. It follows from Kottwitz's theorem (see B.3.1 (c)) that generalized function $O^{\underline{\kappa}}_{\underline{x}}$ depends only on the stable conjugacy class of $\underline{x}$.

The main technical tool for the proof of Theorem B.1.4 is the following simple version of the trace formula. Let θ be a generalized function on $\underline{\mathcal{G}}(\mathbb{A})$ defined by the rule $\theta(f dx) = \sum_{x \in \mathcal{G}(F)} f(x)$. For each $g \in \underline{G}(\mathbb{A})$, put $\theta^g := (\mathrm{Ad}\, g)^*(\theta)$.

Proposition B.3.5. *(a) Let $\phi = \otimes_v \phi_v$ be an element of $\mathcal{S}(\underline{\mathcal{G}}(\mathbb{A}))$ such that $Supp(\phi) \cap$ $\mathrm{Ad}\, \underline{G}(\mathbb{A})(\underline{\mathcal{G}}(\underline{E}))$ consist of regular elliptic elements. Then the integral*

$$\Theta(\phi) := \int_{\underline{G}(\mathbb{A})/\underline{G}(\underline{E})} \theta^g(\phi) dg$$

converges absolutely. Furthermore,

$$\Theta(\phi) = \sum_{\underline{x}} \sum_{\underline{\kappa} \in \widehat{\underline{G}_{\underline{x}}}^{\Gamma}} O^{\underline{\kappa}}_{\underline{x}}(\phi),$$

where $\underline{x}$ runs over the set of regular elliptic stably conjugacy classes of $\underline{\mathcal{G}}(\underline{E})$.
(b) If $\mathcal{F}(\phi)$ also satisfies the support assumption of (a), then $\Theta(\phi) = \Theta(\mathcal{F}(\phi))$.

Proof. (a) The first assertion (see [Wa1, 10.8]) is a direct analog of the corresponding result of Arthur, while the second one (see, for example, [KP, Thm 3.2.1]) is an analog of a result of Kottwitz.

(b) By the Poisson summation formula, we have $\mathcal{F}(\theta) = \theta$. Therefore the assertion follows from the absolute convergence of $\Theta(\phi)$ and $\Theta(\mathcal{F}(\phi))$. □

To apply the trace formula, we will embed our local data into global ones.

Claim B.3.6. *There exist a totally imaginary number field $\underline{E}$, two finite places w and u of $\underline{E}$, a semisimple simply connected group $\underline{G}$ over $\underline{E}$, an inner twisting $\underline{\varphi} : \underline{G} \to \underline{G}'$, an endoscopic triple $\underline{\mathcal{E}} = (\underline{H}, [\underline{\eta}], \underline{\bar{s}})$ for $\underline{G}$, a tori $\underline{T}$ over $\underline{E}$, a pair of stably conjugate embeddings of maximal tori $\underline{a} : \underline{T} \hookrightarrow \underline{H}$ and $\underline{a}' : \underline{T} \hookrightarrow \underline{G}'$, and an embedding $\underline{b} : \underline{T} \hookrightarrow \underline{G}$ compatible with $\underline{a}$ satisfying the following conditions:*

(a) $\underline{E}_w \cong E$, $\underline{G}_w \cong G$, $\underline{\varphi}_w \cong \varphi$, $\underline{\mathcal{E}}_w \cong \mathcal{E}$, $\underline{T}_w \cong T$, $[\underline{b}]_w = [b]$, while $\underline{a}_w$ are $\underline{a}'_w$ are conjugate to a and a', respectively.

(b) For each $v \neq u, w$, the groups $\underline{G}_v$ and $\underline{G}'_v$ are quasi-split, and $\underline{\varphi}_v$ is trivial. Moreover, after we identify $\underline{G}_v$ with $\underline{G}'_v$ by means of some $\underline{G}(\overline{\underline{E}_v})$-conjugate of $\underline{\varphi}_v$, embeddings $\underline{a}_v$ and $\underline{a}'_v$ are conjugate.

(c) u is inert in $\underline{E}[\underline{T}]$, and $\underline{\mathcal{E}}_u$ is elliptic.

Proof. (I) Put $E' := E[T]$, and set $\Gamma' := \mathrm{Gal}(E'/E)$. Choose a dense subfield F' of E', which is a finite extension of $\mathbb{Q}$. Increasing F', we may assume that F' is Γ'-invariant. Set $F := (F')^{\Gamma'}$, and let w_0 be the prime of F, corresponding to the embedding $F \hookrightarrow (E')^{\Gamma'} = E$. In particular, $F_{w_0} \cong E$. Choose a totally imaginary quadratic extension $\underline{E}$ of F_0 such that w_0 splits in $\underline{E}$, and let w and u be the primes of $\underline{E}$ lying over w_0. Finally, let $\underline{E}'$ be the composite field $\underline{E} \cdot F'$. We have natural identifications $\underline{E}_w \cong \underline{E}_u \cong E$ and $\mathrm{Gal}(\underline{E}'/\underline{E}) \cong \mathrm{Gal}(E'/E)$, and both w and u are inert in $\underline{E}'$.

(II) Let $\varphi^* : G \to G^*$ be the inner twisting such that G^* is quasi-split. Since G, H and T split over E', the homomorphisms ρ_G, ρ_H and ρ_T factor through $\mathrm{Gal}(E'/E)$. We denote by $\underline{G}^*$ (resp. $\underline{H}$, resp. $\underline{T}$) the quasi-split group over $\underline{E}$ such that $\widehat{\underline{G}^*} = \widehat{G}$ (resp. $\widehat{\underline{H}} = \widehat{H}$, resp. $\widehat{\underline{T}} = \widehat{T}$) and $\rho_{\underline{G}^*}$ (resp. $\rho_{\underline{H}}$, resp. $\rho_{\underline{T}}$) is the composition of the projection $\underline{\Gamma} \to \mathrm{Gal}(\underline{E}'/\underline{E}) \cong \mathrm{Gal}(E'/E)$ and the homomorphism $\rho_G : \mathrm{Gal}(E'/E) \to \mathrm{Out}(\widehat{G})$ (resp. $\rho_H : \mathrm{Gal}(E'/E) \to \mathrm{Out}(\widehat{H})$, resp. $\rho_T : \mathrm{Gal}(E'/E) \to \mathrm{Out}(\widehat{T})$).

By construction, we have $\underline{G}^*_w \cong G^*$, $\underline{H}_w \cong H$ and $\underline{T}_w \cong T$. Moreover, the conjugacy classes of embeddings $\widehat{\underline{T}} \hookrightarrow \widehat{\underline{G}}$ and $\widehat{\underline{T}} \hookrightarrow \widehat{\underline{H}}$ corresponding to $a : T \hookrightarrow G$ and $[b]$ are $\underline{\Gamma}$-invariant. Therefore they come from stable conjugacy classes of embeddings $\underline{a}^* : \underline{T} \hookrightarrow \underline{G}^*$ and $\underline{b} : \underline{T} \hookrightarrow \underline{H}$. Furthermore, $(\underline{a}^*)_w$ is stably conjugate to a, and $[\underline{b}]_w = [b]$.

(III) Since u is inert in $\underline{E}' = \underline{E}[\underline{T}] \supset \underline{E}[\underline{G}^*]$, the canonical map $H^1(\underline{E}, (\underline{G}^*)^{\mathrm{ad}}) \to \bigoplus_{v \neq u} H^1(\underline{E}_v, (\underline{G}^*)^{\mathrm{ad}})$ is an isomorphism (by Lemma B.3.2 (b)). Hence there exist unique inner twistings $\underline{\varphi}^* : \underline{G}^* \to \underline{G}$ and $\underline{\varphi} : \underline{G} \to \underline{G}'$ such that $\underline{\varphi}^*_w \cong (\varphi^*)^{-1}$, $\underline{\varphi}_w \cong \varphi$, while $\underline{\varphi}^*_v$ and $\underline{\varphi}_v$ are trivial for all $v \neq w, u$.

Applying Lemma B.3.2 (c) for the embedding $\underline{a}^* : \underline{T}/Z(\underline{G}) \hookrightarrow (\underline{G}^*)^{\mathrm{ad}}$ we conclude from 1.1.4 (b) that there exist embeddings $\underline{a} : \underline{T} \hookrightarrow \underline{G}$ and $\underline{a}' : \underline{T} \hookrightarrow \underline{G}'$ stably conjugate to $\underline{a}^*$. Applying now Lemma B.3.2 (c) for $\underline{T}$, we can further replace $\underline{a}$ and $\underline{a}'$ so that $\underline{a}_w$ is conjugate to a, $\underline{a}'_w$ is conjugate to a', while $\underline{a}_v$ and $\underline{a}'_v$ are conjugate for all $v \neq w, u$.

(IV) Since w is inert in $\underline{E}[\underline{T}] \supset \underline{E}[\underline{H}]$, we get that $Z(\widehat{\underline{H}})^{\underline{\Gamma}} = Z(\widehat{H})^{\Gamma}$, and the conjugacy class $[\eta]$ of embeddings $\widehat{\underline{H}} = \widehat{H} \hookrightarrow \widehat{G} = \widehat{\underline{G}}$ is $\underline{\Gamma}$-invariant. Hence the triple $\underline{\mathcal{E}} := (\underline{H}, [\eta], \overline{s})$ is an endoscopic triple for $\underline{G}$. Moreover, $\underline{\mathcal{E}}_w \cong \mathcal{E}$ and $\underline{\mathcal{E}}_u$ is elliptic. Indeed, u is inert in $\underline{E}[\underline{H}]$, therefore $Z(\widehat{\underline{H}})^{\Gamma_u} = Z(\widehat{\underline{H}})^{\underline{\Gamma}} = Z(\widehat{H})^{\Gamma}$ is finite. □

B.4. Proof of Theorem B.1.4

B.4.1. Fix $(a, a'; [b])$-indistinguishable $\phi \in \mathcal{S}(\mathcal{G}(E))$ and $\phi' \in \mathcal{S}(\mathcal{G}'(E))$. We want to show that $e'(G)\mathcal{F}(\phi)$ and $e'(G')\mathcal{F}(\phi')$ are $(a, a'; [b])$-indistinguishable. Using Lemma 1.6.11 and the symmetry between G and G', it will suffice to check that for each compatible $x \in \mathcal{G}^{\mathrm{sr}}(E)$ and $y \in \mathcal{H}^{\mathrm{sr}}(E)$ we have:

(i) $O_x^{[y]}(\mathcal{F}(\phi)) = 0$, if x does not have a stable conjugate in $\mathcal{G}'(E)$;

(ii) $e'(G)O_x^{[y]}(\mathcal{F}(\phi)) = e'(G')\left\langle \frac{x, x'; [y]}{a, a'; [b]} \right\rangle O_{x'}^{[y]}(\mathcal{F}(\phi'))$, for each stable conjugate $x' \in \mathcal{G}'^{\mathrm{sr}}(E)$ of x.

B.4.2. Fix x and y as in B.4.1. Let $\Omega_x \subset \mathcal{G}_x^{\mathrm{sr}}(E)$ be an open neighborhood of x such that $O_{\widetilde{x}}^{\overline{\kappa}[y]}(\mathcal{F}(\phi)) = O_x^{\overline{\kappa}[y]}(\mathcal{F}(\phi))$ for each $\widetilde{x} \in \Omega_x$, and $O_{\widetilde{x}'}^{\overline{\kappa}[y]}(\mathcal{F}(\phi')) = O_{x'}^{\overline{\kappa}[y]}(\mathcal{F}(\phi'))$ for each stable conjugate $x' \in \mathcal{G}'(E)$ of x and each $\widetilde{x}' \in \varphi_{x,x'}(\Omega_x) \subset \mathcal{G}'^{\mathrm{sr}}_{x'}(E)$. Denote by $\Omega_y \subset \mathcal{H}_y^{\mathrm{sr}}(E)$ the image of Ω_x under the natural isomorphism $\mathcal{G}_x \xrightarrow{\sim} \mathcal{H}_y$, sending x to y, and choose an open neighborhood $\Omega \subset \mathcal{H}^{\mathrm{sr}}(E)$ of y contained in $\mathrm{Ad}\, H(E)(\Omega_y)$.

By construction, for each $\widetilde{y} \in \Omega$ there exists $\widetilde{x} \in \Omega_x \subset \mathcal{G}^{\mathrm{sr}}(E)$ compatible with y. Moreover, since conditions (i), (ii) from B.4.1 do not change when y and x are replaced by stable conjugates, it will suffice to show (i), (ii) for some pair of compatible elements $\widetilde{y} \in \Omega \subset \mathcal{H}^{\mathrm{sr}}(E)$ and $\widetilde{x} \in \mathcal{G}^{\mathrm{sr}}(E)$.

B.4.3. Strategy of the proof Choose $\underline{E}, u, w, \underline{G}, \underline{\varphi}, \underline{\mathcal{E}}, \underline{T}, \underline{a}, \underline{a}', [\underline{b}]$ as in Claim B.3.6, and identify $\underline{G}_v$ with $\underline{G}'_v$ for each $v \neq w, u$ as in Claim B.3.6 (b). Finally, we fix a non trivial character $\psi : \mathbb{A}_F/F \to \mathbb{C}^\times$.

Our strategy will be to construct measures $\underline{\phi} = \otimes_{v\in V}\phi_v \in \mathcal{S}(\mathcal{G}(\mathbb{A}))$ and $\underline{\phi}' = \otimes_{v\in V}\phi'_v \in \mathcal{S}(\mathcal{G}'(\mathbb{A}))$ and compatible elements $\underline{y} \in \underline{\mathcal{H}}^{\mathrm{sr}}(\underline{E})$ and $\underline{x} \in \underline{\mathcal{G}}^{\mathrm{sr}}(\underline{E})$ satisfying the following properties:

(A) $\underline{y}_w \in \Omega$, $\phi_w = \phi$ and $\phi'_w = \phi'$.

(B) $\underline{x}$ has a stable conjugate in $\underline{\mathcal{G}}'(\underline{E})$ if $\underline{x}_w$ has a stable conjugate in $\mathcal{G}'(E)$.

(C) both $\underline{\phi}$ and $\underline{\phi}'$ satisfy the support assumption of Proposition B.3.5 (a), and we have $\Theta(\underline{\phi}) = \Theta(\underline{\phi}')$.

(D) $\mathcal{F}(\underline{\phi})$ and $\mathcal{F}(\underline{\phi}')$ satisfy the support assumption of Proposition B.3.5 (a), and we have $\Theta(\mathcal{F}(\underline{\phi})) = O_{\underline{x}}^{[\underline{y}]}(\mathcal{F}(\underline{\phi}))$ and $\Theta(\mathcal{F}(\underline{\phi}')) = O_{\underline{x}}^{[\underline{y}]}(\mathcal{F}(\underline{\phi}'))$. (We define $O_{\underline{x}}^{[\underline{y}]}(\mathcal{F}(\underline{\phi}'))$ to be zero unless there exists a stable conjugate $\underline{x}' \in \underline{\mathcal{G}}'(\underline{E})$ of $\underline{x}$, in which case we define $O_{\underline{x}}^{[\underline{y}]}(\mathcal{F}(\underline{\phi}'))$ to be $O_{\underline{x}'}^{[\underline{y}]}(\mathcal{F}(\underline{\phi}'))$.)

(E) For each $v \neq w$, we have $O_{\underline{x}_v}^{[\underline{y}]_v}(\mathcal{F}(\phi_v)) \neq 0$.

(F) For each $v \in V_f \smallsetminus w$, there exists a stable conjugate $x'_v \in \underline{\mathcal{G}}'(\underline{E}_v)$ of $\underline{x}_v$, and we have

$$e'(\underline{G}_v) O_{\underline{x}_v}^{[\underline{y}]_v}(\mathcal{F}(\phi_v)) = e'(\underline{G}'_v) \left\langle \frac{\underline{x}_v, x'_v; [\underline{y}]_v}{\underline{a}_v, \underline{a}'_v; [\underline{b}]_v} \right\rangle_{\underline{\mathcal{E}}_v} O_{x'_v}^{[\underline{y}]_v}(\mathcal{F}(\phi'_v)).$$

Once these data are constructed, the result follows. Indeed, by (A) and the observation at the end of B.4.2, it will suffice to check that $\underline{y}_w$ and $\underline{x}_w$ satisfy conditions (i), (ii) of B.4.1. Next (C), (D) and Proposition B.3.5 (b) imply that $O_{\underline{x}}^{[\underline{y}]}(\mathcal{F}(\underline{\phi})) = O_{\underline{x}}^{[\underline{y}]}(\mathcal{F}(\underline{\phi}'))$. Assume first that $\underline{x}_w$ does not have a stable conjugate in $\mathcal{G}'(E)$. Then $\underline{x}$ does not have a stable conjugate in $\underline{\mathcal{G}}'(\underline{E})$, thus $O_{\underline{x}}^{[\underline{y}]}(\mathcal{F}(\underline{\phi})) = \prod_v O_{\underline{x}_v}^{[\underline{y}]_v}(\mathcal{F}(\phi_v)) = 0$. Hence the vanishing of $O_{\underline{x}_w}^{[\underline{y}]_w}(\mathcal{F}(\phi_w))$ follows from (E).

Assume now that that $\underline{x}_w$ has a stable conjugate in $\mathcal{G}'(E)$, then by (B), there exists a stable conjugate $\underline{x}' \in \underline{\mathcal{G}}'(\underline{E})$ of $\underline{x}$. Then $O_{\underline{x}}^{[\underline{y}]}(\mathcal{F}(\underline{\phi})) = O_{\underline{x}'}^{[\underline{y}]}(\mathcal{F}(\underline{\phi}'))$ and each $\underline{x}'_v \in \underline{\mathcal{G}}(\underline{E})$ is a stable conjugate of $\underline{x}_v$. Using product formulas $\prod_{v\in V} e'(\underline{G}_v) =$

$\prod_{v\in V} e'(\underline{G}'_v) = 1$ (see [Ko5]) and $\prod_{v\in V_f} \left\langle \frac{\underline{x}_v, \underline{x}'_v; [\underline{y}]_v}{\underline{a}_v, \underline{a}'_v; [\underline{b}]_v} \right\rangle_{\underline{\mathcal{E}}_v} = 1$ (B.3.3 (c)), the required equality $e'(G) O^{[\underline{y}]_w}_{\underline{x}_w}(\mathcal{F}(\phi)) = e'(G') \left\langle \frac{\underline{x}_w, \underline{x}'_w; [\underline{y}]_w}{a, a'; [b]} \right\rangle O^{[\underline{y}]_w}_{\underline{x}'_w}(\mathcal{F}(\phi'))$ follows from (E) and (F).

B.4.4. Construction of $\underline{\phi}, \underline{\phi}', \underline{y}$ and $\underline{x}$ (a) Choose an $\underline{\mathcal{O}}$-subalgebra $\mathcal{K} \subset \underline{\mathcal{G}}(\underline{E})$, and let $S_1 \subset V$ be a finite subset containing $V_\infty \cup \{w, u\}$ such that for each $v \notin S_1$ we have

- $\underline{H}_v$ and $\underline{\mathcal{G}}_v$ are unramified;
- the $\underline{\mathcal{O}}_v$-subalgebra $\mathcal{K}_v \subset \underline{\mathcal{G}}(\underline{E}_v)$, spanned by $\mathcal{K}$, is hyperspecial and satisfies $\mathcal{F}(1_{\mathcal{K}_v}) = 1_{\mathcal{K}_v}$.

(b) Let $\mathcal{A}$ be the set of isomorphisms classes of those endoscopic triples for $\underline{G}$, which are unramified outside of S_1. Then $\mathcal{A}$ is finite (see [La1, Lem. 8.12]), and $\underline{\mathcal{E}} \in \mathcal{A}$. Let $\mathcal{A}'$ be the subset of $\mathcal{A}$ consisting of triples $(H_a, [\eta_a], \overline{s}_a)$ such that $\underline{E}[H_a]$ is not contained in $\underline{E}[\underline{H}]$. For each $a \in \mathcal{A}'$, we fix a prime $v_a \in V_f \smallsetminus S_1$ which splits in $\underline{E}[\underline{H}]$ but does not split in $\underline{E}[H_a]$. Put $S_2 := \{v_a \,|\, a \in \mathcal{A}'\}$, and set $S := S_1 \cup S_2$.

(c) Choose $\underline{y} \in \underline{\mathcal{H}}^{\mathrm{sr}}(\underline{E})$ such that $\underline{y}_w \in \Omega$, $\underline{y}_u \in \mathcal{H}^{\mathrm{sr}}(\underline{E}_u)$ is elliptic (that is, $\underline{H}_{\underline{y}_u} \subset \underline{H}_u$ is elliptic) and $\underline{y}_v \in \mathcal{H}^{\mathrm{sr}}(\underline{E}_v)$ is split (that is, $\underline{H}_{\underline{y}_v} \subset \underline{H}_v$ is split) for each $v \in S_2$.

Choose an element $\underline{x}^* \in \underline{\mathcal{G}}^*(\underline{E})$ compatible with $\underline{y}$ (exists by Lemma 1.1.10 (b)). Then $(\underline{G}^*_{\underline{x}^*})_u \subset \underline{G}^*_u$ is an elliptic torus. Since $\underline{y}_w \in \overline{\Omega}$, element $\underline{x}^*_w$ has a stable conjugate in $\Omega_x \subset \mathcal{G}^{\mathrm{sr}}_x(E)$. Since $\underline{\varphi}^*_v$ is trivial for each $v \neq w, u$, it follows from Lemma B.3.2 (c) (ii) (as in the proof Claim B.3.6 (III)) that there exists a stably conjugate $\underline{x} \in \underline{\mathcal{G}}(\underline{E})$ of $\underline{x}^*$.

(d) Choose a stably conjugate $x'_u \in \underline{\mathcal{G}}'(\underline{E}_u)$ of $\underline{x}_u$ (which exists by Lemma 1.5.3), and set $x'_v := \underline{x}_v$ for each $v \neq w, u$. For each $v \in S \smallsetminus (V_\infty \cup w)$, choose measures $\phi_v \in \mathcal{S}(\underline{G}(\underline{E}_v))$ and $\phi'_v \in \mathcal{S}(\underline{G}'(\underline{E}_v))$ constructed in Corollary B.2.6 for the triple $(\underline{y}_v, \underline{x}_v, x'_v)$. In particular, $\phi_v = \phi'_v$ for each $v \in S \smallsetminus (V_\infty \cup \{u, w\})$.

For every $v \in S \smallsetminus V_\infty$, let $\omega_v \subset \underline{E}^\times_v$ be an open neighborhood of the identity such that Ω is invariant under the multiplication by ω_w, while $\mathcal{F}(\phi_v)$ and $\mathcal{F}(\phi'_v)$ are invariant under the multiplication by ω_v if $v \neq w$.

For each $v \in V \smallsetminus S$, put $\phi_v = \phi'_v = 1_{\mathcal{K}_v} dx_v$. Finally, put $\phi_w := \phi$ and $\phi'_w := \phi'$.

(e) Choose a finite set $S_3 \supset S$ such that for each $v \notin S_3$, we have $\underline{x}_v \in \mathcal{K}_v$, and the reduction of $\underline{x}_v$ modulo v is regular. Choose $\lambda \in \underline{E}^\times$ such that

(i) $\mathrm{val}_v(\lambda) = 0$ for every $v \notin S_3$;

(ii) $\lambda \in \omega_v$ for every $v \in S \smallsetminus V_\infty$;

(iii) For every $v \in S_3 \smallsetminus S$, $\mathrm{val}_v(\lambda)$ is so large that $(\lambda \overline{y})_v$ belongs to the open neighborhood of zero prescribed in Lemma B.2.7 (b).

Finally, we replace $\underline{y}$ and $\underline{x}$ constructed in (c) by $\lambda\underline{y}$ and $\lambda\underline{x}$, respectively.

(f) Recall that $\mathcal{F}(\phi_v)$ and $\mathcal{F}(\phi'_v)$ are compactly supported for each $v \in V_f$ and $\mathcal{F}(\phi_v) = \mathcal{F}(\phi'_v) = 1_{\mathcal{K}_v} dx_v$ for each $v \notin S$. Since $\underline{E} \subset \mathbb{A}$ is discrete, one can choose a compact neighborhood $C_v \subset \underline{\mathcal{G}}(\underline{E}_v) = \underline{\mathcal{G}}'(\underline{E}_v)$ of $\underline{x}_v$ for each $v \in V_\infty$

such that all elements of $\underline{\mathcal{G}}(\underline{E}) \cap (\prod_{v \in V_\infty} C_v \times \prod_{v \in V_f} \mathrm{Supp}(\mathcal{F}(\phi_v)))$ and $\underline{\mathcal{G}}'(\underline{E}) \cap (\prod_{v \in V_\infty} C_v \times \prod_{v \in V_f} \mathrm{Supp}(\mathcal{F}(\phi'_v)))$ are stable conjugate to $\underline{x}$.

For every $v \in V_\infty$, choose a measure $\phi_v = \phi'_v$ of $\mathcal{S}(\underline{\mathcal{G}}(\underline{E}_v))$ such that $\mathcal{F}(\phi_v) = f_v dx_v$ for a smooth non-negative function f_v on $\underline{\mathcal{G}}(\underline{E}_v)$ supported on C_v such that $f_v(\underline{x}_v) \neq 0$. Put $\underline{\phi} := \otimes_{v \in V} \phi_v$ and $\underline{\phi}' := \otimes_{v \in V} \phi'_v$.

To complete the proof of Theorem B.1.4, it remains to show that the constructed above $\underline{\phi}, \underline{\phi}', \underline{y}$ and $\underline{x}$ satisfy conditions (A)–(F) of B.4.3.

B.4.5. Proof of conditions (A)–(F) (A) is clear (see B.4.4 (c) and (e)).

(B) Since $\underline{x}_u$ is elliptic, the assertion follows from Lemma B.3.2 (c) (ii) (as in the proof of Claim B.3.6 (III)).

(C) Since the support of ϕ_u and ϕ'_u is regular elliptic, both $\underline{\phi}$ and $\underline{\phi}'$ satisfy the support assumption of Proposition B.3.5 (a). Because of symmetry between $\underline{\phi}$ and $\underline{\phi}'$, it will therefore suffice to check that for every $\underline{z} \in \underline{\mathcal{G}}(\underline{E})$ and $\underline{\kappa} \in \widehat{\underline{G}_{\underline{z}}}^{\Gamma}$ such that $O_{\underline{z}}^{\underline{\kappa}}(\underline{\phi}) \neq 0$, there exists a stable conjugate $\underline{z}' \in \underline{\mathcal{G}}'(\underline{E})$ of $\underline{z}$, and we have $O_{\underline{z}}^{\underline{\kappa}}(\underline{\phi}) = O_{\underline{z}'}^{\underline{\kappa}}(\underline{\phi}')$.

Fix $\underline{z} \in \underline{\mathcal{G}}(\underline{E})$ and $\underline{\kappa} \in \widehat{\underline{G}_{\underline{z}}}^{\Gamma}$ such that $O_{\underline{z}}^{\underline{\kappa}}(\underline{\phi}) \neq 0$. Consider the endoscopic triple $\mathcal{E}' := \mathcal{E}_{([a_{\underline{z}}], \underline{\kappa})} = (H', [\eta'], \overline{s}')$ for $\underline{G}$. Following [Wa1, 10.9], we will show that $\mathcal{E}' \cong \underline{\mathcal{E}}$.

By the assumption, $O_{\underline{z}_v}^{\underline{\kappa}_v}(\phi_v) \neq 0$ for each $v \in V$. Since $\mathcal{K}_v \subset \underline{\mathcal{G}}(\underline{E}_v)$ is a hyperspecial subalgebra and $\phi_v = 1_{\mathcal{K}_v} dx_v$ for each $v \in V \setminus S$, we conclude from Lemma B.2.7 (a) that $\mathcal{E}'$ is unramified outside of S.

For each $v \in S_2$, measure ϕ_v is supported on split elements (by Corollary B.2.6 (i)) and satisfies $O_{\underline{z}_v}^{\underline{\kappa}_v}(\phi_v) \neq 0$. Therefore $\underline{G}_{\underline{z}_v}$ is split, hence $\mathcal{E}'_v \cong \mathcal{E}_{([a_{\underline{z}_v}], \underline{\kappa}_v)}$ is split. In particular, $\mathcal{E}'$ is unramified outside of S_1. Moreover, H' splits at each $v \in S_2$, hence $\underline{E}[H'] \subset \underline{E}[\underline{H}]$.

Since $O_{\underline{z}_u}^{\underline{\kappa}_u}(\phi_u) \neq 0$, we get by Corollary B.2.6 (i), (iii) that $\underline{z}_u$ is stable conjugate to an element of $\underline{G}_{\underline{x}_u}^{\mathrm{sr}}(\underline{E}_u)$, and the class of $\underline{\kappa}_u \in \widehat{\underline{G}_{\underline{z}}}^{\Gamma_u} \cong \widehat{\underline{G}_{\underline{x}}}^{\Gamma_u}$ equals $\underline{\kappa}_{[\underline{y}]_u}$. Since $\underline{x}_u$ is elliptic, we get that $\mathcal{E}'_u \cong \mathcal{E}_{([a_{\underline{x}_u}], \overline{\kappa}_{[\underline{y}]_u})}$ is isomorphic to $\underline{\mathcal{E}}_u$ (use Lemma 1.3.9 (a)). As u is inert in $\underline{E}[\underline{H}] \supset \underline{E}[H']$, we conclude from this that $\mathcal{E}' \cong \underline{\mathcal{E}}$ (use B.3.3 (b)), as claimed.

By the above, there exists $\underline{y}' \in \underline{\mathcal{H}}^{\mathrm{sr}}(\underline{E})$ compatible with $\underline{z}$ such that $\underline{\kappa} = \overline{\kappa}_{[\underline{y}']}$, thus $O_{\underline{z}}^{\underline{\kappa}}(\underline{\phi}) = O_{\underline{z}}^{[\underline{y}']}(\underline{\phi}) \neq 0$. In particular, $O_{\underline{z}_w}^{[\underline{y}']_w}(\phi_w) \neq 0$. Since ϕ_w and ϕ'_w are $(a, a'; [b])$-indistinguishable, it follows from Lemma 1.6.11 that there exists a stable conjugate $z'_w \in \mathcal{G}'(E)$ of $\underline{z}_w$. Since $\underline{G}_{\underline{z}_u} \subset \underline{G}_u$ is elliptic, there exists $\underline{z}' \in \underline{G}'(\underline{E})$ stably conjugate to $\underline{z}$ such that $\underline{z}'_v$ is conjugate to $\underline{z}_v$ for each $v \neq u, w$ (use Lemma B.3.2 (c) and 1.1.4).

It now remains to show that $O_{\underline{z}}^{[\underline{y}']}(\underline{\phi}) = O_{\underline{z}'}^{[\underline{y}']}(\underline{\phi}')$. By the product formula (B.3.3 (c)), it will suffice to check that for each $v \in V_f$, we have

$$O_{\underline{z}_v}^{[\underline{y}']_v}(\phi_v) = \left\langle \frac{\underline{z}_v, \underline{z}'_v; y'_v}{\underline{a}_v, \underline{a}'_v; [\underline{b}]_v} \right\rangle_{\underline{\mathcal{E}}_v} O_{\underline{z}'_v}^{[\underline{y}']_v}(\phi'_v).$$

The assertion for $v = w$ follows from Lemma 1.6.11, while the assertion for $v = u$ follows from Corollary B.2.6 (ii). Finally, the assertion for $v \neq u, w$ follows from the fact that under the identification $\underline{G}'_v = \underline{G}_v$, we have $\phi'_v = \phi_v$, $\underline{z}'_v$ is conjugate to $\underline{z}_v$, and $\underline{a}'_v$ is conjugate to $\underline{a}_v$.

(D) By B.4.4 (f), $\mathrm{Supp}(\mathcal{F}(\underline{\phi})) \cap \mathrm{Ad}\,\underline{G}(\mathbb{A})(\mathcal{G}(\underline{E}))$ consists of elements stably conjugate to $\underline{x}$. Since $\underline{x}_u$ is elliptic, $\underline{x}$ is elliptic, thus $\mathcal{F}(\underline{\phi})$ satisfies the support assumption of Proposition B.3.5 (a). Therefore $\Theta(\mathcal{F}(\underline{\phi})) = \sum_{\underline{\kappa} \in \widehat{\underline{G}_{\underline{x}}}^{\Gamma}} O_{\underline{x}}^{\underline{\kappa}}(\mathcal{F}(\underline{\phi}))$. Let $\underline{\kappa} \in \widehat{\underline{G}_{\underline{x}}}^{\Gamma}$ be such that $O_{\underline{x}}^{\underline{\kappa}}(\mathcal{F}(\underline{\phi})) \neq 0$. Then $O_{\underline{x}_u}^{\underline{\kappa}_u}(\mathcal{F}(\phi_u)) \neq 0$. Since $\underline{x}_u$ is elliptic, it follows from Corollary B.2.6 (iv) that $\underline{\kappa}_u = \overline{\kappa}_{[\underline{y}]_u}$. But the map $\widehat{\underline{G}_{\underline{x}}}^{\Gamma} \hookrightarrow \widehat{\underline{G}_{\underline{x}}}^{\Gamma_u} = \pi_0(\widehat{\underline{G}_{\underline{x}}}^{\Gamma_u})$ is injective, therefore $\underline{\kappa} = \overline{\kappa}_{[\underline{y}]}$. This shows that $\Theta(\mathcal{F}(\underline{\phi})) = O_{\underline{x}}^{[\underline{y}]}(\mathcal{F}(\underline{\phi}))$. The proof for $\underline{\phi}'$ is similar.

(E), (F) For $v \in S \smallsetminus (V_\infty \cup w)$, the assertions follows from Corollary B.2.6 (v). For $v \neq u, w$, (F) follows from the fact that under the identification $\underline{G}_v = \underline{G}'_v$, we have $\phi'_v = \phi_v$, and $\underline{a}'_v$ is conjugate to $\underline{a}_v$. It remains to show (E) for $v \in V_\infty \cup (V \smallsetminus S)$. If $v \in V_\infty$, the assertion follows from the fact that $\mathcal{F}(\phi_v) = f_v dx_v$, while f_v is non-negative and satisfies $f_v(\underline{x}_v) \neq 0$. Assume now that $v \notin S$. Then $\mathcal{F}(\phi_v) = 1_{\mathcal{K}_v} dx_v$. The assertion now follows from Lemma B.2.7 (c) if $v \notin S_3$ and from the choice of λ in B.4.4 (e) (and Lemma B.2.7 (b)) if $v \in S_3$.

List of main terms and symbols

References

[Bo] M. Borovoi, Abelian Galois Cohomology of Reductive Groups, *Mem. Amer. Math. Soc.* **132** (1998).

[BP] A. Braverman and A. Polishchuk, Kazhdan–Laumon representations of finite Chevalley groups, character sheaves and some generalization of the Lefschetz–Verdier trace formula, preprint, math.AG/9810006.

[BR] P. Bardsley and R. W. Richardson, Étale slices for algebraic transformation groups in characteristic p, *Proc. London Math. Soc.* (3) **51** (1985), no. 2, 295–317.

[BS] A. Borel and T.A. Springer, Rationality properties of linear algebraic groups. II, *Tohoku Math. J.* (2), **20** (1968) 443–497.

[BT] F. Bruhat and J, Tits, Groupes réductifs sur un corps local. II. Schémas en groupes. Existence d'une donnée radicielle valuée, *Inst. Hautes Études Sci. Publ. Math.* **60** (1984), 197–376.

[Ch] P.-H. Chaudouard, Sur certanes identités endoscopiques entre transformées de Fourier, preprint, 2004.

[DB] S. DeBacker, Parametrizing conjugacy classes of maximal unramified tori via Bruhat–Tits Theory, preprint, available at http://www/math.lsa.umich.edu/ smdbackr.

[DBR] —— and M. Reeder, Depth-zero supercuspidal L-packets and their stability, preprint, available at http://www/math.lsa.umich.edu/ smdbackr.

[De] P. Deligne, Les corps locaux de caractéristique p, limites de corps locaux de caractéristique 0, in *Représentations des groupes réductifs sur un corps local*, 119–157, Hermann, Paris, 1984.

[DL] P. Deligne and G. Lusztig, Representations of reductive groups over finite fields, *Ann. Math.* (2), **103** (1976), 103–161.

[HC1] Harish-Chandra, Harmonic Analysis on Reductive p-adic Groups (Notes by G. van Dijk), *Lecture Notes in Mathematics* **162**, Springer-Verlag, Berlin, New York, 1970.

[HC2] ——, Admissible invariant distributions on reductive p-adic groups, Preface and notes by S. DeBacker and P. J. Sally, Jr., University Lecture Series, **16**, *Amer. Math. Soc.*, Providence, 1999.

[Ka1] D. Kazhdan, Proof of Springer's hypothesis, *Israel J. Math.* **28** (1977), 272–286.

[Ka2] ——, On lifting, in Lie group representations, II (College Park, Md., 1982/1983), 209–249, *Lecture Notes in Mathematics* **1041**, Springer, Berlin, 1984.

[Ka3] ——, Representations of groups over close local fields, *J. Analyse Math.* **47** (1986) 175–179.

[KP] —— and A. Polishchuk, Generalization of a theorem of Waldspurger to nice representations in *The Orbit Method in Geometry and Physics* (Marseille, 2000), 197–242, Prog. Math., Vol. **213**, Birkhäuser, Boston, MA, 2003.

[KV1] D. Kazhdan and Y. Varshavsky, Endoscopic decomposition of characters of certain cuspidal representations, *Electron. Res. Announc. Amer. Math. Soc.* **10** (2004), 11–20.

[KV2] ——, *On endoscopic decomposition of certain depth zero decomposition II*, in preparation.

[Ko1] R. E. Kottwitz, Stable trace formula: cuspidal tempered terms, *Duke Math. J.* **51** (1984), 611–650.

[Ko2] ——, Stable trace formula: elliptic singular terms, *Math. Ann.* **275** (1986), 365–399.

[Ko3] ——, Rational conjugacy classes in reductive groups, *Duke Math. J.* **49** (1982), no. 4, 785–806.

[Ko4] ——, Tamagawa numbers, *Ann. Math.* (2) **127** (1988), no. 3, 629–646.

[Ko5] ——, Sign changes in harmonic analysis on reductive groups, *Trans. Amer. Math. Soc.* **278** (1983), no. 1, 289–297.

[La1] R. P. Langlands, Les débuts d'une formule des traces stables, *Publ. Math. Univ. Paris VII* **13**, Paris, 1983.

[La2] ——, Representations of abelian algebraic groups, Olga Taussky-Todd: in memoriam, *Pacific J. Math.* (1997), Special Issue, 231–250.

[LS] —— and D. Shelstad, On the definition of transfer factors, *Math. Ann.* **278** (1987), no. 1–4, 219–271.

[Lu] G. Lusztig, Green functions and character sheaves, *Ann. Math.* (2) **131** (1990), no. 2, 355–408.

[MP1] A. Moy and G. Prasad, Unrefined minimal K-types for p-adic groups, *Invent. Math.* **116** (1994), no. 1–3, 393–408.

[MP2] ——, Jacquet functors and unrefined minimal K-types, *Comment. Math. Helv.* **71** (1996), no. 1, 98–121.

[MS] J. S. Milne and K.-Y. Shih, Conjugates of Shimura Varieties in Hodge Cycles, Motives, and Shimura Varieties *Lecture Notes in Mathematics* **900**, 280–356, Springer-Verlag, Berlin, New York, 1982.

[MW] C. Mœglin and J.-L. Waldspurger, Paquets stables de représentations tempérées et de réduction unipotente pour $SO(2n+1)$, *Invent. Math.* **152** (2003), 461–623.

[PY] G. Prasad and J.-K. Yu, On finite group actions on reductive groups and buildings, *Invent. Math.* **147** (2002), 545–560.

[Sp] T. A. Springer, Trigonometric sums, Green functions of finite groups and representations of Weyl groups, *Invent. Math.* **36** (1976), 173–207.

[SS] —— and R. Steinberg, Conjugacy classes, in Seminar on Algebraic Groups and Related Finite Groups, 167–266, *Lecture Notes in Mathematics*, Vol. **131** Springer, Berlin, 1970.

[St] R. Steinberg, Endomorphisms of linear algebraic groups, *Mem. of the Amer. Math. Soc.* **80**, Amer. Math. Soc., Providence, 1968.

[Wa1] J.-L. Waldspurger, Le lemme fondamental implique le transfert, *Compositio Math.* **105** (1997), 153–236.

[Wa2] ——, Transformation de Fourier et endoscopie, *J. Lie Theory* **10** (2000), 195–206.

[We] A. Weil, Sur certains groupes d'opérateurs unitaires, *Acta Math.* **111** (1964), 143–211.

Odd family algebras

A. A. Kirillov[1] and L. G. Rybnikov[2*]

[1] Department of Mathematics
The University of Pennsylvania
Philadelphia, PA 19104-6395
USA
and
Institute for Problems of Information Transmission of Russian Academy of Sciences
B. Karetnyi 19
Moscow 101 477, GSP-4
Russia
kirillov@math.upenn.edu
[2] Moscow State University
leo_rybnikov@mtu-net.ru

Summary. A new class of associative algebras related to simple complex Lie algebras (or root systems) was introduced and studied in [K1] and [K2]. They were named classical and quantum family algebras. The aim of this paper is to introduce the odd analogue of these algebras and expose some results about their structure. In particular, we describe the structure of $\mathfrak{g}$-module $\Lambda(\mathfrak{g})$ and compute the odd exponents for some cases.

Key words: Semisimple Lie algebra, exterior algebra, Clifford algebra, universal enveloping algebra, irreducible representation, classical and quantum family algebras.

Subject Classification: 17B35

1. Generalities about odd family algebras

We assume that the reader is acquainted with the general background of the theory of semisimple Lie algebras (see e.g., [OV]).

1.1. Basic definitions

First, we recall some standard notations used in [K2].

Let $\mathfrak{g}$ be a simple complex Lie algebra with the canonical decomposition

$$\mathfrak{g} = \mathfrak{n}_- \oplus \mathfrak{h} \oplus \mathfrak{n}_+.$$

* Supported by CRDF grant RM1-2543.

We denote by P (respectively by Q) the weight (resp. root) lattice in $\mathfrak{h}^*$ and by P_+ (resp. Q_+) the semigroup generated by fundamental weights $\omega_1, \ldots, \omega_l$ (resp. by simple roots $\alpha_1, \ldots, \alpha_l$). By ρ we denote the weight $\frac{1}{2}\sum_{\alpha\in R_+}\alpha = \sum_{k=1}^{l}\omega_k$.

For every $\lambda \in P_+$ let (π_λ, V_λ) be an irreducible representation of $\mathfrak{g}$ with highest weight λ. We denote by $d(\lambda)$ the dimension of V_λ.

Let us denote by $Wt\,(\lambda)$ the multiset of all weights of (π_λ, V_λ) and by $m_\lambda(\mu)$ the multiplicity of the weight $\mu \in Wt\,(\lambda)$. In particular, if θ is the maximal root of $\mathfrak{g}$, then (π_θ, V_θ) is equivalent to the adjoint representation $(\mathrm{ad}, \mathfrak{g})$ of $\mathfrak{g}$ on itself.

By λ^* we denote the highest weight of the dual (or contragradient) representation which acts in V_λ^* by $\pi_{\lambda^*}(X) = -(\pi_\lambda(X))^*$. It is clear that $d(\lambda) = d(\lambda^*)$ and $Wt\,(\lambda^*) = -Wt\,(\lambda)$.

Choosing a basis in V_λ, we can identify the $\mathfrak{g}$-module $\mathrm{End}\, V_\lambda \simeq V_\lambda \otimes V_\lambda^*$ with the matrix space $\mathrm{Mat}_{d(\lambda)}(\mathbb{C})$ where $\mathfrak{g}$ acts by the formula

$$X \cdot A = [\pi_\lambda(X), A].$$

Now we introduce the odd analogues of the symmetric algebra $S(\mathfrak{g})$ and the universal enveloping algebra $U(\mathfrak{g})$. The first is the exterior algebra $\Lambda(\mathfrak{g})$ while the second is the Clifford algebra $\mathrm{Cl}(\mathfrak{g}, K)$, where K is the Killing form on $\mathfrak{g}$. Recall that $\mathrm{Cl}(\mathfrak{g}, K)$ is a quotient of the tensor algebra $T(\mathfrak{g})$ by the ideal generated by expressions $X^2 - (X, X)_K \cdot 1$, $X \in \mathfrak{g}$. The algebra $\Lambda(\mathfrak{g})$ is naturally graded by degree of odd polynomials while $\mathrm{Cl}(\mathfrak{g}, K)$ has a natural filtration such that associated graded algebra is isomorphic to $\Lambda(\mathfrak{g})$. Therefore, $\Lambda(\mathfrak{g})$ and $\mathrm{Cl}(\mathfrak{g}, K)$ have isomorphic $\mathfrak{g}$-module structures. The spectrum and multiplicities of this module are still not explicitly known. The following important result is due to B. Kostant [Ko2].

Theorem 1 (Kostant). *There is an isomorphism of $\mathfrak{g}$-modules*

$$\Lambda(\mathfrak{g}) \cong 2^l\, V_\rho \otimes V_\rho.$$

Let $\widetilde{G}$ be a connected and simply connected Lie group with $\mathrm{Lie}(\widetilde{G}) = \mathfrak{g}$. The action of $\mathfrak{g}$ on any finite-dimensional $\mathfrak{g}$-module gives rise to the corresponding action of $\widetilde{G}$. In particular, $\widetilde{G}$ acts on $\Lambda(\mathfrak{g})$, $\mathrm{Cl}(\mathfrak{g}, K)$ and $\mathrm{End}\, V_\lambda$ by automorphisms of these algebras. Actually, this action is trivial for $g \in C$, the center of $\widetilde{G}$. So, we can consider it as the action of the adjoint group $G = \widetilde{G}/C$. Note that the irreducible representation (π_λ, V_λ) of $\widetilde{G}$ is trivial on C iff $\lambda \in P_+ \cap Q$.

In [K1] the following two algebras were introduced: the classical family algebra $\mathcal{C}_\lambda(\mathfrak{g}) = (\mathrm{End}\, V_\lambda \otimes S(\mathfrak{g}))^G$ and the quantum family algebra $\mathcal{Q}_\lambda(\mathfrak{g}) = (\mathrm{End}\, V_\lambda \otimes U(\mathfrak{g}))^G$. We define here the odd analogues of these algebras.

Namely, let

$$\mathcal{C}_\lambda^{\mathrm{odd}}(\mathfrak{g}) := (\mathrm{End}\, V_\lambda \otimes \Lambda(\mathfrak{g}))^G, \qquad \mathcal{Q}_\lambda^{\mathrm{odd}}(\mathfrak{g}) := (\mathrm{End}\, V_\lambda \otimes \mathrm{Cl}(\mathfrak{g}, K))^G. \quad (1)$$

Note that there is a natural filtration on $\mathcal{Q}_\lambda^{\mathrm{odd}}(\mathfrak{g})$ and a natural grading on $\mathcal{C}_\lambda^{\mathrm{odd}}(\mathfrak{g})$ coming from the filtration on $\mathrm{Cl}(\mathfrak{g}, K)$ and the grading on $\Lambda(\mathfrak{g})$, respectively. We have $\mathrm{gr}\, \mathcal{Q}_\lambda^{\mathrm{odd}}(\mathfrak{g}) = \mathcal{C}_\lambda^{\mathrm{odd}}(\mathfrak{g})$.

We hope that some important questions in the theory of semisimple Lie algebras and their representations can be studied using these family algebras.

One of these questions is the computation of odd exponents. Recall (see [Ko1]) that the algebra $S(\mathfrak{g})$ is a free module over the algebra $I(\mathfrak{g}) := S(\mathfrak{g})^G$ and the graded character of the isotypic component of type λ is

$$\chi_\lambda(q) = \frac{\sum_{k=1}^{K(\lambda)} q^{e_k(\lambda)}}{\prod_{i=1}^{\mathrm{rk}\,\mathfrak{g}} 1 - q^{e_i}},$$

where the numbers e_l are ordinary exponent of $\mathfrak{g}$ and $e_k(\lambda)$ are so-called generalized exponents of the simple module (π_λ, V_λ). It is also known that the number $K(\lambda)$ is equal to the multiplicity of zero weight in $Wt(\lambda)$.

In the odd situation the algebra $\Lambda(\mathfrak{g})$ is still a module over $I^{\mathrm{odd}}(\mathfrak{g}) = \Lambda(\mathfrak{g})^G$ but this module is no longer free. Nevertheless, every isotypic component of $\Lambda(\mathfrak{g})$ has a system of homogeneous generators of degrees $e_1^{\mathrm{odd}}(\lambda), \ldots, e_{L(\lambda)}^{\mathrm{odd}}(\lambda)$ which we call *odd exponents* corresponding to the simple module (π_λ, V_λ). In other words, the polynomial $\sum_{i=1}^{L(\lambda)} t^{e_i^{\mathrm{odd}}(\lambda)}$ is a Poincaré series for the graded $\mathfrak{g}$-module $\Lambda(\mathfrak{g})_\lambda / I^{\mathrm{odd}}(\mathfrak{g})_+ \Lambda(\mathfrak{g})_\lambda$, where $I^{\mathrm{odd}}(\mathfrak{g})_+ \subset I^{\mathrm{odd}}(\mathfrak{g})$ is the ideal consisting of all elements of positive degree. It turns out that the isotypic component of the adjoint representation is in many cases an "almost free" module over $I^{\mathrm{odd}}(\mathfrak{g})$, i.e., there is only one (very simple) relation. We will show this in Section 4.

2. The character of $\mathfrak{g}$-module $\Lambda(\mathfrak{g})$

2.1. The general formula

Let (π, V) be any finite-dimensional $\mathfrak{g}$-module and let $Wt(V)$ denote the multiset of weights of V. A simple combinatorial argument shows that the graded $\mathfrak{g}$-module $\Lambda(V)$ has the character

$$\chi_{\Lambda(V)}(q) = \prod_{\mu \in Wt(V)} (1 + q e^{\mu}). \tag{2}$$

This is the odd analogue of the fact that the character of the graded module $S(V)$ is

$$\chi_{S(V)}(q) = \prod_{\mu \in Wt(V)} (1 - q e^{\mu})^{-1}.$$

The right-hand side of (2) for the $\mathfrak{g}$-module $\mathfrak{g}$ can be rewritten as

$$\chi_{\Lambda(\mathfrak{g})}(q) = (1+q)^l \cdot \prod_{\alpha \in R_+} (1 + q e^{\alpha})(1 + q e^{-\alpha})$$

where $l = \mathrm{rk}\ g$ and R_+ is the set of positive roots for $\mathfrak{g}$.

Recall that the character of the irreducible $\mathfrak{g}$-module with the highest weight $\lambda \in P_+$ is

$$\chi_\lambda = \frac{E_{\lambda+\rho}}{E_\rho} \qquad \text{where} \qquad E_\lambda = \sum_{w \in W} (-1)^{l(w)} e^{w(\lambda)}.$$

We use also the identity

$$E_\rho = e^\rho \prod_{\alpha \in R_+} (1 - e^{-\alpha}).$$

Then the multiplicity of π_λ in $\Lambda(\mathfrak{g})$ is the coefficient of $e^{\lambda+\rho}$ in $\chi_{\Lambda(\mathfrak{g})} \cdot E_\rho$. The latter quantity can be written as

$$(1+q)^l e^\rho \prod_{\alpha>0} \left((1+qe^\alpha)(1+qe^{-\alpha})(1-e^{-\alpha})\right) \tag{3}$$

or

$$q^r (1+q)^l e^{3\rho} \prod_{\alpha>0} \left(1 - (q-1+q^{-1})e^{-\alpha} - (q-1+q^{-1})e^{-2\alpha} + e^{-3\alpha}\right), \tag{4}$$

where $l = \operatorname{rk} \mathfrak{g}$, $r = |R_+| = \frac{1}{2}(\dim \mathfrak{g} - \operatorname{rk} \mathfrak{g})$.

We derive from (4) the following.

Theorem 2. *The multiplicity of π_λ in $\Lambda(\mathfrak{g})$ depends on how the weight $2\rho - \lambda$ can be expressed as a linear combination of distinct positive roots with coefficients* 1, 2 *or* 3.

Namely, each linear combination which contains k_1 roots with the coefficient 1, k_2 roots with the coefficient 2 and k_3 roots with the coefficient 3, contributes to the multiplicity in question the summand $(-1)^{k_2+k_3} P_{k_1+k_2}(q)$ where

$$P_k(q) = q^r(1+q)^l(q-1+q^{-1})^k = q^{r-k}(1+q)^{l-k}(1+q^3)^k. \tag{5}$$

Proof. Consider the general term in the expansion of the product (4). To get it, we have to take from each parentesis in (4) one of four summands and multiply the chosen terms.

If we take k_0 times the zeroth summand, k_1 times the first summand, k_2 times the second summand and k_3 times the last summand (so that $k_0 + k_1 + k_2 + k_3 = r$), we get the product of the form

$$(q-1+q^{-1})^{k_1+k_2} \exp\left(-\sum_{i=1}^{k_1} \alpha_i - 2\sum_{i=1}^{k_2} \beta_i - 3\sum_{i=1}^{k_3} \gamma_i\right). \tag{6}$$

The corresponding term in (4) will be (6) multiplied by $q^r(1+q)^l e^{3\rho}$. Now the statement of the theorem becomes evident. □

E.g., the term $e^{2\rho}$ enters with coefficient $P_0(q) = q^r(1+q)^l$ since the zero weight has a unique expression $0 = 0$.

For the same reason, the terms $e^{2\rho-\alpha}$ for a simple root α enter with coefficient $P_1(q) = (1+q)^{l-1}(1+q^3)$.

Theorem 1 gives the method to compute the graded character of $\Lambda(\mathfrak{g})$ for any semisimple Lie algebra $\mathfrak{g}$. So, we have a tool to calculate the odd analogues of exponents for any weight λ between 0 and 2ρ. But this method is practical only for weights which are not far from the maximal weight 2ρ, or for algebras of low rank.

2.2. Example 1: the classical algebra $\mathbf{A}_n \simeq \mathfrak{sl}_{n+1}$

In the case of $\mathfrak{g} = \mathfrak{sl}_{n+1}$ we have $l = n$, $r = \frac{n(n+1)}{2}$. The explicit formulae for multiplicities of simple modules (π_λ, V_λ) in Λ^k are known for $n = 1, 2, 3$. We collect these results in the tables below.

Table 1. Case $g = \mathfrak{sl}_2$

λ	$d(\lambda)$	Λ^0	Λ^1	Λ^2	Λ^3
(0)	1	1	0	0	1
(2)	3	0	1	1	0

Table 2. Case $g = \mathfrak{sl}_3$

λ	$d(\lambda)$	Λ^0	Λ^1	Λ^2	Λ^3	Λ^4	Λ^5	Λ^6	Λ^7	Λ^8
(0,0)	1	1	0	0	1	0	1	0	0	1
(1,1)	8	0	1	1	1	2	1	1	1	0
(3,0)	10	0	0	1	1	0	1	1	0	0
(0,3)	10	0	0	1	1	0	1	1	0	0
(2,2)	27	0	0	0	1	2	1	0	0	0

Table 3. Case $g = \mathfrak{sl}_4$

λ	$d(\lambda)$	Λ^0	Λ^1	Λ^2	Λ^3	Λ^4	Λ^5	Λ^6	Λ^7	Λ^8	Λ^9	...
(0,0,0)	1	1	0	0	1	0	1	0	1	1	0	...
(1,0,1)	15	0	1	1	1	2	2	3	2	2	3	...
(0,1,2)	45	0	0	1	1	1	3	3	3	3	3	...
(2,1,0)	45	0	0	1	1	1	3	3	3	3	3	...
(0,2,0)	20	0	0	0	1	2	1	1	3	3	1	...
(2,0,2)	84	0	0	0	1	2	2	3	4	4	3	...
(1,2,1)	175	0	0	0	1	2	3	5	5	5	5	...
(0,0,4)	35	0	0	0	1	1	0	1	1	1	1	...
(4,0,0)	35	0	0	0	1	1	0	1	1	1	1	...
(0,4,0)	105	0	0	0	0	1	1	0	2	2	0	...
(1,1,3)	256	0	0	0	0	1	2	2	3	3	2	...
(3,1,1)	256	0	0	0	0	1	2	2	3	3	2	...
(3,0,3)	300	0	0	0	0	0	1	2	1	1	2	...
(2,3,0)	280	0	0	0	0	0	1	2	1	1	2	...
(0,3,2)	280	0	0	0	0	0	1	2	1	1	2	...
(2,2,2)	729	0	0	0	0	0	0	1	3	3	1	...

We list here the odd exponents of the adjoint representation:

For $\mathfrak{sl}_2$: 1, 2.
For $\mathfrak{sl}_3$: 1, 2, 3, 4.
For $\mathfrak{sl}_4$: 1, 2, 3, 4, 5, 6.

2.3. Example 2: The classical algebra $\mathbf{B}_2 \simeq \mathfrak{so}_5 \simeq \mathfrak{sp}_4 \simeq \mathbf{C}_2$

In this case we have the following table.

Table 4. Case $\mathfrak{g} = \mathfrak{so}_5 = \mathfrak{sp}_4$.

λ	$d(\lambda)$	Λ^0	Λ^1	Λ^2	Λ^3	Λ^4	Λ^5	Λ^6	Λ^7	Λ^8	Λ^9	Λ^{10}
(0,0)	1	1	0	0	1	0	0	0	0	0	0	1
(0,1)	5	0	0	0	1	1	0	1	1	0	0	0
(2,0)	10	0	1	1	0	1	2	1	0	1	1	0
(4,0)	35	0	0	0	1	1	0	1	1	0	0	0
(2,1)	35	0	0	1	1	1	2	1	1	1	0	0
(0,2)	14	0	0	0	1	1	0	1	1	0	0	0
(2,2)	81	0	0	0	0	1	2	1	0	0	0	0

The odd exponents of the adjoint representation are: 1, 2, 5, 6.

2.4. The exceptional algebra $\mathfrak{g}_2$

In this case we have the following table.

Table 5. Case $\mathfrak{g} = \mathfrak{g}_2$.

λ	$d(\lambda)$	Λ^0	Λ^1	Λ^2	Λ^3	Λ^4	Λ^5	Λ^6	Λ^7	Λ^8	Λ^9	...
(0,0)	1	1	0	0	1	0	0	0	0	0	0	...
(0,1)	7	0	0	0	0	0	1	1	0	1	1	...
(1,0)	14	0	1	1	0	1	1	0	0	0	1	...
(0,2)	27	0	0	0	1	1	0	1	2	1	0	...
(1,1)	64	0	0	0	0	1	1	1	2	1	1	...
(0,3)	77	0	0	1	1	0	2	2	0	2	2	...
(2,0)	77	0	0	0	1	1	0	1	2	1	0	...
(0,4)	182	0	0	0	1	1	0	1	2	1	0	...
(1,2)	189	0	0	0	0	1	2	2	2	2	2	...
(1,3)	448	0	0	0	0	1	1	1	2	1	1	...
(2,1)	286	0	0	0	0	0	1	1	0	1	1	...
(0,5)	378	0	0	0	0	0	1	1	0	1	1	...
(2,2)	729	0	0	0	0	0	0	1	2	1	0	...

The odd exponents of the adjoint representation are: 1, 2, 9, 10.

3. Structure of odd family algebras

3.1. Kostant decomposition of quantum odd family algebras

There is a distinguished $\mathfrak{g}$-module isomorphism between $\Lambda(\mathfrak{g})$ and $\mathrm{Cl}(\mathfrak{g}, K)$, namely, the anti-symmetrization map

$$\sigma : \Lambda(\mathfrak{g}) \to \mathrm{Cl}(\mathfrak{g}, K),$$

sending any element

$$x_1 \wedge \ldots \wedge x_n \in \Lambda(\mathfrak{g})$$

to the element

$$\frac{1}{n!} \sum_{s \in S_n} \mathrm{sign}(s) x_{s(1)} \cdot \ldots \cdot x_{s(n)} \in \mathrm{Cl}(\mathfrak{g}, K).$$

The subspace $\sigma(\Lambda^2(\mathfrak{g})) \subset \mathrm{Cl}(\mathfrak{g}, K)$ is closed with respect to the commutator operation on $\mathrm{Cl}(\mathfrak{g}, K)$. This subspace can be naturally identified with Lie algebra $\mathfrak{so}(\mathfrak{g}, K)$. Namely, for $h \in \sigma(\Lambda^2(\mathfrak{g})) \subset \mathrm{Cl}(\mathfrak{g}, K)$ the operator $\mathrm{ad}\, h$ acting on $\mathrm{Cl}(\mathfrak{g}, K)$ preserves $\sigma(\Lambda^1(\mathfrak{g})) = \mathfrak{g}$. The action of $\mathrm{ad}\, h$ on $\mathfrak{g}$ preserves the Killing form, so the map $h \mapsto \mathrm{ad}\, h$ induces a Lie algebra isomorphism $\sigma(\Lambda^2(\mathfrak{g})) \to \mathfrak{so}(\mathfrak{g}, K)$.

The adjoint representation defines a Lie algebra homomorphism

$$\varphi_0 : \mathfrak{g} \to \mathfrak{so}(\mathfrak{g}, K) = \sigma(\Lambda^2(\mathfrak{g})).$$

Let us write this homomorphism in coordinates. Let $\{x_i\}$ be an orthonormal basis of $\mathfrak{g}$ with respect to the Killing form, and let c_{ijk} be the structure constants of $\mathfrak{g}$ (i.e., $[x_i, x_j] = \sum_k c_{ijk} x_k$). Note that c_{ijk} is a totally antisymmetric tensor. Then

$$\varphi_0(x_i) = -\frac{1}{2} \sum_{j,k} \sigma(c_{ijk} x_j \wedge x_k) = \sum_{j<k} c_{ijk} x_k x_j. \tag{7}$$

This homomorphism extends to a filtered associative algebra homomorphism

$$\varphi : U(\mathfrak{g}) \to \mathrm{Cl}(\mathfrak{g}, K).$$

The following result is due to Kostant [Ko2].

Theorem 3 (Kostant Decomposition Theorem). *There is a $\mathfrak{g}$-equivariant isomorphism of associative algebras*

$$\mathrm{Cl}(\mathfrak{g}, K) \cong \mathrm{End}\, V_\rho \otimes \mathrm{Cl}(\mathfrak{g}, K)^G.$$

The image of the homomorphism φ_0 is $\mathrm{End}\, V_\rho \otimes 1$.

The following result can be easily deduced from the Kostant decomposition Theorem.

Proposition 1. *The quantum odd family algebra* $\mathcal{Q}_\lambda^{\mathrm{odd}}(\mathfrak{g})$ *decomposes as follows*

$$\mathcal{Q}_\lambda^{\mathrm{odd}}(\mathfrak{g}) \cong [\mathrm{End}(V_\lambda \otimes V_\rho)]^G \otimes \mathrm{Cl}(\mathfrak{g}, K)^G.$$

Note that the algebra $[\mathrm{End}(V_\lambda \otimes V_\rho)]^G$ is the image of the even quantum family algebra $\mathcal{Q}_\lambda^{\mathrm{even}}(\mathfrak{g}) = [\mathrm{End}(V_\lambda) \otimes U(\mathfrak{g})]^G$ under the homomorphism π_ρ sending $U(\mathfrak{g})$ to $\mathrm{End}\, V_\rho$.

Recall that the center $Z(\mathfrak{g})$ of the universal enveloping algebra $U(\mathfrak{g})$ is contained in $\mathcal{Q}_\lambda^{\mathrm{even}}(\mathfrak{g}) = [\mathrm{End}(V_\lambda) \otimes U(\mathfrak{g})]^G$ as the subalgebra of scalar matrices, i.e., matrices of the form $1_{V_\lambda} \otimes A$, $A \in Z(\mathfrak{g})$. Since V_ρ is irreducible, elements of $Z(\mathfrak{g})$ go to constants under the homomorphism $\pi_\rho : U(\mathfrak{g}) \to \mathrm{End}(V_\rho)$. Thus we have the following

Proposition 2. *There is a filtered algebra homomorphism*

$$\varphi = 1 \otimes \varphi_0 : \mathcal{Q}_\lambda^{\mathrm{even}}(\mathfrak{g}) \to \mathcal{Q}_\lambda^{\mathrm{odd}}(\mathfrak{g}),$$

which maps the elements of $Z(\mathfrak{g}) \subset \mathcal{Q}_\lambda^{\mathrm{even}}(\mathfrak{g})$ *to constants. The odd family algebra decomposes as follows:*

$$\mathcal{Q}_\lambda^{\mathrm{odd}}(\mathfrak{g}) \cong \varphi(\mathcal{Q}_\lambda^{\mathrm{even}}(\mathfrak{g})) \otimes \mathrm{Cl}(\mathfrak{g}, K)^G.$$

Since the homomorphism φ maps the elements of $Z(\mathfrak{g}) \subset \mathcal{Q}_\lambda^{\mathrm{even}}(\mathfrak{g})$ to constants, the associated graded homomorphism

$$\Phi = 1 \otimes \Phi_0 : \mathcal{C}_\lambda^{\mathrm{even}}(\mathfrak{g}) \to \mathcal{C}_\lambda^{\mathrm{odd}}(\mathfrak{g}),$$

maps the ideal $I_+^{\mathrm{even}}(\mathfrak{g})$, generated by all homogeneous elements of positive degree in $I^{\mathrm{even}}(\mathfrak{g}) \subset \mathcal{Q}_\lambda^{\mathrm{even}}(\mathfrak{g})$, to zero.

Corollary 1.

$$\begin{aligned}\dim\, \mathcal{C}_\lambda^{\mathrm{odd}}(\mathfrak{g}) &= \dim\, \mathcal{Q}_\lambda^{\mathrm{odd}}(\mathfrak{g}) = 2^{\mathrm{rk}\,\mathfrak{g}} \dim\, \varphi(\mathcal{Q}_\lambda^{\mathrm{even}}(\mathfrak{g}))\\ &= 2^{\mathrm{rk}\,\mathfrak{g}} \dim\, \Phi(\mathcal{C}_\lambda^{\mathrm{even}}(\mathfrak{g})) \le 2^{\mathrm{rk}\,\mathfrak{g}} \dim\, (\mathcal{C}_\lambda^{\mathrm{even}}(\mathfrak{g})/I_+^{\mathrm{even}}(\mathfrak{g})).\end{aligned}$$

Remark. Unfortunately, there is no decomposition of the classical odd family algebra into tensor product of $I^{\mathrm{odd}}(\mathfrak{g})$ and some complement subalgebra.

3.2. Differential structures on odd family algebras

We define a differential d_0 on $\Lambda(\mathfrak{g})$ as follows. Put $d_0(x) = \Phi_0(x)$ for $x \in \mathfrak{g} = \Lambda^1(\mathfrak{g})$ and extend d_0 to the whole algebra $\Lambda(\mathfrak{g})$ by the (super-) Leibniz rule. In coordinates we have

$$d_0 = -\frac{1}{2} c_{ijk} x_j \wedge x_k \partial_i, \tag{8}$$

where ∂_i denotes the operator of super-differentiation by x_i.

The cohomology of the differential d_0 is $I^{\mathrm{odd}}(\mathfrak{g}) = \Lambda(\mathfrak{g})^G$ (see e.g., [F]). It is well known that the algebra $I^{\mathrm{odd}}(\mathfrak{g}) = \Lambda(\mathfrak{g})^G$ of $\mathfrak{g}$-invariant elements in $\Lambda(\mathfrak{g})$ is the

Grassmann algebra generated by the elements $\xi_1, \ldots, \xi_l$ whose degrees are very simply related to the (even) exponents of the adjoint representation: $\deg \xi_i = 2e_i(\theta) + 1$. More precisely, let $I_1, \ldots, I_l$ be the homogeneous generators of the algebra $I^{\text{even}}(\mathfrak{g})$ (they have degrees $\deg I_i = e_i(\theta) + 1$). The elements I_i can be considered as symmetric elements of $\mathfrak{g}^{\otimes(e_i+1)}$. The elements ξ_i can be considered in the same way as antisymmetric elements of $\mathfrak{g}^{\otimes(2e_i+1)}$. Let $\text{Alt} : \mathfrak{g}^{\otimes(2e_i+1)} \twoheadrightarrow \Lambda^{2e_i+1}(\mathfrak{g})$ be the alternation operator. In other words, Alt is the unique $GL(\mathfrak{g})$-invariant projector $\mathfrak{g}^{\otimes(2e_i+1)} \twoheadrightarrow \Lambda^{2e_i+1}(\mathfrak{g})$. The following explicit formula is well known (see e.g., [Ko2], Theorem 64).

Theorem 4 (Trangression formula). $\xi_i = \text{Alt} \circ (d_0^{\otimes e_i} \otimes 1)(I_i)$.

One can define the differential d on $\mathcal{C}_\lambda^{\text{odd}}(\mathfrak{g}) = (\text{End}\, V_\lambda \otimes \Lambda(\mathfrak{g}))^G$ as the restriction of $1 \otimes d_0$. Since the differential d_0 is $\mathfrak{g}$-invariant, the definition is correct. The cohomology of this differential is, clearly, $1 \otimes I^{\text{odd}}(\mathfrak{g})$.

Proposition 3. $\Phi(\mathcal{C}_\lambda^{\text{even}}(\mathfrak{g})) \subset d(\mathcal{C}_\lambda^{\text{odd}}(\mathfrak{g})) \subset \mathcal{C}_\lambda^{\text{odd}}(\mathfrak{g})$. *In particular, the differential d is zero on $\Phi(\mathcal{C}_\lambda^{\text{even}}(\mathfrak{g}))$.*

Proof. Observe that for the generators of $S(\mathfrak{g})$ we have

$$\Phi_0(x) = d_0(x).$$

Therefore $\Phi_0(S(\mathfrak{g})) \subset d_0(\Lambda(\mathfrak{g}))$, and the proposition is proved. □

The quantum odd family algebra also has a differential structure. The differential d is defined as follows:

$$d = \text{ad}\,\gamma, \qquad \gamma = -\frac{1}{6}\left(1 \otimes \sum_{ijk} c_{ijk} x_i x_j x_k\right). \tag{9}$$

This quantum differential deforms the classical one and the homology of this differential is zero as well (see [AM]).

3.3. Some distinguished elements in family algebras

Recall that a special element M was introduced in [K1] both in quantum and classical family algebras. Such an element exists also in the odd analogues.

Namely let again $\{x_i\}$ be an orthonormal basis in $\mathfrak{g}$ with respect to the Killing form. Then the element

$$M_{\text{odd}} := \sum_i \pi_\lambda(x_i) \otimes x_i$$

belongs to $\mathcal{C}_\lambda^{\text{odd}}(\mathfrak{g})$ (resp. to $\mathcal{Q}_\lambda^{\text{odd}}(\mathfrak{g})$) if we interpret x_i as an element of $\Lambda(\mathfrak{g})$ (resp. as an element of $\text{Cl}(\mathfrak{g}, K)$).

There is the following relation between the elements M of the odd and even classical family algebras.

Proposition 4.

$$-M_{\text{odd}}^2 = d(M_{\text{odd}}) = \Phi(M_{\text{even}}). \tag{10}$$

Proof. The second equality is obvious. Let us prove the first.

$$M_{\text{odd}}^2 = \sum_i \pi_\lambda(x_i) \otimes x_i \cdot \sum_j \pi_\lambda(x_j) \otimes x_j = \frac{1}{2} \sum_{i,j} [\pi_\lambda(x_i), \pi_\lambda(x_j)] \otimes x_i \wedge x_j$$

$$= \frac{1}{2} \sum_{i,j,k} c_{ijk} \pi_\lambda(x_k) \otimes x_i \wedge x_j = - \sum_k \pi_\lambda(x_k) \otimes d_0(x_k) = -d(M_{\text{odd}}). \qquad \square$$

Remark. Note that the equation (10) is known as the *Maurer–Cartan equation*. It means that the operator $D := d + \operatorname{ad} M$ is also a differential on $\mathcal{C}_\lambda^{\text{odd}}(\mathfrak{g})$. This is a particular case of the general construction of a differential computing $\mathfrak{g}$-module-valued cohomology of $\mathfrak{g}$. The cohomology of D is the same as of d: $1 \otimes I^{\text{odd}}(\mathfrak{g})$ (see e.g., [F]).

Now let us recall the definition of more general elements of $\mathcal{C}_\lambda^{\text{even}}(\mathfrak{g})$ (resp. to $\mathcal{Q}_\lambda^{\text{even}}(\mathfrak{g})$) which correspond to the elements of $I(\mathfrak{g})$ in the classical case and of $Z(\mathfrak{g})$ in the quantum case. We call them *M-type elements*.

I. Classical case. Let again $\{x_i\}$ be an orthonormal basis in $\mathfrak{g}$ and $\{\partial_i\}$ be the corresponding partial derivatives. For any $P \in I(\mathfrak{g})$ we put

$$M_P = \sum_i \pi_\lambda(x_i) \otimes \partial_i(P).$$

An easy check shows that $M_P \in \mathcal{C}_\lambda^{\text{even}}(\mathfrak{g})$. For a quadratic invariant polynomial $C = \frac{1}{2}\sum_i x_i^2$ the element M_C coincides with M_{even}.

II. Quantum case. In this case we have no partial derivatives but we can use the Hopf structure on $U(\mathfrak{g})$ to define M-type elements in $\mathcal{Q}_\lambda^{\text{odd}}(\mathfrak{g})$.

Let $\Delta : U(\mathfrak{g}) \to U(\mathfrak{g}) \otimes U(\mathfrak{g})$ be the comultiplication homomorphism given by

$$\Delta(x) = x \otimes 1 + 1 \otimes x \qquad \text{for} \quad x \in \mathfrak{g}.$$

Define the homomorphism $\delta : U(\mathfrak{g}) \to \operatorname{Mat}_{d(\lambda)}(U(\mathfrak{g}))$ as the composition:

$$\delta : U(\mathfrak{g}) \overset{\Delta}{\to} U(\mathfrak{g}) \otimes U(\mathfrak{g}) \overset{\pi_\lambda \otimes 1}{\to} \operatorname{End} V_\lambda \otimes U(g) \cong \operatorname{Mat}_{d(\lambda)}(U(\mathfrak{g})).$$

Example. For the quadratic element $C = \frac{1}{2}\sum_i x_i^2 \in Z(\mathfrak{g})$ we have

$$\delta(C) = \pi_\lambda(C) \otimes 1 + M_{\text{even}} + 1 \otimes C.$$

We call the elements $\delta(A)$ where $A \in Z(\mathfrak{g})$ the *M-type elements* of the quantum family algebra. It is justified by the fact that $\operatorname{gr}(\delta(A) - 1 \otimes A) = M_{\operatorname{gr} A} \in \mathcal{C}_\lambda^{\text{even}}(\mathfrak{g})$.

Note that the quantum family algebra is the image of the bigger algebra $[U(\mathfrak{g}) \otimes U(\mathfrak{g})]^G$ of diagonal invariants in $U(\mathfrak{g}) \otimes U(\mathfrak{g})$ under the homomorphism $\pi_\lambda \otimes 1$. The algebra $[U(\mathfrak{g}) \otimes U(\mathfrak{g})]^G$ can be considered as the commutant of the algebra $\Delta(U(\mathfrak{g}))$ in $U(\mathfrak{g}) \otimes U(\mathfrak{g})$. For any $A \in Z(\mathfrak{g})$ we have $\Delta(A) \in [U(\mathfrak{g}) \otimes U(\mathfrak{g})]^G$. On the other hand, $\Delta(A) \in \Delta(U(\mathfrak{g}))$. This proves the following.

Proposition 5. *For any $A \in Z(\mathfrak{g})$ the element $\delta(A) = (\pi_\lambda \otimes 1) \circ \Delta(A)$ belongs to the center of $\mathcal{Q}_\lambda^{\text{even}}(\mathfrak{g})$.*

The next assertion follows directly from Propositions 2 and 5.

Proposition 6. *The images of all M-type elements of $\mathcal{Q}_\lambda^{\text{even}}(\mathfrak{g})$ under the homomorphism φ belong to the center of the algebra $\mathcal{Q}_\lambda^{\text{odd}}(\mathfrak{g})$.*

Since the classical family algebras are associated graded of the corresponding quantum family algebras, the classical analog of Proposition 6 is also true.

3.4. Odd family algebras and odd exponents

Let us consider the decomposition of the $\mathfrak{g}$-module End V_λ into irreducible components:

$$\text{End } V_\lambda = \bigoplus_{i=1}^{p} W_i, \qquad W_i \cong V_{\lambda_i} \text{ as } \mathfrak{g}\text{-modules.}$$

In [K1] the highest weights λ_i, $1 \leq i \leq p$ are called (as well as the corresponding irreducible representations $(\pi_{\lambda_i}, V_{\lambda_i})$) the *children* of λ.

When π_λ is self-dual (i.e., $\lambda = \lambda^*$), the space End $V_\lambda \cong V_\lambda \otimes V_\lambda$ naturally splits into the sum of the symmetric part $S^2 V_\lambda$ and the antisymmetric part $\Lambda^2 V_\lambda$. Correspondingly, the set of children splits into *girls* and *boys*.

Note that any $\lambda \in P_+$ has two "obligatory" children: the zero weight 0 (trivial representation) and the maximal root θ (adjoint representation). In a self-dual case 0 is a girl and θ is a boy.

We can construct an element of $C_\lambda^{\text{odd}}(\mathfrak{g})$ by filling the subspace $W_i \subset \text{End } V_\lambda$ by elements of some subspace $U \subset \Lambda(\mathfrak{g})$ which transforms according to the representation $\pi_{\lambda_i^*}$, dual to π_{λ_i}. Namely, let $A^{(i)}(U)$ be a nonzero element of the one-dimensional space $(W_i \otimes U)^G$. Clearly, the odd family algebra $C_\lambda^{\text{odd}}(\mathfrak{g})$ is spanned by the elements of the form $A^{(i)}(U)$.

Let $\{e_1^{(i)}, \ldots, e_{k_i}^{(i)}\}$ be the set of odd exponents for λ_i. To each $e_m^{(i)}$ we have a subspace $H_m^{(i)} \subset \Lambda^{e_m^{(i)}}(\mathfrak{g})_{\lambda^*} / I_+^{\text{odd}}(\mathfrak{g}) \Lambda^{e_m^{(i)}}(\mathfrak{g})_{\lambda^*}$, which transforms according to the representation $\pi_{\lambda_i^*}$, dual to π_{λ_i} (since for all m the $\mathfrak{g}$-module $\Lambda^m(\mathfrak{g})$ is self-dual, λ and λ^* have the same set of exponents).

Setting $U = H_m^{(i)}$ we can construct an element $A_m^{(i)}$ of $C_\lambda^{\text{odd}}(\mathfrak{g}) / I_+^{\text{odd}}$. The following obvious assertion is an odd analog of Proposition 1.3 in [K2].

Proposition 7. *The elements $A_m^{(i)}$, $1 \leq m \leq k_i$, $1 \leq i \leq p$, form a basis of the space $C_\lambda^{\text{odd}}(\mathfrak{g}) / I_+^{\text{odd}}(\mathfrak{g})$.*

4. Odd family algebras for standard representations of Classical Lie algebras (types A, B, C)

The following fact is useful for the further consideration.

Theorem 5 (Amitsur–Levitski). *Let M be an $n \times n$-matrix whose entries are odd elements of any supercommutative algebra. Then $M^{2n} = 0$.*

Proof. For $k = 1, 2, \ldots$, we have

$$\operatorname{tr} M^{2k} = \operatorname{tr}(M \cdot M^{2k-1}) = -\operatorname{tr}(M^{2k-1} \cdot M) = -\operatorname{tr} M^{2k}.$$

Therefore $\operatorname{tr} M^{2k} = 0$ for $k = 1, 2, \ldots$. Since the matrix M^2 has even entries, it satisfies the Hamilton–Cayley identity. Thus $M^{2n} = 0$. □

4.1. The case $\mathfrak{g} = \mathbf{A}_n = \mathfrak{sl}_{n+1}$

Consider the standard representation V_{ω_1} of $\mathfrak{sl}_{n+1}$.

It is known from [K1] that the even family algebra $C^{\mathrm{even}}_{\omega_1}(\mathfrak{sl}_{n+1})$ is a free $I^{\mathrm{even}}(\mathfrak{sl}_{n+1})$-module with the generators $1, M_{\mathrm{even}}, \ldots, M^n_{\mathrm{even}}$. This means that

$$\dim(C^{\mathrm{even}}_{\omega_1}(\mathfrak{g})/I^{\mathrm{even}}_+(\mathfrak{g})) = n + 1. \tag{11}$$

The algebra $I^{\mathrm{even}}(\mathfrak{sl}_{n+1})$ is generated by $I_k = \operatorname{tr} M^{k+1}_{\mathrm{even}}$, $k = 1, \ldots, n$. By the equation (10) and the transgression formula we obtain that the algebra $\Lambda(\mathfrak{sl}_{n+1})^G$ is a Grassmann algebra with the generators $\xi_k = \operatorname{tr} M^{2k+1}_{\mathrm{odd}}$, $k = 1, \ldots, n$. By the Amitsur–Levitski theorem we have $M^{2n+2}_{\mathrm{odd}} = 0$.

We now consider the algebra $C^{\mathrm{odd}}_{\omega_1}(\mathfrak{sl}_{n+1})$ as a $\Lambda(\xi_1, \ldots, \xi_{n-1})$-module, and let $B \subset C^{\mathrm{odd}}_{\omega_1}(\mathfrak{sl}_{n+1})$ be the $\Lambda(\xi_1, \ldots, \xi_{n-1})$-submodule generated by $1, M_{\mathrm{odd}}, M^2_{\mathrm{odd}}, \ldots, M^{2n+1}_{\mathrm{odd}}$.

Lemma 1. *B is a free $\Lambda(\xi_1, \ldots, \xi_{n-1})$-module.*

Proof. It suffices to check that $\xi_1 \wedge \cdots \wedge \xi_{n-1} M^{2n+1}_{\mathrm{odd}} \neq 0$. We have

$$\operatorname{tr} \xi_1 \wedge \cdots \wedge \xi_{n-1} M^{2n+1}_{\mathrm{odd}} = \xi_1 \wedge \cdots \wedge \xi_{n-1} \wedge \xi_n \neq 0.$$ □

Lemma 2. $B = C^{\mathrm{odd}}_{\omega_1}(\mathfrak{sl}_{n+1})$.

Proof. Indeed, $\dim C^{\mathrm{odd}}_{\omega_1}(\mathfrak{g}) \leq 2^n(n+1) = \dim B$. □

Corollary 2. *The algebra $C^{\mathrm{odd}}_{\omega_1}(\mathfrak{sl}_{n+1})$ is generated by the odd super-cummuting elements $\xi_1, \ldots, \xi_{n-1}$ of degrees $3, \ldots, 2n-1$, and the element M_{odd} of degree 1, with the defining relation $M^{2n+2}_{\mathrm{odd}} = 0$. In particular, the Poincaré series of $C^{\mathrm{odd}}_{\omega_1}(\mathfrak{g})$ is*

$$P_{C^{\mathrm{odd}}_{\omega_1}(\mathfrak{sl}_{n+1})}(q) = \frac{(1 - q^{2n+2})(1 + q^3) \ldots (1 + q^{2n-1})}{1 - q}.$$

The standard representation V_{ω_1} of $\mathfrak{sl}_{n+1}$ has two children: $\lambda_0 = 0$ (the trivial representation) and $\lambda_1 = \omega_1 + \omega_n$ (the adjoint representation). The space W_0 consists of scalar matrices and W_1 of traceless matrices. The Poincaré series for the trivial representation is $(1 + q^3) \ldots (1 + q^{2n+1})$.

Corollary 3. *The Poincaré series for the adjoint representation is*

$$P_{\mathrm{ad}}(q) = \frac{q(1 - q^{2n})(1 + q^3) \ldots (1 + q^{2n-1})}{1 - q}.$$

The odd exponents of the adjoint representation are $1, \ldots, 2n$.

4.2. The case $\mathfrak{g} = \mathbf{B}_n = \mathfrak{so}_{2n+1}$

The standard representation V_{ω_1} of $\mathfrak{so}_{2n+1}$ has dimension $2n+1$ and is orthogonal.

It is known from [K1] that the even family algebra $\mathcal{C}^{\text{even}}_{\omega_1}(\mathfrak{so}_{2n+1})$ is a free $I^{\text{even}}(\mathfrak{so}_{2n+1})$-module with generators $1, M_{\text{even}}, \dots, M^{2n}_{\text{even}}$. This means that

$$\dim(\mathcal{C}^{\text{even}}_{\omega_1}(\mathfrak{so}_{2n+1})/I^{\text{even}}_{+}(\mathfrak{so}_{2n+1})) = 2n+1. \tag{12}$$

The algebra $I^{\text{even}}(\mathfrak{so}_{2n+1})$ is generated by the elements $I_k = \operatorname{tr} M^{2k}_{\text{even}}$, $k = 1, \dots, n$. By the equation (10) and the transgression formula we obtain that the algebra $\Lambda(\mathfrak{so}_{2n+1})^G$ is a Grassmann algebra with the generators $\xi_k = \operatorname{tr} M^{4k-1}_{\text{odd}}$, $k = 1, \dots, n$.

Lemma 3. $\dim \varphi(\mathcal{Q}^{\text{even}}_{\omega_1}(\mathfrak{so}_{2n+1})) \leq 2n$.

Proof. Since the the representation V_{ω_1} has simple spectrum, the dimension of $\varphi(\mathcal{Q}^{\text{even}}_{\omega_1}(\mathfrak{so}_{2n+1})) = [\operatorname{End} V_{\omega_1} \otimes V_\rho]^G$ is the number of irreducible components of the representation $V_{\omega_1} \otimes V_\rho$. The highest weights of these components are $\rho + \lambda_i$, $\lambda_i \in Wt(\omega_1)$. Note that $-\alpha_n \in Wt(\omega_1)$ and the weight $\rho - \alpha_n$ is not dominant. Therefore, $\dim \varphi(\mathcal{Q}^{\text{even}}_{\omega_1}(\mathfrak{so}_{2n+1})) \leq \dim V_{\omega_1} - 1 = 2n$. □

Consider the algebra $\mathcal{C}^{\text{odd}}_{\omega_1}(\mathfrak{so}_{2n+1})$ as a $\Lambda(\xi_1, \dots, \xi_{n-1})$-module. Let $B \subset \mathcal{C}^{\text{odd}}_{\omega_1}(\mathfrak{so}_{2n+1})$ be the $\Lambda(\xi_1, \dots, \xi_{n-1})$-submodule generated by $1, M_{\text{odd}}, M^2_{\text{odd}}, \dots, M^{4n-1}_{\text{odd}}$.

The following fact are proved as above.

Lemma 4. *B is a free $\Lambda(\xi_1, \dots, \xi_{n-1})$-module.*

Lemma 5. $B = \mathcal{C}^{\text{odd}}_{\omega_1}(\mathfrak{g})$.

Corollary 4. *The algebra $\mathcal{C}^{\text{odd}}_{\omega_1}(\mathfrak{so}_{2n+1})$ is generated by the odd super-cummuting elements $\xi_1, \dots, \xi_{n-1}$ of degrees $3, \dots, 4n-5$, and the element M_{odd} of degree 1, with the defining relation of degree $4n$. In particular, the Poincaré series of $\mathcal{C}^{\text{odd}}_{\omega_1}(\mathfrak{so}_{2n+1})$ is*

$$P_{\mathcal{C}^{\text{odd}}_{\omega_1}(\mathfrak{so}_{2n+1})}(q) = \frac{(1-q^{4n})(1+q^3)\dots(1+q^{4n-1})}{1-q}.$$

The standard representation V_{ω_1} of $\mathbf{B}_n$ is orthogonal, and therefore the space $\operatorname{End}(V_{\omega_1})$ splits into symmetric and antisymmetric parts. There is one boy $\Lambda^2(V_{\omega_1})$ (isomorphic to the adjoint representation) and two girls: the space of scalar matrices E and the space of traceless symmetric matrices $S^2_0(V_{\omega_1})$.

The matrix $M^2_{\text{odd}} = -\Phi(M_{\text{even}})$ is antisymmetric and its k-th power is antisymmetric for odd k and symmetric for even k. Since $d(M_{\text{odd}}) = -M^2_{\text{odd}}$ we have $d(M^{2k}_{\text{odd}}) = 0$ and $d(M^{2k-1}_{\text{odd}}) = -M^{2k}_{\text{odd}}$. Since the cohomology of the differential d is $I^{\text{odd}}(\mathfrak{g})$, nontrivial irreducible components of M^{2k}_{odd} and M^{2k-1}_{odd} are the same.

Corollary 5. *The Poincaré series for the adjoint representation of $\mathfrak{so}_{2n+1}$ is*

$$P_{\text{ad}}(q) = \frac{q(1-q^{4n-4})(1+q)(1+q^3)\dots(1+q^{4n-5})}{1-q^4}.$$

The odd exponents of the adjoint representation are

$$1,\ 2,\ 5,\ 6, \dots,\ 4n-3,\ 4n-2.$$

The Poincaré series for the representation $S_0(V_{\omega_1})$ *of* $\mathfrak{so}_{2n+1}$ *is*

$$P_{S_0^2}(q) = \frac{q^3(1-q^{4n-8})(1+q)(1+q^3)\dots(1+q^{4n-5})}{1-q^4}.$$

The odd exponents of the representation $S_0^2(V_{\omega_1})$ *of* $\mathfrak{so}_{2n+1}$ *are*

$$3,\ 4,\ 7,\ 8, \dots,\ 4n-5,\ 4n-4.$$

4.3. The case $\mathfrak{g} = \mathbf{C}_n = \mathfrak{sp}_{2n}$

The standard representation V_{ω_1} of $\mathfrak{sp}_{2n}$ has dimension $2n$ and is symplectic.

It is known from [K1] that the even family algebra $C_{\omega_1}^{\text{even}}(\mathfrak{sp}_{2n})$ is a free $I^{\text{even}}(\mathfrak{sp}_{2n})$-module with the generators $1,\ M_{\text{even}}, \dots,\ M_{\text{even}}^{2n-1}$. This means that

$$\dim(C_{\omega_1}^{\text{even}}(\mathfrak{sp}_{2n})/I_+^{\text{even}}(\mathfrak{g})) = 2n. \tag{13}$$

The algebra $I^{\text{even}}(\mathfrak{sp}_{2n})$ is generated by $I_k = \operatorname{tr} M_{\text{even}}^{2k}$, $k = 1, \dots, n$. By the equation (10) and the transgression formula we obtain that the algebra $\Lambda(\mathfrak{sp}_{2n})^G$ is a Grassmann algebra with the generators $\xi_k = \operatorname{tr} M_{\text{odd}}^{4k-1}$, $k = 1, \dots, n$. By the Amitsur–Levitski theorem we have $M_{\text{odd}}^{4n} = 0$.

Let us now consider the algebra $C_{\omega_1}^{\text{odd}}(\mathfrak{sp}_{2n})$ as a $\Lambda(\xi_1, \dots, \xi_{n-1})$-module and let $B \subset C_{\omega_1}^{\text{odd}}(\mathfrak{sp}_{2n})$ be the $\Lambda(\xi_1, \dots, \xi_{n-1})$-submodule generated by $1,\ M_{\text{odd}},\ M_{\text{odd}}^2, \dots,\ M_{\text{odd}}^{4n-1}$.

The following statements can be proved exactly as in the A_n case.

Lemma 6. *B is a free $\Lambda(\xi_1, \dots, \xi_{n-1})$-module.*

Lemma 7. $B = C_{\omega_1}^{\text{odd}}(\mathfrak{g})$.

Corollary 6. *The algebra $C_{\omega_1}^{\text{odd}}(\mathfrak{g})$ is generated by the odd super-cummuting elements $\xi_1, \dots, \xi_{n-1}$ of degrees $3, \dots, 4n-5$, and the element M_{odd} of degree 1, with the defining relation $M_{\text{odd}}^{4n} = 0$. In particular, the Poincaré series of $C_{\omega_1}^{\text{odd}}(\mathfrak{g})$ is*

$$P_{C_{\omega_1}^{\text{odd}}(\mathfrak{g})}(q) = \frac{(1-q^{4n})(1+q^3)\dots(1+q^{4n-5})}{1-q}.$$

The standard representation V_{ω_1} of $\mathbf{C}_n$ is symplectic, and therefore the space $\operatorname{End}(V_{\omega_1})$ splits into symmetric and antisymmetric parts. There is one girl $S^2(V_{\omega_1})$ (isomorphic to the adjoint representation) and two boys: the trivial representation E and its complement $\Lambda_0^2(V_{\omega_1})$ in $\Lambda^2(V_{\omega_1})$.

The matrix $M_{\text{odd}}^2 = -\Phi(M_{\text{even}})$ is antisymmetric and its k-th power is antisymmetric for odd k and symmetric for even k. Since $d(M_{\text{odd}}) = -M_{\text{odd}}^2$ we have $d(M_{\text{odd}}^{2k}) = 0$ and $d(M_{\text{odd}}^{2k-1}) = -M_{\text{odd}}^{2k}$. Since the cohomology of the differential d is $I^{\text{odd}}(\mathfrak{g})$, non-trivial irreducible components of M_{odd}^{2k} and M_{odd}^{2k-1} are the same.

Corollary 7. *The Poincaré series for the adjoint representation is*

$$P_{\mathrm{ad}}(q) = \frac{q(1-q^{4n-4})(1+q)(1+q^3)\dots(1+q^{4n-5})}{1-q^4}.$$

The odd exponents of the adjoint representation are

$$1,\ 2,\ 5,\ 6,\dots,\ 4n-3,\ 4n-2.$$

The Poincaré series for the representation $\Lambda_0^2(V_{\omega_1})$ is

$$P_{\Lambda_0^2}(q) = \frac{q^3(1-q^{4n-8})(1+q)(1+q^3)\dots(1+q^{4n-5})}{1-q^4}.$$

The odd exponents of the representation $\Lambda_0^2(V_{\omega_1})$ are

$$3,\ 4,\ 7,\ 8,\dots,\ 4n-5,\ 4n-4.$$

5. Other examples

5.1. The case of a 7-dimensional representation of $\mathfrak{g} = \mathbf{G}_2$

The representation V_{ω_2} of the exceptional Lie algebra $\mathbf{G}_2$ has dimension 7. This representation has 4 children: the trivial representation, the 7-dimensional representation ω_2, the 14-dimensional representation $\theta = \omega_1$, and the 27-dimensional representation $2\omega_2$. Let us denote by p_1, p_7, p_{14}, and p_{27} the corresponding projectors. We compute here the algebra $C^{\mathrm{odd}}_{\omega_2}(\mathbf{G}_2)$ and the odd exponents for the children of ω_2, using the following result concerning the even family algebra $C^{\mathrm{even}}_{\omega_2}(\mathbf{G}_2)$.

Lemma 8 (Theorem G in [K1]).

1. *The algebra $I^{\mathrm{even}}(\mathbf{G}_2)$ is generated by elements* $\operatorname{tr} M^2_{\mathrm{even}}$, $\operatorname{tr} M^6_{\mathrm{even}}$.
2. *The $I^{\mathrm{even}}(\mathbf{G}_2)$-module $p_{27}(C^{\mathrm{even}}_{\omega_2}(\mathfrak{g}))$ is spanned by elements $p_{27}(M^{2k}_{\mathrm{even}})$ for $k = 1, 2, 3$.*
3. *The $I^{\mathrm{even}}(\mathbf{G}_2)$-module $p_{14}(C^{\mathrm{even}}_{\omega_2}(\mathbf{G}_2))$ is spanned by elements $p_{14}(M^k_{\mathrm{even}})$ for $k = 1, 5$.*
4. *The $I^{\mathrm{even}}(\mathbf{G}_2)$-module $p_7(C^{\mathrm{even}}_{\omega_2}(\mathfrak{g}))$ is spanned by $p_7(M^3_{\mathrm{even}})$.*

By (10) and the transgression formula we obtain that the algebra $\Lambda(\mathbf{G}_2)^G$ is a Grassmann algebra with the generators $\xi_1 = \operatorname{tr} M^3_{\mathrm{odd}}$, $\xi_2 = \operatorname{tr} M^{11}_{\mathrm{odd}}$. Since $p_{27}(\Lambda^{12}(\mathbf{G}_2)) \simeq p_{27}(\Lambda^2(\mathfrak{g})) = 0$ (see Table 5), we have $M^{12}_{\mathrm{odd}} = \Phi(M^6_{\mathrm{even}}) = 0$. Therefore $\dim \Phi(C^{\mathrm{even}}_{\omega_2}) \le 6$.

Consider the algebra $C^{\mathrm{odd}}_{\omega_2}(\mathbf{G}_2)$ as a $\Lambda(\xi_1)$-module. Let $B \subset C^{\mathrm{odd}}_{\omega_2}(\mathbf{G}_2)$ be the $\Lambda(\xi_1)$-submodule generated by $1, M_{\mathrm{odd}}, M^2_{\mathrm{odd}}, \dots, M^{11}_{\mathrm{odd}}$.

Lemma 9. *B is a free $\Lambda(\xi_1)$-module.*

Proof. It suffices to check that $\xi_1 M_{\rm odd}^{11} \neq 0$. We have

$$\operatorname{tr} \xi_1 M_{\rm odd}^{11} = \xi_1 \wedge \xi_2 \neq 0. \qquad \square$$

Lemma 10. $B = C_{\omega_2}^{\rm odd}(\mathbf{G}_2)$.

Proof. Indeed, $\dim C_{\omega_2}^{\rm odd}(\mathbf{G}_2) \leq 24 = \dim B$. $\square$

Corollary 8. *The algebra $C_{\omega_2}^{\rm odd}(\mathbf{G}_2)$ is generated by ξ_1 of degree 3, and the element $M_{\rm odd}$ of degree 1, with the defining relation $M_{\rm odd}^{12} = 0$. In particular, the Poincaré series of $C_{\omega_2}^{\rm odd}(\mathbf{G}_2)$ is*

$$P_{C_{\omega_2}^{\rm odd}(\mathfrak{g})}(t) = \frac{(1-t^{12})(1+t^3)}{1-t}.$$

Corollary 9. 1. *The Poincaré series for the representation V_{ω_2} is*

$$P_{\omega_2}(q) = q^5(1+q)(1+q^3).$$

The odd exponents of this representation are 5, 6.

2. *The Poincaré series for the representation V_{ω_1} is*

$$P_{\omega_1}(q) = q(1+q^9)(1+q)(1+q^3).$$

The odd exponents of this representation are 1, 2, 9, 10.

3. *The Poincaré series for the representation $V_{2\omega_2}$ is*

$$P_{2\omega_2}(q) = q^3(1+q^4)(1+q)(1+q^3).$$

The odd exponents of this representation are 3, 4, 7, 8.

References

[AM] A. Alekseev and E. Meinrenken, The non-commutative Weil algebra, *Invent. Math.*, **139** (2000), No 1, 135–172.

[F] D. B. Fuchs, *Cohomology of infinite-dimensional Lie algebras*, Translated from the Russian by A. B. Sosinskii. Contemporary Soviet Mathematics. Consultants Bureau, New York, 1986, xii+328 pp.

[K1] A. A. Kirillov, Family algebras, Electronic Research Announcements AMS, **6** (2000), No 1.

[K2] A. A. Kirillov, Introduction to family algebras, *Moscow Math. J.*, **1** (2001), No 1, 49–63.

[Ko1] B. Kostant, Lie group representations on polynomial ring, *Amer. J. Math.*, **85** (1963), 327–404.

[Ko2] B. Kostant, Clifford algebra analogue of the Hopf–Kozul–Samelson theorem, the ρ-decomposition $C(\mathfrak{g}) = \operatorname{End} V_\rho \otimes C(P)$, and the $\mathfrak{g}$-module structure of $\Lambda\mathfrak{g}$, *Adv. Math.*, **125** (1997), 275–350.

[OV] A. Onishchik and E. Vinberg, *A seminar on Lie groups and algebraic groups*, Nauka, Moscow, 1988, 1995 (second edition) 44 pp, In Russian. English translation: Series in Soviet Mathematics, Springer-Verlag, 1990, xx+328 pp.

Gelfand–Zeitlin theory from the perspective of classical mechanics. I

Bertram Kostant[1]* and Nolan Wallach[2]†

[1] Dept. of Math.
MIT
Cambridge, MA 02139
USA
kostant@math.mit.edu

[2] Dept. of Math.
UCSD
San Diego, CA 92093
USA
nwallach@ucsd.edu

With admiration,
To a dear friend and brilliant colleague.
B.K.

To Tony on his 60th birthday:
We honor your past accomplishments
and anticipate your future successes.
N.W.

Subject Classifications: Primary: 14L30, 14R20, 33C45, 53D17

Summary. Let $M(n)$ be the algebra (both Lie and associative) of $n \times n$ matrices over $\mathbb{C}$. Then $M(n)$ inherits a Poisson structure from its dual using the bilinear form $(x, y) = -\mathrm{tr}\, xy$. The $Gl(n)$ adjoint orbits are the symplectic leaves and the algebra, $P(n)$, of polynomial functions on $M(n)$ is a Poisson algebra. In particular, if $f \in P(n)$, then there is a corresponding vector field ξ_f on $M(n)$. If $m \leq n$, then $M(m)$ embeds as a Lie subalgebra of $M(n)$ (upper left hand block) and $P(m)$ embeds as a Poisson subalgebra of $P(n)$. Then, as an analogue of the Gelfand–Zeitlin algebra in the enveloping algebra of $M(n)$, let $J(n)$ be the subalgebra of $P(n)$ generated by $P(m)^{Gl(m)}$ for $m = 1, \ldots, n$. One observes that

$$J(n) \cong P(1)^{Gl(1)} \otimes \cdots \otimes P(n)^{Gl(n)}.$$

* Research supported in part by NSF grant DMS-0209473 and in part by the KG & G Foundation.

† Research supported in part by NSF grant DMS-0200305.

We prove that $J(n)$ is a maximal Poisson commutative subalgebra of $P(n)$ and that for any $p \in J(n)$ the holomorphic vector field ξ_p is integrable and generates a global one-parameter group $\sigma_p(z)$ of holomorphic transformations of $M(n)$. If $d(n) = n(n+1)/2$, then $J(n)$ is a polynomial ring $\mathbb{C}[p_1, \ldots, p_{d(n)}]$ and the vector fields ξ_{p_i}, $i = 1, \ldots, d(n-1)$, span a commutative Lie algebra of dimension $d(n-1)$. Let A be a corresponding simply-connected Lie group so that $A \cong \mathbb{C}^{d(n-1)}$. Then A operates on $M(n)$ by an action σ so that if $a \in A$, then

$$\sigma(a) = \sigma_{p_1}(z_1) \cdots \sigma_{p_{d(n-1)}}(z_{d(n-1)})$$

where a is the product of exp $z_i \xi_{p_i}$ for $i = 1, \ldots, d(n-1)$. We prove that the orbits of A are independent of the choice of the generators p_i. Furthermore, for any matrix the orbit $A \cdot x$ may be explicitly given in terms of the adjoint action of a $n-1$ abelian groups determined by x. In addition we prove the following results about this rather remarkable group action.

(1) Let $x \in M(n)$. Then $A \cdot x$ is an orbit of maximal dimension $(d(n-1))$ if and only if the differentials $(dp_i)_x$, $i = 1, \ldots, d(n)$, are linearly independent.
(2) The orbits, O_x, of the adjoint action of $Gl(n)$ on $M(n)$ are A-stable, and if O_x is an orbit of maximal dimension $(n(n-1))$, that is, if x is regular, then the A-orbits of dimension $d(n-1)$ in O_x are the leaves of a polarization of a Zariski open dense subset of the symplectic manifold O_x.
The results of the paper are related to the theory of orthogonal polynomials. Motivated by the interlacing property of the zeros of neighboring orthogonal polynomials on $\mathbb{R}$, we introduce a certain Zariski open subset $M_\Omega(n)$ of $M(n)$ and prove
(3) $M_\Omega(n)$ has the structure of $(\mathbb{C}^\times)^{d(n-1)}$ bundle over a ($d(n)$-dimensional) variety of Hessenberg matrices. Moreover, the fibers are maximal A-orbits. The variety of Hessenberg matrices plays a major role in this paper.

In Part II of this two-part paper, we deal with a commutative analogue of the Gelfand–Kirillov theorem. The fibration in (3) leads to the construction of n^2+1 functions (including a constant function) in an algebraic extension of the function field of $M(n)$ which, under Poisson bracket, satisfies the commutation relations of the direct sum of a $2d(n-1)+1$ dimensional Heisenberg Lie algebra and an n-dimensional commutative Lie algebra.

0. Introduction

0.1.

Let $M(n)$, for any positive integer n, denote the Lie (and asssociative) algebra of all complex $n \times n$ complex matrices. Let $P(n)$ be the graded commutative algebra of all polynomial functions on $M(n)$. The symmetric algebra over $M(n)$, as one knows, is a Poisson algebra. Using the bilinear form $(x, y) = -\mathrm{tr}\, xy$ on $M(n)$ this may be carried over to $P(n)$, defining on $P(n)$ the structure of a Poisson algebra and hence the structure of a Poisson manifold on $M(n)$. Consequently, to each $p \in P(n)$ there is associated a holomorphic vector field ξ_p on $M(n)$ such that $\xi_p q = [p, q]$ where $q \in P(n)$ and $[p, q]$ is Poisson bracket.

For any positive integer k, put $d(k) = k(k+1)/2$ and let I_k be the set $\{1, \ldots, k\}$. If $m \in I_n$ we will regard $M(m)$ (upper left hand $m \times m$ corner) as a Lie subalgebra of $M(n)$. As a "classical mechanics" analogue to the Gelfand–Zeitlin commutative subalgebra of the univeral enveloping algebra of $M(n)$, let $J(n)$ be the subalgebra of $P(n)$ generated by $P(m)^{Gl(m)}$ for all $m \in I_n$. Then

$$J(n) = P(1)^{Gl(1)} \otimes \cdots \times P(n)^{Gl(n)}.$$

In addition, $J(n)$ is a Poisson commutative polynomial subalgebra of $P(n)$ with $d(n)$ generators. In fact we can write $J(n) = \mathbb{C}[p_1, \ldots, p_{d(n)}]$ where, for $x \in M(n)$, $p_i(x), i \in I_{d(n)}$, "run over" the elementary symmetric functions of the roots of the characteristic polynomial of x_m, $m \in I_n$. Here and throughout, $x_m \in M(m)$ is the upper left $m \times m$ minor of x. The algebraic morphism

$$\Phi_n : M(n) \to \mathbb{C}^{d(n)} \text{ where } \Phi_n(x) = (p_1(x), \ldots, p_{d(n)}(x)) \tag{0.1}$$

plays a key role in this paper. Let $\mathfrak{b}_e$ be the $d(n)$-dimensional affine space of all $x \in M(n)$ of the form

$$x = \begin{pmatrix} a_{1\,1} & a_{1\,2} & \cdots & a_{1\,n-1} & a_{1\,n} \\ 1 & a_{2\,2} & \cdots & a_{2\,n-1} & a_{2\,n} \\ 0 & 1 & \cdots & a_{3\,n-1} & a_{3\,n} \\ \vdots & \vdots & \ddots & \vdots & \vdots \\ 0 & 0 & \cdots & 1 & a_{n\,n} \end{pmatrix} \tag{0.2}$$

where $a_{i\,j} \in \mathbb{C}$ are arbitrary. Elements $x \in M(n)$ of the form (0.2) are called Hessenberg matrices. As a generalization of classical facts about companion matrices we prove

Theorem 0.1. *The restriction*

$$\mathfrak{b}_e \to \mathbb{C}^{d(n)} \tag{0.3}$$

of the map Φ_n is an algebraic isomorphism.

See Theorem 2.3 and Remark 2.4. The real and imaginary parts of a complex number define a lexicographical order in $\mathbb{C}$. For any $x \in M(n)$ and $m \in I_n$, let $E_x(m) = \{\mu_{1\,m}(x), \ldots, \mu_{m\,m}(x)\}$ be the (increasing) ordered m-tuple of eigenvalues of x_m, with the multiplicity as roots of the characteristic polynomial. As a corollary of Theorem 0.1, one has the following independence (with respect to m) of the eigenvalue sequences $E_x(m)$.

Theorem 0.2. *For all $m \in I_n$ let $E(m) = \{\mu_{1\,m}, \ldots, \mu_{m\,m}\}$ be an arbitrary m-tuple with values in $\mathbb{C}$. Then there exists a unique $x \in \mathfrak{b}_e$ such that $E(m) = E_x(m)$, up to ordering, for all $m \in I_n$.*

See Theorem 2.5. For any $c \in \mathbb{C}^{d(n)}$ let $M_c(n) = \Phi_n^{-1}(c)$ be the "fiber" of Φ_n over c. If $x, y \in M(n)$, then x and y lie in the same fiber if and only if $E_x(m) = E_y(m)$ for all $m \in I_n$.

Remark 0.3. Theorem 0.1 implies that Φ_n is surjective and asserts that $\mathfrak{b}_e$ is a cross-section of Φ_n. That is, to any $c \in \mathbb{C}^{d(n)}$ the intersection $M_c(n) \cap \mathfrak{b}_e$ consists of exactly one matrix.

0.2.

One of the main results of the present paper, Part I, of a two-part paper, concerns the properties of a complex analytic abelian group A of dimension $d(n-1)$ that operates on $M(n)$. One has

Theorem 0.4. *The algebra $J(n)$ is a maximal Poisson commutative subalgebra of $P(n)$. Furthermore the vector field ξ_p, for any $p \in J(n)$, is globally integrable on $M(n)$, defining an analytic action of $\mathbb{C}$ on $M(n)$. Moreover, the fiber $M_c(n)$ is stable under this action, for any $c \in \mathbb{C}^{d(n)}$.*

See Theorem 3.4, Proposition 3.5 and Theorem 3.25. The generators p_i of $J(n)$ are replaced by a more convenient set of $d(n)$ generators $p_{(i)}$, $i \in I_{d(n)}$. See Section 3.1 and (3.20). The span of $\xi_{p_{(i)}}$, $i \in I_{d(n)}$, is a commutative $d(n-1)$-dimensional Lie algebra of analytic vector fields on $M(n)$. The Lie algebra $\mathfrak{a}$ integrates to an action of a complex analytic group $A \cong \mathbb{C}^{d(n-1)}$ on $M(n)$. In a sense A is a very extensive enlargement of a group, for the case where $\mathbb{R}$ replaces $\mathbb{C}$, introduced in §4 of [GS]. However, no diagonalizability, compactness or eigenvalue interlacing is required for the existence of A. In the complex setting the second statement of Theorem 0.4 and the existence of the action of A can be deduced from an iteration of Theorem 4.1 in [Kn]. The proof given in this paper is independent of the theory supporting Theorem 4.1 in [Kn] and leads to an explicit description of an arbitrary orbit $A \cdot x$ is terms of the adjoint action of $n-1$ abelian groups defined by $x \in M(n)$. (See Theorem 0.6) below.

Now let

$$x \mapsto V_x \tag{0.4}$$

be the nonconstant dimensional tangent space "distribution" on $M(n)$ defined by putting $V_x = \{(\xi_p)_x \mid p \in J(n)\}$. Let $D(n)$ be the group of all diagonal matrices in $Gl(n)$ and let $\operatorname{Ad} D(n)$ be the group of automorphisms of $M(n)$ defined by the adjoint action of $D(n)$ on $M(n)$.

Theorem 0.5. *The orbits of A are leaves of the distribution (0.4). Furthermore $M_c(n)$ is stable under the action of A, for any $c \in \mathbb{C}^{d(n)}$, so that $M_c(n)$ is a union of A-orbits. Finally, the abelian $d(n-1)$-dimensional group of analytic isomorphisms of $M(n)$ (as an analytic manifold) defined by A, contains* $\operatorname{Ad} D(n)$ *as an $n-1$-dimensional subgroup.*

See Theorem 3.4 and Proposition 3.5. For the last statement in the Theorem 0.5, see Theorem 3.27.

For any $x \in M(n)$ and $m \in I_n$, let $Z_{x,m} \subset M(m)$ be the (obviously commutative) associative subalgebra generated by x_m and the identity of $M(m)$. Let $G_{x,m} \subset Gl(m)$ be the commutative algebraic subgroup of $Gl(n)$ corresponding to $Z_{x,m}$ when the latter is regarded as a Lie algebra. The orbits of A are described in

Theorem 0.6. *Let $x \in M(n)$. Consider the following morphism of nonsingular irreducible affine varieties*

$$G_{x,1} \times \cdots \times G_{x,n-1} \to M(n) \tag{0.5}$$

where for $g(m) \in G_{x,m}$, $m \in I_{n-1}$,

$$(g(1), \ldots, g(n-1)) \mapsto \mathrm{Ad}\,(g(1) \cdots g(n-1))(x). \tag{0.6}$$

Then the image of (0.5) is exactly the A-orbit $A \cdot x$. *In particular* $A \cdot x$ *is an irreducible, constructible (in the sense of Chevalley) subset of* $M(n)$. *Furthermore the Zariski closure of* $A \cdot x$ *is the same as its closure in the Euclidean topology. In addition* $A \cdot x$ *contains a Zariski open subset of its closure.*

See Theorem 3.6. One should note that even though the groups $G_{x,m}$, $m \in I_n$, are commutative (and A is commutative) they do not commute with each other, so the order in (0.6) is important.

Let $x \in M(n)$. Then, by definition, x is regular (in $M(n)$) if the $Gl(n)$ adjoint orbit of x is of maximal dimension, $2\,d(n)$. One knows that x is regular if and only if $\dim Z_{x,n} = n$ or if and only if the differentials $(dp_{d(n-1)+k})_x$, $k \in I_n$, are linearly independent where the indexing of the p_i is such that the ring of invariants, $P(m)^{Gl(m)}$, is generated by $p_{d(m-1)+k}$, $k \in I_m$. We will now say that x is strongly regular if $(dp_i)_x$, $i \in I_{d(n)}$, are linearly independent.

One has that $\dim A \cdot x \leq d(n-1)$.

Theorem 0.7. *Let* $x \in M(n)$. *Then the following conditions are equivalent:*

(a) x *is strongly regular,*
(b) $A \cdot x$ *is an orbit of maximal dimension,* $d(n-1)$,
(c) $\dim Z_{x,m} = m$, $\forall m \in I_n$, *and* $Z_{x,m} \cap Z_{x,m+1} = 0$, $\forall m \in I_{n-1}$.

See Theorem 2.14 and (3.29).

Let $M^{\mathrm{sreg}}(n) \subset M(n)$ be the Zariski open set of all strongly regular matrices. Note that $M^{\mathrm{sreg}}(n)$ is not empty since in fact $\mathfrak{b}_e \subset M^{\mathrm{sreg}}(n)$. Theorem 0.6 for the case where $x \in M^{\mathrm{sreg}}(n)$ is especially nice.

Theorem 0.8. *Let* $x \in M^{\mathrm{sreg}}(n)$. *Then the morphism (0.5) is an algebraic isomorphism onto its image, the maximal orbit* $A \cdot x$. *In particular* $A \cdot x$ *is a nonsingular variety and as such*

$$A \cdot x \cong G_{x,1} \times \cdots \times G_{x,n-1}. \tag{0.7}$$

See Theorem 3.14.

0.3.

Let $x \in M(n)$. Motivated by the Jacobi matrices that arise in the theory of orthogonal polynomials on $\mathbb{R}$, we will say that x satisfies the eigenvalue disjointness condition if, for any $m \in I_n$, the eigenvalues of x_m have multiplicity one (so that x_m is regular semisimple in $M(m)$) and, as a set, $E_x(m) \cap E_x(m+1) = \emptyset$ for any $m \in I_{n-1}$. Let $M_\Omega(n)$ be the dense Zariski open set of such $x \in M(n)$. One readily has that $M_\Omega(n) = \Phi_n^{-1}(\Omega(n))$ where $\Omega(n)$ is a dense Zariski open set in $\mathbb{C}^{d(n)}$.

Theorem 0.9. *One has $M_{\Omega}(n) \subset M^{\text{sreg}}(n)$. In fact, if $c \in \Omega(n)$, then the entire fiber $M_c(n)$ is a single maximal A-orbit. Moreover, if $c \in \Omega(n)$ and $x \in M_c(n)$, then $G_{x,m}$ is a maximal (complex) torus in $Gl(m)$, for any $m \in I_{n-1}$, so that $M_c(n) = A \cdot x$ is a closed nonsingular subvariety of $M(n)$ and as such*

$$M_c(n) \cong (\mathbb{C}^\times)^{d(n-1)}. \tag{0.8}$$

See (2.55) and Theorem 3.23. Let $\mathfrak{u}$ be the $d(n-1)$-dimensional nilpotent Lie algebra of all strictly upper triangular matrices. One has a natural projection

$$M(n) \to \mathfrak{u}, \qquad x \mapsto x_{\mathfrak{u}}, \tag{0.9}$$

where $x_{\mathfrak{u}}$ is such that $x - x_{\mathfrak{u}}$ is lower triangular. Another "snapshot" picture of $M_c(n)$, when $c \in \Omega(n)$, is given in

Theorem 0.10. *Let $c \in \Omega(n)$ and let $\mathfrak{u}_c$ be the image of $M_c(n)$ by the projection (0.9). Then $\mathfrak{u}_c$ is a dense Zariski open subset of $\mathfrak{u}$ and the restriction*

$$M_c(n) \to \mathfrak{u}_c \tag{0.10}$$

of (0.9) to $M_c(n)$ is an algebraic isomorphism.

See Theorem 3.26.

Let $c \in \Omega(n)$. Then it follows from Theorem 0.9 that the subgroup $D_c = \{a \in A \mid a \cdot x = x,\ \forall x \in M_c(n)\}$ is closed and discrete in A. Let $A_c = A/D_c$ so that the action of A on $M_c(n)$ descends to an action of A_c. Furthermore, the latter action is simple and transitive so that A_c has the structure of an algebraic group and as such

$$A_c \cong (\mathbb{C}^\times)^{d(n-1)}. \tag{0.11}$$

In particular, if $F_c = \{a \in A_c \mid a^2 = 1\}$, then F_c is a finite group of order $2^{d(n-1)}$.

It is suggestive from the example of symmetric Jacobi matrices that if x is a symmetric matrix, then x is close to being determined by knowing the spectrum of x_m for all $m \in I_n$. Let $c \in \Omega(n)$ and let $M_c^{(\text{sym})}(n)$ be the set of all symmetric matrices in $M_c(n)$. The following theorem appears in the paper as Theorem 3.32.

Theorem 0.11. *Let $c \in \Omega(n)$ (see (2.53)). Then $M_c^{(\text{sym})}(n)$ is a finite set of cardinality $2^{d(n-1)}$. In fact $M_c^{(\text{sym})}(n)$ is an orbit of F_c.*

If $c \in \mathbb{C}^{d(n)}$, put $E_c(m) = E_x(m)$, for $m \in I_n$, and any (and hence all) $x \in M_c(n)$. Also put $\mu_{k\,m}(c) = \mu_{k\,m}(x)$ for any $k \in I_m$. We will say that $c \in \mathbb{C}^{d(n)}$ satisfies the eigenvalue interlacing condition if one has $\mu_{k\,m}(c) \in \mathbb{R}$ for all $m \in I_n$, $k \in I_m$ and, writing $\mu_{k\,m} = \mu_{k\,m}(c)$,

$$\mu_{1\,m+1} < \mu_{1\,m} < \mu_{2\,m+1} < \ldots < \mu_{m\,m+1} < \mu_{m\,m} < \mu_{m+1\,m+1} \tag{0.12}$$

for any $m \in I_{n-1}$. If c satisfies the eigenvalue interlacing condition, then obviously $c \in \Omega(n)$. The following result is established in the paper as Theorem 3.34.

Theorem 0.12. *Let $c \in \Omega(n)$. Then the following conditions are equivalent:*

(a) c satisfies the eigenvalue interlacing condition,
(b) there exists a real symmetric matrix in $M_c^{(\mathrm{sym})}(n)$,
(c) all $2^{d(n-1)}$ matrices in $M_c^{(\mathrm{sym})}(n)$ are real symmetric.

Example 0.13. Consider the case where $n = 3$ so that there are 8 symmetric matrices in $M_c(3)$ for any $c \in \Omega(n)$. Let c be defined so that $E_c(1) = \{0\}$, $E_c(2) = \{1, -1\}$, $E_c(3) = \{\sqrt{2}, 0, -\sqrt{2}\}$ so that c is eigenvalue interlacing. Then 2 of the 8 real symmetric matrices in $M_c(3)$ are

$$x = \begin{pmatrix} 0 & 1 & 0 \\ 1 & 0 & 1 \\ 0 & 1 & 0 \end{pmatrix}, \qquad y = \begin{pmatrix} 0 & 1 & 1 \\ 1 & 0 & 0 \\ 1 & 0 & 0 \end{pmatrix}$$

noting that x, but not y, is Jacobi. The remaining 6 are obtained by sign changes in x and y.

Let $\phi_k(t)$, $k \in \mathbb{Z}_+$, be an orthonormal sequence of polynomials in $L_2(\mathbb{R}, \nu)$ (for a suitable measure ν on $\mathbb{R}$) obtained by applying the Gram–Schmidt process to the monomial functions $\{t^k\}, m \in \mathbb{Z}_+$. The ϕ_k are uniquely determined up to sign. Let W_k be the span of $1, \ldots, t^{k-1}$ and let Λ_k be the orthogonal projection of $L_2(\mathbb{R}, \nu)$ on W_k so that $\{\phi_{j-1}(t)\}$, $j \in I_k$, is an orthonormal basis of W_k. Let $\tilde{t}$ be the operator on $L_2(\mathbb{R}, \nu)$ of multiplication by t.

Theorem 0.14. *Let $x \in M(n)$ be the matrix of $\Lambda_n \tilde{t}$ with respect to the basis ϕ_k, $k = 0, \ldots, n-1$ and let $c = \Phi_n(x)$ (see (0.1)). Then c satisfies the eigenvalue interlacing condition and $x \in M_c^{(\mathrm{sym})}$. Moreover x is Jacobi (see §2.5) and in the $2^{d(n-1)}$-element set $M_c^{(\mathrm{sym})}$ there are precisely 2^{n-1} Jacobi matrices. The latter represents $\Lambda_n \tilde{t}$ when the basis ϕ_k is replaced, using sign changes, by 2^{n-1} different choices of the orthonormal polynomials.*

Finally, the characteristic polynomial of x_m, for $m \in I_n$, is the monic polynomial corresponding to ϕ_m. In particular, the numbers in $E_x(m)$ are the zeros of the orthogonal polynomial ϕ_m.

See Theorems 2.20, 2.21 and (2.67) where the $a_k > 0$.

0.4.

For any $x \in M(n)$ let O_x be the $Gl(n)$ adjoint orbit generated by x. The symplectic leaves of the Poisson structure on $M(n)$ are, of course, all the adjoint orbits and any adjoint orbit stable under the action of A. For any $x \in M(n)$ let $O_x^{\mathrm{sreg}} = O_x \cap M^{\mathrm{sreg}}(n)$ so that O_x^{sreg} is an A-stable Zariski open set in O_x.

Theorem 0.15. *Let $x \in M(n)$. Then O_x^{sreg} is not empty if and only if x is regular. In such a case O_x^{sreg} is a $2\,d(n-1)$-dimensional symplectic (in the complex sense) open and dense submanifold of O_x. Furthermore, the orbits of A in O_x^{sreg} (necessarily of dimension $d(n-1)$) are the leaves of a polarization of O_x^{sreg}.*

See Theorem 3.36.

Part II of this paper, [K-W], is devoted to a "classical" analogue of the Gelfand–Kirillov theorem. The main result there will use results in the present paper. The eigenvalue functions $\mu_{k\,m}(x)$, for $m \in I_n$ and $k \in I_m$, can only be defined locally on $M_\Omega(n)$. However, they can be defined globally on a suitable covering, $M_\Omega(n, \mathfrak{e})$ of $M_\Omega(n)$ and they Poisson commute on this covering. A second set of $d(n-1)$ Poisson commuting functions on $M_\Omega(n, \mathfrak{e})$ is then defined using the action of an algebraic $A_\mathfrak{r}$ on $M_\Omega(n, \mathfrak{e})$. A "Lagrangian" property of Hessenberg matrices is then used to show that these two sets of Poisson commuting elements generate a phase space coordinate system. The coordinate functions are in an algebraic extension of the function field of $M(n)$.

0.5.

We wish to thank Pavel Etingof for fruitful conversations.

1. Preliminaries

1.1.

For any positive integer n let $M(n)$ be the algebra (both Lie and associative) of all $n \times n$ complex matrices. With respect to its Lie algebra structure let $U(n)$ be the universal enveloping algebra of $M(n)$. Let $S(n) = \oplus_{k=0}^{\infty} S^k(n)$ be the (graded) symmetric algebra over $M(n)$. Then $S(n) = \operatorname{Gr} U(n)$ with respect to the usual filtration in $U(n)$. Commutation in $U(n)$ induces the structure of a Poisson algebra $[u, v]$ on $S(n)$ where

$$[S^j(n), S^k(n)] \subset S^{j+k-1}(n). \tag{1.1}$$

If one identifies $M(n)$ with $S^1(n)$, then the Poisson bracket on $M(n)$ induced by (1.1) is of course the same as the given Lie algebra bracket.

Let $P(n) = \oplus_{k=0}^{\infty} P^k(n)$ be the graded algebra of polynomial functions on $M(n)$. We identify $P^1(n)$ with the dual space $M(n)^*$ to $M(n)$. For any positive integer k, let $I_k = \{1, \ldots, k\}$. For $i, j \in I_n$ we will, throughout the paper, let $\alpha_{ij} \in P^1(n)$ be the linear function defined so that, for $x \in M(n)$,

$$\alpha_{ij}(x) \text{ is the } ij^{th} \text{ entry of } x. \tag{1.2}$$

Of course $M(n)$ is the Lie algebra of the general linear group $Gl(n)$. The adjoint action of $Gl(n)$ on $M(n)$ induces a $GL(n)$-module structure on the algebras $U(n)$, $S(n)$ and $P(n)$. It will be convenient in this paper to transfer the coadjoint orbit theory of $GL(n)$ to adjoint orbits and to transfer the Poisson algebra structure on $S(n)$ to $P(n)$. Let $\gamma : M(n) \to P^1(n)$ be the linear isomorphism defined by

$$\langle \gamma(x), y \rangle = -B(x, y) \tag{1.3}$$

where B is the bilinear $B(x, y) = \operatorname{tr} xy$ on $M(n)$. Then γ extends to an algebra $Gl(n)$-isomorphism

$$\gamma : S(n) \to P(n). \tag{1.4}$$

Consequently $P(n)$ becomes a $Gl(n)$-Poisson algebra where $[\gamma(u), \gamma(v)] = \gamma([u, v])$ for $u, v \in S(n)$. If $e_{ij} \in M(n)$ is the ij^{th} matrix unit of $M(n)$, one notes that

$$\alpha_{ij} = \gamma(-e_{ji}). \tag{1.5}$$

Since negative transpose defines a Lie algebra automorphism on $M(n)$, the bracket structure for the linear functionals α_{ij} is the same as for the matrix units. That is,

$$\begin{aligned} [\alpha_{ij}, \alpha_{st}] &= 0 \text{ if } i \neq t \text{ and } j \neq s \\ [\alpha_{ij}, \alpha_{jk}] &= \alpha_{ik} \text{ if } i \neq k \\ [\alpha_{ij}, \alpha_{ji}] &= \alpha_{ii} - \alpha_{jj}. \end{aligned} \tag{1.6}$$

1.2.

Let $T(M(n))$ and $T^*(M(n))$ be, respectively, the holomorphic tangent and cotangent bundles of $M(n)$. All vector fields and 1-forms considered here will be, respectively, holomorphic sections of $T(M(n))$ and $T^*(M(n))$ defined on some open subset (most often on $M(n)$ itself) of $M(n)$. The Poisson algebra structure on $P(n)$ defines a Poisson structure, in the holomorphic sense, on $M(n)$. See Chapter 1 in [CG]. If $\varphi \in P(n)$ the map $\psi \mapsto [\varphi, \psi]$ is a derivation of $P(n)$ and hence there exists a vector field ξ_φ on $M(n)$ such that

$$\xi_\varphi \psi = [\varphi, \psi].$$

Since $\xi_\varphi \psi = -\xi_\psi \varphi$ it is clear that $(\xi_\varphi)_x$, for any $x \in M(n)$, depends only on $(d\varphi)_x$ so that in fact one has a bundle map

$$\Gamma : T^*(M(n)) \to T(M(n)) \tag{1.7}$$

and $(\xi_\varphi)_x = \Gamma((d\varphi)_x)$.

If $W \subset M(n)$ is any open set, let $\mathcal{H}(W)$ be the algebra of holomorphic functions on W and let $\mathrm{Vec}(W)$ be the Lie algebra of (holomorphic) vector fields on W. For any $y \in M(n)$ let $\partial^y \in \mathrm{Vec}(M(n))$ be the translation invariant vector field on $M(n)$ defined so that if $\psi \in \mathcal{H}(M(n))$ and $x \in M(n)$, then

$$\begin{aligned} (\partial^y \psi)(x) &= (\partial^y)_x \psi \\ &= d/dt(\psi(x + ty))_{t=0}. \end{aligned}$$

Let

$$M(n) \to \mathrm{Vec}(M(n)), \qquad y \mapsto \eta^y$$

be the corresponding Lie algebra homomorphism corresponding to the adjoint action of $Gl(n)$. If O is an adjoint orbit, then obviously $\eta^y|O$ is tangent to O. Explicitly it is immediate that if $x, y \in M(n)$, then

$$(\eta^y)_x = (\partial^{-[y,x]})_x \tag{1.8}$$

so that if O_x is the adjoint orbit containing x, then

$$T_x(O_x) = \{(\partial^{-[y,x]})_x \mid y \in M(n)\}. \tag{1.9}$$

Using γ we may carry over the symplectic structure (in the holomorphic sense, see [CG], Chapter 1) on coadjoint orbits to adjoint orbits. If $x \in M(n)$, let ω_x be the value of the symplectic form ω_O on $O = O_x$ at x. Then for $y, z \in M(n)$ (taking into account the minus sign in the definition of γ, and (5.3.1) and (5.3.3) in [K]) one has

$$\omega_x(\eta^y, \eta^z) = B(x, [y, z]) \tag{1.10}$$

in the notation (1.3).

Lemma 1.1. *One has*

$$\xi_{\gamma(y)} = \eta^y \tag{1.11}$$

for any $y \in M(n)$.

Proof. For any $v \in P^1(n)$ and $x \in M(n)$ one has

$$(\eta^y v)(x) = -\langle v, [y, x]\rangle \tag{1.12}$$

by (1.8). But then if $w \in M(n)$, one has

$$(\eta^y \gamma(w))(x) = B(w, [y, x]). \tag{1.13}$$

On the other hand,

$$\begin{aligned}(\xi_{\gamma(y)}\gamma(w))(x) &= [\gamma(y), \gamma(w)](x)\\ &= \gamma([y, w])(x)\\ &= B([w, y], x).\end{aligned}$$

But then $(\eta^y \gamma(w))(x) = (\xi_{\gamma(y)}\gamma(w))(x)$. This proves (1.11). □

We may enlarge the family of functions φ for which the vector field ξ_φ is defined. Let $W \subset M(n)$ be any open set. For any $\varphi \in \mathcal{H}$ let ξ_φ be the vector field on W defined so that if $x \in W$, then

$$(\xi_\varphi)_x = \Gamma((d\varphi)_x). \tag{1.14}$$

Let y_j, $j = 1, \ldots, n^2$, be a basis of $M(n)$ and let $v_j \in P^1(M(n))$ be the dual basis so that, in the notation of (1.14),

$$d\varphi = \sum_j \partial^{y_j}(f)\, dv_j \tag{1.15}$$

on W. But this implies

$$\xi_\varphi = \sum_j \partial^{y_j}(f)\, \xi_{v_j} \tag{1.16}$$

on W.

Remark 1.2. Using (1.8),(1.9) and (1.11) and the notation of (1.14), note that for any $x \in W$ the tangent vectors $(\xi_{v_j})_x$, $j = 1, \ldots, n^2$, span $T_x(O_x)$.

The following proposition essentially recovers the fact that the adjoint orbits are the symplectic leaves of the Poisson structure on $M(n)$. It also removes a possible ambiguity in the definition of the vector fields ξ_φ. If $x \in M(n)$ and ψ is a holomorphic function defined in a neighborhood of x in O_x, then the Hamiltonian vector field ξ_ψ defined in this neighborhood is such that

$$\omega_x((\xi_\psi)_x, u) = u\ \psi \tag{1.17}$$

for any $u \in T_x(O_x)$ (see (4.1.3) in [K]).

Proposition 1.3. *Let $W \subset M(n)$ be an open set and let $x \in W$. Let $\varphi \in \mathcal{H}(W)$. Then*

$$(\xi_\varphi)_x \in T_x(O_x). \tag{1.18}$$

Furthermore, if $V = W \cap O_x$, then

$$\xi_\varphi | V = \xi_{\varphi|V}. \tag{1.19}$$

Proof. One has (1.18) by Remark 1.2 and (1.16). Let $\psi = \varphi|V$. Since ω_x is nonsingular, to prove (1.19) it suffices, by Remark 1.2 and (1.18), to prove that

$$\omega_x((\xi_\varphi)_x, (\xi_v)_x) = \omega_x((\xi_\psi)_x, (\xi_v)_x) \tag{1.20}$$

for any $v \in P^1(n)$. But now it follows from (1.15), (1.16) and (1.17) that to prove (1.20) it suffices to assume $W = M(n)$ and to take $\varphi \in P^1(n)$. Write $v = \gamma(y)$ and $\varphi = \gamma(w)$ for $y, w \in M(n)$. Then the right side of (1.20) is $(\xi_{\gamma(y)}\gamma(w))(x)$ by (1.17). But the end of the proof of Lemma 1 implies that this equals $\operatorname{tr}[w, y]\, x$. On the other hand the left-hand side of (1.20) equals $\operatorname{tr} x\,[w, y]$ by (1.10). This proves (1.19). □

Classical properties (see e.g., (4.1.4) in [K] for the real case) of Poisson bracket of functions and Hamiltonian vector fields on symplectic manifolds remain valid for the Poisson manifold $M(n)$.

Proposition 1.4. *Let $W \subset M(n)$ be any open set. Then $\mathcal{H}(W)$ is a Lie algebra under Poisson bracket and the map*

$$\mathcal{H}(W) \to \operatorname{Vec}(W), \qquad \varphi \mapsto \xi_\varphi \tag{1.21}$$

is a Lie algebra homomorphism.

Proof. This is immediate from (1.19) since any adjoint orbit is a symplectic manifold. □

2. Commuting vector fields arising from Gelfand–Zeitlin theory

2.1.

Let $m \in I_n$ and let L_m be the set of all $n^2 - m^2$ pairs (i, j) where $\{i, j\} \subset I_n$ but $\{i, j\} \not\subset I_m$. We identify ("upper left hand corner") as

$$M(m) = \{x \in M(n) \mid \alpha_{ij}(x) = 0,\ \forall (i, j) \in L_m\}.$$

We will regard $M(m)$ as a Lie subalgebra of $M(n)$. The corresponding subgroup $\{g \in Gl(n) \mid \alpha_{ij}(g) = \delta_{ij}, \forall (i, j) \in L_m\}$ naturally identifies with $Gl(m)$. One has the direct sum

$$M(n) = M(m) \oplus M(m)^{\perp}, \tag{2.1}$$

where $M(m)^{\perp}$ is the B-orthocomplement of $M(m)$ in $M(n)$. For any $x \in M(n)$ let $x_m \in M(m)$ be the component of x in $M(m)$ relative to (2.1). That is, $x_m \in M(m)$ is defined so that $\alpha_{ij}(x_m) = \alpha_{ij}(x)$ for all $(i, j) \in I_m$. We will also refer to x_m as the $m \times m$ cutoff of x. The surjective map $M(n) \to M(m)$ defined by (2.1) induces an injection $P^1(m) \to P^1(n)$. The latter extends to an injective homomorphism

$$P(m) \to P(n). \tag{2.2}$$

It follows immediately from (1.6) that (2.2) is an injective homomorphism of Poisson algebras and hence $M(m)$ embeds as a Poisson submanifold of $M(n)$. Henceforth we will identify $P(m)$ with its image in $P(n)$ under (2.2).

Let $m \in I_n$ and let Id_m be the identity element of the associative algebra $M(m)$. For any $k \in I_m$, let $f_{k,m} \in P(n)$ be the coefficient of the cutoff characteristic polynomial defined by

$$\det(\lambda\ \mathrm{Id}_m - x_m) = \lambda^m + \sum_{k=1}^{m} (-1)^{m-k+1} f_{k,m}(x) \lambda^{k-1}. \tag{2.3}$$

One readily notes that if $\{\mu_j\}$, $j = 1, \ldots, m$, are the eigenvalues (always with multiplicity as roots of the characteristic polynomial) of x_m, then

$$f_{k,m}(x) \text{ is the elementary symmetric polynomial of degree } m - k + 1 \text{ in the } \mu_j. \tag{2.4}$$

For any nonnegative integer k let $d(k) = k(k + 1)/2$. Note that

$$d(n - 1) + d(n) = n^2.$$

Since k, m are arbitrary subject to the condition $1 \le k \le m \le n$, the number of polynomials $f_{k,m}$ defined by (2.3) is $d(n)$. It will be convenient to simply order these polynomials, defining p_j, for $j \in I_{d(n)}$, by putting

$$p_{d(m-1)+k} = f_{k,m}. \tag{2.5}$$

One readily deduces the following proposition from the commutativity of the Gelfand–Zeitlin ring in the enveloping algebra $U(M(n))$. The proof given here is self-contained.

Proposition 2.1. *One has the Poisson commutativity*

$$[p_i, p_j] = 0 \tag{2.6}$$

for any $\{i, j\} \subset I_{d(n)}$.

Proof. One of course knows that the ring of invariants $P(n)^{Gl(n)}$ is the polynomial ring in $p_{d(n-1)+k}$, $k \in I_n$. Hence for these values of k it follows that $p_{d(n-1)+k}$ is constant on any adjoint orbit O_x. This implies that

$$\xi_{p_{d(n-1)+k}} = 0 \tag{2.7}$$

by (1.19). Thus one has (2.6) if $j = d(n-1)+k$ and $k \in I_n$. But then, replacing n by any $m \leq n$, (2.6) follows for all $\{i, j\} \subset I_{d(n)}$. □

2.2.

Let Φ_n be the regular algebraic map

$$\Phi_n : M(n) \to \mathbb{C}^{d(n)}, \qquad \Phi_n(x) = (p_1(x), \ldots, p_{d(n)}(x)), \tag{2.8}$$

and for any $c = (c_1, \ldots, c_{d(n)}) \in \mathbb{C}^{d(n)}$, let $M_c(n) = \Phi_n^{-1}(c)$ so that $M_c(n)$ is a typical fiber of Φ_n and hence

$$M(n) = \sqcup_{c \in \mathbb{C}^{d(n)}} M_c(n). \tag{2.9}$$

From the definition we note that if $x, y \in M(n)$, then x, y lie in the same fiber of Φ_n if and only if the characteristic polynomial of the cutoffs x_m and y_m are equal for all $m \in I_n$. Expressed otherwise, introduce a lexicographical order in $\mathbb{C}$ so that if $z_1, z_2 \in \mathbb{C}$, then $z_1 < z_2$ if $\Re z_1 < \Re z_2$ and $\Im z_1 < \Im z_2$ in case $\Re z_1 = \Re z_2$. Let $c \in \mathbb{C}^{d(n)}$ and for $m \in I_n$, let

$$E_c(m) = \{\mu_{1\,m}(c), \ldots, \mu_{m\,m}(c)\} \tag{2.10}$$

be the roots, in increasing order, of the polynomial

$$\lambda^m + \sum_{k=1}^{m} (-1)^{m-k+1} c_{d(m-1)+k} \lambda^{k-1}. \tag{2.11}$$

Then $x \in M_c(n)$ if and only if $E_c(m)$ is an m-tuple of the eigenvalues of x_m, for all $m \in I_n$, where the multiplicity is as a root of the characteristic polynomial of x_m.

Remark 2.2. Of course given any set of n sequences, $E(m) = \{\mu_{1\,m}, \ldots, \mu_{m\,m}\}$, $m \in I_n$, there exists a unique $c \in \mathbb{C}^{d(n)}$ such that $E(m) = E_c(m)$, up to an ordering, for all $m \in I_n$.

Let $e \in M(n)$ be the principal nilpotent element $e = -\sum_{i=1}^{n-1} e_{i+1,i}$ and let $\mathfrak{b} \subset M(n)$ be the Borel subalgebra of all upper triangular matrices. Thus $x \in e + \mathfrak{b}$ if and only if x is of the form

$$x = \begin{pmatrix} a_{1\,1} & a_{1\,2} & \cdots & a_{1\,n-1} & a_{1\,n} \\ -1 & a_{2\,2} & \cdots & a_{2\,n-1} & a_{2\,n} \\ 0 & -1 & \cdots & a_{3\,n-1} & a_{3\,n} \\ \vdots & \vdots & \ddots & \vdots & \vdots \\ 0 & 0 & \cdots & -1 & a_{n\,n} \end{pmatrix} \tag{2.12}$$

where $a_{ij} \in \mathbb{C}$ is arbitrary. That is $\alpha_{ij}(x) = a_{ij}$ for $i \le j$. If one considers the n-dimensional subvariety $\mathfrak{s} = \{x \in e + \mathfrak{b} \mid a_{ij} = 0 \text{ for } 1 \le i \le j \le n-1\}$, then from well-known facts about companion matrices, for all $k \in I_n$ and $x \in \mathfrak{s}$,

$$\alpha_{k\,n}(x) = p_{d(n-1)+k}(x). \tag{2.13}$$

In particular the restriction of the $Gl(n)$-invariants, $p_{d(n-1)+k}$ to $\mathfrak{s}$ defines a coordinate system on $\mathfrak{s}$ and the map

$$\mathfrak{s} \to \mathbb{C}^n, \qquad x \mapsto (p_{d(n-1)+1}(x), \ldots, p_{d(n-1)+n}(x)) \tag{2.14}$$

is an algebraic isomorphism. We now generalize this statement to the entire affine variety $e + \mathfrak{b}$.

Theorem 2.3. *Let $q_i = p_i | e + \mathfrak{b}$ for $i \in I_{d(n)}$. Then the q_i are a coordinate system on $e + \mathfrak{b}$. In particular for any $x \in e + \mathfrak{b}$ the differentials*

$$(dq_i)_x,\ i \in I_{d(n)},\ \text{are linearly independent} \tag{2.15}$$

and hence the differentials

$$(dp_i)_x,\ i \in I_{d(n)},\ \text{are linearly independent.} \tag{2.16}$$

Furthermore the map

$$e + \mathfrak{b} \to \mathbb{C}^{d(n)}, \qquad x \mapsto (p_1(x), \ldots, p_{d(n)}(x)) \tag{2.17}$$

is an algebraic isomorphism.

Proof. We will prove Theorem 2.3 by induction on n. If $n = 1$, the proof is immediate. Assume that the theorem is true for $n-1$. Let $(e + \mathfrak{b})_{n-1}$ be equal to $e + \mathfrak{b}$ when $n-1$ replaces n. Then $x \mapsto x_{n-1}$ defines a surjection

$$e + \mathfrak{b} \to (e + \mathfrak{b})_{n-1}. \tag{2.18}$$

If $z \in (e + \mathfrak{b})_{n-1}$ and $F(z)$ is the fiber of (2.18) over z, then

$$F(z) \to \mathbb{C}^n, \qquad x \mapsto (\alpha_{1n}(x), \ldots, \alpha_{nn}(x))$$

is an algebraic isomorphism. If $x \in F(z)$, then $x_j = z_j$ for $j = 1, \ldots, n-1$. Now expanding the determinant $\det(\lambda I_n - x)$ using the last column of $x \in F(z)$ one has

$$\det(\lambda I_n - x) = \lambda^n + \sum_{k=1}^{n}(-1)^{n-k+1} p_{d(n-1)+k}(x)\lambda^{k-1}$$
$$= (\lambda - a_{nn}(x)) \det(\lambda\, I_{n-1} - z)$$
$$+ \sum_{k=1}^{n-1}(-1)^{n-k+1}\alpha_{kn}(x) \det(\lambda\, I_{k-1} - z_{k-1})$$
$$= (\lambda - \alpha_{nn}(x))g_{n-1}(\lambda) + \sum_{k=1}^{n-1}(-1)^{n-k+1}\alpha_{kn}(x)\, g_{k-1}(\lambda) \tag{2.19}$$

where $\det(\lambda\, I_0 - z_0)$ is interpreted as 1 and $g_j(\lambda)$ is a monic polynomial of degree j in λ whose coefficients are fixed elements of $P(n-1)$ evaluated at z. Comparing coefficients there exists a upper trianglar $n \times n$ matrix $b_{i,j}$ of fixed elements in $P(n-1)$ evaluated at z such that

$$p_{d(n-1)+n}(x) = \alpha_{n\,n}(x) + b_{n,n}$$
$$p_{d(n-1)+n-1}(x) = \alpha_{n-1\,n}(x) + b_{n-1,n-1} + b_{n-1,n}\alpha_{n\,n}(x)$$
$$p_{d(n-1)+n-2}(x) = \alpha_{n-2\,n}(x) + b_{n-2,n-2} + b_{n-2,n-1}\,\alpha_{n-1\,n}(x) + b_{n-2,n}\,\alpha_{n\,n}(x)$$
$$\vdots = \vdots$$
$$p_{d(n-1)+1}(x) = \alpha_{1\,n}(x) + b_{1,1} + b_{1,2}\,\alpha_{2\,n}(x) + \cdots + b_{1,n}\,\alpha_{n\,n}(x). \tag{2.20}$$

By induction the $b_{i,j}$ are polynomials in $p_i(z) = p_i(x)$, for $i \le d(n-1)$. Given the triangular nature of (2.20) we can solve for the $\alpha_{k\,n}(x)$ in terms of $p_j(x)$, $j \in I_{d(n)}$, and hence we can write any $\alpha_{i\,j}|e + \mathfrak{b}$, $i \le j$, as a polynomial in q_j for $j \in I_{d(n)}$. But since $\alpha_{i\,j}|e + \mathfrak{b}$, $i \le j$ is a coordinate sysytem for $e + \mathfrak{b}$, this proves (2.15) (and hence also (2.16)) and the injectivity of (2.17). The surjectivity of (2.17) follows from induction and the fact that the $\alpha_{k\,n}(x)$ in (2.20) are arbitrary. □

Let $\mathfrak{b}_e = -e + \mathfrak{b}$. The elements of $\mathfrak{b}_e$ are called Hessenberg matrices (We thank Gil Strang for this information).

Remark 2.4. Note that (2.16) in Theorem 2.3 implies that the functions p_j, $j \in I_{d(n)}$, are algebraically independent in $P(n)$. Note also, upon conjugating $e + \mathfrak{b}$ by an invertible diagonal matrix g, that Theorem 2.3 is still true if e is replaced by any sum $\sum_{i=1}^{n-1} z_i\, e_{i+1,i}$ where $z_i \in \mathbb{C}^\times$. This is clear, since g is diagonal,

$$p_j(g\,x g^{-1}) = p_j(x)$$

for any $j \in I_{d(n)}$ and $x \in M(n)$. In particular Theorem 2.3 is true for the space $\mathfrak{b}_e$ of Hessenberg matrices.

Let $x \in M(n)$, and for $m \in I_n$, let $E_x(m) = \{\mu_{1\,m}(x), \ldots, \mu_{m\,m}(x)\}$ be the m-tuple of the eigenvalues of x_m, in increasing (lexicographical) order, with multiplicity as roots of the characteristic polynomial. In the notation of (2.10), one has

$$E_x(m) = E_c(m), \quad \forall m \in I_n, \iff x \in M_c(n). \tag{2.21}$$

Theorem 2.3 has a number of nice consequences. For one thing it implies that $\mathfrak{b}_e$ is a cross-section of the surjection Φ_n. For another, for $x \in \mathfrak{b}_e$, the spectrum of x_m over all $m \in I_n$ can be chosen arbitrarily and (independently) and in so doing uniquely determines x. That is,

Theorem 2.5. *The restriction of the map $\Phi_n : M(n) \to \mathbb{C}^{d(n)}$ (see (2.8)) to $\mathfrak{b}_e$ is an algebraic isomorphism*

$$\mathfrak{b}_e \to \mathbb{C}^{d(n)}. \tag{2.22}$$

Furthermore given any set of n sequences $E(m) = \{\mu_{1\,m}, \ldots, \mu_{m\,m}\}$, $m \in I_n$, there exists a unique (Hessenberg matrix) $x \in \mathfrak{b}_e$ such that $E(m) = E_x(m)$, up to a reordering, for all $m \in I_n$.

Proof. One has only to note that if $c \in \mathbb{C}^{d(n)}$, then

$$M_c(n) \cap \mathfrak{b}_e = \{x\} \tag{2.23}$$

where $x \in \mathfrak{b}_e$ is the unique point such that $p_i(x) = c_i$ for all $i \in I_{d(n)}$. The final statement follows from Remark 2.2. □

2.3.

For any $x \in M(n)$ let $M(n)_x$ be the centralizer of x in $M(n)$ and let $Z_{x,n}$ be the (necessarily commutative) associative subalgebra generated by x and Id_n. One knows that

$$\begin{aligned} Z_{x,n} &= \text{cent } M(n)_x \\ \dim Z_{x,n} &\leq n \leq \dim M(n)_x. \end{aligned} \tag{2.24}$$

One says that x is regular if $\dim M(n)_x = n$, and as one knows

$$x \text{ is regular} \iff \dim Z_{x,n} = n \iff Z_{x,n} = M(n)_x. \tag{2.25}$$

If $x \in M(n)$, it follows from (2.24) that $\dim O_x \leq 2\,d(n)$ and

$$x \text{ is regular} \iff \dim O_x = 2\,d(n). \tag{2.26}$$

Let $T_x(O_x)^{\perp}$ be the orthocomplement of $T_x(O_x)$ in $T_x^*(M(n)$.

Proposition 2.6. *Let $x \in M(n)$. Then x is regular if and only if $\{(df_{k,n})_x\}$, $k \in I_n$, are linearly independent. In fact x is regular if and only if $\{(df_{k,n})_x\}$, $k \in I_n$, is a basis of $T_x(O_x)^{\perp}$.*

Proof. The first statement follows from Theorem 9, p. 382 in [K-1]. If $\{(df_{k,n})_x\}$, $k \in I_n$, is a basis of $T_x(O_x)^{\perp}$, then necessarily $\dim O_x = 2\,d(n)$ so that x is regular by (2.26). Conversely, $(df_{k,n})_x \in T_x(O_x)^{\perp}$ for any $k \in I_n$ and any $x \in M(n)$ since $f_{k,n}$ is constant on O_x. But then $\{(df_{k,n})_x\}$, $k \in I_n$, is a basis of $T_x(O_x)^{\perp}$ by the first statement and (2.26). □

We will say that $x \in M(n)$ is strongly regular if $(dp_i)_x$ for all $i \in I_{d(n)}$ are linearly independent. Let $M^{\text{sreg}}(n)$ be the set of all strongly regular elements in $M(n)$. By (2.16) one has

$$\mathfrak{b}_e \subset M^{\text{sreg}}(n) \tag{2.27}$$

It is then clear that

$$M^{\text{sreg}}(n) \text{ is a nonempty Zariski open subset of } M(n). \tag{2.28}$$

Theorem 2.7. *Let $x \in M(n)$ so that $(\xi_p)_x \in T_x(O_x)$ by (1.18) for any $p \in P(n)$. Then x is strongly regular if and only if the tangent vectors $(\xi_{p_i})_x \in T_x(O_x)$, $i \in I_{d(n-1)}$, are linearly independent.*

Proof. Assume that x is strongly regular. Then x is regular by Proposition 2.6. But if $q_i = p_i \mid O_x$, then $(dq_i)_x$, for all $i \in I_{d(n-1)}$ is a linearly independent set since otherwise there is a nontrivial linear combination of $(dp_i)_x$ for $i \in I_{d(n-1)}$ which lies in $T_x(O_x)^{\perp}$. But this and Proposition 2.6 would contradict the strong regularity of x. But now

$$\xi_{q_i} = \xi_{p_i} | O_x \tag{2.29}$$

by (1.19). But this proves the tangent vectors $(\xi_{p_i})_x \in T_x(O_x)$, $i \in I_{d(n-1)}$, are linearly independent since O_x is symplectic. Conversely, assume that the tangent vectors $(\xi_{p_i})_x \in T_x(O_x)$, $i \in I_{d(n-1)}$, are linearly independent. Then recalling (1.7) it follows that the covectors $(dp_i)_x$ for $i \in I_{d(n-1)}$ are linearly independent. But then, by (2.29), the same statement is true if we replace the p_i by the restrictions q_i (using the notation of (2.29)). Let $W_x \subset T_x(O_x)$ be the span of $(\xi_{p_i})_x \in T_x(O_x)$, $i \in I_{d(n-1)}$, so that $\dim W_x = d(n-1)$. But (2.6) and (1.19) imply that $\omega_x((\xi_{p_i})_x, (\xi_{p_j})_x) = 0$ for $i, j \in I_{d(n-1)}$. Thus W_x is a Lagrangian subspace of the symplectic vector space $T_x(O_x)$. Thus $\dim O_x \geq 2\,d(n-1)$. Hence x is regular. Thus $(dp_{d(n-1)+k})_x$ for $k \in I_n$ are linearly independent by Proposition 2.6. But then the full set $(dp_i)_x$, for $i \in I_{d(n)}$, is linearly independent by Proposition 2.6 and the fact that the $(dq_i)_x$, for $i \in I_{d(n-1)}$, are linearly independent. □

2.4.

For $m \in I_n$ one knows that $P(m)^{Gl(m)}$ is a polynomial ring with the m generators $p_{d(m-1)+k}$, $k \in I_m$. Let $J(n) \subset P(n)$ be the algebra generated by $P(m)^{Gl(m)}$, over all $m \in I_n$. By Remark 2.4

$$J(n) \cong P(1)^{Gl(1)} \otimes \cdots \otimes P(n)^{Gl(n)} \tag{2.30}$$

and $J(n)$ is the polynomial ring with the $d(n)$ generators p_i, $i \in I_{d(n)}$. It then follows from (2.6) that

$$[p, q] = 0, \text{ for all } p, q \in J(n), \tag{2.31}$$

and hence for the corresponding vector fields

$$[\xi_p, \xi_q] = 0, \text{ for all } p, q \in J(n) \tag{2.32}$$

(see Proposition 1.4). Now for any $x \in M(n)$ one defines a subspace of $T_x(M(n))$ by putting

$$V_x = \{(\xi_p)_x \mid \forall p \in J(n)\}.$$

Remark 2.8. If $x \in M(n)$, note that it follows from Proposition 1.3, (1.19) and (2.32) that $V_x \subset T_x(O_x)$ and V_x is an isotropic subspace of $T_x(O_x)$ with respect to the symplectic form ω_x. In particular

$$\dim V_x \leq \frac{1}{2} \dim O_x \tag{2.33}$$

and hence

$$\dim V_x \leq d(n-1). \tag{2.34}$$

Moreover note that if $J(n)$ is generated by $\{q_i\}$, $i \in I_k$, for some integer k, then V_x is spanned by $(\xi_{q_i})_x$, $i \in I_k$, since, clearly, $(dp)_x$ is in the span of $(dq_i)_x$, $i \in I_k$ (see (1.7)) for any $p \in J(n)$. Recalling (2.7) note then that Theorem 2.7 may be expressed by the statement

$$\text{One has equality in (2.34)} \iff x \text{ is strongly regular.} \tag{2.35}$$

Remark 2.9. Even though the function $x \mapsto \dim V_x$ on $M(n)$ is nonconstant, by abuse of terminology, we will refer to $x \mapsto V_x$ as a distribution on $M(n)$ (one readily shows that the two other conditions in the definition of an involutory distribution are satisfied). An analytic submanifold X of $M(n)$ will be said to be a leaf of this distribution if $T_x(X) = V_x$ for all $x \in X$.

One of the main results of the paper will be to show that that there exists a connected analytic group A, operating on $M(n)$, all of whose orbits are leaves of the distribution $x \mapsto V_x$. In addition, it will be shown that ξ_p, for any $p \in J(n)$, integrates to a flow on $M(n)$. Furthermore the flow commutes with A and stabilizes the orbits of A. In order to do this we will first determine V_x very explicitly. For any $m \in I_n$ and $x \in M(n)$ let $V_x(m) = \{(\xi_p)_x \mid p \in P(m)^{Gl(m)}\}$. It is immediate (2.7) that $V_x(n) = 0$ and from Remark 2.8 one has (choosing $k = d(n)$ and $q_i = p_i$) that

$$V_x = \sum_{m \in I_{n-1}} V_x(m). \tag{2.36}$$

For $m \in I_n$ let $M(m)^\perp$ be defined as in (2.1). Clearly if $u \in M(m)$, then

$$u + M(m)^\perp = \{y \in M(n) \mid y_m = u\}. \tag{2.37}$$

To determine $V_x(m)$ we replace the generators $f_{k,m}$ (see (2.3)) of $P(m)^{Gl(m)}$ by the generators $f_{(k,m)}$ of $P(m)^{Gl(m)}$ (recalling the theory of symmetric functions) where for $k \in I_m$ and $x \in M(n)$,

$$f_{(k,m)}(x) = \frac{1}{m+1-k} \operatorname{tr}(x_m)^{m+1-k}. \tag{2.38}$$

Lemma 2.10. *Let $x \in M(n)$ and let $m \in I_n$, $k \in I_m$. Then for any $v \in M(n)$ and any $w \in x_m + M(m)^\perp$ one has*

$$(\partial^v)_w f_{(k,m)} = \operatorname{tr}(x_m)^{m-k}\, v. \tag{2.39}$$

Proof. By definition

$$\begin{aligned}(\partial^v)_w f_{(k,m)} &= \frac{1}{m+1-k}\, d/dt(\operatorname{tr}((w+tv)_m)^{m+1-k})_{t=0} \\ &= \frac{1}{m+1-k}\, d/dt(\operatorname{tr}((x_m+tv_m)^{m+1-k})_{t=0} \\ &= \operatorname{tr}(x_m)^{m-k}\, v_m \\ &= \operatorname{tr}(x_m)^{m-k} v\end{aligned}$$

since $(x_m)^{m-k} \in M(m)$, and hence $(x_m)^{m-k}$ is B-othogonal to the component of v in $M(m)^\perp$ relative to the decomposition $M(n) = M(m) + M(m)^\perp$. □

Let $x \in M(n)$ and let

$$T_x^*(M(n)) \to P^1(n), \qquad \rho \mapsto \widetilde{\rho} \tag{2.40}$$

be the isomorphism defined so that if $v \in M(n)$ and $\rho \in T_x^*(M(n))$, then $\langle \rho, (\partial^v)_x \rangle = \langle \widetilde{\rho}, v \rangle$. Clearly $(d\widetilde{\rho})_x = \rho$ so that if $q \in P(n)$ and we put $\rho = (dq)_x$, then

$$(\xi_{\widetilde{\rho}})_x = (\xi_q)_x \tag{2.41}$$

by (1.7).

Proposition 2.11. *Let $x \in M(n)$. Let $m \in I_n$ and $k \in I_m$. Let γ be defined as in (1.3). Then*

$$\gamma((x_m)^{m-k}) = -(\widetilde{df_{(k,m)}})_x. \tag{2.42}$$

Proof. This is immediate from (2.39) since we can put $w = x$ in (2.39). □

For notational convenience put $\xi_{(k,m)} = \xi_{f_{(k,m)}}$.

Theorem 2.12. *Let $m \in I_n$ and $k \in I_m$. Then for any $x \in M(n)$ one has*

$$(\xi_{(k,m)})_x = (\partial^{[(x_m)^{m-k}, x]})_x. \tag{2.43}$$

Proof. Put $y = -(x_m)^{m-k}$. Then by (1.11) and (1.8) one has

$$(\xi_{\gamma(y)})_x = (\partial^{-[y,x]})_x. \tag{2.44}$$

But then if $\rho = (df_{(k,m)})_x$ one has

$$\gamma(y) = \widetilde{\rho} \tag{2.45}$$

by (2.42). But $(\xi_{\widetilde{\rho}})_x = (\xi_{(k,m)})_x$ by (2.41) where $q = f_{(k,m)}$. But then (2.43) follows from (2.44) and (2.45). □

For any $x \in M(n)$ and $m \in I_n$, let $Z_{x,m}$ be the associative subalgebra generated by Id_m and x_m so that $Z_{x,m}$ is spanned by $(x_m)^{m-k}$, $k \in I_m$. Here $(x_m)^0 = \mathrm{Id}_m$. Also let $M(m)_{x_m}$ be the centralizer of x_m in $M(m)$. One has that dim $Z_{x,m} \leq m$ and x_m is regular in $M(m)$ if and only if dim $Z_{x,m} = m$ which is the case if and only if

$$Z_{x,m} = M(m)_{x_m}. \tag{2.46}$$

Also put

$$Z_x = \sum_{m \in I_n} Z_{x,m}. \tag{2.47}$$

As a corollary of Theorem 2.12 one has

Theorem 2.13. *Let $x \in M(n)$. Then for any $m \in I_m$,*

$$\begin{aligned} V_x(m) &= \{(\partial^{[z,x]})_x \mid z \in Z_{x,m}\} \text{ and} \\ V_x &= \{(\partial^{[z,x]})_x \mid z \in Z_x\}. \end{aligned} \tag{2.48}$$

Moreover x is strongly regular if and only if (1) x_m is regular in $M(m)$ for all $m \in I_n$ and (2) the sum (2.47) is a direct sum. Equivalently, x is strongly regular if and only if the elements $(x_m)^{m-k}$, over all $m \in I_n$ and $k \in I_m$, are linearly independent in $M(n)$.

Proof. The equalities (2.48) follow from (2.36) and of course Theorem 2.12. See also the last part of Remark 2.8. The statement about strong regularity follows from Proposition 2.11 and the isomorphism (2.41). □

The following result is a simpler criterion for strong regularity.

Theorem 2.14. *Let $x \in M(n)$. Then x is strongly regular if and only if (a) x_m is regular in $M(m)$ for all $m \in I_n$ and (b)*

$$Z_{x,m} \cap Z_{x,m+1} = 0, \quad \forall m \in I_{n-1}. \tag{2.49}$$

Proof. If x is strongly regular, then (a) and (b) are satisfied by (1) and (2) in Theorem 2.13. Conversely, assume (a) and (b) are satisfied. We have only to show that (2) in Theorem 2.13 is satisfied. Assume (2) is not satisfied and so there exists $0 \neq z_i \in Z_{x,m_i}$, $i \in I_k$, for $k > 1$, and m_i a strictly increasing sequence with values in I_n, such that

$$\sum_{i=1}^{k} z_i = 0. \tag{2.50}$$

Let $m = m_1$ so that $m \in I_{n-1}$. But now if $i > 1$, then note that

$$[z_i, x]_{m+1} = 0. \tag{2.51}$$

Indeed

$$M(n) = M(m_i) \oplus M(m_i)^{\perp} \tag{2.52}$$

by (2.1). But $z_i \in M(m_i)$ and clearly both components in (2.52) are stable under $\operatorname{ad} M(m_i)$. However the component of x in $M(m_i)$ is x_{m_i}. But $z_i \in Z_{x,m_i}$ so that z commutes with x_{m_i}. Thus $[z_i, x] \in M_{m_i}^{\perp}$. But $m_i \geq m+1$ and hence $M_{m_i}^{\perp} \subset M_{m+1}^{\perp}$. This proves (2.51). But then (2.50) implies that $[z_1, x]_{m+1} = 0$. But $\operatorname{ad} M(m)$ clearly stabilizes both components of the decomposition $M(n) = M(m+1) \oplus M(m+1)^{\perp}$. Since the component of x in $M(m+1)$ is x_{m+1} one has $[z_1, x_{m+1}] = 0$. Thus $z_1 \in M(m+1)_{x_{m+1}} = Z_{x,m+1}$. But then $z_1 \in Z_{x,m} \cap Z_{x,m+1}$. This is a contradiction since $z_1 \neq 0$. □

Recall equation (2.10). We will say that $c \in \mathbb{C}^{d(n)}$ satisfies the eigenvalue disjointness condition if (I) the numbers in $E_c(m)$ are distinct for all $m \in I_n$ and (II), as a set, one has $E_c(m) \cap E_c(m+1) = \emptyset$ for all $m \in I_{n-1}$. Similarly, we will say x satisfies the eigenvalue disjointness condition if (I) and (II) are satisfied where c is replaced by x. See (2.21). As explained later conditions (I) and (II) are motivated by the theory of orthogonal polynomials. If x satisfies the eigenvalue disjointness condition, by abuse of notation, we will, on occasion, regard $E_x(m)$ as a set, rather than as a sequence.

Remark 2.15. Note that if $x \in M(n)$ satisfies the eigenvalue disjointness condition, then x_m is a regular semisimple element of $M(m)$ for all $m \in I_n$.

Let $\Omega(n)$ be the set of all $c \in \mathbb{C}^{d(n)}$ which satisfy the eigenvalue disjointness condition and let $M_\Omega(n)$ be the set of all $x \in M(n)$ which satisfy this condition so that in the notation of (2.8),

$$M_\Omega(n) = \Phi_n^{-1}(\Omega(n)). \tag{2.53}$$

Remark 2.16. We note here that $\Omega(n)$ and $M_\Omega(n)$ are (obviously nonempty) Zariski open subsets of $\mathbb{C}^{d(n)}$ and $M(n)$, respectively. Indeed if $c \in \mathbb{C}^{d(n)}$, then condition I is just the condition that the discriminant of the polynomial (2.11) is nonzero for all $m \in I_n$. Condition II is that

$$\prod_{i\in I_m, j\in I_{m+1}} \mu_{i,m}(c) - \mu_{j,m+1}(c) \tag{2.54}$$

not vanish for $m \in I_{n-1}$. But the polynomial functions (2.54) on the roots $E_c(m)$, $m \in I_n$, are clearly invariant under the product of the root permutation groups $\Sigma(m)$, $m \in I_n$. Consequently (2.54) can be expressed as polynomial functions on $\mathbb{C}^{d(n)}$. Thus conditions I and II are the nonvanishing of certain polynomial functions on $\mathbb{C}^{d(n)}$. A similar statement clearly then applies to $M(n)$, thereby establishing the assertion of Remark 2.16.

Although with a different application in mind, the constructions in the proof of the following theorem already appear in Section 4 of [GS].

Theorem 2.17. *If $x \in M(n)$ satisfies the eigenvalue disjointness condition, then x is strongly regular. That is,*

$$M_\Omega(n) \subset M^{\mathrm{sreg}}(n). \tag{2.55}$$

Proof. Assume $x \in M(n)$ satisfies the eigenvalue disjointness condition. Let $m \in I_{n-1}$. It suffices by (2.46) and Theorem 2.14 to prove that

$$Z_{x,m} \cap M(m+1)_{x_{m+1}} = 0. \tag{2.56}$$

But now since x_m is regular semisimple it can be diagonalized. That is, there exists $g \in Gl(m)$ such that

$$g\, x_m\, g^{-1} = \operatorname{diag}(\mu_{1\,m}(x), \ldots, \mu_{m\,m}(x)). \tag{2.57}$$

Then, writing $\mu_{i\,m} = \mu_{i\,m}(x)$, $g\, x_{m+1}\, g^{-1}$ is of the form

$$g\, x_{m+1}\, g^{-1} = \begin{pmatrix} \mu_{1\,m} & 0 & \cdots & 0 & a_{1\,m+1} \\ 0 & \mu_{2\,m} & \cdots & 0 & a_{2\,m+1} \\ \vdots & \vdots & \ddots & \vdots & \vdots \\ 0 & 0 & \cdots & \mu_{m\,m} & a_{m\,m+1} \\ b_{m+1\,1} & b_{m+1\,2} & \cdots & b_{m+1\,m} & d \end{pmatrix}. \tag{2.58}$$

But now $\{\mu_{1\,m+1}(x), \ldots, \mu_{m+1\,m+1}(x)\}$ are roots of the characteristic polynomial $p(\lambda)$ of x_{m+1}. On the other hand, by (2.58)

$$p(\lambda) = (\lambda - d)(\prod_{i=1}^{m}(\lambda - \mu_{i\,m}) - \sum_{i=1}^{m} a_{i\,m+1} b_{m+1\,i} \prod_{j \in (I_m - \{i\})} (\lambda - \mu_{j\,m}). \tag{2.59}$$

But $\mu_{i\,m}$ is a root of all the $m+1$ polynomial summands of (2.59) except for the summand

$$a_{i\,m+1} b_{m+1\,i} \prod_{j \in (I_m - \{i\})} (\lambda - \mu_{j\,m}).$$

Since, as sets, $E_x(m) \cap E_x(m+1) = \emptyset$ one must have that

$$a_{i\,m+1} b_{m+1\,i} \neq 0, \;\; \forall i \in I_m. \tag{2.60}$$

But now the subalgebra of $M(m)$ generated by Id_m and $g\, x_m g^{-1}$ is the algebra $\mathfrak{d}(m)$ of all diagonal matrices in $M(m)$. But now if $y = g\, x_{m+1}\, g^{-1}$ and $M(m+1)_y$ is the centralizer of y in $M(m)$, to prove (2.56) it suffices, by conjugation, to prove

$$\mathfrak{d}(m) \cap M(m+1)_y = 0. \tag{2.61}$$

But if $0 \neq d \in \mathfrak{d}(m)$ and $d = \operatorname{diag}(d_1, \ldots, d_m)$, then $d_j \neq 0$ for some $j \in I_m$. But

$$a_{j\,m+1}([d, y]) = d_j\, a_{j\,m+1}. \tag{2.62}$$

Thus $[d, y] \neq 0$ by (2.60). This proves (2.61). □

2.5.

If $x \in M(n)$ and x is a real symmetric matix, then x_m, for any $m \in I_n$, is real symmetric so that the eigenvalues of x_m are real. Hence, for the m-tuple, $E_x(m)$, one has

$$\mu_{1\,m} \leq \cdots \leq \mu_{m\,m} \tag{2.63}$$

where we have written $\mu_{j,m}$ for $\mu_{j,m}(x)$. Under certain conditions, strong regularity implies the eigenvalue disjointness condition. A key argument in the proof of Proposition 2.18 is well known.

Proposition 2.18. *Assume $x \in M(n)$ is strongly regular and x is a real symmetric matrix, so that in the notation of (2.63), the inequality $\leq$ is replaced by the strict inequality $<$. Then x satisfies the eigenvalue disjointness condition. In fact if $m \in I_{n-1}$, then using the notation of (2.63) one has the interlacing condition*

$$\mu_{1\,m+1} < \mu_{1\,m} < \mu_{2\,m+1} < \ldots < \mu_{m\,m+1} < \mu_{m\,m} < \mu_{m+1\,m+1}. \tag{2.64}$$

Proof. We use the notation in the proof of Theorem 2.17. But now instead of assuming that x satisfies the eigenvalue disjointness condition, assume that x is strongly regular so that one has (2.56) and hence also (2.61). Since x is symmetric, the element $g \in Gl(m)$ can be chosen to be orthogonal. Thus one has $a_{i\,m+1} = b_{m+1,i}$ for all $i \in I_m$. But (2.61) implies that $[d, y] \neq 0$. On the other hand, clearly $\alpha_{ik}([d, y]) = 0$ for all $i, k \in I_n$ unless, as an ordered pair, $\{i, k\} = \{j, m+1\}$ or $\{i, k\} = \{m+1, j\}$. But $\alpha_{j\,m+1}([d, y]) = d_j\, a_{j\,m+1}$ and $\alpha_{m+1\,j}([d, y]) = -d_j\, a_{j\,m+1}$. Thus $a_{j\,m+1} \neq 0$ for all $j \in I_m$. But as a rational function of the parameter λ of $\mathbb{R}$ one has

$$p(\lambda)/\prod_{i=1}^{m}(\lambda - \mu_{i\,m}) = \lambda - d - \prod_{j\in(I_m-\{i\})} a^2_{j\,m+1}/(\lambda - \mu_{i\,m}). \tag{2.65}$$

Consider λ in the open interval $(\mu_{i\,m}, \mu_{i+1\,m})$ for $i \in I_{m-1}$. As λ approaches $\mu_{i\,m}$ the function (2.65) clearly approaches $-\infty$ and (2.65) approaches $+\infty$ as λ approaches $\mu_{i+1\,m}$. Thus there exists $\mu_{i+1\,m+1} \in (\mu_{i\,m}, \mu_{i+1\,m})$. On the other hand, if we consider λ in the open interval $(-\infty, \mu_{1\,m})$, one notes that (2.65) approaches $-\infty$ as λ approaches $-\infty$ and (2.65) approaches $+\infty$ as λ approaches $\mu_{1\,m}$. Thus there exists $\mu_{1\,m+1} \in (-\infty, \mu_{1\,m})$. Similarly there exists $\mu_{m+1,m+1} \in (\mu_{m\,m}, +\infty)$. This accounts for $m+1$ distinct roots of $p(\lambda)$ not only establishing $E_x(m) \cap E_x(m+1) = \emptyset$ (so that x satisfies the eigenvalue disjointness condition) but also (2.64). □

In this article $x \in M(n)$ is called a Jacobi matrix if $\alpha_{ij}(x) = 0$ if $|i - j| > 1$ and $\alpha_{ij}(x) \neq 0$ if $|i - j| = 1$. That is, x is "tridiagonal" except that the main diagonal is arbitrary and the entries of x are nonzero along the two diagonals adjacent to the main diagonal. If these nonzero entries are all positive, we will say that x has positive adjacent diagonals. Among other statements the following result says that if x is real symmetric and Jacobi, the assumption of strong regularity in Proposition 2.18 is unnecessary since it is automatically satisfied. In particular the conclusions of that proposition hold.

It is clear that $M_c(n)$ is stable under conjugation by any diagonal matrix in $Gl(n)$ for any $c \in \mathbb{C}^{d(n)}$.

Theorem 2.19. *Assume $x \in M(n)$ is a Jacobi matrix. Then x is strongly regular. Furthermore if $y \in M(n)$ is also a Jacobi matrix, then x and y are conjugate by a diagonal* *matrix in $Gl(n)$ if and only if $E_x(m) = E_y(m)$ for all $m \in I_n$ (i.e., if and only if x and y lie in the same fiber $M_c(n)$ of Φ_n — see (2.21)). Moreover if x is real symmetric, then x satisfies the eigenvalue disjointness condition and one has the eigenvalue interlacing*

$$\mu_{1\,m+1} < \mu_{1\,m} < \mu_{2\,m+1} < \ldots < \mu_{m\,m+1} < \mu_{m\,m} < \mu_{m+1\,m+1} \tag{2.66}$$

where we have written $\mu_{j\,k} = \mu_{j\,k}(x)$.

Proof. The first statement follows from (2.16) and Remark 2.4. Given Jacobi matrices x, y it follows from Remark 2.5 that there exists diagonal matrices g_x and g_v in $Gl(n)$ such that $g_x\,x\,g_x^{-1}$ and $g_y\,y\,g_y^{-1}$ both lie in $e + \mathfrak{b}$. But then $g_x\,x\,g_x^{-1} = g_y\,y\,g_y^{-1}$ by Theorem 2.5 if x and y lie in the same fiber $M_c(n)$. Thus the same fiber condition implies that x and y are conjugate by a diagonal matrix in $Gl(n)$. The converse direction is obvious. But now if x is real symmetric, then the remaining statements follow from Proposition 2.18. □

Theorem 2.19 has an immediate application to the theory of orthogonal polynomials. Let $\rho(t)$ be a nonnegative integrable (with respect to Lebesgue measure) function on a (finite or infinite) interval $(a, b) \subset \mathbb{R}$. Assume that the measure $\nu = \rho(t)dt$ on (a, b) has a positive integral and $\phi(t)$ is integrable with respect to ν for any polynomial function $\phi(t)$. Let W_k for $k \in \mathbb{Z}_+$ be the k-dimensional subspace of the Hilbert space $L_2(\mathbb{R}, \nu)$ spanned by $1, \ldots, t^{k-1}$. Let $\phi_k(t)$, $k \in \mathbb{Z}_+$, be the orthonormal sequence in $L_2(\mathbb{R}, \nu)$ obtained by applying the Gram–Schmidt process to the functions $\{t^k\}$, $k \in \mathbb{Z}_+$, so that, for any $k \in \mathbb{Z}_+$, $\{\phi_{j-1}(t)\}$, $j \in I_k$, is an orthonormal basis of W_k. This basis may be normalized so that $a_k > 0$ where $a_k\,t^k$ is the leading term of $\phi_k(t)$. Then one knows that there is a 3-term formula (see e.g., (10), p. 157 in [J])

$$t\,\phi_k(t) = \frac{a_k}{a_{k+1}}\,\phi_{k+1}(t) + c_{kk}\,\phi_k(t) + \frac{a_{k-1}}{a_k}\phi_{k-1}(t) \tag{2.67}$$

where $c_{kk} \in \mathbb{R}$. Let Λ_k be the orthogonal projection of $L_2(\mathbb{R}, \nu)$ onto W_k. Let $\tilde{t}$ be the multiplication operator on $L_2(\mathbb{R}, \nu)$ by the function t. Then if $x \in M(n)$ is the matrix of $\Lambda_n\,\tilde{t}|W_n$ with respect to the basis $\phi_{m-1}(t)$, $m \in I_n$, of W_n, it follows from (2.67) that x is the real symmetric Jacobi matrix given by

$$x = \begin{pmatrix} c_{00} & a_0/a_1 & 0 & 0 & \cdots & 0 \\ a_0/a_1 & c_{11} & a_1/a_2 & 0 & \cdots & 0 \\ 0 & a_1/a_2 & c_{22} & a_2/a_3 & \cdots & 0 \\ \vdots & \vdots & \vdots & \ddots & \vdots & \vdots \\ 0 & 0 & 0 & \cdots & a_{n-1}/a_n & c_{nn} \end{pmatrix}. \tag{2.68}$$

By Theorem 2.19 x is strongly regular and satisfies the eigenvalue disjointness condition. In addition one has the eigenvalue interlacing

$$\mu_{1\,m+1} < \mu_{1\,m} < \mu_{2\,m+1} < \ldots < \mu_{m\,m+1} < \mu_{m\,m} < \mu_{m+1\,m+1} \tag{2.69}$$

where we have put $\mu_{jk} = \mu_{jk}(x)$. An important object of mathematical study is the zero set of the polynomials $\phi_k(t)$. The following result, which is no doubt well known, but proved here for completeness, asserts that one can recover the orthogonal polynomial $\phi_m(t)$, $m \in I_n$, from the characteristic polynomial of x_m. In particular $E_x(m) = \{\mu_{1m}, \ldots, \mu_{mm}\}$ is the zero set of the polynomial $\phi_m(t)$.

Theorem 2.20. *Let the notation be as above. In particular, let $x \in M(n)$ be given by (2.68). For $m \in I_n$, let $\phi_m^{(1)}(t) = \frac{1}{a_m}\phi_m(t)$ so that $\phi_m^{(1)}(t)$ is the monic polynomial corresponding to $\phi_m(t)$. Then $\phi_m^{(1)}(\lambda)$ is the characteristic polynomial of x_m so that*

$$E_x(m) \text{ is the set of zeros of } \phi_m(t) \tag{2.70}$$

and one has the interlacing (2.69).

Proof. Let $m \in I_n$ so that, by (2.67), x_m is the matrix of $\Lambda_m\,\tilde{t}\,|W_m$ with respect to the basis $\{\phi_{j-1}\}$, $j \in I_m$, of W_m. Let $y \in M(m)$ be the matrix of $\Lambda_m\,\tilde{t}\,|W_m$ with respect to the basis $\{t^{j-1}\}$, $j \in I_m$, of W_m. If $j \in I_{m-1}$, then of course $\Lambda_m\,\tilde{t}(t^{j-1}) = t^j \in W_m$. But now let $b_i \in \mathbb{R}$ be such that

$$\phi_m^{(1)}(t) = t^m + \sum_{j\in I_m} b_j\, t^{j-1}. \tag{2.71}$$

But $\Lambda_m(\phi_m^{(1)}(t)) = 0$ by orthogonality. Thus

$$\begin{aligned}\Lambda_m\,\tilde{t}(t^{m-1}) &= \Lambda_m(t^m)\\ &= -\sum_{j\in I_m} b_j\, t^{j-1}.\end{aligned} \tag{2.72}$$

Hence y is the companion matrix

$$\begin{pmatrix} 0 & 0 & \cdots & 0 & -b_1\\ 1 & 0 & \cdots & 0 & -b_2\\ 0 & 1 & \cdots & 0 & -b_3\\ \vdots & \vdots & \ddots & \vdots & \vdots\\ 0 & 0 & \cdots & 1 & -b_m\end{pmatrix}.$$

But then $\phi_m^{(1)}(\lambda)$ is the characteristic polynomial of y. Consequently, $\phi_m^{(1)}(\lambda)$ is also the characteristic polynomial of x_m. □

Applying Theorem 2.19 we can say something about the uniqueness of $x \in M(n)$ (see (2.68)) in yielding the orthogonal polynomials $\phi_m(t)$, $m \in I_n$.

Theorem 2.21. *Let $\{\phi_m(t)\}$, $m \in I_n$, be the set of normalized orthogonal polynomials, where $\deg \phi_m(t) = m$, for a general measure ν on $\mathbb{R}$ specified as above. Then the matrix $x \in M(n)$, given by (2.68), is the unique symmetric Jacobi matrix with positive adjacent diagonals such that $E_x(m)$, for any $m \in I_n$, as a set, is the set of zeros of the polynomial $\phi_m(t)$.*

Proof. If $y \in M(n)$ is a Jacobi matrix such that $E_y(m) = E_x(m)$ for any $m \in I_n$ then, by Theorem 2.19, y is a conjugate of x by a diagonal matrix $g \in Gl(n)$. But if conjugation by g is to preserve symmetry and the positivity condition, it is immediate that g must be a constant matrix. In such a case of course $x = y$. □

3. The group A and its orbit structure on $M(n)$

3.1.

Let $m \in I_n$. Recall that $P(m)^{Gl(m)}$ (see (2.30)) is the polynomial ring with the m generators $f_{(k,m)}$, $k \in I_m$ (see (2.38)). Let $\mathfrak{a}(m)$ be the commutative (see (2.32)) Lie algebra of analytic vector fields on $M(n)$ spanned by $\xi_{(k,m)}$ for $k \in I_m$. Here we are retaining the notation (2.43).

If $m = n$, then $\mathfrak{a}(n) = 0$ by the argument which implies (2.7). Now assume $m \in I_{n-1}$. Note then that dim $\mathfrak{a}(m) = m$ by (2.43), (2.46), (2.49) and the existence of strongly regular elements. Let $A(m)$ be a simply connected complex analytic group where $\mathfrak{a}(m) = \text{Lie}\, A(m)$. Of course $A(m) \cong \mathbb{C}^m$. We wish to prove that $\mathfrak{a}(m)$ integrates to an analytic action of $A(m)$ on $M(n)$.

For any $x \in M(n)$ we recall the subspace $V_x(m) \subset T_x(M(n))$ defined as in (2.36). Since $\{f_{(k,m)}\}$, $k \in I_m$, generate $P(m)^{Gl(m)}$,

$$V_x(m),\ \text{for any } x \in M(n), \text{ is spanned by } \{(\xi_{(k,m)})_x\},\ k \in I_m. \tag{3.1}$$

Consequently, if $x \in M(n)$,

$$\dim V_x(m) \le m. \tag{3.2}$$

However

$$\dim V_x(m) = m \text{ if } x \text{ is strongly regular, by Theorem 2.7.}$$

For any $w \in M(m)$ clearly

$$w + M(m)^{\perp} = \{x \in M(n) \mid x_m = w\} \tag{3.3}$$

using the notation of (2.37). But obviously one has the fibration

$$M(n) = \sqcup_{w \in M(m)}\ \ w + M(m)^{\perp}. \tag{3.4}$$

Lemma 3.1. *For any $\xi \in \mathfrak{a}$ and any $w \in M(m)$ the vector field ξ is tangent to the fiber $w + M(m)^{\perp}$. That is,*

$$V_x(m) \subset T_x(x_m + M(m)^{\perp}) \tag{3.5}$$

for any $x \in M(n)$ (noting of course that $x \in x_m + M(m)^{\perp}$).

Proof. Obviously $w + M(m)^{\perp}$ is the nonsingular affine subvariety of $M(n)$ defined by the equations $\alpha_{ij} - \alpha_{ij}(w) = 0$ for all $\{i, j\} \subset I_m$. A vector field η on $M(n)$ is then clearly tangent to $w + M(m)^{\perp}$ if $\eta(\alpha_{ij}) = 0$ for all $\{i, j\} \subset I_m$. But this is certainly the case if $\eta = \xi$. Indeed $f_{(k,m)} \in P(m)^{Gl(m)}$, for $k \in I_m$, and $\alpha_{ij} \in P(m)$. Hence $f_{(k,m)}$ Poisson commutes with α_{ij}. □

Recalling the notation of (2.46) let $x \in M(n)$ and let $G_{x,m} \subset Gl(m)$ be the (commutative) subgroup of all $g \in Gl(m)$ such that g_m is an invertible element in $Z_{x,m}$ (where Id_m is taken as the identity). It is clear that $G_{x,m}$ is an algebraic subgroup of $Gl(n)$ and $Z_{x,m}$ (as a Lie algebra) is the Lie algebra of $G_{x,m}$. Since $A(m)$ is simply-connected there clearly exists a homomorphism of commutative complex analytic groups

$$\rho_{x,m} : A(m) \to G_{x,m} \tag{3.6}$$

whose differential $(\rho_{x,m})_*$ is given by

$$(\rho_{x,m})_*(\xi_{(k,m)}) = -(x_m)^{m-k} \tag{3.7}$$

for any $k \in I_m$.

Remark 3.2. Note that $G_{x,m}$ is connected since first of all $G_{x,m} \cong \{g_m \mid g \in G_{x,m}\}$ and clearly

$$\{g_m \mid g \in G_{x,m}\} = \{y \in Z_{x,m} \mid f_{1,m}(y) \neq 0\}.$$

It follows that

$$\rho_{x,m} : A(m) \to G_{x,m}, \qquad \text{is surjective} \tag{3.8}$$

since, clearly, $(\rho_{x,m})_*$ is a surjective Lie algebra homomorphism of $\mathfrak{a}(m)$ to $Z_{x,m}$.

Let $x \in M(n)$. Now obviously $M(m)$ and $M(m)^\perp$ are both stable under ad $M(m)$ and Ad $Gl(m)$. But obviously x_m is fixed by Ad $G_{x,m}$. Consequently,

$$x_m + M(m)^\perp \text{ is stable under Ad } G_{x,m}. \tag{3.9}$$

Theorem 3.3. *Let $m \in I_{n-1}$. Then $\mathfrak{a}(m)$ integrates to a (complex analytic) action of $A(m)$ on $M(n)$,*

$$A(m) \times M(n) \to M(n), \qquad (a, y) \mapsto a \cdot y. \tag{3.10}$$

More explicitly, if $x \in M(n)$ and $y \in x_m + M(m)^\perp$ (noting that any $y \in M(n)$ is of this form by (3.3) and (3.4)) one has

$$a \cdot y = \mathrm{Ad}\,(\rho_{x,m}(a))(y) \tag{3.11}$$

for any $a \in A(m)$.

Proof. For the first statement it suffices, by Lemma 3.1, to show that $\mathfrak{a}(m)|(x_m + M(m)^\perp)$ integrates to an action of $A(m)$ on $x_m + M(m)^\perp$.

Using the notation of §2.1 let $Y \subset P^1(n)$ be the span of α_{ij} for $(i, j) \in L_m$ so that dim $Y = n^2 - m^2$ and Y is the orthocomplement of $M(m)$ in $P^1(n)$. Let $\mathcal{Y}$ be the affine algebra of functions on $x_m + M(m)^\perp$. For any $\alpha \in Y$ let $\alpha_o \in \mathcal{Y}$ be defined so that if $u \in M(m)^\perp$, then

$$\alpha_o(x_m + u) = \alpha(u). \tag{3.12}$$

Let $Y_o = \{\alpha_o \mid \alpha \in Y\}$ so that clearly the subspace Y_o of $\mathcal{Y}$ generates $\mathcal{Y}$. Let

$$r : Y \to Y_o \tag{3.13}$$

be the linear isomorphism defined by putting $r(\alpha) = \alpha_o$. Since any $\alpha \in Y$ vanishes on x_m, note that

$$r(\alpha) = \alpha|(x_m + M(m)^{\perp}) \quad \forall \alpha \in Y. \tag{3.14}$$

Since $Z_{x,m} \subset M(m)$ it is clear that Y is stable under the coadjoint representation of $Z_{x,m}$ and $G_{x,m}$ (resp. coad and Coad) on $P^1(n)$. We now wish to prove that Y_o is stable under $\mathfrak{a}(m)$ and in fact, on Y, one has

$$\xi \circ r = r \circ \text{coad}\,((\rho_{x,m})_*(\xi)) \tag{3.15}$$

for any $\xi \in \mathfrak{a}(m)$. To prove (3.15) we first note that if $k \in I_m$ and $y \in x_m + M(m)^{\perp}$, then

$$(\xi_{(k,m)})_y = (\partial^{[(x_m)^{m-k},y]})_y. \tag{3.16}$$

Indeed (3.16) follows from Theorem 2.12 where we have replaced x by y in (2.43) and recognize that $x_m = y_m$ in our present notation. Applying both sides of (3.16) to $\alpha \in Y$ one has, by (3.14),

$$\begin{aligned}(\xi_{(k,m)}\alpha_o)(y) &= (\xi_{(k,m)}\alpha)(y)\\ &= \langle \alpha, [(x_m)^{m-k}, y]\rangle\\ &= (-\text{coad }(x_m)^{m-k}(\alpha))(y).\end{aligned}$$

But this proves (3.15) (and hence also the stability of Y_o under $\mathfrak{a}(m)$) since $(\rho_{x,m})_*(\xi_{(k,m)}) = -(x_m)^{m-k}$ for $k \in I_m$.

But now the action of $\mathfrak{a}(m)$ on Y_o is a linear representation of $\mathfrak{a}(m)$ on a finite-dimensional vector space. Hence it exponentiates to a representation π of $A(m)$ on Y. But then the commutative diagram (3.15) yields the commutative diagram

$$\pi(a) \circ r = r \circ \text{Coad }(\rho_{x,m}(a)) \tag{3.17}$$

on Y for any $a \in A(m)$. But now if one defines an action of $A(m)$ on $x_m + M(m)^{\perp}$ by putting $a \cdot y = \text{Ad}\,(\rho_{x,m}(a))(y)$ we observe that, for any $\alpha \in Y$,

$$\alpha_o(a^{-1} \cdot y) = (\pi(a)(\alpha_o))(y). \tag{3.18}$$

Indeed this follows from (3.17) since

$$\alpha(\text{Ad }\rho_{x,m}(a^{-1})(y)) = (\text{Coad }(\rho_{x,m}(a))(\alpha))(y).$$

But now replacing a in (3.18) by $\exp\ t\xi$ for $\xi \in \mathfrak{a}(m)$ and differentiating it follows from the definition of π that the action of $A(m)$ on $x_m + M(m)^{\perp}$ given by (3.11) integrates the Lie algebra $\mathfrak{a}(m)|(x_m + M(m)^{\perp})$. □

3.2.

By Theorem 2.7, and the existence of strongly regular elements (see e.g., (2.27)) the sum $\mathfrak{a}$ of the $\mathfrak{a}(m)$ for $m \in I_n$ is a direct sum of the $\mathfrak{a}(m)$ where $m \in I_{n-1}$ so that we can write

$$\mathfrak{a} = \oplus_{m \in I_{n-1}} \mathfrak{a}(m). \tag{3.19}$$

We note then that $\mathfrak{a}$ is a commutative Lie algebra of analytic vector fields on $M(n)$ of dimension $d(n-1)$ with basis $\xi_{p_{(i)}}$ $i \in d(n-1)$, where for $m \in I_n$, and $k \in I_m$,

$$p_{(d(m-1)+k)} = f_{(k,m)} \tag{3.20}$$

(see (2.38)). Let $A = A(1) \times \cdots \times A(n-1)$ so that we can regard $\mathfrak{a} = \operatorname{Lie} A$ and note that as an analytic group

$$A \cong \mathbb{C}^{d(n-1)}. \tag{3.21}$$

As a consequence of Theorem 3.3 one has

Theorem 3.4. *The Lie algebra $\mathfrak{a}$ of vector fields on $M(n)$ integrates to an action of A on $M(n)$ where if $a \in A$ and $a = (a(1), \ldots, a(n-1))$ for $a(m) \in A(m)$, then*

$$a \cdot y = a(1) \cdot (\cdots (a(n-1) \cdot y) \cdots) \tag{3.22}$$

for any $y \in M(n)$. Furthermore the orbits of A on $M(n)$ are leaves of the distribution, $x \mapsto V_x$, on $M(n)$. See §2.4, Remark 2.9 and Theorem 2.13.

Proof. Since $\mathfrak{a}$ is commutative the actions of $A(m)$ for all $m \in I_{n-1}$ clearly commute with one another. Thus (3.22) defines an action of A on $M(n)$ which, by Theorem 3.3, integrates $\mathfrak{a}$. The final statement of the theorem follows from Theorem 2.12 and (2.48). □

Recall that $J(n) \subset P(n)$ is the Poissson commutative associative subalgebra defined in §2.4. See (2.30).

Theorem 3.5. *Let $p \in J(n)$ be arbitrary. Then the vector field ξ_p is globally integrable defining an action of $\mathbb{C}$ on $M(n)$. Furthermore this action commutes with A and stabilizes any A-orbit.*

Let $\{p_i'\}$, $i \in I_d$, for some integer d, be an arbitrary set of generators of $J(n)$. Let $\mathfrak{a}'$ be the (commutative) Lie algebra spanned by $\xi_{p_i'}$, $i \in I_d$, and let A' be a corresponding simply connected Lie group. Then $\mathfrak{a}'$ integrates to an action of A' on $M(n)$. Furthermore the action of A' commutes with the action of A and the orbits of A' are the same as the orbits of A. In particular any A'-orbit is a leaf of the distribution $x \mapsto V_x$ on $M(n)$.

Proof. Let $y \in M(n)$ and let Q be the orbit $A \cdot y$ of A so that Q is an analytic submanifold of $M(n)$. By Theorem 3.4 one has

$$V_x = T_x(Q) \quad \forall x \in Q. \tag{3.23}$$

Let $\mathfrak{p} = \{f \in \mathfrak{a} \mid \xi_f | Q = 0\}$ and let $\mathfrak{q}$ be a linear complement of $\mathfrak{p}$ in $\mathfrak{q}$. Then by the commutativity of A it follows that if $\dim Q = \ell$, then $\dim \mathfrak{q} = \ell$ and if q_j, $j \in I_\ell$, is a basis of $\mathfrak{q}$, then $\{(\xi_{q_j})_x\}$, $j \in I_\ell$, is a basis of $T_x(Q)$ for all $x \in Q$. But now $\xi_p | Q$ is tangent to Q by (3.23). Thus there exists functions h_j, $j \in I_\ell$ on Q so that

$$\xi_p = \sum_{j \in I_\ell} h_j \xi_{q_j} \tag{3.24}$$

on Q. But the commutativity $[\xi_p, \xi_{q_j}] = 0$ implies that the functions h_j are constant on Q. By Theorem 3.3 the vector field $\xi_{q_j}|Q$ integrates to an action, $t \mapsto \exp\ t\,\xi_{q_j}|Q$ of $\mathbb{C}$ on Q. But then one gets another action of $\mathbb{C}$ on Q by putting

$$b(t) = \exp\ t\,h_1\,\xi_{q_1} \cdots \exp\ t\,h_\ell\,\xi_{q_\ell}|Q.$$

But then it is immediate from (3.24) that $\xi_p|Q$ integrates to the action of $b(t)$ on Q. Thus ξ_p integrates to an action of $\mathbb{C}$ on $M(n)$. Clearly this action of commutes with the action of A on $M(n)$ by (2.32). In addition the argument above implies that any A-orbit is stable under the action of $\mathbb{C}$.

It then follows that $\mathfrak{a}'$ integrates to an action of A' on $M(n)$ and the action of A' commutes with the action of A. In addition the argument above implies that any A-orbit Q is stable under the action of A'. But since the p_i' generate $J(n)$ the span of $(dp_i')_x$, for $i \in I_d$, is the same as the span of $(dp_{(j)})_x$, for $j \in I_{d(n)}$, for any $x \in M(n)$. Thus $(\xi_{p_i'})_x$, for $i \in I_d$, spans V_x by (1.14). Consequently all orbits of A' on Q are open. But Q is connected since A is connected. But then A' is obviously transitive on Q. □

3.3.

In this section we will give some properties of the group A. We first observe the elementary property that A operates "vertically" with respect to the "fibration" (2.9) of $M(n)$. Recall (see (2.8)) the morphism $\Phi_n : M(n) \to \mathbb{C}^{d(n)}$. Regarding c_j as a coordinate function on $\mathbb{C}^{d(n)}$ note that, by (2.8), for $j \in I_{d(n)}$,

$$c_j \circ \Phi_n = p_j. \tag{3.25}$$

Proposition 3.6. *The group A stabilizes the "fiber" $M_c(n)$ of $M(n)$ (see (2.9)) for any $c \in \mathbb{C}^{d(n)}$. In fact $M_c(n)$ is stabilized by the action of $\mathbb{C}$ defined in Theorem 3.5.*

Proof. For $i, j \in I_{d(n)}$ one has $\xi_{p_{(i)}} p_j = 0$ by (2.31) (see (3.20)). But then the function p_j on $M(n)$ is invariant under A. Hence $M_c(n)$ is stabilized by A, for any $c \in \mathbb{C}^{d(n)}$, by (3.25). The final statement follows from Theorem 3.5. □

Note: At a later point Knop determined that Theorems 3.4, 3.5 and Proposition 3.6 could be deduced from Theorem 4.1 in [Kn].

Let $x \in M(n)$. We wish to give an explicit description of the A-orbit $A \cdot x$. Obviously $G_{x,m}$ is an algebraic subgroup of $Gl(n)$, for any $m \in I_n$, and the adjoint representation of $G_{x,m}$ on $M(n)$ is an algebraic representation. In the following theorem we use the term "constructible set". This is a concept in algebraic geometry due to C. Chevalley. For its precise definition and properties see §5, Chapter II in [C] or Exercise 1.9.6, p. 19 in [S], or §1.3, p. 3 in [B].

Theorem 3.7. *Let $x \in M(n)$. Consider the following morphism of nonsingular irreducible (see Remark 3.2) affine varieties*

$$G_{x,1} \times \cdots \times G_{x,n-1} \to M(n) \tag{3.26}$$

where for $g(m) \in G_{x,m},\ m \in I_{n-1}$,

$$(g(1), \ldots, g(n-1)) \mapsto \mathrm{Ad}\,(g(1)) \cdots \mathrm{Ad}\,(g(n-1))(x). \tag{3.27}$$

Then the image of (3.26) is exactly the A-orbit $A \cdot x$. In fact if $a \in A$ and $a = a(1) \cdots a(n-1)$ where $a(m) \in A(m)$, then $a \cdot x$ is given by the right side of (3.27) where $g(m) = \rho_{x,m}(a(m))$. In particular $A \cdot x$ is an irreducible, constructible (in the sense of Chevalley) subset of $M(n)$. Furthermore, the Zariski closure of $A \cdot x$ is the same as its usual Hausdorff closure. In addition $A \cdot x$ contains a Zariski open subset of its closure.

Proof. Let $m \in I_{n-1}$. For $g(m) \in G_{x,m}$ there exists, by (3.8) and Theorem 3.2, $a(m) \in A(m)$ such that $a(m) = \mathrm{Ad}\, g(m)$ on $x_m + M(m)^{\perp}$. Let $b(m) = a(m) \cdots a(n-1)$ and $h(m) = g(m) \cdots g(n-1)$. Obviously the right side of (3.27) can be written $\mathrm{Ad}\, h(1)(x)$. Inductively (downwards) assume that $b(m+1) \cdot x = \mathrm{Ad}\, h(m+1)(x)$ for $m < n-1$. Note that the induction assumption is satisfied if $m+1 = n-1$ by Theorem 3.2. But since $p_{(i)}$ is invariant under $Gl(m)$ for $i > d(m-1)$, note (key observation) that $(b(m+1) \cdot x)_m = x_m$ so that $b(m+1) \cdot x \in x_m + M(m)^{\perp}$. Thus $b(m) \cdot x = \mathrm{Ad}\, h(m)(x)$ by Theorem 3.2. Hence, by induction, $\mathrm{Ad}\, h(1)(x) \in A \cdot x$. Conversely let $a \in A$ so that we can write $a = a(1) \cdots a(n-1)$ for $a(m) \in A(m)$. Let $g(m) = \rho_{x,m}(a(m))$ so that $g(m) \in G_{x,m}$. Then the argument above establishes that $a \cdot x = \mathrm{Ad}\, h(1)(x)$. Hence the image of (3.26) is $A \cdot x$. The image of an irreducible set under a morphism is obviously irreducible. But the image $A \cdot x$ is constructible by Proposition 5 in Chapter V, p. 95–96 in [C] or by Proposition, p. 4 in [B]. See also Corollary 2, §8, p.51 in [M]. But then the Euclidean closure of $A \cdot x$ is the same as its Zariski closure by Corollary 1, §9, p. 60 in [M]. On the other hand the irreducible constructible set $A \cdot x$ contains a Zariski open subset of its closure by Proposition 4 in Chapter 5, p. 95 in [C] or by Proposition, p. 4 in [B]. □

Remark 3.8. In (3.22) the ordering of the terms $a(m)$ is immaterial since these elements commute with one another. However, the ordering of the terms $g(m)$ in (3.27) cannot be permuted in general. The groups $G_{x,m}$ do not, in general, commute with one another. The point is that $A(m)$ is given by $\mathrm{Ad}\, G_{x,m}$ only on $x_m + M(m)^{\perp}$.

By the final statement of Theorem 3.4 and (2.34) one has, for any $x \in M(n)$,

$$\dim\, A \cdot x \leq d(n-1) \tag{3.28}$$

and, by (2.35),

$$\dim\, A \cdot x = d(n-1) \iff x \text{ is strongly regular (see §2.3).} \tag{3.29}$$

That is, (see §2.3) the nonempty Zariski open subset $M^{\mathrm{sreg}}(n)$ of $M(n)$ is stable under the action of A and the A-orbits in $M^{\text{sreg}}(n)$ are exactly all the A-orbits of maximal dimension $(d(n-1))$. (3.30)

We will now see that we can make a much stronger statement about $A \cdot x$ than that in Theorem 3.6 in case x is strongly regular. We first need some preliminary results.

Let $c \in \mathbb{C}^{d(n)}$. One has $M_c(n) = (\Phi_n)^{-1}(c)$ (see (2.8)) so that $M_c(n)$ is a Zariski closed subset of $M(n)$. Now let $M_c^{\text{sreg}}(n) = M_c(n) \cap M^{\text{sreg}}(n)$.

Remark 3.9. Note that $M_c^{\text{sreg}}(n)$ is a Zariski open subset of $M_c(n)$. Furthermore $M_c^{\text{sreg}}(n)$ is not empty since $M_c(n) \cap (e + \mathfrak{b})$ is a one point subset of $M_c^{\text{sreg}}(n)$, by Theorem 2.5 and (2.27). However $M_c^{\text{sreg}}(n)$ may not be dense in $M_c(n)$. In fact one can show that it is not dense if $n = 3$ and $c = 0$.

Let $N(c)$ be the number of irreducible components of $M_c^{\text{sreg}}(n)$ and let $M_{c,i}^{\text{sreg}}(n)$, $i = 1, \ldots, N(c)$, be some indexing of these components. In the following proposition overline is Zariski closure.

Proposition 3.10. *The irreducible component decomposition of* $\overline{M_c^{\text{sreg}}(n)}$ *is*

$$\overline{M_c^{\text{sreg}}(n)} = \bigcup_{i=1}^{N(c)} \overline{M_{c,i}^{\text{sreg}}(n)}, \tag{3.31}$$

noting that $\overline{M_{c,i}^{\text{sreg}}(n)}$ *is a closed subvariety of* $M_c(n)$*. Furthermore one recovers* $M_{c,i}^{\text{sreg}}(n)$ *from its closure by*

$$M_{c,i}^{\text{sreg}}(n) = (\overline{M_{c,i}^{\text{sreg}}(n)}) \cap M^{\text{sreg}}(n) \tag{3.32}$$

so that $M_{c,i}^{\text{sreg}}(n)$ *is an Zariski open subvariety of its Zariski closure. Furthermore,* $M_{c,i}^{\text{sreg}}(n)$ *is a constructible set so that its Zariski closure is the same as its usual Hausdorff closure (see p. 60 in [M]).*

Proof. The equality (3.31) is immediate. The statement that (3.31) is an irreducible component decomposition is an easy consequence of (3.32). In fact it is a general result. See the last paragraph in §1.1, p.3 in [B]. Clearly the left side of (3.32) is contained in the right-hand side. But the right-hand side is an irreducible set in $M_c^{\text{sreg}}(n)$. Thus one has (3.32) by the irreducible maximality of $M_{c,i}^{\text{sreg}}(n)$ in $M_c^{\text{sreg}}(n)$. We have used Proposition 2 in Chapter II of [C], p. 35 throughout. $M_{c,i}^{\text{sreg}}(n)$ is a constructible set by (3.32) and Proposition 3 in Chapter II of [C], p. 94. The last statement of the proposition follows from Corollary 1, §9, p. 60 in [M]. □

Proposition 3.11. *Let the assumptions and notations be as in Proposition 3.10. Then* $M_{c,i}^{\text{sreg}}(n)$ *and hence* $\overline{M_{c,i}^{\text{sreg}}(n)}$ *have dimension* $d(n-1)$ *for all* i*. Furthermore* $M_{c,i}^{\text{sreg}}(n)$ *is nonsingular in* $\overline{M_c^{\text{sreg}}(n)}$ *and a fortiori* $M_{c,i}^{\text{sreg}}(n)$ *is nonsingular in* $\overline{M_{c,i}^{\text{sreg}}(n)}$.

Proof. The proof is an application of Theorem 4 in §4 of Chapter 3 in [M], p. 172, where in the notation of that reference, $X = M(n)$, $Y = \overline{M_c^{\text{sreg}}(n)}$, U runs through all open affine subvarieties of a finite affine cover of $M^{\text{sreg}}(n)$, $k = d(n)$, $f_i = p_i - c_i$. □

Let $c \in \mathbb{C}^{d(n)}$ be arbitrary. Then since $M_c(n)$ is stable under the action of A it follows from (3.30) that $M_c^{\mathrm{sreg}}(n)$ is stable under A.

Theorem 3.12. *Let the assumptions and notations be as in Proposition 3.10. Then there are exactly $N(c)$ orbits of A in $M_c^{\mathrm{sreg}}(n)$ and these orbits are identical to the irreducible components $M_{c,i}^{\mathrm{sreg}}(n)$ of $M_c^{\mathrm{sreg}}(n)$. In particular all the orbits are nonsingular algebraic varieties of dimension $d(n-1)$.*

Proof. Let $x \in M_c^{\mathrm{sreg}}(n)$ so that $A \cdot x \subset M_c^{\mathrm{sreg}}(n)$. But by the irreducibility of $A \cdot x$ (see Theorem 3.6) the Zariski closure $\overline{A \cdot x}$ is irreducible. Thus, by Proposition 3.10, there exists i such that

$$\overline{A \cdot x} \subset \overline{M_{c,i}^{\mathrm{sreg}}(n)}. \tag{3.33}$$

But then if $k = \dim \overline{A \cdot x}$ one has $k \le d(n-1)$ and one has equality in (3.33) if $k = d(n-1)$. But $A \cdot x$ contains a Zariski open set U of $\overline{A \cdot x}$ by the last statement in Theorem 3.7. Let $W \subset U$ be the nonsingular Zariski open subvariety of simple points in U. Then W is an analytic submanifold of dimension k in $M(n)$. But, as a nonempty Zariski open subvariety of a closed subvariety, $W = V \cap \overline{A \cdot x}$ where V is a Zariski open subset of $M(n)$. Thus W is open in $A \cdot x$. But $A \cdot x$ is a $d(n-1)$-dimensional analytic submanifold of $M(n)$. Thus $k = d(n-1)$, by invariance of dimension of submanifolds, so that

$$\overline{A \cdot x} = \overline{M_{c,i}^{\mathrm{sreg}}(n)}. \tag{3.34}$$

But now (3.34) implies that $\overline{M_{c,i}^{\mathrm{sreg}}(n)}$ is stable under the action of A since $\overline{A \cdot x}$ is the closure of $A \cdot x$ in the usual Euclidean topology. But then $M_{c,i}^{\mathrm{sreg}}(n)$ must be stable under the action of A by (3.32). In particular $A \cdot x \subset M_{c,i}^{\mathrm{sreg}}(n)$ since, clearly, $x \in M_{c,i}^{\mathrm{sreg}}(n)$ by (3.32) and (3.33). Assume $A \cdot x \neq M_{c,i}^{\mathrm{sreg}}(n)$ and let $y \in M_{c,i}^{\mathrm{sreg}}(n) - A \cdot x$. Then clearly we may replace x by y in (3.34) so that $\overline{A \cdot y} = \overline{M_{c,i}^{\mathrm{sreg}}(n)}$. But then $A \cdot x$ and $A \cdot y$ contain disjoint nonempty Zariski open subsets of $\overline{M_{c,i}^{\mathrm{sreg}}(n)}$. This contradicts the irreducibility of $\overline{M_{c,i}^{\mathrm{sreg}}(n)}$. Thus we have proved that $M_{c,i}^{\mathrm{sreg}}(n)$ is an A-orbit. If $j \neq i$ we may find $z \in M_{c,j}^{\mathrm{sreg}}(n)$ which does not lie in any other component of $M_c^{\mathrm{sreg}}(n)$. The argument above then implies that $M_{c,j}^{\mathrm{sreg}}(n) = A \cdot z$. □

3.4.

Let $x \in M^{\mathrm{sreg}}(n)$. Then by Theorems 3.6 and 3.12 one has the following surjective morphism of nonsingular varieties:

$$\nu_x : G_{x,1} \times \cdots \times G_{x,n-1} \to A \cdot x \tag{3.35}$$

where

$$\nu_x(g(1), \ldots, g(n-1)) = \mathrm{Ad}(g(1)) \cdots \mathrm{Ad}(g(n-1))(x). \tag{3.36}$$

We now prove

Lemma 3.13. *The map (3.35) is injective so that in fact ν_x is bijective for any $x \in M^{\text{sreg}}(n)$.*

Proof. Assume $g(m), g(m)' \in G_{x,m}$ for $m = 1, \ldots, n-1$, and

$$\text{Ad}(g(1)) \cdots \text{Ad}(g(n-1))(x) = \text{Ad}(g(1)') \cdots \text{Ad}(g(n-1)')(x).$$

By upward induction we will prove that $g(m) = g(m)'$ for all m. The inductive argument to follow also establishes that the result is true for $m = 1$. Assume we have proved that $g(i) = g(i)'$ for $i < m$. Then if $y = \text{Ad}(g(m)) \cdots \text{Ad}(g(n-1))(x)$ one has $y = \text{Ad}(g(m)') \cdots \text{Ad}(g(n-1)')(x)$. Let $z = \text{Ad}(g(m))^{-1}(y)$ and $z' = \text{Ad}(g(m)')^{-1}(y)$. Then $z_{m+1} = z'_{m+1} = x_{m+1}$. But $z = \text{Ad}(h(m))(z')$ where $h(m) = g(m)^{-1}g(m)'$. But then $h(m)$ commutes with x_{m+1}. Thus $h(m) - \text{Id}_n \in Z_{x,m} \cap Z_{x,m+1}$ by (2.46). But $Z_{x,m} \cap Z_{x,m+1} = 0$ by (2.49). Thus $g(m) = g(m)'$ and hence ν_x is injective. □

We can now describe the structure of all A-orbits of maximal (i.e., $d(n-1)$) dimension.

Theorem 3.14. *Let $x \in M^{\text{sreg}}(n)$ and for $m = 1, \ldots, n-1$, let $G_{x,m}$ be the centralizer of x_m in $Gl(m)$ so that $G_{x,m}$ is a connected commutative algebraic group of dimension m. Then the morphism (3.35) is an algebraic isomorphism of nonsingular varieties so that as a variety*

$$A \cdot x \cong G_{x,1} \times \cdots \times G_{x,n-1}. \tag{3.37}$$

Proof. Since ν_x is a bijective morphism of varieties of dimension $d(n-1)$ it follows that ν_x is birational by Theorem 3 of §IV in [C], p. 115 ($n = 0$ in the notation of this reference). But then ν_x is an isomorphism of algebraic varieties by Theorem, p. 78, §18 of Chapter AG in [B] and also Theorem 5.2.8, p. 85 in [S] since the range and domain of ν_x are nonsingular by Theorem 3.12. □

Example 3.15. Since $e \in e + \mathfrak{b}$ note that the principal nilpotent element e is strongly regular. If $x = e$, then one readily sees that $A \cdot x$ is the set of all principal nilpotent elements in the nilpotent Lie algebra $\mathfrak{u}'$ of all strictly lower triangular matrices and $\overline{A \cdot x} = \mathfrak{u}'$. Writing the set of all principal nilpotent elements in $\mathfrak{u}'$ uniquely, using the right-hand side of (3.36), when $x = e$, may be new.

3.5.

In this section we will assume x satisfies the eigenvalue disjointness condition. That is, $x \in M_\Omega(n)$. See §2.4 and more specifically (2.53). We recall that $M_\Omega(n)$ is a Zariski open subset of $M^{\text{sreg}}(n)$ so that, in particular, Theorem 3.14 applies to x.

Theorem 3.16. *Let $x \in M(n)$ satisfy the eigenvalue disjointness condition. Then $Z_{x,m}$ is a Cartan subalgebra of $M(m)$, for any $m \in I_n$, and $G_{x,m}$ is a maximal (complex) torus in $Gl(m)$ so that*

$$G_{x,m} \cong (\mathbb{C}^{\times})^m. \tag{3.38}$$

In addition the orbit $A \cdot x$ is an algebraic subvariety of $M(n)$ and as such

$$A \cdot x \cong (\mathbb{C}^{\times})^{d(n-1)}. \tag{3.39}$$

Proof. $Z_{x,m}$ is a Cartan subalgebra of $M(m)$ since x_m is a regular semisimple element of $M(m)$. But this of course implies that $G_{x,m}$ is a maximal (complex) torus in $Gl(m)$ since Lie $G_{x,m} = Z_{x,m}$. One then has (3.38). Theorem 3.14 then implies the remaining statements of Theorem 3.16. □

For $m \in I_{n-1}$ let $Y(m)$ be the span of $e_{i\,m+1}$, $i \in I_m$, so that $Y(m)$ is an m-dimensional subspace of $M(m)^{\perp}$. It is clear that $Y(m)$ is stable under $\mathrm{Ad}\, Gl(m)$ and $\mathrm{ad}\, M(m)$ and hence $Y(m)$ inherits, respectively, the structure of a $Gl(m)$ and a $M(m)$-module. On the other hand $\mathbb{C}^m$ is a module for $Gl(m)$ and $M(m)$ with respect to the natural action of $Gl(m)$ and $M(m)$. One immediately observes that the linear isomorphism

$$Y(m) \to \mathbb{C}^m, \quad y \mapsto (\alpha_{1\,m+1}(y), \ldots, \alpha_{m\,m+1}(y)) \tag{3.40}$$

is an isomorphism of $Gl(m)$ and $M(m)$-modules.

Let $H(m)$ be a maximal torus of $Gl(m)$ and let $\mathfrak{h}(m) = \mathrm{Lie}\, H(m)$ be the corresponding Cartan subalgebra of $M(m)$. Then, by restriction, $Y(m)$ is a $H(m)$-module with respect to Ad and an $\mathfrak{h}(m)$-module with respect to ad. One notes in fact that $Y(m)$ is a cyclic $H(m)$-module and a cyclic $\mathfrak{h}(m)$-module. This is transparent, from the module isomorphism (3.40), in case we choose $H(m) = \mathrm{Diag}(m)$ where $\mathrm{Diag}(m)$ is the group of diagonal matrices in $Gl(m)$. The general case follows by conjugation. An element $y \in Y(m)$ will be called a cyclic $H(m)$-generator in case it generates $Y(m)$ under the action of $H(m)$. A similar terminology will be used for $\mathfrak{h}(m)$. By going to the diagonal case it is immediate that y is a cyclic generator for $H(m)$ if and only if it is a cyclic generator for $\mathfrak{h}(m)$. Indeed let

$$Y^{\times}(m) = \{y \in Y(m) \mid \alpha_{i\,m+1}(y) \neq 0, \ \forall i \in I_m\}. \tag{3.41}$$

Then clearly y is a cyclic generator for $\mathrm{Diag}(m)$ and a cyclic generator for Lie $\mathrm{Diag}(m)$ if and only if $y \in Y^{\times}(m)$. The following proposition is established by again going to the diagonal case and conjugating (using of course the module isomorphism (3.40)).

Proposition 3.17. *Let $m \in I_{n-1}$. Let $H(m)$ be a maximal torus of $GL(m)$ and let $\mathfrak{h}(m) = \mathrm{Lie}\, H(m)$. Let $y \in Y(m)$. Then the following conditions are all equivalent:*

(a) y is a cyclic generator for $H(m)$ (or $\mathfrak{h}(m)$),
(b) $\mathrm{Ad}\, H(m)(y)$ is the unique $\mathrm{Ad}\, H(m)$-orbit in $Y(m)$ of maximal dimension (m),
(c) $\mathrm{Ad}\, h(y) = y$ for $h \in H(m)$ implies $h = \mathrm{Id}_n$,
(d) If $w \in Y(m)$, then $w \in \mathrm{Ad}\, H(m)(y) \iff w$ is a cyclic generator for $H(m)$,
(e) $\{[z_{(i)}, y]\}$, $i \in I_m$, is a basis of $Y(m)$ if $\{z_{(i)}\}$, $i \in I_m$, is a basis of $\mathfrak{h}(m)$.

For any $x \in M_\Omega$ and $m \in I_{n-1}$ let

$$Y^x(m) = \{y \in Y(m) \mid y \text{ is a cyclic generator for } G_{x,m}\}. \tag{3.42}$$

We recall that x_m is the "cutoff" of x in $M(m)$ for any $x \in M(n)$ where $m \in I_n$. Henceforth, if $m \in I_{n-1}$, let $x_{\{m\}} \in Y(m)$ be the "component" of x in $Y(m)$. That is, $x_{\{m\}} = \sum_{i=1}^{m} \alpha_{i\,m+1}(x) e_{i\,m+1}$.

Proposition 3.18. *Let $x \in M_\Omega(n)$. Then $x_{\{m\}} \in Y^x(m)$.*

Proof. Let $g \in Gl(m)$ be such that $g\,x_m\,g^{-1} \in \mathrm{Diag}(m)$. Then clearly $g\,G_{x,m}\,g^{-1} = \mathrm{Diag}(m)$. It suffices then to show that $g\,x_{\{m\}}\,g^{-1} \in Y^\times(m)$. To prove this we use the notation and the argument in the proof of Theorem 2.17. But then the result follows from (2.60). □

Proposition 3.17 lists a number of criteria for $y \in Y(m)$ to be an element of $Y^x(m)$. It is convenient to add the following to this list.

Proposition 3.19. *Let $x \in M_\Omega(n)$ and let $m \in I_{n-1}$. Let $y \in Y(m)$. Then $y \in Y^x(m)$ if and only if $\{(\mathrm{ad}\, x_m)^j(y)\}$, $j = 0, \ldots, m-1$, is a basis of $Y(m)$.*

Proof. As noted in Proposition 3.17 one has $y \in Y^x(m)$ if and only if y is a cyclic $Z_{x,m}$ generator. But x_m is a regular semisimple element of $M(m)$. But then $\mathrm{ad}\, x_m | Y(m)$ is diagonalizable with m distinct eigenvalues, by the module isomorphism (3.40). But this readily establishes the proposition by standard linear algebra. □

Remark 3.20. If $x \in M_\Omega(n)$, note that $Y^x(m)$ is a Zariski open $\mathrm{Ad}\, G_{x,m}$-orbit in $Y(m)$ by Proposition 3.17. However, if $m > 1$ the maximal torii $G_{x,m}$ of $Gl(m)$ do not run through all maximal torii of $Gl(m)$ as x runs through $M_\Omega(n)$. For example one easily has $G_{x,m} \neq \mathrm{Diag}(m)$ for any $x \in M_\Omega(n)$. This restricts the possible sets $Y^x(m)$. In fact, for example, the following theorem implies that $Y^x(m) \neq Y^\times(m)$ for all $x \in M_\Omega(n)$.

Theorem 3.21. *Let $m \in I_{n-1}$. One has $e_{m\,m+1} \in Y^x(m)$ for all $x \in M_\Omega(n)$.*

Proof. Let $x \in M_\Omega(n)$. If $m = 1$, the result is obvious. Assume $m > 1$. Let W be the subspace of $Y(m)$ spanned by $(\mathrm{ad}\, x_m)^j(e_{m,m+1})$ for $j = 0, \ldots, m-1$. Using the characteristic polynomial of $\mathrm{ad}\, x_m$ it is clear that W is stable under $\mathrm{ad}\, x_m$. To show that $e_{m,m+1}$ is a cyclic $G_{x,m}$ generator (= cyclic $Z_{x,m}$ generator) it suffices, by Proposition 3.19, to show that $W = Y(m)$. For this, of course, it suffices to show

$$\dim\, W = m. \tag{3.43}$$

Let $Y_o(m)$ be the span of $e_{i\,m+1}$, $i = 1, \ldots, m-1$, so that $\dim\, Y_o(m) = m - 1$. Let $Q : Y(m) \to Y_o(m)$ be the projection with respect to the decomposition $Y(m) = Y_o(m) \oplus \mathbb{C}\, e_{m\,m+1}$. Let $W_o = W \cap Y_o(m)$. Since $e_{m\,m+1} \in W$ one has $W = W_o \oplus \mathbb{C}\, e_{m\,m+1}$. To prove (3.43) it suffices to show that $W_o = Y_o(m)$. Clearly $Q(W) = W_o$. Let $v \in W_o$. Then, since $\alpha_{m\,m+1}(v) = 0$, note that

$$Q[x_m, v] = [x_{m-1}, v] \tag{3.44}$$

so that W_o is stable under $\operatorname{ad} x_{m-1}$. But now $[x_m, e_{m\,m+1}] \in W$ and hence $Q[x_m, e_{m\,m+1}] \in W_o$. Thus if $w_j = (\operatorname{ad} x_{m-1})^j (Q[x_m, e_{m\,m+1}])$, then

$$w_j \in W_o \tag{3.45}$$

for $j \in \{0, \dots, m-2\}$. But note that

$$[Q[x_m, e_{m\,m+1}], e_{m+1\,m}] = x_{\{m-1\}}. \tag{3.46}$$

But $\operatorname{ad} x_{m-1}$ commutes with $\operatorname{ad} e_{m+1\,m}$. Thus

$$[w_j, e_{m+1\,m}] = (\operatorname{ad} x_{m-1})^j x_{\{m-1\}}. \tag{3.47}$$

But $x_{\{m-1\}} \in Y^x(m-1)$ by Proposition 3.18. Hence the dimension of the subspace spanned by the vectors on the right-hand side of (3.47) for $j \in \{0, \dots, m-2\}$ is $m-1$ by Proposition 3.19. Thus the space spanned by the w_j must have dimension $m-1$. This proves that $W_o = Y_o(m)$. □

For any $x \in M(n)$ let x^T be the transpose matrix and for any subset $X \subset M(n)$ let $X^T = \{x^T \mid x \in X\}$. Obviously

$$p_i(x) = p_i(x^T) \tag{3.48}$$

for all $x \in M(n)$ and $i \in I_{d(n)}$ so that

$$M_c(n) = M_c(n)^T \tag{3.49}$$

for any $c \in \mathbb{C}^{d(n)}$. Note that (see Sections 2.1 and (2.12))

$$-e^T = \sum_{m \in I_{n-1}} e_{m\,m+1}. \tag{3.50}$$

Let $\mathfrak{u} \subset M(n)$ be the Lie algebra of strictly upper triangular matrices. Then clearly

$$\mathfrak{u} = \oplus_{m \in I_{n-1}} Y(m). \tag{3.51}$$

Lemma 3.22. *Let $x \in M_\Omega(n)$. Then there exists $b_x \in A$ such that*

$$b_x \cdot x \in (-e + \mathfrak{b})^T. \tag{3.52}$$

Proof. Let $m \in I_{n-1}$. Assume inductively (downward) we have found $g(k) \in G_{x,k}$ for $k = m, \dots, n-1$ such that if $h(m) = g(m) \cdots g(n-1)$ and $y(m) = (\operatorname{Ad} h(m))(x)$, then

$$y(m)_{\{j\}} = e_{j\,j+1} \text{ for } j = m, \dots, n-1.$$

If $m = n-1$, this induction assumption is satisfied, by Theorem 3.21, in that we can choose $g(n-1) \in G_{x,n-1}$ such that $\operatorname{Ad} g(n-1)(x_{\{n-1\}}) = e_{n-1\,n}$. To advance the induction assume $m > 1$ and note that, since x_k is fixed by $\operatorname{Ad} G_{x,k}$ for all k, one has $(y(m))_m = x_m$. Hence $y(m)_{\{m-1\}} = x_{\{m-1\}}$. By Theorem 3.21 we can choose $g(m-1) \in G_{x,m-1}$ so that

$$\operatorname{Ad} g(m-1)(x_{\{m-1\}}) = e_{m-1\,m}. \tag{3.53}$$

Let $h(m-1) = g(m-1)h(m)$ and $y(m-1) = (\operatorname{Ad} h(m-1))(x)$. The point is that the induction assumption is satisfied for $y(m-1)$ since the element $e_{j\,j+1} \in Y(j)$ is clearly fixed under the adjoint action of $Gl(m-1)$ on $Y(j)$ for $j = m, \ldots, n-1$. By Theorem 3.6 there exists $b_x \in A$ such that $b_x \cdot x = (\operatorname{Ad} h(1))(x)$. But then $(b_x \cdot x)_{\{k\}} = e_{k\,k+1}$ for $k = 1, \ldots, n-1$. But then by (3.50) one has $b_x \cdot x + e^T \in \mathfrak{b}^T$. □

Let $c \in \Omega(n)$ where we recall $\Omega(n) \subset \mathbb{C}^{d(n)}$ is the Zariski open set defined by the eigenvalue disjointness condition. See Sections 2.4 and (2.53). We now have the following neat description of the fiber $M_c(n)$ of (2.8).

Theorem 3.23. *Let $c \in \Omega(n)$. See (2.53). Then the group A operates transitively on the fiber $M_c(n)$ of (2.8). Furthermore $M_c(n)$ is an A-orbit of maximal dimension in $M(n)$. Moreover $M_c(n)$ is a nonsingular Zariski closed subvariety of $M(n)$ of dimension $d(n-1)$. As an algebraic variety*

$$M_c(n) \cong (\mathbb{C}^\times)^{d(n-1)}. \tag{3.54}$$

Proof. Since $M_c(n) \subset M^{\text{sreg}}(n)$ (see (2.55)) one has $M_c^{\text{sreg}}(n) = M_c(n)$ in the notation of Remark 3.9. But obviously $M_c(n)$ is Zariski closed in $M(n)$. But, by Theorem 3.12, the irreducible components of $M_c(n)$ are then the A-orbits in $M_c(n)$, each of which is a maximal orbit and a nonsingular variety of dimension $d(n-1)$. We now show that there is only one orbit (i.e., $N(c) = 1$ in the notation of Theorem 3.12). By Remark 2.4 we may replace e by $-e$ in Theorem 2.5. We have put $\mathfrak{b}_e = -e + \mathfrak{b}$ (see Section 2.2) and, hence recalling (3.48), $\mathfrak{b}_e$ by $\mathfrak{b}_e^T$ so that the restriction

$$\Phi_n : \mathfrak{b}_e^T \to \mathbb{C}^{d(n)} \tag{3.55}$$

is an algebraic isomorphism. But now if $x, y \in M_c(n)$ there exists, by Lemma 3.22, $b_x, b_y \in A$ such that both $b_x \cdot x$ and $b_y \cdot y$ are in $\mathfrak{b}_e^T$. But since $M_c(n)$ is stabilized by A, it follows from (3.55) that $b_x \cdot x = b_y \cdot y$. Thus x and y are A conjugate and hence $N(c) = 1$. The isomorphism (3.54) follows from the isomorphism (3.39). □

The following two results could have been proved at an earlier point. However they are manifestly transparent as a consequence of Theorem 3.23.

Proposition 3.24. *Let $c \in \Omega(n)$ and let $x \in M_c(n)$. Then the tangent vectors $(\xi_{p(i)})_x$, $i \in I_{d(n-1)}$, are a basis of $T_x(M_c(n))$ (see (3.20)).*

Proof. The tangent vectors in the statement of the theorem are obviously a basis of $T_x(A \cdot x)$ since $A \cdot x$ is a maximal orbit. But then the proposition follows from the equality $A \cdot x = M_c(n)$. □

The subalgebra $J(n) \subset P(n)$ (see (2.30)) is Poisson commutative by (2.31). Since $P(n)$ is a Poisson algebra, the set of all $f \in P(n)$, which Poisson commutes with all polynomials in $J(n)$, is an algebra containing $J(n)$. We now assert that this set is equal to $J(n)$.

Theorem 3.25. *Let $J(n) \subset P(n)$ be the subalgebra (and polynomial ring in $d(n)$ generators) defined by (2.30). Then $J(n)$ is a maximal Poisson commutative subalgebra of $P(n)$.*

Proof. Since $e + \mathfrak{b}$ is a translate of linear subspace of $M(n)$ it follows from Theorem 2.3 that the affine ring of $e+\mathfrak{b}$ is the polynomial ring in the restrictions $p_i|(e+\mathfrak{b})$, $i \in I_{d(n)}$. Thus for any $f \in P(n)$ there exists $p \in J(n)$ such that $f = p$ on $e + \mathfrak{b}$. But now if f Poisson commutes with all elements in $J(n)$, then $f|M_c(n)$ is a constant by Proposition 3.24 and the connectivity of $M_c(n)$, implied by Theorem 3.23. Thus $f = p$ on $M_\Omega(n)$. But $M_\Omega(n)$ is Zariski dense in $M(n)$. Hence $f = p$. □

We recall (see (3.51)) $\mathfrak{u}$ is the Lie algebra of all lower triangular matrices. For any $x \in M(n)$ let $x_\mathfrak{u}$ be the "component" of x in $\mathfrak{u}$. That is $x_\mathfrak{u} \in \mathfrak{u}$ is such that $x - x_\mathfrak{u} \in \mathfrak{b}^T$. Noting Proposition 3.18 one has

$$x_\mathfrak{u} = \sum_{m \in I_{n-1}} x_{\{m\}}. \tag{3.56}$$

Clearly (3.56) is just the decomposition of $x_\mathfrak{u}$ defined by the direct sum (3.51). Now let $c \in \mathbb{C}^{d(n)}$ and let

$$\beta_c : M_c(n) \to \mathfrak{u} \tag{3.57}$$

be the regular morphism defined by putting $\beta_c(x) = x_\mathfrak{u}$. Of course $\mathfrak{u}$ is a nonsingular variety of dimension $d(n-1)$. If $c \in \Omega(n)$, then $M_c(n)$ is also a nonsingular variety of dimension $d(n-1)$ by Theorem 3.23. For such c we can establish the following description of $M_c(n)$.

Theorem 3.26. *Let $c \in \Omega(n)$ (see (2.53)). Let $\mathfrak{u}_c \subset \mathfrak{u}$ be the image of β_c. Then $\mathfrak{u}_c$ is a nonempty Zariski open subset of $\mathfrak{u}$ and*

$$\beta_c : M_c(n) \to \mathfrak{u}_c \tag{3.58}$$

is an algebraic isomorphism.

Proof. We first prove that β_c is injective. Let $x, y \in M_c(n)$ and assume $x_\mathfrak{u} = y_\mathfrak{u}$. But then $x_{\{m\}} = y_{\{m\}}$ for all $m \in I_n$. To prove that $x = y$ we will inductively (upward) prove that $x_m = y_m$ for all $m \in I_n$. Since x_m and y_m have the same spectrum one has $\operatorname{tr} x_m = \operatorname{tr} y_m$ for all $m \in I_n$ and hence $\alpha_{jj}(x) = \alpha_{jj}(y)$ for all $j \in I_n$. In particular this holds for $j = 1$. Thus $x_1 = y_1$. Assume $m \in I_{n-1}$ and $x_m = y_m$. Let $g \in GL(m)$ be such that $g x_m g^{-1}$ is diagonal in $M(m)$. We use the notation and arguments in the proof of Theorem 2.17. The characteristic polynomial of x_{m+1} and also of y_{m+1} is given by (2.59). But, from (2.59), the residue of $p(\lambda)/\prod_{i=1}^m(\lambda - \mu_{i\,m})$ at $\lambda = \mu_{i\,m}$, where $i \in I_m$, is $-a_{i\,m+1}b_{m+1\,i}$.

This is nonzero by (2.60) and is unchanged if y replaces x. But $x_{\{m\}} = y_{\{m\}}$ implies that $\alpha_{i\,m+1}(g\,x_{m+1}g^{-1}) = \alpha_{i\,m+1}(g\,y_{m+1}g^{-1})$ (recalling that Y_m is stable under $\operatorname{Ad} Gl(m)$). Thus $\alpha_{m+1\,i}(g\,x_{m+1}g^{-1}) = \alpha_{m+1\,i}(g\,y_{m+1}g^{-1})$. But then $g\,x_{m+1}g^{-1} = g\,y_{m+1}g^{-1}$. Hence $x_{m+1} = y_{m+1}$. This proves the injectivity of (3.57).

We next wish to prove that the differential of (3.57) is an isomorphism at all points of $M_c(n)$. Assume not, so there exists $x \in M_c(n)$, $0 \neq \xi \in \mathfrak{a}$ such that $(\beta_c)_*(\xi_x) = 0$. Recalling (3.19) let ξ_j be the component of ξ in $\mathfrak{a}(j)$ (so that $j \in I_{n-1}$). Let m be minimal such that $\xi_m \neq 0$. But now $\xi(\alpha_{k\,m+1})(x) = 0$ for $k \in I_m$. On the other hand $\xi_j(\alpha_{k\,m+1}) = 0$ for $j \geq m+1$ since $f_{(i,j)} \in P(j)^{Gl(j)}$ (see (2.38)) Poisson commutes with $\alpha_{k\,m+1}$ for all $i \in I_j$. Thus $\xi_m(\alpha_{k\,m+1})(x) = 0$. But then, by (2.43) there exists $0 \neq z \in Z_{x,m}$ (see (2.46)) such that $\alpha_{k\,m+1}([z,x]) = 0$ for all $k \in I_m$. Let $W(m)$ be the B-orthocomplement of $Y(m)^T$ in $M(n)$ so that one has

$$M(n) = Y(m) \oplus W(m) \tag{3.59}$$

(see (3.40)). It is clear that both summands in (3.59) are stable under $\operatorname{ad} M(m)$. But the component of x in $Y(m)$ relative to (3.59) is $x_{\{m\}}$. But $\alpha_{k\,m+1}$ clearly vanishes on $W(m)$. Thus $\alpha_{k\,m+1}([z, x_{\{m\}}]) = 0$ for all $k \in I_m$. But this implies $[z, x_{\{m\}}] = 0$. However $x_{\{m\}} \in Y^x(m)$ by Proposition 3.18. But then $[z, x_{\{m\}}] = 0$ clearly contradicts (e) of Proposition 3.17. This proves that the differential of (3.57) is everywhere a linear isomorphism. The fact that $\mathfrak{u}_c$ is Zariski open in $\mathfrak{u}$ follows by combining III, Proposition 10.4, p. 270 in [H] with III, Exercise 9.1, p. 266 in [H]. (We thank P. Etingof for this reference). But then (3.58) is an isomorphism by Theorem 5.2.8, p. 85 in [S]. □

3.6.

Let $\mathfrak{a}^{\mathrm{diag}}$ be the $n-1$ dimensional subalgebra spanned by $\xi_{(m,m)}$, $m \in I_{n-1}$ and let $A^{\mathrm{diag}} \subset A$ be the corresponding $n-1$ dimensional subgroup. In Remark 2.4 it was noted that the p_j, $j \in I_n$, are invariant under $\operatorname{Ad}\operatorname{Diag}(n)$. Consequently the same statement is true for any $p \in J(n)$ and hence $\operatorname{Ad}\operatorname{Diag}(n)$ necessarily commutes with the action of A. We now observe that, in fact, $A^{\mathrm{diag}}(n)$ operates on $M(n)$ as $\operatorname{Ad}\operatorname{Diag}(n)$ so that A can be thought of as an $d(n-1)$-dimensional extension of $\operatorname{Ad}\operatorname{Diag}(n)$. Let $\mathfrak{d}(n)$ be the Lie algebra of all diagonal matrices in $M(n)$ so that $\mathfrak{d}(n) = \operatorname{Lie}\operatorname{Diag}(n)$. Actually we have already, analogously defined $\mathfrak{d}(m) \subset M(m)$ for $m \in I_{n-1}$. See (2.61). Clearly I_m, $m \in I_n$, is a basis of $\mathfrak{d}(n)$. But $\operatorname{ad}\mathrm{Id}_n$ operates trivially on $M(n)$ so that $\operatorname{Ad}\operatorname{Diag}(n)$ is an $n-1$-dimensional group. Let

$$\rho^{\mathrm{diag}} : A^{\mathrm{diag}} \to \operatorname{Diag}(n) \tag{3.60}$$

be the homomorphism whose differential, $(\rho^{\mathrm{diag}})_*$, is such that

$$(\rho^{\mathrm{diag}})_*(\xi_{(m,m)}) = -I_m \tag{3.61}$$

for $m \in I_{n-1}$. One notes then that $A^{\mathrm{diag}} \to \operatorname{Ad}\operatorname{Diag}(n)$ where $a \mapsto \operatorname{Ad}\rho^{\mathrm{diag}}(a)$ is an epimorphism.

Theorem 3.27. *Let $a \in A^{\rm diag}$ and $x \in M(n)$. Then one has*

$$a \cdot x = \mathrm{Ad}\ \rho^{\rm diag}(a)(x) \tag{3.62}$$

so that the action of $\mathrm{Ad\,Diag}(n)$ *on* $M(n)$ *is realized by the subgroup* $A^{\rm diag}$ *of* A *on* $M(n)$.

Proof. Write $a = a(1)\cdots a(n-1)$ where $a(m) \in A(m)$ (see (3.21)). Then there exists $b(m) \in \mathbb{C}$ such that $a(m) = \exp\ b(m)\,\xi_{(m,m)}$. Let $g(m) = \rho_{x,m}(a(m))$ (see (3.7)). Then, by Theorem 3.6, $a\cdot x = \mathrm{Ad}\,h(x)$ where $h = g(1)\cdots g(m-1)$. Of course $a(m) \in A^{\rm diag}$ and hence it suffices to show that $\rho_{x,m}(a(m)) = \rho^{\rm diag}(a(m))$ (which implies $g(m)$, in this case, is independent of x). But to prove this it suffices to show that

$$(\rho_{x,m})_*(\xi_{(m,m)}) = (\rho^{\rm diag})_*(\xi_{(m,m)}). \tag{3.63}$$

But indeed both sides of (3.63) are equal to $-\mathrm{Id}_m$ by (3.8) and (3.61). □

Let $c \in \Omega(n)$ (see (2.53)). Let

$$D_c = \{a \in A \mid a \text{ operates as the identity map on } M_c(n)\}.$$

Since A is abelian and operates transitively on $M_c(n)$ (see Theorem 3.23) it follows that

$$D_c = \{a \in A \mid \text{there exists } x \in M_c(n) \text{ such that } a\cdot x = x\}. \tag{3.64}$$

For $m \in I_{n-1}$ let $D_c(m) = D_c \cap A(m)$. Let $A_c = A/D_c$.

Theorem 3.28. *Let $c \in \Omega(n)$ (see (2.53)) and let $x \in M_c(n)$. Then (see (3.7))* $D_c(m) = \mathrm{Ker}\,\rho_{x,m}$. *Moreover $D_c(m)$ is a closed discrete subgroup of $A(m)$ and*

$$\begin{aligned} A(m)/D_c(m) &\cong G_{x,m} \\ &\cong (C^\times)^m \end{aligned} \tag{3.65}$$

giving $A(m)/D_c(m)$ the structure of an abelian reductive algebraic group (i.e., a complex torus). In addition $D_c = D_c(1) \times \cdots \times D_c(n-1)$ so that

$$A_c \cong (A(1)/D_c(1)) \times \cdots \times (A(n-1)/D_c(n-1)) \tag{3.66}$$

so that A_c has the structure of a complex torus of dimension $d(n-1)$ which operates simply and transitively on $M_c(n)$.

Proof. One has $\mathrm{Ker}\,\rho_{x,m} \subset D_c$ by (3.11) where we put $y = x$. On the other hand let $a \in A(m)$ and let $g = \rho_{x,m}(a)$ so that $a\cdot x = \mathrm{Ad}\,g(x)$ by (3.11). But if $\mathrm{Ad}\,g(x) = x$, then $g = 1$ by Theorem 3.14 where we put $g = g(m)$ and $g(k) = 1$ for $k \neq m$ (see (3.35)). This proves $D_c(m) = \mathrm{Ker}\,\rho_{x,m}$, using (3.64). Of course $D_c(m)$ is closed and discrete in $A(m)$ since $(\rho_{x,m})_*$ is an isomorphism (see (3.8)). The equalities (3.65) then follow from (3.38). Now let $a \in D_c$ so that $a\cdot x = 1$. Write $a = a(1)\cdots a(n-1)$ where $a(m) \in A(m)$. Let $g(m) = \rho_{x,m}(a(m))$. Then $g(m) = 1$ for all $m \in I_{n-1}$ by Theorem 3.14 (see (3.35)). Hence $a(m) \in D_c(m)$ for all $m \in I_{n-1}$. But this readily implies (3.66). The last statement then follows from Theorem 3.23. □

Let $c \in \Omega(n)$ and let $D_c^{\text{diag}} = D_c \cap A^{\text{diag}}$.

Theorem 3.29. *Let $c \in \Omega(n)$. Then independent of c one has (see (3.60))*

$$D_c^{\text{diag}} = \operatorname{Ker} \rho^{\text{diag}} \tag{3.67}$$

and ρ^{diag} induces an isomorphism

$$\begin{aligned} A^{\text{diag}}/D_c^{\text{diag}} &\cong \operatorname{Ad} \operatorname{Diag}(n) \\ &\cong (\mathbb{C}^\times)^{n-1}. \end{aligned} \tag{3.68}$$

In particular Ad Diag(n) *operates faithfully and without fixed point on $M_c(n)$.*

Proof. Let $x \in M_c(n)$. Obviously $\operatorname{Ker} \rho^{\text{diag}} \subset D_c^{\text{diag}}$ by (3.62). Conversely, let $a \in D_c^{\text{diag}}$. We will use the notation and arguments in the proof of Theorem 3.27. One has $a \cdot x = x$. But then $g(m) = 1$ for all $m \in I_{n-1}$ by Theorem 3.14. But, as established in the proof of Theorem 3.27, $g(m) = \rho^{\text{diag}}(a(m))$. Thus $a(m) \in \operatorname{Ker} \rho^{\text{diag}}$ for all $m \in I_{n-1}$. Hence $a \in \operatorname{Ker} \rho^{\text{diag}}$. But it is immediate from the definition of ρ^{diag} that

$$\operatorname{Diag}(n) = \operatorname{Cent} Gl(n) \times \operatorname{Im} \rho^{\text{diag}}. \tag{3.69}$$

But this readily implies (3.68). □

3.7.

We recall that the operation of transpose ($x \mapsto x^T$) in $M(n)$ stabilizes $M_c(n)$ for any $c \in \mathbb{C}^{d(n)}$. The relation of this operation to the action of A is given in

Proposition 3.30. *Let $a \in A$ and let $x \in M(n)$. Then*

$$(a \cdot x)^T = a^{-1} \cdot x^T. \tag{3.70}$$

Proof. Write $a = a(1) \cdots a(n-1)$ where $a(m) \in A(m)$ for $m \in I_{n-1}$. Let $g(m) = \rho_{x,m}(a(m))$. Then for $y \in x_m + M(m)^\perp$ one has $a(m) \cdot y = g(m)\, y\, g(m)^{-1}$, by (3.11), so that

$$(a(m) \cdot y)^T = (g(m)^T)^{-1}\, y^T\, g(m)^T \tag{3.71}$$

since $(g(m)^T)^{-1} = (g(m)^{-1})^T$. Let $g_T(m) = \rho_{x^T,m}(a(m))$. On the other hand $(x_m)^T = (x^T)_m$ and hence $y^T \in (x^T)_m + M(m)^\perp$. Thus

$$(a(m))^{-1} \cdot y^T = (g_T(m))^{-1}\, y^T\, g_T(m). \tag{3.72}$$

But we now assert that

$$g(m)^T = g_T(m). \tag{3.73}$$

Indeed, by (3.8), $(\rho_{x^T,m})_*(\xi_{(k,m)}) = -((x^T)_m)^{m-k}$. But clearly $((x^T)_m)^{m-k} = ((x_m)^{m-k})^T$. Thus

$$(\rho_{x^T,m})_*(\xi) = (\rho_{x,m})_*)(\xi)^T \tag{3.74}$$

for any $\xi \in \mathfrak{a}(m)$. But this clearly implies (3.73). But then (3.71) and (3.72) yield

$$(a(m) \cdot y)^T = a(m)^{-1} \cdot y^T \tag{3.75}$$

for any $y \in x_m + M(m)^\perp$. But since x is arbitrary one has (3.75) for all $y \in M(n)$ and all $m \in I_{n-1}$. However since A is commutative this immediately implies (3.70). □

Remark 3.31. If $c \in \Omega(n)$ the action of A on $M_c(n)$ descends to an action of the $d(n-1)$-dimensional torus A_c on $M_c(n)$, by Theorem 3.28. If, in Proposition 3.30, one has $x \in M_c(n)$, where $c \in \Omega(n)$, one clearly has (3.70) where we can regard $a \in A_c$.

It is suggestive from Theorems 2.19 and 2.21 that if $x \in M(n)$ satisfies the eigenvalue disjointness condition (i.e., $x \in M_\Omega(n)$) and x is symmetric (i.e., $x = x^T$) then x is "essentially" uniquely determined by the set of eigenvalues $E_x(m)$ of x_m for all $m \in I_n$. The precise statement is given in Theorem 3.32 below. If $c \in \Omega(n)$ let F_c be the group of elements of order ≤ 2 in A_c so that F_c is a group of order $2^{d(n-1)}$ and let $M_c^{(\mathrm{sym})}(n)$ be the set of symmetric matrices in $M_c(n)$.

Theorem 3.32. *Let $c \in \Omega(n)$ (see (2.53)). Then $M_c^{(\mathrm{sym})}(n)$ is a finite set of cardinality $2^{d(n-1)}$. In fact $M_c^{(\mathrm{sym})}(n)$ is an orbit of F_c (see Remark 3.31).*

Proof. Let $x \in M_c(n)$ so that the most general element $y \in M_c(n)$ is uniquely of the form $y = a \cdot x$ for $a \in A_c$ by Theorem 3.28. But then the condition that y be symmetric is that $a \cdot x = a^{-1} \cdot x^T$ by (3.70) and Remark 3.31. But this is just the condition that

$$a^2 \cdot x = x^T. \tag{3.76}$$

But now by (3.49) and Theorem 3.28 there exists $b \in A_c$ such that $b \cdot x = x^T$. But since $A_c \cong (C^\times)^{d(n-1)}$ there exists $a \in A_c$ such that $b = a^2$. But then $y \in M_c^{(\mathrm{sym})}(n)$ where $y = a \cdot x$. Thus $M_c^{(\mathrm{sym})}(n) \neq \emptyset$. Now choose $x \in M_c^{(\mathrm{sym})}(n)$. But then, by (3.76), a necessary and sufficient conditon that $y \in M_c^{(\mathrm{sym})}(n)$ is that $a \in F_c$. □

An important special case of Theorem 3.32 is when $M_c(n)$ contains a real symmetric matrix. Let $c \in \mathbb{C}^{d(n)}$. The using the notation of (2.10) we will say that c satisfies the eigenvalue interlacing condition if $\mu_{k\,m}(c)$ is real for all $m \in I_n$, $k \in I_m$, and (2.64) is satisfied where $\mu_{k\,m} = \mu_{k\,m}(c)$. In such a case of course $c \in \Omega(n)$.

For any $c \in \mathbb{C}^{d(n)}$ and $m \in I_n$ let $p_{c,m}(\lambda)$ be the polynomial defined by putting

$$p_{c,m}(\lambda) = \prod_{k=1}^{m} (\lambda - \mu_{k\,m}). \tag{3.77}$$

Remark 3.33. Assume that $c \in \Omega(n)$ and that $\mu_{k\,m}(c)$ is real for all $m \in I_n$, $k \in I_m$. The lexicographical ordering implies

$$\mu_{k\,m}(c) < \mu_{k+1\,m}(c) \quad \forall m \in I_n,\ k \in I_{m-1}. \tag{3.78}$$

Note that if c satisfies the eigenvalue interlacing condition, then for $m \in I_{n-1}$,

$$\mathrm{sign}(p_{c,m+1}(\mu_{i\,m}(c))) = (-1)^{m+1-i} \quad \forall i \in I_m. \tag{3.79}$$

Theorem 3.34. *Let $c \in \Omega(n)$. Then the following conditions are equivalent:*

$$\begin{aligned} &(1)\ c \text{ satisfies the eigenvalue interlacing condition,}\\ &(2)\ \text{there exists a real symmetric matrix in } M_c^{(\mathrm{sym})}(n),\\ &(3)\ \text{all } 2^{d(n-1)} \text{ matrices in } M_c^{(\mathrm{sym})}(n) \text{ are real symmetric.}\end{aligned} \tag{3.80}$$

Proof. Obviously (3) implies (2). Assume (2) and let $x \in M_c^{(\mathrm{sym})}(n)$ be real symmetric. Then x is strongly regular by (2.55). Thus (2) implies (1) by Proposition 2.18. Now assume (1) and let $x \in M_c^{(\mathrm{sym})}(n)$. We wish to show that x is real. But now (3.78) is satisfied. Obviously x_1 is real since $x_1 = \mu_{1\,1}(c)$. Assume inductively that $m \in I_{n-1}$ and that x_m is real. We use the notation and argument in the proof of Theorem 2.17. We can take g to be an orthogonal matrix. Then $g\,x_{m+1}g^{-1}$ is still symmetric so that $a_{j\,m+1} = b_{m+1\,j}$. It suffices then to show that $a_{i\,m+1}$ is real for $i \in I_m$. (One has $a_{m+1\,m+1}(x)$ is real since the reality of $E_x(m+1)$ implies that $\operatorname{tr} x_{m+1}$ is real.) But $p(\lambda) = p_{c,m+1}(\lambda)$. See (2.59). But then

$$p_{c,m+1}(\mu_{i,m}(c)) = -a_{i\,m+1}^2 \prod_{j \in I_m - \{i\}} (\mu_{i\,m} - \mu_{j\,m}). \tag{3.81}$$

But clearly

$$\operatorname{sign}(- \prod_{j \in I_m - \{i\}} (\mu_{i\,m} - \mu_{j\,m}) = (-1)^{m+1-i}.$$

Thus $a_{i\,m+1}^2 > 0$ by (3.79). Hence $a_{i\,m+1}$ is real for all $i \in I_m$. □

Let $c \in \Omega(n)$. Recalling Theorem 3.29 put $A_c^{\mathrm{diag}} = A^{\mathrm{diag}}/D_c^{\mathrm{diag}}$ so that A_c^{diag} is an $n-1$-dimensional subtorus of A_c, by (3.68). Now let F_c^{diag} be the group of all elements $a \in A_c^{\mathrm{diag}}$ of order ≤ 2 so that F_c^{diag} is a subgroup of F_c of order 2^{n-1}. The determination of $a \cdot x$ for $a \in F_c$ and $x \in M_c(n)$ seems quite nontransparent to us. However if $a \in F_c^{\mathrm{diag}}$, then it is easy to exhibit $a \cdot x$. Indeed it follows from Theorem 3.29 (see also (3.69)) that for each $a \in F_c^{\mathrm{diag}}$ and $j \in I_{n-1}$ there exists $\varepsilon_j(a) \in \{1, -1\}$ such that for the faithful descent of ρ^{diag} to F_c^{diag} one has

$$\rho^{\mathrm{diag}}(a) = \operatorname{diag}(\varepsilon_1(a), \ldots, \varepsilon_{n-1}(a), 1) \tag{3.82}$$

and

$$a \cdot x = \rho^{\mathrm{diag}}(a)\, x\, \rho^{\mathrm{diag}}(a). \tag{3.83}$$

Remark 3.35. Let $x \in M(n)$ be a real symmetric Jacobi matrix (e.g., arising say from orthogonal polynomials on $\mathbb{R}$—see Theorem 2.20). Let $c \in \mathbb{C}^{d(n)}$ be such that $x \in M_c(n)$ so that c is eigenvalue interlacing by Theorem 2.19. Then note that

$$\{a \cdot x \mid a \in F_c^{\mathrm{diag}}\} \text{ is the set of all symmetric Jacobi matrices in } M_c^{(\mathrm{sym})}, \tag{3.84}$$

by Theorem 2.19. It is interesting to note that if $a \in F_c - F_c^{\mathrm{diag}}$, then $a \cdot x \in M_c^{(\mathrm{sym})}$ is non-Jacobi.

Example. Consider the case where $n = 3$ so that there are 8 symmetric matrices in $M_c(3)$ for any $c \in \Omega(n)$. Let c be defined so that $E_c(1) = \{0\}$, $E_c(2) = \{1, -1\}$, $E_c(3) = \{\sqrt{2}, 0, -\sqrt{2}\}$ so that c is eigenvalue interlacing. Then 2 of the 8 real symmetric matrices in $M_c(3)$ are

$$x = \begin{pmatrix} 0 & 1 & 0 \\ 1 & 0 & 1 \\ 0 & 1 & 0 \end{pmatrix}, \qquad y = \begin{pmatrix} 0 & 1 & 1 \\ 1 & 0 & 0 \\ 1 & 0 & 0 \end{pmatrix}$$

noting that x, but not y, is Jacobi. The remaining 6 are obtained by sign changes in x and y. That is, by applying the three nontrivial elements in F_c^{diag} to x and y.

3.8.

In Part II we will need the following result on polarizations of regular adjoint orbits and maximal A-orbits. Let $x \in M(n)$. It is clear from (1.18) that the adjoint O_x is stable under the action of A. Let $O_x^{\text{sreg}} = M^{\text{sreg}} \cap O_x$ so that O_x^{sreg} is Zariski open subset of O_x.

Theorem 3.36. *Let $x \in M(n)$. Then O_x^{sreg} is non-empty (and hence dense in O_x) if and only if x is regular, in $M(n)$. In particular if x is regular, then O_x^{sreg} is a symplectic $2\,d(n-1)$-dimensional manifold (in the complex sense). Furthermore O_x^{sreg} is stable under the action of A and the orbits of A in O_x^{sreg} (necessarily of dimension $d(n-1)$) are the leaves of a polarization of O_x^{sreg}.*

Proof. If x is not regular, then all elements in O_x are not regular so that O_x^{sreg} is empty (strongly regular implies regular — see Theorem 2.13). Now assume that x is regular. But now by Theorem 2.5 there exists $y \in e + \mathfrak{b}$ such that $E_x(n) = E_y(n)$. But y is regular by (2.27). But, as one knows, any two regular matrices with the same spectrum are conjugate (they are both conjugate to the same companion matrix). Thus $y \in O_x$. But then $y \in O_x^{\text{sreg}}$ by (2.27). Hence O_x^{sreg} is not empty. But then O_x^{sreg} is a union of maximal A-orbits by (3.30). But each such orbit is a Lagrangian submanifold of O_x by dimension (see (3.29)) and the vanishing of (1.17), by (2.31), where $\psi = p_{(i)}$ and $u = (\xi_{p_{(j)}})_x$ (see (1.18)) for $i, j \in I_{d(n-1)}$. □

References

[B] A. Borel, *Linear Algebraic Groups*, W. A. Benjamin, Inc., 1969.

[C] C. Chevalley, *Fondements de la Géométrie Algébrique*, Faculté des Sciences de Paris, Mathématiques approfondies, 1957/1958.

[CG] N. Chris and V. Ginzburg, *Representation Theory and Complex Geometry*, Birkhäuser, 1997.

[GS] V. Guillemin and S. Sternberg, On the collective integrability according to the method of Thimm, *Ergod. Th. & Dynam. Sys.*, **3** (1983), 219–230.

[H] R. Hartshorne, *Algebraic Geometry*, Grad. Texts in Math. Vol. 52, Springer-Verlag, 1977.

[J] D. Jackson, *Fourier Series and Orthogonal Polynomials*, The Carus Mathematical Monographs, no. 6, Math. Assoc. of America, 1948.

[K] B. Kostant, Quantization and Unitary Representations, Lecture Notes in Math, Vol. 170, Springer-Verlag, 1970.

[K-1] B. Kostant, Lie group representations on polynomial rings, *AJM*, **85** (1963), 327–404.

[K-W] B. Kostant and N. Wallach, Gelfand–Zeitlin from the Perspective of Classical Mechanics. II, Prog. Math., Vol. 244 (2005), pp. 387–420.

[Kn] F. Knop, Automorphisms, Root Systems and Compactifications of Homogeneous Varieties, *JAMS*, **9** (1996), no. 1, 153–174.

[M] D. Mumford, *The Red Book of Varieties and Schemes*, Lecture Notes in Math., Vol. 1358, Springer, 1995.

[S] T. Springer, *Linear Algebraic Groups*, 2nd edition, Prog. Math., Vol. 9, Birkhäuser Boston, 1998.

Extensions of algebraic groups

Shrawan Kumar[1] and Karl-Hermann Neeb[2]

[1] Department of Mathematics
University of North Carolina
Chapel Hill, NC 27599-3250
USA
shrawan@email.unc.edu

[2] Fachbereich Mathematik
Darmstadt University of Technology
Schloßgartenstr. 7
D-64289, Darmstadt
Germany
neeb@mathematik.tu-darmstadt.de

Dedicated to Professor Anthony Joseph on his sixtieth birthday

Summary. Let G be a connected complex algebraic group and A an abelian connected algebraic group, together with an algebraic action of G on A via group automorphisms. The aim of this article is to study the group of isomorphism classes of extensions of G by A in the algebraic group category. We describe this group as a direct sum of the group $\mathrm{Hom}(\pi_1([G, G]), A)$ and a relative Lie algebra cohomology space. We also prove a version of Van Est's theorem for algebraic groups, identifying the cohomology of G with values in a G-module $\mathfrak{a}$ in terms of relative Lie algebra cohomology.

Subject Classification: 20G10

Introduction

Let G be a connected complex algebraic group and A an abelian connected algebraic group, together with an algebraic action of G on A via group automorphisms. The aim of this article is to study the set of isomorphism classes $\mathrm{Ext}_{\mathrm{alg}}(G, A)$ of extensions of G by A in the algebraic group category. The following is our main result (cf. Theorem 1.8).

0.1. Theorem. *For G and A as above, there exists a split exact sequence of abelian groups:*

$$0 \to \mathrm{Hom}(\pi_1([G, G]), A) \to \mathrm{Ext}_{\mathrm{alg}}(G, A) \xrightarrow{\pi} H^2(\mathfrak{g}, \mathfrak{g}_{\mathrm{red}}, \mathfrak{a}_u) \to 0,$$

where A_u is the unipotent radical of A, G_{red} is a Levi subgroup of G, $\mathfrak{g}_{\text{red}}, \mathfrak{g}, \mathfrak{a}_u$ are the Lie algebras of G_{red}, G, A_u, respectively, and $H^*(\mathfrak{g}, \mathfrak{g}_{\text{red}}, \mathfrak{a}_u)$ is the Lie algebra cohomology of the pair $(\mathfrak{g}, \mathfrak{g}_{\text{red}})$ with coefficients in the $\mathfrak{g}$-module $\mathfrak{a}_u$.

Our next main result is the following analogue of the Van-Est Theorem for the algebraic group cohomology (cf. Theorem 2.2).

0.2. Theorem. *Let G be a connected algebraic group and let $\mathfrak{a}$ be a finite-dimensional algebraic G-module. Then, for any $p \geq 0$,*

$$H^p_{\text{alg}}(G, \mathfrak{a}) \simeq H^p(\mathfrak{g}, \mathfrak{g}_{\text{red}}, \mathfrak{a}).$$

By an algebraic group G we mean an affine algebraic group over the field of complex numbers $\mathbb{C}$ and the varieties are considered over $\mathbb{C}$. The Lie algebra of G is denoted by $L(G)$.

1. Extensions of Algebraic Groups

1.1. Definition. Let G be an algebraic group and A an abelian algebraic group, together with an algebraic action of G on A via group automorphisms, i.e., a morphism of varieties $\rho : G \times A \to A$ such that the induced map $G \to \operatorname{Aut} A$ is a group homomorphism. Such an A is called an *algebraic group with G-action*.

By $\operatorname{Ext}_{\text{alg}}(G, A)$ we mean the set of isomorphism classes of extensions of G by A in the algebraic group category, i.e., surjective morphisms $q : \widehat{G} \to G$ with kernel isomorphic to A as an algebraic group with G-action. We obtain on $\operatorname{Ext}_{\text{alg}}(G, A)$ the structure of an abelian group by assigning to two extensions $q_i : \widehat{G}_i \to G$ of G by A the fiber product extension $\widehat{G}_1 \times_G \widehat{G}_2$ of G by $A \times A$ and then applying the group morphism $m_A : A \times A \to A$ fiberwise to obtain an A-extension of G (this is the Baer sum of two extensions). Then, $\operatorname{Ext}_{\text{alg}}$ assigns to a pair of an algebraic group G and an abelian algebraic group A with G-action, an abelian group, and this assignment is contravariant in G (via pulling back the action of G and the extension) and if G is fixed, $\operatorname{Ext}_{\text{alg}}(G, \cdot)$ is a covariant functor from the category of abelian algebraic groups with G-actions to the category of abelian groups. Here we assign to a G-equivariant morphism $\gamma : A_1 \to A_2$ of abelian algebraic groups and an extension $q : \widehat{G} \to G$ of G by A_1 the extension

$$\gamma_*\widehat{G} := (A_2 \rtimes \widehat{G})/\Gamma(\gamma) \to G, \quad [(a, g)] \mapsto q(g),$$

where $\Gamma(\gamma)$ is the graph of γ in $A_2 \times A_1$ and the semidirect product refers to the action of $\widehat{G}$ on A_2 obtained by pulling back the action of G on A_2 to $\widehat{G}$. In view of the equivariance of γ, its graph is a normal algebraic subgroup of $A_2 \rtimes \widehat{G}$ so that we can form the quotient $\gamma_*\widehat{G}$.

We define a map

$$D : \operatorname{Ext}_{\text{alg}}(G, A) \to \operatorname{Ext}(L(G), L(A))$$

by assigning to an extension

$$1 \to A \xrightarrow{i} \widehat{G} \xrightarrow{q} G \to 1$$

of algebraic groups the corresponding extension

$$0 \to L(A) \xrightarrow{di} L(\widehat{G}) \xrightarrow{dq} L(G) \to 0$$

of Lie algebras. Since i is injective, di is injective. Similarly, dq is surjective. Moreover, $\dim G = \dim L(G)$ and hence the above sequence of Lie algebras is indeed exact.

It is clear from the definition of D that it is a homomorphism of abelian groups. If $\mathfrak{g}$ is the Lie algebra of G and $\mathfrak{a}$ the Lie algebra of A, then the group $\mathrm{Ext}(\mathfrak{g}, \mathfrak{a})$ is isomorphic to the second Lie algebra cohomology space $H^2(\mathfrak{g}, \mathfrak{a})$ of $\mathfrak{g}$ with coefficients in the $\mathfrak{g}$-module $\mathfrak{a}$ (with respect to the derived action) ([CE]). Therefore, the description of the group $\mathrm{Ext}_{\mathrm{alg}}(G, A)$ depends on a "good" description of kernel and cokernel of D which will be obtained below in terms of an exact sequence involving D.

In the following G is always assumed to be connected. The following lemma reduces the extension theory for connected algebraic groups A with G-actions to the two cases of a torus A_s and the case of a unipotent group A_u.

1.2. Lemma. *Let G be connected and let A be a connected algebraic group with G-action. Further, let $A = A_u A_s$ denote the decomposition of A into its unipotent and reductive factors. Then, $A \cong A_u \times A_s$ as a G-module, where G acts trivially on A_s and G acts on A_u as a G-stable subgroup of A. Thus, we have*

$$\mathrm{Ext}_{\mathrm{alg}}(G, A) \cong \mathrm{Ext}_{\mathrm{alg}}(G, A_u) \oplus \mathrm{Ext}_{\mathrm{alg}}(G, A_s). \tag{1}$$

Proof. Decompose

$$A = A_u A_s, \tag{2}$$

where A_s is the set of semisimple elements of A and A_u is the set of unipotent elements of A. Then, A_s and A_u are closed subgroups of A and (2) is a direct product decomposition (see [H, Theorem 15.5]). The action of G on A clearly keeps A_s and A_u stable separately. Also, G acts trivially on A_s since $\mathrm{Aut}(A_s)$ is discrete and G is connected (by assumption). Thus, the action of G on A decomposes as the product of actions on A_s and A_u with the trivial action on A_s. Hence, the isomorphism (1) follows from the functoriality of $\mathrm{Ext}_{\mathrm{alg}}(G, \cdot)$. □

If $G = G_u \rtimes G_{\mathrm{red}}$ is a Levi decomposition of G, then G_u being simply-connected,

$$\pi_1(G) \cong \pi_1(G_{\mathrm{red}}),$$

where G_u is the unipotent radical of G, G_{red} is a Levi subgroup of G and π_1 denotes the fundamental group. The connected reductive group G_{red} is a product of its connected center $Z := Z(G_{\mathrm{red}})_0$ and its commutator group $G'_{\mathrm{red}} := [G_{\mathrm{red}}, G_{\mathrm{red}}]$ which is a connected semisimple group. Thus, $G'_{\rm red}$ has an algebraic universal covering group $\tilde{G}'_{\rm red}$, with the finite abelian group $\pi_1(G'_{\rm red})$ as its fiber. We write $\tilde{G}_{\rm red} := Z \times \tilde{G}'_{\rm red}$, which is an algebraic covering group of $G_{\rm red}$; denote its kernel by Π_G and observe that

$$\tilde{G} := G_u \rtimes \tilde{G}_{\rm red}$$

is a covering of G with Π_G as its fiber. We write $q_G : \tilde{G} \to G$ for the corresponding covering map.

1.3. Lemma. *If G and A are tori, then* $\mathrm{Ext}_{\rm alg}(G, A) = 0$.

Proof. Let $q : \widehat{G} \to G$ be an extension of the torus G by A. Then, as is well known, $\widehat{G}$ is again a torus (cf. [B, §11.5]). Since any character of a subtorus of a torus extends to a character of the whole torus ([B, §8.2]), the identity $I_A : A \to A$ extends to a morphism $f : \widehat{G} \to A$. Now, $\ker f$ yields a splitting of the above extension. □

The following proposition deals with the case $A = A_s$.

1.4. Proposition. *If $A = A_s$, then $D = 0$ and we obtain an exact sequence*

$$\mathrm{Hom}(\tilde{G}, A_s) \xrightarrow{res} \mathrm{Hom}(\Pi_G, A_s) \xrightarrow{\Phi} \mathrm{Ext}_{\rm alg}(G, A_s),$$

where Φ assigns to any $\gamma \in \mathrm{Hom}(\Pi_G, A_s)$ the extension $\gamma_* \tilde{G}$. The kernel of Φ consists of those homomorphisms vanishing on the fundamental group $\pi_1(G'_{\rm red})$ of $G'_{\rm red}$ and Φ factors through an isomorphism

$$\Phi' : \mathrm{Hom}(\pi_1(G'_{\rm red}), A_s) \simeq \mathrm{Ext}_{\rm alg}(G, A_s).$$

Proof. Consider an extension

$$1 \to A_s \to \widehat{G} \to G \to 1.$$

Since A_s is a central torus in $\widehat{G}$, the unipotent radical $\widehat{G}_u$ of $\widehat{G}$ maps isomorphically on G_u. Also

$$1 \to A_s \to \widehat{G}_{\rm red} \to G_{\rm red} \to 1$$

is an extension whose restriction to Z splits by the preceding lemma. On the other hand the commutator group of $\widehat{G}_{\rm red}$ has the same Lie algebra as $G'_{\rm red}$, hence is a quotient of $\tilde{G}'_{\rm red}$. Thus, $\widehat{G}_{\rm red}$ is a quotient of $A_s \times Z \times \tilde{G}'_{\rm red}$, which implies that $\widehat{G}$ is a quotient of $A_s \times \tilde{G}$. Hence, $\widehat{G}$ is obtained from $A_s \times \tilde{G}$ by taking its quotient by the graph of a homomorphism $\Pi_G \to A_s$. This proves that Φ is surjective. In particular, the pullback $q_G^* \widehat{G}$ of $\widehat{G}$ to $\tilde{G}$ always splits.

We next show that $\ker \Phi$ coincides with the image of the restriction map from $\mathrm{Hom}(\tilde{G}, A_s)$ to $\mathrm{Hom}(\Pi_G, A_s)$. Assume that the extension $\widehat{G}_\gamma = \gamma_* \tilde{G}$ defined by $\gamma \in \mathrm{Hom}(\Pi_G, A_s)$ splits. Let $\sigma : G \to \widehat{G}_\gamma$ be a splitting morphism. Pulling σ back via q_G, we obtain a splitting morphism

$$\tilde{\sigma} : \tilde{G} \to q_G^* \widehat{G}_\gamma \cong A_s \times \tilde{G}.$$

Thus, there exists a morphism $\delta : \tilde{G} \to A_s$ of algebraic groups such that σ satisfies $\sigma(q_G(g)) = \beta(\delta(g), g)$ for all $g \in \tilde{G}$, where $\beta : A_s \times \tilde{G} \to \widehat{G}_\gamma = (A_s \times \tilde{G})/\Gamma(\gamma)$ is the standard quotient map. For $g \in \Pi_G = \ker q_G$ we have $\beta(\delta(g), g) = 1$, and therefore $\delta(g) = \gamma(g)$ for all $g \in \Pi_G$. This shows that δ is an extension of γ to $\tilde{G}$. Conversely, if γ extends to $\tilde{G}$, $\widehat{G}_\gamma$ is a trivial extension of G.

That $D = 0$ follows from the fact that $\widehat{G}$ and $q_G^*\widehat{G}$ have the same Lie algebras, which is a split extension of $\mathfrak{g}$ by $\mathfrak{a}_s$.

We recall that $\tilde{G} = G_u \rtimes (Z \times \tilde{G}'_{\rm red})$. If a homomorphism $\gamma : \Pi_G \to A_s$ extends to $\tilde{G}$, then it must vanish on the subgroup $\pi_1(G'_{\rm red})$ of Π_G since, $\tilde{G}'_{\rm red}$ being a semisimple group, there are no nonconstant homomorphisms from $\tilde{G}'_{\rm red} \to A_s$. Conversely, if a homomorphism $\gamma : \Pi_G \to A_s$ vanishes on $\pi_1(G'_{\rm red})$, then γ defines a homomorphism

$$Z \cap G'_{\rm red} \cong \Pi_G/\pi_1(G'_{\rm red}) \to A_s.$$

But since A_s is a torus, this extends to a morphism $f : Z \to A_s$ ([B, §8.2]) which in turn can be pulled back via $Z \cong \tilde{G}/(G_u \rtimes \tilde{G}'_{\rm red})$ to a morphism $\tilde{f} : \tilde{G} \to A_s$ extending γ. This proves that the image of $\mathrm{Hom}(\tilde{G}, A_s)$ under the restriction map in $\mathrm{Hom}(\Pi_G, A_s)$ is the annihilator of $\pi_1(G'_{\rm red})$ so that

$$\Phi : \mathrm{Hom}(\Pi_G, A_s) \to \mathrm{Ext}_{\rm alg}(G, A_s)$$

factors through an isomorphism

$$\Phi' : \mathrm{Hom}(\pi_1(G'_{\rm red}), A_s) \simeq \mathrm{Ext}_{\rm alg}(G, A_s). \qquad \square$$

1.5. Remark. A unipotent group A_u over $\mathbb{C}$ has no nontrivial finite subgroups, so that

$$\mathrm{Hom}(\pi_1(G'_{\rm red}), A_s) \cong \mathrm{Hom}(\pi_1(G'_{\rm red}), A).$$

Now, we turn to the study of extensions by unipotent groups. In contrast to the situation for tori, we shall see that these extensions are faithfully represented by the corresponding Lie algebra extensions.

1.6. Lemma. *The canonical restriction map*

$$H^2(\mathfrak{g}, \mathfrak{g}_{\rm red}, \mathfrak{a}_u) \longrightarrow H^2(\mathfrak{g}, \mathfrak{a}_u)$$

is injective.

Proof. Let $\omega \in Z^2(\mathfrak{g}, \mathfrak{a}_u)$ be a Lie algebra cocycle representing an element of $H^2(\mathfrak{g}, \mathfrak{g}_{\rm red}, \mathfrak{a}_u)$ and suppose that the class $[\omega] \in H^2(\mathfrak{g}, \mathfrak{a}_u)$ vanishes so that the extension

$$\widehat{\mathfrak{g}} := \mathfrak{a}_u \oplus_\omega \mathfrak{g} \to \mathfrak{g}, \quad (a, x) \mapsto x$$

with the bracket $[(a, x), (a', x')] = (x.a' - x'.a + \omega(x, x'), [x, x'])$ splits. We have to find a $\mathfrak{g}_{\rm red}$-module map $f : \mathfrak{g} \to \mathfrak{a}_u$ vanishing on $\mathfrak{g}_{\rm red}$ with

$$\omega(x, x') = (d_{\mathfrak{g}} f)(x, x') := x.f(x') - x'.f(x) - f([x, x']), \quad x, x' \in \mathfrak{g}.$$

Since the space $C^1(\mathfrak{g}, \mathfrak{a}_u)$ of linear maps $\mathfrak{g} \to \mathfrak{a}_u$ is a semisimple $\mathfrak{g}_{\mathrm{red}}$-module ($\mathfrak{a}_u$ being a G-module; in particular, a G_{red}-module), we have

$$C^1(\mathfrak{g}, \mathfrak{a}_u) = C^1(\mathfrak{g}, \mathfrak{a}_u)^{\mathfrak{g}_{\mathrm{red}}} \oplus \mathfrak{g}_{\mathrm{red}}.C^1(\mathfrak{g}, \mathfrak{a}_u);$$

and similarly for the space $Z^2(\mathfrak{g}, \mathfrak{a}_u)$ of 2-cocycles. As the Lie algebra differential $d_{\mathfrak{g}} : C^1(\mathfrak{g}, \mathfrak{a}_u) \to Z^2(\mathfrak{g}, \mathfrak{a}_u)$ is a $\mathfrak{g}_{\mathrm{red}}$-module map, each $\mathfrak{g}_{\mathrm{red}}$-invariant coboundary is the image of a $\mathfrak{g}_{\mathrm{red}}$-invariant cochain in $C^1(\mathfrak{g}, \mathfrak{a}_u)$. We conclude, in particular, that $\omega = d_{\mathfrak{g}} h$ for some $\mathfrak{g}_{\mathrm{red}}$-module map $h : \mathfrak{g} \to \mathfrak{a}_u$. For $x \in \mathfrak{g}_{\mathrm{red}}$ and $x' \in \mathfrak{g}$, it follows that

$$\begin{aligned} 0 &= \omega(x, x') = x.h(x') - x'.h(x) - h([x, x']) \\ &= h([x, x']) - x'.h(x) - h([x, x']) = -x'.h(x), \end{aligned}$$

showing that $h(\mathfrak{g}_{\mathrm{red}}) \subseteq \mathfrak{a}_u^{\mathfrak{g}}$, which in turn leads to $[\mathfrak{g}_{\mathrm{red}}, \mathfrak{g}_{\mathrm{red}}] \subseteq \ker h$. As $\mathfrak{z}(\mathfrak{g}_{\mathrm{red}}) \cap [\mathfrak{g}, \mathfrak{g}] = \{0\}$, the map $h|_{\mathfrak{z}(\mathfrak{g}_{\mathrm{red}})}$ extends to a linear map $f : \mathfrak{g} \to \mathfrak{a}_u^{\mathfrak{g}}$ vanishing on $[\mathfrak{g}, \mathfrak{g}]$. Moreover, since f vanishes on $[\mathfrak{g}, \mathfrak{g}]$, f is clearly a $\mathfrak{g}$-module map; in particular, a $\mathfrak{g}_{\mathrm{red}}$-module map. Then, $d_{\mathfrak{g}} f = 0$, so that $d_{\mathfrak{g}}(h - f) = \omega$, and $h - f$ vanishes on $\mathfrak{g}_{\mathrm{red}}$. □

1.7. Proposition. *For $A = A_u$ the map $D : \mathrm{Ext}_{\mathrm{alg}}(G, A_u) \to H^2(\mathfrak{g}, \mathfrak{a}_u)$ induces a bijection*

$$D : \mathrm{Ext}_{\mathrm{alg}}(G, A_u) \to H^2(\mathfrak{g}, \mathfrak{g}_{\mathrm{red}}, \mathfrak{a}_u).$$

Proof. In view of the preceding lemma, we may identify $H^2(\mathfrak{g}, \mathfrak{g}_{\mathrm{red}}, \mathfrak{a}_u)$ with a subspace of $H^2(\mathfrak{g}, \mathfrak{a}_u)$. First we claim that $\mathrm{im}(D)$ is contained in this subspace. For any extension

$$1 \to A_u \to \widehat{G} \to G \to 1, \tag{3}$$

we choose a Levi subgroup $\widehat{G}_{\mathrm{red}} \subset \widehat{G}$ mapping to G_{red} under the above map $\widehat{G} \to G$. Then,

$$\widehat{G}_{\mathrm{red}} \cap A_u = \{1\}.$$

Moreover, $\widehat{G}_{\mathrm{red}} \to G_{\mathrm{red}}$ is surjective and hence an isomorphism. This shows that the extension (3) restricted to G_{red} is trivial and that $\widehat{\mathfrak{g}}_u$ contains a $\widehat{\mathfrak{g}}_{\mathrm{red}}$-invariant complement to $\mathfrak{a}_u$. Therefore, $\widehat{\mathfrak{g}}$ can be described by a cocycle $\omega \in Z^2(\mathfrak{g}, \mathfrak{g}_{\mathrm{red}}, \mathfrak{a}_u)$; in particular, ω vanishes on $\mathfrak{g} \times \mathfrak{g}_{\mathrm{red}}$. This shows that $\mathrm{Im}\, D \subset H^2(\mathfrak{g}, \mathfrak{g}_{\mathrm{red}}, \mathfrak{a}_u)$.

If the image of the extension (3) under D vanishes, then the extension $\mathfrak{a}_u \hookrightarrow \widehat{\mathfrak{g}}_u \twoheadrightarrow \mathfrak{g}_u$ splits, which implies that the corresponding extension of unipotent groups $A_u \hookrightarrow \widehat{G}_u \twoheadrightarrow G_u$ splits. Moreover, the splitting map can be chosen to be G_{red}-equivariant since ω is G_{red}-invariant. This means that we have a morphism $G_u \rtimes G_{\mathrm{red}} \to \widehat{G} \cong \widehat{G}_u \rtimes G_{\mathrm{red}}$ splitting the extension (3). This proves that D is injective.

To see that D is surjective, let $\omega \in Z^2(\mathfrak{g}, \mathfrak{g}_{\mathrm{red}}, \mathfrak{a}_u)$. Let $q : \widehat{\mathfrak{g}} := \mathfrak{a}_u \oplus_\omega \mathfrak{g} \to \mathfrak{g}$ denote the corresponding Lie algebra extension. Since $\mathfrak{a}_u$ is a nilpotent module of $\mathfrak{g}_u$, the subalgebra $\widehat{\mathfrak{g}}_u := \mathfrak{a}_u \oplus_\omega \mathfrak{g}_u$ of $\widehat{\mathfrak{g}}$ is nilpotent, hence corresponds to a unipotent algebraic group $\widehat{G}_u$ which is an extension of G_u by A_u. Further, the G_{red}-invariance of the decomposition $\widehat{\mathfrak{g}} = \mathfrak{a}_u \oplus \mathfrak{g}$ implies that G_{red} acts algebraically on $\widehat{\mathfrak{g}}_u$ and hence on $\widehat{G}_u$, so that we can form the semidirect product $\widehat{G} := \widehat{G}_u \rtimes G_{\mathrm{red}}$ which is an extension of G by A_u mapped by D onto $\widehat{\mathfrak{g}}$. □

Combining the previous results, we get the following.

1.8. Theorem. *For a connected algebraic group G and a connected abelian algebraic group A with G-action, we have the following isomorphisms of abelian groups induced respectively by the decomposition $A = A_u A_s$, the map Φ of Proposition 1.4 and the map D.*

(a) $\mathrm{Ext}_{\mathrm{alg}}(G, A) \simeq \mathrm{Ext}_{\mathrm{alg}}(G, A_u) \oplus \mathrm{Ext}_{\mathrm{alg}}(G, A_s)$,
(b) $\mathrm{Ext}_{\mathrm{alg}}(G, A_s) \simeq \mathrm{Hom}(\pi_1([G, G]), A_s)$, *and*
(c) $\mathrm{Ext}_{\mathrm{alg}}(G, A_u) \simeq H^2(\mathfrak{g}, \mathfrak{g}_{\mathrm{red}}, \mathfrak{a}_u) \simeq H^2(\mathfrak{g}_u, \mathfrak{a}_u)^{\mathfrak{g}}$,

where $\mathfrak{a} = L(A)$, G_{red} is a Levi subgroup of G, $\mathfrak{g}_{\mathrm{red}} = L(G_{\mathrm{red}})$, $\mathfrak{g}_u = L(G_u)$, $\mathfrak{g} = L(G)$ and $\mathfrak{a}_u = L(A_u)$.

(Observe that, by the following proof, the fundamental group $\pi_1([G, G])$ is a finite group.)

Moreover, the isomorphisms in (a) and (b) are functorial in both A and G. Also, the first isomorphism in (c) is functorial in A as well as in the category of pairs (G, G_{red}), where G is a connected algebraic group and G_{red} is a Levi subgroup. (Recall that given any algebraic group morphism $f : G \to H$, we can always choose Levi subgroups $G_{\mathrm{red}} \subset G$ and $H_{\mathrm{red}} \subset H$ such that $f(G_{\mathrm{red}}) \subset H_{\mathrm{red}}$.)

Proof. In view of the Levi decomposition of the commutator $[G, G] = [G, G]_u \rtimes G'_{\mathrm{red}}$, we have $\pi_1([G, G]) = \pi_1(G'_{\mathrm{red}})$. Now, we only have to use Lemma 1.2 and the preceding results Propositions 1.4 and 1.7 on extensions by A_s and A_u to complete the proof of (a), (b) and the first isomorphism of (c). The restriction map $\gamma : C^*(\mathfrak{g}, \mathfrak{g}_{\mathrm{red}}, \mathfrak{a}) \to C^*(\mathfrak{g}_u, \mathfrak{a})^{\mathfrak{g}_{\mathrm{red}}}$ between the standard cochain complexes is clearly an isomorphism. Thus, the last isomorphism in (*c*) follows since the action of $\mathfrak{g}_{\mathrm{red}}$ on $\mathfrak{g}$ and $\mathfrak{a}_u$ is completely reducible and, moreover, the standard action of any Lie algebra $\mathfrak{s}$ on $H^*(\mathfrak{s}, M)$ is trivial (for any $\mathfrak{s}$-module M).

The functoriality of the isomorphisms follows from their proofs. □

2. Analogue of Van-Est Theorem for algebraic group cohomology

2.1. Definition. Let G be an algebraic group and A an abelian algebraic group with G-action. For any $n \geq 0$, let $C^n_{\mathrm{alg}}(G, A)$ be the abelian group consisting of all the variety morphisms $f : G^n \to A$ under the pointwise addition. Define the differential

$$\delta : C^n_{\mathrm{alg}}(G, A) \to C^{n+1}_{\mathrm{alg}}(G, A) \qquad \text{by}$$

$$(\delta f)(g_0, \ldots, g_n) = g_0 \cdot f(g_1, \ldots, g_n) + (-1)^{n+1} f(g_0, \ldots, g_{n-1}) + \sum_{i=0}^{n-1} (-1)^{i+1} f(g_0, g_1, \ldots, g_i g_{i+1}, \ldots, g_n).$$

Then, as is well known (and easy to see),

$$\delta^2 = 0. \tag{4}$$

The *algebraic group cohomology* $H^*_{\text{alg}}(G, A)$ *of* G *with coefficients in* A is defined as the cohomology of the complex

$$0 \to C^0_{\text{alg}}(G, A) \xrightarrow{\delta} C^1_{\text{alg}}(G, A) \xrightarrow{\delta} \cdots .$$

We have the following analogue of the Van-Est Theorem [V] for the algebraic group cohomology.

2.2. Theorem. *Let* G *be a connected algebraic group and let* $\mathfrak{a}$ *be a finite-dimensional algebraic* G*-module. Then, for any* $p \geq 0$,

$$H^p_{\text{alg}}(G, \mathfrak{a}) \simeq H^p(\mathfrak{g}, \mathfrak{g}_{\text{red}}, \mathfrak{a}) \simeq H^p(\mathfrak{g}_u, \mathfrak{a})^{\mathfrak{g}},$$

where $\mathfrak{g}$ *is the Lie algebra of* G, $\mathfrak{g}_u$ *is the Lie algebra of the unipotent radical* G_u *of* G*, and* $\mathfrak{g}_{\text{red}}$ *is the Lie algebra of a Levi subgroup* G_{red} *of* G *as in Section 1.*

Proof. Consider the homogeneous affine variety $X := G/G_{\text{red}}$ and let $\{\Omega^q = \Omega^q(X, \mathfrak{a})\}_q$ denote the complex vector space of algebraic de Rham forms on X with values in the vector space $\mathfrak{a}$. Since X is a G-variety under the left multiplication of G and $\mathfrak{a}$ is a G-module, Ω^q has a natural locally-finite algebraic G-module structure. Define a double cochain complex $A = \bigoplus_{p,q \geq 0} A^{p,q}$, where

$$A^{p,q} := C^p_{\text{alg}}(G, \Omega^q)$$

and $C^p_{\text{alg}}(G, \Omega^q)$ consists of all the maps $f : G^p \to \Omega^q$ such that $\operatorname{im} f \subset M_f$, for some finite-dimensional G-stable subspace $M_f \subset \Omega^q$ and, moreover, the map $f : G^p \to M_f$ is algebraic. Let $\delta : A^{p,q} \to A^{p+1,q}$ be the group cohomology differential as in Section 2.1 and let $d : A^{p,q} \to A^{p,q+1}$ be induced from the standard de Rham differential $\Omega^q \to \Omega^{q+1}$, which is a G-module map. It is easy to see that $d\delta - \delta d = 0$ and, of course, $d^2 = \delta^2 = 0$. Thus, (A, δ, d) is a double cochain complex. This gives rise to two spectral sequences both converging to the cohomology of the associated single complex $(C, \delta + d)$ with their E_1-terms given as follows:

$$^{\backslash}E_1^{p,q} = H^q_d(A^{p,*}), \qquad \text{and} \tag{5}$$

$$^{\backslash\backslash}E_1^{p,q} = H^q_\delta(A^{*,p}). \tag{6}$$

We now determine $^{\backslash}E_1$ and $^{\backslash\backslash}E_1$ more explicitly in our case.

Since X is a contractible variety, by the algebraic de Rham theorem [GH, Chap. 3, §5], the algebraic de Rham cohomology

$$H^q_{\text{dR}}(X, \mathfrak{a}) \begin{cases} \simeq \mathfrak{a}, & \text{if } q = 0 \\ = 0, & \text{otherwise.} \end{cases}$$

Thus,

$$
{}'E_1^{p,q} \begin{cases} \simeq C^p_{\mathrm{alg}}(G, \mathfrak{a}), & \text{if } q = 0 \\ = 0, & \text{otherwise.} \end{cases}
$$

Therefore,

$$
{}'E_2^{p,q} = H^p_\delta(H^q_d(A)) = \begin{cases} H^p_{\mathrm{alg}}(G, \mathfrak{a}), & \text{if } q = 0 \\ 0, & \text{otherwise.} \end{cases} \tag{7}
$$

In particular, the spectral sequence ${}'E_*$ collapses at ${}'E_2$. From this we see that there is a canonical isomorphism

$$
H^p_{\mathrm{alg}}(G, \mathfrak{a}) \simeq H^p(C, \delta + d). \tag{8}
$$

We next determine ${}''E_1$ and ${}''E_2$. But first we need the following two lemmas.

2.3. Lemma. *For any* $p \geq 0$,

$$
H^q_{\mathrm{alg}}(G, \Omega^p) = \begin{cases} \Omega^{pG}, & \textit{if } q = 0 \\ 0, & \textit{otherwise,} \end{cases}
$$

where Ω^{pG} *denotes the subspace of* G*-invariants in* $\Omega^p = \Omega^p(X, \mathfrak{a})$.

Proof. The assertion for $q = 0$ follows from the general properties of group cohomology. So we need to consider the case $q > 0$ now.

Since $L := G_{\mathrm{red}}$ is reductive, any algebraic L-module M is completely reducible. Let

$$
\pi^M : M \to M^L
$$

be the unique L-module projection onto the space of L-module invariants M^L of M. Taking M to be the ring of regular functions $\mathbb{C}[L]$ on L under the left regular representation, i.e., under the action

$$
(k \cdot f)(k') = f(k^{-1}k'), \quad \text{for } f \in \mathbb{C}[L], k, k' \in L,
$$

we get the L-module projection $\pi = \pi^{\mathbb{C}[L]} : \mathbb{C}[L] \to \mathbb{C}$. Thus, for any complex vextor space V, we get the projection $\pi \otimes I_V : \mathbb{C}[L] \otimes V \to V$, which we abbreviate simply by π, where I_V is the identity map of V. We define a "homotopy operator" H, for any $q \geq 0$,

$$
H : C^{q+1}_{\mathrm{alg}}(G, \Omega^p) \to C^q_{\mathrm{alg}}(G, \Omega^p)
$$

by

$$
\Big((Hf)(g_1, \ldots, g_q)\Big)_{g_0 L} = \pi\left(\Theta^f_{(g_0, \ldots, g_q)}\right),
$$

for $f \in C^{q+1}_{\mathrm{alg}}(G, \Omega^p)$ and $g_0, \ldots, g_q \in G$, where $\Theta^f_{(g_0, \ldots, g_q)} : L \to \Omega^p_{g_0 L}$ is defined by

$$
\Theta^f_{(g_0, \ldots, g_q)}(k) = \Big((g_0 k) \cdot f(k^{-1} g_0^{-1}, g_1, g_2, \ldots, g_q)\Big)_{g_0 L},
$$

for $k \in L$. (Here $\Omega^p_{g_0L}$ denotes the fiber at g_0L of the vector bundle of p-forms in X with values in $\mathfrak{a}$ and, for a form ω, ω_{g_0L} denotes the value of the form ω at g_0L.) It is easy to see that on $C^q_{\mathrm{alg}}(G, \Omega^p)$, for any $q \geq 1$,

$$H\delta + \delta H = I. \tag{9}$$

To prove this, take any $f \in C^q_{\mathrm{alg}}(G, \Omega^p)$ and $g_0, \dots, g_q \in G$. Then,

$$\begin{aligned}
&\Big((H\delta f)(g_1, \dots, g_q)\Big)_{g_0L} \\
&\quad = \pi\left(\Theta^{\delta f}_{(g_0,\dots,g_q)}\right) \\
&\quad = \left(f(g_1, \dots, g_q)\right)_{g_0L} \\
&\qquad + (-1)^{q+1}\pi\Big(\left((g_0k)\cdot f(k^{-1}g_0^{-1}, g_1, \dots, g_{q-1})\right)_{g_0L}\Big) \\
&\qquad + \sum_{i=1}^{q-1}(-1)^{i+1}\pi\Big(\left((g_0k)\cdot f(k^{-1}g_0^{-1}, g_1, \dots, g_i g_{i+1}, \dots, g_q)\right)_{g_0L}\Big) \\
&\qquad - \pi\Big(\left((g_0k)\cdot f(k^{-1}g_0^{-1}g_1, g_2, \dots, g_q)\right)_{g_0L}\Big),
\end{aligned} \tag{10}$$

where $\left((g_0k)\cdot f(k^{-1}g_0^{-1}, g_1, \dots, g_{q-1})\right)_{g_0L}$ means the function from L to $\Omega^p_{g_0L}$ defined as $k \mapsto \left((g_0k)\cdot f(k^{-1}g_0^{-1}, g_1, \dots, g_{q-1})\right)_{g_0L}$. Similarly,

$$\begin{aligned}
&\left((\delta H f)(g_1, \dots, g_q)\right)_{g_0L} \\
&\quad = \Big(g_1 \cdot \left((Hf)(g_2, \dots, g_q)\right)\Big)_{g_0L} \\
&\qquad + (-1)^q\left((Hf)(g_1, \dots, g_{q-1})\right)_{g_0L} \\
&\qquad + \sum_{i=1}^{q-1}(-1)^i\left((Hf)(g_1, \dots, g_i g_{i+1}, \dots, g_q)\right)_{g_0L} \\
&\quad = \Big(g_1 \cdot \left((Hf)(g_2, \dots, g_q)\right)\Big)_{g_0L} \\
&\qquad + (-1)^q\pi\Big(\left((g_0k)\cdot f(k^{-1}g_0^{-1}, g_1, \dots, g_{q-1})\right)_{g_0L}\Big) \\
&\qquad + \sum_{i=1}^{q-1}(-1)^i\pi\Big(\left((g_0k)\cdot f(k^{-1}g_0^{-1}, g_1, \dots, g_i g_{i+1}, \dots, g_q)\right)_{g_0L}\Big).
\end{aligned} \tag{11}$$

From the definition of the G-action on Ω^p, it is easy to see that

$$\pi\Big(\left((g_0k)\cdot f(k^{-1}g_0^{-1}g_1, g_2, \dots, g_q)\right)_{g_0L}\Big) = \Big(g_1 \cdot \left((Hf)(g_2, \dots, g_q)\right)\Big)_{g_0L}. \tag{12}$$

Combining (10)–(12), we clearly get (9).

From the above identity (9), we see, of course, that any cocycle in $C^q_{\mathrm{alg}}(G, \Omega^p)$ (for any $q \geq 1$) is a coboundary, proving the lemma. □

2.4. Lemma. *The restriction map* $\gamma : \Omega^{pG} \to C^p(\mathfrak{g}, \mathfrak{g}_{\mathrm{red}}, \mathfrak{a})$ *(defined below in the proof) is an isomorphism for all* $p \geq 0$*, where* $C^*(\mathfrak{g}, \mathfrak{g}_{\mathrm{red}}, \mathfrak{a})$ *is the standard cochain complex for the Lie algebra pair* $(\mathfrak{g}, \mathfrak{g}_{\mathrm{red}})$ *with coefficient in the* $\mathfrak{g}$*-module* $\mathfrak{a}$*. Moreover,* γ *commutes with differentials. Thus,* γ *induces an isomorphism in cohomology*

$$H^*(\Omega^G) \xrightarrow{\sim} H^*(\mathfrak{g}, \mathfrak{g}_{\mathrm{red}}, \mathfrak{a}).$$

Proof. For any $\omega \in \Omega^{pG}$, define $\gamma(\omega)$ as the value of ω at eL. Since G acts transitively on X, and ω is G-invariant, γ is injective.

Since any $\omega_o \in C^p(\mathfrak{g}, \mathfrak{g}_{\mathrm{red}}, \mathfrak{a})$ can be extended (uniquely) to a G-invariant form on X with values in $\mathfrak{a}$, γ is surjective. Further, from the definition of differentials on the two sides, it is easy to see that γ commutes with differentials. □

2.5. Continuation of the proof of Theorem 2.2

We now determine ${}''E$. First, by (6) of (2.2),

$${}''E_1^{p,q} = H_\delta^q(A^{*,p}) = H_{\mathrm{alg}}^q(G, \Omega^p).$$

Thus, by Lemma 2.3,

$${}''E_1^{p,0} = H_{\mathrm{alg}}^0(G, \Omega^p) = \Omega^{pG}$$

and

$${}''E_1^{p,q} = 0, \qquad \text{if } q > 0.$$

Moreover, under the above equality, the differential of the spectral sequence $d_1 : {}''E_1^{p,0} \to {}''E_1^{p+1,0}$ can be identified with the restriction of the de Rham differential

$$\Omega^{pG} \to \Omega^{p+1G}.$$

Thus, by Lemma 2.4,

$${}''E_2^{p,q} = \begin{cases} H^p(\mathfrak{g}, \mathfrak{g}_{\mathrm{red}}, \mathfrak{a}), & \text{if } q = 0 \\ 0, & \text{otherwise.} \end{cases} \tag{13}$$

In particular, the spectral sequence ${}''E$ as well degenerates at the ${}''E_2$-term. Moreover, we have a canonical isomorphism

$$H^p(\mathfrak{g}, \mathfrak{g}_{\mathrm{red}}, \mathfrak{a}) \simeq H^p(C, \delta + d). \tag{14}$$

Comparing the above isomorphism with the isomorphism (8) of §2.2, we get a canonical isomorphism:

$$H_{\mathrm{alg}}^p(G, \mathfrak{a}) \simeq H^p(\mathfrak{g}, \mathfrak{g}_{\mathrm{red}}, \mathfrak{a}).$$

For the isomorphism $H^p(\mathfrak{g}, \mathfrak{g}_{\mathrm{red}}, \mathfrak{a}) \simeq H^p(\mathfrak{g}_u, \mathfrak{a})^{\mathfrak{g}}$, see Theorem 1.8 and its proof. This proves Theorem 2.2. □

2.6. Remark. Even though we took the field $\mathbb{C}$ as our base field, all the results of this paper hold (by the same proofs) over any algebraically closed field of characteristic 0, if we replace the fundamental group π_1 by the algebraic fundamental group.

Acknowledgments

This work was done while the authors were visiting the Fields Institute, Toronto (Canada) in July, 2003, hospitality of which is gratefully acknowledged. The first author was partially supported from NSF. We thank J-P. Serre for a helpful advice regarding exposition.

References

[B] A. Borel, *Linear Algebraic Groups*, Graduate Texts in Math. Vol. 126, second ed., Springer–Verlag, 1991.

[CE] C. Chevalley and S. Eilenberg, Cohomology theory of Lie groups and Lie algebras, *Transactions Amer. Math. Soc.* **63** (1948), 85–124.

[GH] P. Griffiths and J. Harris, *Principles of Algebraic Geometry*, John Wiley and Sons., Inc., 1978.

[H] J. Humphreys, *Linear Algebraic Groups*, Graduate Texts in Math. Vol. 21, Springer–Verlag 1995.

[HS] G. Hochschild and J-P. Serre, Cohomology of Lie algebras, *Annals of Math.* **57** (1953), 591–603.

[K] S. Kumar, *Kac-Moody Groups, their Flag Varieties and Representation Theory*, Progress in Math. Vol. 204, Birkhäuser, 2002.

[V] W.T. Van-Est, Une application d'une méthode de Cartan–Leray, *Indag. Math.* **18** (1955), 542–544.

Differential operators and cohomology groups on the basic affine space

Thierry Levasseur[1] and J. T. Stafford[2]

[1] Département de Mathématiques
Université de Brest
29238 Brest cedex 3
France
Thierry.Levasseur@univ-brest.fr

[2] Department of Mathematics
University of Michigan
Ann Arbor, MI 48109-1043
USA
jts@umich.edu

This paper is dedicated to Tony Joseph on the occasion of his 60th *birthday.*

Summary. We study the ring of differential operators $\mathcal{D}(\mathbf{X})$ on the basic affine space $\mathbf{X} = G/U$ of a complex semisimple group G with maximal unipotent subgroup U. One of the main results shows that the cohomology group $\mathrm{H}^*(\mathbf{X}, \mathcal{O}_\mathbf{X})$ decomposes as a finite direct sum of non-isomorphic simple $\mathcal{D}(\mathbf{X})$-modules, each of which is isomorphic to a twist of $\mathcal{O}(\mathbf{X})$ by an automorphism of $\mathcal{D}(\mathbf{X})$.

We also use $\mathcal{D}(\mathbf{X})$ to study the properties of $\mathcal{D}(\mathbf{Z})$ for highest weight varieties $\mathbf{Z}$. For example, we prove that $\mathbf{Z}$ is $\mathcal{D}$-simple in the sense that $\mathcal{O}(\mathbf{Z})$ is a simple $\mathcal{D}(\mathbf{Z})$-module and produce an irreducible G-module of differential operators on $\mathbf{Z}$ of degree -1 and specified order.

Key words: semisimple Lie group, basic affine space, highest weight variety, rings of differential operators, D-simplicity

Subject Classifications: 13N10, 14L30, 16S32, 17B56, 20G10

1. Introduction

Fix a complex semisimple, connected and simply connected Lie group G with maximal unipotent subgroup U and Lie algebra $\mathfrak{g}$. Then the *basic affine space* is the quasi-

The second author was supported in part by the NSF through the grants DMS-9801148 and DMS-0245320. Part of this work was done while he was visiting the Mittag–Leffler Institute and he would like to thank the Institute for its financial support and hospitality.

affine variety $\mathbf{X} = G/U$. The ring of global differential operators $\mathcal{D}(\mathbf{X})$ has a long history, going back to the work [GK] of Gelfand and Kirillov in the late 1960s who used this space to formulate and partially solve their conjecture that the quotient division ring of the enveloping algebra $U(\mathfrak{g})$ should be isomorphic to a Weyl skew field.

The variety $\mathbf{X}$ is only quasi-affine and, when $\mathfrak{g}$ is not isomorphic to a direct sum of copies of $\mathfrak{sl}(2)$, the affine closure $\overline{\mathbf{X}}$ of $\mathbf{X}$ is singular. In general, rings of differential operators on a singular variety $\mathbf{Z}$ can be quite unpleasant; for example, and in contrast to the case of a smooth affine variety, $\mathcal{D}(\mathbf{Z})$ need not be noetherian, finitely generated or simple and (conjecturally) it will not be generated by the derivations $\mathrm{Der}_{\mathbb{C}}(\mathbf{Z})$. Moreover, the canonical module $\mathcal{O}(\mathbf{Z})$ need not be a simple $\mathcal{D}(\mathbf{Z})$-module. Recently, Bezrukavnikov, Braverman and Positselskii [BBP] proved a remarkable result on the structure of $\mathcal{D}(\mathbf{X})$ which shows that it actually has very pleasant properties. Specifically, they proved that there exist automorphisms $\{F_w\}_w$, indexed by the Weyl group W of G, such that for any nonzero $\mathcal{D}(\mathbf{X})$-module M there exists $w \in W$ such that $\mathcal{D}_{\mathbf{X}} \otimes_{\mathcal{D}(\mathbf{X})} M^w \neq 0$, where $M^w = M^{F_w}$ is the twist of M by F_w. (The F_w should be thought of as analogues of partial Fourier transforms.) Since $\mathbf{X}$ is smooth this implies that, for any finite open affine cover $\{\mathbf{X}_i\}_i$ of $\mathbf{X}$, the ring $\bigoplus_{i,w} \mathcal{D}(\mathbf{X}_i)^w$ is a noetherian, faithfully flat overring of $\mathcal{D}(\mathbf{X})$. As is shown in [BBP], it follows easily that $\mathcal{D}(\mathbf{X})$ is a noetherian domain of finite injective dimension.

The aim of this paper is to extend and apply the results of [BBP]. Our first result, which combines Proposition 3.1, Theorem 3.3 and Theorem 3.8, further elucidates the structure of $\mathcal{D}(\mathbf{X})$.

Proposition 1.1. *Let $\mathbf{X} = G/U$ denote the basic affine space of G. Then:*

(1) *$\mathcal{D}(\mathbf{X})$ is a simple ring satisfying the Auslander–Gorenstein and Cohen–Macaulay conditions (see Section 3 for the definitions);*
(2) *$\mathcal{D}(\mathbf{X})$ is (finitely) generated, as a $\mathbb{C}$-algebra, by $\{\mathcal{O}(\mathbf{X})^w : w \in W\} \cup \widehat{\mathfrak{h}}$.*

Note that $\mathcal{D}(\mathbf{X})$ is quite a subtle ring; for example, if $G = \mathrm{SL}(3, \mathbb{C})$, then $\mathcal{D}(\mathbf{X}) \cong U(\mathfrak{so}(8))/J$, where J is the *Joseph ideal* as defined in [Jo1] (see Example 2.4).

The variety $\mathbf{X} = G/U$ has a natural left action of G and a right action of the maximal torus H for which $B = HU$ is a Borel subgroup. Differentiating these actions gives embeddings of $\mathfrak{g} = \mathrm{Lie}(G)$, respectively $\widehat{\mathfrak{h}} = \mathrm{Lie}(H)$ into $\mathrm{Der}_{\mathbb{C}}(\mathbf{X})$. It follows easily from the simplicity of $\mathcal{D}(\mathbf{X})$ that $\mathcal{O}(\mathbf{X}) = \mathrm{H}^0(\mathbf{X}, \mathcal{O}_{\mathbf{X}})$ is a simple $\mathcal{D}(\mathbf{X})$-module. One of the main results of this paper extends this to describe the $\mathcal{D}(\mathbf{X})$-module structure of the full cohomology group $\mathrm{H}^*(\mathbf{X}, \mathcal{O}_{\mathbf{X}})$:

Theorem 1.2 (Theorem 4.8). *Let $\mathbf{X} = G/U$ and set $\mathcal{M} = \mathcal{D}(\mathbf{X})/\mathcal{D}(\mathbf{X})\mathfrak{g}$.*

(1) $\mathcal{M} \cong \mathrm{H}^*(\mathbf{X}, \mathcal{O}_{\mathbf{X}})$.
(2) *For each i,* $\mathrm{H}^i(\mathbf{X}, \mathcal{O}_{\mathbf{X}}) \cong \bigoplus \Big\{ \mathcal{O}(\mathbf{X})^w : w \in W : \ell(w) = i \Big\}$.
(3) *For $w \neq v \in W$, the $\mathcal{D}(\mathbf{X})$-modules $\mathcal{O}(\mathbf{X})^w$ and $\mathcal{O}(\mathbf{X})^v$ are simple and nonisomorphic.*

A result analogous to Theorem 1.2, but in the l-adic setting, has been proved in [Po, Lemma 12.0.1]. It is not clear to us what is the relationship between the two results.

A key point in the proof of Theorem 1.2 is that by the Borel–Weil–Bott theorem one has an explicit description of the G-module structure of the $\mathrm{H}^*(\mathbf{X}, \mathcal{O}_\mathbf{X})$ and one then proves Theorem 1.2 by comparing that structure with the G-module structure of the $\mathcal{O}(\mathbf{X})^w$. Theorem 1.2 is rather satisfying since it relates the left ideal of $\mathcal{D}(\mathbf{X})$ generated by the derivations coming from $\mathfrak{g}$ to the only "obvious" simple $\mathcal{D}(\mathbf{X})$-modules $\mathcal{O}(\mathbf{X})^w$. In contrast, if one considers the left ideal generated by all the "obvious" derivations, $\mathcal{D}(\mathbf{X})\mathfrak{g} + \mathcal{D}(\mathbf{X})\widehat{\mathfrak{h}}$, then one obtains:

Proposition 1.3 (Theorem 4.5). *As left $\mathcal{D}(\mathbf{X})$-modules,*

$$\frac{\mathcal{D}(\mathbf{X})}{\mathcal{D}(\mathbf{X})\mathfrak{g} + \mathcal{D}(\mathbf{X})\widehat{\mathfrak{h}}} = \frac{\mathcal{D}(\mathbf{X})}{\mathcal{D}(\mathbf{X})\mathrm{Der}_\mathbb{C}(\mathbf{X})} \cong \mathcal{O}(\mathbf{X}).$$

This proposition is somewhat surprising since, at least when $\mathfrak{g}$ is not a direct sum of copies of $\mathfrak{sl}(2)$, one can show that $\mathcal{D}(\mathbf{X})$ is *not* generated by $\mathcal{O}(\mathbf{X})$ and $\mathrm{Der}_\mathbb{C}(\mathbf{X})$ as a $\mathbb{C}$-algebra (see Corollary 5.10). Of course, the analogue of Proposition 1.3 for smooth varieties is standard.

A natural class of varieties associated to G are *S-varieties*: closures $\overline{\mathbf{Y}}$ of a G-orbit $\mathbf{Y} = \mathbf{Y}_\Gamma = G.v_\Gamma$ where v_Γ is a sum of highest weight vectors in some finite-dimensional G-module. In such a case there exists a natural surjection $\mathbf{X} \twoheadrightarrow \mathbf{Y} = G/S_\Gamma$, for the isotropy group S_Γ of v_Γ. This induces, by restriction of operators, a map $\psi_\Gamma : \mathcal{D}(\mathbf{X})^{S_\Gamma} \to \mathcal{D}(\overline{\mathbf{Y}})$ and allows us to use our structure results on $\mathcal{D}(\mathbf{X})$ to give information on $\mathcal{D}(\mathbf{Y})$. For this to be effective we need the mild assumption that $\overline{\mathbf{Y}}$ is normal and $\mathrm{codim}_{\overline{\mathbf{Y}}}(\overline{\mathbf{Y}} \smallsetminus \mathbf{Y}) \geq 2$ or, equivalently, that $\mathcal{O}(\mathbf{Y}) = \mathcal{O}(\overline{\mathbf{Y}})$ (see Theorem 5.3 for further equivalent conditions).

Corollary 1.4 (Proposition 5.6). *Let $\overline{\mathbf{Y}}$ be an S-variety such that $\mathcal{O}(\mathbf{Y}) = \mathcal{O}(\overline{\mathbf{Y}})$. Then $\mathcal{O}(\mathbf{Y})$ is a simple $\mathcal{D}(\mathbf{Y})$-module.*

When $\overline{\mathbf{Y}} = \overline{\mathbf{Y}}_\Gamma$ is singular, this result says that there exists operators $D \in \mathcal{D}(\mathbf{Y})$ that cannot be constructed from derivations; these are "exotic" operators in the terminology of [AB]). As the name suggests, exotic operators can be hard to construct—see [AB] or [BK2], for example—but in our context their construction is easy; they arise as $\psi_\Gamma F_{w_0}(\mathcal{O}(\mathbf{Y}_{\Gamma^*}))$. In the special case of a highest weight variety (this is just an S-variety $\overline{\mathbf{Y}}_\gamma = \overline{\mathbf{Y}}_\Gamma$ for $\Gamma = \mathbb{N}\gamma$) we can be more precise about these operators.

Corollary 1.5 (Corollary 6.6). *Suppose that $\overline{\mathbf{Y}} = \overline{\mathbf{Y}}_\gamma$ is a highest weight variety. Then there exists an irreducible G-module $E \cong V(\gamma)$ of differential operators on $\mathbf{Y}$ of degree -1 and order $\langle \gamma, 2\varrho^\vee \rangle$.*

Corollary 1.4 also proves the Nakai conjecture for S-varieties satisfying the hypotheses of that result and this covers most of the known cases of normal, singular varieties for which the conjecture is known. See Section 5 for the details.

A fundamental question in the theory of differential operators on invariant rings asks the following. Suppose that Q is a (reductive) Lie group acting on a finite-dimensional vector space V such that the fixed ring $\mathcal{O}(V)^Q$ is singular. Then, is the natural map $\mathcal{D}(V)^Q \to \mathcal{D}(\mathcal{O}(V)^Q)$ surjective? A positive answer to this question is known in a number of cases and this has had significant applications to representation theory; see, for example, [Jo2, LS1, LS2, Sc] and Remark 6.8. It is therefore natural to ask when the analogous map $\psi_\Gamma : \mathcal{D}(\mathbf{X})^{S_\Gamma} \to \mathcal{D}(\mathbf{Y}_\Gamma)$ is surjective. Although we do not have a general answer to this question, we suspect that ψ_Γ will usually not be surjective. As evidence for this we prove that this is the case for one of the fundamental examples of an S-variety: the closure of the minimal orbit in a simple Lie algebra $\mathfrak{g}$.

Proposition 1.6 (Theorem 6.7). *If $\mathbf{Y}_\gamma$ is the minimal (nonzero) nilpotent orbit of a simple classical Lie algebra $\mathfrak{g}$, then ψ_γ is surjective if and only if $\mathfrak{g} = \mathfrak{sl}(2)$ or $\mathfrak{g} = \mathfrak{sl}(3)$.*

2. Preliminaries

In this section we describe the basic results and notation we need from the literature, notably the relevant results from [BBP]. The reader is also referred to [GK, HV, Sh1, Sh2] for the interrelationship between differential operators on the base affine space and the corresponding enveloping algebra.

We begin with some necessary notation. The base field will always be the field $\mathbb{C}$ of complex numbers. Let G be a connected simply-connected semi-simple algebraic group of rank ℓ, B a Borel subgroup, $H \subset B$ a maximal torus, $U \subset B$ a maximal unipotent subgroup of G. The Lie algebra of an algebraic group is denoted by the corresponding gothic character; thus $\mathfrak{g} = \mathrm{Lie}(G)$, $\mathfrak{h} = \mathrm{Lie}(H)$ and $\mathfrak{u} = \mathrm{Lie}(U)$.

We will use standard Lie theoretic notation, as for example given in [Bo]. In particular, let Δ denote the root system of $(\mathfrak{g}, \mathfrak{h})$ and fix a set of positive roots Δ_+ such that $\mathfrak{u} = \bigoplus_{\alpha\in\Delta_+} \mathfrak{u}_\alpha$. Denote by W the Weyl group of Δ. Let Λ be the weight lattice of Δ which we identify with the character group of H. The set of dominant weights is denoted by Λ^+. Fix a basis $\Sigma = \{\alpha_1, \dots, \alpha_\ell\}$ of Δ_+ and write $\{\varpi_1, \dots, \varpi_\ell\}$ for the fundamental weights; thus $\langle \varpi_i, \alpha_j^\vee\rangle = \delta_{ij}$. Denote by ϱ, respectively $\varrho^\vee$, the half sum of the positive roots, respectively coroots. Let w_0 be the longest element of W and set $\lambda^* = -w_0(\lambda)$ for all $\lambda \in \Lambda$. Then $w_0(\varrho) = -\varrho$, $w_0(\varrho^\vee) = -\varrho^\vee$ and, by [Bo, Chapter 6, 1.10, Corollaire],

$$\langle \lambda^*, \varrho^\vee\rangle = \langle \lambda, \varrho^\vee\rangle = \sum_{i=1}^{\ell} m_i \qquad \text{if} \quad \lambda = \sum_{i=1}^{\ell} m_i\alpha_i .$$

For each $\omega \in \Lambda^+$ let $V(\omega)$ be the irreducible G-module of highest weight ω and let $V(\omega)_\mu$ be the subspace of elements of weight $\mu \in \Lambda$; we will denote by v_ω a highest weight vector of $V(\omega)$. Recall that the G-module $V(\omega)^*$ identifies naturally with $V(\omega^*)$. If E is a locally finite $(G \times H)$-module, we denote by $E[\lambda]$ the isotypic G-component of type $\lambda \in \Lambda^+$ and by E^μ the μ-weight space for the action of $H \equiv \{1\} \times H$. Hence,

$$E = \bigoplus \big\{E[\lambda]^\mu : \mu \in \Lambda,\ \lambda \in \Lambda^+\big\}.$$

Let $\mathbf{Y}$ be an algebraic variety. We denote by $\mathcal{O}_\mathbf{Y}$ the structural sheaf of $\mathbf{Y}$ and by $\mathcal{O}(\mathbf{Y})$ its algebra of regular functions. The sheaf of differential operators on $\mathbf{Y}$ is denoted by $\mathcal{D}_{\mathbf{Y}}$ with global sections $\mathcal{D}(\mathbf{Y}) = \mathrm{H}^0(\mathbf{Y}, \mathcal{D}_{\mathbf{Y}})$. The $\mathcal{O}(\mathbf{Y})$-module of elements of order $\leq k$ in $\mathcal{D}(\mathbf{Y})$ is denoted by $\mathcal{D}_k(\mathbf{Y})$, and the order of $D \in \mathcal{D}(\mathbf{Y})$ will be written ord D. Now suppose that $\mathbf{Y}$ is an irreducible quasi-affine algebraic variety, embedded as an open subvariety of an affine variety $\overline{\mathbf{Y}}$. We will frequently use the fact that if $\overline{\mathbf{Y}}$ is normal with $\operatorname{codim}_{\overline{\mathbf{Y}}}(\overline{\mathbf{Y}} \smallsetminus \mathbf{Y}) \geq 2$, then $\mathcal{O}(\mathbf{Y}) = \mathcal{O}(\overline{\mathbf{Y}})$, and so $\mathcal{D}(\mathbf{Y}) = \mathcal{D}(\overline{\mathbf{Y}})$.

Let Q be an affine algebraic group. We say that an algebraic variety $\mathbf{Y}$ is a Q-variety if it is equipped with a rational action of Q. For such a variety, $\mathcal{O}(\mathbf{Y})$, $\mathcal{D}_k(\mathbf{Y})$ and $\mathcal{D}(\mathbf{Y})$ are locally finite Q-modules, with the action of $a \in Q$ on $\varphi \in \mathcal{O}(\mathbf{Y})$ and $D \in \mathcal{D}(\mathbf{Y})$ being defined by $a.\varphi(y) = \varphi(a^{-1}.y)$ for $y \in \mathbf{Y}$, respectively $(a.D)(\varphi) = a.D(a^{-1}.\varphi)$. If $\mathbf{Y}$ is also an affine variety such that $\mathcal{O}(\mathbf{Y})^Q$ is an affine algebra, we define the *categorical quotient* $\mathbf{Y}/\!/Q$ by $\mathcal{O}(\mathbf{Y}/\!/Q) = \mathcal{O}(\mathbf{Y})^Q$. We then have a restriction morphism

$$\psi : \mathcal{D}(\mathbf{Y})^Q \longrightarrow \mathcal{D}(\mathbf{Y}/\!/Q), \qquad \psi(D)(f) = D(f) \text{ for } f \in \mathcal{O}(\mathbf{Y})^Q.$$

Notice that $\psi(\mathcal{D}_k(\mathbf{Y})^Q) \subseteq \mathcal{D}_k(\mathbf{Y}/\!/Q)$ for all k. In many cases $\mathcal{O}(\mathbf{Y})$ will be a $\mathbb{Z}$-graded algebra, $\mathcal{O}(\mathbf{Y}) = \bigoplus_{n\in\mathbb{Z}} \mathcal{O}^n$, in which case $\mathcal{D}(\mathbf{Y})$ has an induced $\mathbb{Z}$-graded structure, with the nth graded piece being

$$\mathcal{D}(\mathbf{Y})^n = \{\theta \in \mathcal{D}(\mathbf{Y}) : \theta(\mathcal{O}^r) \subseteq \mathcal{O}^{r+n} \text{ for all } r \in \mathbb{Z}\}. \tag{2.1}$$

In this situation, $\mathcal{D}(\mathbf{Y})^n$ will be called operators of *degree n*.

Assume now that V is a $(G \times H)$-module and that $\mathbf{Y}$ is a $(G \times H)$-subvariety of V. Then

$$\mathcal{O}(\mathbf{Y}) = \bigoplus_{\substack{\mu\in\Lambda \\ \lambda\in\Lambda^+}} \mathcal{O}(\mathbf{Y})[\lambda]^\mu \subset \mathcal{D}(\mathbf{Y}) = \bigoplus_{\substack{\mu\in\Lambda \\ \lambda\in\Lambda^+}} \mathcal{D}(\mathbf{Y})[\lambda]^\mu.$$

It follows easily from the definitions that

$$\mathcal{D}(\mathbf{Y})^\mu = \left\{d \in \mathcal{D}(\mathbf{Y}) : d(\mathcal{O}(\mathbf{Y})^\lambda) \subseteq \mathcal{O}(\mathbf{Y})^{\lambda+\mu} \text{ for all } \lambda \in \Lambda\right\}. \tag{2.2}$$

One clearly has a surjection $S(V^*)[\lambda]^\mu \twoheadrightarrow \mathcal{O}(\mathbf{Y})[\lambda]^\mu$. Furthermore, if $V = V^{-\gamma^*}$ for some $0 \neq \gamma^* \in \Lambda$, we will identify $S^m(V^*)$ with $S(V^*)^{m\gamma^*}$ and obtain the surjective G-morphism $S^m(V^*)[\lambda] \twoheadrightarrow \mathcal{O}(\mathbf{Y})[\lambda]^{m\gamma^*}$. In this case, $\mathcal{O}(\mathbf{Y}) = \bigoplus_{m\in\mathbb{N}} \mathcal{O}(\mathbf{Y})^{m\gamma^*}$ and $\mathcal{D}(\mathbf{Y}) = \bigoplus_{m\in\mathbb{Z}} \mathcal{D}(\mathbf{Y})^{m\gamma^*}$ are $\mathbb{Z}$-graded and (2.2) can be interpreted as saying that $\mathcal{D}(\mathbf{Y})^{m\gamma^*}$ is the set of differential operators of degree m on $\mathbf{Y}$.

The previous results apply in particular to the basic affine space $\mathbf{X} = G/U$. We need to collect here a few facts about the $(G \times H)$-variety $\mathbf{X}$ and its canonical affine closure $\overline{\mathbf{X}}$. For these assertions, see, for example, [GK, Gr, VP]. Set $V = V(\varpi_1) \bigoplus \cdots \bigoplus V(\varpi_\ell)$ and recall that there is an isomorphism

$$\mathbf{X} \xrightarrow{\sim} G.(v_{\varpi_1} \oplus \cdots \oplus v_{\varpi_\ell}) \subset V, \qquad \bar{g} \mapsto g.(v_{\varpi_1} \oplus \cdots \oplus v_{\varpi_\ell}),$$

where $\bar{g}$ denotes the class of $g \in G$ modulo U. We identify $\mathbf{X}$ with this G-orbit and write

$$\mathcal{O} = \mathcal{O}(\mathbf{X}) \quad \text{and} \quad \mathcal{D} = \mathcal{D}(\mathbf{X}).$$

Then the Zariski closure $\overline{\mathbf{X}}$ of $\mathbf{X}$ in V is a normal irreducible affine variety that satisfies $\operatorname{codim}_{\overline{\mathbf{X}}}(\overline{\mathbf{X}} \smallsetminus \mathbf{X}) \geq 2$; thus $\mathcal{O} = \mathcal{O}(\overline{\mathbf{X}})$, etc.

Identify w_0 with an automorphism of H; thus $\lambda^*(h) = \lambda(w_0(h^{-1}))$ for all $\lambda \in \Lambda$ and $h \in H$. Define the twisted (right) action of H on $\mathbf{X}$ by $r_h.\bar{g} = \overline{g w_0(h)}$. Endow the G-module V with the action of $H \equiv \{1\} \times H$ defined by

$$r_h.(u_1 \oplus \cdots \oplus u_\ell) = \varpi_1^*(h^{-1})u_1 \oplus \cdots \oplus \varpi_\ell^*(h^{-1})u_\ell.$$

Then the embedding $\mathbf{X} \hookrightarrow V$ is a morphism of $(G \times H)$-varieties. The induced (left) action of H on $\mathcal{O}$ is then given by $(r_h.f)(\bar{g}) = f(r_{h^{-1}}.\bar{g})$ for all $f \in \mathcal{O}, h \in H, \bar{g} \in \mathbf{X}$. It follows that $\mathcal{O}[\nu^*] = \mathcal{O}^{\nu^*}$, $\mathcal{O}^\mu = 0$ if $\mu \notin \Lambda^+$, and we can decompose the $(G \times H)$-module $\mathcal{O}$ as

$$\mathcal{O} = \bigoplus_{\lambda \in \Lambda^+} \mathcal{O}^\lambda, \qquad \mathcal{O}^\lambda = \mathcal{O}[\lambda] \cong V(\lambda). \tag{2.3}$$

The final isomorphism in (2.3) comes from the fact that the G-action on $\mathcal{O}$ is multiplicity free [VP, Theorem 2]. Notice that the algebra $\mathcal{O}$ is (finitely) generated by the G-modules $\mathcal{O}^{\varpi_j}$, $1 \leq j \leq \ell$. Also, (2.2) implies that $\mathcal{D}^\mu(\mathcal{O}^\lambda) = 0$ when $\lambda + \mu$ is not dominant.

Notation 2.1. *The differentials of the actions of G and H (via $h \mapsto r_h$) on $\mathbf{X}$ yield morphisms of algebras $\imath : U(\mathfrak{g}) \to \mathcal{D}$ and $\jmath : U(\mathfrak{h}) \to \mathcal{D}$. By [GK, Corollary 8.1 and Lemma 9.1], $\imath$ and $\jmath$ are injective; from now on we will identify $U(\mathfrak{g})$ with $\imath(U(\mathfrak{g}))$ but write $\widehat{\mathfrak{h}} = \jmath(\mathfrak{h})$ and $U(\widehat{\mathfrak{h}}) = \jmath(U(\mathfrak{h}))$, to distinguish these objects from their images under $\imath$.*

Let $0 \neq M$ be a left $\mathcal{D}$-module and $\tau \in \operatorname{Aut}(\mathcal{D})$, the $\mathbb{C}$-algebra automorphism group of $\mathcal{D}$. Then the *twist of M by τ* is the $\mathcal{D}$-module M^τ defined as follows: $M^\tau = M$ as an abelian group but $a \cdot x = \tau(a)x$ for all $a \in \mathcal{D}, x \in M$. Recall also that the localization of M on $\mathbf{X}$ is

$$L(M) = \mathcal{D}_{\mathbf{X}} \otimes_{\mathcal{D}} M \cong \mathcal{O}_{\mathbf{X}} \otimes_{\mathcal{O}} M.$$

Thus $L(M)$ is a quasi-coherent left $\mathcal{D}_{\mathbf{X}}$-module. Clearly $L(M) = 0$ when M is supported on $\overline{\mathbf{X}} \smallsetminus \mathbf{X}$ but, remarkably, one can obtain a nonzero localization by first twisting the module M:

Theorem 2.2 ([BBP, Proposition 3.1 and Theorem 3.4]). *Let $\mathcal{D} = \mathcal{D}(\mathbf{X})$. Then:*

(1) *There exists an injection of groups $F : W \hookrightarrow \operatorname{Aut}(\mathcal{D})$ written $w \mapsto F_w$.*
(2) *For each $\mathcal{D}$-module $M \neq 0$ there exists $w \in W$ such that $L(M^{F_w}) \neq 0$.* □

This theorem is also valid for right modules. The morphisms F_w can be regarded as variants of partial Fourier transforms, and more details on their structure and properties can be found in [BBP].

It will be convenient to reformulate Theorem 2.2, for which we need some notation. Since $\mathbf{X}$ is an open subset of the nonsingular locus of $\overline{\mathbf{X}}$, we can fix an open (and smooth) affine cover of $\mathbf{X}$:

$$\mathbf{X} = \bigcup_{j=1}^{k} \mathbf{X}_j, \quad \text{where} \quad \mathbf{X}_j = \{x \in \overline{\mathbf{X}} : f_j(x) \neq 0\}, \tag{2.4}$$

for the appropriate $f_i \in \mathcal{O}(\mathbf{X})$. Notice that $\mathcal{D}(\mathbf{X}_j) = \mathcal{D}[f_j^{-1}]$ for each j and so $\mathcal{D}(\mathbf{X}_j)$ is a flat $\mathcal{D}$-module. If M is $\mathcal{D}$-module and $w \in W$, let M^w denote the twist of M by F_w.

For each pair (w, j) with $w \in W$ and $1 \leq j \leq k$, we have an injective morphism of algebras $\phi_{wj} : \mathcal{D} \hookrightarrow \mathcal{D}(\mathbf{X}_j)$ given by $\phi_{wj}(d) = F_w^{-1}(d)$. We write $\mathcal{D}(\mathbf{X}_j)_w$ for the algebra $\mathcal{D}(\mathbf{X}_j)$ regarded as an overring of $\mathcal{D}$ under this embedding. The significance of this construction is that, for any left $\mathcal{D}$-module M, the map $d \otimes v \mapsto d \otimes v$ induces an isomorphism $\mathcal{D}(\mathbf{X}_j) \otimes_{\mathcal{D}} M^w \cong \mathcal{D}(\mathbf{X}_j)_w \otimes_{\mathcal{D}} M$ of left $\mathcal{D}(\mathbf{X}_j)$-modules. Theorem 2.2 can now be rewritten as follows.

Corollary 2.3. *Let $0 \neq M$ be a left $\mathcal{D}$-module. Then:*

(1) *There exists a pair (w, j) such that $\mathcal{D}(\mathbf{X}_j)_w \otimes_{\mathcal{D}} M \neq 0$.*
(2) *Set $R_w = \bigoplus_{j=1}^{k} \mathcal{D}(\mathbf{X}_j)_w$ and $R = \bigoplus_{w \in W} R_w$. Then R is a faithfully flat (left or right) overring of $\mathcal{D}$.*

Proof. Part (1) is just a reformulation of Theorem 2.2. This in turn implies that R is a faithful right $\mathcal{D}$-module. It is flat since $\mathcal{D}(\mathbf{X}_j)_w$ is isomorphic, as a $\mathcal{D}$-module, to the localization of $\mathcal{D}$ at the powers of $F_w(f_j)$. Since Theorem 2.2 also holds for right modules, the same argument shows that R is a faithful flat left $\mathcal{D}$-module. □

When $\mathfrak{g} = \mathfrak{sl}(2)$ it is easy to check that $\overline{\mathbf{X}} = \mathbb{C}^2$, and so there is no subtlety to the structure of either $\overline{\mathbf{X}}$ or $\mathcal{D}$. However, when $\mathfrak{g} \neq \mathfrak{sl}(2)^m$, $\overline{\mathbf{X}}$ will be singular and $\mathcal{D}$ will be rather subtle. The simplest example is:

Example 2.4. Assume that $\mathfrak{g} = \mathfrak{sl}(3)$ and set $\mathbf{X} = \mathrm{SL}(3)/U$. Then $\overline{\mathbf{X}}$ is the quadric $\sum_{i=1}^{3} u_i y_i = 0$ inside $\mathbb{C}^6$. Moreover, $\mathcal{D} = \mathcal{D}(\mathbf{X}) \cong U(\mathfrak{so}(8))/J$, where J is the Joseph ideal.

An explicit set of generators of $\mathcal{D}$ is given in [LS3, (2.2)]. The algebra $F_{w_0}(\mathcal{O})$ is generated by the operators $\{\Phi_j, \Theta_j\}$ of order 2 from [LS3, (2.2)].

Proof. The proof of the first assertion is an elementary and classical computation, which is left to the reader. The second assertion then follows from [LSS, Remark 3.2(v) and Corollary A]. The claim in the second paragraph will not be used in this paper and so is left to the interested reader (the next proposition may prove useful). A second way of interpreting this example is given in Remark 6.8. □

In the main body of the paper we will need some more technical results from [BBP] about the automorphisms F_w and for the reader's convenience we record them in the following proposition. This summarizes Lemma 3.3, Corollary 3.10, Proposition 3.11 and the proof of Lemma 3.12 of that paper.

Proposition 2.5. *Let $\eta \in \Lambda^+$, $\mu \in \Lambda$ and $w \in W$. Then:*

(1) *F_w is G-linear and $F_w(\mathcal{D}^\mu) = \mathcal{D}^{w(\mu)}$. Thus $F_w(\mathcal{O}^\eta) \cong V(\eta)$ as G-modules.*
(2) *$F_w(h) = w(h) + \langle w(h) - h, \varrho\rangle$ for all $h \in \widehat{\mathfrak{h}}$.*
(3) ord $F_w(d)$ = ord $d + \langle \mu - w(\mu), \varrho^\vee\rangle$ *for all $0 \neq d \in \mathcal{D}^\mu$. In particular, if $0 \neq f \in \mathcal{O}^\eta$, then* ord $F_{w_0}(f) = \langle \eta, 2\varrho^\vee\rangle$.
(4) *Let $\{f_i\}_{1\leq i\leq n} \subset \mathcal{O}^\eta$ and $\{g_i\}_{1\leq i\leq n} \subset \mathcal{O}^{\eta^*} \cong (\mathcal{O}^\eta)^*$ be dual bases such that (up to a constant) $\sum_i f_i \otimes g_i$ is the unique G-invariant element in $\mathcal{O}^\eta \otimes \mathcal{O}^{\eta^*}$. Then the $(G \times H)$-invariant operator $P_\eta = \sum_{i=1}^n f_i F_{w_0}(g_i) \in U(\widehat{\mathfrak{h}})$ is given by*

$$P_\eta = c_\eta \prod_{\alpha^\vee \in \Delta_+^\vee} \prod_{i=1}^{\langle \eta, \alpha^\vee\rangle} (\alpha^\vee + \langle \alpha^\vee, \varrho\rangle - i), \tag{2.5}$$

for some $c_\eta \in \mathbb{C} \smallsetminus \{0\}$. Moreover, ord $P_\eta = \langle \eta, 2\varrho^\vee\rangle$.
(5) *$U(\widehat{\mathfrak{h}}) = \sum_{w\in W} U(\widehat{\mathfrak{h}}) F_w(P_\eta)$.* □

3. The structure of $\mathcal{D}(G/U)$.

In [BBP], the authors use Theorem 2.2 to prove that $\mathcal{D}$ is a noetherian ring of finite injective dimension. In this section we investigate other consequences of that result to the structure of $\mathcal{D}$. The notation from the last section will be retained; in particular, G is a connected, simply connected semi-simple algebraic group over $\mathbb{C}$, with basic affine space $\mathbf{X} = G/U$ and $\mathcal{D} = \mathcal{D}(\mathbf{X})$.

We begin with an easy application of the faithful flatness of the ring $R = \bigoplus_{w,j} \mathcal{D}(\mathbf{X}_j)_w$ defined in Corollary 2.3.

Proposition 3.1. *The ring $\mathcal{D}$ is simple and $\mathcal{O}$ is a simple left $\mathcal{D}$-module.*

Proof. Let J be a nonzero ideal of $\mathcal{D}$. As in the proof of Corollary 2.3, $\mathcal{D}(\mathbf{X}_j)$ is a noetherian localization of $\mathcal{D}$ and so, by [MR, Proposition 2.1.16(vi)], each $\mathcal{D}(\mathbf{X}_j)_w \otimes_{\mathcal{D}} J \cong \mathcal{D}(\mathbf{X}_j) \otimes_{\mathcal{D}} J^w$ is an ideal of $\mathcal{D}(\mathbf{X}_j)$ and it is nonzero since $\mathcal{D}$ is a domain. But $\mathbf{X}_j$ is a smooth affine variety and so $\mathcal{D}(\mathbf{X}_j)$ is a simple ring. Thus $\mathcal{D}(\mathbf{X}_j)_w \otimes_{\mathcal{D}} J = \mathcal{D}(\mathbf{X}_j)_w$ for all $w \in W$ and $1 \leq j \leq k$. This means that the module $M = \mathcal{D}/J$ satisfies $R \otimes_{\mathcal{D}} M = 0$, hence $M = 0$ by Corollary 2.3(2). In other words, $J = \mathcal{D}$.

If $\mathcal{O}$ is not a simple $\mathcal{D}$-module, pick a proper factor module $\mathcal{O}/K$ and note that K is then an ideal of $\mathcal{O}$. But now the annihilator $\mathrm{ann}_{\mathcal{D}}(\mathcal{O}/K)$ of $\mathcal{O}/K$ as a $\mathcal{D}$-module is an ideal of $\mathcal{D}$ that contains K. This contradicts the simplicity of $\mathcal{D}$. □

We now turn to the homological properties of $\mathcal{D}$. Two conditions that have proved very useful in applying homological techniques (see for example [Bj] or [LS1]) are the Auslander and Cohen–Macaulay conditions. These are defined as follows. A noetherian algebra A of finite injective dimension is called *Auslander–Gorenstein* if, for any finitely generated (left) A-module M and any right submodule $N \subseteq \mathrm{Ext}^i_A(M, A)$, one has $\mathrm{Ext}^j_A(N, A) = 0$, for $j < i$. The *grade* of M is $\mathrm{j}_A(M) = \inf\{j : \mathrm{Ext}^j_A(M, A) \neq 0\}$ (with the convention that $\mathrm{j}_A(0) = +\infty$). The

Gelfand–Kirillov dimension of M will be denoted $\mathrm{GKdim}_A M$. We say that the algebra A is *Cohen–Macaulay* if

$$\mathrm{GKdim}_A M + \mathrm{j}_A(M) = \mathrm{GKdim}\, A \quad \text{for all } M \neq 0.$$

If $\mathbf{Z}$ is a smooth affine variety, then $\mathcal{D}(\mathbf{Z})$ is Auslander–Gorenstein and Cohen–Macaulay with $\mathrm{GKdim}\, \mathcal{D}(\mathbf{Z}) = 2 \dim \mathbf{Z}$ (see [Bj, Chapter 2, Section 7]). This applies, of course, when $\mathbf{Z} = \mathbf{X}_j$ for $1 \le j \le k$ and so $R = \bigoplus_{w,j} \mathcal{D}(\mathbf{X}_j)_w$ also satisfies these properties. As we show in Theorem 3.3, these properties descend to $\mathcal{D} = \mathcal{D}(\mathbf{X})$.

If M is a (finitely-generated) $\mathcal{D}$-module, write

$$M_j^w = \mathcal{D}(\mathbf{X}_j) \otimes_{\mathcal{D}} M^w, \qquad \text{for} \quad 1 \le j \le k \text{ and } w \in W.$$

Lemma 3.2. *If M is a finitely generated left $\mathcal{D}$-module, then*

$$\mathrm{GKdim}_{\mathcal{D}} M = \max\{\mathrm{GKdim}_{\mathcal{D}(\mathbf{X}_j)} M_j^w : 1 \le j \le k,\, w \in W\}.$$

Proof. By definition, $R \otimes_{\mathcal{D}} M = \bigoplus_{j=1}^k \bigoplus_{w \in W} M_j^w$ and so

$$\mathrm{GKdim}_R\, R \otimes_{\mathcal{D}} M = \max\{\mathrm{GKdim}_{\mathcal{D}(\mathbf{X}_j)} M_j^w : 1 \le j \le k,\, w \in W\}.$$

By faithfully flatness (Corollary 2.3), the natural map $M \to R \otimes_{\mathcal{D}} M$ is injective and it follows that $\mathrm{GKdim}_{\mathcal{D}} M \le \mathrm{GKdim}_R\, R \otimes_{\mathcal{D}} M$. Conversely, since M_j^w is the localization of M^w at the Ore subset $\{f_j^s : s \in \mathbb{N}\}$, it follows from [Lo, Theorem 3.2] that $\mathrm{GKdim}_{\mathcal{D}(\mathbf{X}_j)} M_j^w \le \mathrm{GKdim}_{\mathcal{D}} M^w$. Since $\mathrm{GKdim}_{\mathcal{D}} M^w = \mathrm{GKdim}_{\mathcal{D}} M$, this implies that $\mathrm{GKdim}_R\, R \otimes_{\mathcal{D}} M \le \mathrm{GKdim}_{\mathcal{D}} M$ and the lemma is proved. □

Theorem 3.3. *The algebra $\mathcal{D}$ is Auslander–Gorenstein and Cohen–Macaulay.*

Proof. Let M be a finitely-generated (left) $\mathcal{D}$-module and N is a (right) submodule of $\mathrm{Ext}^p_{\mathcal{D}}(M, \mathcal{D})$. As R is a flat $\mathcal{D}$-module,

$$R \otimes_{\mathcal{D}} \mathrm{Ext}^i_{\mathcal{D}}(N, \mathcal{D}) \cong \mathrm{Ext}^i_R(N \otimes_{\mathcal{D}} R, R)$$

and $N \otimes_{\mathcal{D}} R$ is a submodule of $\mathrm{Ext}^p_R(R \otimes_{\mathcal{D}} M, R)$. Since R is Auslander–Gorenstein, this implies that $R \otimes_{\mathcal{D}} \mathrm{Ext}^q_{\mathcal{D}}(N, \mathcal{D}) = 0$, for any $q < p$. Since R is a faithful $\mathcal{D}$-module, it follows that $\mathrm{Ext}^q_{\mathcal{D}}(N, \mathcal{D}) = 0$. Thus $\mathcal{D}$ is Auslander–Gorenstein.

From $\mathrm{Ext}^i_R(R \otimes_{\mathcal{D}} M, R) \cong \mathrm{Ext}^i_{\mathcal{D}}(M, \mathcal{D}) \otimes_{\mathcal{D}} R$ and the faithful flatness of $R_{\mathcal{D}}$ we get that $\mathrm{Ext}^i_R(R \otimes_{\mathcal{D}} M, R) = 0 \iff \mathrm{Ext}^i_{\mathcal{D}}(M, \mathcal{D}) = 0$. Thus, $\mathrm{j}_R(R \otimes_{\mathcal{D}} M) = \mathrm{j}_{\mathcal{D}}(M)$. But,

$$\mathrm{Ext}^p_R(R \otimes_{\mathcal{D}} M, R) \cong \bigoplus_{j,w} \mathrm{Ext}^p_{\mathcal{D}(\mathbf{X}_j)}(M_j^w, \mathcal{D}(\mathbf{X}_j))$$

and so $\mathrm{j}_{\mathcal{D}}(M) = \mathrm{j}_R(R \otimes_{\mathcal{D}} M) = \min\{\mathrm{j}_{\mathcal{D}(\mathbf{X}_j)}(M_j^w) : 1 \le j \le k,\, w \in W\}$.

Each $\mathbf{X}_j$ is a smooth affine variety of dimension $\dim \mathbf{X}$ and so each $\mathcal{D}(\mathbf{X}_j)$ is Cohen–Macaulay. It therefore follows from Lemma 3.2 that

$$\begin{aligned}\mathrm{GKdim}_{\mathcal{D}} M &= \max\{\mathrm{GKdim}_{\mathcal{D}(\mathbf{X}_j)} M_j^w : 1 \le j \le k, w \in W, M_j^w \neq 0\} \\ &= \max\{2\dim \mathbf{X} - \mathrm{j}_{\mathcal{D}(\mathbf{X}_j)}(M_j^w) : 1 \le j \le k, w \in W, M_j^w \neq 0\} \\ &= 2\dim \mathbf{X} - \min\{\mathrm{j}_{\mathcal{D}(\mathbf{X}_j)}(M_j^w) : 1 \le j \le k, w \in W\} \\ &= 2\dim \mathbf{X} - \mathrm{j}_{\mathcal{D}}(M),\end{aligned}$$

as required. □

Remark 3.4. The last result should be compared with [YZ, Corollary 0.3] which shows that a simple noetherian $\mathbb{C}$-algebra with an affine commutative associated graded ring is automatically Auslander–Gorenstein and Cohen–Macaulay. It is not clear whether that result applies to $\mathcal{D}$, since we do not have a good description of the associated graded ring of $\mathcal{D}$.

When $\mathbf{Z}$ is a smooth affine variety, $\mathcal{D}(\mathbf{Z})$ is a finitely generated $\mathbb{C}$-algebra, simply because it is generated by $\mathcal{O}(\mathbf{Z})$ and $\mathrm{Der}_{\mathbb{C}}(\mathbf{Z})$. When $\mathbf{Z}$ is singular, it can easily happen that $\mathcal{D}(\mathbf{Z})$ is not affine (see, for example [BGG]). However, as we will show in Theorem 3.8, the $\mathbb{C}$-algebra $\mathcal{D}$ is finitely generated. We begin with some lemmas.

Lemma 3.5. *Let $\gamma \in \Lambda^+$. There exists a nondegenerate pairing of G-modules*

$$\tau : F_{w_0}(\mathcal{O}^\gamma) \times \mathcal{O}^{\gamma^*} \longrightarrow \mathbb{C}, \quad (F_{w_0}(f), g) \mapsto F_{w_0}(f)(g).$$

Proof. By Proposition 2.5 and (2.2), $F_{w_0}(\mathcal{O}^\gamma)(\mathcal{O}^{\gamma^*}) \subseteq \mathcal{O}^{w_0(\gamma)+\gamma^*} = \mathcal{O}^0 = \mathbb{C}$. Also,

$$(a.F_{w_0}(f))(a.g) = a.F_{w_0}(f)(a^{-1}.(a.g)) = a.F_{w_0}(f)(g) = F_{w_0}(f)(g),$$

for all $a \in G$. Therefore, τ is a well-defined pairing of G-modules and it induces a G-linear map $F_{w_0}(\mathcal{O}^\gamma) \to (\mathcal{O}^{\gamma^*})^*$. Since $F_{w_0}(\mathcal{O}^\gamma) \cong (\mathcal{O}^{\gamma^*})^* \cong V(\gamma)$, it suffices to show that τ is nonzero in order to show that it is nondegenerate. In the notation of Proposition 2.5(4), we will show that $F_{w_0}(f_i)(g_i) \neq 0$ for some i.

By Proposition 2.5(2),

$$F_{w_0}(\alpha^\vee) = w_0(\alpha^\vee) + \langle w_0(\alpha^\vee) - \alpha^\vee, \varrho\rangle = w_0(\alpha^\vee) - 2\langle\alpha^\vee, \varrho\rangle$$

and so Proposition 2.5(4) implies that

$$F_{w_0}(P_\gamma) = c_\gamma \prod_{\alpha^\vee \in \Delta_+^\vee} \prod_{i=1}^{\langle\gamma,\alpha^\vee\rangle} (w_0(\alpha^\vee) - \langle\alpha^\vee, \varrho\rangle - i).$$

Since $w_0(\alpha^\vee) \in \widehat{\mathfrak{h}}$ is a vector field on $\mathbf{X}$, we have $w_0(\alpha^\vee)(1) = 0$ and so

$$F_{w_0}(P_\gamma)(1) = c_\gamma \prod_{\alpha^\vee \in \Delta_+^\vee} (-1)^{\langle\gamma,\alpha^\vee\rangle} \prod_{i=1}^{\langle\gamma,\alpha^\vee\rangle} (\langle\alpha^\vee, \varrho\rangle + i) \neq 0,$$

since $\langle\alpha^\vee, \varrho\rangle > 0$ for all $\alpha^\vee$. By Proposition 2.5(4), $\sum_i F_{w_0}(f_i)(g_i) = F_{w_0}(P_\gamma)(1)$ and so $F_{w_0}(f_i)(g_i) \neq 0$ for some $1 \le i \le n$. □

Lemma 3.6. *The $\mathcal{O}(\mathbf{X})$-module map* $\mathsf{m} : \mathcal{O}(\mathbf{X}) \otimes F_{w_0}(\mathcal{O}(\mathbf{X})) \to \mathcal{D}(\mathbf{X})$ *given by* $g \otimes F_{w_0}(f) \mapsto gF_{w_0}(f)$ *is injective.*

Proof. Let $t \in \operatorname{Ker}(\mathsf{m})$ and write $t = \sum_{i,j} h_{ij} \otimes F_{w_0}(f_{ij})$ where $\{f_{ij}\}_j$ is a basis of $\mathcal{O}^{\mu_i}$, for some $\mu_1, \dots, \mu_s \in \Lambda^+$, and $h_{ij} \in \mathcal{O}$ for all i, j. We may assume that s is minimal; thus, for each $1 \le i \le s$, some $h_{ij} \neq 0$. Partially order Λ by $\omega^* \geqslant \lambda^*$ if $\omega^* - \lambda^* \in \Lambda^+$ and assume that μ_1^* is minimal among the μ_i^*'s for this ordering. By Lemma 3.5, for each i there exists a basis $\{g_{ij}\}_j$ of $\mathcal{O}^{\mu_i^*}$ such that $F_{w_0}(f_{ij})(g_{ik}) = \delta_{jk}$. When $i > 1$ we have $F_{w_0}(f_{ij})(g_{1k}) \subseteq \mathcal{O}^{\mu_1^* - \mu_i^*} = 0$, since $\mu_1^* - \mu_i^* \notin \Lambda^+$. Therefore, for each k, we have

$$0 = \mathsf{m}(t)(g_{1k}) = \sum_{i,j} h_{ij} F_{w_0}(f_{ij})(g_{1k}) = \sum_j h_{1j} F_{w_0}(f_{1j})(g_{1k}) = h_{1k},$$

contradicting the minimality of s. □

Define a finitely generated subalgebra of $\mathcal{D}(\mathbf{X})$ by

$$\mathcal{S} = \mathbb{C}\langle \mathcal{O}, F_{w_0}(\mathcal{O})\rangle = \mathbb{C}\langle \mathcal{O}^{\varpi_i}, F_{w_0}(\mathcal{O}^{\varpi_j})\,;\ 1 \le i, j \le \ell\rangle,$$

where $\ell = \operatorname{rank}(\mathfrak{g})$. The elements of $\mathcal{O}$ act locally nilpotently on $\mathcal{D}$, and therefore on $\mathcal{S}$. Thus $C = \mathcal{O} \smallsetminus \{0\}$ is an Ore subset of $\mathcal{S}$. Let $\mathbb{K} = C^{-1}\mathcal{O}$ denote the field of fractions of $\mathcal{O}$. Recall that $C^{-1}\mathcal{D}(\mathbf{X}) = \mathcal{D}(\mathbb{K})$ is the ring of differential operators on $\mathbb{K}$ and that $\mathcal{D}_r(\mathbf{X}) = \mathcal{D}_r(\mathbb{K}) \cap \mathcal{D}(\mathbf{X})$ for all $r \in \mathbb{N}$ (see, for example, [MR, Theorem 15.5.5]).

Lemma 3.7. *We have* $C^{-1}\mathcal{S} = \mathcal{D}(\mathbb{K})$. *In particular, for any finite dimensional subspace* $E \subset \mathcal{D}(\mathbf{X})$, *there exists* $0 \neq f \in \mathcal{O}(\mathbf{X})$ *such that* $fE \subset \mathcal{S}$.

Proof. The aim of the proof is to apply [LS2, Lemma 8].

Applying the exact functor $\mathbb{K} \otimes_{\mathcal{O}} -$ to the injective map m of Lemma 3.6 yields the $\mathbb{K}$-linear injection $\mathsf{m} : \mathbb{K} \otimes F_{w_0}(\mathcal{O}) \hookrightarrow A = C^{-1}\mathcal{S}$. Since $F_{w_0}(\mathcal{O})$ is a commutative algebra of dimension $n = \dim \mathbf{X} = \operatorname{trdeg}_{\mathbb{C}} \mathbb{K}$, we may pick $u_1, \dots, u_n \in F_{w_0}(\mathcal{O})$ algebraically independent over $\mathbb{C}$ and set $P = \mathbb{C}[u_1, \dots, u_n]$. For any $q \in \mathbb{N}$, denote by P_q the subspace of polynomials of degree at most q and define

$$\Theta_q = \{d \in \mathbb{K} \otimes P : \operatorname{ord} \mathsf{m}(d) \le q\} \quad \text{and} \quad k = \max\{\operatorname{ord} \mathsf{m}(u_i) : 1 \le i \le n\}.$$

Observe that $\{\Theta_q\}_q$ and $\{\mathbb{K} \otimes P_q\}_q$ are two increasing filtrations on the $\mathbb{K}$-vector space $\mathbb{K} \otimes P$ and that the map $q \mapsto \dim_{\mathbb{K}}(\mathbb{K} \otimes P_q)$ is a polynomial function of degree n. Furthermore, since $\mathsf{m}(u_i u_j) = \mathsf{m}(u_i)\mathsf{m}(u_j)$, we have $\mathbb{K} \otimes P_q \subseteq \Theta_{kq}$ for all q. Hence, $\dim_{\mathbb{K}} \Theta_q \ge p(q)$ for some polynomial p of degree n. If $A_q = \mathcal{D}_q(\mathbb{K}) \cap A$, then $\mathsf{m}(\Theta_q) \subseteq A_q$ and so $\dim_{\mathbb{K}} A_q \ge p(q)$. It follows that

$$\limsup_{q \to \infty} \{\log_q(\dim_{\mathbb{K}} A_q / A_{q-1})\} \ge n - 1.$$

The hypotheses of [LS2, Lemma 8] are now satisfied by the pair $A \subseteq \mathcal{D}(\mathbb{K})$ and, by that result, $\mathcal{D}(\mathbb{K}) = A = C^{-1}\mathcal{S}$. The final assertion of the lemma follows by clearing denominators. □

We can now describe a generating set for the $\mathbb{C}$-algebra $\mathcal{D}$, for which we recall the definition of $\widehat{\mathfrak{h}}$ from Notation 2.1.

Theorem 3.8. *As a $\mathbb{C}$-algebra, $\mathcal{D}$ is generated by $\widehat{\mathfrak{h}}$ and the $F_w(\mathcal{O})$, for all $w \in W$.*

Proof. Set $\mathcal{B} = \mathbb{C}\langle \widehat{\mathfrak{h}}, F_w(\mathcal{O})\,;\ w \in W\rangle$; thus $\mathcal{B}$ is a G-submodule of $\mathcal{D}$ containing both $\mathcal{S}$ and $U(\widehat{\mathfrak{h}})$. Moreover, as $F_w(U(\widehat{\mathfrak{h}})) = U(\widehat{\mathfrak{h}})$ (see Proposition 2.5(2)), $F_w(\mathcal{B}) = \mathcal{B}$ for all $w \in W$. As $\mathcal{D}$ is a locally finite G-module, it suffices to show that $E \subset \mathcal{B}$ for all finite-dimensional G-submodules E of $\mathcal{D}$. For such a module E, set $L = \{b \in \mathcal{B} : bE \subset \mathcal{B}\}$ and $I_w(E) = \{g \in \mathcal{O} : F_w(g)E \subset \mathcal{B}\}$ for $w \in W$. We aim to show that the left ideal L of $\mathcal{B}$ contains 1.

Clearly, $I_w(E)$ is an ideal of $\mathcal{O}$. It is also a G-submodule since

$$F_w(a.g)E = F_w(a.g)a.E = a.(F_w(g)E) \subset a.\mathcal{B} = \mathcal{B}$$

for all $a \in G$ and $g \in I_w(E)$. Since $\mathcal{S} \subseteq \mathcal{B}$, Lemma 3.7 implies that $I_1(E) \neq 0$. For each w, $F_{w^{-1}}(E)$ is a G-submodule of $\mathcal{D}$ (isomorphic to E) and

$$g \in I_w(E) \iff gF_{w^{-1}}(E) \subset \mathcal{B} \iff g \in I_1(F_{w^{-1}}(E)).$$

Thus, $I_w(E) \neq 0$ for all $w \in W$. Since $\mathcal{O}$ is a domain, it follows that $I = \bigcap_{w\in W} I_w(E)$ is a nonzero G-submodule of $\mathcal{O}$. Now, $\mathcal{O} = \bigoplus_{\lambda\in\Lambda^+} \mathcal{O}^\lambda$ and so $\mathcal{O}^\gamma \subset I$ for some $\gamma \in \Lambda^+$. By Proposition 2.5(4), $F_w(P_{\gamma^*}) = \sum_i F_w(f_i)F_{ww_0}(g_i)$ for some $f_i \in \mathcal{O}^{\gamma^*}$ and $g_i \in \mathcal{O}^\gamma$. By the definition of I and the choice of γ, we have $F_{ww_0}(g_i)E \subset \mathcal{B}$; that is, $F_{ww_0}(g_i) \in L$, for all $w \in W$. Since $F_w(f_i) \in F_w(\mathcal{O}) \subset \mathcal{B}$ we obtain that $F_w(P_{\gamma^*}) \in L$ for all $w \in W$. Finally, as $U(\widehat{\mathfrak{h}}) \subset \mathcal{B}$, Proposition 2.5(5) implies that $1 \in L$ and hence that $E \subset \mathcal{B}$. □

We conjecture that $\widehat{\mathfrak{h}}$ is unnecessary in the last theorem; i.e., we conjecture that

$$\mathcal{D} = \mathbb{C}\langle F_w(\mathcal{O}) : w \in W\rangle.$$

Using Example 2.4 the authors can prove this for $\mathfrak{g} = \mathfrak{sl}(3)$; indeed we can even show that $\mathcal{D} = \mathbb{C}\langle \mathcal{O}, F_{w_0}(\mathcal{O})\rangle$ in this case. However, the argument heavily uses facts about $\mathfrak{so}(8)$ and so it may not be a good guide to the general case.

4. The $\mathcal{D}(\mathbf{X})$-module $\mathrm{H}^*(\mathbf{X}, \mathcal{O}_\mathbf{X})$

As before, set $\mathbf{X} = G/U$ with associated rings $\mathcal{O} = \mathcal{O}(\mathbf{X})$ and $\mathcal{D} = \mathcal{D}(\mathbf{X})$. In this section we give a complete description of $\mathrm{H}^*(\mathbf{X}, \mathcal{O}_\mathbf{X}) = \bigoplus_{i\in\mathbb{N}} \mathrm{H}^i(\mathbf{X}, \mathcal{O}_\mathbf{X})$ as a $\mathcal{D}$-module. Specifically, we will show that $\mathrm{H}^*(\mathbf{X}, \mathcal{O}_\mathbf{X})$ is simply the direct sum of the twists $\mathcal{O}^w = \mathcal{O}^{F_w}$ of $\mathcal{O}$ by $w \in W$.

The first cohomology group to consider is $\mathrm{H}^0(\mathbf{X}, \mathcal{O}_\mathbf{X}) = \mathcal{O}$. As a $\mathcal{D}$-module, $\mathcal{O} \cong \mathcal{D}/I$ where $I = \{d \in \mathcal{D} : d(1) = 0\}$ is a maximal left ideal of $\mathcal{D}$ (Proposition 3.1). Clearly, $I \supseteq \mathcal{D}\mathfrak{g} + \mathcal{D}\widehat{\mathfrak{h}}$, in the terminology of Notation 2.1. The first main result of this section, Theorem 4.5, shows that this is actually an equality and, further, that $\mathcal{D}/\mathcal{D}\mathfrak{g} \cong \bigoplus_{w\in W} \mathcal{O}^w$.

Before proving this theorem, we need some preliminary notation and lemmas. As in (2.4), we cover $\mathbf{X}$ by affine open subsets $\mathbf{X}_j = \{x \in \overline{\mathbf{X}} : f_j(x) \neq 0\}$ and let $\mathcal{C}_j = \{f_j^s : s \in \mathbb{N}\}$, for $1 \leq j \leq k$, denote the associated Ore subsets in $\mathcal{D}$. If M is a left $\mathcal{D}$-module, the kernel of the localization map $M \to M_{f_j} = \mathcal{D}(\mathbf{X}_j) \otimes_{\mathcal{D}} M$ is

$$T_j(M) = \{v \in M : f_j^s v = 0 \text{ for some } s > 0\}.$$

Lemma 4.1. *For all $w \in W$ and $x \in \mathfrak{g}$, one has $F_w(x) = x$.*

Proof. By Proposition 2.5(1), F_w is a G-linear automorphism of $\mathcal{D}$ and hence is $\mathfrak{g}$-linear, where $\mathfrak{g}$ acts by the adjoint action. Therefore, for any $\theta \in \mathcal{D}$ and $x \in \mathfrak{g}$,

$$[F_w(x), F_w(\theta)] = F_w([x, \theta]) = [x, F_w(\theta)].$$

Thus $[F_w(x), y] = [x, y]$ for all $y \in \mathcal{D}$. If $y \in \mathfrak{g}$, then $\mathfrak{g}$-linearity also implies that $[F_w(x), y] = F_w([x, y])$ and hence that $[x, y] = [F_w(x), y] = F_w([x, y])$. As $\mathfrak{g}$ is semisimple, $[\mathfrak{g}, \mathfrak{g}] = \mathfrak{g}$, and so this implies $F_w(z) = z$ for all $z \in \mathfrak{g}$. □

The next lemma shows why we should expect all the $\mathcal{O}^w$ to appear in a decomposition of $\mathcal{D}/\mathcal{D}\mathfrak{g}$.

Lemma 4.2. (1) *Each $\mathcal{D}$-module $\mathcal{O}^w$ is a factor of $\mathcal{M} = \mathcal{D}/\mathcal{D}\mathfrak{g}$.*
(2) *Set $\mathcal{N} = I/\mathcal{D}\mathfrak{g}$. Then, $\mathcal{N} = T_j(\mathcal{M})$ for $1 \leq j \leq k$ and $\mathcal{D}/I$ is the unique factor of $\mathcal{M}$ isomorphic to $\mathcal{O}$ as a $\mathcal{D}$-module.*

Proof. (1) Since the action of $d \in \mathcal{D}$ on $\mathcal{O}^w$ is given by $d \cdot f = F_w(d)(f)$ for $f \in \mathcal{O}$, we have $\mathcal{O}^w \cong \mathcal{D}/F_w^{-1}(I)$. As we noted above, $I \supseteq \mathcal{D}\mathfrak{g} + \mathcal{D}\widehat{\mathfrak{h}}$ and so, by Lemma 4.1, $F_w^{-1}(I) \supseteq \mathcal{D}\mathfrak{g}$.
(2) Let $T_x\mathbf{X}$ denote the tangent space of $\mathbf{X}$ at $x \in \mathbf{X}$. Observe that, if $e = U/U \in \mathbf{X}$, then $\operatorname{Stab}_G(e) = U$ and so $\mathfrak{g}/\mathfrak{u}$ identifies naturally with $T_e\mathbf{X}$. Since $\mathbf{X}$ is a homogeneous space, the map $\iota : \mathfrak{g} \to \operatorname{Der}_{\mathbb{C}} \mathcal{O}$ induces an isomorphism $\mathfrak{g}/\mathfrak{u} \cong T_e\mathbf{X} \cong T_x\mathbf{X}$. As $\mathbf{X}_j \subseteq \mathbf{X}$ is affine, it follows that $\operatorname{Der} \mathcal{O}(\mathbf{X}_j) = \mathcal{O}(\mathbf{X}_j)\mathfrak{g}$ for each $j = 1, \ldots, k$. Furthermore, since $\mathbf{X}_j$ is smooth, the $\mathcal{D}(\mathbf{X}_j)$-module $\mathcal{O}(\mathbf{X}_j)$ is simple and isomorphic to $\mathcal{D}(\mathbf{X}_j)/\mathcal{D}(\mathbf{X}_j) \operatorname{Der} \mathcal{O}(\mathbf{X}_j)$. Since $\mathcal{O} = \mathcal{M}/\mathcal{N}$, this implies that

$$\mathcal{O}(\mathbf{X}_j) = \mathcal{C}_j^{-1}(\mathcal{M}/\mathcal{N}) \cong \mathcal{D}(\mathbf{X}_j)/\mathcal{D}(\mathbf{X}_j)\mathfrak{g} = \mathcal{C}_j^{-1}\mathcal{M} \tag{4.1}$$

for all j. Therefore $\mathcal{C}_j^{-1}\mathcal{N} = 0$, and so $\mathcal{N} \subseteq T_j(\mathcal{M})$. The equality $\mathcal{N} = T_j(\mathcal{M})$ then follows from $T_j(\mathcal{O}) = 0$.

Now suppose that $\mathcal{O} \cong \mathcal{D}/L$ for some maximal left ideal $L \supseteq \mathcal{D}\mathfrak{g}$. Then $\mathcal{C}_j^{-1}L \supseteq \mathcal{D}(\mathbf{X}_j)\mathfrak{g}$ and $\mathcal{C}_j^{-1}\mathcal{O} = \mathcal{O}(\mathbf{X}_j) \cong \mathcal{D}(\mathbf{X}_j)/\mathcal{C}_j^{-1}L$ for all j. By (4.1), the left ideal $\mathcal{D}(\mathbf{X}_j)\mathfrak{g}$ is maximal and so $\mathcal{C}_j^{-1}L = \mathcal{D}(\mathbf{X}_j)\mathfrak{g}$ for all j. Therefore $L/\mathcal{D}\mathfrak{g} \subseteq T_j(\mathcal{M}) = \mathcal{N} = I/\mathcal{D}\mathfrak{g}$, which implies that $L \subseteq I$. Hence $L = I$. □

As in [Bo, Théorème 2, Section VI.1.5], $\Sigma^\vee = \{\alpha_1^\vee, \dots, \alpha_\ell^\vee\}$ defines a dominant chamber

$$C(\Sigma^\vee) = \left\{y \in \widehat{\mathfrak{h}}_{\mathbb{R}} : \langle y, \alpha_i\rangle > 0 \text{ for all } i = 1, \dots, \ell\right\}$$

in the root system $\Delta^\vee = \jmath(\Delta^\vee) \subset \widehat{\mathfrak{h}}_{\mathbb{R}} = \bigoplus_{i=1}^{\ell} \mathbb{R}\alpha_i^\vee \subset \widehat{\mathfrak{h}}$.

Lemma 4.3. *If $y \in C(\Sigma^\vee)$ and $w \in W \smallsetminus \{1\}$, then $F_w(y) \notin I$.*

Proof. Using [Bo, Ch. VI, Section 1.6,Corollaire de la Proposition 18], we have $0 \neq y - w(y) = \sum_{i=1}^{\ell} n_i \alpha_i^\vee$ with $n_i \in \mathbb{R}_+$. Hence, Proposition 2.5(2) implies that

$$F_w(y) - w(y) \;=\; \langle w(y) - y, \varrho\rangle \;=\; -\sum_i n_i \;\in\; \mathbb{C} \smallsetminus \{0\}.$$

Thus $F_w(y) - w(y) \notin I$. Since $w(y) \in \widehat{\mathfrak{h}} \subset I$, this implies that $F_w(y) \notin I$. □

Lemma 4.4. *As $\mathcal{D}$-modules, $\mathcal{O}^w \cong \mathcal{O}^{w'}$ if and only if $w = w'$.*

Proof. Since $(\mathcal{O}^w)^v \cong \mathcal{O}^{wv}$ it suffices to prove the result when $w' = 1$. So, assume that $\mathcal{O}^w \cong \mathcal{O}$ for some $w \neq 1$ and set $\mathcal{N}_w = F_w^{-1}(I)/\mathcal{D}\mathfrak{g}$; thus $\mathcal{N}_1 = \mathcal{N}$. Then $\mathcal{O}^w = \mathcal{D}/F_w^{-1}(I) = \mathcal{M}/\mathcal{N}_w$ and Lemma 4.2 implies that $\mathcal{N}_w = \mathcal{N}_1$; equivalently, $F_{w^{-1}}(I) = I$. Now pick $y \in C(\Sigma^\vee)$. Since $y \in I$, this implies that $F_{w^{-1}}(y) \in I$, contradicting Lemma 4.3. □

The following theorem gives a precise description of the $\mathcal{D}$-modules $\mathcal{O}$ and $\mathcal{M}$; it shows in particular that $\mathcal{M}$ is a multiplicity free, semisimple module of length $|W|$.

Theorem 4.5. *Write $\mathcal{O}(\mathbf{X}) \cong \mathcal{D}(\mathbf{X})/I$ for $I = \{d \in \mathcal{D}(\mathbf{X}) : d(1) = 0\}$ and define $\mathcal{M} = \mathcal{D}(\mathbf{X})/\mathcal{D}(\mathbf{X})\mathfrak{g}$. Then:*

(1) *$\mathcal{M} \cong \bigoplus_{w\in W} \mathcal{O}(\mathbf{X})^w$ as a $\mathcal{D}(\mathbf{X})$-module;*
(2) *$I = \mathcal{D}(\mathbf{X})\mathfrak{g} + \mathcal{D}(\mathbf{X})\widehat{\mathfrak{h}} = \mathcal{D}(\mathbf{X})\mathfrak{g} + \mathcal{D}(\mathbf{X})y$ for all $y \in C(\Sigma^\vee)$.*

Proof. (1) Set $\mathcal{N}_w = F_w^{-1}(I)/\mathcal{D}\mathfrak{g}$, for $w \in W$. By Lemmas 4.2(2) and 4.4, the $\mathcal{D}$-modules $\mathcal{O}^w \cong \mathcal{M}/\mathcal{N}_w$ are nonisomorphic, so the natural map $\mathcal{M} \to \bigoplus_{w\in W} \mathcal{O}^w$ is surjective with kernel $N = \bigcap_{w\in W} \mathcal{N}_w$. It therefore remains to prove that $N = 0$. By Lemma 4.2, if $x \in I$, then there exists $t \in \mathbb{N}$ such that $f_j^t x \in \mathcal{D}\mathfrak{g}$ for $j = 1, \dots, k$. Therefore $F_{w^{-1}}(f_j)^t F_{w^{-1}}(x) \in \mathcal{D}\mathfrak{g}$ for all j; equivalently, each element $[F_{w^{-1}}(x) + \mathcal{D}\mathfrak{g}] \in \mathcal{N}_w$ is torsion for $F_{w^{-1}}(\mathcal{C}_j)$. Consequently, if $v \in N$, then there exists $s \in \mathbb{N}$ such that $F_w(f_j)^s v = 0$ for all $1 \leq j \leq k$ and $w \in W$. In other words, $N_{f_j}^w = (0)$, for all such j and w. By faithful flatness (Corollary 2.3) this implies that $N = 0$.
(2) It suffices to prove that $I = \mathcal{D}\mathfrak{g} + \mathcal{D}y$ for $y \in C(\Sigma^\vee)$. Set $\mathcal{J} = (\mathcal{D}\mathfrak{g} + \mathcal{D}y)/\mathcal{D}\mathfrak{g}$. By part (1) and Lemma 4.2, the $\mathcal{N}_w$, $w \in W$, are the only maximal submodules of $\mathcal{M}$. Therefore, if $\mathcal{J} \subsetneq \mathcal{N}_1$ we must have $\mathcal{J} \subseteq \mathcal{N}_w$ for some $w \neq 1$. This implies that $F_w(\mathcal{D}\mathfrak{g} + \mathcal{D}y) \subseteq I$, hence $F_w(y) - w(y) \in I$, in contradiction with Lemma 4.3. Therefore $\mathcal{N}_1 = \mathcal{J}$ and $I = \mathcal{D}\mathfrak{g} + \mathcal{D}y$. □

We will prove in Theorem 4.8 that $\mathrm{H}^*(\mathbf{X}, \mathcal{O}_{\mathbf{X}})$ is isomorphic to $\mathcal{M}$ as a $\mathcal{D}$-module. In order to prove this, we need to recall some results on the cohomology of line bundles over the flag variety $\mathbf{B} = G/B$.

We begin with some general remarks. Let $\mathbf{Z}$ be a smooth G-variety and write $\tau : \mathfrak{g} \to \operatorname{Der} \mathcal{O}_{\mathbf{Z}}$ for the differential of the G-action. Let $\mathcal{F}$ be a G-equivariant $\mathcal{O}_{\mathbf{Z}}$-module as defined, for example, in [Ka, Section 4.4]. This implies, in particular, that $\mathcal{F}$ is a compatible $(\mathfrak{g}, \mathcal{O}_{\mathbf{Z}})$-module in the sense that for any open subset $\Omega \subseteq \mathbf{Z}$, one has $\xi.(fv) = \tau(\xi)(f)v + f(\xi.v)$ for all $\xi \in \mathfrak{g}$, $f \in \mathcal{O}_{\mathbf{Z}}(\Omega)$ and $v \in \mathcal{F}(\Omega)$. In this setting, the cohomology group $\mathrm{H}^i(\mathbf{Z}, \mathcal{F})$ inherits a structure of compatible $(G, \mathcal{O}(\mathbf{Z}))$-module, see [Ke, Theorem 11.6]. If $\mathcal{F}$ is a coherent G-equivariant $\mathcal{D}_{\mathbf{Z}}$-module, then it follows from [Ka, Sections 4.10 and 4.11] that $\mathrm{H}^i(\mathbf{Z}, \mathcal{F})$ is endowed with a $\mathcal{D}(\mathbf{Z})$-module structure such that $\tau(\xi)v = \frac{d}{dt}_{|t=0}(e^{t\xi}.v)$ for all $v \in \mathrm{H}^i(\mathbf{Z}, \mathcal{F})$ and $\xi \in \mathfrak{g}$. We will apply these observations in two cases: one is when $\mathbf{Z} = \mathbf{X} = G/U$ and $\mathcal{F} = \mathcal{O}_{\mathbf{X}}$ is a G-equivariant $\mathcal{D}_{\mathbf{X}}$-module under left translation; the other is described next.

Let $\pi : G \twoheadrightarrow \mathbf{B} = G/B$, $\phi : G \twoheadrightarrow \mathbf{X}$ and $\varphi : \mathbf{X} \twoheadrightarrow \mathbf{B}$ be the natural projections, thus $\pi = \varphi \circ \phi$. For each $\lambda \in \Lambda$, the one-dimensional H-module $\mathbb{C}_{-\lambda}$ can be viewed as a B-module with trivial action of U. As in [Ke, pp. 333–335], define the G-equivariant $\mathcal{O}_{\mathbf{B}}$-module $\mathcal{L}(\lambda)$ to be the sections of the line bundle $G \times^B \mathbb{C}_{-\lambda}$. Since $G \times^B \mathbb{C}_{-\lambda} \cong \mathbf{X} \times^H \mathbb{C}_{-\lambda}$, one has

$$\begin{aligned}\Gamma(\Omega, \mathcal{L}(\lambda)) = \{f : \varphi^{-1}\Omega \to \mathbb{C} : f(\bar{g}h) = \lambda(h) f(\bar{g})\\ \text{for all } h \in H \text{ and } \bar{g} \in \varphi^{-1}\Omega\}.\end{aligned} \tag{4.2}$$

where $\Omega \subseteq \mathbf{B}$ is any open subset.

Recall from (2.3) that the decomposition $\mathcal{O} = \mathcal{O}(\mathbf{X}) = \bigoplus_{\gamma \in \Lambda^+} \mathcal{O}^\gamma$ is induced by the twisted action of H on $\mathbf{X}$. Hence, $h \in H$ acts on $f \in \Gamma(\Omega, \mathcal{L}(\lambda))$ via

$$(r_h.f)(\bar{g}) = f(\bar{g}w_0(h^{-1})) = \lambda(w_0(h)^{-1})f(\bar{g}) = \lambda^*(h)f(\bar{g}). \tag{4.3}$$

Passing to Čech cohomology, H therefore acts on $\mathrm{H}^i(\mathbf{B}, \mathcal{L}(\lambda))$ with weight λ^*.

The cohomology groups of the line bundle $\mathcal{L}(\lambda)$ are described as G-modules by the Borel–Weil–Bott theorem, that we now recall (see [Ja, Corollaries II.5.5 and II.5.6], up to a switch from B to the opposite Borel). The length of $w \in W$ is denoted by $\ell(w)$ and we will write

$$W(i) = \{w \in W : \ell(w) = i\}.$$

The "dot action" of $w \in W$ on $\xi \in \mathfrak{h}^*$ is defined by $w \cdot \xi = w(\xi + \varrho) - \varrho$.

Theorem 4.6 (Borel–Weil–Bott). *The G-module $\mathrm{H}^i(\mathbf{B}, \mathcal{L}(\lambda))$ is isomorphic to*

$$\begin{cases} V(\mu)^* & \text{if } \exists\,(w, \mu) \in W(i) \times \Lambda^+ \text{ such that } \lambda = w \cdot \mu; \\ 0 & \text{otherwise.} \end{cases}$$

□

The following standard proposition reduces the computation of the G-module $\mathrm{H}^*(\mathbf{X}, \mathcal{O}_{\mathbf{X}})$ to the Borel–Weil–Bott theorem. We include a proof since we could not find an appropriate reference.

Proposition 4.7. *The morphism $\varphi : \mathbf{X} \to \mathbf{B}$ is affine and one has $\varphi_* \mathcal{O}_{\mathbf{X}} \cong \bigoplus_{\nu \in \Lambda} \mathcal{L}(\nu)$ as $(\mathfrak{g}, \mathcal{O}_{\mathbf{B}})$-modules. In particular, for each $i \in \mathbb{N}$, there is a G-module isomorphism*

$$\mathrm{H}^i(\mathbf{X}, \mathcal{O}_{\mathbf{X}}) \cong \bigoplus_{\nu \in \Lambda} \mathrm{H}^i(\mathbf{B}, \mathcal{L}(\nu)).$$

Proof. By the Bruhat decomposition, $\mathbf{B}$ is covered by the affine open subsets $\Omega_w = \pi(\dot{w}U^- B)$ where $\dot{w} \in N_G(H)$ is a representative of $w \in W$ and U^- is the opposite maximal unipotent subgroup of G, see [Ja, (II.1.10)]. As $\dot{w}U^- B \cong U^- \times H \times U$, the subset $\varphi^{-1}\Omega_w = \phi(\dot{w}U^- B)$ is affine and isomorphic to $U^- \times H$ as an H-variety. Thus φ is an affine morphism and $\mathrm{H}^i(\mathbf{X}, \mathcal{O}_{\mathbf{X}}) \cong \mathrm{H}^i(\mathbf{B}, \varphi_* \mathcal{O}_{\mathbf{X}})$ by [Ha, III, Exercise 4.1].

The affine algebra $\Gamma(\Omega_w, \varphi_* \mathcal{O}_{\mathbf{X}}) = \mathcal{O}_{\mathbf{X}}(\varphi^{-1}\Omega_w)$ is endowed with a regular action of H. Hence, $\Gamma(\Omega_w, \varphi_* \mathcal{O}_{\mathbf{X}})$ decomposes as $\bigoplus_{\nu \in \Lambda} \mathcal{O}_{\mathbf{X}}(\varphi^{-1}\Omega_w)_\nu$ with

$$\mathcal{O}_{\mathbf{X}}(\varphi^{-1}\Omega_w)_\nu = \{f : \varphi^{-1}\Omega_w \to \mathbb{C} : f(\bar{g}h) = \nu(h) f(\bar{g}) \quad \text{for all } h \in H \text{ and } \bar{g} \in \varphi^{-1}\Omega_w\}.$$

Therefore, by (4.2), $\Gamma(\Omega_w, \varphi_* \mathcal{O}_{\mathbf{X}}) \cong \bigoplus_{\nu \in \Lambda} \Gamma(\Omega_w, \mathcal{L}(\nu))$ as $(\mathfrak{g}, \mathcal{O}_{\mathbf{B}}(\Omega_w))$-modules, and it follows that $\varphi_* \mathcal{O}_{\mathbf{X}} \cong \bigoplus_{\nu \in \Lambda} \mathcal{L}(\nu)$ as $(\mathfrak{g}, \mathcal{O}_{\mathbf{B}})$-modules. □

One consequence of this proposition is that $\mathrm{H}^i(\mathbf{X}, \mathcal{O}_{\mathbf{X}}) = 0$ for $i > \dim \mathbf{B} = \ell(w_0) = |R^+|$. We can now give the promised description of $\mathrm{H}^*(\mathbf{X}, \mathcal{O}_{\mathbf{X}})$.

Theorem 4.8. *As $\mathcal{D}(\mathbf{X})$ modules, $\mathrm{H}^i(\mathbf{X}, \mathcal{O}_{\mathbf{X}}) \cong \bigoplus_{w \in W(i)} \mathcal{O}(\mathbf{X})^w$ for $0 \le i \le \dim B$. Moreover*

$$\mathrm{H}^*(\mathbf{X}, \mathcal{O}_{\mathbf{X}}) \cong \mathcal{D}(\mathbf{X})/\mathcal{D}(\mathbf{X})\mathfrak{g} \cong \bigoplus_{w \in W} \mathcal{O}(\mathbf{X})^w$$

is a direct sum of nonisomorphic simple $\mathcal{D}$-modules.

Proof. By Proposition 3.1 and Lemma 4.4, the $\mathcal{O}^w$ are nonisomorphic simple modules. Combining Theorem 4.6 with Proposition 4.7 gives

$$\mathrm{H}^i(\mathbf{X}, \mathcal{O}_{\mathbf{X}}) \cong \bigoplus_{\nu \in \Lambda} \mathrm{H}^i(\mathbf{B}, \mathcal{L}(\nu)) \cong \bigoplus_{\substack{\mu \in \Lambda^+ \\ w \in W(i)}} \mathrm{H}^i(\mathbf{B}, \mathcal{L}(w \cdot \mu)).$$

Since $\mathrm{H}^{\ell(w)}(\mathbf{B}, \mathcal{L}(w \cdot \mu)) \cong V(\mu^*)$, the multiplicity $[\mathrm{H}^i(\mathbf{X}, \mathcal{O}_{\mathbf{X}}) : V(\lambda)]$ is equal to $|W(i)|$ for any $\lambda \in \Lambda^+$.

Fix $w \in W(i)$ and pick $0 \neq e_w$ in the trivial G-module $\mathrm{H}^{\ell(w)}(\mathbf{B}, \mathcal{L}(w \cdot 0))$. As $\mathcal{O}_{\mathbf{X}}$ is a $(G \times H)$-equivariant $\mathcal{D}_{\mathbf{X}}$-module, $xe_w = \frac{d}{dt}_{|t=0}(\mathrm{e}^{tx}.e_w) = 0$ for $x \in \mathfrak{g}$ and, by (4.3), $y \in \widehat{\mathfrak{h}}$ acts on e_w with weight

$$(w \cdot 0)^* = -w_0(w(\varrho) - \varrho) = w_0 w w_0(\varrho) - \varrho.$$

Hence $ye_w = \langle w_0 w w_0(\varrho) - \varrho, y \rangle e_w = \langle \varrho, w_0 w^{-1} w_0(y) - y \rangle e_w$, for all $y \in \widehat{\mathfrak{h}}$. Substituting this into the formula from Proposition 2.5(2) shows that $F_{w_0 w w_0}(y) e_w = 0$.

Thus $(\mathcal{D}\mathfrak{g} + \mathcal{D}F_{w_0 w w_0}(\widehat{\mathfrak{h}}))e_w = 0$ and so, by Theorem 4.5(2), $\mathcal{D}e_w \cong \mathcal{O}^{w_0 w^{-1} w_0}$. Since the map $w \mapsto w_0 w^{-1} w_0$ permutes $W(i)$ and the $\mathcal{O}^w$ are nonisomorphic simple modules, we conclude that $\mathrm{H}^i(\mathbf{X}, \mathcal{O}_{\mathbf{X}}) \supseteq \sum_{w \in W(i)} \mathcal{D}e_w \cong \bigoplus_{w \in W(i)} \mathcal{O}^w$. By Lemma 4.1, $\mathcal{O}^w \cong \mathcal{O} \cong \bigoplus_{\lambda \in \Lambda^+} V(\lambda)$ as G-modules. Thus, for each $\lambda \in \Lambda^+$, the first paragraph of the proof implies that

$$\left[\bigoplus\{\mathcal{O}^w : w \in W(i)\} : V(\lambda)\right] = |W(i)| = \left[\mathrm{H}^i(\mathbf{X}, \mathcal{O}_{\mathbf{X}}) : V(\lambda)\right].$$

Consequently, $\mathrm{H}^i(\mathbf{X}, \mathcal{O}_{\mathbf{X}}) \cong \bigoplus_{w \in W(i)} \mathcal{O}^w$. Theorem 4.5 then implies that

$$\mathrm{H}^*(\mathbf{X}, \mathcal{O}_{\mathbf{X}}) = \bigoplus_{i=0}^{\ell(w_0)} \mathrm{H}^i(\mathbf{X}, \mathcal{O}_{\mathbf{X}}) \cong \bigoplus_{w \in W} \mathcal{O}^w \cong \mathcal{D}/\mathcal{D}\mathfrak{g},$$

which completes the proof. □

5. Differential operators on S-varieties

In this section we consider highest weight varieties and, more generally, S-varieties $\overline{\mathbf{Y}}$ in the sense of [VP]. These are natural generalizations of the closure $\overline{\mathbf{X}}$ of the basic affine space $\mathbf{X}$ and there is a natural map from $\mathcal{D}(\mathbf{X})$ to the ring of differential operators $\mathcal{D}(\overline{\mathbf{Y}})$ over such a variety. Although this map need not be surjective (see Theorem 6.7) it does carry enough information to prove, under mild assumptions, that $\overline{\mathbf{Y}}$ is $\mathcal{D}$-simple in the sense that $\mathcal{O}(\overline{\mathbf{Y}})$ is a simple $\mathcal{D}(\overline{\mathbf{Y}})$-module. We will continue to write $\mathcal{O} = \mathcal{O}(\mathbf{X})$ and $\mathcal{D} = \mathcal{D}(\mathbf{X})$.

Definition 5.1. An irreducible affine G-variety $\overline{\mathbf{Y}}$ is called an *S-variety* if it contains a dense orbit $\mathbf{Y} = G.v$ such that $U \subseteq \mathrm{Stab}_G(v)$, the stabilizer of v in G.

Remark 5.2. One important feature of S-varieties is that any affine spherical variety (i.e., an irreducible affine G-variety having a dense B-orbit) is a flat deformation of an S-variety (see, for example, [Gr, Theorem 22.3]).

The S-varieties have been completely described in [VP]. We begin with the relevant notation. Set $\Gamma = \sum_{j=1}^{s} \mathbb{N}\gamma_j$, where $\gamma_1, \ldots, \gamma_s \in \Lambda^+$ are distinct dominant weights. Write $V_\Gamma = \bigoplus_{j=1}^{s} V(\gamma_j) \ni v_\Gamma = v_{\gamma_1} \oplus \cdots \oplus v_{\gamma_s}$ and define $\mathbf{Y}_\Gamma = G.v_\Gamma$. The following theorem summarizes the results of [VP, Section 3] that we need.

Theorem 5.3. (1) *The closures* $\overline{\mathbf{Y}}_\Gamma$ *give all the S-varieties.*
(2) *One has* $\mathcal{O}(\overline{\mathbf{Y}}_\Gamma) = \bigoplus_{\gamma \in \Gamma^*} \mathcal{O}^\gamma \subseteq \mathcal{O}(\mathbf{Y}_\Gamma) = \bigoplus_{\mu \in \mathbb{Z}\Gamma^* \cap \Lambda^+} \mathcal{O}^\mu$.
(3) *The following assertions are equivalent:*
 (i) $\mathcal{O}(\overline{\mathbf{Y}}_\Gamma) = \mathcal{O}(\mathbf{Y}_\Gamma)$;
 (ii) $\Gamma = \mathbb{Z}\Gamma \cap \Lambda^+$;
 (iii) $\overline{\mathbf{Y}}_\Gamma$ *is normal and* $\mathrm{codim}_{\overline{\mathbf{Y}}_\Gamma}(\overline{\mathbf{Y}}_\Gamma \smallsetminus \mathbf{Y}_\Gamma) \geq 2$. □

We will always assume that Γ satisfies the equivalent conditions from Theorem 5.3(3); that is,

$$\mathbb{Z}\Gamma \cap \Lambda^+ = \Gamma. \tag{5.1}$$

Hence, for an S-variety $\overline{\mathbf{Y}}_\Gamma$ satisfying (5.1) we have $\mathcal{D}(\mathbf{Y}_\Gamma) = \mathcal{D}(\overline{\mathbf{Y}}_\Gamma)$. Notice also that $\Gamma^* = \sum_{j=1}^{s} \mathbb{N}\gamma_j^*$ satisfies (5.1) if and only if Γ does. Natural examples of S-varieties satisfying (5.1) are the following and more examples can be found in [Gr, Ch. 2, §11]).

Examples 5.4. (1) For $\Gamma = \Lambda^+$ we obtain the basic affine space $\mathbf{X} = \mathbf{Y}_{\Lambda^+}$.

(2) Let $\gamma \in \Lambda^+$ and $\Gamma = \mathbb{N}\gamma$. Then, $\mathbf{Y}_\Gamma$ will be denoted by $\mathbf{Y}_\gamma$ and is the orbit of a highest weight vector $v_\gamma \in V(\gamma)$. Its closure $\overline{\mathbf{Y}}_\gamma$ is called a *highest weight* or HV-variety [VP, Section 1].

(3) An important example of highest weight variety is the closure of the minimal (nonzero) orbit in a simple Lie algebra $\mathfrak{g}$: in this case $\gamma = \tilde{\alpha}$ is the highest root and $V(\tilde{\alpha}) \cong \mathfrak{g}$ is the adjoint representation.

Set $S_\Gamma = \mathrm{Stab}_G(v_\Gamma)$; thus $\mathbf{Y}_\Gamma \cong G/S_\Gamma$. Since $S_\Gamma = \bigcap_{j=1}^{s} \mathrm{Stab}_G(v_{\gamma_j})$, we see that $\mathfrak{h}$ normalizes $\mathfrak{s}_\Gamma = \mathrm{Lie}(S_\Gamma)$. Let $\Delta(\mathfrak{h}, \mathfrak{s}_\Gamma)$ denote the set of roots of $\mathfrak{h}$ in the Lie algebra $\mathfrak{s}_\Gamma$. Then $S_\Gamma = S'_\Gamma Q_\Gamma$ where $Q_\Gamma = H \cap S_\Gamma$ and S'_Γ is generated by the one parameter groups U_α, $\alpha \in \Delta(\mathfrak{h}, \mathfrak{s}_\Gamma)$ (see [VP, p. 753] or [Gr, Corollary 3.5]). Clearly, $Q_\Gamma = \{h \in H : \forall \gamma \in \Gamma,\ \gamma(h) = 1\}$ is a diagonalizable subgroup of H with character group $\Lambda/\mathbb{Z}\Gamma$. Under the right action of the given groups on $\mathcal{O}(\mathbf{Y}_\Gamma)$, the proof of [Gr, Lemma 17.1(b)] shows that

$$\mathcal{O}(\mathbf{Y}_\Gamma) = \mathcal{O}(G)^{S_\Gamma} = \mathcal{O}(G)^{UQ_\Gamma} = \mathcal{O}^{Q_\Gamma}.$$

When (5.1) holds, this implies that $\mathcal{O}(\overline{\mathbf{Y}}_\Gamma) = \mathcal{O}(\mathbf{Y}_\Gamma) = \mathcal{O}^{Q_\Gamma}$ and $\overline{\mathbf{Y}}_\Gamma = \overline{\mathbf{X}}/\!\!/Q_\Gamma$.

If an algebraic group L acts on the right on some variety $\mathbf{Z}$ we denote by $\delta_g : z \mapsto z.g$ the right translation by $g \in L$ on $\mathbf{Z}$. It induces a right action on $f \in \mathcal{O}(\mathbf{Z})$ by $\delta_g.f(z) = f(z.g)$. This applies for example to S_Γ acting on G and Q_Γ acting on $\overline{\mathbf{X}}$. In this notation, $\mathcal{D}(\mathbf{X})^\mu = \{d \in \mathcal{D}(\mathbf{X}) : \forall h \in H,\ \delta_h.d = \mu^*(h)d\}$ for any $\mu \in \Lambda$.

When $\overline{\mathbf{Y}}_\Gamma = \overline{\mathbf{Y}}_\gamma$ is an HV-variety we will set $S_\gamma = S_\Gamma$ and $Q_\gamma = Q_\Gamma$, etc.

Proposition 5.5. *Suppose that* $\overline{\mathbf{Y}}_\Gamma$ *is an S-variety satisfying* (5.1). *Then* $\mathcal{D}^{Q_\Gamma} = \bigoplus_{\mu \in \mathbb{Z}\Gamma^*} \mathcal{D}^\mu$ *and there is a natural morphism of algebras* $\psi_\Gamma : \mathcal{D}^{Q_\Gamma} \to \mathcal{D}(\mathbf{Y}_\Gamma)$.

Proof. Let $d = \sum_{\mu \in \Lambda} d_{\mu^*}$ with $d_{\mu^*} \in \mathcal{D}(\mathbf{X})^{\mu^*}$. It is clear that $d \in \mathcal{D}(\mathbf{X})^{Q_\Gamma}$ if and only if $\delta_h.d_{\mu^*} = d_{\mu^*}$, for all $h \in Q_\Gamma$. This condition is equivalent to $\mu(h)d_{\mu^*} = d_{\mu^*}$; that is, $\mu(h) = 1$ when $d_{\mu^*} \neq 0$. Since Q_Γ is a diagonalizable group, [Sp, Proposition 2.5.7(iii)] implies that $\mathbb{Z}\Gamma = \{\mu \in \Lambda : \forall h \in Q_\Gamma,\ \mu(h) = 1\}$. It follows that $\mathcal{D}(\mathbf{X})^{Q_\Gamma} = \bigoplus_{\mu \in \mathbb{Z}\Gamma^*} \mathcal{D}(\mathbf{X})^\mu$. The morphism ψ_Γ is simply the restriction morphism coming from the identification $\overline{\mathbf{Y}}_\Gamma = \overline{\mathbf{X}}/\!\!/Q_\Gamma$. □

An affine variety $\mathbf{Z}$ (respectively an algebra R) is called $\mathcal{D}$*-simple* if $\mathcal{O}(\mathbf{Z})$ (respectively R) is a simple $\mathcal{D}(\mathbf{Z})$-module (respectively $\mathcal{D}(R)$-module). This does not hold for arbitrary varieties; for example when $\mathbf{Z}$ is the cubic cone in $\mathbb{C}^3$, $\mathcal{O}(\mathbf{Z})$ does not even have finite length as a $\mathcal{D}(\mathbf{Z})$-module [BGG]. It is, however, important in many situations to know that a variety is $\mathcal{D}$-simple. We first note that for S-varieties satisfying (5.1) this is an easy consequence of Proposition 3.1:

Proposition 5.6. *Let $\overline{\mathbf{Y}}_\Gamma$ be an S-variety satisfying* (5.1). *Then $\mathcal{O}(\mathbf{Y}_\Gamma)$ is a simple left $\mathcal{D}(\mathbf{Y}_\Gamma)$-module.*

Proof. Since $\mathcal{O}(\mathbf{Y}_\Gamma) = \mathcal{O}(\mathbf{X})^{Q_\Gamma}$ with Q_Γ reductive, $\mathcal{O}(\mathbf{Y}_\Gamma)$ is an $\mathcal{O}(\mathbf{Y}_\Gamma)$-module summand of $\mathcal{O}(\mathbf{X})$. The proposition is now a consequence of Proposition 3.1 and the following result. □

Proposition 5.7 ([Sm, Proposition 3.1]). *Let $R \hookrightarrow T$ be an inclusion of commutative $\mathbb{C}$-algebras and suppose that R is a direct summand of the R-module T. If T is $\mathcal{D}$-simple, then R is $\mathcal{D}$-simple.* □

We next refine Proposition 5.6 by showing that $\mathcal{O}(\mathbf{Y}_\Gamma)$ is a simple module over a rather explicit subring of $\mathcal{D}(\mathbf{Y}_\Gamma)$. Let $\mathcal{S}_\Gamma$ be the subalgebra of $\mathcal{D}(\mathbf{X})$ generated by the two finitely generated commutative subalgebras $\mathcal{O}(\mathbf{Y}_\Gamma)$ and $F_{w_0}(\mathcal{O}(\mathbf{Y}_{\Gamma^*}))$. Observe that $F_{w_0}(\mathcal{O}(\mathbf{Y}_{\Gamma^*}))$ is isomorphic to $\mathcal{O}(\mathbf{Y}_{\Gamma^*})$ (as both an algebra and a G-module) and so

$$\mathcal{S}_\Gamma = \mathbb{C}\langle\, \mathcal{O}^{\gamma_i^*}, F_{w_0}(\mathcal{O}^{\gamma_j})\,;\ 1 \le i, j \le s\rangle.$$

By Proposition 2.5(1), $F_{w_0}(\mathcal{O}^{\gamma}) \subseteq \mathcal{D}^{-\gamma^*}$ and so $\mathcal{S}_\Gamma \subseteq \mathcal{D}(\mathbf{X})^{Q_\Gamma}$, by Proposition 5.5. We will consider $\mathcal{O}(\mathbf{Y}_\Gamma)$ as an $\mathcal{S}_\Gamma$-module through the map ψ_Γ defined in the latter result.

Proposition 5.8. *Let $\overline{\mathbf{Y}}_\Gamma$ be an S-variety satisfying* (5.1). *Then $\mathcal{O}(\mathbf{Y}_\Gamma)$ is a simple $\mathcal{S}_\Gamma$-module.*

Proof. As in Lemma 3.6, partially order Λ by $\omega^* \geqslant \lambda^*$ if $\omega^* - \lambda^* \in \Lambda^+$. Let $0 \neq g \in \mathcal{O}(\mathbf{Y}_\Gamma)$ and write $g = g_{\lambda_0^*} + \sum_{\lambda_j^* \ngeqslant \lambda_0^*} g_{\lambda_j^*}$ with $g_{\lambda_0^*} \neq 0$ and $g_{\lambda_i^*} \in \mathcal{O}^{\lambda_i^*} \subset \mathcal{O}(\mathbf{Y}_\Gamma)$ for all i. By Lemma 3.5 there exists $f \in \mathcal{O}^{\lambda_0}$ such that $F_{w_0}(f)(g_{\lambda_0^*}) = 1$. Hence Proposition 2.5(1) implies that $F_{w_0}(f)(g) = 1 + \sum_{\lambda_j^* \ngeqslant \lambda_0^*} F_{w_0}(f)(g_{\lambda_j^*})$ with $F_{w_0}(f)(g_{\lambda_j^*}) \in \mathcal{O}^{-\lambda_0^*+\lambda_j^*}$. But $\mathcal{O}^{-\lambda_0^*+\lambda_j^*} = 0$ when $\lambda_j^* \ngeqslant \lambda_0^*$. Thus $F_{w_0}(f)(g) = 1$ and $\mathcal{O}(\mathbf{Y}_\Gamma)$ is simple over $\mathcal{S}_\Gamma$. □

Let $\mathbf{Z}$ be an affine variety with $R = \mathcal{O}(\mathbf{Z})$ and consider the subalgebra

$$\Delta(R) = \Delta(\mathbf{Z}) = \mathbb{C}\langle\, \mathcal{O}(\mathbf{Z}), \mathrm{Der}_{\mathbb{C}}(\mathcal{O}(\mathbf{Z})\,\rangle \subseteq \mathcal{D}(\mathbf{Z}).$$

It is known [MR, Corollary 15.5.6] that $\Delta(\mathbf{Z}) = \mathcal{D}(\mathbf{Z})$ when $\mathbf{Z}$ is smooth. The Nakai conjecture [Na] says that the converse should be true:

$$\Delta(\mathbf{Z}) = \mathcal{D}(\mathbf{Z}) \overset{?}{\Longrightarrow} R \text{ is regular.}$$

The reader can consult [Be, Tr], [Sc, Section 12.3] and the references therein for work related to this conjecture.

The following observation, which is implicit in [Tr], implies that many singular varieties, notably S-varieties satisfying (5.1), do satisfy the conclusion of Nakai's conjecture.

Lemma 5.9. *Assume that* $\mathbf{Z}$ *is irreducible and* $\mathcal{O}(\mathbf{Z})$ *is* $\mathcal{D}$*-simple. Then the Nakai conjecture holds for* $\mathbf{Z}$.

Proof. Suppose that $\mathcal{D}(R) = \Delta(R)$. Then R is simple as a $\Delta(R)$-module and the result follows from [MR, Theorem 15.3.8]. □

Corollary 5.10. (1) *The Nakai conjecture holds for S-varieties which satisfy* (5.1).
(2) *More generally, suppose that* $R \subseteq T$ *are finitely generated* $\mathbb{C}$*-algebras such that* R *is a summand of the* R*-module* T *and that* T *is* $\mathcal{D}$*-simple. Then the Nakai conjecture holds for* R.

Proof. Part (1) is immediate from Lemma 5.9 combined with Proposition 5.6. Similarly, part (2) follows from Lemma 5.9 and Proposition 5.7. □

One significance of the Nakai conjecture is that it implies the Zariski–Lipman conjecture: $\mathbf{Z}$ is smooth whenever $\operatorname{Der}_{\mathbb{C}} \mathcal{O}(\mathbf{Z})$ is a projective $\mathcal{O}(\mathbf{Z})$-module. Part (1) of Corollary 5.10 does not give new information about that conjecture; indeed since S-varieties always have graded coordinate rings, the Zariski–Lipman conjecture for these varieties already follows from [Ho]. It is not clear whether part (2) of Corollary 5.10 has significant applications in this direction.

A natural situation where Corollary 5.10 applies is for invariant rings:

Corollary 5.11. *Let* Q *be an affine algebraic group and* $\mathbf{V}$ *be an irreducible affine* Q*-variety. Suppose that* $\mathcal{O}(\mathbf{V})$ *is* $\mathcal{D}$*-simple (for example, when* $\mathbf{V}$ *is smooth). Then the Nakai conjecture holds for* $\mathbf{V}/\!\!/Q$ *in the following two cases:*

(a) Q *is reductive;*
(b) $\mathbf{V}$ *is a* G*-variety and* $Q = U$.

Proof. (a) As in the proof of Proposition 5.6, $\mathcal{O}(\mathbf{V}/\!\!/Q)$ is a summand of $\mathcal{O}(\mathbf{V})$. Thus Corollary 5.10(2) applies.
(b) Observe that $\mathcal{D}(\overline{\mathbf{X}} \times \mathbf{V}) \cong \mathcal{D}(\overline{\mathbf{X}}) \otimes \mathcal{D}(\mathbf{V})$; therefore, by hypothesis and Proposition 3.1 (or Proposition 5.8), $\mathcal{O}(\overline{\mathbf{X}} \times \mathbf{V}) \cong \mathcal{O}(\overline{\mathbf{X}}) \otimes \mathcal{O}(\mathbf{V})$ is $\mathcal{D}$-simple. By [Kr, III.3.2, p. 191]), $(\overline{\mathbf{X}} \times \mathbf{V})/\!\!/G \cong \mathbf{V}/\!\!/U$ where G acts componentwise on $\overline{\mathbf{X}} \times \mathbf{V}$. Thus the result follows from (a) applied to G acting on $\overline{\mathbf{X}} \times \mathbf{V}$. □

Surprisingly, and despite the simplicity of its proof, only special cases of Corollary 5.11 have appeared before in the literature and these have typically required substantially harder proofs. See, for example, [Is, Theorem 2.3] and [Sc, Section 12.3].

6. Exotic differential operators

The results from the last section raise two questions which we study in this section. First, can one say more about the structure, notably the order, of the exotic differential operators in $\mathcal{D}(\mathbf{Y}_\Gamma)$ induced from the ring $\mathcal{S}_\Gamma$ of Proposition 5.8? This is answered by Theorem 6.5 and Corollary 6.6 and proves Corollary 1.5 from the introduction.

The second question concerns the following basic question in the theory of rings of differential operators. If V is a finite dimensional representation of a reductive group K, then restriction of operators induces a ring homomorphism $\mathcal{D}(V)^K \to \mathcal{D}(V /\!/ K)$. When is this map surjective? The conjectural answer is that this happens if and only if $V /\!/ K$ is singular. Positive answers to this question have been found in many cases and these solutions had significant applications to Lie theory (see, for example, [Jo2, Jo3, LS1, Sc]). These results have almost always been in situations where $V /\!/ K$ is a highest weight variety. Now let $\overline{\mathbf{Y}} = \overline{\mathbf{Y}}_\Gamma$ be any HV-variety, or even S-variety satisfying (5.1). It is natural to ask whether the resulting map $\psi_\Gamma : \mathcal{D}(\mathbf{X})^{Q_\Gamma} \to \mathcal{D}(\mathbf{Y})$ is surjective. As we will show in Theorem 6.7, this even fails for the minimal orbit $\mathbf{O}_{\min} = \mathbf{Y}_{\tilde{\alpha}}$ of a simple classical Lie algebra $\mathfrak{g}$. This proves Proposition 1.6 from the introduction.

The idea behind the proof of Theorem 6.7 is as follows. Let $\overline{\mathbf{Y}} = \overline{\mathbf{Y}}_\gamma$ be an HV-variety. Then $\mathcal{O}(\mathbf{Y}) = \bigoplus_{p\in\mathbb{N}} \mathcal{O}^{p\gamma^*}$ is an $\mathbb{N}$-graded algebra and so $\mathcal{D}(\mathbf{Y})$ is $\mathbb{Z}$-graded by (2.1). There exist examples of HV-varieties $\overline{\mathbf{Y}}$, notably the closures of minimal orbits $\mathbf{Y}_{\tilde{\alpha}}$, for which one can find an irreducible G-module E consisting of "exotic" operators of degree -1 and order at most 4, see [AB, BK2, LS3]. On the other hand, by combining (2.2) with Propositions 2.5(1,3) and 5.5 the only obvious operators of order -1 in $\mathrm{Im}(\psi_\gamma)$ and small order are those in $\psi_\gamma(F_{w_0}(\mathcal{O}^\gamma))$. But their order is only bounded above by $k(\gamma)$, where we write

$$k(\lambda) = \langle \lambda, 2\varrho^\vee \rangle = 2\textstyle\sum_i m_i \qquad \text{for} \quad \lambda = \sum_i m_i\alpha_i \in \Lambda^+. \tag{6.1}$$

Typically $k(\gamma)$ is significantly larger than 4 (see the table at the end of the section). The aim of the proof is therefore to show that for the minimal orbit the bound $k(\gamma)$ is attained and hence that the G-module E cannot lie in $\mathrm{Im}(\psi_\gamma)$.

We begin with some technical lemmas, the first of which is a mild generalization of [BBP, Lemma 3.8]. The subalgebra $U(\widehat{\mathfrak{h}}) \cong S(\mathfrak{h})$ of $\mathcal{D}$ can be identified with $\mathcal{O}(\mathfrak{h}^*)$ and it follows from the definitions that $u(g) = u(\lambda)g$ for $u \in U(\widehat{\mathfrak{h}})$ and $g \in \mathcal{O}^\lambda$. We denote by $\{U_m(\widehat{\mathfrak{h}})\}_{m\in\mathbb{N}}$ the standard filtration on the enveloping algebra $U(\widehat{\mathfrak{h}})$.

Lemma 6.1. *Let $f_1, \dots, f_n \in \mathcal{O}$ be linearly independent and pick $D \in \mathcal{D}_p$. Suppose that there exist functions $c_i : \Lambda^+ \to \mathbb{C}$ such that $D(g) = \sum_{i=1}^n c_i(\mu) f_i g$ for all $g \in \mathcal{O}^\mu$ and $\mu \in \Lambda^+$. Then $D = \sum_{i=1}^n f_i\tilde{c}_i$ for some $\tilde{c}_1, \dots, \tilde{c}_n \in U_p(\widehat{\mathfrak{h}})$.*

Proof. For any function $c : \Lambda^+ \to \mathbb{C}$ and $\lambda \in \Lambda^+$, define $T_\lambda(c) : \Lambda^+ \to \mathbb{C}$ by $T_\lambda(c)(\mu) = c(\lambda+\mu) - c(\mu)$. As $\mathrm{ord}\, D \le p$, we have $[g_{p+1}, [g_p, [\dots, [g_1, D]] \dots] = 0$ for all $0 \neq g_j \in \mathcal{O}^{\lambda_j}$. This easily implies that $T_{\lambda_1} \circ T_{\lambda_2} \circ \cdots \circ T_{\lambda_p} \circ T_{\lambda_{p+1}}(c_i) = 0$ for all $\lambda_i \in \Lambda^+$. Since T_λ is a difference operator, it follows that c_i is a polynomial function of degree $\le p$ on Λ^+ and it is clear that there exist $\tilde{c}_i \in U(\mathfrak{h})$ (of degree $\le p$) such that $\tilde{c}_i(\lambda) = c_i(\lambda)$ for all $\lambda \in \Lambda^+$. Obviously, $D(g) = \left(\sum_{i=1}^n f_i\tilde{c}_i\right)(g)$ for all $g \in \mathcal{O}^\lambda$, hence the result. □

[Sh2, Theorem 1] shows that, for all $\gamma \in \Lambda^+$ and $w \in W$, one has $\mathcal{D}[\gamma]^{w(\gamma)} \cong U(\widehat{\mathfrak{h}}) \otimes E$, where E is a $G \times H$-module of dimension $\dim V(\gamma)$. The next lemma provides an explicit $G \times H$-module E with this property.

Lemma 6.2. *Let $\gamma \in \Lambda^+$. Then, via the multiplication map $\mathsf{m} : \mathcal{D} \otimes \mathcal{D} \to \mathcal{D}$, we have*

$$\mathcal{D}_p[\gamma]^{w(\gamma)} = F_w(\mathcal{O}^\gamma) \otimes U_p(\widehat{\mathfrak{h}}) = U_p(\widehat{\mathfrak{h}}) \otimes F_w(\mathcal{O}^\gamma)$$

for all $w \in W$ and $p \in \mathbb{N}$. In particular, $\mathcal{D}[\gamma]^{w(\gamma)}$ is a free $U(\widehat{\mathfrak{h}})$-module with basis being any $\mathbb{C}$-basis of $F_w(\mathcal{O}^\gamma)$.

Proof. It is sufficient to prove the lemma for $w = 1$; indeed, applying F_w to the equalities $\mathcal{D}_p[\gamma]^\gamma = \mathcal{O}^\gamma \otimes U_p(\widehat{\mathfrak{h}}) = U_p(\widehat{\mathfrak{h}}) \otimes \mathcal{O}^\gamma$ and appealing to Proposition 2.5(1) gives the general result.

By [Sh2, Theorem 1], $n = \mathrm{rk}_{U(\widehat{\mathfrak{h}})} \mathcal{D}[\gamma]^\gamma = \dim V(\gamma)$. Let $\{D_1, \dots, D_n\}$ be a basis of the right $U(\widehat{\mathfrak{h}})$-module $\mathcal{D}[\gamma]^\gamma$, and fix a basis $\{f_1, \dots, f_n\}$ of the G-module $\mathcal{O}^\gamma$. It is easy to see that $U_p(\widehat{\mathfrak{h}}) \otimes \mathcal{O}^\gamma \cong U_p(\widehat{\mathfrak{h}})\mathcal{O}^\gamma = \mathcal{O}^\gamma U_p(\widehat{\mathfrak{h}}) \cong \mathcal{O}^\gamma \otimes U_p(\widehat{\mathfrak{h}})$, and that this space is contained in $\mathcal{D}[\gamma]^\gamma$. For the converse, write $f_j = \sum_i D_i a_{ij}$ for $a_{ij} \in U(\widehat{\mathfrak{h}})$. Then, $f_j g = \sum_i a_{ij}(\mu) D_i(g)$ for all $0 \neq g \in \mathcal{O}^\mu$. Thus $\bigoplus_{j=1}^n \mathbb{C} f_j g \subseteq \sum_{j=1}^n \mathbb{C} D_j(g)$. Since $\dim\bigl(\bigoplus_{j=1}^n \mathbb{C} f_j g\bigr) = n \geq \dim\bigl(\sum_{j=1}^n \mathbb{C} D_j(g)\bigr)$, we obtain that $\bigoplus_{j=1}^n \mathbb{C} f_j g = \bigoplus_{j=1}^n \mathbb{C} D_j(g)$. Therefore, for all $g \in \mathcal{O}^\mu$, there exist unique elements $c_{ij}(g, \mu) \in \mathbb{C}$ such that $D_j(g) = \sum_{i=1}^n c_{ij}(g, \mu) f_i g$.

We next show that the $c_{ij}(g, \mu)$'s depend only on μ. Indeed, it follows from

$$f_j g = \sum_k a_{kj}(\mu) D_k(g) = \sum_i \sum_k a_{kj}(\mu) c_{ik}(g, \mu) f_i g$$

that $\sum_k c_{ik}(g, \mu) a_{kj}(\mu) = \delta_{ij}$ for all $1 \leq i, j \leq n$. In other words, the matrix $[c_{ik}(g, \mu)]_{ik}$ is the inverse of the matrix $[a_{kj}(\mu)]_{kj}$. Since the a_{kj}'s do not depend on g, nor do the $c_{ik}(g, \mu)$.

We can therefore write $D_j(g) = \sum_{i=1}^n c_{ij}(\mu) f_i g$ for any $g \in \mathcal{O}^\mu$. By Lemma 6.1 we deduce that $D_j = \sum_{i=1}^n f_i \tilde{c}_{ij}$ with $\tilde{c}_{ij} \in U_p(\widehat{\mathfrak{h}})$. □

Return to an S-variety $\overline{\mathbf{Y}} = \overline{\mathbf{Y}}_\Gamma$ satisfying (5.1) and define $\psi = \psi_\Gamma : \mathcal{D}(\mathbf{X})^{Q_\Gamma} \to \mathcal{D}(\mathbf{Y})$ by Proposition 5.5. Recall that $\Gamma = \sum_{i=1}^s \mathbb{N}\gamma_i$ for some $\gamma_i \in \Lambda^+$ and order the γ_i so that $\mathbb{Q}\Gamma = \bigoplus_{i=1}^r \mathbb{Q}\gamma_i$. Set $\mathfrak{t} = \{h \in \widehat{\mathfrak{h}} : \langle \Gamma^*, h \rangle = 0\}$; thus $\mathfrak{t} \cong \mathrm{Lie}(Q_\Gamma)$ has dimension $\mathrm{rank}(\mathfrak{g}) - r$. Pick $x_1, \dots, x_r \in \widehat{\mathfrak{h}}$ such that $\widehat{\mathfrak{h}} = \bigl(\bigoplus_{i=1}^r \mathbb{C}x_i\bigr) \oplus \mathfrak{t}$ and $\langle \gamma_i^*, x_j \rangle = \delta_{ij}$ for $1 \leq i, j \leq r$. Set $y_j = \psi(x_j) \in \mathcal{D}(\mathbf{Y})$ and let $u \in U(\widehat{\mathfrak{h}})$. From the identity $u(f) = \psi(u)(f) = u(\mu) f$ for $\mu \in \Gamma^*$ and $f \in \mathcal{O}^\mu$, one deduces easily that $\psi(u) = 0$ when $u \in \mathfrak{t}U(\widehat{\mathfrak{h}})$ and that ψ induces an isomorphism of polynomial algebras:

$$\psi : \mathbb{C}[\mathbf{x}] = \mathbb{C}[x_1, \dots, x_r] \xrightarrow{\sim} \mathbb{C}[\mathbf{y}] = \mathbb{C}[y_1, \dots, y_r].$$

For $\mathbf{m} = (m_1, \dots, m_r) \in \mathbb{N}^r$ we set $x^{\mathbf{m}} = \prod_{i=1}^r x_i^{m_i}$ and $y^{\mathbf{m}} = \prod_{i=1}^r y_i^{m_i}$. Note that if $u(y) \in \mathbb{C}[\mathbf{y}]$, the (total) degree of $u(y)$ coincides with its order as a differential operator on $\mathbf{Y}$. We recall the definition of $k(\lambda)$ from (6.1).

Proposition 6.3. *Let $\overline{\mathbf{Y}} = \overline{\mathbf{Y}}_\Gamma$ satisfy* (5.1) *and set $\psi = \psi_\Gamma$. Let $\gamma \in \Gamma$, $0 \neq f \in \mathcal{O}^\gamma$ and $u(y) \in \mathbb{C}[\mathbf{y}]$. Then,*

$$\mathrm{ord}\, \psi(F_{w_0}(f)) u(y) = k(\gamma) + \deg u(y).$$

In particular, the elements $\{\psi(F_{w_0}(f_{\mathbf{m}}))y^{\mathbf{m}} : \mathbf{m} \in \mathbb{N}^r\}$ *are linearly independent for any* $f_{\mathbf{m}} \in \mathcal{O}^\gamma \smallsetminus \{0\}$.

Proof. As the variety $\overline{\mathbf{Y}}$ is irreducible, the associated graded ring $\operatorname{gr}\mathcal{D}(\mathbf{Y}) = \bigoplus_k \mathcal{D}_k(\mathbf{Y})/\mathcal{D}_{k-1}(\mathbf{Y})$ is a domain and so $\operatorname{ord} ab = \operatorname{ord} a + \operatorname{ord} b$ for $a, b \in \mathcal{D}(\mathbf{Y})$. Since $\operatorname{ord} u(y) = \deg u(y)$, it therefore suffices to show that $\operatorname{ord} \psi(F_{w_0}(f)) = k(\gamma)$. By Proposition 2.5(3), we do have $\operatorname{ord} \psi(F_{w_0}(f)) \leq \operatorname{ord} F_{w_0}(f) = k(\gamma)$.

Consider the operator $P_{\gamma^*} = \sum_i g_i F_{w_0}(f_i)$ (where $g_i \in \mathcal{O}^{\gamma^*}$, $f_i \in \mathcal{O}^\gamma$) defined by Proposition 2.5(4). For each $\alpha^\vee \in \Delta_+^\vee$ we have $\alpha^\vee = \sum_{j=1}^r \langle \alpha^\vee, \gamma_j^* \rangle x_j + z$ with $z \in \mathfrak{t}$. Thus, applying ψ to (2.5) gives

$$\psi(P_{\gamma^*}) = c_{\gamma^*} \prod_{\alpha^\vee \in \Delta_+^\vee} \prod_{i=1}^{\langle \gamma^*, \alpha^\vee \rangle} \Big(\sum_j \langle \alpha^\vee, \gamma_j^* \rangle y_j + \langle \alpha^\vee, \varrho \rangle - i \Big). \tag{6.2}$$

Write $\gamma^* = \sum_{j=1}^r m_j \gamma_j^*$ for some $m_j \in \mathbb{Q}$ and pick α such that $\langle \gamma^*, \alpha^\vee \rangle \neq 0$. Then $\sum_j m_j \langle \alpha^\vee, \gamma_j^* \rangle \neq 0$ and so $\langle \alpha^\vee, \gamma_j^* \rangle \neq 0$ for some $1 \leq j \leq r$. Hence $\deg \psi(\alpha^\vee) = \deg\big(\sum_j \langle \alpha^\vee, \gamma_j^* \rangle y_j\big) = 1$. Thus (6.2) implies that $\psi(P_{\gamma^*}) = \sum_i g_i \psi(F_{w_0}(f_i))$ has order $\sum_{\alpha^\vee \in \Delta_+^\vee} \langle \gamma^*, \alpha^\vee \rangle = k(\gamma)$. Therefore, there exists $i \in \{1, \ldots, n\}$ such that $\operatorname{ord} \psi(F_{w_0}(f_i)) = k(\gamma)$. In particular, $\psi(F_{w_0}(\mathcal{O}^\gamma))$ is nonzero.

Consider the symbol map $\operatorname{gr}_{k(\gamma)} : \mathcal{D}_{k(\gamma)}(\mathbf{Y}) \to \mathcal{D}_{k(\gamma)}(\mathbf{Y})/\mathcal{D}_{k(\gamma)-1}(\mathbf{Y})$. Notice that $\operatorname{gr}_{k(\gamma)}$ is a morphism of G-modules and that, by the previous paragraph, $\operatorname{gr}_{k(\gamma)}(\psi(F_{w_0}(f_i))) \neq 0$ for some $f_i \in \mathcal{O}^\gamma$. Since $\psi(F_{w_0}(\mathcal{O}^\gamma)) \cong V(\gamma)$ is an irreducible G-module, we deduce that $\operatorname{gr}_{k(\gamma)}(\psi(F_{w_0}(f))) \neq 0$ for all $0 \neq f \in \mathcal{O}^\gamma$; that is to say, $\operatorname{ord} \psi(F_{w_0}(f)) = k(\gamma)$. This proves the first assertion, from which the second follows immediately. □

Corollary 6.4. *Let* $\overline{\mathbf{Y}}_\Gamma$ *satisfy* (5.1), *write* $\psi = \psi_\Gamma$ *and pick* $\gamma \in \Gamma$. *Then* ψ *induces an isomorphism of* G*-modules:*

$$F_{w_0}(\mathcal{O}^\gamma) \otimes \mathbb{C}[\mathbf{x}] \xrightarrow{\ \sim\ } \psi(\mathcal{D}[\gamma]^{-\gamma^*}) = \psi(F_{w_0}(\mathcal{O}^\gamma)) \otimes \mathbb{C}[\mathbf{y}].$$

Proof. We claim that $\mathcal{D}[\gamma]^{-\gamma^*} \cap \operatorname{Ker} \psi = F_{w_0}(\mathcal{O}^\gamma) \otimes \mathfrak{t}U(\widehat{\mathfrak{h}})$. Let $Q \in \mathcal{D}[\gamma]^{-\gamma^*} \cap \operatorname{Ker} \psi$. By Lemma 6.2, write $Q = P + T$, where $P = \sum_{\mathbf{m} \in \mathbb{N}^r} F_{w_0}(f_{\mathbf{m}}) x^{\mathbf{m}}$ for some $f_{\mathbf{m}} \in \mathcal{O}^\gamma$, and $T \in F_{w_0}(\mathcal{O}^\gamma) \otimes \mathfrak{t}U(\widehat{\mathfrak{h}})$. Since $T \in \operatorname{Ker} \psi$, certainly $0 = \psi(P) = \sum_{\mathbf{m}} \psi(F_{w_0}(f_{\mathbf{m}}))y^{\mathbf{m}}$. By Proposition 6.3, this implies that $f_{\mathbf{m}} = 0$ for all $\mathbf{m}$. Hence $P = 0$ and $Q \in F_{w_0}(\mathcal{O}^\gamma) \otimes \mathfrak{t}U(\widehat{\mathfrak{h}})$. Since the opposite inclusion is clear, the claim is proven. As $U(\widehat{\mathfrak{h}}) = \mathbb{C}[\mathbf{x}] \bigoplus \mathfrak{t}U(\widehat{\mathfrak{h}})$, it follows that ψ induces the required isomorphism. □

Theorem 6.5. *Let* $\overline{\mathbf{Y}}_\Gamma$ *be an S-variety satisfying* (5.1), *set* $\psi = \psi_\Gamma$ *and pick* $\gamma \in \Gamma$. *Let* $E \subset \psi(\mathcal{D}[\gamma]^{-\gamma^*})$ *be an irreducible* G*-module. Then* $E = \psi(F_{w_0}(\mathcal{O}^\gamma))u(y)$ *for some* $0 \neq u(y) \in \mathbb{C}[\mathbf{y}]$ *and* $\operatorname{ord} q = k(\gamma) + \deg u(y)$ *for all* $0 \neq q \in E$.

In particular, if ψ *is surjective then* $\operatorname{ord} q \geq k(\gamma)$ *for all* $0 \neq q \in \mathcal{D}(\mathbf{Y})[\gamma]^{-\gamma^*}$.

Proof. Let $q \in E \smallsetminus \{0\}$; by Corollary 6.4 we may write $q = \sum_i p_i u_i(y)$ where the $p_i \in \psi(F_{w_0}(\mathcal{O}^\gamma))$ are linearly independent and $0 \neq u_i(y) \in \mathbb{C}[\mathbf{y}]$. Since $\psi(F_{w_0}(\mathcal{O}^\gamma))$ is an irreducible G-module, the Jacobson Density Theorem produces an element $a \in \mathbb{C}G$ such that $a.p_i = \delta_{i,1} p_1$ for all i. Recall that G acts trivially on $\widehat{\mathfrak{h}}$, hence $a.q = p_1 u_1(y) \in E \smallsetminus \{0\}$. Therefore, $E = \mathbb{C}G.(a.q) = \psi(F_{w_0}(\mathcal{O}^\gamma))u_1(y)$. This proves the first claim and the second is then a consequence of Proposition 6.3.

Suppose that ψ is surjective. Since ψ is a $(G \times H)$-module map, we must have $\mathcal{D}(\mathbf{Y})[\gamma]^{-\gamma^*} = \psi(\mathcal{D}[\gamma]^{-\gamma^*})$ and the final claim follows from the previous ones. $\square$

Assume now that $\overline{\mathbf{Y}}_\Gamma = \overline{\mathbf{Y}}_\gamma$ is an HV-variety and recall that, by Example 5.4(2), $\overline{\mathbf{Y}}_\gamma$ automatically satisfies (5.1). Then $\mathcal{O}(\mathbf{Y}_\gamma)$ is graded and from the discussion after (2.2) we know that $\mathcal{D}(\mathbf{Y}_\gamma)^{-m\gamma^*}$ identifies with the space of differential operators of degree $-m$ on the graded ring $\mathcal{O}(\mathbf{Y}_\gamma)$. In particular, we obtain the following explicit module of exotic differential operators:

Corollary 6.6. *Let $\overline{\mathbf{Y}} = \overline{\mathbf{Y}}_\gamma$ be an HV-variety. Then $\mathcal{D}(\mathbf{Y})$ contains an irreducible G-module $E = \psi_\gamma(F_{w_0}(\mathcal{O}^\gamma)) \cong V(\gamma)$ of differential operators of degree -1 and order $k(\gamma) = \langle \gamma, 2\varrho^\vee \rangle$.* $\square$

For a number of important HV-varieties, G-modules of differential operators of degree -1 have been constructed, but these constructions are typically quite subtle (see, for example, [AB, BK1, BK2]). The results of [AB] apply to the minimal nilpotent orbit. In our final result we will show that their operators almost never appear in the image of ψ and therefore that ψ is not surjective for those varieties.

Assume that G is simple. Then the minimal (nonzero) nilpotent orbit of $\mathfrak{g} = \mathrm{Lie}(G)$ is $\mathbf{O}_{\min} = \mathbf{Y}_{\tilde{\alpha}}$, where $\tilde{\alpha} = \tilde{\alpha}^*$ is the highest root. In this case $k(\tilde{\alpha})$ is easy to compute. Indeed, by [Bo, Ch. VI, 1.11, Proposition 31], $k(\tilde{\alpha}) = 2(h-1)$, where h is the Coxeter number of the root system Δ. These Coxeter numbers are described, for example, in [Bo, Planche I–IX] and we therefore obtain the following values for $k(\tilde{\alpha})$.

Type of $\mathfrak{g}$	A_ℓ	$\mathsf{B}_\ell, \ell \geq 2$	$\mathsf{C}_\ell, \ell \geq 2$	$\mathsf{D}_\ell, \ell \geq 3$	E_6	E_7	E_8	F_4	G_2
$k(\tilde{\alpha})$	2ℓ	$2(2\ell-1)$	$2(2\ell-1)$	$2(2\ell-3)$	22	34	58	22	10

It is now easy to prove Proposition 1.6 from the introduction.

Theorem 6.7. *Let $\mathbf{O}_{\min} = \mathbf{Y}_{\tilde{\alpha}}$ be the minimal nonzero orbit in a simple classical Lie algebra $\mathfrak{g}$. Then the restriction map $\psi : \mathcal{D}(\mathbf{X})^{Q_{\tilde{\alpha}}} \to \mathcal{D}(\mathbf{Y}_{\tilde{\alpha}})$ is surjective if and only if $\mathfrak{g} = \mathfrak{sl}(2)$ or $\mathfrak{g} = \mathfrak{sl}(3)$.*

Proof. By Example 5.4(3), $\overline{\mathbf{Y}}_{\tilde{\alpha}}$ is an HV-variety and so it satisfies (5.1).

Suppose that ψ is surjective. As $\mathfrak{g}$ is classical, it follows from [AB, Theorem 3.2.3 and Equation 3] that $\mathcal{D}(\mathbf{Y}_{\tilde{\alpha}})$ contains a G-module $E \cong V(\tilde{\alpha})$ of differential operators of degree -1 and order ≤ 4. As ψ is a $G \times H$-module map, and $\mathcal{D}(\mathbf{Y}_{\tilde{\alpha}})^{-\tilde{\alpha}^*}$ is the space of operators of degree -1, this forces $E \subseteq \mathcal{D}(\mathbf{Y}_{\tilde{\alpha}})[\tilde{\alpha}]^{-\tilde{\alpha}^*} = \psi(\mathcal{D}[\tilde{\alpha}]^{-\tilde{\alpha}^*})$. Therefore, Theorem 6.5 says the operators in E have order $4 \geq k(\tilde{\alpha})$. The table shows that this is only possible when $\mathfrak{g} = \mathfrak{sl}(2)$ or $\mathfrak{sl}(3)$.

Conversely, if $\mathfrak{g} = \mathfrak{sl}(2)$, then $\mathcal{D} = \mathbb{C}[u, v, \partial_u, \partial_v]$ is the second Weyl algebra, $Q_{\tilde{\alpha}} = \{\pm \mathrm{id}\} \cong \mathbb{Z}/2\mathbb{Z}$ and $\mathcal{O}(\mathbf{Y}_{\tilde{\alpha}}) = \mathcal{O}^{Q_{\tilde{\alpha}}} = \mathbb{C}[u^2, v^2, uv]$. Thus ψ is just the isomorphism $\mathcal{D}^{Q_{\tilde{\alpha}}} = \mathbb{C}[u^2, v^2, uv, \partial_u^2, \partial_v^2, \partial_u\partial_v] \xrightarrow{\sim} \mathcal{D}(\mathbf{Y}_{\tilde{\alpha}})$.

Now suppose that $\mathfrak{g} = \mathfrak{sl}(3)$. Then $\overline{\mathbf{X}}$ is the quadratic cone $\{\sum_{i=1}^3 u_i y_i = 0\}$ in $\mathbb{C}^3 \times \mathbb{C}^3$, $Q_{\tilde{\alpha}} \cong \mathbb{C}^*$ and the natural map $\psi : \mathcal{D}^{Q_{\tilde{\alpha}}} \to \mathcal{D}(\mathbf{Y}_{\tilde{\alpha}})$ is surjective by [LS3, Lemma 1.1 and Theorem 2.14]. □

Differential operators have also been extensively studied for lagrangian subvarieties of minimal orbits in [BK1, BK2, LS3]. The varieties discussed in those papers are HV-varieties for an appropriate Lie algebra (see, for example, [BK1, Table 1]) and they again have differential operators of order ≤ 4 and degree -1. As might be expected by analogy with Theorem 6.7, the corresponding map ψ does produce operators of the required order for Lie algebras of small rank but in large rank the map ψ is definitely not surjective. We omit the details of these assertions since they are rather technical.

The differential operators constructed in [BK1, BK2, AB] have a number of interesting properties, as is explained in those papers. It would be interesting to know whether the operators constructed for arbitrary HV-varieties by Corollary 6.6 also have distinctive properties.

Remark 6.8. It is instructive to compare the results of this section with those from [LSS, LS1]. One of the main aims of those papers was to construct $\mathcal{D}(\mathbf{Z})$ for certain specific singular affine varieties $\mathbf{Z}$. The typical situation is that $\mathbf{Z}$ is an irreducible component of $\overline{\mathbf{O} \cap \mathfrak{n}^+}$, where $\mathbf{O}$ is a nilpotent orbit of a simple Lie algebra $\widetilde{\mathfrak{g}}$ with triangular decomposition $\widetilde{\mathfrak{g}} = \mathfrak{n}^- \oplus \mathfrak{l} \oplus \mathfrak{n}^+$. Those papers then show that $\mathcal{D}(\mathbf{Z}) = U(\widetilde{\mathfrak{g}})/P$ for some primitive ideal P. However, $\mathbf{Z}$ will almost never be an S-variety for $\widetilde{\mathfrak{g}}$. Rather, $\mathbf{Z}$ will be contained in the nilradical of a parabolic subalgebra $\mathfrak{p}$ of $\widetilde{\mathfrak{g}}$ and, at least when $\mathbf{O}$ is the minimal orbit, $\mathbf{Z}$ then will be an HV-variety for a smaller Lie algebra $\mathfrak{g}$ contained in the Levi factor of $\mathfrak{p}$. For these examples one would not expect to find a group Q and a surjective map $\psi : \mathcal{D}(G/U)^Q \to \mathcal{D}(\mathbf{Z})$, simply because it is unlikely for the big Lie algebra $\widetilde{\mathfrak{g}}$ to lie in the image of such a map.

The reader is referred to [LSS, Theorem 5.2] and [LS1, Introduction] for explicit examples of this behaviour and to [Jo2] for a more general framework. One example is provided by Example 2.4, for which we take $\widetilde{\mathfrak{g}} = \mathfrak{so}(8)$. The parabolic $\mathfrak{p}$ is described explicitly in [LSS, Table 3.1 and Remark 3.2(v)], so suffice it to say that $\mathfrak{p}$ has radical $\mathfrak{r} \cong \mathbb{C}^6$ and Levi factor $\mathfrak{a} \oplus \mathbb{C}$, where $\mathfrak{a} \cong \mathfrak{so}(6)$. The variety $\mathbf{Z}$ is the quadric $\sum_{i=1}^3 x_i y_i = 0$ inside $\mathfrak{r}$. Since $\mathfrak{r}$ is the natural representation for $\mathfrak{a}$ it follows easily that $\mathbf{Z}$ is an HV-variety for $\mathfrak{a}$. Example 2.4 follows from this discussion by the lucky coincidence that the closure $\overline{\mathbf{X}}$ of the basic affine space for $\mathfrak{sl}(3)$ identifies with $\mathbf{Z}$ under an embedding of $\mathfrak{sl}(3)$ into $\mathfrak{a}$.

Acknowledgement

We would like to thank the referee for some helpful comments.

References

[AB] A. Astashkevich and R. Brylinski, Exotic differential operators on complex minimal nilpotent orbits, in *Advances in Geometry*, Prog. Math., Vol. 172, Birkhäuser, Boston, 1999, 19–51.

[Be] J. Becker, Higher derivations and integral closure, *Amer. J. Math.*, **100** (1978), 495–521.

[BGG] J. N. Bernstein, I. M. Gelfand and S. I. Gelfand, Differential operators on the cubic cone, *Russian Math. Surveys*, **27** (1972), 466–488.

[BBP] R. Bezrukavnikov, A. Braverman and L. Positselskii, Gluing of abelian categories and differential operators on the basic affine space, *J. Inst. Math. Jussieu*, **1** (2002), 543–557.

[Bj] J.-E. Björk, *Rings of Differential Operators*, North-Holland, Amsterdam, 1979.

[Bo] N. Bourbaki, *Groupes et Algèbres de Lie, Chapitres 4,5 et 6*, Masson, Paris, 1981.

[BK1] R. Brylinski and B. Kostant, Minimal representations, geometrical quantization and unitarity, *Proc. Natl. Acad. Sci.*, **91** (1994), 6026–6029.

[BK2] ———, Differential operators on conical Lagrangian manifolds, in *Lie Theory and Geometry: in honor of Bertram Kostant on the occasion of his 65th birthday*, Prog. Math. Vol. 123, Birkhäuser, Boston, 1994, 65–96.

[GK] I. M. Gel'fand and A. A. Kirillov, The structure of the Lie field connected with a split semisimple Lie algebra, *Funct. Anal. Appl.*, **3** (1969), 6–21.

[Gr] F. D. Grosshans, *Algebraic Homogeneous Spaces and Invariant Theory*, Lecture Notes in Mathematics, Vol. 1673, Springer-Verlag, Berlin Heidelberg, 1997.

[Ha] R. Hartshorne, *Algebraic Geometry*, Springer-Verlag, New York Heidelberg Berlin, 1977.

[Ho] M. Hochster, The Zariski–Lipman conjecture in the graded case, *J. Algebra*, **47** (1977), 411–424.

[HV] A. van den Hombergh and H. de Vries, On the differential operators on the quasi-affine variety G/N, *Indag. Math.*, **40** (1978), 460–466.

[Is] Y. Ishibashi, Nakai's conjecture for invariant subrings, *Hiroshima Math. J.*, **15** (1985), 429–436.

[Ja] J. C. Jantzen, *Representations of Algebraic Groups, Second Ed.*, Math. Surveys and Monographs Vol. 107, Amer. Math. Soc., Providence, RI, 2003.

[Jo1] A. Joseph, The minimal orbit in a simple Lie algebra and its associated maximal ideal, *Ann. Sci. Éc. Norm. Sup.*, **9** (1976), 1–29.

[Jo2] ———, A surjectivity theorem for rigid highest weight modules, *Invent. Math.*, **92** (1988), 567–596.

[Jo3] ———, Annihilators and associated varieties of unitary highest weight modules, *Ann. Sci. Éc. Norm. Sup.*, **25** (1992), 1–45.

[Ka] M. Kashiwara, Representation theory and D-modules on flag varieties, *Astérisque*, **173–174** (1989), 55–109.

[Ke] G. Kempf, The Grothendieck–Cousin complex of an induced representation, *Advances in Mathematics*, **29** (1978), 310–396.

[Kr] H. Kraft, *Geometrische Methoden in der Invariantentheorie*, Aspekte der Mathematik, Vieweg Verlag, 1985.

[LSS] T. Levasseur, S. P. Smith and J. T. Stafford, The minimal nilpotent orbit, the Joseph ideal, and differential operators, *J. Algebra*, **116** (1988), 480–501.

[LS1] T. Levasseur and J. T. Stafford, *Rings of Differential Operators on Classical Rings of Invariants*, Mem. Amer. Math. Soc. 412, 1989.

[LS2] ———, Invariant differential operators and an homomorphism of Harish-Chandra, *J. Amer. Math. Soc.*, **8** (1995), 365–372.

[LS3] ———, Differential operators on some nilpotent orbits, *Representation Theory*, **3** (1999), 457–473.

[Lo] M. Lorenz, *Gelfand–Kirillov Dimension*, Cuadernos de Algebra, No. 7 (Grenada, Spain), 1988.

[MR] J. C. McConnell and J. C. Robson, *Noncommutative Noetherian Rings*, Grad. Texts in Math., Vol. 30, Amer. Math. Soc., Providence, RI, 2000.

[Na] Y. Nakai, On the theory of differentials in commutative rings, *J. Math. Soc. Japan*, **13** (1961), 63–84.

[Po] A. Polishchuk, Gluing of perverse sheaves on the basic affine space. With an appendix by R. Bezrukavnikov and the author, *Selecta Math. (N.S.)*, **7** (2001), 83–147.

[Sc] G. W. Schwarz, Lifting differential operators from orbit spaces, *Ann. Sci. Éc. Norm. Sup.*, **28** (1995), 253–306.

[Sh1] N. N. Shapovalov, On a conjecture of Gel'fand–Kirillov, *Funct. Anal. Appl.*, **7** (1973), 165–6.

[Sh2] ———, Structure of the algebra of regular differential operators on a basic affine space, *Funct. Anal. Appl.*, **8** (1974), 37–46.

[Sm] K. E. Smith, The D-module structure of F-split rings, *Math. Research Letters*, **2** (1995), 377–386.

[Sp] T. A. Springer, *Linear Algebraic Groups*, Progress in Math., Vol. 9, Birkhäuser, Boston, MA., 1998.

[Tr] W. N. Traves, Nakai's conjecture for varieties smoothed by normalization, *Proc. Amer. Math. Soc.*, **127** (1999), 2245–2248.

[VP] E. B. Vinberg and V. L. Popov, On a class of quasi-homogeneous affine varieties, *Math. USSR Izvestija*, **6** (1972), 743–758.

[YZ] A. Yekutieli and J. J. Zhang, Dualizing complexes and tilting complexes over simple rings, *J. Algebra*, **256** (2002), 556–567.

A q-analogue of an identity of N. Wallach

G. Lusztig

Department of Mathematics
M.I.T.
Cambridge, MA 02139
USA
gyuri@math.mit.edu

To Tony Joseph on the occasion of his 60th birthday

Summary. N. Wallach has considered an element of the group algebra of the symmetric group S_n which is the sum of an n-cycle, an $(n-1)$-cycle, ..., a 2-cycle and the identity. He showed that multiplication by this element has eigenvalues $0, 1, 2, \dots, n-2, n$. We prove an analogous result in which the group algebra of S_n is replaced by the corresponding Hecke algebra.

Subject Classification: 20G99

1.

Let n be an integer ≥ 2. For $i \in [1, n-1]$ let s_i be the transposition $(i, i+1)$ in the group S_n of permutations of $[1, n]$. (Given two integers a, b we denote by $[a, b]$ the set of all integers c such that $a \leq c \leq b$.) Consider the following element of $\mathbf{C}[S_n]$ (the group algebra of S_n):

$$\mathbf{t} = s_1 s_2 \dots s_{n-1} + s_2 s_3 \dots s_{n-1} + \dots + s_{n-2} s_{n-1} + s_{n-1} + 1.$$

(The sum of an n-cycle,an $n-1$-cycle, ..., a 2-cycle and the identity.) Wallach [WA] proved the remarkable identity

$$\mathbf{t} \prod_{k \in [1,n], k \neq n-1} (\mathbf{t} - k) = 0 \tag{a}$$

in $\mathbf{C}[S_n]$ and used it to establish a vanishing result for some Lie algebra cohomologies. In particular, left multiplication by $\mathbf{t}$ in $\mathbf{C}[S_n]$ has eigenvalues in $\{0, 1, 2, \dots, n-2, n\}$. A closely related result appeared later in connection with a problem concerning shuffling of cards in Diaconis, Fill and Pitman [DFP] and also in Phatarfod [PH].

Supported in part by the National Science Foundation.

Let $\mathbf{q}$ be an indeterminate. Let H be the $\mathbf{Z}[\mathbf{q}]$-algebra with generators $T_1, T_2, \dots, T_{n-1}$ and relations $(T_i + 1)(T_i - \mathbf{q}) = 0$ for all i, $T_i T_j T_i = T_j T_i T_j$ for $|i-j| = 1$, $T_i T_j = T_j T_i$ for $|i-j| \geq 2$, a Hecke algebra of type A_{n-1}. Set

$$\tau = T_1 T_2 \dots T_{n-1} + T_2 T_3 \dots T_{n-1} + \dots + T_{n-2} T_{n-1} + T_{n-1} + 1 \in H.$$

Under the specialization $\mathbf{q} = 1$, τ becomes the element $\mathbf{t}$ of $\mathbf{C}[S_n]$. Our main result is the following q-analogue of (a):

Proposition 2. *The following equality in H holds:*

$$\tau \prod_{k \in [1,n], k \neq n-1} (\tau - 1 - \mathbf{q} - \mathbf{q}^2 - \dots - \mathbf{q}^{k-1}) = 0.$$

The proof will be given in Section 4. The proof of the Proposition is a generalization of the proof of 1(a) given in [GW]. However, there is a new difficulty due to the fact that the product of two standard basis elements of H is not a standard basis element (as for S_n) but a complicated linear combination of basis elements. To overcome this difficulty we will work in a model of H as a space of functions on a product of two flag manifolds over a finite field.

Let V be a vector space of dimension n over a finite field $\mathbf{F}_q$ of cardinal q. Let $\mathcal{F}$ be the set of complete flags

$$V_* = (V_0 \subset V_1 \subset V_2 \subset \dots \subset V_n)$$

in V where V_k is a subspace of V of dimension k for $k \in [0, n]$. Now $GL(V)$ acts on $\mathcal{F}$ by

$$g : V_* \mapsto gV_* = (gV_0 \subset gV_1 \subset gV_2 \subset \dots \subset gV_n)$$

and on $\mathcal{F} \times \mathcal{F}$ by $g : (V_*, V'_*) \mapsto (gV_*, gV'_*)$. Let $\mathcal{H}$ be the $\mathbf{C}$-vector space of all functions $f : \mathcal{F} \times \mathcal{F} \to \mathbf{C}$ that are constant on the orbits of $GL(V)$. This is an associative algebra with multiplication

$$f, f' \mapsto f * f', \quad (f * f')(W_*, V_*) = \sum_{V'_* \in \mathcal{F}} f(W_*, V'_*) f'(V'_*, V_*).$$

Define $f_1 \in \mathcal{H}$ by

$f_1(W_*, V'_*) = 1$ if there exists $g \in [1, n]$ (necessarily unique) with $W_r = V'_r$ for $r \in [1, g-1]$, $V'_r \neq W_r \subset V'_{r+1}$ for $r \in [g, n-1]$;

$f_1(W_*, V'_*) = 0$, otherwise.

For $t \in [0, n]$ and any sequence $1 \leq i_1 < i_2 < \dots < i_{n-t} \leq n$ let $X_t^{i_1, i_2, \dots, i_{n-t}}$ be the set of all pairs $(V'_*, V_*) \in \mathcal{F} \times \mathcal{F}$ such that $V'_r \subset V_{i_r}$, $V'_r \not\subset V_{i_r - 1}$ for $r \in [1, n-t]$. For $t \in [0, n]$ let $X_t = \cup X_t^{i_1, i_2, \dots, i_{n-t}} \subset \mathcal{F} \times \mathcal{F}$ where the union is taken over all sequences $1 \leq i_1 < i_2 < \dots < i_{n-t} \leq n$. Clearly, this union is disjoint and $X_0 \subset X_1 \subset X_2 \subset \dots \subset X_n = \mathcal{F} \times \mathcal{F}$. Also, X_0 is the diagonal in $\mathcal{F} \times \mathcal{F}$. Define $f_t \in \mathcal{H}$ by

$f_t(V'_*, V_*) = 1$ if $(V'_*, V_*) \in X_t$, $f_t(V'_*, V_*) = 0$, otherwise.

For $t = 1$ this agrees with the earlier definition of f_1. Note that f_0 is the unit element of $\mathcal{H}$. The following result is a q-analogue of a result in [DFP].

Lemma 3. *For $t \in [1, n-1]$ we have $f_1 * f_t = (1+q+q^2+\cdots+q^{t-1})f_t + q^t f_{t+1}$.*

Let $f = f_1 * f_t$. From the definitions we have $f = \sum_{g=1}^{n} \phi_g$ where $\phi_g \in \mathcal{H}$ is defined as follows: for $(W_*, V_*) \in \mathcal{F} \times \mathcal{F}$, $\phi_g(W_*, V_*)$ is the number of $V'_* \in \mathcal{F}$ such that

$V'_r = W_r$ for $r \in [1, g-1]$, $V'_r \neq W_r \subset V'_{r+1}$ for $r \in [g, n-1]$ and there exists $1 \le i_1 < i_2 < \cdots < i_{n-t} \le n$ with $V'_r \subset V_{i_r}$, $V'_r \not\subset V_{i_r-1}$ for $r \in [1, n-t]$.

Here V'_r is uniquely determined for $r \in [1, g-1]$ (we have $V'_r = W_r$) while for $r \in [g+1, n-1]$, V'_r is equal to $V'_g + W_{r-1}$ (this follows by induction from $V'_r = V'_{r-1} + W_{r-1}$ which holds since V'_{r-1}, W_{r-1} must be distinct hyperplanes of V'_r). Hence $\phi_g(W_*, V_*)$ is the cardinal of the set Y_g consisting of all g-dimensional subspaces V'_g of V such that

$W_{g-1} \subset V'_g$,

$V'_g + W_{r-1} \neq W_r$ for $r \in [g, n-1]$ (or equivalently $V'_g \not\subset W_{n-1}$),

and there exists $1 \le i_1 < i_2 < \cdots < i_{n-t} \le n$ (necessarily unique) with

$W_r \subset V_{i_r}$, $W_r \not\subset V_{i_r-1}$ if $r \in [1, n-t] \cap [1, g-1]$,

$V'_g \subset V_{i_g}$, $V'_g \not\subset V_{i_g-1}$ if $g \in [1, n-t]$,

$V'_g + W_{r-1} \subset V_{i_r}$, $V'_g + W_{r-1} \not\subset V_{i_r-1}$ if $r \in [1, n-t] \cap [g+1, n-1]$.

Assume first that $g \in [1, n-t]$. If a $V'_g \in Y_g$ exists and if $1 \le i_1 < i_2 < \cdots < i_{n-t} \le n$ is as above then, setting $j_r = i_r$ for $r \in [1, g-1]$ and $j_r = i_{r+1}$ for $r \in [g, n-t-1]$, we have $1 \le j_1 < j_2 < \cdots < j_{n-t-1} \le n$ and

(a) $W_r \subset V_{j_r}$, $W_r \not\subset V_{j_r-1}$ for $r \in [1, n-t-1]$.

(For $r \in [1, g-1]$ this is clear. Assume now that $r \in [g, n-t-1]$. Since $V'_g + W_r \subset V_{j_r}$, we have $W_r \subset V_{j_r}$. If $W_r \subset V_{j_r-1}$ then, since $V'_g \subset V_{i_g} \subset V_{j_r-1}$ and $j_r = i_{r+1}$, we would have $V'_g + W_r \subset V_{i_{r+1}-1}$, contradiction.) We see that $\phi_g(W_*, V_*) = 0$ if $(W_*, V_*) \notin X_{t+1}$. We now assume that $(W_*, V_*) \in X_{t+1}$. Let $1 \le j_1 < j_2 < \cdots < j_{n-t-1} \le n$ be such that (a) holds. Then $\phi_g(W_*, V_*)$ is the number of g-dimensional subspaces V'_g of V such that

(b) $W_{g-1} \subset V'_g \not\subset W_{n-1}$,

and

(c) if $g = 1 \le n-t-1$, then $V'_g \subset V_i$, $V'_g \not\subset V_{i-1}$ for some i with $1 \le i < j_g$;

(d) if $g \in [2, n-t-1]$, then $V'_g \subset V_i$, $V'_g \not\subset V_{i-1}$ for some i with $j_{g-1} < i < j_g$;

(e) if $g = n-t \ge 2$, then $V'_g \subset V_i$, $V'_g \not\subset V_{i-1}$ for some i with $j_{g-1} < i \le n$.

Now conditions (c), (d), (e) can be replaced by:

(c′) if $g = 1 \le n-t-1$, then $V'_g \subset V_{j_g-1}$;

(d′) if $g \in [2, n-t-1]$, then $V'_g \subset V_{j_g-1}$, $V'_g \not\subset V_{j_{g-1}}$;

(e′) if $g = n-t \ge 2$, then $V'_g \not\subset V_{j_{g-1}}$.

Setting $L = V'_g / W_{g-1}$ we see that $\phi_g(W_*, V_*)$ is the number of lines L in V/W_{g-1} such that $L \not\subset W_{n-1}/W_{g-1}$ and

if $g = 1 \le n-t-1$, then $L \subset V_{j_g-1}/W_{g-1}$;

if $g \in [2, n-t-1]$, then $L \subset V_{j_g-1}/W_{g-1}$, $L \not\subset V_{j_{g-1}}/W_{g-1}$;

if $g = n-t \ge 2$, then $L \not\subset V_{j_{g-1}}/W_{g-1}$.

Since W_{n-1}/W_{g-1} is a hyperplane in V/W_{g-1}, we see that $\phi_g(W_*, V_*)$ is given by

$(q^{j_g-g} - q^{j_g-g-1})/(q-1) = q^{j_g-g-1}$ if $g = 1 \leq n-t-1$ and $V_{j_g-1} \not\subset W_{n-1}$,

0 if $g = 1 \leq n-t-1$ and $V_{j_g-1} \subset W_{n-1}$,

$(q^{j_g-g} - q^{j_g-g-1} - q^{j_{g-1}-g+1} + q^{j_{g-1}-g})/(q-1) = q^{j_g-g-1} - q^{j_{g1}-g}$ if $g \in [2, n-t-1]$ and $V_{j_{g-1}} \not\subset W_{n-1}$,

$(q^{j_g-g} - q^{j_g-g-1})/(q-1) = q^{j_g-g-1}$ if $g \in [2, n-t-1]$ and $V_{j_{g-1}} \subset W_{n-1}$, $V_{j_g-1} \not\subset W_{n-1}$,

0 if $g \in [2, n-t-1]$ and $V_{j_g-1} \subset W_{n-1}$,

$(q^{n-g+1} - q^{n-g} - q^{j_{g-1}-g+1} + q^{j_{g-1}-g})/(q-1) = q^{n-g} - q^{j_{g-1}-g}$ if $g = n-t \geq 2$ and $V_{j_{g-1}} \not\subset W_{n-1}$,

q^{n-g} if $g = n-t \geq 2$ and $V_{j_{g-1}} \subset W_{n-1}$,

q^{n-g} if $g = 1 = n-t$.

Now there is a unique $u \in [1, n]$ such that $V_{u-1} \subset W_{n-1}$, $V_u \not\subset W_{n-1}$.

From (a) we see that $u \notin \{j_1, j_2, \dots, j_{n-t-1}\}$ (we use that $n-t-1 < n-1$). Using the formulas above, we can now compute $N = \sum_{g \in [1, n-t]} \phi_g(W_*, V_*)$.

If $u < j_1$ and $n-t \geq 2$ (so that $V_{j_1-1} \not\subset W_{n-1}$) we have

$N = q^{j_1-2} + \sum_{g=2}^{n-t-1}(q^{j_g-g-1} - q^{j_{g-1}-g}) + (q^t - q^{j_{n-t-1}-1})q^t$.

If $j_{h-1} < u < j_h$ for some $h \in [2, n-t-1]$ (so that $V_{j_{h-1}} \subset W_{n-1}$, $V_{j_h-1} \not\subset W_{n-1}$) we have

$$N = q^{j_h-h-1} + \sum_{g=h+1}^{n-t-1}(q^{j_g-g-1} - q^{j_{g-1}-g}) + (q^t - q^{j_{n-t-1}1}) = q^t.$$

If $j_{n-t-1} < u$ and $n-t \geq 2$ (so that $V_{j_{n-t-1}} \subset W_{n-1}$) we have $N = q^t$.

If $n-t = 1$, we have $N = q^t$.

We see that in any case we have $N = q^t$.

Assume next that $g \in [n-t+1, n]$. If a $V'_g \in Y_g$ exists then there exists $1 \leq i_1 < i_2 < \cdots < i_{n-t} \leq n$ such that

(f) $W_r \subset V_{i_r}$, $W_r \not\subset V_{i_r-1}$ if $r \in [1, n-t]$.

We see that $\phi_g(W_*, V_*) = 0$ if $(W_*, V_*) \notin X_t$. We now assume that $(W_*, V_*) \in X_t$ and that $1 \leq i_1 < i_2 < \cdots < i_{n-t} \leq n$ is such that (f) holds. Then $\phi_g(W_*, V_*)$ is the number of g-dimensional subspaces V'_g of V such that

$W_{g-1} \subset V'_g \not\subset W_{n-1}$

that is, the number of lines L in V/W_{g-1} such that $L \not\subset W_{n-1}/W_{g-1}$. We see that $\phi_g(W_*, V_*) = q^{n-g}$. Hence $\sum_{g \in [n-t+1, n]} \phi_g(W_*, V_*) = 1 + q + q^2 + \cdots + q^{t-1}$.

Summarizing, we see that for $(W_*, V_*) \in \mathcal{F} \times \mathcal{F}$, $f(W_*, V_*) = \sum_{g=1}^{n} \phi_g(W_*, V_*)$ is equal to

$1 + q + q^2 + \cdots + q^t$ if $(W_*, V_*) \in X_t$,

q^t if $(W_*, V_*) \in X_{t+1} - X_t$,

0 if $(W_*, V_*) \notin X_{t+1}$.

The lemma follows immediately.

4.

We show that

(a) $q^{1+2+\cdots+(t-1)} f_t = f_1 * (f_1 - 1) * (f_1 - 1 - q) * \cdots * (f_1 - 1 - q - q^2 - \cdots - q^{t-2})$
for $t \in [1, n-1]$ by induction on t. For $t = 1$ this is clear. Assume that $t \in [2, n-1]$ and that (a) holds when t is replaced by $t-1$. Using Lemma 3 we have $q^{t-1} f_t = (f_1 - 1 - q - q^2 - \cdots - q^{t-2}) * f_{t-1}$. Using this and the induction hypothesis we have

$$q^{1+2+\cdots+(t-1)} f_t = (f_1 - 1 - q - q^2 - \cdots - q^{t-2}) * f_1 * (f_1 - 1) * \\ * (f_1 - 1 - q) * \cdots * (f_1 - 1 - q - q^2 - \cdots - q^{t-3}).$$

This proves (a).

Next we note that X_{n-1} is the set of all $(V'_*, V_*) \in \mathcal{F} \times \mathcal{F}$ such that for some $i \in [1, n]$ we have $V'_1 \subset V_i, V'_1 \not\subset V_{i-1}$. Thus, $X_{n-1} = \mathcal{F} \times \mathcal{F} = X_n$ so that $f_{n-1} = f_n$. Using this and Lemma 3 we see that $f_1 * f_{n-1} = (1 + q + q^2 + \cdots + q^{n-1}) f_{n-1}$ that is $(f_1 - 1 - q - q^2 - \cdots - q^{n-1}) f_{n-1} = 0$. Hence multiplying both sides of (a) (for $t = n-1$) by $(f_1 - 1 - q - q^2 - \cdots - q^{n-1})$ we obtain

$$f_1 * (f_1 - 1) * (f_1 - 1 - q) * \cdots * \\ * (f_1 - 1 - q - q^2 - \cdots - q^{n-3}) * (f_1 - 1 - q - q^2 - \cdots - q^{n-1}) = 0.$$

Thus an identity like that in Proposition 2 holds in $\mathcal{H}$ instead of H (with f_1, q instead of $\tau, \mathbf{q}$). It is known that the algebra $\mathcal{H}$ may be identified with $\mathbf{C} \otimes_{\mathbf{Z}} [\mathbf{q}] H$ (where $\mathbf{C}$ is regarded as a $\mathbf{Z}[\mathbf{q}]$-algebra via the specialization $\mathbf{q} \mapsto q$) in such a way that $1 \otimes \tau$ is identified with f_1. Since q can take infinitely many values, the identity in Proposition 2 follows.

5.

Setting $f_t = 0$ for $t > n$, we see that the identity in Lemma 3 remains valid for any $t \geq 0$. We see that the subspace of $\mathcal{H}$ spanned by $\{f_t; t \geq 0\}$ coincides with the subspace spanned by $\{f_1^t; t \geq 0\}$; in particular it is a commutative subring.

6.

Consider the endomorphism of $\mathbf{Q}(\mathbf{q}) \otimes_{\mathbf{Z}[\mathbf{q}]} H$ given by left multiplication by τ. Proposition 2 shows that the eigenvalues of this endomorphism are in

$$\{0, 1, 1 + \mathbf{q}, 1 + \mathbf{q} + \mathbf{q}^2, \ldots, 1 + \mathbf{q} + \cdots + \mathbf{q}^{n-3}, 1 + \mathbf{q} + \cdots + \mathbf{q}^{n-1}\}.$$

The multiplicity of the eigenvalue $1 + \mathbf{q} + \cdots + \mathbf{q}^{k-1}$ is preserved by the specialization $\mathbf{q} = 1$ hence it is the same as the multiplicity of the eigenvalue k for the left multiplication by $\mathbf{t}$ on $\mathbf{C}[S_n]$, which by [DFP] is the number of permutations of $[1, n]$ with exactly k fixed points.

References

[DFP] P. Diaconis, J. A. Fill and J. Pitman, Analysis of top to random shuffles, *Combinatorics, probability and computing* **1** (1992), 135–155.

[GW] A. M. Garsia and N. Wallach, Qsym over Sym is free, *J. Comb. Theory, Ser. A.* **104** (2003), 217–263.

[PH] R. M. Phatarfod, On the matrix occuring in a linear search problem, *Jour. Appl. Prob.* **28** (1991), 336–346.

[WA] N. Wallach, Lie algebra cohomology and holomorphic continuation of generalized Jacquet integrals, *Advanced Studies in Pure Math.* **14** (1988), 123–151.

Centralizers in the quantum plane algebra

L. Makar-Limanov*

Department of Mathematics & Computer Science
Bar-Ilan University
52900 Ramat-Gan
Israel
lml@macs.biu.ac.il
Department of Mathematics
Wayne State University
Detroit, MI 48202
USA
lml@math.wayne.edu

To Tony Joseph who is certainly an inspiration.

Summary. Dixmier discovered that the centralizers of elements of the first Weyl algebra have some unexpected properties. Sometimes a centralizer is not integrally closed. Also there are cases when the field of fractions of a centralizer is not a purely transcendental field. In this article I am going to discuss what happens if the Weyl algebra is replaced by the quantum plane algebra or a quantum space algebra of any dimension. I became interested in this question after a conversation with L. Small and J. Zhang during a meeting in Taiwan in June of 2001. To my great surprise it turns out that though the centralizers (of non-central elements) are not necessarily integrally closed, the fields of fractions of centralizers of non-constants are always purely transcendental fields of dimension 1 for a "general position" situation.

Subject Classifications: 16S36, 16W35, 17B37, 16W50

Introduction

Let A_1 be the first Weyl algebra over $\mathbb{C}$, i.e., an algebra over the field of complex numbers $\mathbb{C}$ generated by p and q subject to the relation $pq - qp = 1$. In the paper [Di] Dixmier gave the following example of peculiar behavior of centralizers of elements of this algebra. If $u = p^3 + q^2 - \alpha$ where $\alpha \in \mathbb{C}$, $v = \frac{1}{2}p$, $U = u^2 + 4v$, $V = u^3 + 3(uv + vu)$, then $V^2 - U^3 = \alpha$. This equality implies that U and V commute, and so when $\alpha \neq 0$, the centralizer $C(U)$ of U is isomorphic to the ring of regular

* The author was supported by an NSA grant while working on this project.

functions on an elliptic curve and its field of fractions is not isomorphic to the field of rational functions $\mathbb{C}(z)$.

On the other hand, when $\alpha = 0$, the field of fractions of $C(U)$ is isomorphic to $\mathbb{C}(z)$, but $C(U)$ is not integrally closed. Here is a simpler example of the last kind: if $t = pq$ and $h = pt(t-1)^{-1}(t-2)$, then both $h^2, h^3 \in A_1$ although $h \in D_1 \backslash A_1$, where D_1 is the skew-field of fractions of A_1. I cannot properly attribute the latter example save that it was constructed by someone in Moscow, Russia about 1968.

Another example of this kind was found by Bergman (see [Be]) that is, of $h = p^{-1}(t-1)(t+1)$. These two examples are actually very similar. Very close examples were found in 1922 by Burchnall and Chaundy (see [BC]). These examples are of a bit different kind because Burchnall and Chaundy were looking only at monic differential operators, i.e., at elements that are monic polynomials in p and worked in an algebra larger than A_1. Burchnall and Chaundy observed, e.g., that $P = p^2 - 2q^{-2}$ and $Q = p^3 - 3q^{-2} + 3q^{-3}$ commute without being polynomials of an element of the form $p + f(q)$. Since $P = q^{-2}t(t-3)$ and $Q = q^{-3}t(t-2)(t-4)$, they are the square and cube of $h = q^{-1}t(t-1)^{-1}(t-2)$ correspondingly. Burchnall and Chaundy also gave an example of a centralizer isomorphic to the ring of regular functions on an elliptic curve but the elements involved are not in D_1.

The effect discovered by Dixmier is somewhat surprising because A_1 may be looked at as a deformation of the polynomial algebra $\mathbb{C}[x, y]$, and in $\mathbb{C}[x, y]$ any maximal subalgebra of transcendence degree one is isomorphic to a polynomial ring $\mathbb{C}[z]$ (see [Za]).

It is reasonable to compare the centralizers of non-scalar elements of A_1 with the maximal subalgebras of transcendence degree one of $\mathbb{C}[x, y]$ because it is known that the centralizer of a non-scalar element of A_1 is a maximal subalgebra of A_1 of transcendence degree one. Also if $\mathbb{C}[x, y]$ is embedded into a Poisson algebra, then centralizers of non-scalar elements of $\mathbb{C}[x, y]$ are maximal subalgebras of $\mathbb{C}[x, y]$ of transcendence degree one (see [SU]).

The theorem about centralizers in A_1 has a somewhat entertaining history. It is usually attributed to Amitsur (see [Am]) who attributes it to Flanders (see [Fl]) who attributes it to Schur (see [Sc]). See also the paper [Go] which contains a lot of information about centralizers of A_1 and its generalizations.

Another popular deformation of $\mathbb{C}[x, y]$ is the coordinate algebra of the quantum plane $\mathbb{C}_q[x, y]$, i.e., the $\mathbb{C}$-algebra generated by x and y subject to the relation $yx = qxy$. So it seems rather natural to compare the structure of the centralizers of elements of this algebra with the structure of the centralizers of elements of A_1 and with the maximal subalgebras of transcendence degree one of polynomial rings.

Centralizers in $\mathbf{C_q}$

It is known that the centralizers of non-scalar elements of $\mathbb{C}_q[x, y]$ when q is not a root of 1 are commutative algebras of transcendence degree one (see [AC] and [BS]). Actually in [AC] this is shown for the field of fractions $\mathbb{C}_q(x, y)$ of $\mathbb{C}_q[x, y]$. See also [HS] where it is independently proved that any two commuting elements of the q-deformed Heisenberg algebra $H_q[x, y]$, which is a subalgebra of $\mathbb{C}_q(x, y)$, are algebraically dependent and [LS] where some examples of centralizers of $H_q[x, y]$ are provided.

In this section we check that, unlike the Weyl algebra setting, these centralizers are always subalgebras of a polynomial ring in one variable. In the next section we extend this result to quantum space algebras, i.e., algebras with n generators $x_1, x_2, \ldots, x_n$ subject to the relations $x_j x_i = q_{ij} x_i x_j$.

Actually this result can be deduced from Theorem 1.1 of [BS] which implies that a centralizer of a non-scalar element of $\mathbb{C}_q[x, y]$ has transcendence degree one and Proposition 6.1 of [Be] which implies that any subalgebra of $\mathbb{C}_q[x, y]$ has a non-trivial mapping into $\mathbb{C}[z]$. On the other hand, the proofs here are shorter and more straightforward than in these papers which deal with much more general settings.

It is tempting to conjecture that the centralizers of non-central elements of $\mathbb{C}_q[x, y]$ when q is a root of 1 are subalgebras of a polynomial ring in two variables. But this is not the case, as we can see from the following example.

Let $q = -1$. The center of $\mathbb{C}_{-1}[x, y]$ is $\mathbb{C}[x^2, y^2]$. Let us take $z = x^3 + xyt^3$ where $t = x^2y^2 + 1$. It is easy to check by a straightforward computation that the centralizer $C(z) = \mathbb{C}[x^2, y^2, z]$ and that $z^2 = x^6 - (t - 1)t^6$. The claim is that $\mathbb{C}[u, v][\sqrt{u^3 - (t - 1)t^6}]$ where $u = x^2$, $v = y^2$, and $t = uv + 1$ cannot be embedded into a polynomial ring with two variables. If we would have such an embedding, say into $\mathbb{C}[a, b]$, then the images U and V of u and v should be algebraically independent polynomials and the image T of t should be relatively prime with both U and V. Assume that $U^3 - (T - 1)T^6 = Z^2$. Then T and Z are also relatively prime polynomials. By taking the Jacobian (relative to a, b) of this equality with U, we get $T^5(6 - 7T)\mathrm{J}(U, T) = 2Z\mathrm{J}(U, Z)$. So T^5 divides $\mathrm{J}(U, Z)$, while Z divides $(6 - 7T)\mathrm{J}(U, T)$ since $(T, Z) = 1$. Now $\deg \mathrm{J}(f, g) < \deg f + \deg g$ where deg denotes the total degree of the corresponding polynomial relative to a, b. Therefore $\deg U + \deg Z > 5 \deg T$ and $2 \deg T + \deg U > \deg Z$. So $\deg U + \deg Z + 2 \deg T + \deg U > 5 \deg T + \deg Z$ and $2 \deg U > 3 \deg T$, which is impossible since $\deg T = \deg U + \deg V$.

So let us go back to the case when q is not a root of 1. The next examples show that we should not expect a centralizer to be integrally closed. Take $h = x(1 - y)(1 - qy)^{-1}(1 - q^2y)$ in the skew-field $\mathbb{C}_q(x, y)$ of fractions of $\mathbb{C}_q[x, y]$. Then both $h^2, h^3 \in \mathbb{C}_q[x, y]$ and $C(h^2) = \mathbb{C}[h^2, h^3]$. Similarly, if we take $h = x(1 - y)^n(1 - qy)^{-1}(1 - q^2y)^{-1} \ldots (1 - q^ny)^{-1}(1 - q^{n+1}y)^n$, then $h^i \notin \mathbb{C}_q[x, y]$ for $i < n + 1$, $h^{n+1} \in \mathbb{C}_q[x, y]$, and $C(h^{n+1}) = \mathbb{C}[h^{n+1}, \ldots, h^{2n+1}]$. So the centralizers of elements of $\mathbb{C}_q[x, y]$ are neither integrally closed nor is there a bound on the size of a set of generators of a centralizer.

Let us start the proof of the main claim also by an example. Let $h = \sum_{i=0}^{n} h_i(x)y^i$ and assume that h_0 is not a constant. If $g \in C(h)\backslash 0$ and $g = \sum_{i=0}^{m} g_i(x)y^i$, then g_0 is not zero since otherwise g and h cannot commute. So the restriction of the substitution map $f(x, y) \to f(x, 0)$ on $C(h)$ is an embedding of $C(h)$ into $\mathbb{C}[x]$.

Now the general case. Define the weights $w(x) = \rho$, $w(y) = \sigma$ and $w(x^i y^j) = i\rho + j\sigma$ where ρ and σ are integers. It is clear that $w(M_1 M_2) = w(M_1) + w(M_2)$ for any monomials.

Let $h(x, y) \in \mathbb{C}_q[x, y]\backslash\mathbb{C}$. We can always find a non-zero pair of integers ρ and σ, so that the weights of all monomials of h are non-negative and that at least one non-constant monomial of h has zero weight. If we take $\rho \leq 0 \leq \sigma$ and relatively prime, then $u = x^\sigma y^{-\rho} \in \mathbb{C}_q[x, y]$ generates the semigroup of all monomials of weight zero. It is clear that $C(u) = \mathbb{C}[u]$ since $x^{i_2} y^{j_2} x^{i_1} y^{j_1} = q^D x^{i_1} y^{j_1} x^{i_2} y^{j_2}$ where $D = i_1 j_2 - j_1 i_2$. It is also clear that $C(p(u)) = \mathbb{C}[u]$ if $p(u)$ is a non-constant polynomial in u. Indeed, any $g \in \mathbb{C}_q[x, y]$ can be presented as $g = \sum_{i=-k}^{m} g_i(u) G_i$ where $w(G_i) = i$ and $g_i(u)$ are polynomials in u. So if $[p(u), g] = 0$, then $\sum_{i=-k}^{m} g_i(u)[p(u), G_i] = 0$. Since $w([p(u), G_i]) = i$ if $[p(u), G_i] \neq 0$, we have $[p(u), G_i] = 0$ for all i. But if $w(G_i) \neq 0$, then from $M_2 M_1 = q^D M_1 M_2$ we see that G_i cannot commute with $p(u)$. So $g \in \mathbb{C}[u]$.

Let us look at $h = \sum_{i=0}^{n} h_i(u) H_i$ where h_i are polynomials in u and H_i are monomials with $w(H_i) = i$. Since the choice of H_i may be not unique let us choose H_i of the smallest total degree possible. With our choice of weights, $H_0 = 1$ and h_0 is a non-trivial polynomial in u. For $g \in C(h)\backslash\mathbb{C}$, we have $g = \sum_{i=-k}^{m} g_i(u) G_i$ where $w(G_i) = i$. If $gh = hg$, then $g_{-k}(u) G_{-k} h_0(u) = h_0 g_{-k}(u) G_{-k}$ and monomial G_{-k} must commute with h_0. We observed that this implies $G_{-k} \in \mathbb{C}[u]$. So $k = 0$, and the mapping of the elements of $C(h)$ given by $g \to g_0$ defines an embedding of $C(h)$ into $C(u) = \mathbb{C}[u]$.

Centralizers in quantum spaces

Let $\mathbb{C}_{\overline{q}}$ be an algebra over $\mathbb{C}$ with n generators $x_1, x_2, \ldots, x_n$ subject to the relations $x_j x_i = q_{ij} x_i x_j$ where $i < j$. Assume that q_{ij} where $i < j$ is a free basis of an abelian group Q.

In this section we check that the centralizers of non-constant elements of $\mathbb{C}_{\overline{q}}$ are subalgebras of a polynomial ring in one variable. The proof is very similar to the one in the previous section.

For any integers $\rho_1, \rho_2, \ldots, \rho_n$, we can define the weight $w(x_1^{i_1} x_2^{i_2} \ldots x_n^{i_n}) = \sum i_k \rho_k$ and, of course, $w(M_1 M_2) = w(M_1) + w(M_2)$ for any two monomials. If $h \in \mathbb{C}_{\overline{q}}\backslash\mathbb{C}$, we can choose integers $\rho_1, \rho_2, \ldots, \rho_n$ so that the weights of all monomials of h are non-negative, at least one non-constant monomial of h has zero weight, and all monomials of h of zero weight are proportional to the powers of a monomial M. Indeed, take the Newton polyhedron $\mathcal{H}$ of h, that is, mark the origin in n-dimensional space and all integer points which correspond to the monomials of h, and then take the convex hull of this set of points. Clearly the origin is one of the vertices of the polyhedron. Take an edge e of $\mathcal{H}$ which contains the origin and choose a hyperplane P which intersects $\mathcal{H}$ only at e and has a normal vector with integer coefficients. Take a normal vector to P in the direction of $\mathcal{H}$ and use the values of its components as the weights $\{\rho_i\}$. Then all conditions on the weight are satisfied with monomial M corresponding to the shortest vector with integer coordinates belonging to e.

As in the previous section it is easy to check that $C(M) = \mathbb{C}[M]$. Indeed, if N is a monomial, then $MN = (\prod g_{ij}^{d_{ij}})NM$ where d_{ij} is the determinant of a two by two matrix formed by powers of x_i and x_j in the monomials N and M and, since q_{ij} form a free group, commutation means that the power-vectors of M and N are proportional.

It is again clear that if $g \in C(h)\backslash\mathbb{C}$, then the weights of all monomials of g are non-negative and that the zero-weight component of g is a non-zero polynomial of M. So we have an embedding of the $C(h)$ into the algebra $\mathbb{C}[M]$.

Conclusion and remarks

The lesson here is that as far as the centralizers are concerned, $\mathbb{C}_{\overline{q}}$ is closer to the polynomial algebra with n generators or to the free algebra of rank n (just recall the result of G. Bergman that the centralizer of a non-scalar element of a free algebra is isomorphic to the polynomial ring with one generator (see [Be])).

On the other hand, consider, instead of $\mathbb{C}_{\overline{q}}$, its field of fractions or even the analog of the Laurent polynomial ring. It is still possible to show that the centralizers of non-central elements are commutative by looking at appropriately defined deficit functions on centralizers (see [ML]; also in [AC] it is proved that the centralizers are commutative in the case of the field of fractions of $\mathbb{C}_q[x, y]$). But the structure of the centralizers may become more complicated since it is not always possible to introduce weights giving "good" components of weight zero and our technique fails in general. Of course, it is still entirely possible that the centralizers of non-central elements of the field of fractions of $\mathbb{C}_{\overline{q}}$ with appropriate restrictions on the group Q are isomorphic to the field of rational functions in one variable, and it would be very interesting to research this question.

If we consider the subalgebra L of $\mathbb{C}_q(x, y)$ generated by x, y, and y^{-1}, then it is still possible to find "good" zero components and show that the centralizer of a non-scalar element is always isomorphic to a subalgebra of the Laurent polynomial ring $\mathbb{C}[t, t^{-1}]$; so the corresponding fields of fractions are isomorphic to the field of rational functions in one variable. Since the q-deformed Heisenberg algebra $H_q[x, y]$ can be embedded in L when $q \neq 1$, we can conclude that the same is true for the centralizers of H_q.

Lastly, all of these centralizers are finitely generated since any subalgebra of $\mathbb{C}[t, t^{-1}]$ is finitely generated. Let $A \subset \mathbb{C}[t, t^{-1}]$. Assume that $A \not\subset \mathbb{C}[t]$. Choose an $a \in A$ for which order $\mathrm{o}(a) = n$ of a is negative and the largest possible under this condition. (The order of an element is the smallest degree of monomials of this element.) For any remainder $i \pmod n$ take an element a_i for which $\mathrm{o}(a_i) \equiv i \pmod n$, $\mathrm{o}(a_i) < 0$, and $\mathrm{o}(a_i)$ is the largest possible under this conditions. Of course, for some i we may have no elements at all. Then for any element $b \in A$ with negative order we can find a_i and a non-negative integer k so that $\mathrm{o}(b) = \mathrm{o}(a_i a^k)$. Hence we can find $c \in \mathbb{C}$ for which $\mathrm{o}(b - c a_i a^k) > \mathrm{o}(b)$. After several steps like that we will get an element of $R = A \cap \mathbb{C}[t]$. We will be done if we show that any subalgebra R of $\mathbb{C}[t]$ is finitely generated. Choose an $r \in R$ for which degree $\deg(r) = m$ of r is positive and the smallest possible under this condition. For any remainder $i \pmod{m}$ take an element r_i for which $\deg(r_i) \equiv i \pmod{m}$ and $\deg(r_i)$ is the smallest possible under this condition. It is clear that R is generated by r and $r_1, r_2, \dots, r_{m-1}$ and therefore A is generated by $a, a_1, a_2, \dots, a_{n-1}$ and $r, r_1, r_2, \dots, r_{m-1}$.

References

[Am] S. A. Amitsur, Commutative linear differential operators, *Pacific J. Math.* **8** (1958), 1–10.

[AC] V. A. Artamonov, P. M. Cohn, The skew field of rational functions on the quantum plane, Algebra, 11. *J. Math. Sci.* (New York) **93** (1999), no. 6, 824–829.

[BS] J. P. Bell, L.W. Small, Centralizers in domains of GK dimension two, *Bull. London Math. Soc.* **36** (2004), no. 6, 779–785.

[Be] G. M. Bergman, Centralizers in free associative algebras, *Trans. Amer. Math. Soc.* **137** (1969), 327–344.

[BC] J. L. Burchnall, T. W. Chaundy, *Commutative ordinary differential operators*, Proc. London Math. Soc. (Ser. 2), **21** (1922), 420–440.

[Di] J. Dixmier, Sur les algèbres de Weyl, *Bull. Soc Math. France*, **96** (1968), 209–242.

[Fl] H. Flanders, *Commutative linear differential operators*, Department of Mathematics, University of California, Berkeley (1955), Technical Report No. 1.

[Go] K. R. Goodearl, Centralizers in differential, pseudodifferential, and fractional differential operator rings, *Rocky Mountain J. Math.* **13** (1983), no. 4, 573–618.

[HS] L. Hellström, S. D. Silvestrov, *Commuting elements in q-deformed Heisenberg algebras*, (English. English summary) World Scientific Publishing Co., Inc., River Edge, NJ, 2000. xiv+257, pp. ISBN 981-02-4403-7.

[LS] D. Larsson, S. D. Silvestrov, Burchnall-Chaundy Theory for q-Difference Operators and q-Deformed Heisenberg Algebras, *Journal of Nonlinear Mathematical Physics* **10**, Supplement 2 (2003), 95–106 SIDE V.

[ML] L. G. Makar-Limanov, Commutativity of certain centralizers in the rings $R_{n,k}$, (Russian) *Funkcional. Anal. i Priložen.* **4** (1970), no. 4, 332–333.

[Sc] I. Schur, Über vertauschbare lineare Differentialausdrücke, Berlin Math. Gesellschaft, Sitzungsbericht 3 (*Archiv der Math.*, Beilage (3), 8 (1904), 2–8.

[SU] I. P. Shestakov, U. U. Umirbaev, Poisson brackets and two-generated subalgebras of rings of polynomials, *J. Amer. Math. Soc.* **17** (2004), no. 1, 181–196 (electronic).

[Za] A. Zaks, Dedekind subrings of $k[x_1, x_2, \dots, x_n]$ are rings of polynomials, *Israel J. Math.* **9** (1971), 285–289.

Centralizer construction of the Yangian of the queer Lie superalgebra

Maxim Nazarov[1] and Alexander Sergeev[2]

[1] Department of Mathematics
University of York
Heslington
York YO10 5DD
England
mln1@york.ac.uk

[2] Steklov Institute of Mathematics
Fontanka 27
St. Petersburg 191011
Russia
sergeev@pdmi.ras.ru

To Professor Anthony Joseph on the occasion of his 60th birthday

Summary. Consider the complex matrix Lie superalgebra $\mathfrak{gl}_{N|N}$ with the standard generators E_{ij} where $i\,,j=\pm 1\,,\,\ldots\,,\,\pm\,N$. Define an involutive automorphism η of $\mathfrak{gl}_{N|N}$ by $\eta\,(E_{ij})=E_{-i,-j}$. The queer Lie superalgebra $\mathfrak{q}_N$ is the fixed point subalgebra in $\mathfrak{gl}_{N|N}$ relative to η. Consider the twisted polynomial current Lie superalgebra

$$\mathfrak{g}=\{\,X(t)\in\mathfrak{gl}_{N|N}[t]\;:\;\eta\,(X(t))=X(-t)\,\}\,.$$

The enveloping algebra $\mathrm{U}(\mathfrak{g})$ of the Lie superalgebra $\mathfrak{g}$ has a deformation, called the Yangian of $\mathfrak{q}_N$. For each $M=1,2,\,\ldots\,,$ denote by A_N^M the centralizer of $\mathfrak{q}_M\subset\mathfrak{q}_{N+M}$ in the associative superalgebra $\mathrm{U}(\mathfrak{q}_{N+M})$. In this article we construct a sequence of surjective homomorphisms $\mathrm{U}(\mathfrak{q}_N)\leftarrow\mathrm{A}_N^1\leftarrow\mathrm{A}_N^2\leftarrow\ldots\,.$ We describe the inverse limit of the sequence of centralizer algebras $\mathrm{A}_N^1,\mathrm{A}_N^2,\,\ldots$ in terms of the Yangian of $\mathfrak{q}_N$.

Subject Classifications: 16S30, 16S40, 16W35, 17B35, 17B37, 17B65

1. Main results

In this article we work with the queer Lie superalgebra $\mathfrak{q}_N$. This is perhaps the most interesting super-analogue of the general linear Lie algebra $\mathfrak{gl}_N$, see for instance [S2]. We will realize $\mathfrak{q}_N$ as a subalgebra in the general linear Lie superalgebra $\mathfrak{gl}_{N|N}$ over the complex field $\mathbb{C}$. Let the indices $i\,,j$ run through $-N\,,\,\ldots\,,\,-1,1,\,\ldots\,,\,N$. Put

$\bar{\imath} = 0$ if $i > 0$ and $\bar{\imath} = 1$ if $i < 0$. Take the $\mathbb{Z}_2$-graded vector space $\mathbb{C}^{N|N}$. Let $e_i \in \mathbb{C}^{N|N}$ be the standard basis vectors. The $\mathbb{Z}_2$-gradation on $\mathbb{C}^{N|N}$ is defined so that $\deg e_i = \bar{\imath}$. Let $E_{ij} \in \mathrm{End}(\mathbb{C}^{N|N})$ be the matrix units: $E_{ij}\, e_k = \delta_{jk}\, e_i$. The algebra $\mathrm{End}(\mathbb{C}^{N|N})$ is $\mathbb{Z}_2$-graded so that $\deg E_{ij} = \bar{\imath} + \bar{\jmath}$. We will also regard E_{ij} as basis elements of the Lie superalgebra $\mathfrak{gl}_{N|N}$. The *queer* Lie superalgebra $\mathfrak{q}_N$ is the fixed point subalgebra in $\mathfrak{gl}_{N|N}$ with respect to the involutive automorphism η defined by

$$\eta : E_{ij} \mapsto E_{-i,-j}\,. \tag{1.1}$$

Thus as a vector subspace, $\mathfrak{q}_N \subset \mathfrak{gl}_{N|N}$ is spanned by the elements

$$F_{ij} = E_{ij} + E_{-i,-j}\,.$$

Note that $F_{-i,-j} = F_{ij}$. The elements F_{ij} with $i > 0$ form a basis of $\mathfrak{q}_N$.

The vector subspace of $\mathrm{End}(\mathbb{C}^{N|N})$ spanned by the elements F_{ij} is closed with respect to the usual matrix multiplication. Hence we can also regard it as an associative algebra. Denote this associative algebra by Q_N to distinguish its structure from that of the Lie superalgebra $\mathfrak{q}_N$. Both $\mathrm{End}(\mathbb{C}^{N|N})$ and Q_N are simple as associative $\mathbb{Z}_2$-graded algebras, see [J, Theorem 2.6].

The enveloping algebra $\mathrm{U}(\mathfrak{q}_N)$ of the Lie superalgebra $\mathfrak{q}_N$ is a $\mathbb{Z}_2$-graded associative unital algebra. In this article we will always keep to the following convention. Let A and B be any two associative $\mathbb{Z}_2$-graded algebras. Their tensor product $\mathrm{A} \otimes \mathrm{B}$ is a $\mathbb{Z}_2$-graded algebra such that for any homogeneous elements $X, X' \in \mathrm{A}$ and $Y, Y' \in \mathrm{B}$

$$(X \otimes Y)(X' \otimes Y') = (-1)^{\deg X' \deg Y}\, X X' \otimes Y Y', \tag{1.2}$$

$$\deg\,(X \otimes Y) = \deg X + \deg Y\,. \tag{1.3}$$

By definition, an anti-homomorphism $\omega : \mathrm{A} \to \mathrm{B}$ is any linear map which preserves the $\mathbb{Z}_2$-gradation and satisfies any for homogeneous $X, X' \in \mathrm{A}$

$$\omega(X X') = (-1)^{\deg X \deg X'}\, \omega(X')\, \omega(X)\,. \tag{1.4}$$

For any Lie superalgebra $\mathfrak{a}$, the *principal* anti-automorphism of the enveloping $\mathbb{Z}_2$-graded algebra $\mathrm{U}(\mathfrak{a})$ is determined by the assignment $X \mapsto -X$ for $X \in \mathfrak{a}$.

The *supercommutator* of any two homogeneous elements $X, Y \in \mathrm{A}$ is by definition

$$[\,X\,,Y\,] = XY - (-1)^{\deg X \deg Y}\, YX. \tag{1.5}$$

This definition extends to arbitrary elements $X, Y \in \mathrm{A}$ by linearity. It is the bracket (1.5) that defines the Lie superalgebra structure on the vector space $\mathrm{A} = \mathrm{End}(\mathbb{C}^{N|N})$. Thus, for any indices $i\,,j\,,k\,,l = \pm 1, \ldots, \pm N$ we have

$$[\,E_{ij}\,, E_{kl}\,] = \delta_{kj}\, E_{il} - (-1)^{(\bar{\imath}+\bar{\jmath})(\bar{k}+\bar{l})}\, \delta_{il}\, E_{kj}\,;$$

$$\begin{aligned}[\,F_{ij}\,, F_{kl}\,] = {} & \delta_{kj}\, F_{il} - (-1)^{(\bar{\imath}+\bar{\jmath})(\bar{k}+\bar{l})}\, \delta_{il}\, F_{kj} \\ & + \delta_{-k,j}\, F_{-i,l} - (-1)^{(\bar{\imath}+\bar{\jmath})(\bar{k}+\bar{l})}\, \delta_{i,-l}\, F_{k,-j}\,. \end{aligned} \tag{1.6}$$

For any A and any subset $\mathrm{C} \subset \mathrm{A}$, by the *centralizer* of C in A we mean the collection of all elements $X \in \mathrm{A}$ such that $[X, Y] = 0$ for any $Y \in \mathrm{C}$. To remind the reader about this convention, we shall then refer to the $\mathbb{Z}_2$-graded algebra A as a superalgebra. The centre $\mathrm{Z}(\mathfrak{q}_N)$ of the enveloping algebra $\mathrm{U}(\mathfrak{q}_N)$ will be always taken in the superalgebra sense. A set of generators of the algebra $\mathrm{Z}(\mathfrak{q}_N)$ was given in [S1]. In particular, all central elements of $\mathrm{U}(\mathfrak{q}_N)$ were shown to have $\mathbb{Z}_2$-degree 0. A distinguished basis of the vector space $\mathrm{Z}(\mathfrak{q}_N)$ was constructed in [N2].

Let us recall the principal results of [S1] here. For any indices $n \geqslant 1$ and $i, j = \pm 1, \ldots, \pm N$, denote by $F^{(n)}_{ij\,|\,N}$ the element of the algebra $\mathrm{U}(\mathfrak{q}_N)$

$$\sum_{k_1,\ldots,k_{n-1}} (-1)^{\bar{k}_1 + \ldots + \bar{k}_{n-1}}\, F_{i k_1} F_{k_1 k_2} \ldots F_{k_{n-2} k_{n-1}} F_{k_{n-1} j} \tag{1.7}$$

where each of the indices $k_1, \ldots, k_{n-1}$ runs through $\pm 1, \ldots, \pm N$. Note that

$$F^{(n)}_{-i,-j\,|\,N} = (-1)^{n-1}\, F^{(n)}_{ij\,|\,N}\,. \tag{1.8}$$

Of course, here $F^{(1)}_{ij\,|\,N} = F_{ij}$. Observe that if $n > 1$, then by the definition (1.7)

$$F^{(n)}_{ij\,|\,N} = \sum_k (-1)^{\bar{k}}\, F_{ik}\, F^{(n-1)}_{kj\,|\,N} \tag{1.9}$$

where the index k runs through $\pm 1, \ldots, \pm N$. Using this observation one proves by induction on $n = 1, 2, \ldots$ the following generalization of (1.6): in the $\mathbb{Z}_2$-graded algebra $\mathrm{U}(\mathfrak{q}_N)$ the supercommutator

$$\begin{aligned} [\, F_{ij}\,, F^{(n)}_{kl\,|\,N}\,] = {} & \delta_{kj}\, F^{(n)}_{il\,|\,N} - (-1)^{(\bar{\imath}+\bar{\jmath})(\bar{k}+\bar{l})}\, \delta_{il}\, F^{(n)}_{kj\,|\,N} \\ & + \delta_{-k,j}\, F^{(n)}_{-i,l\,|\,N} - (-1)^{(\bar{\imath}+\bar{\jmath})(\bar{k}+\bar{l})}\, \delta_{i,-l}\, F^{(n)}_{k,-j\,|\,N}\,. \end{aligned} \tag{1.10}$$

For a more general formula, expressing the supercommutator $[\, F^{(m)}_{ij\,|\,N}\,, F^{(n)}_{kl\,|\,N}\,]$ for any m and n, see Proposition 3.1 and the remark after its proof. Now put

$$C^{(n)}_N = \sum_k F^{(n)}_{kk\,|\,N} \tag{1.11}$$

where the index k runs through $\pm 1, \ldots, \pm N$. The relations (1.10) immediately imply that $C^{(n)}_N \in \mathrm{Z}(\mathfrak{q}_N)$. Note that $C^{(2)}_N = C^{(4)}_N = \ldots = 0$ due to (1.8). The following proposition has been stated in [S1] without proof.

Proposition 1.1. *The elements* $C^{(1)}_N, C^{(3)}_N, \ldots$ *generate the centre* $\mathrm{Z}(\mathfrak{q}_N)$.

The dependence of the elements $C^{(n)}_N$ and $F^{(n)}_{ij\,|\,N}$ of $\mathrm{U}(\mathfrak{q}_N)$ on the index N has been indicated for the purposes of the next argument, which extends [S1].

For any integers $N \geqslant 0$ and $M \geqslant 1$ consider the Lie superalgebra $\mathfrak{q}_{N+M}$. Now let the indices i, j run through $-N-M, \ldots, -1, 1, \ldots, N+M$. Regard the

Lie superalgebras $\mathfrak{q}_N$ and $\mathfrak{q}_M$ as the subalgebras of $\mathfrak{q}_{N+M}$ spanned by the elements $F_{ij} = F^{(1)}_{ij\,|\,N+M}$ where $|i|,|j| \leqslant N$ and $|i|,|j| > N$, respectively. Denote by A^M_N the centralizer of $\mathfrak{q}_M$ in the associative superalgebra $\mathrm{U}(\mathfrak{q}_{N+M})$.

By definition, the centralizer A^M_N contains the centre $\mathrm{Z}(\mathfrak{q}_{N+M})$ of the $\mathrm{U}(\mathfrak{q}_{N+M})$. It also contains the subalgebra $\mathrm{U}(\mathfrak{q}_N) \subset \mathrm{U}(\mathfrak{q}_{N+M})$. Moreover, the relations (1.10) imply that the centralizer A^M_N contains the elements

$$F^{(1)}_{ij\,|\,N+M}\,,\ F^{(2)}_{ij\,|\,N+M}\,,\ \ldots \quad \text{where } |i|,|j| \leqslant N. \tag{1.12}$$

Theorem 1.2. *The elements* $C^{(1)}_{N+M}\,, C^{(3)}_{N+M}\,, \ldots$ *and* (1.12) *generate* A^M_N.

We prove this theorem in Section 2 of the present article. In the particular case $N=0$, we will then obtain Proposition 1.1.

Now take the Lie superalgebra $\mathfrak{q}_{N+M-1}$. As a subalgebra of the Lie superalgebra $\mathfrak{q}_{N+M}$, it is spanned by the elements F_{ij} where $|i|,|j| < N+M$. In particular, the subalgebras $\mathfrak{q}_N$ and $\mathfrak{q}_{M-1}$ of $\mathfrak{q}_{N+M-1}$ are spanned by the elements F_{ij} where $|i|,|j| \leqslant N$ and $N < |i|,|j| < N+M$, respectively. The enveloping algebra $\mathrm{U}(\mathfrak{q}_{N+M-1})$ and its subalgebra A^{M-1}_N will be also regarded as subalgebras in the associative algebra $\mathrm{U}(\mathfrak{q}_{N+M})$. We assume that $M \geqslant 1$ and $\mathrm{A}^0_N = \mathrm{U}(\mathfrak{q}_N)$.

Denote by I_{N+M} the right ideal in the algebra $\mathrm{U}(\mathfrak{q}_{N+M})$ generated by the elements

$$F_{N+M,\pm 1}\,,\ \ldots,\ F_{N+M,\pm(N+M)}\,. \tag{1.13}$$

Lemma 1.3. *(a) the intersection* $\mathrm{I}_{N+M} \cap \mathrm{A}^M_N$ *is a two-sided ideal of* A^M_N;

(b) there is a decomposition $\mathrm{A}^M_N = \mathrm{A}^{M-1}_N \oplus (\,\mathrm{I}_{N+M} \cap \mathrm{A}^M_N\,)$.

We prove this lemma in Section 3. Using Part (b) of the lemma, denote by α_M the projection of A^M_N to its direct summand A^{M-1}_N. By (a), the map $\alpha_M : \mathrm{A}^M_N \to \mathrm{A}^{M-1}_N$ is a homomorphism of associative algebras. The proof of the next proposition will also be given in Section 3.

Proposition 1.4. *For any* $n \geqslant 1$ *and any* $i\,, j$ *such that* $|i|,|j| \leqslant N$ *we have*

$$\alpha_M(F^{(n)}_{ij\,|\,N+M}) = F^{(n)}_{ij\,|\,N+M-1} \quad \text{and} \quad \alpha_M(C^{(n)}_{N+M}) = C^{(n)}_{N+M-1}\,.$$

The standard filtration (2.32) on the enveloping algebra $\mathrm{U}(\mathfrak{q}_{N+M})$ defines a filtration on its subalgebra A^M_N. By definition, the map $\alpha_M : \mathrm{A}^M_N \to \mathrm{A}^{M-1}_N$ preserves that filtration. It also preserves the $\mathbb{Z}_2$-gradation, inherited from $\mathrm{U}(\mathfrak{q}_{N+M})$. Using the homomorphisms $\alpha_1, \alpha_2, \ldots$ define an algebra A_N as the inverse limit of the sequence $\mathrm{A}^0_N, \mathrm{A}^1_N, \mathrm{A}^2_N, \ldots$ in the category of associative filtered algebras. The main result of this article is an explicit description of the algebra A_N in terms of generators and relations.

By definition, an element of A_N is any sequence of elements $Z_0, Z_1, Z_2, \ldots$ of the algebras $\mathrm{A}^0_N, \mathrm{A}^1_N, \mathrm{A}^2_N, \ldots$, respectively, such that $\alpha_M(Z_M) = Z_{M-1}$ for each $M \geqslant 1$, and the filtration degrees of the elements in the sequence are bounded. Utilising

Proposition 1.4, for any indices $n = 1, 2, \ldots$ and any $i, j = \pm 1, \ldots, \pm N$ define an element $F_{ij}^{(n)} \in \mathrm{A}_N$ as the sequence

$$F_{ij\,|\,N}^{(n)}, \; F_{ij\,|\,N+1}^{(n)}, \; F_{ij\,|\,N+2}^{(n)}, \; \ldots . \tag{1.14}$$

Further, for any $n = 1, 3, \ldots$ define an element $C^{(n)} \in \mathrm{A}_N$ as the sequence

$$C_N^{(n)}, \; C_{N+1}^{(n)}, \; C_{N+2}^{(n)}, \; \ldots . \tag{1.15}$$

The filtration degree of every element in (1.14) and (1.15) does not exceed n.

Note that the algebra A_N is unital and comes with a $\mathbb{Z}_2$-gradation, such that for all possible indices n and i, j we have

$$\deg C^{(n)} = 0 \quad \text{and} \quad \deg F_{ij}^{(n)} = \bar{\imath} + \bar{\jmath} \, . \tag{1.16}$$

By their definition, the elements $C^{(1)}, C^{(3)}, \ldots \in \mathrm{A}_N$ are central. Due to (1.8)

$$F_{-i,-j}^{(n)} = (-1)^{n-1} \, F_{ij}^{(n)} . \tag{1.17}$$

Theorem 1.5. *(a) The algebra A_N is generated by the elements $C^{(1)}, C^{(3)}, \ldots$ and $F_{ij}^{(1)}, F_{ij}^{(2)}, \ldots$.*

(b) The central elements $C^{(1)}, C^{(3)}, \ldots$ of A_N are algebraically independent.

(c) Together with the centrality and algebraic independence of $C^{(1)}, C^{(3)}, \ldots$, the defining relations of the $\mathbb{Z}_2$-graded algebra A_N are (1.17) and

$$\begin{aligned}
&[\, F_{ij}^{(m)}, F_{kl}^{(n)} \,] \\
&\quad = F_{il}^{(m+n-1)} \, \delta_{kj} - (-1)^{(\bar{\imath}+\bar{\jmath})(\bar{k}+\bar{l})} \, \delta_{il} \, F_{kj}^{(m+n-1)} \\
&\qquad + (-1)^{m-1} \, (\, F_{-i,l}^{(m+n-1)} \, \delta_{-k,j} - (-1)^{(\bar{\imath}+\bar{\jmath})(\bar{k}+\bar{l})} \, \delta_{i,-l} \, F_{k,-j}^{(m+n-1)} \,) \\
&\qquad + (-1)^{\bar{\jmath}\bar{k}+\bar{\jmath}\bar{l}+\bar{k}\bar{l}} \sum_{r=1}^{\min(m,n)-1} (\, F_{il}^{(m+n-r-1)} \, F_{kj}^{(r)} - F_{il}^{(r)} \, F_{kj}^{(m+n-r-1)} \,) \\
&\qquad + (-1)^{\bar{\jmath}\bar{k}+\bar{\jmath}\bar{l}+\bar{k}\bar{l}+\bar{k}+\bar{l}} \\
&\qquad \times \sum_{r=1}^{\min(m,n)-1} (-1)^{m+r} (\, F_{-i,l}^{(m+n-r-1)} \, F_{-k,j}^{(r)} - F_{i,-l}^{(r)} \, F_{k,-j}^{(m+n-r-1)} \,)
\end{aligned} \tag{1.18}$$

where $m, n = 1, 2, \ldots$ and $i, j, k, l = \pm 1, \ldots, \pm N$.

The proof will be given in Section 3. In particular, Theorem 1.5 shows that the algebra A_N is isomorphic to the tensor product of its two subalgebras, generated by the elements $C^{(1)}, C^{(3)}, \ldots$ and by the elements $F_{ij}^{(1)}, F_{ij}^{(2)}, \ldots$ respectively. Denote the latter subalgebra by B_N, it is a $\mathbb{Z}_2$-graded subalgebra.

The algebra B_N appeared in [N3] in the following guise. Let us consider the associative unital $\mathbb{Z}_2$-graded algebra $\mathrm{Y}(\mathfrak{q}_N)$ over the field $\mathbb{C}$ with the countable set of generators $T_{ij}^{(n)}$ where $n = 1, 2, \ldots$ and $i, j = \pm 1, \ldots, \pm N$. The $\mathbb{Z}_2$-gradation on the algebra $\mathrm{Y}(\mathfrak{q}_N)$ is determined by setting $\deg T_{ij}^{(n)} = \bar{\imath} + \bar{\jmath}$ for any $n \geqslant 1$. To write down the defining relations for these generators, put

$$T_{ij}(x) = \delta_{ij} \cdot 1 + T_{ij}^{(1)} x^{-1} + T_{ij}^{(2)} x^{-2} + \ldots$$

where x is a formal parameter, so that $T_{ij}(x) \in \mathrm{Y}(\mathfrak{q}_N)[[x^{-1}]]$. Then for all possible indices i, j, k, l we have the relations

$$T_{-i,-j}(x) = T_{ij}(-x), \tag{1.19}$$

$$\begin{aligned} &(x^2 - y^2) \cdot [\, T_{ij}(x), T_{kl}(y) \,] \cdot (-1)^{\bar{\imath}\bar{k} + \bar{\imath}\bar{l} + \bar{k}\bar{l}} \\ &\quad = (x + y) \cdot (T_{kj}(x)\, T_{il}(y) - T_{kj}(y)\, T_{il}(x)) \\ &\qquad - (x - y) \cdot (T_{-k,j}(x)\, T_{-i,l}(y) - T_{k,-j}(y)\, T_{i,-l}(x)) \cdot (-1)^{\bar{k} + \bar{l}} \end{aligned} \tag{1.20}$$

where y is a formal paramater independent of x, so that (1.20) is an equality in the algebra of formal Laurent series in x^{-1}, y^{-1} with coefficients in $\mathrm{Y}(\mathfrak{q}_N)$.

The algebra $\mathrm{Y}(\mathfrak{q}_N)$ is called the *Yangian* of the Lie superalgebra $\mathfrak{q}_N$. Note that the centre of the associative superalgebra $\mathrm{Y}(\mathfrak{q}_N)$ with $N \geqslant 1$ is not trivial. For a description of the centre of $\mathrm{Y}(\mathfrak{q}_N)$ see [N3, Section 3]. In particular, all central elements of $\mathrm{Y}(\mathfrak{q}_N)$ have $\mathbb{Z}_2$-degree 0. In our Section 3 we will prove

Proposition 1.6. *The assignment $F_{ij}^{(n)} \mapsto (-1)^{\bar{\imath}}\, T_{ji}^{(n)}$ for any $n = 1, 2, \ldots$ and $i, j = \pm 1, \ldots, \pm N$ extends to an anti-isomorphism of $\mathbb{Z}_2$-graded algebras*

$$\omega : \mathrm{B}_N \to \mathrm{Y}(\mathfrak{q}_N).$$

Now denote by ω_{N+M} the principal anti-automorphism of the enveloping algebra $\mathrm{U}(\mathfrak{q}_{N+M})$. It preserves the subalgebra $\mathrm{U}(\mathfrak{q}_M) \subset \mathrm{U}(\mathfrak{q}_{N+M})$. So it also preserves the centralizer $\mathrm{A}_N^M \subset \mathrm{U}(\mathfrak{q}_{N+M})$ of that subalgebra. For any $M \geqslant 0$ let $\pi_M : \mathrm{A}_N \to \mathrm{A}_N^M$ be the canonical homomorphism. By definition,

$$\pi_M(F_{ij}^{(n)}) = F_{ij\,|\,N+M}^{(n)} \tag{1.21}$$

for any $n = 1, 2, \ldots$ and $i, j = \pm 1, \ldots, \pm N$. Using Proposition 1.6, we can define a homomorphism $\tau_M : \mathrm{Y}(\mathfrak{q}_N) \to \mathrm{A}_N^M$ by the equality

$$\tau_M \circ \omega = \omega_{N+M} \circ (\pi_M \,|\, \mathrm{B}_N).$$

By (1.21),

$$\tau_M(T_{ij}^{(n)}) = (-1)^{\bar{\jmath}}\, \omega_{N+M}(F_{ji\,|\,N+M}^{(n)}).$$

Using the homomorphisms τ_M for all $M = 0, 1, 2, \ldots$, one can define a family of irreducible finite-dimensional $\mathrm{Y}(\mathfrak{q}_N)$-modules; see [N4, Section 1] and [P]. Another family of irreducible finite-dimensional $\mathrm{Y}(\mathfrak{q}_N)$-modules can be defined by using the results of [N1] and [N3, Section 5]. It should be possible to give a parametrization of all irreducible finite-dimensional $\mathrm{Y}(\mathfrak{q}_N)$-modules, similarly to the parametrization of the irreducible finite-dimensional $\mathrm{Y}(\mathfrak{gl}_N)$-modules as given by V. Drinfeld; see [D2, Theorem 2] and [M].

It was shown in [N3] that the associative $\mathbb{Z}_2$-graded algebra $\mathrm{Y}(\mathfrak{q}_N)$ has a natural Hopf superalgebra structure. In particular, the homomorphism of comultiplication $\mathrm{Y}(\mathfrak{q}_N) \to \mathrm{Y}(\mathfrak{q}_N) \otimes \mathrm{Y}(\mathfrak{q}_N)$ can be defined by

$$T_{ij}(x) \mapsto \sum_k T_{ik}(x) \otimes T_{kj}(x) \cdot (-1)^{(\bar\imath+\bar k)(\bar\jmath+\bar k)} \tag{1.22}$$

where the tensor product is over the subalgebra $\mathbb{C}[[x^{-1}]]$ of $\mathrm{Y}(\mathfrak{q}_N)[[x^{-1}]]$, and k runs through $\pm 1, \ldots, \pm N$. See [N3, Section 2] for the definitions of the the counit map $\mathrm{Y}(\mathfrak{q}_N) \to \mathbb{C}$ and the antipodal map $\mathrm{Y}(\mathfrak{q}_N) \to \mathrm{Y}(\mathfrak{q}_N)$.

There is a distinguished ascending $\mathbb{Z}$-filtration on the associative algebra $\mathrm{Y}(\mathfrak{q}_N)$. It is obtained by assigning to every generator $F_{ij}^{(n)}$ the degree $n-1$. The corresponding $\mathbb{Z}$-graded algebra will be denoted by $\mathrm{gr}\,\mathrm{Y}(\mathfrak{q}_N)$. Let $G_{ij}^{(n)}$ be the element of $\mathrm{gr}\,\mathrm{Y}(\mathfrak{q}_N)$ corresponding to the generator $F_{ij}^{(n)} \in \mathrm{Y}(\mathfrak{q}_N)$. The algebra $\mathrm{gr}\,\mathrm{Y}(\mathfrak{q}_N)$ inherits the $\mathbb{Z}_2$-gradation from the algebra $\mathrm{Y}(\mathfrak{q}_N)$, so that

$$\deg G_{ij}^{(n)} = \bar\imath + \bar\jmath\,.$$

Moreover, $\mathrm{gr}\,\mathrm{Y}(\mathfrak{q}_N)$ inherits from $\mathrm{Y}(\mathfrak{q}_N)$ the Hopf superalgebra structure. It follows from the definition (1.22) that with respect to the homomorphism of comultiplication $\mathrm{gr}\,\mathrm{Y}(\mathfrak{q}_N) \to \mathrm{gr}\,\mathrm{Y}(\mathfrak{q}_N) \otimes \mathrm{gr}\,\mathrm{Y}(\mathfrak{q}_N)$, for any $n \geqslant 1$ we have

$$G_{ij}^{(n)} \mapsto G_{ij}^{(n)} \otimes 1 + 1 \otimes G_{ij}^{(n)}.$$

On the other hand, for arbitrary Lie superalgebra $\mathfrak{a}$, a comultiplication map $\mathrm{U}(\mathfrak{a}) \to \mathrm{U}(\mathfrak{a}) \otimes \mathrm{U}(\mathfrak{a})$ can be defined for $X \in \mathfrak{a}$ by $X \mapsto X \otimes 1 + 1 \otimes X$, and then extended to a homomorphism of $\mathbb{Z}_2$-graded associative algebras by using the convention (1.2). Let us now consider the enveloping algebra $\mathrm{U}(\mathfrak{g})$ of the *twisted polynomial current* Lie superalgebra

$$\mathfrak{g} = \{\, X(t) \in \mathfrak{gl}_{N|N}[t] \;:\; \eta(X(t)) = X(-t)\,\}\,.$$

Here we employ the automorphism (1.1) of the Lie superalgebra $\mathfrak{gl}_{N|N}$. As a vector space, $\mathfrak{g}$ is spanned by the elements

$$E_{ij}\, t^n + E_{-i,-j}\,(-t)^n \tag{1.23}$$

where $n = 0, 1, 2 \ldots$ and $i, j = \pm 1, \ldots, \pm N$. Note that the $\mathbb{Z}_2$-degree of the element (1.23) equals $\bar\imath + \bar\jmath$. The algebra $\mathrm{U}(\mathfrak{g})$ also has a natural $\mathbb{Z}$-gradation, such that the degree of the element (1.23) is n.

It turns out that $\mathrm{U}(\mathfrak{g})$ and $\mathrm{gr}\,\mathrm{Y}(\mathfrak{q}_N)$ are isomorphic as Hopf superalgebras. By [N3, Theorem 2.3] their isomorphism $\mathrm{U}(\mathfrak{g}) \to \mathrm{gr}\,\mathrm{Y}(\mathfrak{q}_N)$ can be defined by mapping the element (1.23) of the algebra $\mathrm{U}(\mathfrak{g})$ to the element $(-1)^{\bar{\imath}+1}\,G_{ji}^{(n+1)}$ of the algebra $\operatorname{gr}\mathrm{Y}(\mathfrak{q}_N)$. Moreover, $\mathrm{Y}(\mathfrak{q}_N)$ is a deformation of $\mathrm{U}(\mathfrak{g})$ as a Hopf superalgebra; see the end of [N3, Section 2] for details.

Let us finish this introductory section with a few remarks of an historical nature. Our construction of the algebra A_N follows a similar construction for the general linear Lie algebra $\mathfrak{gl}_N$ instead of the queer Lie superalgebra $\mathfrak{q}_N$, due to G. Olshanski [O1, O2]. It was he who first considered the inverse limit of the sequence of centralizers of $\mathfrak{gl}_M$ in the enveloping algebras $\mathrm{U}(\mathfrak{gl}_{N+M})$ for $M = 1, 2, \ldots$. Following a suggestion of B. Feigin, he then described the inverse limit in terms of the Yangian $\mathrm{Y}(\mathfrak{gl}_N)$ of the Lie algebra $\mathfrak{gl}_N$. The latter Yangian is a deformation of the enveloping algebra of the polynomial current Lie algebra $\mathfrak{gl}_N[t]$ in the class of Hopf algebras [D1].

The elements (1.7) of $\mathrm{U}(\mathfrak{q}_N)$ were initially considered by A. Sergeev [S1], in order to describe the centre of the superalgebra $\mathrm{U}(\mathfrak{q}_N)$. The homomorphisms $\alpha_M : \mathrm{A}_N^M \to \mathrm{A}_N^{M-1}$ for $M = 1, 2, \ldots$ and the elements $F_{ij}^{(1)}, F_{ij}^{(2)}, \ldots$ of the inverse limit algebra A_N were introduced by M. Nazarov following [O1, O2]. He then identified the algebra *defined* by the relations (1.19) and (1.20), as a deformation of the enveloping algebra $\mathrm{U}(\mathfrak{g})$ in the class of Hopf superalgebras [N3]. It was also explained in [N3] why this deformation should be called the Yangian of $\mathfrak{q}_N$. Discovery of this deformation led to the construction [N1] of analogues of the classical Young symmetrizers for projective representations of the symmetric groups. However, Theorems 1.2 and 1.5 were only conjectured by M. Nazarov. The purpose of the present article is to prove these conjectures.

We hope these remarks indicate importance of the role that G. Olshanski played at various stages of our work. We are very grateful to him for friendly advice. This work was finished while M. Nazarov stayed at the Max Planck Institute of Mathematics in Bonn. He is grateful to the Institute for hospitality. M. Nazarov has been also supported by the EC grant MRTN-CT-2003-505078.

2. Proof of Theorem 1.2

We will use basic properties of complex semisimple associative superalgebras and their modules [J]. Most of these properties were first established in [W], in a generality greater than we need in the present article. We will also use the following simple lemma. Its proof carries over almost verbatim from the ungraded case, but we shall include the proof for the sake of completeness. Let A be any finite-dimensional $\mathbb{Z}_2$-graded associative algebra over the complex field $\mathbb{C}$. Let G be a finite group of automorphisms of A. The crossed product algebra $G \ltimes \mathrm{A}$ is also $\mathbb{Z}_2$-graded: for any $g \in G$ we have $\deg g = 0$ in $G \ltimes \mathrm{A}$.

Lemma 2.1. *Suppose the superalgebra* A *is semisimple. Then the superalgebra* $G \ltimes \mathrm{A}$ *is also semisimple.*

Proof. We will write $G \ltimes \mathrm{A} = \mathrm{B}$. Let V be any module over the superalgebra B, and let $\rho : \mathrm{B} \to \operatorname{End}(V)$ be the corresponding homomorphism. Here we assume that the homomorphism ρ preserves $\mathbb{Z}_2$-gradation. Let $U \subset V$ be any B-submodule. Since A is semisimple, we have the decomposition $V = U \oplus U'$ into a direct sum of A-modules for some $U' \subset V$. Let $P \in \mathrm{End}(V)$ be the projection onto U along U'. Put

$$S = \frac{1}{|G|} \sum_{g \in G} \rho(g)\, P\, \rho(g)^{-1} \in \mathrm{End}(V)\,.$$

For any element $X \in \mathrm{A}$ let X^g be its image under the automorphism g. Then

$$\begin{aligned} \rho(X)\, S &= \frac{1}{|G|} \sum_{g \in G} \rho(X)\rho(g)\, P\, \rho(g)^{-1} = \frac{1}{|G|} \sum_{g \in G} \rho(g)\rho(X^g)\, P\, \rho(g)^{-1} \\ &= \frac{1}{|G|} \sum_{g \in G} \rho(g)\, P\, \rho(X^g)\rho(g)^{-1} = \frac{1}{|G|} \sum_{g \in G} \rho(g)\, P\, \rho(g)^{-1} \rho(X) = S\, \rho(X)\,. \end{aligned}$$

For any $h \in G$ we also have $\rho(h)\, S = S\, \rho(h)$ by the definition of S. Note that $S \in \mathrm{End}(V)$ is of $\mathbb{Z}_2$-degree 0, as is P. So $\ker S \subset V$ is a B-submodule.

Since U is a B-submodule, we have the equalities $P\rho(g)P = \rho(g)P$ for all $g \in G$. Using the definition of S, these equalities imply that $SP = P$ and $PS = S$. The latter pair of equalities guarantees that $\mathrm{im}\, S = \mathrm{im}\, P = U$ and $S^2 = S$. So $V = U \oplus \ker S$. Since V is an arbitrary module over the superalgebra B, this superalgebra is semisimple by [J, Proposition 2.4]. □

We will also need the following general "double centralizer theorem". Let V be a $\mathbb{Z}_2$-graded complex vector space. The associative algebra $\mathrm{End}(V)$ is then also $\mathbb{Z}_2$-graded. Take any subalgebra B in the superalgebra $\mathrm{End}(V)$. Here we assume that B as a vector space splits into the direct sum of its subspaces of $\mathbb{Z}_2$-degrees 0 and 1. Denote by B' the centralizer of B in the superalgebra $\mathrm{End}(V)$.

Proposition 2.2. *Suppose that the superalgebra* B *is finite dimensional and semisimple. Also suppose that the* B'*-module* V *is finitely generated. Then* $\mathrm{B} = \mathrm{B}''$ *in* $\mathrm{End}(V)$.

Proof. First, we will establish an analogue of the Jacobson density theorem [L, Theorem XVII.1] for the superalgebra B. Namely, we shall prove that for any homogeneous $v_1, \ldots, v_n \in V$ and $X \in \mathrm{B}''$, there exists $Y \in \mathrm{B}$ such that $Xv_r = Yv_r$ for any index $r = 1, \ldots, n$. Then we will choose $v_1, \ldots, v_n$ to be homogeneous generators of V over B'. By writing any vector $v \in V$ as the sum $Z_1 v_1 + \ldots + Z_n v_n$ for some homogeneous $Z_1, \ldots, Z_n \in \mathrm{B}'$, we will get

$$\begin{aligned} Xv &= \sum_{r=1}^{n} XZ_r v_r = \sum_{r=1}^{n} (-1)^{\deg X \deg Z_r} Z_r X v_r \\ &= \sum_{r=1}^{n} (-1)^{\deg X \deg Z_r} Z_r Y v_r = \sum_{r=1}^{n} YZ_r v_r = Yv\,. \end{aligned}$$

Along with the obvious embedding $\mathrm{B} \subset \mathrm{B}''$, this will prove Proposition 2.2.

Recall that the $\mathbb{Z}_2$-graded vector space $\overline{V}$ *opposite* to V is obtained from V by changing the $\mathbb{Z}_2$-gradation deg to deg+1. Define the action of the algebra B in $\overline{V}$ as the pullback of its action in V via the involutive automorphism $Y \mapsto (-1)^{\deg Y}\, Y$, where Y is any homogeneous element of the $\mathbb{Z}_2$-graded algebra B. Now consider the direct sum of B-modules

$$W = \bigoplus_{r=1}^{n} V_r\,,$$

where the B-module V_r equals V or $\overline{V}$ depending on whether $\deg v_r$ in V is 0 or 1. For each index $r = 1, \ldots, n$, we have an embedding of vector spaces $A_r : V \to W$, and a projection $B_r : W \to V$. Note that the $\mathbb{Z}_2$-degrees of the linear maps A_r and B_r coincide with that of the vector v_r. We also have the equality $B_p\, A_q = \delta_{pq} \cdot \mathrm{id}$ in $\mathrm{End}(V)$, and the equality

$$\sum_{r=1}^{n} A_r\, B_r = \mathrm{id}$$

in $\mathrm{End}(W)$. Any homogeneous element $Y \in \mathrm{B}$ acts in W as the linear operator

$$\widetilde{Y} = \sum_{r=1}^{n} (-1)^{\deg v_r \deg Y} A_r\, Y\, B_r\,.$$

Given $X \in \mathrm{B}''$, put

$$\widetilde{X} = \sum_{r=1}^{n} (-1)^{\deg v_r \deg X} A_r\, X\, B_r\,.$$

Any homogeneous element $Z \in \mathrm{End}(W)$ can be written as

$$Z = \sum_{p,q=1}^{n} A_p\, Z_{pq}\, B_q$$

for some homogeneous $Z_{pq} \in \mathrm{End}(V)$, where $\deg Z_{pq} = \deg Z + \deg v_p + \deg v_q$. Then

$$\widetilde{Y} Z = \sum_{r=1}^{n} (-1)^{\deg v_r \deg Y} A_r\, Y\, B_r \cdot \sum_{p,q=1}^{n} A_p\, Z_{pq}\, B_q \qquad (2.24)$$

$$= \sum_{p,q=1}^{n} (-1)^{\deg v_p \deg Y} A_p\, Y Z_{pq}\, B_q\,,$$

$$Z\, \widetilde{Y} = \sum_{p,q=1}^{n} A_p\, Z_{pq}\, B_q \cdot \sum_{r=1}^{n} (-1)^{\deg v_r \deg Y} A_r\, Y\, B_r \qquad (2.25)$$

$$= \sum_{p,q=1}^{n} (-1)^{\deg v_q \deg Y} A_p\, Z_{pq}\, Y\, B_q\,.$$

Suppose that the element $Z \in \mathrm{End}(W)$ belongs to the centralizer of the image of the superalgebra B in $\mathrm{End}(W)$. Due to (2.24) and (2.25), the assumption $[\widetilde{Y}, Z] = 0$ is equivalent to the collection of equalities $[Y, Z_{pq}] = 0$ for all $p, q = 1, \dots, n$. Since here $Y \in \mathrm{B}$ is arbitrary and $X \in \mathrm{B}''$, we then have $[X, Z_{pq}] = 0$ for all $p, q = 1, \dots, n$. A calculation similar to (2.24) and (2.25) then shows that $[\widetilde{X}, Z] = 0$ in $\mathrm{End}(W)$.

Now take the vector

$$w = \sum_{r=1}^{n} A_r v_r \in W\,,$$

it has $\mathbb{Z}_2$-degree 0. Let U be the cyclic span of the vector w under the action of B. Since the superalgebra B is finite-dimensional semisimple, we have the decomposition $W = U \oplus U'$ into a direct sum of B-modules for some $U' \subset W$; see [J, Proposition 2.4]. Choose the element $Z \in \mathrm{End}(W)$ from the centralizer of the image of B to be the projector onto U' along U. Here $\deg Z = 0$. Then $Z\widetilde{X}w = \widetilde{X}Zw = 0$, and $\widetilde{X}w \in U$. So there exists $Y \in \mathrm{B}$ such that $\widetilde{X}w = \widetilde{Y}w$. The last equality means that $Xv_r = Yv_r$ for each $r = 1, \dots, n$. □

In the notation of Proposition 2.2, we have the following corollary.

Corollary 2.3. *Suppose that the vector space V is finite dimensional, and that the superalgebra* $\mathrm{B} \subset \mathrm{End}(V)$ *is semisimple. Then* $\mathrm{B} = \mathrm{B}''$ *in* $\mathrm{End}(V)$.

Now for any integer $n \geqslant 1$ consider the tensor product $(\mathrm{End}(V))^{\otimes n}$ of n copies of the $\mathbb{Z}_2$-graded algebra $\mathrm{End}(V)$. This tensor product is a $\mathbb{Z}_2$-graded associative algebra defined using the conventions (1.2) and (1.3). The proof of the following lemma is also included for the sake of completeness.

Lemma 2.4. *Suppose that* B *contains the identity* $1 \in \mathrm{End}(V)$. *Then the centralizer of* $\mathrm{B}^{\otimes n}$ *in the superalgebra* $(\mathrm{End}(V))^{\otimes n}$ *coincides with* $(\mathrm{B}')^{\otimes n}$.

Proof. We will use the induction on n. If $n = 1$, the statement of Lemma 2.4 is tautological. Suppose that $n > 1$. The centralizer of $\mathrm{B}^{\otimes n}$ contains $(\mathrm{B}')^{\otimes n}$ due to the conventions (1.2) and (1.3). Now suppose that for some homogeneous elements $X_1, \dots, X_l \in \mathrm{End}(V)$ and $Y_1, \dots, Y_l \in (\mathrm{End}(V))^{\otimes(n-1)}$ the sum of the products $X_1 Y_1 + \dots + X_l Y_l$ belongs to the centralizer of $\mathrm{B}^{\otimes n}$. In particular, then for any homogeneous $Y \in \mathrm{B}^{\otimes(n-1)}$ we have

$$\begin{aligned}
\sum_{k=1}^{l} X_k \otimes (Y_k Y) &= \sum_{k=1}^{l} (X_k \otimes Y_k)(1 \otimes Y) \\
&= \sum_{k=1}^{l} (-1)^{(\deg X_k + \deg Y_k)\deg Y} (1 \otimes Y)(X_k \otimes Y_k) \\
&= \sum_{k=1}^{l} (-1)^{\deg Y_k \deg Y} X_k \otimes (Y Y_k)\,.
\end{aligned}$$

We may assume that the elements $X_1, \ldots, X_l$ are linearly independent. Then the above equalities imply that for any k the element $Y_k \in (\mathrm{End}\,(V))^{\otimes\,(n-1)}$ belongs to the centralizer of $\mathrm{B}^{\otimes\,(n-1)}$. Then $Y_k \in (\mathrm{B}')^{\otimes\,(n-1)}$ by the induction assumption. A similar argument shows that $X_k \in \mathrm{B}'$ for any index k. □

For any $N \geqslant 0$ and $M \geqslant 1$ take the $\mathbb{Z}_2$-graded vector space $\mathbb{C}^{\,N+M\,|\,N+M}$. Identify the $\mathbb{Z}_2$-graded vector spaces $\mathbb{C}^{N|N}$ and $\mathbb{C}^{M|M}$ with the subspaces in $\mathbb{C}^{\,N+M\,|\,N+M}$ spanned by the vectors e_i where respectively $i = -N, \ldots, -1, 1, \ldots, N$ and $i = -N-M, \ldots, -N-1, N+1, \ldots, N+M$. The decomposition

$$\mathbb{C}^{N|N} = \mathbb{C}^{N|N} \oplus \mathbb{C}^{M|M} \tag{2.26}$$

determines the embeddings of the Lie superalgebras $\mathfrak{gl}_{N|N}$ and $\mathfrak{gl}_{M|M}$ into $\mathfrak{gl}_{\,N+M\,|\,N+M}$, and of their subalgebras $\mathfrak{q}_N$ and $\mathfrak{q}_M$ into the Lie algebra $\mathfrak{q}_{\,N+M}$.

Let us now regard $\mathrm{End}\,(\mathbb{C}^{N|N})$ and Q_M as subalgebras in the associative superalgebra $\mathrm{End}\,(\mathbb{C}^{\,N+M\,|\,N+M})$, using the decomposition (2.26). The elements

$$\sum_{|i|>N} E_{ii} \quad \text{and} \quad \sum_{|i|>N} (-1)^{\bar{\imath}}\, E_{i,-i}$$

of $\mathrm{End}\,(\mathbb{C}^{\,N+M\,|\,N+M})$ span a subalgebra, isomorphic to the associative algebra Q_1. Using this isomorphism, a direct calculation shows that the centralizer of Q_M in $\mathrm{End}\,(\mathbb{C}^{\,N+M\,|\,N+M})$ coincides with $\mathrm{End}\,(\mathbb{C}^{N|N}) \oplus \mathrm{Q}_1$. This centralizer is a semisimple associative $\mathbb{Z}_2$-graded algebra, denote it by C.

For any integer $n \geqslant 1$ consider the tensor product $V = (\mathbb{C}^{\,N+M\,|\,N+M})^{\otimes n}$ of n copies of the $\mathbb{Z}_2$-graded vector space $\mathbb{C}^{\,N+M\,|\,N+M}$. Identify the algebras

$$(\mathrm{End}\,(\mathbb{C}^{\,N+M\,|\,N+M}))^{\otimes n} \quad \text{and} \quad \mathrm{End}\,((\mathbb{C}^{\,N+M\,|\,N+M})^{\otimes n}) \tag{2.27}$$

so that for any homogeneous elements $X_1, \ldots, X_n \in \mathrm{End}\,(\mathbb{C}^{\,N+M\,|\,N+M})$ and any homogeneous vectors $u_1, \ldots, u_n \in \mathbb{C}^{\,N+M\,|\,N+M}$

$$(X_1 \otimes \ldots \otimes X_n)\,(u_1 \otimes \ldots \otimes u_n) = (-1)^d\, X_1 u_1 \otimes \ldots \otimes X_n u_n$$

where

$$d = \sum_{1 \leqslant p < q \leqslant n} \deg u_p \, \deg X_q \,.$$

By identifying the two algebras we determine an action on V of the subalgebra

$$\mathrm{C}^{\otimes n} \subset (\mathrm{End}\,(\mathbb{C}^{\,N+M\,|\,N+M}))^{\otimes n}. \tag{2.28}$$

The symmetric group S_n acts on V so that for any adjacent transposition $\sigma_p = (p\,, p+1)$

$$\begin{aligned} &\sigma_p\,(u_1 \otimes \ldots \otimes u_p \otimes u_{p+1} \otimes \ldots \otimes u_n) \\ &\quad = (-1)^{\deg u_p \deg u_{p+1}}\, u_1 \otimes \ldots \otimes u_{p+1} \otimes u_p \otimes \ldots \otimes u_n \,. \end{aligned}$$

The action of S_n and $\mathrm{C}^{\otimes n}$ on V extends to that of $S_n \ltimes \mathrm{C}^{\otimes n}$. Here the group S_n acts by automorphisms of the algebra $(\mathrm{End}(\mathbb{C}^{N+M\,|\,N+M}))^{\otimes n}$ so that

$$\begin{aligned}\sigma_p\,(X_1 \otimes \ldots \otimes X_p \otimes X_{p+1} \otimes \ldots \otimes X_n)\\ = (-1)^{\deg X_p \deg X_{p+1}}\, X_1 \otimes \ldots \otimes X_{p+1} \otimes X_p \otimes \ldots \otimes X_n\,.\end{aligned} \tag{2.29}$$

Now consider the action of enveloping algebra $\mathrm{U}(\mathfrak{q}_N)$ on the vector space V, as of a subalgebra of the $\mathbb{Z}_2$-graded associative algebra $\mathrm{U}(\mathfrak{gl}_{N+M\,|\,N+M})$. We use the comultiplication

$$\mathrm{U}(\mathfrak{gl}_{N+M\,|\,N+M}) \to \mathrm{U}(\mathfrak{gl}_{N+M\,|\,N+M})^{\otimes n} \tag{2.30}$$

along with the identification of the two algebras in (2.27).

Proposition 2.5. *The centralizer of the image of* $\mathrm{U}(\mathfrak{q}_M)$ *in the associative superalgebra* $\mathrm{End}(V)$ *coincides with the image of* $S_n \ltimes \mathrm{C}^{\otimes n}$.

Proof. The $\mathbb{Z}_2$-graded algebra $\mathrm{C}^{\otimes n}$ is semisimple, see [J, Proposition 2.10]. So is the crossed product $S_n \ltimes \mathrm{C}^{\otimes n}$, see our Lemma 2.1. Let B be the image of the crossed product in $\mathrm{End}(V)$. The $\mathbb{Z}_2$-graded algebra B is semisimple too. By Corollary 2.3 it suffices to prove that the centralizer of B in $\mathrm{End}(V)$ coincides with the image of $\mathrm{U}(\mathfrak{q}_M)$.

Let us describe the latter image. Consider the span in $\mathrm{End}(\mathbb{C}^{N+M\,|\,N+M})$ of $\mathfrak{q}_M$ and of the identity element 1. It is a subalgebra in $\mathrm{End}(\mathbb{C}^{N+M\,|\,N+M})$, denote this subalgebra by D. We will prove that the invariant subalgebra

$$(\mathrm{D}^{\otimes n})^{S_n} \subset (\mathrm{End}(\mathbb{C}^{N+M\,|\,N+M}))^{\otimes n}$$

coincides with the image of $\mathrm{U}(\mathfrak{q}_M)$. Here we use the definition (2.29), and the comultiplication (2.30). There is no need to identify the two algebras (2.27) here.

Due to the convention (1.2) and the definition of the comultiplication (2.30), the image of $\mathrm{U}(\mathfrak{q}_M)$ in $(\mathrm{End}(\mathbb{C}^{N+M\,|\,N+M}))^{\otimes n}$ is contained in $(\mathrm{D}^{\otimes n})^{S_n}$. Now for any n elements $X_1, \ldots, X_n \in \mathrm{D}$ consider their *symmetrized* tensor product

$$\langle X_1, \ldots, X_n \rangle = \frac{1}{n!} \sum_{\sigma \in S_n} \sigma\,(X_1 \otimes \ldots \otimes X_n)\,. \tag{2.31}$$

Suppose that for some $p \in \{1, \ldots, n\}$ we have $X_q \in \mathbb{C}\,1$ if and only if $p < q$. By induction on p, let us prove that $\langle X_1, \ldots, X_p, 1, \ldots, 1\rangle$ belongs to the image of $\mathrm{U}(\mathfrak{q}_M)$ in $(\mathrm{End}(\mathbb{C}^{N+M\,|\,N+M}))^{\otimes n}$. This is evident if $p = 1$. If $p > 1$, then

$$\langle X_1, \ldots, X_{p-1}, 1, \ldots, 1\rangle\,\langle X_p, 1, \ldots, 1\rangle$$

equals

$$\frac{n-p+1}{n}\,\langle X_1, \ldots, X_p, 1, \ldots, 1\rangle$$

plus a sum of certain symmetrized tensor products in $(\mathrm{End}(\mathbb{C}^{N+M\,|\,N+M}))^{\otimes n}$ which belong to the image of $\mathrm{U}(\mathfrak{q}_M)$ by the induction assumption.

To prove that the centralizer of B in End(V) coincides with the image of $\mathrm{U}(\mathfrak{q}_M)$, it now suffices to show that the centralizer of the subalgebra (2.28) coincides with $\mathrm{D}^{\otimes n}$. But we have $\mathrm{C}' = \mathrm{D}$ in $\mathrm{End}(\mathbb{C}^{N+M\,|\,N+M})$ by definition. Using Lemma 2.4 we now complete the proof of Proposition 2.5. □

Consider the symmetric algebra $\mathrm{S}(\mathfrak{q}_{N+M})$ of the Lie superalgebra $\mathfrak{q}_{N+M}$. The standard filtration

$$\mathbb{C} = \mathrm{U}^0(\mathfrak{q}_{N+M}) \subset \mathrm{U}^1(\mathfrak{q}_{N+M}) \subset \mathrm{U}^2(\mathfrak{q}_{N+M}) \subset \ldots \tag{2.32}$$

on the algebra $\mathrm{U}(\mathfrak{q}_{N+M})$ determines for each $n = 1, 2, \ldots$ a linear map

$$\mathrm{U}^n(\mathfrak{q}_{N+M}) \,/\, \mathrm{U}^{n-1}(\mathfrak{q}_{N+M}) \;\to\; \mathrm{S}^n(\mathfrak{q}_{N+M})\,, \tag{2.33}$$

which is bijective by the Poincaré–Birkhoff–Witt theorem for Lie superalgebras [MM, Theorem 5.15]. Using (2.33), for all indices $i, j = \pm 1, \ldots, \pm(N+M)$ define an element $f^{(n)}_{ij\,|\,N+M} \in \mathrm{S}^n(\mathfrak{q}_{N+M})$ as the image of the element

$$F^{(n)}_{ij\,|\,N+M} \in \mathrm{U}^n(\mathfrak{q}_{N+M})\,;$$

the latter element is the sum (1.7) where each of the indices $k_1, \ldots, k_{n-1}$ runs through $\pm 1, \ldots, \pm(N+M)$. Define $c^{(n)}_{N+M} \in \mathrm{S}^n(\mathfrak{q}_{N+M})$ as the image of

$$C^{(n)}_{N+M} \in \mathrm{U}^n(\mathfrak{q}_{N+M})\,;$$

see (1.11). Note that if n is even, then $C^{(n)}_{N+M} = 0$ and hence $c^{(n)}_{N+M} = 0$.

Now consider adjoint action of the Lie superalgebra $\mathfrak{q}_{N+M}$ on $\mathrm{S}(\mathfrak{q}_{N+M})$. In particular, consider the action of $\mathfrak{q}_M$ on $\mathrm{S}(\mathfrak{q}_{N+M})$ as that of a subalgebra of $\mathfrak{q}_{N+M}$. Then take the subalgebra of invariants $\mathrm{S}(\mathfrak{q}_{N+M})^{\mathfrak{q}_M} \subset \mathrm{S}(\mathfrak{q}_{N+M})$.

Proposition 2.6. *The subalgebra* $\mathrm{S}(\mathfrak{q}_{N+M})^{\mathfrak{q}_M}$ *is generated by the elements* $c^{(1)}_{N+M}$, $c^{(3)}_{N+M}, \ldots$ *and* $f^{(1)}_{ij\,|\,N+M}, f^{(2)}_{ij\,|\,N+M}, \ldots$ *where* $|i|, |j| \leqslant N$.

Proof. Consider $(\mathrm{End}(\mathbb{C}^{N+M\,|\,N+M}))^{\otimes n}$ as a vector space. Define a linear map φ_n from this space to the nth symmetric power $\mathrm{S}^n(\mathfrak{q}_{N+M})$ by setting

$$\varphi_n(E_{i_1 j_1} \otimes \ldots \otimes E_{i_n j_n}) = F_{i_1 j_1} \ldots F_{i_n j_n} \tag{2.34}$$

for any indices $i, j = \pm 1, \ldots, \pm(N+M)$. The map φ_n commutes with the natural action of the Lie superalgebra $\mathfrak{q}_{N+M}$ on $(\mathrm{End}(\mathbb{C}^{N+M\,|\,N+M}))^{\otimes n}$ and $\mathrm{S}^n(\mathfrak{q}_{N+M})$. The map φ_n has a right inverse linear map ψ_n which commutes with the action of $\mathfrak{q}_{N+M}$ as well. Namely, using (2.31) we set

$$\psi_n(F_{i_1 j_1} \ldots F_{i_n j_n}) = 2^{-n}\, \langle F_{i_1 j_1}, \ldots, F_{i_n j_n} \rangle\,.$$

It follows that the subspace of $\mathfrak{q}_M$-invariants of the $\mathbb{Z}$-degree n in $\mathrm{S}(\mathfrak{q}_{N+M})$,

$$\mathrm{S}^n(\mathfrak{q}_{N+M})^{\mathfrak{q}_M} = \varphi_n(((\mathrm{End}(\mathbb{C}^{N+M\,|\,N+M}))^{\otimes n})^{\mathfrak{q}_M}) = \varphi_n(\mathrm{B})$$

by Proposition 2.5; here B is the image of the crossed product $S_n \ltimes \mathrm{C}^{\otimes n}$ in

$$(\mathrm{End}(\mathbb{C}^{N+M\,|\,N+M}))^{\otimes n} = \mathrm{End}((\mathbb{C}^{N+M\,|\,N+M})^{\otimes n}).$$

The vector subspace $\mathrm{C} \subset \mathrm{End}(\mathbb{C}^{N+M\,|\,N+M})$ is spanned by the identity element 1, the elements E_{ij} where $|i|, |j| \leqslant N$ and by the element

$$J = \sum_i (-1)^{\bar{\imath}}\, E_{i,-i}$$

where the summation index i runs through $\pm 1, \ldots, \pm(N+M)$. For each $p = 1, \ldots, n$ introduce the elements of the algebra $(\mathrm{End}(\mathbb{C}^{N+M\,|\,N+M}))^{\otimes n}$,

$$E_{ij}^{(p)} = 1^{\otimes(p-1)} \otimes E_{ij} \otimes 1^{\otimes(n-p)} \quad \text{and} \quad J_p = 1^{\otimes(p-1)} \otimes J \otimes 1^{\otimes(n-p)}.$$

The vector subspace $\mathrm{B} \subset (\mathrm{End}(\mathbb{C}^{N+M\,|\,N+M}))^{\otimes n}$ is spanned by the products of the form

$$J_{q_1} \ldots J_{q_b}\, E_{i_1 j_1}^{(p_1)} \ldots E_{i_a j_a}^{(p_a)}\, H \tag{2.35}$$

where

$$1 \leqslant p_1 < \ldots < p_a \leqslant n, \quad 1 \leqslant q_1 < \ldots < q_b \leqslant n,$$
$$\{p_1, \ldots, p_a\} \cap \{q_1, \ldots, q_b\} = \varnothing,$$
$$|i_1|, |j_1|, \ldots, |i_a|, |j_a| \leqslant N$$

and H is the image in $(\mathrm{End}(\mathbb{C}^{N+M\,|\,N+M}))^{\otimes n}$ of some permutation from S_n.

By the definition of the symmetric algebra $\mathrm{S}(\mathfrak{q}_{N+M})$ and due to (2.29), we have the identity $\varphi_n \circ \sigma = \varphi_n$ for any $\sigma \in S_n$. Therefore it suffices to compute the φ_n-image of the element (2.35) where the factor H corresponds to a permutation of the form

$$(1, 2, \ldots, r_1)\,(r_1+1, r_1+2, \ldots, r_2) \ldots (r_c+1, r_c+2, \ldots, n)$$

where $c \geqslant 0$ and $0 < r_1 < r_2 < \ldots < r_c < n$. But for any $p = 1, \ldots, n-1$ we have the relation

$$\varphi_p(X)\, \varphi_{n-p}(Y) = \varphi_n(X \otimes Y), \tag{2.36}$$
$$X \in (\mathrm{End}(\mathbb{C}^{N+M\,|\,N+M}))^{\otimes p} \quad \text{and} \quad Y \in (\mathrm{End}(\mathbb{C}^{N+M\,|\,N+M}))^{\otimes(n-p)}.$$

Hence it suffices to consider only the case where H corresponds to the cyclic permutation $(1, 2, \ldots, n)$. Suppose this is the case. Then we may assume that $p_1 = 1$ or $a = 0$. Here we use the identity $\varphi_n \circ \sigma = \varphi_n$ for $\sigma = (1, 2, \ldots, n)$.

For the cyclic permutation $(1, 2, \ldots, n)$ we have

$$H = \sum_{k_1, \ldots, k_n} (-1)^{\bar{k}_1 + \ldots + \bar{k}_n - 1}\, E_{k_n k_1}^{(1)} E_{k_1 k_2}^{(2)} \ldots E_{k_{n-1} k_n}^{(n)} \tag{2.37}$$

where each of the indices $k_1, \dots, k_n$ runs through $\pm 1, \dots, \pm(N+M)$. Hence

$$E^{(1)}_{ij} H = \sum_{k_1,\dots,k_{n-1}} (-1)^{\bar{k}_1 + \dots + \bar{k}_{n-1}} E^{(1)}_{k_n k_1} E^{(2)}_{k_1 k_2} \dots E^{(n)}_{k_{n-1} k_n} . \tag{2.38}$$

Further, for any $p = 1, \dots, n-1$ we have the equality

$$E^{(1)}_{ij} E^{(p+1)}_{kl} H = (-1)^{\bar{j}\bar{k} + \bar{j}\bar{l} + \bar{k}\bar{l}} E^{(1)}_{il} E^{(p+1)}_{kj} H' H''$$

where the factors H' and H'' are the images in $(\mathrm{End}(\mathbb{C}^{N+M\,|\,N+M}))^{\otimes n}$ of the cyclic permutations $(1, \dots, p)$ and $(p+1, \dots, n)$. Using this equality together with the relation (2.36), we reduce the case $a > 1$ to the case $a = 1$.

Suppose that $a = 1$ and $p_1 = 1$. In this case, the φ_n-image of (2.35) equals

$$\sum_{k_1,\dots,k_{n-1}} (-1)^{\bar{k}_1 + \dots + \bar{k}_{n-1} + b\bar{\imath} + q_1 + \dots + q_b} F_{(-1)^b i, k_1} F_{k_1 k_2} \dots F_{k_{n-1} j}$$
$$= (-1)^{b\bar{\imath} + q_1 + \dots + q_b} f^{(n)}_{(-1)^b i,\, j\,|\,N+M} ;$$

we use the equality (2.38), the definition (2.34) and the identity $F_{-k,-l} = F_{kl}$.

It remains to consider the case $a = 0$. Then the φ_n-image of (2.35) equals

$$(-1)^{q_1 + \dots + q_b} \sum_{k_1,\dots,k_n} (-1)^{\bar{k}_1 + \dots + \bar{k}_{n-1} + b\bar{k}_n} F_{(-1)^b k_n, k_1} F_{k_1 k_2} \dots F_{k_{n-1} k_n} ;$$

we use the equality (2.37), the definition (2.34) and the identity $F_{-k,-l} = F_{kl}$. Denote by f the sum over $k_1, \dots, k_n$ in the above display. If the number b is even, then $f = c^{(n)}_{N+M}$. Now suppose that the number b is odd, so that f is

$$\sum_{k_1,\dots,k_n} (-1)^{\bar{k}_1 + \dots + \bar{k}_{n-1} + \bar{k}_n} F_{-k_n, k_1} F_{k_1 k_2} \dots F_{k_{n-1} k_n} . \tag{2.39}$$

By changing the signs of the indices $k_1, \dots, k_n$ in (2.39) and then using the identity $F_{-k,-l} = F_{kl}$ one shows that $f = (-1)^n f$. So $f = 0$ if n is odd. Since

$$\deg F_{-k_n, k_1} = \bar{k}_1 + \bar{k}_n + 1 = \deg(F_{k_1 k_2} \dots F_{k_{n-1} k_n}) + 1 ,$$

the element (2.39) of the symmetric algebra $\mathrm{S}(\mathfrak{q}_{N+M})$ can be also written as

$$\sum_{k_1,\dots,k_n} (-1)^{\bar{k}_1 + \dots + \bar{k}_{n-1} + \bar{k}_n} F_{k_1 k_2} \dots F_{k_{n-1} k_n} F_{-k_n, k_1} . \tag{2.40}$$

By changing the signs of the indices $k_2, \dots, k_n$ in (2.40) and then using the identity $F_{-k,-l} = F_{kl}$ one shows that $f = (-1)^{n-1} f$. So $f = 0$ if n is even. □

By using the fact that for each $n \geqslant 1$ the linear map (2.33) commutes with the adjoint action of the Lie superalgebra $\mathfrak{q}_{N+M}$ on $\mathrm{U}^n(\mathfrak{q}_{N+M})$ and $\mathrm{S}^n(\mathfrak{q}_{N+M})$, we can now complete the proof of Theorem 1.2. Take any element $X \in \mathrm{A}^M_N$. We have

$X \in \mathrm{U}^n(\mathfrak{q}_{N+M})$ for some $n \geqslant 0$. Let us demonstrate by induction on n that X belongs to the subalgebra in $\mathrm{U}(\mathfrak{q}_{N+M})$ generated by the elements $C^{(1)}_{N+M}, C^{(3)}_{N+M}, \ldots$ and (1.12). Since $\mathrm{U}^0(\mathfrak{q}_{N+M}) = \mathbb{C}$, here we can take $n \geqslant 1$ and make the induction assumption. By Proposition 2.6, the image of X in $\mathrm{S}^n(\mathfrak{q}_{N+M})$ under the map (2.33) is a linear combination of the products of the elements $c^{(1)}_{N+M}, c^{(3)}_{N+M}, \ldots$ and $f^{(1)}_{ij\,|\,N+M}, f^{(2)}_{ij\,|\,N+M}, \ldots$ where $|i|, |j| \leqslant N$. By replacing these elements of $\mathrm{S}(\mathfrak{q}_{N+M})$ respectively by $C^{(1)}_{N+M}, C^{(3)}_{N+M}, \ldots$ and (1.12) in the linear combination, we obtain a certain element $Y \in \mathrm{U}(\mathfrak{q}_{N+M})$ such that $X - Y \in \mathrm{U}^{(n-1)}(\mathfrak{q}_{N+M})$. We also have $Y \in \mathrm{A}^M_N$. By applying the induction assumption to the difference $X - Y$, we complete the the proof.

3. Proof of Theorem 1.5

Here we prove Theorem 1.5 along with Lemma 1.3 and Propositions 1.4 and 1.6.

Proof of Lemma 1.3. Together with the right ideal I_{N+M} generated by the elements (1.13), consider the left ideal J_{N+M} in $\mathrm{U}(\mathfrak{q}_{N+M})$ generated by the elements

$$F_{\pm 1,\, N+M}\,, \ \ldots\,, \ F_{\pm(N+M),\, N+M}\,.$$

By the Poincaré–Birkhoff–Witt theorem for Lie superalgebras, every element $X \in \mathrm{U}(\mathfrak{q}_{N+M})$ can be uniquely written as a sum of the products of the form

$$\begin{aligned} &F^{\,p_1}_{N+M,\,1}\, F^{\,p_{-1}}_{N+M,\,-1} \cdots F^{\,p_{N+M-1}}_{N+M,\,N+M-1}\, F^{\,p_{-N-M+1}}_{N+M,\,-N-M+1} \\ &\quad\times F^{\,p_{N+M}}_{N+M,\,N+M}\, F^{\,p_{-N-M}}_{N+M,\,-N-M}\; Y \\ &\quad\times F^{\,q_1}_{1,\,N+M}\, F^{\,q_{-1}}_{-1,\,N+M} \cdots F^{\,q_{N+M-1}}_{N+M-1,\,N+M}\, F^{\,q_{-N-M+1}}_{-N-M+1,\,N+M} \end{aligned} \tag{3.1}$$

where each of the exponents p_k and q_k runs through $0, 1, 2, \ldots$ or through $0, 1$ if $k > 0$ or $k < 0$ respectively, whereas the factor $Y \in \mathrm{U}(\mathfrak{q}_{N+M-1})$ depends on these exponents. Note that here

$$[\, F_{N+M,\,N+M}\,, Y\,] = 0 \quad \text{and} \quad [\, F_{-N-M,\,N+M}\,, Y\,] = 0\,. \tag{3.2}$$

Now suppose that $X \in \mathrm{A}^M_N$. In particular, then we have

$$[\, F_{N+M,\,N+M}\,, X\,] = 0 \tag{3.3}$$

since $M \geqslant 1$ by our assumption. Note that if $|\,k\,| < N + M$, then due to (1.6)

$$\begin{aligned} [\, F_{N+M,\,N+M}\,, F_{N+M,\,k}\,] &= F_{N+M,k}\,, \\ [\, F_{N+M,\,N+M}\,, F_{k,\,N+M}\,] &= -\,F_{N+M,k}\,. \end{aligned}$$

If $|k| = N + M$, then

$$[\, F_{N+M,\,N+M}\,,\, F_{k,\,N+M}\,] = 0\,.$$

Hence the condition (3.3) implies that X is a sum of the products (3.1) where

$$\begin{aligned} &p_1 + p_{-1} + \ldots + p_{N+M-1} + p_{-N-M+1} \\ &\quad = q_1 + q_{-1} + \ldots + q_{N+M-1} + q_{-N-M+1}\,. \end{aligned} \tag{3.4}$$

The intersection $\mathrm{I}_{N+M} \cap \mathrm{A}_N^M$ consists of those elements $X \in \mathrm{A}_N^M$ which are sums of the products (3.1) where

$$p_1 + p_{-1} + \ldots + p_{N+M} + p_{-N-M} > 0\,.$$

Due to the equality (3.4), the latter inequality is equivalent to

$$q_1 + q_{-1} + \ldots + q_{N+M-1} + q_{-N-M+1} + p_{N+M} + p_{-N-M} > 0\,.$$

So by using (3.2),

$$\mathrm{I}_{N+M} \cap \mathrm{A}_N^M = \mathrm{J}_{N+M} \cap \mathrm{A}_N^M\,. \tag{3.5}$$

In particular, the intersection $\mathrm{I}_{N+M} \cap \mathrm{A}_N^M$ is a two-sided ideal of A_N^M. Thus we get Part (a) of Lemma 1.3.

Furthemore, due to (3.4) there is only one summand (3.1) of $X \in \mathrm{A}_N^M$ with

$$p_1 + p_{-1} + \ldots + p_{N+M} + p_{-N-M} = 0\,, \tag{3.6}$$

this summand has the form of $Y \in \mathrm{U}(\mathfrak{q}_{N+M-1})$. Note that the right ideal I_{N+M} of $\mathrm{U}(\mathfrak{q}_{N+M})$ is stable under the adjoint action of the subalgebra $\mathfrak{q}_{N+M-1} \subset \mathfrak{q}_{N+M}$. Indeed, if $|i|, |j| < N + M$, then by (1.6) we have

$$[\, F_{ij}\,,\, F_{N+M,l}\,] = -(-1)^{\,\bar{\imath}\,\bar{l} + \bar{\jmath}\,\bar{l}}\; \delta_{il}\, F_{N+M,\,j} \;-\; (-1)^{\,\bar{\imath}\,\bar{l} + \bar{\jmath}\,\bar{l}}\; \delta_{i,-l}\, F_{N+M,\,-j}$$

for any index $l = \pm 1\,, \ldots\,, \pm (N+M)$. In particular, I_{N+M} is stable under the adjoint action of $\mathfrak{q}_{M-1}$. So the condition $[\,Z\,,\,X\,] = 0$ on X for any $Z \in \mathfrak{q}_{M-1}$ implies the condition $[\,Z\,,\,Y\,] = 0$ on the summand Y of X corresponding to (3.6). Therefore $Y \in \mathrm{A}_N^{M-1}$, and we get Part (b) of Lemma 1.3. □

Proof of Proposition 1.4. Suppose that $|i|, |j| \leqslant N$. Let us prove by induction on $n = 1, 2, \ldots$ that the differences

$$F_{ij\,|\,N+M}^{(n)} - F_{ij\,|\,N+M-1}^{(n)} \quad \text{and} \quad C_{N+M}^{(n)} - C_{N+M-1}^{(n)} \tag{3.7}$$

belong to the left ideal J_{N+M} in $\mathrm{U}(\mathfrak{q}_{N+M})$. Due to (3.5), Proposition 1.4 will then follow. Neither of the elements $F_{ij\,|\,N+M}^{(n)}$ and $C_{N+M}^{(n)}$, nor the ideal J_{N+M} depend on the partition of the number $N + M$ into N and M. Hence it suffices to consider the case $M = 1$. Note that according to the definition (1.11)

$$C_{N+1}^{(n)} - C_N^{(n)} = \sum_{|k|\leqslant N} (F_{kk\,|\,N+1}^{(n)} - F_{kk\,|\,N}^{(n)}) + F_{N+1,N+1\,|\,N+1}^{(n)} + F_{-N-1,-N-1\,|\,N+1}^{(n)}$$

where the last two summands belong to J_{N+1}, by their definition. Therefore it suffices to consider only the first of the differences (3.7), where $M = 1$.

If $n = 1$, that difference is zero. Now suppose that $n > 1$, and make the induction assumption. Using the relation (1.9),

$$F_{ij\,|\,N+1}^{(n)} - F_{ij\,|\,N}^{(n)} = \sum_{|k|\leqslant N} (-1)^{\bar{k}} F_{ik} (F_{kj\,|\,N+1}^{(n-1)} - F_{kj\,|\,N}^{(n-1)}) + F_{i\,,N+1} F_{N+1,\,j\,|\,N+1}^{(n-1)} - F_{i\,,-N-1} F_{-N-1,\,j\,|\,N+1}^{(n-1)} \,.$$

At the right-hand side of the last equality, the summands corresponding to $|k| \leqslant N$ belong to the left ideal J_{N+1} by the induction assumption. Using (1.10) and (1.8), the remainder of the right-hand side is equal to the sum

$$(-1)^{\bar{\imath}\bar{\jmath}} F_{N+1,\,j\,|\,N+1}^{(n-1)} F_{i\,,N+1} - (-1)^{(\bar{\imath}+1)(\bar{\jmath}+1)} F_{-N-1,\,j\,|\,N+1}^{(n-1)} F_{i\,,-N-1}$$
$$+ (-1)^{\bar{\imath}\bar{\jmath}+1} (1 + (-1)^{\bar{\imath}+\bar{\jmath}+n}) \, \delta_{ij} \, F_{N+1,\,N+1\,|\,N+1}^{(n-1)}$$
$$+ (-1)^{(\bar{\imath}+1)(\bar{\jmath}+1)} (1 + (-1)^{\bar{\imath}+\bar{\jmath}+n}) \, \delta_{i,-j} \, F_{-N-1,\,N+1\,|\,N+1}^{(n-1)} \,.$$

But in this sum, every summand evidently belongs to the left ideal J_{N+1}. □

Let us now prove Theorem 1.5. First, we will verify the formula (1.18) for the supercommutator $[\, F_{ij}^{(m)},\, F_{kl}^{(n)} \,]$ in the algebra A_N. We will use

Proposition 3.1. *In* $\mathrm{U}(\mathfrak{q}_N)$ *for* $m\,,n = 1,2,\,\ldots$ *and* $i\,,j\,,k\,,l = \pm 1,\,\ldots,\,\pm N$

$$[\, F_{ij\,|\,N}^{(m)}, F_{kl\,|\,N}^{(n)} \,] = F_{il\,|\,N}^{(m+n-1)} \delta_{kj} - (-1)^{(\bar{\imath}+\bar{\jmath})(\bar{k}+\bar{l})} \delta_{il} F_{kj\,|\,N}^{(m+n-1)}$$
$$+ (-1)^{m-1} (F_{-i,l\,|\,N}^{(m+n-1)} \delta_{-k,j}$$
$$- (-1)^{(\bar{\imath}+\bar{\jmath})(\bar{k}+\bar{l})} \delta_{i,-l} F_{k,-j\,|\,N}^{(m+n-1)}) + (-1)^{\bar{\jmath}\bar{k}+\bar{\jmath}\bar{l}+\bar{k}\bar{l}}$$
$$\times \sum_{r=1}^{m-1} (F_{il\,|\,N}^{(n+r-1)} F_{kj\,|\,N}^{(m-r)} - F_{il\,|\,N}^{(m-r)} F_{kj\,|\,N}^{(n+r-1)})$$
$$+ (-1)^{\bar{\jmath}\bar{k}+\bar{\jmath}\bar{l}+\bar{k}\bar{l}+\bar{k}+\bar{l}}$$
$$\times \sum_{r=1}^{m-1} (-1)^r (F_{-i,l\,|\,N}^{(n+r-1)} F_{-k,j\,|\,N}^{(m-r)} - F_{i,-l\,|\,N}^{(m-r)} F_{k,-j\,|\,N}^{(n+r-1)}) \,.$$

Proof. The formula for the supercommutator $[\, F_{ij\,|\,N}^{(m)},\, F_{kl\,|\,N}^{(n)} \,]$ in the $\mathbb{Z}_2$-graded algebra $\mathrm{U}(\mathfrak{q}_N)$ displayed above is easy to verify by using the induction on m. When $m = 1$, this formula coincides with (1.10). Now make the induction assumption. Let the index h run through $\pm 1, \ldots, \pm N$. Then due to (1.9),

$$
\begin{aligned}
[\, F^{(m+1)}_{ij\,|\,N}, F^{(n)}_{kl\,|\,N} \,] &= \sum_h (-1)^{\bar h}\, [\, F_{ih}\, F^{(m)}_{hj\,|\,N}, F^{(n)}_{kl\,|\,N} \,] \\
&= \sum_h (-1)^{\bar h} F_{ih} [F^{(m)}_{hj\,|\,N}, F^{(n)}_{kl\,|\,N}\,] + \sum_h (-1)^{\bar h + (\bar h + \bar\jmath)(\bar k + \bar l)}\, [\, F_{ih}, F^{(n)}_{kl\,|\,N}\,]\, F^{(m)}_{hj\,|\,N} \\
&= \sum_h (-1)^{\bar h} F_{ih}\, (\, F^{(m+n-1)}_{hl\,|\,N}\, \delta_{kj} - (-1)^{(\bar h + \bar\jmath)(\bar k + \bar l)}\, \delta_{hl}\, F^{(m+n-1)}_{kj\,|\,N} \\
&\quad + (-1)^{m-1} (\, F^{(m+n-1)}_{-h,l\,|\,N}\, \delta_{-k,j} - (-1)^{(\bar h + \bar\jmath)(\bar k + \bar l)}\, \delta_{h,-l}\, F^{(m+n-1)}_{k,-j\,|\,N} \,) \\
&\quad + (-1)^{\bar\jmath \bar k + \bar\jmath \bar l + \bar k \bar l} \sum_{r=1}^{m-1} (\, F^{(n+r-1)}_{hl\,|\,N}\, F^{(m-r)}_{kj\,|\,N} - F^{(m-r)}_{hl\,|\,N}\, F^{(n+r-1)}_{kj\,|\,N} \,) \\
&\quad + (-1)^{\bar\jmath \bar k + \bar\jmath \bar l + \bar k \bar l + \bar k + \bar l} \sum_{r=1}^{m-1} (-1)^r (F^{(n+r-1)}_{-h,l\,|\,N}\, F^{(m-r)}_{-k,j\,|\,N} - F^{(m-r)}_{h,-l\,|\,N}\, F^{(n+r-1)}_{k,-j\,|\,N} \,)) \\
&\quad + \sum_h (-1)^{\bar h + (\bar h + \bar\jmath)(\bar k + \bar l)} (\, \delta_{kh}\, F^{(n)}_{il\,|\,N} - (-1)^{(\bar\imath + \bar h)(\bar k + \bar l)}\, \delta_{il}\, F^{(n)}_{kh\,|\,N} \\
&\quad + \delta_{-k,h}\, F^{(n)}_{-i,l\,|\,N} - (-1)^{(\bar\imath + \bar h)(\bar k + \bar l)}\, \delta_{i,-l}\, F^{(n)}_{k,-h\,|\,N} \,)\, F^{(m)}_{hj\,|\,N} .
\end{aligned}
$$

Here we used the induction assumption with the index i replaced by h, and the relation (1.10) with the index j replaced by h. Using the relation (1.9) repeatedly, the right-hand side of the above displayed equalities equals

$$
\begin{aligned}
& F^{(m+n)}_{il\,|\,N}\, \delta_{kj} - (-1)^{(\bar\imath + \bar\jmath)(\bar k + \bar l)}\, \delta_{il}\, F^{(m+n)}_{kj\,|\,N} \\
&\quad + (-1)^m (\, F^{(m+n)}_{-i,l\,|\,N}\, \delta_{-k,j} - (-1)^{(\bar\imath + \bar\jmath)(\bar k + \bar l)}\, \delta_{i,-l}\, F^{(m+n)}_{k,-j\,|\,N}) \\
&\quad + (-1)^{\bar\jmath \bar k + \bar\jmath \bar l + \bar k \bar l} (\, F^{(n)}_{il\,|\,N}\, F^{(m)}_{kj\,|\,N} - F^{(1)}_{il\,|\,N}\, F^{(m+n-1)}_{kj\,|\,N}) \\
&\quad + (-1)^{\bar\jmath \bar k + \bar\jmath \bar l + \bar k \bar l + \bar k + \bar l} ((-1)^{m+1}\, F^{(1)}_{i,-l\,|\,N}\, F^{(m+n-1)}_{k,-j\,|\,N} - F^{(n)}_{-i,l\,|\,N}\, F^{(m)}_{-k,j\,|\,N}) \\
&\quad + (-1)^{\bar\jmath \bar k + \bar\jmath \bar l + \bar k \bar l} \sum_{r=1}^{m-1} (\, F^{(n+r)}_{il\,|\,N}\, F^{(m-r)}_{kj\,|\,N} - F^{(m-r+1)}_{il\,|\,N}\, F^{(n+r-1)}_{kj\,|\,N}) \\
&\quad + (-1)^{\bar\jmath \bar k + \bar\jmath \bar l + \bar k \bar l + \bar k + \bar l} \\
&\quad \times \sum_{r=1}^{m-1} (-1)^{r+1} (\, F^{(n+r)}_{-i,l\,|\,N}\, F^{(m-r)}_{-k,j\,|\,N} + F^{(m-r+1)}_{i,-l\,|\,N}\, F^{(n+r-1)}_{k,-j\,|\,N}).
\end{aligned}
$$

But the last displayed sum can also be obtained by replacing m by $m+1$ on the right-hand side of the equality in Proposition 3.1. Thus we have made the induction step. □

If $m > n$, then on the right-hand side of the equality in Proposition 3.1, the summands corresponding to the indices $r = 1, \ldots, m-n$ cancel in each of the two sums over $r = 1, \ldots, m-1$. In the first of the two sums, this is obvious. To cancel these summands in the second sum, one utilizes the relations (1.8). Hence if $m > n$, the summation over $r = 1, \ldots, m-1$ in Proposition 3.1 can be replaced by the summation over $r = m-n+1, \ldots, m-1$. Thus if we change the running index r to $m-r$, the latter index should run through $1, \ldots, \min(m,n)-1$. Using this remark, the relation (1.18) in Theorem 1.5 follows from Proposition 3.1.

In the remainder of the proof of Theorem 1.5, we will also make use of the next proposition. For any integers $M \geqslant 0$ and $n \geqslant 1$ consider the elements

$$c_{N+M}^{(n)} \quad \text{and} \quad f_{ij\,|\,N+M}^{(n)} \tag{3.8}$$

of the algebra $\mathrm{S}(\mathfrak{q}_{N+M})^{\mathfrak{q}_M}$, see Proposition 2.6. Fix any positive integer s.

Proposition 3.2. *Take the elements* $f_{ij\,|\,N+M}^{(n)}$ *where*

$$1 \leqslant n \leqslant s, \quad 1 \leqslant i \leqslant N, \quad 1 \leqslant |j| \leqslant N. \tag{3.9}$$

Along with these elements, take the elements $c_{N+M}^{(n)}$ *where* $1 \leqslant n \leqslant s$ *and* n *is odd. For any sufficiently large number* M*, all these elements are algebraically independent in the supercommutative algebra* $\mathrm{S}(\mathfrak{q}_{N+M})$.

Proof. We will use arguments from [MO, Subsection 2.11]. By the Poincaré–Birkhoff–Witt theorem for Lie superalgebras, the elements

$$F_{kl} = f_{kl\,|\,N+M}^{(1)}$$

where $k = 1, \ldots, N+M$ and $l = \pm 1, \ldots, \pm N+M$ are free generators of the supercommutative algebra $\mathrm{S}(\mathfrak{q}_{N+M})$. Let X_s be the quotient algebra of $\mathrm{S}(\mathfrak{q}_{N+M})$, defined by imposing the following relations on these free generators.

For every triple (i,j,n) satisfying the conditions (3.9), choose a subset

$$\mathcal{O}_{ij}^{(n)} \subset \{N+1, N+2, \ldots\}$$

of cardinality $n-1$ in such a way that all these subsets are disjoint. Let M be so large that all these subsets are contained in $\{N+1, \ldots, N+M-s\}$. If

$$\mathcal{O}_{ij}^{(n)} = \{l_1, \ldots, l_{n-1}\},$$

then put

$$F_{i\,l_1} = F_{l_1 l_2} = \ldots = F_{l_{n-2}\,l_{n-1}} = 1.$$

Denote by $x_{ij}^{(n)}$ the image of the element $F_{l_{n-1}\,j} \in \mathrm{S}(\mathfrak{q}_{N+M})$ in the algebra X_s. Having done this for every triple (i,j,n) satisfying the conditions (3.9), for every $r = 1, \ldots, s$ denote by x_r the image in X_s of the element

$$F_{N+M-s+r,\,N+M-s+r} \in \mathrm{S}(\mathfrak{q}_{N+M}).$$

Finally, put $F_{kl}=0$ if $k>0$ and $(k\,,l)$ is not one of the pairs

$$(i\,,l_1)\,,\,(l_1\,,l_2)\,,\,\ldots\,,\,(l_{n-2}\,,l_{n-1})\,,\,(l_{n-1}\,,j)$$

for any triple $(i\,,j\,,n)$ satisfying (3.9), and not one of the pairs

$$(N+M-s+r\,,N+M-s+r)\ \text{ where } r=1\,,\,\ldots\,,\,s\,.$$

The elements $x_1\,,\,\ldots\,,\,x_s$ and the elements $x_{ij}^{(n)}$ for all the triples $(i\,,j\,,n)$ satisfying (3.9) are free generators of the algebra X_s. For any of these triples, the image in X_s of

$$f_{ij\,|\,N+M}^{(n)}\in\mathrm{S}(\mathfrak{q}_{N+M})$$

equals $x_{ij}^{(n)}$ plus a certain linear combination of products of the elements $x_{kl}^{(m)}$ where $1\leqslant m<n$. For any odd n, the image in X_s of the element

$$c_{N+M}^{(n)}\in\mathrm{S}(\mathfrak{q}_{N+M})$$

equals

$$2^n(x_1^n+\ldots+x_s^n)$$

plus a linear combination of products of elements $x_{kl}^{(m)}$ where $1\leqslant m\leqslant n$. Hence all these images are algebraically independent in the quotient X_s of the supercommutative algebra $\mathrm{S}(\mathfrak{q}_{N+M})$. □

Let us show that the associative algebra A_N is generated by the elements $C^{(1)}$, $C^{(3)},\ldots$ and $F_{ij}^{(1)},F_{ij}^{(2)},\ldots$. Take any element $Z\in\mathrm{A}_N$, and consider its canonical image $Z_M=\pi_M(Z)\in\mathrm{A}_N^M$ for any $M\geqslant 0$. By Theorem 1.2, the element Z_M is a linear combination of the products of the elements $C_{N+M}^{(n)}$ where $n=1,3,\ldots$ and of the elements $F_{ij\,|\,N+M}^{(n)}$ where $n=1,2,\ldots,$ whereas $i=1,\ldots,N$ and $j=\pm1,\ldots,\pm N$; see (1.8). Choose any linear ordering of all these elements. Applying Proposition 3.1 to the algebra $\mathrm{U}(\mathfrak{q}_{N+M})$ instead of $\mathrm{U}(\mathfrak{q}_N)$, we will assume that any of the products in the linear combination Z_M is an ordered monomial in these elements. If $\bar\imath+\bar\jmath=1$, then the element $F_{ij\,|\,N+M}^{(n)}$ may appear in any of these monomials only with degree 1.

We will assume that for any $M\geqslant 1$, the map α_M preserves the ordering. Then for every monomial Y_M appearing in the linear combination Z_M, the monomial $\alpha_M(Y_M)$ may appear in the linear combination $\alpha_M(Z_M)=Z_{M-1}$.

The filtration degrees of the elements in the sequence $Z_0,Z_1,Z_2,\ldots$ are bounded. Hence for any factor $C_{N+M}^{(n)}$ or $F_{ij\,|\,N+M}^{(n)}$ of the monomials appearing in linear combination Z_M, we have $n\leqslant s$ for a certain integer s which does not depend on M. Then for a sufficiently large number M, the coefficients of the monomials appearing in the linear combinations $Z_M,Z_{M+1},Z_{M+2},\ldots$ are determined uniquely. The uniqueness follows from Proposition 3.2.

Now fix a sufficiently large number M, as above. Let Y_M be any monomial appearing in the linear combination Z_M, say with a coefficient $z\in\mathbb{C}$. We assume that

$z \neq 0$. Then for any integer $L > M$, the linear combination Z_L contains the summand $z\,Y_L$ where Y_L is a monomial and

$$(\alpha_{M+1} \circ \ldots \circ \alpha_L)(Y_L) = Y_M\,.$$

For any nonnegative integer $L \leqslant M$, define $Y_L = (\alpha_L \circ \ldots \circ \alpha_M)(Y_M)$. The sequence $Y_0, Y_1, Y_2, \ldots$ determines an element $Y \in \mathrm{A}_N$, which is a monomial in $C^{(1)}, C^{(3)}, \ldots$ and $F_{ij}^{(1)}, F_{ij}^{(2)}, \ldots$. Here $1 \leqslant i \leqslant N$ and $1 \leqslant |j| \leqslant N$. The element $Z \in \mathrm{A}_N$ is then a sum of the products of the form $z\,Y$. This sum is finite because any such product corresponds to a summand $z\,Y_M$ in the linear combination Z_M. Thus we have proved Part (a) of Theorem 1.5.

Let us now prove Parts (b) and (c). By definition, the algebra A_N comes with an ascending $\mathbb{Z}$-filtration, such that the generators $C^{(n)}$ and $F_{ij}^{(n)}$ of A_N have the degree n. Denote the corresponding $\mathbb{Z}$-graded algebra by $\mathrm{gr}\,\mathrm{A}_N$. Let $c^{(n)}$ and $f_{ij}^{(n)}$ be the generators of the algebra $\mathrm{gr}\,\mathrm{A}_N$ corresponding to $C^{(n)}$ and $F_{ij}^{(n)}$. We always assume that the index n in $C^{(n)}$ and $c^{(n)}$ is odd. We also assume that $|i|, |j| \leqslant N$ in $F_{ij}^{(n)}$ and $f_{ij}^{(n)}$. It follows from (1.17) that for any $n = 1, 2, \ldots$

$$f_{-i,-j}^{(n)} = (-1)^{n-1} f_{ij}^{(n)}.$$

The algebra $\mathrm{gr}\,\mathrm{A}_N$ also inherits from A_N a $\mathbb{Z}_2$-gradation, such that

$$\deg c^{(n)} = 0 \quad \text{and} \quad \deg f_{ij}^{(n)} = \bar{\imath} + \bar{\jmath}\,,$$

see (1.16). The relation (1.18) demonstrates that the $\mathbb{Z}_2$-graded algebra $\mathrm{gr}\,\mathrm{A}_N$ is supercommutative. To complete the proof of Theorem 1.5, it suffices to show that the elements $c^{(1)}, c^{(3)}, \ldots$ together with the elements $f_{ij}^{(1)}, f_{ij}^{(2)}, \ldots$ where $i = 1, \ldots, N$ and $j = \pm 1, \ldots, \pm N$ are algebraically independent in the supercommutative algebra $\mathrm{gr}\,\mathrm{A}_N$.

The algebra $\mathrm{gr}\,\mathrm{A}_N$ can also be obtained as an inverse limit of the sequence of the supercommutative algebras $\mathrm{S}(\mathfrak{q}_{N+M})^{\mathfrak{q}_M}$ where $M = 0, 1, 2, \ldots$. The limit is taken in the category of $\mathbb{Z}$-graded algebras. We assume that if $M = 0$, then $\mathrm{S}(\mathfrak{q}_{N+M})^{\mathfrak{q}_M} = \mathrm{S}(\mathfrak{q}_N)$. The definition of the surjective homomorphism

$$\mathrm{S}(\mathfrak{q}_{N+M})^{\mathfrak{q}_M} \to \mathrm{S}(\mathfrak{q}_{N+M-1})^{\mathfrak{q}_{M-1}}$$

for any $M \geqslant 1$ is similar to the definition of the surjective homomorphism $\alpha_M : \mathrm{A}_N^M \to \mathrm{A}_N^{M-1}$, see Lemma 1.3. Here we omit the details, but notice that the elements $c^{(n)}$ and $f_{ij}^{(n)}$ of $\mathrm{gr}\,\mathrm{A}_N$ correspond to the sequences of elements (3.8) of the algebras $\mathrm{S}(\mathfrak{q}_{N+M})^{\mathfrak{q}_M}$ where $M = 0, 1, 2, \ldots$. Proposition 3.2 now guarantees the algebraic independence of the elements $c^{(1)}, c^{(3)}, \ldots$ together with the elements $f_{ij}^{(1)}, f_{ij}^{(2)}, \ldots$ where $i = 1, \ldots, N$ and $j = \pm 1, \ldots, \pm N$.

Proof of Proposition 1.6. Under the correspondence $F_{ij}^{(n)} \mapsto (-1)^{\bar{\imath}}\, T_{ji}^{(n)}$, the collection of relations (1.17) in the algebra B_N for the indices $n = 1, 2, \ldots$ corresponds to the equality (1.19) in $\mathrm{Y}(\mathfrak{q}_N)[[u^{-1}]]$. Put $T_{ij}^{(0)} = \delta_{ij}$ for any $i, j = 1, \ldots, \pm N$. Using (1.4), the relation (1.18) in B_N then corresponds to

$$
\begin{aligned}
&(-1)^{\bar{\imath}\bar{\jmath}+\bar{\imath}\bar{k}+\bar{\jmath}\bar{k}}\,[\,T_{ji}^{(m)}, T_{lk}^{(n)}\,] \\
&\quad = \sum_{r=0}^{m-1} (\,T_{jk}^{(m+n-r-1)}\,T_{li}^{(r)} - T_{jk}^{(r)}\,T_{li}^{(m+n-r-1)}\,) + (-1)^{\bar{\imath}+\bar{\jmath}+1} \\
&\quad \times \sum_{r=0}^{m-1} (-1)^{m+r}\,(\,T_{-j,k}^{(m+n-r-1)}\,T_{-l,i}^{(r)} - T_{j,-k}^{(r)}\,T_{l,-i}^{(m+n-r-1)}\,). \qquad (3.10)
\end{aligned}
$$

Here we also used a remark on the summation over $r = 1, \ldots, m-1$ similar to that made immediately after the proof of Proposition 3.1.

Put $T_{ij}^{(-1)} = 0$ for any $i, j = 1, \ldots, \pm N$. The collection of relations (3.10) for $m, n = 1, 2, \ldots$ is equivalent to the collection of relations

$$
\begin{aligned}
&(-1)^{\bar{\imath}\bar{\jmath}+\bar{\imath}\bar{k}+\bar{\jmath}\bar{k}}\,(\,[\,T_{ji}^{(m+1)}, T_{lk}^{(n-1)}\,] - [\,T_{ji}^{(m-1)}, T_{lk}^{(n+1)}\,]\,) \\
&\quad = T_{jk}^{(n-1)}\,T_{li}^{(m)} - T_{jk}^{(m)}\,T_{li}^{(n-1)} + T_{jk}^{(n)}\,T_{li}^{(m-1)} - T_{jk}^{(m-1)}\,T_{li}^{(n)} + (-1)^{\bar{\imath}+\bar{\jmath}} \\
&\quad \times (T_{-j,k}^{(n-1)}\,T_{-l,i}^{(m)} - T_{j,-k}^{(m)}\,T_{l,-i}^{(n-1)} - T_{-j,k}^{(n)}\,T_{-l,i}^{(m-1)} + T_{j,-k}^{(m-1)}\,T_{l,-i}^{(n)}) \qquad (3.11)
\end{aligned}
$$

for $m, n = 0, 1, 2, \ldots$. Multiplying the relation (3.11) by $x^{1-m}\,y^{1-n}$ and taking the sum of resulting relations over $m, n = 0, 1, 2, \ldots$ we get the relation

$$
\begin{aligned}
&(x^2 - y^2) \cdot [\,T_{ji}(x), T_{lk}(y)\,] \cdot (-1)^{\bar{\imath}\bar{\jmath}+\bar{\imath}\bar{k}+\bar{\jmath}\bar{k}} \\
&\quad = (x + y) \cdot (T_{jk}(y)\,T_{li}(x) - T_{jk}(x)\,T_{li}(y)) \\
&\quad + (x - y) \cdot (T_{-j,k}(y)\,T_{-l,i}(x) - T_{j,-k}(x)\,T_{l,-i}(y)) \cdot (-1)^{\bar{\imath}+\bar{\jmath}}. \qquad (3.12)
\end{aligned}
$$

Using (1.5), we can rewrite the left-hand side of the relation (3.12) as

$$
(y^2 - x^2) \cdot [\,T_{lk}(y), T_{ji}(x)\,] \cdot (-1)^{\bar{\imath}\bar{\jmath}+\bar{\imath}\bar{l}+\bar{\jmath}\bar{l}}\,.
$$

Replacing in the resulting relation the indices i, j, k, l and the parameters x, y by l, k, j, i and y, x respectively, we obtain exactly the relation (1.20). Thus the defining relations of the subalgebra $\mathrm{B}_N \subset \mathrm{A}_N$ correspond to the defining relations of the algebra $\mathrm{Y}(\mathfrak{q}_N)$, see Theorem 1.5. □

References

[D1] V. Drinfeld, Hopf algebras and the quantum Yang-Baxter equation, *Soviet Math. Dokl.* **32** (1985), 254–258.

[D2] V. Drinfeld, A new realization of Yangians and quantized affine algebras, *Soviet Math. Dokl.* **36** (1988), 212–216.

[J] T. Józefiak, *Semisimple superalgebras*, Lecture Notes Math., Vol. 1352 (1988), 96–113.

[L] S. Lang, *Algebra*, Addison-Wesley, Reading MA, 1965.

[M] A. Molev, Finite-dimensional irreducible representations of twisted Yangians, *J. Math. Phys.* **39** (1998), 5559–5600.

[MM] J. Milnor and J. Moore, On the structure of Hopf algebras, *Ann. of Math.* **81** (1965), 211–264.

[MO] A. Molev and G. Olshanski, Centralizer construction for twisted Yangians, *Selecta Math.* **6** (2000), 269–317.

[N1] M. Nazarov, Young's symmetrizers for projective representations of the symmetric group, *Adv. Math.* **127** (1997), 190–257.

[N2] M. Nazarov, Capelli identities for Lie superalgebras, *Ann. Scient. Éc. Norm. Sup.* **30** (1997), 847–872.

[N3] M. Nazarov, Yangian of the queer Lie superalgebra, *Commun. Math. Phys.* **208** (1999), 195–223.

[N4] M Nazarov, Representations of twisted Yangians associated with skew Young diagrams, *Selecta Math.* **10** (2004), 71–129.

[O1] G. Olshanski, Extension of the algebra $U(g)$ for infinite-dimensional classical Lie algebras g, and the Yangians $Y(gl(m))$, *Soviet Math. Dokl.* **36** (1988), 569–573.

[O2] G. Olshanski, Representations of infinite-dimensional classical groups, limits of enveloping algebras, and Yangians, *Adv. Soviet Math.* **2** (1991), 1–66.

[P] I. Penkov, Characters of typical irreducible finite-dimensional $\mathfrak{q}(n)$-modules, *Funct. Anal. Appl.* **20** (1986), 30–37.

[S1] A. Sergeev, The centre of enveloping algebra for Lie superalgebra $Q(n, \mathbb{C})$, *Lett. Math. Phys.* **7** (1983), 177–179.

[S2] A. Sergeev, The tensor algebra of the identity representation as a module over the Lie superalgebras $GL(n, m)$ and $Q(n)$, *Math. Sbornik* **51** (1985), 419–427.

[W] T. Wall, Graded Brauer groups, *J. Reine Angew. Math.* **213** (1964), 187–199.

Definitio nova algebroidis verticiani

Vadim Schechtman

Laboratoire Emile Picard
Université Paul Sabatier
118 route de Narbonne
31062 Toulouse
France
schechtman@math.ups-tlse.fr

Antonio Ioseph sexagesimi anniversarii diei causa

Summary. An algebra of differential operators is the enveloping algebra of a Lie algebroid T of vector fields. Similarly, a vertex algebra of differential operators is the enveloping algebra of a vertex algebroid, which is a Lie algebroid equipped with certain complementary differential operators. These operators should satisfy some complicated identities, these identities being a corollary of the Borcherd's axioms of a vertex algebra.

In this note we attempt to shed some light at the definition of a vertex algebroid, by proposing a new, equivalent definition which has nothing to do with the axioms of a vertex algebra and uses only classical objects such as complexes of De Rham, Hochschild and Koszul. This point of view works nicely for Calabi–Yau structures as well and opens the way to higher dimensional generalisations.

Subject Classification: 17B69

Prooemium

1. Sint $\mathfrak{k}$ anulus commutativus $\mathbb{Q}$ continens, A $\mathfrak{k}$-algebra commutativa et T A-algebroid Lietianus.
Posito $\Omega := \mathrm{Hom}_A(T, A)$, habemus derivatio canonica $d : A \longrightarrow \Omega$, ubi $\langle \tau, da \rangle = \tau(a)$, denotante per $\langle , \rangle : T \otimes_A \Omega \longrightarrow A$ copulationem canonicam.

2. Revocamus (vide commentatione [V. Gorbounov, F. Malikov, V. Schechtman, Gerbes of chiral differenial operators. II. Vertex algebroids, *Inventiones Mathematicae*, **155**, 605–680 (2004)], 1.4), *structura verticiana* super T triplex $\mathcal{A} = (\gamma, \langle , \rangle, c)$ est, ubi elementa $\gamma \in \mathrm{Hom}(A \otimes T, \Omega)$, $\langle , \rangle \in \mathrm{Hom}(S^2 T, A)$ et $c \subset \mathrm{Hom}(\Lambda^2 T, \Omega)$ aequationibus sequentibus satisfacit:

$$\gamma(a, b\tau) - \gamma(ab, \tau) + a\gamma(b, \tau) = -\tau(a)db - \tau(b)da \tag{A1}$$

$$\langle a\tau, \tau'\rangle = a\langle\tau, \tau'\rangle - \tau\tau'(a) + \langle\gamma(a, \tau), \tau'\rangle \tag{A2}$$

$$\begin{aligned} c(a\tau, \tau') &= ac(\tau, \tau') + \gamma(a, [\tau, \tau']) - \gamma(\tau'(a), \tau) \\ &\quad + \tau'\gamma(a, \tau) + \frac{1}{2}ad\langle\tau, \tau'\rangle - \frac{1}{2}d\langle a\tau, \tau'\rangle \end{aligned} \tag{A3}$$

$$\begin{aligned} \langle[\tau, \tau'], \tau''\rangle + \langle\tau', [\tau, \tau'']\rangle &= \tau\langle\tau', \tau''\rangle - \frac{1}{2}\tau'\langle\tau, \tau''\rangle - \frac{1}{2}\tau''\langle\tau, \tau'\rangle \\ &\quad + \langle\tau', c(\tau, \tau'')\rangle + \langle\tau'', c(\tau, \tau')\rangle \end{aligned} \tag{A4}$$

atque

$$\begin{aligned} &\mathrm{Cycle}_{\tau,\tau',\tau''}\left[c([\tau, \tau'], \tau'') - \tau c(\tau', \tau'') + \frac{1}{3}d\langle\tau, c(\tau', \tau'')\rangle\right] \\ &= -\frac{1}{6}\mathrm{Cycle}_{\tau,\tau',\tau''}d\langle\tau, [\tau', \tau'']\rangle, \end{aligned} \tag{A5}$$

ubi denotabimus, brevitatis gratia:

$$\mathrm{Cycle}_{\tau,\tau',\tau''}f(\tau, \tau', \tau'') := f(\tau, \tau', \tau'') + f(\tau', \tau'', \tau) + f(\tau'', \tau, \tau').$$

3. E (A2) prodit:

$$\frac{1}{2}ad\langle\tau, \tau'\rangle - \frac{1}{2}d\langle a\tau, \tau'\rangle = -\frac{1}{2}da\,\langle\tau, \tau'\rangle + \frac{1}{2}d\tau\tau'(a) - \frac{1}{2}d\langle\tau', \gamma(a, \tau)\rangle,$$

ergo (A3) ita exhiberi licet:

$$\begin{aligned} c(a\tau, \tau') &= ac(\tau, \tau') + \gamma(a, [\tau, \tau']) - \gamma(\tau'(a), \tau) + \tau'\gamma(a, \tau) \\ &\quad - \frac{1}{2}d\langle\tau', \gamma(a, \tau)\rangle - \frac{1}{2}da\,\langle\tau, \tau'\rangle + \frac{1}{2}d\tau\tau'(a). \end{aligned} \tag*{(A3)^{bis}}$$

4. Applicatio $h : \mathcal{A} \longrightarrow \mathcal{A}'$ elementum $h \in \mathrm{Hom}(T, \Omega)$ est, axiomatibus sequentibus satisfaciens:

$$h(a\tau) - ah(\tau) = \gamma(a, \tau) - \gamma'(a, \tau) \tag*{$(Mor)_\gamma$}$$

$$\langle\tau, h(\tau')\rangle + \langle\tau', h(\tau)\rangle = \langle\tau, \tau'\rangle - \langle\tau, \tau'\rangle' \tag*{$(Mor)_{\langle,\rangle}$}$$

et

$$\begin{aligned} &h(|\tau, \tau'|) - \tau h(\tau') + \tau' h(\tau) + \frac{1}{2}d\langle\tau, h(\tau')\rangle - \langle\tau', h(\tau)\rangle\} \\ &= c(\tau, \tau') - c'(\tau, \tau'). \end{aligned} \tag*{$(Mor)_c$}$$

5. Posito $\Omega^n := \mathrm{Hom}_A(\Lambda^n_A T, A)$, revocamus differentiale DE RHAMIANUM

$$d : \Omega^{n-1} \longrightarrow \Omega^n$$

ubi

$$d\omega(\tau_1,\tau_2,\dots)=\omega([\tau_1,\tau_2],\tau_3,\dots)-\dots+(-1)^{i+j+1}\omega([\tau_i,\tau_j],\tau_1,\dots)+\dots$$

$$-\tau_1\omega(\tau_2,\dots)+\dots+(-1)^i\tau_i\omega(\tau_1,\dots,\hat{\tau}_i,\dots)+\dots$$

$$=\mathrm{Alt}_{12\dots n}\left\{\frac{1}{2(n-2)!}\omega([\tau_1,\tau_2],\tau_3,\dots)-\frac{1}{(n-1)!}\tau_1\omega(\tau_2,\dots)\right\}.$$

6. Quodque Ω^n in complexum HOCHSCHILDIANUM immergi potest:

$$0\longrightarrow\Omega^n\longrightarrow\mathrm{Hom}(\Lambda^{n-1}T,\Omega)\xrightarrow{d_H}\mathrm{Hom}(A\otimes T\otimes\Lambda^{n-2}T,\Omega)\xrightarrow{d_H}\dots$$

$$\longrightarrow\dots\xrightarrow{d_H}\mathrm{Hom}(A^{\otimes i}\otimes T\otimes\Lambda^{n-2}T,\Omega)\xrightarrow{d_H}\dots,$$

ubi

$$d_H\omega(a,b,c,\dots,e,f,\tau_1,\dots)=a\omega(b,c,\dots,e,f,\tau_1,\dots)$$

$$-\omega(ab,c,\dots,e,f,\tau_1,\dots)+\dots\pm\omega(a,b,\dots,e,f\tau_1,\dots)$$

Manifesto, $\Omega^n=\mathrm{Ker}\,d_H$.

7. Rursus, sit V $\mathfrak{k}$-modulus, potestas extera sua in genum complexus KOSZULIANI immerseri potest:

$$0\longrightarrow\Lambda^nV\longrightarrow\mathrm{Hom}(V^*,\Lambda^{n-1}V)\xrightarrow{Q}\mathrm{Hom}(S^2V^*,\Lambda^{n-2}V)$$

$$\xrightarrow{Q}\dots\xrightarrow{Q}\mathrm{Hom}(S^iV^*,\Lambda^{n-i}V)\xrightarrow{Q}\dots,$$

ubi $V^*:=\mathrm{Hom}(V,\mathfrak{k})$ ac

$$Qc(\tau_1,\tau_2)=\langle\tau_1,c(\tau_2)\rangle+\langle\tau_2,c(\tau_1)\rangle;$$

$$Qc(\tau_1,\tau_2,\tau_3)=\langle\tau_1,c(\tau_2,\tau_3)\rangle+\langle\tau_2,c(\tau_3,\tau_1)\rangle+\langle\tau_3,c(\tau_1,\tau_2)\rangle,$$

etc. Manifesto, $\Lambda^nV=\mathrm{Ker}\,Q$.

8. In hac commentatione solum inspicimus partem complexus de Rhamiani

$$\Omega^{[2,5]}:\ \Omega^2\longrightarrow\dots\longrightarrow\Omega^5$$

Copulatione complexuum Koszulianorum, Hochschildianorum de Rhamianorumque usa, definiamus complexum

$$W^{[2,5]}:\ W^2\longrightarrow\dots\longrightarrow W^5,$$

de inclusione complexuum $\Omega^{[2,5]}\subset W^{[2,5]}$ atque de cocyclo canonico $\mathcal{E}\in W^4$ ornatum. Structura verticiana super T elementum $\mathcal{A}\in W^3$ est, aequationi $D\mathcal{A}=\mathcal{E}$ satisfaciens. Sagittula $\mathcal{A}\longrightarrow\mathcal{A}'$ elementum $h\in W^2$ est, talis ut fit $Dh=\mathcal{A}-\mathcal{A}'$, vide Caput Secundum, Pars Tertia.

Caput primum. Structurae Calabi–Yautianae

1. Revocatio

1.1.

Revocamus (vide op.cit., 11.1), *structure Calabi–Yautianae* super T est applicatio $c : T \longrightarrow A$, duabus proprietatibus sequentibus satisfaciens:

$$c(a\tau) = ac(\tau) + \tau(a) \qquad (CY1)$$

atque

$$c([\tau, \tau']) = \tau c(\tau') - \tau' c(\tau). \qquad (CY2)$$

2. Complexus Hochschild–De Rhamianus

(a)

2.1.

Definimus operator

$$d_{DR} : \mathrm{Hom}(T, A) \longrightarrow \mathrm{Hom}(\Lambda^2 T, A)$$

per formulam:

$$d_{DR}c(\tau_1, \tau_2) = c([\tau_1, \tau_2]) - \tau_1 c(\tau_2) + \tau_2 c(\tau_1).$$

2.2.

Rursus, inspicimus complexum Hochschildianum

$$0 \longrightarrow \mathrm{Hom}(T, A) \xrightarrow{d_H^0} \mathrm{Hom}(A \otimes T, A) \xrightarrow{d_H^1} \mathrm{Hom}(A^{\otimes 2} \otimes T, A),$$

ubi differentialia Hochschildiana per regulas definitur:

$$d_H^0 c(a, \tau) = c(a\tau) - ac(\tau)$$

ac

$$d_H^1 c(a, b, \tau) = ac(b, \tau) - c(ab, \tau) + c(a, b\tau)$$

Liquet quod fit $\Omega = \mathrm{Ker} d_H^0$.

2.3.

Simili modo, consideremus complexum Hochschildianum:

$$0 \longrightarrow \mathrm{Hom}(\Lambda^2 T, A) \xrightarrow{d_H^0} \mathrm{Hom}(A \otimes T^{\otimes 2}, A) \xrightarrow{d_H^1} \mathrm{Hom}(A^{\otimes 2} \otimes T^{\otimes 2}, A),$$

ubi differentialia Hochschildiana per formulas definiuntur:

$$d_H^0 c(a, \tau, \tau') = c(a\tau, \tau') - ac(\tau, \tau')$$

atque

$$d_H^1 c(a, b, \tau, \tau') = ac(b, \tau, \tau') - c(ab, \tau, \tau') + c(a, b\tau, \tau')$$

Nunc erit $\Omega^2 = \mathrm{Ker}\, d_H^0$.

2.4.

Porro, introducamus operator

$$d_{DR} : \mathrm{Hom}(A \otimes T, A) \longrightarrow \mathrm{Hom}(A \otimes T^{\otimes 2}, A)$$

per regulam

$$d_{DR} c(a, \tau, \tau') = c(a, [\tau, \tau']) - c(\tau'(a), \tau) + \tau' c(a, \tau) = \mathrm{Lie}_{\tau'} c(a, \tau).$$

2.5. Lemma. $d_H^0 d_{DR} = d_{DR} d_H^0$.

2.6. Demonstratio. Pro elemento $c \in \mathrm{Hom}(T, A)$, habebimus

$$d_H^0 d_{DR} c(a, \tau, \tau') = d_{DR} c(a\tau, \tau') - a d_{DR} c(\tau, \tau'),$$

ubi

$$d_{DR} c(a\tau, \tau') = c([a\tau, \tau']) - a\tau c(\tau') + \tau' c(a\tau)$$

ac

$$-a d_{DR} c(\tau, \tau') = -ac([\tau, \tau']) + a\tau c(\tau') - a\tau' c(\tau),$$

unde

$$\begin{aligned} d_H^0 d_{DR} c(a, \tau, \tau') &= c(a[\tau, \tau']) - ac([\tau, \tau']) - c(\tau'(a)\tau) + \tau' c(a\tau) - a\tau' c(\tau) \\ &= c(a[\tau, \tau']) - ac([\tau, \tau']) - c(\tau'(a)\tau) + \tau' c(a\tau) \\ &\quad - \tau'\{ac(\tau)\} + \tau'(a) c(\tau) = d_{DR} d_H^0 c(a, \tau, \tau'), \end{aligned}$$

qed

(b)

2.7.

Definimus operator

$$d_{DR}: \mathrm{Hom}(A^{\otimes 2} \otimes T, A) \longrightarrow \mathrm{Hom}(A^{\otimes 2} \otimes T^{\otimes 2}, A)$$

per

$$d_{DR}c(a,b,\tau,\tau') = \mathrm{Lie}_{\tau'}c(a,b,\tau) = \tau'c(a,b,\tau) - c(\tau'(a),b,\tau) \\ - c(a,\tau'(b),\tau) - c(a,b,[\tau',\tau]).$$

2.8. Lemma. $d_H d_{DR} = d_{DR} d_H$.

2.9. Demonstratio. Pro elemento $c \in \mathrm{Hom}(A \otimes T, A)$, fit

$$d_H d_{DR}c(a,b,\tau,\tau') = a d_{DR}c(b,\tau,\tau') - d_{DR}c(ab,\tau,\tau') + d_{DR}c(a,b\tau,\tau'),$$

ubi

$$a d_{DR}c(b,\tau,\tau') = a\tau'c(b,\tau) - ac(\tau'(b),\tau) - ac(b,[\tau',\tau]); -d_{DR}c(ab,\tau,\tau') \\ = -\tau'c(ab,\tau) + c(\tau'(ab),\tau) + c(ab,[\tau',\tau]) \\ = -\tau'c(ab,\tau) + c(\tau'(a)b + a\tau'(b),\tau) + c(ab,[\tau',\tau])$$

atque

$$d_{DR}c(a,b\tau,\tau') = \tau'c(a,b\tau) - c(\tau'(a),b\tau) - c(a,[\tau',b\tau]) \\ = \tau'c(a,b\tau) - c(\tau'(a),b\tau) - c(a,\tau'(b)\tau - b[\tau',\tau])$$

Adde huc

$$0 = \tau'(a)c(b,\tau) - \tau'(a)c(b,\tau).$$

2.10.

Sed

$$-ac(b,[\tau',\tau]) + c(ab,[\tau',\tau]) + c(a,b[\tau',\tau]) = -d_H c(a,b,[\tau',\tau]);$$
$$a\tau'c(b,\tau) - \tau'c(ab,\tau) + \tau'c(a,b\tau) + \tau'(a)c(b,\tau) = \tau' d_H c(a,b,\tau);$$
$$-ac(\tau'(b),\tau) + c(a\tau'(b),\tau) - c(a,\tau'(b)\tau) = -d_H c(a,\tau'(b),\tau)$$

atque

$$c(\tau'(a)b,\tau) - c(\tau'(a),b\tau) - \tau'(a)c(b,\tau) = -d_H(\tau'(a),b,\tau)$$

Qua addendo obtinemus effatum lemmatis.

3. Cocyclus canonicus

3.1.

Inspicimus elementum $\epsilon \in \mathrm{Hom}(A \otimes T, A)$ per formulam definitum:

$$\epsilon(a, \tau) = \tau(a).$$

3.2. Lemma. $d_H\epsilon = d_{DR}\epsilon = 0.$

Habemus enim,

$$d_{DR}\epsilon(a, \tau, \tau') = \mathrm{Lie}_{\tau'}\epsilon(\tau, a) = \tau'\tau(a) - \tau\tau'(a) + [\tau, \tau'](a) = 0$$

(scilicet, ϵ operator invariens est).

Rursus,

$$d_H\epsilon(a, b, \tau) = a\tau(b) - \tau(ab) + b\tau(a) = 0.$$

3.3.

Aliter, ϵ cocyclum (bi)complexus Hochschild–De Rhamiani est.

3.4. Definitio altera. Structura Calabi–Yautiana est elementum $c \in \mathrm{Hom}(T, A)$, satisfaciens equationi $Dc = \epsilon$, denotanti per D differentiale complexus Hochschild–De Rhamiani.

Caput secundum. Structurae verticianae

Pars prima. Aedificium sinistrum

1. Koszul et de Rham

(a)

1.1.

Definimus operatores $Q : \mathrm{Hom}(T, \Omega) \longrightarrow \mathrm{Hom}(S^2T, A)$ per

$$Qh(\tau, \tau') = \langle\tau, h(\tau')\rangle + \langle\tau', h(\tau)\rangle,$$

ergo $\mathrm{Ker}\, Q = \Omega^2$, atque $Q : \mathrm{Hom}(\Lambda^2T, \Omega) \longrightarrow \mathrm{Hom}(S^2T \otimes T, A)$ per

$$Qc(\tau, \tau', \tau'') = \langle\tau, c(\tau', \tau'')\rangle + \langle\tau', c(\tau, \tau'')\rangle,$$

ergo $\mathrm{Ker}\, Q = \Omega^3$.

1.2.

Definimus operatores $d_{DR}: \mathrm{Hom}(T,\Omega) \longrightarrow \mathrm{Hom}(\Lambda^2 T,\Omega)$ per

$$d_{DR}h(\tau,\tau') = h([\tau,\tau']) - \tau\{h(\tau')\} + \tau'\{h(\tau)\} + \frac{1}{2}d\{\langle\tau,h(\tau')\rangle - \langle\tau',h(\tau)\rangle\}.$$

1.3.

atque

$$d_{DR}: \mathrm{Hom}(S^2T,A) \longrightarrow \mathrm{Hom}(S^2T\otimes T,A)$$

per

$$d_{DR}h(\tau,\tau',\tau'') = \mathrm{Sym}_{\tau,\tau'}\left[h(\tau,[\tau',\tau'']) + \frac{1}{2}\tau''\{h(\tau,\tau')\} - \frac{1}{2}\tau\{h(\tau',\tau'')\}\right].$$

1.4. Lemma. $Qd_{DR} = d_{DR}Q$.

Demonstratio. Si $h \in \mathrm{Hom}(T,\Omega)$, habemus

$$\begin{aligned}
Qd_{DR}h(\tau,\tau',\tau'') &= \mathrm{Sym}_{\tau,\tau'}\langle\tau, d_{DR}h(\tau',\tau'')\rangle \\
&= \mathrm{Sym}_{\tau,\tau'}\Big\langle\tau, h([\tau',\tau'']) - \tau'\{h(\tau'')\} + \tau''\{h(\tau')\} \\
&\quad + \frac{1}{2}d\{\langle\tau',h(\tau'')\rangle - \langle\tau'',h(\tau')\rangle\}\Big\rangle \\
&= \mathrm{Sym}_{\tau,\tau'}\Big[\langle\tau,h([\tau',\tau''])\rangle - \tau'(\langle\tau,h(\tau'')\rangle \\
&\quad + \langle[\tau',\tau],h(\tau'')\rangle + \tau''(\langle\tau,h(\tau')\rangle - \langle[\tau'',\tau],h(\tau')\rangle \\
&\quad + \frac{1}{2}\tau(\langle\tau',h(\tau'')\rangle) - \frac{1}{2}\tau(\langle\tau'',h(\tau')\rangle)\Big]
\end{aligned}$$

Sed

$$\begin{aligned}
&\mathrm{Sym}_{\tau,\tau'}[\langle\tau,h([\tau',\tau''])\rangle - \langle[\tau'',\tau],h(\tau')\rangle] \\
&\quad = \mathrm{Sym}_{\tau,\tau'}[\langle\tau',h([\tau,\tau''])\rangle + \langle[\tau,\tau''],h(\tau')\rangle] = \mathrm{Sym}_{\tau,\tau'}Qh(\tau,[\tau',\tau'']); \\
&\mathrm{Sym}_{\tau,\tau'}\langle[\tau',\tau],h(\tau'')\rangle = 0, \\
&\mathrm{Sym}_{\tau,\tau'}[\tau''(\langle\tau,h(\tau')\rangle] = \tau''\{Qh(\tau,\tau')\}
\end{aligned}$$

ac

$$\begin{aligned}
&\mathrm{Sym}_{\tau,\tau'}\left[-\tau'(\langle\tau,h(\tau'')\rangle + \frac{1}{2}\tau(\langle\tau',h(\tau'')\rangle) - \frac{1}{2}\tau(\langle\tau'',h(\tau')\rangle)\right] \\
&\quad = -\frac{1}{2}\mathrm{Sym}_{\tau,\tau'}\left[\tau(\langle\tau',h(\tau'')\rangle + \tau(\langle\tau'',h(\tau')\rangle)\right] = -\frac{1}{2}\mathrm{Sym}_{\tau,\tau'}\tau\{Qh(\tau',\tau'')\},
\end{aligned}$$

unde $Qd_{DR}h(\tau,\tau',\tau'') = d_{DR}Qh(\tau,\tau',\tau'')$, *quod erat demonstrandum.*

1.5.

Axioma (A4) etiam sic exhiberi potest:

$$Qc = d_{DR}\langle , \rangle \tag{A4}$$

(b)

1.6.

Definimus operatores

$$d_{DR}: \operatorname{Hom}(\Lambda^2 T, \Omega) \longrightarrow \operatorname{Hom}(\Lambda^3 T, \Omega)$$

per

$$d_{DR}c(\tau, \tau', \tau'') = \mathrm{Cycle}_{\tau,\tau',\tau''}\left[c([\tau, \tau'], \tau'') - \tau\{c(\tau', \tau'')\} + \frac{1}{3}d\langle \tau, c(\tau', \tau'')\rangle\right]$$

$$= d_{Lie}c(\tau, \tau', \tau'') + d'c(\tau, \tau', \tau''),$$

ubi

$$d'c(\tau, \tau', \tau'') = \frac{1}{3}\mathrm{Cycle}_{\tau,\tau',\tau''}d\langle \tau, c(\tau', \tau'')\rangle$$

1.7.

atque

$$R: \operatorname{Hom}(S^2 T, A) \longrightarrow \operatorname{Hom}(\Lambda^3 T, \Omega)$$

per

$$Rh(\tau, \tau', \tau'') = -\frac{1}{6}\mathrm{Cycle}_{\tau,\tau',\tau''}dh([\tau, \tau'], \tau'').$$

1.8. Lemma. $d^2_{DR} = RQ$.

1.9. Demonstratio. Fit

$$d^2_{DR} = (d_{Lie} + d')^2 = d_{Lie}d' + d'd_{Lie} + d'^2$$

ob $d^2_{Lie} = 0$.

Si $h \in \operatorname{Hom}(T, \Omega)$, habemus

$$d_{Lie}d'h(\tau, \tau', \tau'') = \mathrm{Cycle}_{\tau,\tau',\tau''}[d'h([\tau, \tau'], \tau'') - \tau\{d'h(\tau', \tau'')\}]$$

$$= \frac{1}{2}\mathrm{Cycle}_{\tau,\tau',\tau''}\big[d\{\langle[\tau, \tau'], h(\tau'')\rangle - \langle \tau'', h([\tau, \tau'])\rangle\}$$

$$- \tau d\{\langle \tau', h(\tau'')\rangle - \langle \tau', h(\tau'')\rangle\}\big].$$

Observamus:

$$-\frac{1}{2}\mathrm{Cycle}_{\tau,\tau',\tau''}\big[\tau d\{\langle \tau', h(\tau'')\rangle - \langle \tau', h(\tau'')\rangle\}\big] = -\frac{1}{2}\mathrm{Alt}_{\tau,\tau',\tau''}\tau d\langle \tau', h(\tau'')\rangle.$$

1.10.

Rursus

$$d'd_{Lie}h(\tau,\tau',\tau'') = \frac{1}{3}\mathrm{Cycle}_{\tau,\tau',\tau''}d\langle\tau, d_{Lie}h(\tau',\tau'')\rangle$$
$$= \frac{1}{3}\mathrm{Cycle}_{\tau,\tau',\tau''}d\langle\tau, h([\tau',\tau''])\rangle - \frac{1}{3}\mathrm{Alt}_{\tau,\tau',\tau''}d\langle\tau,\tau'\{h(\tau'')\}\rangle,$$

ubi

$$-\frac{1}{3}\mathrm{Alt}_{\tau,\tau',\tau''}d\langle\tau,\tau'\{h(\tau'')\}\rangle = -\frac{1}{3}\mathrm{Alt}_{\tau,\tau',\tau''}d[\tau'\langle\tau,h(\tau'')\rangle + \langle[\tau,\tau'],h(\tau'')\rangle]$$
$$= -\frac{1}{3}\mathrm{Alt}_{\tau,\tau',\tau''}d\tau'\langle\tau,h(\tau'')\rangle + \frac{2}{3}\mathrm{Cycle}_{\tau,\tau',\tau''}d\langle[\tau,\tau'],h(\tau'')\rangle.$$

1.11.

Denique,

$$d'^2h(\tau,\tau',\tau'') = \frac{1}{6}\mathrm{Alt}_{\tau,\tau',\tau''}d\tau\{\langle\tau',h(\tau'')\rangle\}.$$

1.12.

Post summationem termini $d\tau\{\langle\tau',h(\tau'')\rangle\}$ exeunt, dum termini reliqui praebunt

$$d^2_{DR}h(\tau,\tau',\tau'') = -\frac{1}{6}\mathrm{Cycle}_{\tau,\tau',\tau''}d\big\{\langle[\tau,\tau'],h(\tau'')\rangle + \langle\tau'',h([\tau,\tau'])\rangle\big\}$$
$$= -\frac{1}{6}\mathrm{Cycle}_{\tau,\tau',\tau''}dQh([\tau,\tau'],\tau'') = RQh(\tau,\tau',\tau''), \qquad \text{qed}$$

1.13.

Axioma (A5) sic exhiberi potest:

$$d_{DR}c = R\langle,\rangle \tag{A5}$$

(c)

1.14.

Determinamus operator

$$d_{DR}: \mathrm{Hom}(S^2T\otimes T, A) \longrightarrow \mathrm{Hom}(S^2T\otimes\Lambda^2T, A)$$

per formulam

$$d_{DR}c(\tau_1,\tau_2,\tau_3,\tau_4) = -c(\tau_1,\tau_2,[\tau_3,\tau_4]) - \mathrm{Alt}_{3,4}\tau_4c(\tau_1,\tau_2,\tau_3)$$
$$+ \mathrm{Sym}_{1,2}\mathrm{Alt}_{3,4}\left\{c(\tau_1,[\tau_2,\tau_3],\tau_4) - \frac{1}{3}\tau_1c(\tau_2,\tau_3,\tau_4)\right\}.$$

1.15. Lemma. *Fit* $d_{DR}Q = Qd_{DR}$.

1.16. Démonstratio. Si $c \in \mathrm{Hom}(\Lambda^2 T, \Omega)$, habemus

$$Qd_{DR}c(\tau_1, \tau_2, \tau_3, \tau_4) = \mathrm{Sym}_{1,2}\langle \tau_1, d_{DR}c(\tau_2, \tau_3, \tau_4)\rangle$$

$$= \mathrm{Sym}_{1,2}\mathrm{Cycle}_{2,3,4}\Big\langle \tau_1, c([\tau_2, \tau_3], \tau_4) - \tau_2 c(\tau_3, \tau_4) + \frac{1}{3}d\langle \tau_2, c(\tau_3, \tau_4)\rangle\Big\rangle,$$

ubi

$$-\langle \tau_1, \tau_2 c(\tau_3, \tau_4)\rangle = \langle [\tau_2, \tau_1], c(\tau_3, \tau_4)\rangle - \tau_2\langle \tau_1, c(\tau_3, \tau_4)\rangle$$

atque

$$\langle \tau_1, d\langle \tau_2, c(\tau_3, \tau_4)\rangle\rangle = \tau_1\langle \tau_2, c(\tau_3, \tau_4)\rangle.$$

1.17.

Primo, fit

$$\mathrm{Sym}_{1,2}\mathrm{Cycle}_{2,3,4}\{\langle \tau_1, c([\tau_2, \tau_3], \tau_4)\rangle + \langle [\tau_2, \tau_1], c(\tau_3, \tau_4)\rangle\}$$

$$= \mathrm{Sym}_{1,2}\mathrm{Alt}_{3,4}\Big\{\Big\langle \tau_1, c([\tau_2, \tau_3], \tau_4) + \frac{1}{2}c([\tau_3, \tau_4], \tau_2)\Big\rangle + \frac{1}{2}\langle [\tau_2, \tau_1], c(\tau_3, \tau_4)\rangle + \langle [\tau_3, \tau_1], c(\tau_4, \tau_2)\rangle\Big\}$$

$$= -Qc(\tau_1, \tau_2, [\tau_3, \tau_4]) + \mathrm{Sym}_{1,2}\mathrm{Alt}_{3,4}Qc(\tau_1, [\tau_2, \tau_3], \tau_4).$$

1.18.

Secundo,

$$\mathrm{Sym}_{1,2}\mathrm{Cycle}_{2,3,4}\Big\{-\tau_2\langle \tau_1, c(\tau_3, \tau_4)\rangle + \frac{1}{3}\tau_1\langle \tau_2, c(\tau_3, \tau_4)\rangle\Big\}$$

$$= \mathrm{Sym}_{1,2}\mathrm{Alt}_{3,4}\Big\{-\frac{1}{2}\tau_2\langle \tau_1, c(\tau_3, \tau_4)\rangle - \tau_3\langle \tau_1, c(\tau_4, \tau_2)\rangle + \frac{1}{6}\tau_1\langle \tau_2, c(\tau_3, \tau_4)\rangle + \frac{1}{3}\tau_1\langle \tau_3, c(\tau_4, \tau_2)\rangle\Big\}$$

$$= \mathrm{Alt}_{3,4}\tau_3 Qc(\tau_1, \tau_2, \tau_4) - \frac{1}{3}\mathrm{Sym}_{1,2}\mathrm{Alt}_{3,4}\tau_1 Qc(\tau_2, \tau_3, \tau_4)$$

Hinc lemma nostra sponte sequitur.

(d) *Junctio*

1.19. Lemma. *Compositio*

$$d^2_{DR}: \operatorname{Hom}(S^2T, A) \longrightarrow \operatorname{Hom}(S^2T \otimes T, A) \longrightarrow \operatorname{Hom}(S^2T \otimes \Lambda^2 T, A)$$

aequat QR.

1.20. Demonstratio. Si $c \in \operatorname{Hom}(S^2T, A)$, fit

$$d^2_{DR}c(\tau_1, \tau_2, \tau_3, \tau_4) = -d_{DR}c(\tau_1, \tau_2, [\tau_3, \tau_4]) - \mathrm{Alt}_{3,4}\tau_4 d_{DR}c(\tau_1, \tau_2, \tau_3)$$
$$+ \mathrm{Sym}_{1,2}\mathrm{Alt}_{3,4}\left\{d_{DR}c(\tau_1, [\tau_2, \tau_3], \tau_4) - \frac{1}{3}\tau_1 d_{DR}c(\tau_2, \tau_3, \tau_4)\right\}.$$

1.21.

Primo,

$$-d_{DR}c(\tau_1, \tau_2, [\tau_3, \tau_4]) = -c(\tau_1, [\tau_2, [\tau_3, \tau_4]]) - c(\tau_2, [\tau_1, [\tau_3, \tau_4]])$$
$$- [\tau_3, \tau_4]c(\tau_1, \tau_2) + \frac{1}{2}\tau_1 c(\tau_2, [\tau_3, \tau_4]) + \frac{1}{2}\tau_2 c(\tau_1, [\tau_3, \tau_4])$$

Secundo,

$$\mathrm{Sym}_{1,2}\mathrm{Alt}_{3,4}d_{DR}c(\tau_1, [\tau_2, \tau_3], \tau_4) = \mathrm{Sym}_{1,2}\mathrm{Alt}_{3,4}\Big\{c(\tau_1, [[\tau_2, \tau_3], \tau_4])$$
$$+ c([\tau_2, \tau_3], [\tau_1, \tau_4]) + \tau_4 c(\tau_1, [\tau_2, \tau_3])$$
$$-\frac{1}{2}\tau_1 c([\tau_2, \tau_3], \tau_4) - \frac{1}{2}[\tau_2, \tau_3]c(\tau_1, \tau_4)\Big\}$$

Tertio,

$$-\mathrm{Alt}_{3,4}\tau_4 d_{DR}c(\tau_1, \tau_2, \tau_3) = -\mathrm{Alt}_{3,4}\Big\{\tau_4 c(\tau_1, [\tau_2, \tau_3]) + \tau_4 c(\tau_2, [\tau_1, \tau_3])$$
$$+ \tau_4\tau_3 c(\tau_1, \tau_2) - \frac{1}{2}\tau_4\tau_1 c(\tau_2, \tau_3)$$
$$-\frac{1}{2}\tau_4\tau_2 c(\tau_1, \tau_3)\Big\}$$

et quatro,

$$-\frac{1}{3}\mathrm{Sym}_{1,2}\mathrm{Alt}_{3,4}\tau_1 d_{DR}c(\tau_2, \tau_3, \tau_4) = -\frac{1}{3}\mathrm{Sym}_{1,2}\mathrm{Alt}_{3,4}\Big\{\tau_1 c(\tau_2, [\tau_3, \tau_4])$$
$$+ \tau_1 c(\tau_3, [\tau_2, \tau_4]) + \tau_1\tau_4 c(\tau_2, \tau_3)$$
$$-\frac{1}{2}\tau_1\tau_2 c(\tau_3, \tau_4) - \frac{1}{2}\tau_1\tau_3 c(\tau_2, \tau_4)\Big\}$$

1.22.

Termini formae $c([\tau_i, \tau_j], [\tau_k, \tau_l])$ evanescunt symmetrisatione causa, cum c symmetricos est.

Post summationem, termini cum triplicibus uncinis evanescunt Jacobi causa. Termini formae $\tau_i \tau_j c(\tau_k, \tau_l)$ quoque exire videri possunt.

Tandem termini formae $\tau_i c(\tau_j, [\tau_k, \tau_l])$ praebunt

$$-\frac{1}{6}\mathrm{Sym}_{1,2}\mathrm{Cycle}_{2,3,4}\tau_1 c(\tau_2, [\tau_3, \tau_4]) = -\frac{1}{6}\mathrm{Sym}_{1,2}\mathrm{Cycle}_{2,3,4}\langle \tau_1, dc(\tau_2, [\tau_3, \tau_4])\rangle$$
$$= QRc(\tau_1, \tau_2, \tau_3, \tau_4), \qquad \text{qed}$$

(e) *Differentiale de Rhamianum tertium*

1.23.

Definimus operator

$$d_{DR} : \mathrm{Hom}(\Lambda^3 T, \Omega) \longrightarrow \mathrm{Hom}(\Lambda^4 T, \Omega)$$

ubi

$$\begin{aligned} d_{DR}c(\tau_1, \tau_2, \tau_3, \tau_4) &= c([\tau_1, \tau_2], \tau_3, \tau_4) - c([\tau_1, \tau_3], \tau_2, \tau_4) + \dots \\ &\quad - \tau_1 c(\tau_2, \tau_3, \tau_4) + \tau_2 c(\tau_1, \tau_3, \tau_4) - \dots \\ &\quad + \frac{1}{4} d\{\langle \tau_1, c(\tau_2, \tau_3, \tau_4)\rangle - \langle \tau_2, c(\tau_1, \tau_3, \tau_4)\rangle + \} \\ &= \mathrm{Alt}_{1234}\left[\frac{1}{4}c([\tau_1, \tau_2], \tau_3, \tau_4) - \frac{1}{6}\tau_1 c(\tau_2, \tau_3, \tau_4)\right. \\ &\quad \left. + \frac{1}{24} d\langle \tau_1, c(\tau_2, \tau_3, \tau_4)\rangle\right] = \{d_{Lie} + d'\}c(\tau_1, \tau_2, \tau_3, \tau_4), \end{aligned}$$

ubi

$$d'c(\tau_1, \tau_2, \tau_3, \tau_4) := \frac{1}{24}\mathrm{Alt}_{1234}\, d\langle \tau_1, c(\tau_2, \tau_3, \tau_4)\rangle,$$

confer artt. 1.2 et 1.6.

1.24.

Insuper introducamus operator

$$R : \mathrm{Hom}(S^2 T \otimes T, A) \longrightarrow \mathrm{Hom}(\Lambda^4 T, \Omega),$$

ubi

$$Rc(\tau_1, \tau_2, \tau_3, \tau_4) := -\frac{1}{24}\mathrm{Alt}_{1234}\, dc([\tau_1, \tau_2], \tau_3, \tau_4),$$

confer art. 1.7.

1.25. Lemma. *Fit* $d_{DR}^2 = RQ$.

1.26. Demonstratio. Primo, $d_{Lie}^2 = 0$.
Secundo, ostendetur methodo simili a 1.9.–1.12,

$$\begin{aligned}
&\{d_{Lie}d' + d'd_{Lie} + d'^2\}c(\tau_1, \tau_2, \tau_3, \tau_4) \\
&\quad = -\frac{1}{24}\mathrm{Alt}_{1234}\, d\big\{\langle[\tau_1, \tau_2], c(\tau_3, \tau_4)\rangle + \langle\tau_3, c([\tau_1, \tau_2], \tau_4)\rangle\big\} \\
&\quad = -\frac{1}{24}\mathrm{Alt}_{1234}\, dQc([\tau_1, \tau_2], \tau_3, \tau_4) = RQc(\tau_1, \tau_2, \tau_3, \tau_4),
\end{aligned}$$

qed

1.27.

Consideremus duos operatores R ex artt. 1.7 et 1.24.

1.28. Lemma. *Fit* $d_{DR}R = Rd_{DR}$.

1.29. Demonstratio. Primo, habemus

$$\begin{aligned}
d_{DR}Rc(\tau_1, \tau_2, \tau_3, \tau_4) &= \mathrm{Alt}_{1234}\Bigg[\frac{1}{4}Rc([\tau_1, \tau_2], \tau_3, \tau_4) \\
&\quad -\frac{1}{6}\tau_1 Rc(\tau_2, \tau_3, \tau_4) + \frac{1}{24}d\langle\tau_1, Rc(\tau_2, \tau_3, \tau_4)\rangle\Bigg] \\
&= \mathrm{Alt}_{1234}\Bigg[-\frac{1}{24}d\{c([[\tau_1, \tau_2], \tau_3], \tau_4) + c([\tau_3, \tau_4], [\tau_1, \tau_2]) \\
&\quad +c([\tau_4, [\tau_1, \tau_2]], \tau_3)\} + \frac{1}{36}d\tau_1\{c([\tau_2, \tau_3], \tau_4) \\
&\quad +c([\tau_3, \tau_4], \tau_2) + c([\tau_4, \tau_2], \tau_3)\} - \frac{1}{144}d\langle\tau_1, d\{c([\tau_2, \tau_3], \tau_4) \\
&\quad + c([\tau_3, \tau_4], \tau_2) + c([\tau_4, \tau_2], \tau_3)\}\rangle\Bigg]
\end{aligned}$$

(termini, uncinos triplices continentes, exeunt, relatione identica Jacobiana causa)

$$= \mathrm{Alt}_{1234}\left[-\frac{1}{24}dc([\tau_3, \tau_4], [\tau_1, \tau_2]) + \frac{1}{16}d\tau_1 c([\tau_2, \tau_3], \tau_4)\right].$$

1.30.

Rursus fit

$$Rd_{DR}c(\tau_1,\tau_2,\tau_3,\tau_4) = -\frac{1}{24}\mathrm{Alt}_{1234}\, dd_{DR}c([\tau_1,\tau_2],\tau_3,\tau_4)$$

$$= -\frac{1}{24}\mathrm{Alt}_{1234}\, d\Big[c([\tau_1,\tau_2],[\tau_3,\tau_4]) + c(\tau_3,[[\tau_1,\tau_2],\tau_4])$$

$$+\tau_4 c([\tau_1,\tau_2],\tau_3) - \frac{1}{2}[\tau_1,\tau_2]c(\tau_3,\tau_4) - \frac{1}{2}\tau_3 c([\tau_1,\tau_2],\tau_3)\Big]$$

(termini formae $[\tau_1,\tau_2]c(\tau_3,\tau_4)$ exibunt, quum c symmetricos est)

$$= -\frac{1}{24}\mathrm{Alt}_{1234}\, d\left\{c([\tau_1,\tau_2],[\tau_3,\tau_4]) - \frac{3}{2}\tau_4 c([\tau_1,\tau_2],\tau_3)\right\} = d_{DR}Rc(\tau_1,\tau_2,\tau_3,\tau_4),$$

qed

2. Pede plana

(a) *Paries recessus*

2.1.

Definimus operatores:

$$d_H : \mathrm{Hom}(T,\Omega) \longrightarrow \mathrm{Hom}(A\otimes T,\Omega)$$

per formulam

$$d_H c(a,\tau) = c(a\tau) - ac(\tau)$$

et

$$d_H : \mathrm{Hom}(\Lambda^2 T,\Omega) \longrightarrow \mathrm{Hom}(A\otimes T^{\otimes 2},\Omega)$$

per regulam:

$$d_H c(a,\tau,\tau') = c(a\tau,\tau') - ac(\tau,\tau').$$

2.2.

Introducamus operator:

$$d_{DR} : \mathrm{Hom}(A\otimes T,\Omega) \longrightarrow \mathrm{Hom}(A\otimes T^{\otimes 2},\Omega)$$

per regulam:

$$d_{DR}c(a,\tau,\tau') = c(a,[\tau,\tau']) - c(\tau'(a),\tau) + \tau' c(a,\tau) - \frac{1}{2}d\langle\tau',c(a,\tau)\rangle$$

$$= \mathrm{Lie}_{\tau'}c(a,\iota) - \frac{1}{2}d\langle\tau',c(a,\tau)\rangle.$$

2.3.

Insuper, operator:

$$M : \mathrm{Hom}(S^2T, A) \longrightarrow \mathrm{Hom}(A \otimes T^{\otimes 2}, \Omega)$$

definitur per regulam:

$$Mc(a, \tau, \tau') = -\frac{1}{2}c(\tau, \tau')da.$$

2.4. Lemma. *Fit* $d_H d_{DR} = d_{DR} d_H + MQ$.

2.5. Demonstratio. Elementum $c \in \mathrm{Hom}(T, \Omega)$ datum, habebimus

$$d_H d_{DR} c(a, \tau, \tau') = d_{DR} c(a\tau, \tau') - a d_{DR} c(\tau, \tau')$$

ubi

$$\begin{aligned} d_{DR}c(a\tau, \tau') &= c([a\tau, \tau']) - (a\tau)c(\tau') + \tau'c(a\tau) + \frac{1}{2}d\{\langle a\tau, c(\tau')\rangle - \langle \tau', c(a\tau)\rangle\} \\ &= c(a[\tau, \tau'] - \tau'(a)\tau) - a\tau c(\tau') - da\langle \tau, c(\tau')\rangle + \tau'c(a\tau) \\ &\quad + \frac{1}{2}da\langle \tau, c(\tau')\rangle + \frac{1}{2}ad\langle \tau, c(\tau')\rangle - \frac{1}{2}d\langle \tau', c(a\tau) - ac(\tau)\rangle \\ &\quad - \frac{1}{2}da\langle \tau', c(\tau)\rangle - \frac{1}{2}ad\langle \tau', c(\tau)\rangle, \end{aligned}$$

addemus huc:

$$0 = -\tau'\{ac(\tau)\} + \tau'(a)c(\tau) + a\tau'c(\tau)$$

Rursus,

$$-ad_{DR}c(\tau, \tau') = -ac([\tau, \tau']) + a\tau c(\tau') - a\tau'c(\tau) - \frac{1}{2}ad\{\langle \tau, c(\tau')\rangle - \langle \tau', c(\tau)\rangle\}.$$

2.6.

Sed

$$c(a[\tau, \tau']) - ac([\tau, \tau']) = d_H c(a, [\tau, \tau']);$$

$$-c(\tau'(a)\tau) + \tau'(a)c(\tau) = -d_H c(\tau'(a), \tau);$$

$$\tau'c(a\tau) - \tau'\{ac(\tau)\} = \tau' d_H c(a, \tau);$$

$$-\frac{1}{2}d\langle \tau', c(a\tau) - ac(\tau)\rangle = -\frac{1}{2}d\langle \tau', d_H c(a, \tau)\rangle,$$

terminos reliquos praebendo

$$-\frac{1}{2}da\{\langle \tau, c(\tau')\rangle + \langle \tau', c(\tau)\rangle\} = -\frac{1}{2}da Qc(\tau, \tau') = MQc(a, \tau, \tau'),$$

lemma nostrum sequitur.

(b) *Frons*

2.7.

Revocamus,

$$d_{DR} : \operatorname{Hom}(S^2T, A) \longrightarrow \operatorname{Hom}(S^2T \otimes T, A)$$

definitur per formulam:

$$\begin{aligned} d_{DR}c(\tau_1, \tau_2, \tau_3) &= c(\tau_1, [\tau_2, \tau_3]) + c(\tau_2, [\tau_1, \tau_3]) + \tau_3 c(\tau_1, \tau_2) \\ &\quad - \frac{1}{2}\tau_1 c(\tau_2, \tau_3) - \frac{1}{2}\tau_2 c(\tau_1, \tau_3) \\ &= \mathrm{Lie}_{\tau_3} c(\tau_1, \tau_2) - \frac{1}{2}\tau_1 c(\tau_2, \tau_3) - \frac{1}{2}\tau_2 c(\tau_1, \tau_3) \\ &= \{\mathrm{Lie} + d'_{DR}\} c(\tau_1, \tau_2, \tau_3), \end{aligned}$$

ubi ponamus

$$\mathrm{Lie}\, c(\tau_1, \tau_2, \tau_3) = \mathrm{Lie}_{\tau_3} c(\tau_1, \tau_2)$$

et

$$d'_{DR} c(\tau_1, \tau_2, \tau_3) = -\frac{1}{2}\tau_1 c(\tau_2, \tau_3) - \frac{1}{2}\tau_2 c(\tau_1, \tau_3)$$

Rursus, definimus

$$d_{DR} : \operatorname{Hom}(A \otimes T^{\otimes 2}, A) \longrightarrow \operatorname{Hom}(A \otimes T^{\otimes 3}, A)$$

per formulam:

$$\begin{aligned} d_{DR}c(a, \tau_1, \tau_2, \tau_3) &= \tau_3 c(a, \tau_1, \tau_2) - c(\tau_3(a), \tau_1, \tau_2) + c(a, [\tau_1, \tau_3], \tau_2) \\ &\quad + c(a, \tau_1, [\tau_2, \tau_3]) - \frac{1}{2}\tau_2 c(a, \tau_1, \tau_3) \\ &= \mathrm{Lie}_{\tau_3} c(a, \tau_1, \tau_2) - \frac{1}{2}\tau_2 c(a, \tau_1, \tau_3) \\ &= \{\mathrm{Lie} + d'_{DR}\} c(a, \tau_1, \tau_2, \tau_3), \end{aligned}$$

ubi ponamus

$$\mathrm{Lie}\, c(a, \tau_1, \tau_2, \tau_3) = \mathrm{Lie}_{\tau_3} c(a, \tau_1, \tau_2)$$

et

$$d'_{DR} c(a, \tau_1, \tau_2, \tau_3) = -\frac{1}{2}\tau_2 c(a, \tau_1, \tau_3)$$

Denique, introducamus

$$Q : \operatorname{Hom}(A \otimes T^{\otimes 2}, \Omega) \longrightarrow \operatorname{Hom}(A \otimes T^{\otimes 3}, A)$$

per regulam:

$$Qc(a, \tau_1, \tau_2, \tau_3) = \langle \tau_2, c(a, \tau_1, \tau_3) \rangle.$$

2.8. Lemma. *Fit* $d_H d_{DR} = d_{DR} d_H + QM$.

2.9. Demonstratio. Primo, d_H commutat cum Lie. Si enim $c \in \mathrm{Hom}(S^2T, A)$, habemus:

$$d_H \mathrm{Lie}\, c(a, \tau_1, \tau_2, \tau_3) = \mathrm{Lie}\, c(a\tau_1, \tau_2, \tau_3) - a\mathrm{Lie}\, c(\tau_1, \tau_2, \tau_3)$$

ubi

$$\mathrm{Lie}\, c(a\tau_1, \tau_2, \tau_3) = \tau_3 c(a\tau_1, \tau_2) + c(a[\tau_1, \tau_3] - \tau_3(a)\tau_1, \tau_2) + c(a\tau_1, [\tau_2, \tau_3])$$

et

$$-a\mathrm{Lie}\, c(\tau_1, \tau_2, \tau_3) = -a\tau_3 c(\tau_1, \tau_2) - ac([\tau_1, \tau_3], \tau_2) - ac(\tau_1, [\tau_2, \tau_3])$$

Addemus huc:

$$0 = -\tau_3\{ac(\tau_1, \tau_2)\} + \tau_3(a)c(\tau_1, \tau_2) + a\tau_3 c(\tau_1, \tau_2)$$

Habebimus

$$\tau_3 c(a\tau_1, \tau_2) - \tau_3\{ac(\tau_1, \tau_2)\} = \tau_3 d_H c(a, \tau_1, \tau_2);$$

$$c(a[\tau_1, \tau_3], \tau_2) - ac([\tau_1, \tau_3], \tau_2) = d_H c(a, [\tau_1, \tau_3], \tau_2);$$

$$c(a\tau_1, [\tau_2, \tau_3]) - ac(\tau_1, [\tau_2, \tau_3]) = d_H c(\tau_1, [\tau_2, \tau_3])$$

et

$$-c(\tau_3(a)\tau_1, \tau_2) + \tau_3(a)c(\tau_1, \tau_2) = -d_H c(\tau_3(a), \tau_1, \tau_2)$$

unde

$$d_H \mathrm{Lie}\, c(a, \tau_1, \tau_2, \tau_3) = \mathrm{Lie}_{\tau_3} d_H c(a, \tau_1, \tau_2).$$

2.10.

Secundo,

$$d_H d'_{DR} c(a, \tau_1, \tau_2, \tau_3) = d'_{DR} c(a\tau_1, \tau_2, \tau_3) - a d'_{DR} c(\tau_1, \tau_2, \tau_3)$$

ubi

$$d'_{DR} c(a\tau_1, \tau_2, \tau_3) = -\frac{1}{2} a\tau_1 c(\tau_2, \tau_3) - \frac{1}{2}\tau_2 c(a\tau_1, \tau_3)$$

et

$$-a d'_{DR} c(\tau_1, \tau_2, \tau_3) = \frac{1}{2} a\tau_1 c(\tau_2, \tau_3) + \frac{1}{2} a\tau_2 c(\tau_1, \tau_3).$$

Addemus huc:

$$0 = \frac{1}{2}\tau_2\{ac(\tau_1, \tau_3)\} - \frac{1}{2}\tau_2(a)c(\tau_1, \tau_3) - \frac{1}{2} a\tau_2 c(\tau_1, \tau_3).$$

Nanciscemur:

$$d_H d'_{DR} c(a, \tau_1, \tau_2, \tau_3) = -\frac{1}{2}\tau_2 d_H c(a, \tau_1, \tau_3) - \frac{1}{2}\tau_2(a)c(\tau_1, \tau_3),$$

ubi

$$-\frac{1}{2}\tau_2(a)c(\tau_1, \tau_3) = \left\langle \tau_2, -\frac{1}{2}da\, c(\tau_1, \tau_3)\right\rangle = QMc(a, \tau_1, \tau_2, \tau_3)$$

unde hoc lemma sequitur.

(c) *Camera*

2.11.

Revocamus differentialia de Rhamiana:

$$d_{DR} : \operatorname{Hom}(A \otimes T, \Omega) \longrightarrow \operatorname{Hom}(A \otimes T^{\otimes 2}, \Omega)$$

definiuntur per formulam (vide art. 2.2):

$$d_{DR} c(a, \tau, \tau') = \operatorname{Lie}_{\tau'} c(a, \tau) - \frac{1}{2} d\langle \tau', c(a, \tau)\rangle$$

ac

$$d_{DR} : \operatorname{Hom}(A \otimes T^{\otimes 2}, A) \longrightarrow \operatorname{Hom}(A \otimes T^{\otimes 3}, A)$$

definiuntur per regulam (vide art. 2.7):

$$d_{DR} c(a, \tau_1, \tau_2, \tau_3) = \operatorname{Lie}_{\tau_3} c(a, \tau_1, \tau_2) - \frac{1}{2}\tau_2 c(a, \tau_1, \tau_3).$$

2.12. Lemma. *Fit* $d_{DR} Q = Q d_{DR}$.

2.13. Demonstratio. Primo,

$$\operatorname{Lie} Q = Q \operatorname{Lie}$$

Habemus enim,

$$\begin{aligned}
&\operatorname{Lie}_{\tau_3} Qc(a, \tau_1, \tau_2)\\
&\quad = \tau_3 Qc(a, \tau_1, \tau_2) - Qc(\tau_3(a), \tau_1, \tau_2) + Qc(a, [\tau_1, \tau_3], \tau_2) + Qc(a, \tau_1, [\tau_2, \tau_3])\\
&\quad = \tau_3\langle \tau_2, c(a, \tau_1)\rangle - \langle \tau_2, c(\tau_3(a), \tau_1)\rangle + \langle \tau_2, c(a, [\tau_1, \tau_3]\rangle + \langle [\tau_2, \tau_3], c(a, \tau_1)\rangle\\
&\quad = \langle \tau_2, \operatorname{Lie}_{\tau_3} c(a, \tau_1)\rangle.
\end{aligned}$$

2.14.

Secundo,

$$d'_{DR}Q = Qd'_{DR}$$

Computamus enim,

$$\begin{aligned} d'_{DR}Qc(a,\tau_1,\tau_2,\tau_3) &= -\frac{1}{2}\tau_2 Qc(a,\tau_1,\tau_3) = -\frac{1}{2}\tau_2\langle\tau_3, c(a,\tau_1)\rangle \\ &= \left\langle \tau_2, -\frac{1}{2}d\langle\tau_3, c(a,\tau_1)\rangle\right\rangle \\ &= \langle\tau_2, d'_{DR}c(a,\tau_1,\tau_3)\rangle = Qd'_{DR}c(a,\tau_1,\tau_2,\tau_3), \end{aligned}$$

quod trahit effatum lemmatis.

(d) *Paries rectus*

2.15.

Revocamus operatores:

$$Q:\ \mathrm{Hom}(\Lambda^2 T,\Omega) \longrightarrow \mathrm{Hom}(S^2T\otimes T, A)$$

definitum per formulam:

$$Qc(\tau_1,\tau_2,\tau_3) = \langle\tau_1, c(\tau_2,\tau_3)\rangle + \langle\tau_2, c(\tau_1,\tau_3)\rangle$$

atque

$$Q:\ \mathrm{Hom}(A\otimes T^{\otimes 2},\Omega) \longrightarrow \mathrm{Hom}(A\otimes T^{\otimes 3}, A)$$

definitum per regulam:

$$Qc(a,\tau_1,\tau_2,\tau_3) = \langle\tau_2, c(a,\tau_1,\tau_3)\rangle.$$

2.16. Lemma. $d_H Q = Qd_H$.

Si enim $c\in\mathrm{Hom}(\Lambda^2 T,\Omega)$, habemus:

$$\begin{aligned} d_H Qc(a,\tau_1,\tau_2,\tau_3) &= Qc(a\tau_1,\tau_2,\tau_3) - aQc(a\tau_1,\tau_2,\tau_3) \\ &= \langle a\tau_1, c(\tau_2,\tau_3)\rangle + \langle\tau_2, c(a\tau_1,\tau_3)\rangle - a\langle\tau_1, c(\tau_2,\tau_3)\rangle \\ &\quad - a\langle\tau_2, c(\tau_1,\tau_3)\rangle \\ &= \langle\tau_2, d_H c(a,\tau_1,\tau_3)\rangle = Qd_H c(a,\tau_1,\tau_2,\tau_3) \end{aligned}$$

(e) *Paries sinister*

2.17.

Revocamus operatores:

$$Q : \mathrm{Hom}(T, \Omega) \longrightarrow \mathrm{Hom}(S^2T, A)$$

definitur per regulam:

$$Qc(\tau_1, \tau_2) = \langle \tau_1, c(\tau_2) \rangle + \langle \tau_2, c(\tau_1) \rangle$$

atque

$$Q : \mathrm{Hom}(A \otimes T, \Omega) \longrightarrow \mathrm{Hom}(A \otimes T^{\otimes 2}, A)$$

definitur per formulam:

$$Qc(a, \tau_1, \tau_2) = \langle \tau_2, c(a, \tau_1) \rangle.$$

2.18. Lemma. *Fit* $d_H Q = Q d_H$.

Quod probatur eadem ratione ut in art. 2.16.

3. Tabulatum primum

(a) *Paries recessus*

3.1.

Determinamus sagittulas

$$d_{DR} : \mathrm{Hom}(A^{\otimes 2} \otimes T, \Omega) \longrightarrow \mathrm{Hom}(A^{\otimes 2} \otimes T^{\otimes 2}, \Omega)$$

per formulam

$$\begin{aligned} d_{DR}c(a, b, \tau, \tau') &= \tau' c(a, b, \tau) - c(\tau'(a), b, \tau) - c(a, \tau'(b), \tau) + c(a, b, [\tau, \tau']) \\ &\quad - \frac{1}{2} d \langle \tau', c(a, b, \tau) \rangle \\ &= \mathrm{Lie}_{\tau'} c(a, b, \tau) - \frac{1}{2} d \langle \tau', c(a, b, \tau) \rangle; \end{aligned}$$

(commodum est introducere operatores

$$\mathrm{Lie}\, c(a, b, \tau, \tau') := \mathrm{Lie}_{\tau'} c(a, b, \tau)$$

atque

$$d'_{DR} c(a, b, \tau, \tau') := -\frac{1}{2} d \langle \tau', c(a, b, \tau) \rangle,$$

ergo $d_{DR} = \mathrm{Lie} + d'_{DR}$);

3.2.

ac

$$d_H : \mathrm{Hom}(A \otimes T, \Omega) \longrightarrow \mathrm{Hom}(A^{\otimes 2} \otimes T, \Omega)$$

per regulam

$$d_H c(a, b, \tau) = ac(b, \tau) - c(ab, \tau) + c(a, b\tau);$$

deinde

$$d_H : \mathrm{Hom}(A \otimes T^{\otimes 2}, \Omega) \longrightarrow \mathrm{Hom}(A^{\otimes 2} \otimes T^{\otimes 2}, \Omega)$$

per formulam

$$d_H c(a, b, \tau, \tau') = ac(b, \tau, \tau') - c(ab, \tau, \tau') + c(a, b\tau, \tau');$$

3.3.

$$Q : \mathrm{Hom}(A \otimes T, \Omega) \longrightarrow \mathrm{Hom}(A \otimes T^{\otimes 2}, A)$$

per regulam

$$Qc(a, \tau, \tau') = \langle \tau', c(a, \tau) \rangle,$$

3.4.

denique

$$M : \mathrm{Hom}(A \otimes T^{\otimes 2}, A) \longrightarrow \mathrm{Hom}(A^{\otimes 2} \otimes T^{\otimes 2}, \Omega)$$

per formulam

$$Mc(a, b, \tau, \tau') = \frac{1}{2} da\, c(b, \tau, \tau').$$

3.5. Lemma. *Fit* $d_H d_{DR} = d_{DR} d_H + MQ$.

3.6. Demonstratio. Pro $c \in \mathrm{Hom}(A \otimes T, \Omega)$ habebimus

$$d_H d_{DR} c(a, b, \tau, \tau') = d_H \{\mathrm{Lie} + d'_{DR}\} c(a, b, \tau, \tau').$$

3.7.

Primo, derivatio Lietiana et differentiale Hochschildianum commutant:

$$d_H \mathrm{Lie}\; c(a, b, \tau, \tau') = \mathrm{Lie}\; d_H c(a, b, \tau, \tau').$$

3.8.

Secundo, fit:

$$d_H d'_{DR} c(a, b, \tau, \tau') = -\frac{1}{2}\{ad\langle\tau', c(b, \tau)\rangle - d\langle\tau', c(ab, \tau)\rangle + d\langle\tau', c(a, b\tau)\rangle\}.$$

Addemus huc:

$$0 = \frac{1}{2}\{-d\langle\tau', ac(b, \tau)\rangle + da\ \langle\tau', c(b, \tau)\rangle + ad\langle\tau', c(b, \tau)\}.$$

Adipiscemur:

$$d_H d'_{DR} c(a, b, \tau, \tau') = -\frac{1}{2}d\langle\tau', d_H c(a, b, \tau)\rangle + \frac{1}{2}da\ \langle\tau', c(b, \tau)\rangle$$

ubi manifesto

$$\frac{1}{2}da\ \langle\tau', c(b, \tau)\rangle = MQc(a, b, \tau, \tau')$$

unde lemma sequitur.

(b) *Paries sinister*

3.9.

Contemplemur operatores

$$Q : \mathrm{Hom}(A \otimes T, \Omega) \longrightarrow \mathrm{Hom}(A \otimes T^{\otimes 2}, A),$$

per regulam

$$Qc(a, \tau, \tau') = \langle\tau', c(a, \tau)\rangle,$$

definitur, tamquam in art. 2.17, atque

$$Q : \mathrm{Hom}(A^{\otimes 2} \otimes T, \Omega) \longrightarrow \mathrm{Hom}(A^{\otimes 2} \otimes T^{\otimes 2}, A)$$

per formulam

$$Qc(a, b, \tau, \tau') = \langle\tau', c(a, b, \tau)\rangle$$

definitur.

3.10. Lemma. *Fit* $d_H Q = Q d_H$.

3.11. Demonstratio. Pro $c \in \mathrm{Hom}(A \otimes T, \Omega)$ habeatur

$$\begin{aligned} d_H Qc(a, b, \tau, \tau') &= aQc(b, \tau, \tau') - Qc(ab, \tau, \tau') + Qc(a, b\tau, \tau') \\ &= a\langle\tau', c(b, \tau)\rangle - \langle\tau', c(ab, \tau)\rangle + \langle\tau', c(a, b\tau)\rangle \\ &\quad - \langle\tau', d_H c(a, b, \tau)\rangle = Qd_H c(a, b, \tau, \tau'), \end{aligned}$$

qed

(c) *Paries rectus*

3.12.

Contemplemur operatores:

$$Q: \operatorname{Hom}(A \otimes T^{\otimes 2}, \Omega) \longrightarrow \operatorname{Hom}(A \otimes T^{\otimes 3}, A)$$

per regulam

$$Qc(a, \tau_1, \tau_2, \tau_3) = \langle \tau_2, c(a, \tau_1, \tau_3) \rangle,$$

definitur, et

$$Q: \operatorname{Hom}(A^{\otimes 2} \otimes T^{\otimes 2}, \Omega) \longrightarrow \operatorname{Hom}(A^{\otimes 2} \otimes T^{\otimes 3}, A)$$

per formulam

$$Qc(a, b, \tau_1, \tau_2, \tau_3) = \langle \tau_2, c(a, b, \tau_1, \tau_3) \rangle,$$

definitur.

3.13. Lemma. *Fit* $d_H Q = Q d_H$.

Quod probatur eodem modo ut in art. 3.11.

(d) *Camera*

3.14.

Contemplemur operatores: primo, sagittulam

$$d_{DR}: \operatorname{Hom}(A^{\otimes 2} \otimes T, \Omega) \longrightarrow \operatorname{Hom}(A^{\otimes 2} \otimes T^{\otimes 2}, \Omega)$$

per formulam

$$d_{DR} c(a, b, \tau, \tau') = \operatorname{Lie}_{\tau'} c(a, b, \tau) - \frac{1}{2} d \langle \tau', c(a, b, \tau) \rangle$$

definitam (vide art. 3.1); secundo, sagittulam novam,

$$d_{DR}: \operatorname{Hom}(A^{\otimes 2} \otimes T^{\otimes 2}, A) \longrightarrow \operatorname{Hom}(A^{\otimes 2} \otimes T^{\otimes 3}, A)$$

per formulam

$$d_{DR} c(a, b, \tau_1, \tau_2, \tau_3) = \operatorname{Lie}_{\tau_3} c(a, b, \tau_1, \tau_2) - \frac{1}{2} \tau_2 c(a, b, \tau_1, \tau_3)$$

definitam.

3.15. Lemma. *Fit* $Q d_{DR} = d_{DR} Q$.

3.16.

Primo, Q Lie $=$ Lie Q.

Habeatur enim,

$$
\begin{aligned}
Q\mathrm{Lie}c(a,b,\tau_1,\tau_2,\tau_3) &= \langle \tau_2, \mathrm{Lie}_{tau_3} c(a,b,\tau_1)\rangle \\
&= \langle \tau_2, \tau_3 c(a,b,\tau_1) - c(\tau_3(a),b,\tau_1) - c(a,\tau_3(b),\tau_1) \\
&\quad + c(a,b,[\tau_1,\tau_3])\rangle \\
&= \tau_3\langle \tau_2, c(a,b,\tau_1)\rangle + \langle [\tau_2,\tau_3], c(a,b,\tau_1)\rangle \\
&\quad + \langle \tau_2, -c(\tau_3(a),b,\tau_1) - c(a,\tau_3(b),\tau_1) + c(a,b,[\tau_1,\tau_3])\rangle \\
&= \tau_3 Qc(a,b,\tau_1,\tau_2) + Qc(a,b,\tau_1,[\tau_2,\tau_3]) \\
&\quad - Qc(\tau_3(a),b,\tau_1,\tau_2) - Qc(a,\tau_3(b),\tau_1,\tau_2) \\
&\quad + Qc(a,b,[\tau_1,\tau_3],\tau_2) \\
&= \mathrm{Lie}_{\tau_3} Qc(a,b,\tau_1,\tau_2), \qquad \text{qed}
\end{aligned}
$$

3.17.

Secundo,

$$
\begin{aligned}
Qd'_{DR}c(a,b,\tau_1,\tau_2,\tau_3) &= \left\langle \tau_2, -\frac{1}{2} d\langle \tau_3, c(a,b,\tau_1)\rangle \right\rangle \\
&= -\frac{1}{2}\tau_2\tau_3, c(a,b,\tau_1)\Big\rangle\Big\rangle = -\frac{1}{2}\tau_2 Qc(a,b,\tau_1,\tau_3) \\
&= d'_{DR} Qc(a,b,\tau_1,\tau_2,\tau_3),
\end{aligned}
$$

unde lemma sequitur.

(e) *Frons*

3.18.

Revocamus sagittulas:

$$d_{DR}: \mathrm{Hom}(A\otimes T^{\otimes 2}, A) \longrightarrow \mathrm{Hom}(A\otimes T^{\otimes 3}, A)$$

per formulam

$$d_{DR}c(a,\tau_1,\tau_2,\tau_3) = \mathrm{Lie}_{\tau_3}c(a,\tau_1,\tau_2) - \frac{1}{2}\tau_2 c(a,\tau_1,\tau_3),$$

definitam, vide art. 2.7, atque

$$d_{DR}: \mathrm{Hom}(A^{\otimes 2}\otimes T^{\otimes 2}, A) \longrightarrow \mathrm{Hom}(A^{\otimes 2}\otimes T^{\otimes 3}, A)$$

per regulam

$$d_{DR}c(a,b,\tau_1,\tau_2,\tau_3) = \mathrm{Lie}_{\tau_3}c(a,b,\tau_1,\tau_2) - \frac{1}{2}\tau_2 c(a,b,\tau_1,\tau_3),$$

definitam, vide art. 3.14, ac denique

$$M : \mathrm{Hom}(A \otimes T^{\otimes 2}, A) \longrightarrow \mathrm{Hom}(A^{\otimes 2} \otimes T^{\otimes 2}, \Omega)$$

per formulam

$$Mc(a,b,\tau,\tau') = \frac{1}{2}da\ c(b,\tau,\tau'),$$

definitam, vide art. 3.4.

3.19. Lemma. *Fit* $d_H d_{DR} = d_{DR} d_H + QM$.

3.20. Demonstratio. Primo, fit $d_H\ \mathrm{Lie} = \mathrm{Lie}\ d_H$.
Habeatur enim:

$$d_H\ \mathrm{Lie}c(a,b,\tau_1,\tau_2,\tau_3) = a\mathrm{Lie}_{\tau_3}c(b,\tau_1,\tau_2) - \mathrm{Lie}_{\tau_3}c(ab,\tau_1,\tau_2) + \mathrm{Lie}_{\tau_3}c(a,b\tau_1,\tau_2),$$

ubi

$$\begin{aligned} a\mathrm{Lie}_{\tau_3}c(b,\tau_1,\tau_2) &= a\tau_3 c(b,\tau_1,\tau_2) - ac(\tau_3(b),\tau_1,\tau_2) + ac(b,[\tau_1,\tau_3],\tau_2) \\ &\quad + ac(b,\tau_1,[\tau_2,\tau_3]); \\ -\mathrm{Lie}_{\tau_3}c(ab,\tau_1,\tau_2) &= -\tau_3 c(ab,\tau_1,\tau_2) + c(\tau_3(a)b + a\tau_3(b),\tau_1,\tau_2) \\ &\quad - c(ab,[\tau_1,\tau_3],\tau_2) - c(ab,\tau_1,[\tau_2,\tau_3]) \end{aligned}$$

atque

$$\begin{aligned} \mathrm{Lie}_{\tau_3}c(a,b\tau_1,\tau_2) &= \tau_3 c(a,b\tau_1,\tau_2) - c(\tau_3(a),b\tau_1,\tau_2) + c(a,b[\tau_1,\tau_3] \\ &\quad - \tau_3(b)\tau_1,\tau_2) + c(a,b\tau_1,[\tau_2,\tau_3]). \end{aligned}$$

Addemus huc:

$$0 = \tau_3\{ac(b,\tau_1,\tau_2)\} - \tau_3(a)c(b,\tau_1,\tau_2) + a\tau_3 c(b,\tau_1,\tau_2)$$

Post summationem, videamus statim:

$$\begin{aligned} d_H\ \mathrm{Lie}c(a,b,\tau_1,\tau_2,\tau_3) &= \tau_3 d_H c(a,b,\tau_1,\tau_2) - d_H c(\tau_3(a),b,\tau_1,\tau_2) \\ &\quad - d_H c(a,\tau_3(b),\tau_1,\tau_2) + d_H c(a,b,[\tau_1,\tau_3],\tau_2) \\ &\quad + d_H c(a,b,\tau_1,[\tau_2,\tau_3]) \\ &= \mathrm{Lie}_{\tau_3} d_H c(a,b,\tau_1,\tau_2), \end{aligned}$$

qed

3.21.

Secundo,

$$d_H d'_{DR} c(a, b, \tau_1, \tau_2, \tau_3) = -\frac{1}{2}\{a\tau_2 c(b, \tau_1, \tau_3) - \tau_2 c(ab, \tau_1, \tau_3) + \tau_2 c(a, b\tau_1, \tau_3)\}.$$

Addemus huc:

$$0 = -\frac{1}{2}\{\tau_2\{ac(b, \tau_1, \tau_3)\} - \tau_2(a)c(b, \tau_1, \tau_3) - a\tau_2 c(b, \tau_1, \tau_3)\}$$

Post summationem, adipiscemur:

$$d_H d'_{DR} c(a, b, \tau_1, \tau_2, \tau_3) = d'_{DR} d_H c(a, b, \tau_1, \tau_2, \tau_3) + \frac{1}{2}\tau_2(a)c(b, \tau_1, \tau_3),$$

ubi

$$\frac{1}{2}\tau_2(a)c(b, \tau_1, \tau_3) = \left\langle \tau_2, \frac{1}{2}da\; c(b, \tau_1, \tau_3)\right\rangle$$

$$= \langle \tau_2, Mc(b, \tau_1, \tau_2, \tau_3)\rangle = QMc(a, b, \tau_1, \tau_2, \tau_3) \qquad \text{qed}$$

(f) *Junctio*

3.22.

Revocamus sagittulas:

$$M : \operatorname{Hom}(S^2 T, A) \longrightarrow \operatorname{Hom}(A \otimes T^{\otimes 2}, \Omega)$$

per regulam

$$Mc(a, \tau, \tau') = -\frac{1}{2}da\; c(\tau, \tau')$$

definitam, atque

$$M : \operatorname{Hom}(A \otimes T^{\otimes 2}, A) \longrightarrow \operatorname{Hom}(A^{\otimes 2} \otimes T^{\otimes 2}, \Omega)$$

per formulam

$$Mc(a, b, \tau, \tau') = \frac{1}{2}da\; c(b, \tau, \tau')$$

definitam.

3.23. Lemma. *Fit* $d_H M = -M d_H$.

3.24. Demonstratio. Pro elemento $c \in \operatorname{Hom}(S^2 T, A)$, habemus

$$d_H Mc(a, b, \tau, \tau') = aMc(b, \tau, \tau') - Mc(ab, \tau, \tau') + Mc(a, b\tau, \tau')$$

$$= -\frac{1}{2}\{adb\; c(\tau, \tau') - d(ab)c(\tau, \tau') + da\; c(b\tau, \tau')\}$$

$$-\frac{1}{2}\{-da\; bc(\tau, \tau') + da\; c(b\tau, \tau')\} = -\frac{1}{2}dad_H c(b, \tau, \tau')$$

$$= -Md_H c(a, b, \tau, \tau'), \qquad \text{qed}$$

4. Tabulatum secundum

(a) *Paries recessus*

4.1.

Revocamus sagittulam:

$$d_{DR} : \operatorname{Hom}(A^{\otimes 2} \otimes T, \Omega) \longrightarrow \operatorname{Hom}(A^{\otimes 2} \otimes T^{\otimes 2}, \Omega)$$

per formulam

$$d_{DR}c(a, b, \tau, \tau') = \operatorname{Lie}_{\tau'} c(a, b, \tau) - \frac{1}{2} d\langle \tau', c(a, b, \tau)\rangle,$$

definitam (vide art. 3.1) atque introducamus sagittulam:

$$d_{DR} : \operatorname{Hom}(A^{\otimes 3} \otimes T, \Omega) \longrightarrow \operatorname{Hom}(A^{\otimes 3} \otimes T^{\otimes 2}, \Omega)$$

per regulam

$$d_{DR}c(a, b, c, \tau, \tau') = \operatorname{Lie}_{\tau'} c(a, b, c, \tau) - \frac{1}{2} d\langle \tau', c(a, b, c, \tau)\rangle$$

definitam.

4.2.

Determinamus sagittulas:

$$d_H : \operatorname{Hom}(A^{\otimes 2} \otimes T, \Omega) \longrightarrow \operatorname{Hom}(A^{\otimes 3} \otimes T, \Omega)$$

per formulam:

$$d_H c(a, b, c, \tau) = ac(b, c, \tau) - c(ab, c, \tau) + c(a, bc, \tau) - c(a, b, c\tau);$$

porro

$$d_H : \operatorname{Hom}(A^{\otimes 2} \otimes T^{\otimes 2}, \Omega) \longrightarrow \operatorname{Hom}(A^{\otimes 3} \otimes T^{\otimes 2}, \Omega)$$

per regulam:

$$d_H c(a, b, c, \tau, \tau') = ac(b, c, \tau, \tau') - c(ab, c, \tau, \tau') + c(a, bc, \tau, \tau') - c(a, b, c\tau, \tau').$$

4.3.

atque

$$M : \operatorname{Hom}(A^{\otimes 2} \otimes T^{\otimes 2}, A) \longrightarrow \operatorname{Hom}(A^{\otimes 3} \otimes T^{\otimes 2}, \Omega)$$

per formulam:

$$Mc(a, b, c, \tau, \tau') = \frac{1}{2} da\, c(b, c, \tau, \tau').$$

4.4.

Tandem, revocamus sagittulam:

$$Q : \mathrm{Hom}(A^{\otimes 2} \otimes T, \Omega) \longrightarrow \mathrm{Hom}(A^{\otimes 2} \otimes T^{\otimes 2}, A)$$

per formulam

$$Qc(a, b, \tau, \tau') = \langle \tau', c(a, b, \tau) \rangle,$$

definitam, vide 3.9.

4.5. Lemma. *Fit* $d_H d_{DR} = d_{DR} d_H + MQ$.

Quod probatur eodem modo ut in arts. 3.7, 3.8.

(b) *Paries sinister*

4.6.

Revocamus sagittulam:

$$Q : \mathrm{Hom}(A^{\otimes 2} \otimes T, \Omega) \longrightarrow \mathrm{Hom}(A^{\otimes 2} \otimes T^{\otimes 2}, A)$$

per regulam

$$Qc(a, b, \tau, \tau') = \langle \tau', c(a, b, \tau) \rangle,$$

definitam, vide art. 3.9, atque introducamus sagittulam novam:

$$Q : \mathrm{Hom}(A^{\otimes 3} \otimes T, \Omega) \longrightarrow \mathrm{Hom}(A^{\otimes 3} \otimes T^{\otimes 2}, A)$$

per formulam

$$Qc(a, b, c, \tau, \tau') = \langle \tau', c(a, b, c, \tau) \rangle$$

definitam.

4.7. Lemma. *Fit* $d_H Q = Q d_H$.

Quod probatur eodem modo ut in art. 3.11.

(c) *Paries rectus*

4.8.

Revocamus operatorem:

$$Q : \mathrm{Hom}(A^{\otimes 2} \otimes T^{\otimes 2}, \Omega) \longrightarrow \mathrm{Hom}(A^{\otimes 2} \otimes T^{\otimes 3}, A),$$

definitum per formulam:

$$Qc(a, b, \tau_1, \tau_2, \tau_3) = \langle \tau_2, c(a, b, \tau_1, \tau_3) \rangle,$$

vide art. 3.12, ac determinamus operatorem novum:

$$Q : \mathrm{Hom}(A^{\otimes 3} \otimes T^{\otimes 2}, \Omega) \longrightarrow \mathrm{Hom}(A^{\otimes 3} \otimes T^{\otimes 3}, A)$$

per regulam:

$$Qc(a, b, c, \tau_1, \tau_2, \tau_3) = \langle \tau_2, c(a, b, c, \tau_1, \tau_3) \rangle.$$

4.9. Lemma. *Fit* $d_H Q = Q d_H$.

Quod etiam probatur eadem ratione ut in art. 3.11.

(d) *Camera*

4.10.

Contemplemur sagittulam:

$$d_{DR} : \operatorname{Hom}(A^{\otimes 3} \otimes T, \Omega) \longrightarrow \operatorname{Hom}(A^{\otimes 3} \otimes T^{\otimes 2}, \Omega)$$

per regulam

$$d_{DR}c(a, b, c, \tau, \tau') = \operatorname{Lie}_{\tau'} c(a, b, c, \tau) - \frac{1}{2} d\langle \tau', c(a, b, c, \tau)\rangle,$$

definitam, vide art. 4.1, atque introducamus sagittulam novam:

$$d_{DR} : \operatorname{Hom}(A^{\otimes 3} \otimes T^{\otimes 2}, A) \longrightarrow \operatorname{Hom}(A^{\otimes 3} \otimes T^{\otimes 3}, A)$$

per formulam:

$$d_{DR}c(a, b, c, \tau_1, \tau_2, \tau_3) = \operatorname{Lie}_{\tau_3} c(a, b, c, \tau_1, \tau_3) - \frac{1}{2} \tau_2 c(a, b, c, \tau_1, \tau_3).$$

4.11. Lemma. *Fit* $Q d_{DR} = d_{DR} Q$.

Quod probatur simili calculo ut in artt. 3.16, 3.17.

(e) *Frons*

4.12.

Revocamus sagittulas:

$$d_{DR} : \operatorname{Hom}(A^{\otimes 2} \otimes T^{\otimes 2}, A) \longrightarrow \operatorname{Hom}(A^{\otimes 2} \otimes T^{\otimes 3}, A),$$

definitam per:

$$d_{DR}c(a, b, \tau_1, \tau_2, \tau_3) = \operatorname{Lie}_{\tau_3} c(a, b, \tau_1, \tau_2) - \frac{1}{2} \tau_2 c(a, b, \tau_1, \tau_3),$$

vide art. 3.14, porro:

$$d_{DR} : \operatorname{Hom}(A^{\otimes 3} \otimes T^{\otimes 2}, A) \longrightarrow \operatorname{Hom}(A^{\otimes 3} \otimes T^{\otimes 3}, A)$$

definitam per:

$$d_{DR}c(a, b, c, \tau_1, \tau_2, \tau_3) = \operatorname{Lie}_{\tau_3} c(a, b, c, \tau_1, \tau_2) - \frac{1}{2} \tau_2 c(a, b, c, \tau_1, \tau_3),$$

vide art 4.10, denique:

$$M : \mathrm{Hom}(A^{\otimes 2} \otimes T^{\otimes 2}, A) \longrightarrow \mathrm{Hom}(A^{\otimes 3} \otimes T^{\otimes 2}, \Omega)$$

definitam per:

$$Mc(a, b, c, \tau, \tau') = \frac{1}{2} da\, c(b, c, \tau, \tau'),$$

vide art. 4.3.

4.13. Lemma. *Fit* $d_H d_{DR} = d_{DR} d_H + QM$.

Quod probatur omnino simili modo ut in artt. 3.20, 3.21.

(f) *Junctio*

4.14.

Revocamus operatores:

$$M : \mathrm{Hom}(A \otimes T^{\otimes 2}, A) \longrightarrow \mathrm{Hom}(A^{\otimes 2} \otimes T^{\otimes 2}, \Omega)$$

definitum per:

$$Mc(a, b, \tau, \tau') = \frac{1}{2} da\, c(b, \tau, \tau'),$$

vide art. 3.4, et

$$M : \mathrm{Hom}(A^{\otimes 2} \otimes T^{\otimes 2}, A) \longrightarrow \mathrm{Hom}(A^{\otimes 3} \otimes T^{\otimes 2}, \Omega)$$

definitum per:

$$Mc(a, b, c, \tau, \tau') = \frac{1}{2} da\, c(b, c, \tau, \tau'),$$

vide art. 4.3.

4.15. Lemma. *Fit* $d_H\, M = M\, d_H$.

Quod probatur eodem modo ut in art. 3.24.

Pars secunda. Aedificium rectum

1. Pede plana

(a) *Paries recessus*

1.1.

Revocamus operatores:

$$d_{DR} : \mathrm{Hom}(\Lambda^2 T, \Omega) \longrightarrow \mathrm{Hom}(\Lambda^3 T, \Omega)$$

definitum per formulam:

$$d_{DR}c(\tau,\tau',\tau'') = \mathrm{Cycle}_{\tau,\tau',\tau''}\left[c([\tau,\tau'],\tau'') - \tau\{c(\tau',\tau'')\} + \frac{1}{3}d\langle\tau, c(\tau',\tau'')\rangle\right]$$

$$= \mathrm{Alt}_{\tau',\tau''}\left\{c([\tau,\tau'],\tau'') - \frac{1}{2}c(\tau,[\tau',\tau'']) - \frac{1}{2}\tau c(\tau',\tau'')\right.$$

$$\left. + \tau' c(\tau,\tau'') + \frac{1}{6}d\langle\tau, c(\tau',\tau'')\rangle - \frac{1}{3}d\langle\tau', c(\tau,\tau'')\rangle\right\},$$

vide Pars Prima, art. 1.6, atque

$$Q : \mathrm{Hom}(\Lambda^2 T, \Omega) \longrightarrow \mathrm{Hom}(S^2 T \otimes T, A)$$

definitum per:

$$Qc(\tau,\tau',\tau'') = \mathrm{Sym}_{\tau,\tau'}\langle\tau, c(\tau',\tau'')\rangle.$$

1.2.

Introducamus operatores:

$$d_{DR} : \mathrm{Hom}(A \otimes T^{\otimes 2}, \Omega) \longrightarrow \mathrm{Hom}(A \otimes T \otimes \Lambda^2 T, \Omega)$$

per regulam:

$$d_{DR}c(a,\tau,\tau',\tau'') = \mathrm{Alt}_{\tau',\tau''}\left\{\tau' c(a,\tau,\tau'') - c(\tau'(a),\tau,\tau'') + c(a,[\tau,\tau'],\tau'')\right.$$

$$\left. - \frac{1}{2}c(a,\tau,[\tau',\tau'']) - \frac{1}{3}d\langle\tau', c(a,\tau,\tau'')\rangle\right\}$$

$$= \{\mathrm{Lie} + d'_{DR}\}c(a,\tau,\tau',\tau''),$$

ubi

$$\mathrm{Lie}c(a,\tau,\tau',\tau'') = \mathrm{Alt}_{\tau',\tau''}\left\{\tau' c(a,\tau,\tau'') - c(\tau'(a),\tau,\tau'')\right.$$

$$\left. + c(a,[\tau,\tau'],\tau'') - \frac{1}{2}c(a,\tau,[\tau',\tau''])\right\}$$

et

$$d'_{DR}c(a,\tau,\tau',\tau'') = -\frac{1}{3}\mathrm{Alt}_{\tau',\tau''}d\langle\tau', c(a,\tau,\tau'')\rangle,$$

1.3.

ac

$$M : \mathrm{Hom}(S^2T \otimes T, A) \longrightarrow \mathrm{Hom}(A \otimes T \otimes \Lambda^2 T, \Omega)$$

definitam per:

$$Mc(a,\tau,\tau',\tau'') = -\frac{1}{3} da \ \mathrm{Alt}_{\tau',\tau''} c(\tau,\tau',\tau'').$$

1.4. Lemma. *Fit* $d_H d_{DR} = d_{DR} d_H + MQ$.

1.5. Demonstratio. Habemus

$$d_H d_{DR} c(a,\tau,\tau',\tau'') = d_{DR} c(a\tau,\tau',\tau'') - a d_{DR} c(\tau,\tau',\tau''),$$

ubi

$$\begin{aligned} d_{DR} c(a\tau,\tau',\tau'') = \mathrm{Alt}_{\tau',\tau''} \Big\{ & c([a\tau,\tau'],\tau'') - \frac{1}{2} c(a\tau,[\tau',\tau'']) \\ & - \frac{1}{2}(a\tau) c(\tau',\tau'') + \tau' c(a\tau,\tau'') + \frac{1}{6} d\langle a\tau, c(\tau',\tau'')\rangle \\ & - \frac{1}{3} d\langle \tau', c(a\tau,\tau'')\rangle \Big\} \\ = \mathrm{Alt}_{\tau',\tau''} \Big\{ & c(a[\tau,\tau'] - \tau'(a)\tau, \tau'') \\ & - \frac{1}{2} c(a\tau,[\tau',\tau'']) - \frac{1}{2} a\tau c(\tau',\tau'') - \frac{1}{2} da\langle \tau, c(\tau',\tau'')\rangle \\ & + \tau' c(a\tau,\tau'') + \frac{1}{6} da\langle \tau, c(\tau',\tau'')\rangle + \frac{1}{6} ad\langle \tau, c(\tau',\tau'')\rangle \\ & - \frac{1}{3} d\langle \tau', c(a\tau,\tau'')\rangle \Big\} \end{aligned}$$

ac

$$\begin{aligned} -a d_{DR} c(\tau,\tau',\tau'') = -a \mathrm{Alt}_{\tau',\tau''} \Big\{ & \tau' c(a,\tau,\tau'') - c(\tau'(a),\tau,\tau'') \\ & + c(a,[\tau,\tau'],\tau'') - \frac{1}{2} c(a,\tau,[\tau',\tau'']) \\ & - \frac{1}{3} d\langle \tau', c(a,\tau,\tau'')\rangle \Big\}. \end{aligned}$$

Addemus huc:

$$0 = -\mathrm{Alt}_{\tau',\tau''}\Big[\tau'\{ac(\tau,\tau'')\} - \tau'(a)c(\tau,\tau'') - a\tau'c(\tau,\tau'')$$
$$-\frac{1}{3}\{d\langle\tau', ac(\tau,\tau'')\rangle - da\langle\tau', c(\tau,\tau'')\rangle - ad\langle\tau', c(\tau,\tau'')\rangle\}\Big]$$

Post rationem parvam, effectus proditur.

(b) *Junctio camerae*

1.6.

Determinamus operatorem:

$$R:\ \mathrm{Hom}(A\otimes T^{\otimes 2}, A) \longrightarrow \mathrm{Hom}(A\otimes T\otimes \Lambda^2 T, \Omega)$$

per:

$$Rc(a,\tau,\tau',\tau'') = -\frac{1}{6}\mathrm{Alt}_{\tau',\tau''}d\Big[c(a,[\tau,\tau'],\tau'') + \frac{1}{2}c(a,\tau,[\tau',\tau''])$$
$$+ c(\tau''(a),\tau,\tau')\Big]$$

atque revocamus operatores:

$$d_{DR}:\ \mathrm{Hom}(A\otimes T,\Omega) \longrightarrow \mathrm{Hom}(A\otimes T^{\otimes 2},\Omega)$$

definitum per:

$$d_{DR}c(a,\tau,\tau') = \mathrm{Lie}_{\tau'}c(a,\tau) - \frac{1}{2}d\langle\tau', c(a,\tau)\rangle,$$

vide Pars Prima, art. 2.2, ac

$$Q:\ \mathrm{Hom}(A\otimes T,\Omega) \longrightarrow \mathrm{Hom}(A\otimes T^{\otimes 2}, A)$$

definitum per:

$$Qc(a,\tau,\tau') = \langle\tau', c(a,\tau)\rangle,$$

vide art. 2.17.

1.7. Lemma. *Fit* $d_{DR}^2 = RQ$.

1.8. Demonstratio. Primo, $\mathrm{Lie}^2 = 0$. Vero,

$$\mathrm{Lie}^2 c(a, \tau, \tau', \tau'') = \mathrm{Alt}_{\tau',\tau''}\{\tau' \mathrm{Lie}_{\tau''} c(a, \tau) - \mathrm{Lie}_{\tau''} c(\tau'(a), \tau) + \mathrm{Lie}_{\tau''} c(a, [\tau, \tau'])\} - \mathrm{Lie}_{[\tau',\tau'']} c(a, \tau),$$

ubi

$$\mathrm{Alt}_{\tau',\tau''} \tau' \mathrm{Lie}_{\tau''} c(a, \tau) = \mathrm{Alt}_{\tau',\tau''}\{\tau'\tau'' c(a, \tau) - \tau' c(\tau''(a), \tau) + \tau' c(a, [\tau, \tau''])\};$$

$$-\mathrm{Alt}_{\tau',\tau''} \mathrm{Lie}_{\tau''} c(\tau'(a), \tau) = \mathrm{Alt}_{\tau',\tau''}\{-\tau'' c(\tau'(a), \tau) + c(\tau''\tau'(a), \tau) - c(\tau'(a), [\tau, \tau''])\};$$

$$\mathrm{Alt}_{\tau',\tau''} \mathrm{Lie}_{\tau''} c(a, [\tau, \tau']) = \mathrm{Alt}_{\tau',\tau''}\{\tau'' c(a, [\tau, \tau']) - c(\tau''(a), [\tau, \tau']) + c(a, [[\tau, \tau'], \tau''])\}$$

ac

$$-\mathrm{Lie}_{[\tau',\tau'']} c(a, \tau) = -[\tau', \tau''] c(a, \tau) + c([\tau', \tau''](a), \tau) - c(a, [\tau, [\tau', \tau'']])$$

Addendo adipiscimur protenus 0.

1.9.

Secundo, fit

$$\begin{aligned}\mathrm{Lie} d'_{DR} c(a, \tau, \tau', \tau'') &= \mathrm{Alt}_{\tau',\tau''}\{\tau' d'_{DR} c(a, \tau, \tau'') \\ &\quad - d'_{DR} c(\tau'(a), \tau, \tau'') + d'_{DR} c(a, [\tau, \tau'], \tau'')\} \\ &\quad - d'_{DR} c(a, \tau, [\tau', \tau'']) \\ &= -\frac{1}{2}\big[\mathrm{Alt}_{\tau',\tau''}\{d\tau'\langle\tau'', c(a, \tau)\rangle \\ &\quad - d\langle\tau'', c(\tau'(a), \tau)\rangle + d\langle\tau'', c(a, [\tau, \tau'])\rangle\} \\ &\quad - d\langle[\tau', \tau''], c(a, \tau)\rangle\big]\end{aligned}$$

ubi

$$\tau'\langle\tau'', c(a, \tau)\rangle = \langle[\tau', \tau''], c(a, \tau)\rangle + \langle\tau'', \tau' c(a, \tau)\rangle$$

1.10.

Tertio, fit

$$\begin{aligned}d'_{DR} \mathrm{Lie}\, c(a, \tau, \tau', \tau'') &= -\frac{1}{3}\mathrm{Alt}_{\tau',\tau''} d\langle\tau', \mathrm{Lie}_{\tau''} c(a, \tau)\rangle \\ &= -\frac{1}{3}\mathrm{Alt}_{\tau',\tau''} d\langle\tau', \tau'' c(a, \tau) - c(\tau''(a), \tau) + c(a, [\tau, \tau''])\rangle.\end{aligned}$$

1.11.

Denique fit quarto

$$d'^2_{DR}c(a,\tau,\tau',\tau'') = -\frac{1}{3}\mathrm{Alt}_{\tau',\tau''}d\langle\tau', d'_{DR}c(a,\tau,\tau'')\rangle$$
$$= \frac{1}{6}\mathrm{Alt}_{\tau',\tau''}d\langle\tau', d\langle\tau'', c(a,\tau)\rangle\rangle = \frac{1}{6}\mathrm{Alt}_{\tau',\tau''}d\tau'\langle\tau'', c(a,\tau)\rangle.$$

1.12.

Addendo obtenebimus:

$$d^2_{DR}c(a,\tau,\tau',\tau'') = \{\mathrm{Lie}d'_{DR} + d'_{DR}\mathrm{Lie} + d'^2_{DR}\}c(a,\tau,\tau',\tau'')$$
$$= -\frac{1}{6}d\langle[\tau',\tau''], c(a,\tau)\rangle + \frac{1}{6}\mathrm{Alt}_{\tau',\tau''}\Big\{d\langle\tau', c(a,[\tau,\tau''])\rangle$$
$$- d\langle\tau', c(\tau''(a),\tau)\rangle\Big\}$$
$$= \frac{1}{6}d\big[-Qc(a,\tau,[\tau',\tau''])$$
$$+ \mathrm{Alt}_{\tau',\tau''}\{Qc(a,[\tau,\tau''],\tau') - Qc(\tau''(a),\tau,\tau')\}\big]$$
$$= RQc(a,\tau,\tau',\tau''), \qquad \text{qed}$$

(c) *Junctiones...*

1.13.

Revocamus operatores:

$$R:\ \mathrm{Hom}(S^2T, A) \longrightarrow \mathrm{Hom}(\Lambda^3T, \Omega)$$

definitum per:

$$Rc(\tau,\tau',\tau'') = -\frac{1}{6}\mathrm{Cycle}_{\tau,\tau',\tau''}dc([\tau,\tau'],\tau'')$$
$$= -\frac{1}{6}\big[\mathrm{Alt}_{\tau',\tau''}dc([\tau,\tau'],\tau'') + dc([\tau',\tau''],\tau)\big],$$

vide Pars Prima, 1.7, atque

$$R:\ \mathrm{Hom}(A\otimes T^{\otimes 2}, A) \longrightarrow \mathrm{Hom}(A\otimes T\otimes\Lambda^2T, \Omega)$$

definitum per:

$$Rc(a,\tau,\tau',\tau'') = -\frac{1}{6}d\big[\mathrm{Alt}_{\tau',\tau''}\{c(a,[\tau,\tau'],\tau'') + c(\tau''(a),\tau,\tau')\}$$
$$+ c(a,\tau,[\tau',\tau''])\big],$$

vide art. 1.6.

1.14.

Porro,

$$M:\ \mathrm{Hom}(S^2T,A) \longrightarrow \mathrm{Hom}(A\otimes T^{\otimes 2},\Omega)$$

definitum per:

$$Mc(a,\tau,\tau') = -\frac{1}{2}c(\tau,\tau')da,$$

vide Pars Prima, art. 2.3, ac

$$M:\ \mathrm{Hom}(S^2T\otimes T,A) \longrightarrow \mathrm{Hom}(A\otimes T\otimes\Lambda^2T,\Omega)$$

definitum per

$$Mc(a,\tau,\tau',\tau'') = -\frac{1}{3}da\ \mathrm{Alt}_{\tau',\tau''}c(\tau,\tau',\tau''),$$

vide art. 1.3.

1.15. Lemma. *Fit* $d_H R = Rd_H + Md_{DR} + d_{DR}M$.

1.16. Demonstratio. Primo, habeatur

$$HRc(a,\tau,\tau',\tau'') = Rc(a\tau,\tau',\tau'') - aRc(\tau,\tau',\tau''),$$

ubi

$$Rc(a\tau,\tau',\tau'') = -\frac{1}{6}\big[\mathrm{Alt}_{\tau',\tau''}dc([a\tau,\tau'],\tau'') + dc([\tau',\tau''],a\tau)\big]$$
$$= -\frac{1}{6}\big[\mathrm{Alt}_{\tau',\tau''}dc(a[\tau,\tau'] - \tau'(a)\tau,\tau'') + dc([\tau',\tau''],a\tau)\big]$$

et

$$-aRc(\tau,\tau',\tau'') = -\frac{1}{6}a\big[\mathrm{Alt}_{\tau',\tau''}dc([\tau,\tau'],\tau'') + dc([\tau',\tau''],\tau)\big].$$

Addemus huc:

$$0 = \frac{1}{6}\mathrm{Alt}_{\tau',\tau''}[d\{ac([\tau,\tau'],\tau'')\} - da\,c([\tau,\tau'],\tau'') - adc([\tau,\tau'],\tau'')],$$

$$0 = -\frac{1}{6}\mathrm{Alt}_{\tau',\tau''}[d\{\tau'(a)c(\tau,\tau'')\} - d\tau'(a)\,c(\tau,\tau'') - \tau'(a)dc(\tau,\tau'')]$$

et

$$0 = \frac{1}{6}\mathrm{Alt}_{\tau',\tau''}[d\{ac([\tau',\tau''],\tau)\} - da\, c([\tau',\tau''],\tau) - a\, dc([\tau',\tau''],\tau)],$$

unde post summationem:

$$\begin{aligned} d_H Rc(a,\tau,\tau',\tau'') = -\frac{1}{6}\Big\{ & \mathrm{Alt}_{\tau',\tau''}[dd_H c(a,[\tau,\tau'],\tau'') - dd_H c(\tau'(a),\tau,\tau'')] \\ & + dd_H c(a,\tau,[\tau',\tau'']) + \mathrm{Alt}_{\tau',\tau''}[da\, c([\tau,\tau'],\tau'') \\ & + adc([\tau,\tau'],\tau'') - d\tau'(a)\, c(\tau,\tau'') - \tau'(a)dc(\tau,\tau'')] \\ & + da\, c([\tau',\tau''],\tau) + a\, dc([\tau',\tau''],\tau)\Big\} \\ = {} & Rd_H c(a,\tau,\tau',\tau'') + \mathrm{Alt}_{\tau',\tau''}[da\, c([\tau,\tau'],\tau'') \\ & + adc([\tau,\tau'],\tau'') - d\tau'(a)\, c(\tau,\tau'') - \tau'(a)dc(\tau,\tau'')] \\ & + da\, c([\tau',\tau''],\tau) + a\, dc([\tau',\tau''],\tau). \end{aligned}$$

1.17.

Secundo autem,

$$Md_{DR}c(a,\tau,\tau',\tau'') = -\frac{1}{3}da\ \mathrm{Alt}_{\tau',\tau''}d_{DR}c(\tau,\tau',\tau'')$$

(vide Pars Prima, art. 1.3)

$$= -\frac{1}{3}da\ \mathrm{Alt}_{\tau',\tau''}\{c(\tau,[\tau',\tau'']) + c(\tau',[\tau,\tau'']) - \frac{1}{2}\tau'c(\tau,\tau'') + \tau''c(\tau,\tau')\}$$

et

$$\begin{aligned} d_{DR}Mc(a,\tau,\tau',\tau'') = \mathrm{Alt}_{\tau',\tau''}\Big\{ & \tau'Mc(a,\tau,\tau'') - Mc(\tau'(a),\tau,\tau'') \\ & + Mc(a,[\tau,\tau'],\tau'') - \frac{1}{3}d\langle\tau', Mc(a,\tau,\tau'')\rangle\Big\} \\ & - Mc(a,\tau,[\tau',\tau'']), \end{aligned}$$

ubi

$$\tau'Mc(a,\tau,\tau'') = -\frac{1}{2}\{\tau'da\, c(\tau,\tau'') + da\, \tau'c(\tau,\tau'')\};$$

$$-Mc(\tau'(a),\tau,\tau'') = \frac{1}{2}d\tau'(a)\, c(\tau,\tau'');$$

$$Mc(a, [\tau, \tau'], \tau'') = -\frac{1}{2} da\, c([\tau, \tau'], \tau'');$$

$$-Mc(a, \tau, [\tau', \tau'']) = \frac{1}{2} da\, c(\tau, [\tau', \tau''])$$

et

$$\begin{aligned} -\frac{1}{3} d\langle \tau', Mc(a, \tau, \tau'')\rangle &= \frac{1}{6} d\{\tau'(a) c(\tau, \tau'')\} \\ &= \frac{1}{6}\{d\tau'(a)\, c(\tau, \tau'') + \tau'(a) dc(\tau, \tau'')\}, \end{aligned}$$

unde, post summationem,

$$\begin{aligned} \{M d_{DR} + d_{DR} M\} c(a, \tau, \tau', \tau'') = \mathrm{Alt}_{\tau', \tau''}[&da\, c([\tau, \tau'], \tau'') + adc([\tau, \tau'], \tau'') \\ &- d\tau'(a)\, c(\tau, \tau'') - \tau'(a) dc(\tau, \tau'')] \\ &+ da\, c([\tau', \tau''], \tau) + a\, dc([\tau', \tau''], \tau), \end{aligned}$$

unde lemma nostrum sponte sequitur.

(d) *Paries rectus*

1.18.

Revocamus operatorem:

$$Q : \mathrm{Hom}(\Lambda^3 T, \Omega) \longrightarrow \mathrm{Hom}(S^2 T \otimes \Lambda^2 T, A),$$

definitum per

$$Qc(\tau_1, \tau_2, \tau_3, \tau_4) = \mathrm{Sym}_{1,2}\langle \tau_1, c(\tau_2, \tau_3, \tau_4)\rangle$$

atque introducamus operatorem

$$Q : \mathrm{Hom}(A \otimes T \otimes \Lambda^2 T, \Omega) \longrightarrow \mathrm{Hom}(A \otimes T^{\otimes 2} \otimes \Lambda^2 T, A)$$

per

$$Qc(a, \tau_1, \tau_2, \tau_3, \tau_4) = \langle \tau_2, c(a, \tau_1, \tau_3, \tau_4)\rangle.$$

1.19. Lemma. *Fit* $d_H Q = Q d_H$.

Demonstratio. Exstat

$$d_H Qc(a, \tau_1, \tau_2, \tau_3, \tau_4) = Qc(a\tau_1, \tau_2, \tau_3, \tau_4) - a Qc(\tau_1, \tau_2, \tau_3, \tau_4),$$

ubi

$$Q(a\tau_1, \tau_2, \tau_3, \tau_4) = \langle a\tau_1, c(\tau_2, \tau_3, \tau_4)\rangle + \langle \tau_2, c(a\tau_1, \tau_3, \tau_4)\rangle$$
$$= a\langle \tau_1, c(\tau_2, \tau_3, \tau_4)\rangle + \langle \tau_2, c(a\tau_1, \tau_3, \tau_4)\rangle,$$

unde

$$d_H Qc(a, \tau_1, \tau_2, \tau_3, \tau_4) = \langle \tau_2, d_H c(\tau_1, \tau_3, \tau_4)\rangle$$

Hinc lemma nostra sequitur sponte.

(e) *Camera*

1.20.

Revocamus sagittulam

$$D_{DR} : \operatorname{Hom}(A \otimes T^{\otimes 2}, \Omega) \longrightarrow \operatorname{Hom}(A \otimes T \otimes \Lambda^2 T, \Omega)$$

per formulam

$$d_{DR}c(a, \tau, \tau', \tau'') = \operatorname{Alt}_{\tau',\tau''}\Big\{\tau' c(a, \tau, \tau'') - c(\tau'(a), \tau, \tau'')$$
$$+ c(a, [\tau, \tau'], \tau'') - \frac{1}{2}c(a, \tau, [\tau', \tau'']) - \frac{1}{3}d\langle \tau', c(a, \tau, \tau'')\rangle\Big\}$$

definitam, vide art. 1.2.

Eadem definitio etiam ita scriberi potest:

$$d_{DR}c(a, \tau, \tau', \tau'') = \operatorname{Alt}_{\tau',\tau''}\operatorname{Lie}_{\tau'}c(a, \tau, \tau'') + c(a, \tau, [\tau', \tau''])$$
$$- \frac{1}{3}\operatorname{Alt}_{\tau',\tau''} d\langle \tau', c(a, \tau, \tau'')\rangle.$$

Introducamus autem operatorem:

$$d_{DR} : \operatorname{Hom}(A \otimes T^{\otimes 3}, A) \longrightarrow \operatorname{Hom}(A \otimes T^{\otimes 2} \otimes \Lambda^2 T, A)$$

per formulam:

$$d_{DR}c(a, \tau_1, \tau_2, \tau_3, \tau_4) = \operatorname{Alt}_{3,4}\Big\{\operatorname{Lie}_{\tau_3}c(a, \tau_1, \tau_2, \tau_4) - \frac{1}{3}\tau_2 c(a, \tau_1, \tau_3, \tau_4)\Big\}$$
$$+ c(a, \tau_1, \tau_2, [\tau_3, \tau_4]).$$

1.21. Lemma. *Fit* $Qd_{DR} = d_{DR}Q$.

1.22. Demonstratio. Exstat

$$Qd_{DR}c(a,\tau_1,\tau_2,\tau_3,\tau_4) = \langle \tau_2, d_{DR}c(a,\tau_1,\tau_3,\tau_4)\rangle$$
$$= \mathrm{Alt}_{3,4}\Big\langle \tau_2, \mathrm{Lie}_{\tau_3} c(a,\tau_1,\tau_4) + \frac{1}{2}c(a,\tau_1,[\tau_3,\tau_4])$$
$$- \frac{1}{3}d\langle \tau_3, c(a,\tau_1,\tau_4)\rangle\Big\rangle.$$

Probatur sponte, primo:

$$\langle \tau_2, \mathrm{Lie}_{\tau_3} c(a,\tau_1,\tau_4)\rangle = \mathrm{Lie}_{\tau_3} Qc(a,\tau_1,\tau_2,\tau_4);$$

Secundo, manifesto:

$$\langle \tau_2, c(a,\tau_1,[\tau_3,\tau_4])\rangle = Qc(a,\tau_1,\tau_2,[\tau_3,\tau_4])$$

et

$$\langle \tau_2, d\langle \tau_3, c(a,\tau_1,\tau_4)\rangle\rangle = \tau_2\langle \tau_3, c(a,\tau_1,\tau_4)\rangle\rangle = \tau_2 Qc(a,\tau_1,\tau_3,\tau_4),$$

unde lemma nostrum statim sequitur.

(f) *Frons*

1.23. Lemma. *Fit* $d_H d_{DR} = d_{DR} d_H + QM$.

1.24. Demonstratio. Exstat:

$$d_H d_{DR} c(a,\tau_1,\tau_2,\tau_3,\tau_4) = d_{DR}c(a\tau_1,\tau_2,\tau_3,\tau_4) - a d_{DR}c(\tau_1,\tau_2,\tau_3,\tau_4)$$

Sed (vide Pars Prima, art. 1.14)

$$d_{DR}c(a\tau_1,\tau_2,\tau_3,\tau_4) = -c(a\tau_1,\tau_2,[\tau_3,\tau_4]) + \mathrm{Alt}_{3,4}\Big\{c(a\tau_1,[\tau_2,\tau_3],\tau_4)$$
$$+ c(\tau_2, a[\tau_1,\tau_3] - \tau_3(a)\tau_1, \tau_4) - \tau_4 c(a\tau_1,\tau_2,\tau_3)$$
$$- \frac{1}{3}a\tau_1 c(\tau_2,\tau_3,\tau_4) - \frac{1}{3}\tau_2 c(a\tau_1,\tau_3,\tau_4)\Big\}.$$

Addemus huc:

$$0 = \mathrm{Alt}_{3,4}\{\tau_4\{ac(\tau_1,\tau_2,\tau_3)\} - \tau_4(a)c(\tau_1,\tau_2,\tau_3) - a\tau_4 c(\tau_1,\tau_2,\tau_3)\}$$

cum

$$0 = \frac{1}{3}\mathrm{Alt}_{3,4}\{\tau_2\{ac(\tau_1,\tau_3,\tau_4)\} - \tau_2(a)c(\tau_1,\tau_3,\tau_4) - a\tau_2 c(\tau_1,\tau_3,\tau_4)\}.$$

Post summationem obtinemus statim:

$$d_H d_{DR} c(a, \tau_1, \tau_2, \tau_3, \tau_4) = d_{DR} d_H c(a, \tau_1, \tau_2, \tau_3, \tau_4) - \frac{1}{3}\mathrm{Alt}_{3,4}\tau_2(a)c(\tau_1, \tau_3, \tau_4),$$

ubi

$$-\frac{1}{3}\mathrm{Alt}_{3,4}\tau_2(a)c(\tau_1, \tau_3, \tau_4) = \left\langle \tau_2, -\frac{1}{3}\mathrm{Alt}_{3,4}da\ c(\tau_1, \tau_3, \tau_4)\right\rangle = QMc(a, \tau_1, \tau_2, \tau_3, \tau_4),$$ qed

(g) *Junctio camerae altera*

1.25.

Revocamus operatores:

$$d_{DR} : \mathrm{Hom}(A \otimes T^{\otimes 2}, A) \longrightarrow \mathrm{Hom}(A \otimes T^{\otimes 3}, A),$$

ubi

$$d_{DR} c(a, \tau_1, \tau_2, \tau_3) = \mathrm{Lie}_{\tau_3} c(a, \tau_1, \tau_2) - \frac{1}{2}\tau_2 c(a, \tau_1, \tau_3) =: \{L + d'_{DR}\}c(a, \tau_1, \tau_2, \tau_3)$$

(vide Pars Prima, art. 2.7),

$$d_{DR} : \mathrm{Hom}(A \otimes T^{\otimes 3}, A) \longrightarrow \mathrm{Hom}(A \otimes T^{\otimes 2} \otimes \Lambda^2 T, A),$$

ubi

$$d_{DR} c(a, \tau_1, \tau_2, \tau_3, \tau_4) = \mathrm{Alt}_{3,4}\mathrm{Lie}_{\tau_3} c(a, \tau_1, \tau_2, \tau_4) + c(a, \tau_1, \tau_2, [\tau_3, \tau_4]) - \frac{1}{3}\mathrm{Alt}_{3,4}\tau_2 c(a, \tau_1, \tau_3, \tau_4) =: \{L + d'_{DR}\}c(a, \tau_1, \tau_2, \tau_3),$$

vide art 1.20; tandem,

$$R : \mathrm{Hom}(A \otimes T^{\otimes 2}, A) \longrightarrow \mathrm{Hom}(A \otimes T \otimes \Lambda^2 T, \Omega),$$

ubi

$$Rc(a, \tau, \tau', \tau'') = -\frac{1}{6}\mathrm{Alt}_{\tau', \tau''} d\{c(a, [\tau, \tau'], \tau'') - c(\tau'(a), \tau, \tau'')\} - \frac{1}{6}c(a, \tau, [\tau', \tau'']),$$

vide art. 1.6.

1.26. Lemma. *Fit* $d_{DR}^2 = QR$.

1.27. Demonstratio. Habemus $d_{DR}^2 = \{L + d'_{DR}\}^2$.
Primo, ostendetur, posito

$$\text{Lie } c(a, \tau_1, \tau_2, \tau_3) := \text{Lie}_{\tau_3} c(a, \tau_1, \tau_2)$$

et

$$\text{Lie } c(a, \tau_1, \tau_2, \tau_3, \tau_4) := \text{Lie}_{\tau_4} c(a, \tau_1, \tau_2, \tau_3),$$

habebimus

$$\text{Alt}_{3,4} \text{Lie}^2 c(a, \tau_1, \tau_2) = -\text{Lie}_{[\tau_3, \tau_4]} c(a, \tau_1, \tau_2),$$

Hinc subito fluit $L^2 = 0$.

1.28.

Secundo, videamus post rationem:

$$\begin{aligned}
&\{L d'_{DR} + d'_{DR} L + d'^2_{DR}\} c(a, \tau_1, \tau_2, \tau_3, \tau_4) \\
&= -\frac{1}{6} \text{Alt}_{3,4} \tau_2 \left\{ c(a, [\tau_1, \tau_3], \tau_4) - c(\tau_3(a), \tau_1, \tau_4) + \frac{1}{2} c(a, \tau_1, [\tau_3, \tau_4]) \right\} \\
&= \left\langle \tau_2, -\frac{1}{6} \text{Alt}_{3,4} d \left\{ c(a, [\tau_1, \tau_3], \tau_4) - c(\tau_3(a), \tau_1, \tau_4) + \frac{1}{2} c(a, \tau_1, [\tau_3, \tau_4]) \right\} \right\rangle \\
&= QRc(a, \tau_1, \tau_2, \tau_3, \tau_4),
\end{aligned}$$

qed

2. Tabulatum primum

(a) *Paries recessus*

2.1.

Revocamus operatorem:

$$d_{DR} : \text{Hom}(A \otimes T^{\otimes 2}, \Omega) \longrightarrow \text{Hom}(A \otimes T \otimes \Lambda^2 T, \Omega),$$

ubi

$$\begin{aligned}
d_{DR} c(a, \tau, \tau', \tau'') &= \text{Alt}_{\tau', \tau''} \text{Lie}_{\tau'} c(a, \tau, \tau'') + c(a, \tau, [\tau', \tau'']) \\
&\quad - \frac{1}{3} \text{Alt}_{\tau', \tau''} d \langle \tau', c(a, \tau, \tau'') \rangle,
\end{aligned}$$

sive

$$d_{DR} c(a, \tau, \tau', \tau'') = \{L + d'_{DR}\} c(a, \tau, \tau', \tau'')$$

ubi

$$Lc(a, \tau, \tau', \tau'') = \mathrm{Alt}_{\tau', \tau''}\mathrm{Lie}_{\tau'}c(a, \tau, \tau'') + c(a, \tau, [\tau', \tau'']),$$

vide art. 1.20, et definimus operatores:

$$d_{DR} : \mathrm{Hom}(A^{\otimes 2} \otimes T^{\otimes 2}, \Omega) \longrightarrow \mathrm{Hom}(A^{\otimes 2} \otimes T \otimes \Lambda^2 T, \Omega),$$

ubi

$$d_{DR}c(a, b, \tau, \tau', \tau'') = \mathrm{Alt}_{\tau', \tau''}\mathrm{Lie}_{\tau'}c(a, b, \tau, \tau'')$$

$$+ c(a, b, \tau, [\tau', \tau'']) - \frac{1}{3}\mathrm{Alt}_{\tau', \tau''}d\langle \tau', c(a, b, \tau, \tau'')\rangle,$$

sive

$$d_{DR}c(a, b, \tau, \tau', \tau'') = \{L + d'_{DR}\}c(a, b, \tau, \tau', \tau'')$$

ubi

$$Lc(a, b, \tau, \tau', \tau'') = \mathrm{Alt}_{\tau', \tau''}\mathrm{Lie}_{\tau'}c(a, b, \tau, \tau'') + c(a, b, \tau, [\tau', \tau'']),$$

porro:

$$M : \mathrm{Hom}(A \otimes T^{\otimes 3}, A) \longrightarrow \mathrm{Hom}(A^{\otimes 2} \otimes T \otimes \Lambda^2 T, \Omega),$$

ubi

$$Mc(a, b, \tau, \tau', \tau'') = \frac{1}{3}da \ \mathrm{Alt}_{\tau', \tau''}c(b, \tau, \tau', \tau''),$$

confer Pars Prima, art. 3.4.

2.2. Lemma. *Fit* $d_H d_{DR} = d_{DR} d_H + MQ$.

2.3. Demonstratio. Primo, posito

$$\mathrm{Lie}\ c(a, \tau, \tau', \tau'') = \mathrm{Lie}_{\tau'}c(a, \tau, \tau'')$$

et simili modo

$$\mathrm{Lie}\ c(a, b, \tau, \tau', \tau'') = \mathrm{Lie}_{\tau'}c(a, b, \tau, \tau'')$$

probatur, d_H Lie $=$ Lie d_H, unde subito sequitur $Ld_H = d_H L$.

Secundo, fit

$$d_H d'_{DR}c(a, b, \tau, \tau', \tau'') = d'_{DR}d_H c(a, b, \tau, \tau', \tau'') + \frac{1}{3}da \ \mathrm{Alt}_{\tau', \tau''}\langle \tau', c(b, \tau, \tau'')\rangle,$$

ubi patet

$$\frac{1}{3}da \ \mathrm{Alt}_{\tau', \tau''}\langle \tau', c(b, \tau, \tau'')\rangle = MQc(a, b, \tau, \tau', \tau''),$$

unde manifesto lemma nostrum fluit.

(b) *Junctio camerae*

2.4.

Revocamus operatorem

$$d_{DR} : \operatorname{Hom}(A^{\otimes 2} \otimes T, \Omega) \longrightarrow \operatorname{Hom}(A^{\otimes 2} \otimes T^{\otimes 2}, \Omega)$$

ubi

$$d_{DR}c(a, b, \tau, \tau') = \tau' c(a, b, \tau) - c(\tau'(a), b, \tau) - c(a, \tau'(b), \tau) + c(a, b, [\tau, \tau'])$$
$$- \frac{1}{2} d\langle \tau', c(a, b, \tau)\rangle =:= \{L + d'_{DR}\}c(a, b, \tau, \tau'),$$

vide Pars Prima, art. 3.1, definimusque operatorem

$$R : \operatorname{Hom}(A^{\otimes 2} \otimes T^{\otimes 2}, A) \longrightarrow \operatorname{Hom}(A^{\otimes 2} \otimes T \otimes \Lambda^2 T, \Omega),$$

ubi

$$Rc(a, b, \tau, \tau', \tau'') = -\frac{1}{6} d\big[\mathrm{Alt}_{\tau', \tau''}\{-c(\tau'(a), b, \tau, \tau'') - c(a, \tau'(b), \tau, \tau'')$$
$$+ c(a, b, [\tau, \tau'], \tau'')\} + c(a, b, \tau, [\tau', \tau''])\big],$$

confer 1.6.

2.5. Lemma. *Fit* $d_{DR}^2 = RQ$.

2.6. Demonstratio. Primo, probatur, $L^2 = 0$.
Secundo, computatur:

$$\{L d'_{DR} + d'_{DR} L + d'^2_{DR}\} c(a, b, \tau, \tau', \tau'')$$
$$= -\frac{1}{6} d\big[\langle [\tau', \tau''], c(a, b, \tau)\rangle + \mathrm{Alt}_{\tau', \tau''}\langle \tau'', c(a, b, [\tau, \tau'])$$
$$- c(\tau'(a), b, \tau) - c(a, \tau'(b), \tau)\rangle\big]$$
$$= RQc(a, b, \tau, \tau', \tau''),$$

unde lemma nostrum subito fluit.

(c) *Junctio duarum cellarum tabulati primi*

2.7.

Revocamus operatorem

$$R : \operatorname{Hom}(A \otimes T^{\otimes 2}, A) \longrightarrow \operatorname{Hom}(A \otimes T \otimes \Lambda^2 T, \Omega)$$

ubi

$$Rc(a,\tau,\tau',\tau'') = -\frac{1}{6}d\big[\mathrm{Alt}_{\tau',\tau''}\{c(a,[\tau,\tau'],\tau'') - c(\tau'(a),\tau,\tau'')\} + c(a,\tau,[\tau',\tau''])\big],$$

vide art. 1.6 vel art. 1.13.

2.8. Lemma. *Fit* $d_H R = R d_H + d_{DR} M + M d_{DR}$.

Confer art. 1.15.

2.9. Demonstratio. Fit

$$d_H Rc(a,b,\tau,\tau',\tau'') = aRc(a,b,\tau,\tau',\tau'') - Rc(ab,\tau,\tau',\tau'') + Rc(a,b\tau,\tau',\tau'').$$

Primo, computatur methodo simili ut in art. 1.16:

$$\{d_H R - R d_H\}c(a,b,\tau,\tau',\tau'') = \frac{1}{6}da\; d[\mathrm{Alt}_{\tau',\tau''}\{c(b,[\tau,\tau'],\tau'') - c(\tau'(b),\tau,\tau'')\} - c(b,\tau,[\tau',\tau''])] - \frac{1}{6}\mathrm{Alt}_{\tau',\tau''} d\{\tau'(a)c(b,\tau,\tau'')\}$$

Secundo, ostendetur $\{d_{DR}M + Md_{DR}\}c(a,b,\tau,\tau',\tau'')$ eamdem responsionem praebere, unde lemma fluit.

(d) *Paries rectus*

2.10.

Introducamus operatorem:

$$Q:\ \mathrm{Hom}(A^{\otimes 2}\otimes T\otimes \Lambda^2 T,\Omega) \longrightarrow \mathrm{Hom}(A^{\otimes 2}\otimes T^{\otimes 2}\otimes \Lambda^2 T, A),$$

ubi

$$Qc(a,b,\tau_1,\tau_2,\tau_3,\tau_4) = \langle \tau_2, c(a,b,\tau_1,\tau_3,\tau_4)\rangle.$$

2.11. Lemma. *Fit* $d_H Q = Q d_H$.

(e) *Camera*

2.12.

Introducamus sagittulam:

$$d_{DR}:\ \mathrm{Hom}(A^{\otimes 2}\otimes T^{\otimes 3}, A) \longrightarrow \mathrm{Hom}(A^{\otimes 2}\otimes T^{\otimes 2}\otimes \Lambda^2 T, A),$$

ubi

$$d_{DR}c(a,b,\tau_1,\tau_2,\tau_3,\tau_4) = \mathrm{Alt}_{3,4}\Big\{\mathrm{Lie}_{\tau_3}c(a,b,\tau_1,\tau_2,\tau_4) - \frac{1}{3}\tau_2 c(a,b,\tau_1,\tau_3,\tau_4)\Big\} + c(a,b,\tau_1,\tau_2,[\tau_3,\tau_4]),$$

confer art. 1.20.

2.13. Lemma. *Fit* $Qd_{DR} = d_{DR}Q$.

Demonstratio. procedit ut in art. 1.22.

(f) *Frons*

2.14. Lemma. *Fit* $d_H d_{DR} = d_{DR} d_H + QM$.

Confer art. 1.23.

Demonstratio. Probatur eodem modo ut in art. 1.24:

$$\begin{aligned}\{d_H d_{DR} - d_{DR} d_H\}c(a,b,\tau_1,\tau_2,\tau_3,\tau_4) &= \frac{1}{3}\mathrm{Alt}_{3,4}\tau_2(a)c(b,\tau_1,\tau_3,\tau_4)\\ &= \Big\langle \tau_2, \frac{1}{3}da\ \mathrm{Alt}_{3,4}c(b,\tau_1,\tau_3,\tau_4)\Big\rangle\\ &= \langle \tau_2, Mc(a,b,\tau_1,\tau_3,\tau_4)\rangle\\ &= QMc(a,b,\tau_1,\tau_2,\tau_3,\tau_4),\end{aligned}$$ qed

(g) *Junctio camerae altera*

2.15.

Contemplemur compositio sagittulae

$$d_{DR} : \mathrm{Hom}(A^{\otimes 2} \otimes T^{\otimes 2}, A) \longrightarrow \mathrm{Hom}(A^{\otimes 2} \otimes T^{\otimes 3}, A),$$

ubi

$$d_{DR}c(a,b,\tau_1,\tau_2,\tau_3) = \mathrm{Lie}_{\tau_3}c(a,b,\tau_1,\tau_2) - \frac{1}{2}\tau_2 c(a,b,\tau_1,\tau_3),$$

vide Pars Prima, art. 3.14, cum sagittula d_{DR} ex art. 2.12.

2.16. Lemma. *Fit* $d_{DR}^2 = QR$.

Demonstratio. Eadem ut in art. 1.27.

Pars tertia. Finale

1. Cocyclus canonicus

(a)

1.1.

Contemplemur elementa $\epsilon \in \mathrm{Hom}(A \otimes T^{\otimes 2}, A)$, $\epsilon' \in \mathrm{Hom}(A \otimes T^{\otimes 2}, \Omega)$ atque $\epsilon'' \in \mathrm{Hom}(A^{\otimes 2} \otimes T, \Omega)$ definita per:

$$\epsilon(a, \tau, \tau') = \tau\tau'(a), \ \epsilon'(a, \tau, \tau') = -\frac{1}{2}d\tau\tau'(a)$$

et

$$\epsilon''(a, b, \tau) = -\tau(a)db - \tau(b)da.$$

1.2. Lemma. *Fit* $d_{DR}\epsilon = Q\epsilon'$.

1.3. Demonstratio. Constat:

$$d_{DR}\epsilon(a, \tau_1, \tau_2, \tau_3) = \mathrm{Lie}_{\tau_3}\epsilon(a, \tau_1, \tau_2) - \frac{1}{2}\tau_2\tau_1\tau_3(a),$$

vide Pars Prima, art. 2.11. Terminus primus evadit, quod ϵ operator invariens sit. Hinc

$$d_{DR}\epsilon(a, \tau_1, \tau_2, \tau_3) = -\frac{1}{2}\tau_2\tau_1\tau_3(a) = \langle\tau_2, \epsilon'(a, \tau_1, \tau_3)\rangle = Q\epsilon'(a, \tau_1, \tau_2, \tau_3),$$

1.4. Lemma. *Fit* $d_H\epsilon = Q\epsilon''$.

1.5. Demonstratio. Habemus:

$$\begin{aligned} d_H\epsilon(a, b, \tau, \tau') &= a\tau\tau'(b) - \tau\tau'(ab) + b\tau\tau'(a) \\ &= -\tau(a)\tau'(b) - \tau'(a)\tau(b) = \langle\tau', \epsilon''(a, b, \tau)\rangle = Q\epsilon''(a, b, \tau, \tau'), \end{aligned}$$

qed

1.6. Lemma. *Fit* $d_{DR}\epsilon'' = d_H\epsilon' - M\epsilon$.

1.7. Demonstratio. Habemus (vide Pars Prima, art. 2.7):

$$d_{DR}\epsilon''(a, b, \tau, \tau') = \mathrm{Lie}_{\tau'}\epsilon''(a, b, \tau) - \frac{1}{2}d\langle\tau', \epsilon''(a, b, \tau)\rangle$$

(cum ϵ'' invariens est)

$$= \frac{1}{2}d\{\tau(a)\tau'(b) + \tau'(a)\tau(b)\}$$

Rursus,

$$d_H\epsilon'(a,b,\tau,\tau') = -\frac{1}{2}\big[ad\tau\tau'(b) - d\tau\tau'(ab) + d\{b\tau\tau'(a)\}\big]$$
$$= \frac{1}{2}[d\{\tau(a)\tau'(b) + \tau'(a)\tau(b)\} + da\ \tau\tau'(b)]$$
$$= d_{DR}\epsilon''(a,b,\tau,\tau') + M\epsilon,$$

unde sequitur lemma.

1.8. Lemma. *Fit* $d_H\epsilon'' = 0$.

1.9. Demonstratio. Statuamus:

$$\epsilon_0''(a,b,\tau) := \quad \tau(a)db;\ \epsilon_1''(a,b,\tau) := -\tau(b)da,$$

ergo $\epsilon'' = \epsilon_0'' + \epsilon_1''$. Adipiscimur:

$$d_H\epsilon_0''(a,b,c,\tau) = a\epsilon_0''(b,c,\tau) - \epsilon_0''(ab,c,\tau) + \epsilon_0''(a,bc,\tau) - \epsilon_0''(a,b,c\tau)$$
$$= -a\tau(b)dc + \tau(ab)dc - \tau(a)d(bc) + c\tau(a)db = 0$$

Simili modo probatur, $d_H\epsilon_1'' = 0$, unde lemma fluit.

1.10. Lemma. $d_{DR}\epsilon' = R\epsilon$.

1.11. Demonstratio. Primo observamus, quod definitio sagittulae d_{DR} ex Parte Secunda, art. 1.2, ita exhiberi potest:

$$d_{DR}c(a,\tau,\tau',\tau'') = \mathrm{Alt}_{\tau',\tau''}\mathrm{Lie}_{\tau'}c(a,\tau,\tau'') + c(a,\tau,[\tau',\tau''])$$
$$-\frac{1}{3}\mathrm{Alt}_{\tau',\tau''}d\langle\tau',c(a,\tau,\tau'')\rangle,$$

unde, quia $\mathrm{Lie}_\tau\epsilon' = 0$, sequitur:

$$d_{DR}\epsilon'(a,\tau,\tau',\tau'') = \epsilon'(a,\tau,[\tau',\tau'']) - \frac{1}{3}\mathrm{Alt}_{\tau',\tau''}d\langle\tau',\epsilon'(a,\tau,\tau'')\rangle$$
$$= -\frac{1}{2}d\tau[\tau',\tau''](a) + \frac{1}{6}\mathrm{Alt}_{\tau',\tau''}d\tau'\tau\tau''(a)$$
$$= \mathrm{Alt}_{\tau',\tau''}\left\{\frac{1}{2}d\tau\tau''\tau'(a) + \frac{1}{6}d\tau'\tau\tau''(a)\right\}.$$

1.12.

Secundo, habemus

$$\begin{aligned}R\epsilon(a,\tau,\tau',\tau'') &= -\frac{1}{6}d[\mathrm{Alt}_{\tau',\tau''}\{\epsilon(\tau''(a),\tau,\tau') + \epsilon(a,[\tau,\tau'],\tau'')\} + \epsilon(a,\tau,[\tau'\tau''])] \\ &= -\frac{1}{6}d\mathrm{Alt}_{\tau',\tau''}\{\tau\tau'\tau''(a) + [\tau,\tau']\tau''(a) + \tau\tau'\tau''(a)\} \\ &= -\frac{1}{6}d\mathrm{Alt}_{\tau',\tau''}\{3\tau\tau'\tau''(a) - \tau'\tau\tau''(a)\} = d_{DR}\epsilon'(a,\tau,\tau',\tau''),\end{aligned}$$

qed

2. Definitio altera

(a)

2.1.

Primo, axioma (A1) structurae verticianae ita exhiberi potest:

$$d_H\gamma = \epsilon''. \tag{A1}$$

2.2.

Secundo, axioma (A2) ita scriberi licet:

$$d_H\langle,\rangle - Q\gamma = -\epsilon. \tag{A2}$$

2.3.

Tertio, axioma $(A3)^{\mathrm{bis}}$ ita exhiberi potest:

$$d_Hc - d_{DR}\gamma - M\langle,\rangle = -\epsilon'. \qquad (A3)^{\mathrm{bis}}$$

2.4.

Quatro, axioma (A4) ita quoque exhiberi licet:

$$Qc = d_{DR}\langle,\rangle, \tag{A4}$$

confer Pars Prima, 1.5.

2.5.

Tandem axioma (A5) ita exhiberi licet:

$$d_{DR}c = R\langle,\rangle, \tag{A5}$$

confer Pars Prima, art. 1.13.

(b)

2.6.

Applicatio structurarum verticianarum

$$h:\ \mathcal{A} = (\gamma, \langle,\rangle, c) \longrightarrow \mathcal{A}' = (\gamma', \langle,\rangle', c')$$

elementum $h \in \mathrm{Hom}(T, \Omega)$ est, talis ut:

$$d_H h = \gamma - \gamma';\ \ Qh = \langle,\rangle - \langle,\rangle'$$

atque

$$d_{DR}h = c - c'.$$

3. Complexus de Rham–Koszul–Hochschildianus

3.1.

Introducamus moduli: W^{ijk}, $i \geq 2$; $j = 0, 1$; $k \geq 0$, posito: $W^{200} := \mathrm{Hom}(T, \Omega)$, tractandoque indices: i tamquam gradum DE RHAMIANUM, j tamquam gradum KOSZULIANUM ac k tamquam gradum HOCHSCHILDIANUM, ergo:

$$W^{300} = \mathrm{Hom}(\Lambda^2 T, \Omega),\ \ W^{210} = \mathrm{Hom}(S^2 T, A),\ \ W^{201} = \mathrm{Hom}(A \otimes T, \Omega),$$

etc.

Statuimus $W^n := \oplus_{i+j+k=n} W^{ijk}$, ergo:

$$W^2 = W^{200};$$

$$W^3 = W^{300} \oplus W^{210} \oplus W^{201};$$

$$W^4 = W^{400} \oplus W^{310} \oplus W^{301} \oplus W^{211} \oplus W^{202}$$

et

$$W^5 = W^{500} \oplus W^{410} \oplus W^{401} \oplus W^{311} \oplus W^{302} \oplus W^{212} \oplus W^{203}.$$

3.2.

Introducamus operatores $D = D_{DRQH} + \mathcal{R} + \mathcal{M} : \; W^i \longrightarrow W^{i+1}$, ubi

$$D_{DRQH}c^{ijk} = \{d_{DR} + (-1)^i Q + (-1)^{i+j} d_H\}c^{ijk},$$
$$\mathcal{R}c^3 = -Rc^{210}; \; \mathcal{R}c^4 = Rc^{310} - Rc^{211}$$

atque

$$\mathcal{M}c^3 = Mc^{210}; \; \mathcal{M}c^4 = Mc^{310} + Mc^{211}$$

Partium Primae Secundaeque summa significat, $D^2 = 0$, unde eruimus complexum

$$W^{[2,5]} : \; W^2 \xrightarrow{D} W^3 \xrightarrow{D} W^4 \xrightarrow{D} W^5,$$

de inclusione canonica complexuum: $\Omega^{[2,5]} \longrightarrow W^{[2,5]}$ ornatum.

3.3.

Contemplemur elementum:

$$\mathcal{E} := (\epsilon', \epsilon, \epsilon'') \in W^{301} \oplus W^{211} \oplus W^{202} \subset W^4$$

Sectionis 1 summa significat, $D\mathcal{E} = 0$.

3.4.

Structura verticiana super T est elementum

$$\mathcal{A} = (c, \langle , \rangle, \gamma) \in W^{300} \oplus W^{210} \oplus W^{201} = W^3,$$

talis ut fit $D\mathcal{A} = \mathcal{E}$.

3.5.

Applicatio structurarum verticianarum $h : \; \mathcal{A} \longrightarrow \mathcal{A}'$ elementum $h \in W^{200} = W^2$ est, talis ut fit $Dh = \mathcal{A} - \mathcal{A}'$.